INTRODUCTION TO
SPACE SCIENCE

INTRODUCTION TO
SPACE SCIENCE

Second Edition
Revised and Enlarged

Written by the Staff of the Goddard Space Flight Center
National Aeronautics and Space Administration,
Greenbelt, Maryland

Edited by
WILMOT N. HESS
NASA Manned Spacecraft Center
Houston, Texas

and

GILBERT D. MEAD
NASA Goddard Space Flight Center
Greenbelt, Maryland

GORDON AND BREACH, SCIENCE PUBLISHERS
New York / London / Paris

GORDON AND BREACH, Science Publishers, Inc.
150 Fifth Avenue, New York, N. Y. 10011
Library of Congress Catalog Card Number 65-26600

Editorial Office for Great Britain:
Gordon and Breach, Science Publishers Ltd.
8 Bloomsbury Way
London, W. C. 1

Editorial Office for France:
Gordon & Breach
7–9 rue Emile Dubois
Paris 14e

Distributed in Canada by:
The Ryerson Press
299 Queen Street West
Toronto 2B, Ontario

Printed in the United States of America

PREFACE TO THE SECOND EDITION

It was obvious even before the first edition of this book was available that a revision would soon be needed. In the intervening two and a half years, for example, five Surveyor spacecraft have soft-landed on the moon, sent back thousands of high-resolution photographs of the lunar surface, and tested the surface for its mechanical properties. Five Lunar Orbiters have returned high-quality photographs of possible Apollo landing sites under varying conditions of illumination. The unusual photograph opposite the title page is one of these, with the Earth serving as a backdrop. Synchronous satellites are now regularly transmitting color photographs of the entire earth's surface and cloud cover as seen from an altitude of 35,000 kilometers. Recent Explorer and OGO earth-and lunar-orbiting satellites have returned a wealth of additional information on the magnetic, energetic particle, and radiation environment of the earth, moon, and interplanetary space. These and many other experimental and theoretical studies have provided much new information about the environment in space.

In light of these developments, substantial revisions and additions have been made in the chapters on the Ionosphere, Meterology from Space, the Interplanetary Medium, the Boundary of the Magnetosphere, Interplanetary Dust Particles, Cosmic Chemistry, the Moon, Planetary Atmospheres, Stellar Evolution, and Extragalactic Radio Sources. The chapters on the Shape of the Earth, Man in Space, the Sun, Space Astronomy, and Nucleogenesis have been completely rewritten.

Wilmot N. Hess
Gilbert D. Mead

January, 1968

CONTRIBUTORS

Although all of the first authors of the various chapters were once associated with the Goddard Space Flight Center, a number of them are now located elsewhere. The present affiliations of all the authors and co-authors are listed below:

AIKIN, ARTHUR C., Planetary Ionospheres Branch, Laboratory for Space Sciences, Goddard Space Flight Center

ALEXANDER, WESLEY M., Department of Physics, Baylor University, Waco, Texas

BAUER, SIEGFRIED J., Head, Planetary Ionospheres Branch, Laboratory for Space Sciences, Goddard Space Flight Center

BOGGESS, ALBERT, III, Astrophysics Branch, Laboratory for Space Sciences, Goddard Space Flight Center

BRANDT, JOHN C., Head, Solar Physics Branch, Laboratory for Space Sciences, Goddard Space Flight Center

CAMERON, A. G. W., Belfer Graduate School of Science, Yeshiva University, New York, New York

CAMERON, ROBERT C., Laboratory for Theoretical Studies, Goddard Space Flight Center

CAMERON, WINIFRED S., Laboratory for Theoretical Studies, Goddard Space Flight Center

DAVIS, T. NEIL, Geophysical Institute, University of Alaska, College, Alaska

DONN, BERTRAM D., Astrophysics Branch, Laboratory for Space Sciences, Goddard Space Flight Center

DUNKELMAN, LAWRENCE, Astrophysics Branch, Laboratory for Space Sciences, Goddard Space Flight Center

FICHTEL, CARL E., Energetic Particles Branch, Laboratory for Space Sciences, Goddard Space Flight Center

GILL, JOCELYN R., NASA Headquarters, Washington, D. C.

HARRIS, ISADORE, Laboratory for Theoretical Studies, Goddard Space Flight Center

HEPPNER, JAMES P., Head, Fields and Plasmas Branch, Laboratory for Space Sciences, Goddard Space Flight Center

HESS, WILMOT N., Director of Science and Applications, NASA Manned Spacecraft Center, Houston, Texas

JASTROW, ROBERT D., Director, Goddard Institute for Space Studies, New York, New York

KAULA, WILLIAM M., Institute of Geophysics and Planetary Physics, University of California, Los Angeles, California

LOWMAN, PAUL D., JR., Planetology Branch, Laboratory for Theoretical Studies, Goddard Space Flight Center

MARAN, STEPHEN P., Kitt Peak National Observatory, Tucson, Arizona

McCRACKEN, CURTIS W., Astrophysics Branch, Laboratory for Space Sciences, Goddard Space Flight Center

MEAD, GILBERT D., Laboratory for Theoretical Studies, Goddard Space Flight Center

MUSEN, PETER, Laboratory for Theoretical Studies, Goddard Space Flight Center

NESS, NORMAN F., Ass't Head, Fields and Plasmas Branch, Laboratory for Space Sciences, Goddard Space Flight Center

NORDBERG, WILLIAM, Ass't Chief, Laboratory for Atmospheric and Biological Sciences, Goddard Space Flight Center

O'KEEFE, JOHN A., Ass't Chief, Laboratory for Theoretical Studies, Goddard Space Flight Center

PALMEIRA, R. A. R., Southwest Center for Advanced Studies, Dallas, Texas

PIEPER, GEORGE F., Ass't Director, Space Sciences Directorate, Goddard Space Flight Center

PRIESTER, WOLFGANG, Director, Institut für Astrophysik und Extraterrestrische Forschung, Bonn University, W. Germany

RASOOL, SAIYED I., Goddard Institute for Space Studies, New York, New York

ROSENBERG, JOHAN, University of Bonn, West Germany

SPENCER, NELSON W., Chief, Laboratory for Atmospheric and Biological Sciences, Goddard Space Flight Center

SQUIRES, R. KENNETH, Head, Special Projects Branch, Laboratory for Theoretical Studies, Goddard Space Flight Center

SUGIURA, MASAHISA, Fields and Plasmas Branch, Laboratory for Space Sciences, Goddard Space Flight Center

TRURAN, JAMES W., Belfer Graduate School of Science, Yeshiva University, New York, New York

EDITOR NOTES

FOR THE FIRST EDITION

The subject, Space Science, is not one that has well defined boundaries; so the question of what is included or excluded in this book is clearly a subjective one. As an operational definition we have considered space science to be the synthesis of the interests of scientists working on the national space program. This means, as one might expect, that it is a continuation of sea-level science in all disciplines that can be extended and amplified usefully by going into space. In this sense it includes most of the fields of astronomy, both optical and radio, since by getting above the earth's atmosphere and ionosphere observations are possible in frequency ranges not possible on the ground. It includes and extends the fields of cosmic rays, atmospheric and ionospheric physics. It includes the field of plasma physics with regimes in density and temperature not easily reproduced in the laboratory. It includes the study of the geomagnetic field and charged particles contained in it. It includes the field of geophysics in the study of the moon and planets as well as the earth. We are very fortunate to have specialists in all of the major fields of space science at Goddard Space Flight Center so that the book could be a laboratory project.

An attempt has been made to introduce each subject historically and lead up to a statement of current research problems. Referencing has been kept relatively brief. Appropriate review papers have been referenced but few original research papers are covered. The reader who wants to get more complete coverage of a certain subject should refer to the review papers.

The material covered here goes through the Spring of 1965. The subject is progressing fast enough so that some of the material presented will inevitably be out of date when it is published.

I am very grateful for all the help provided by Goddard Space Flight Center in getting the book prepared. I would especially like to thank Mr. Robert Tanner and Mrs. Margaret Becker and Mrs. Diana Uber for their very able assistance in preparing the manuscript. I would also like to thank Drs. A. Cameron, R. Cameron, F. Jones, K. Maeda, G. Mead, P. Nakada and J. O'Keefe for reading and criticizing parts of the book. Dr. Mead was very helpful in final preparation of the manuscript. Thanks also to Sacramento Peak Observatory, Lick Observatory, Mt. Wilson and Mt. Palomar Observatories and Prof. Hessler of the Geophysical Institute in College, Alaska, for the use of their excellent photographs. Thanks also to the JOURNAL OF GEOPHYSICAL RESEARCH for the use of much of their material. Finally, speaking for all the authors, I would like to thank Dr. Goett, Director of Goddard Space Flight Center, for his support in carrying out this project.

Wilmot N. Hess

Greenbelt, Maryland
June, 1965

PREFACE

TO THE FIRST EDITION

The advent of the artificial satellite as a research tool has opened new frontiers to the traditional scientific disciplines. In the few years of the space age the geodesist, astronomer, particle physicist and meteorologist all have used satellites and space probes to do experimental work beyond the limitations of the ground based laboratories and observatories. The influence of this new research tool has been felt not only in the experimental areas of these various sciences; there also has been a corresponding expanded effort in the theoretical areas to keep pace with the new experimental data that is becoming available.

Some measure of stimulating effect of this research tool is given by the marked increase in space related papers published in scientific journals. In one typical case, the Journal of Geophysical Research, there were 35 space related papers published in 1959; in 1964 the number had grown to about 130. Through papers published in these scientific journals there has been an adequate and remarkably rapid communication of results between researchers in any given discipline. However this medium of communication is primarily designed for exchange of information between specialists in the same scientific discipline. There remains a need for cross-discipline communication, for communication between scientist and engineer, and for summaries useful to graduate students or the well educated technical audience that is not involved in space research. It is for the purpose of filling this need that the group of articles that make up this book has been assembled.

The subject matter of the articles has been chosen to cover the spectrum of research areas that have been made available by the artificial satellite. Each chapter has been written in such manner as to be self contained and does not assume prior knowledge of the subject. They are intended to introduce the subject and to convey an understanding of its basic principles with mathematical developments held to that necessary to explain significant points. There has been an intentional use of pictures and figures to convey an understanding of the key phenomena. The bibliography for each chapter is not intended to be comprehensive; it has been confined to survey articles or major research papers on the topic of the individual article.

The contributors of the various chapters are each actively engaged in research directly related to the subject matter of the article. They are all members of the staff of the Goddard Space Flight Center. In general they have not only made personal contributions of an experimental or theoretical nature in the area of space research on which they write, but they also have played an important role in the planning of the space research program. This book therefore is a report on the present status of our space research program, prepared by some of the architects of and participants in this program.

Dr. Harry Goett

June, 1965 *Director, Goddard Space Flight Center*

CONTENTS

PART I THE EARTH AND ITS ENVIRONMENT

PART II SPACE

Part I
THE EARTH AND ITS ENVIRONMENT

INTRODUCTION

The era of satellites and rockets has allowed us to get up above the dense atmosphere of the earth and have a look to see what is there. There have been some surprises. Probably the largest one was the Van Allen radiation belt found on the first U. S. satellite, Explorer I. One or two components of the radiation belt are rather well understood now, but we still have only vague ideas about the origin of some other parts of the radiation belt. The outer region of the radiation belt appears to be related to auroral activity. Aurorae are now being directly studied from satellites.

The experimental study of the upper atmosphere and upper ionosphere from satellites and rockets had led to a detailed knowledge of the ion and atom distributions in altitude, latitude and type, and their diurnal, seasonal and solar cycle time variations well above the regions previously explored.

We have also been able to study the earth from rockets and satellites in ways not possible from the ground. The Tiros and Nimbus meteorological satellites have studied world-wide cloud patterns, discovered hurricanes, and proved their usefulness in weather forecasting and analysis. The earth's magnetic field is better measured from space than the ground. Complete detailed coverage can be obtained from satellites in a few days much faster than is possible on the ground.

The fact that satellite orbits are not really ellipses as they would be if the earth were a point mass enables us to find the real gravitational potential field of the earth, and from this we can learn something about the interior of the earth from satellites.

This section of the book considers in detail these subjects about the earth and the region of space near the earth extending out to roughly 10 earth radii.

Chapter 1

THE EARTH'S MAGNETIC FIELD

M. Sugiura and J. P. Heppner

INTRODUCTION

The earth possesses two fundamental fields, magnetic and gravitational. The conception of magnetism of the earth as a *planetary* property is due to William Gilbert, who wrote (in Latin) the great treatise *De Magnete* in 1600, eighty-six years before Newton gave his discourse on gravitation and mechanics in his *Principia*. The first scientific paper on magnetism is probably that published in 1269 by Petrus Peregrinus, a Frenchman, who recognized that a spherical lodestone has two poles at which a small magnet orients itself parallel to the axis connecting the poles. However, Peregrinus supposed in the medieval traditions of philosophy that the poles of the magnet received their power from the heavens.

The first recognition of the fact that the direction of a compass needle deviates from the true north—the deviation now being called declination—can be traced back to the middle of the 16th Century when the manufacturing of compass sundials flourished at Nüremberg. By placing a small magnet on a spherical lodestone, which he called a *terrela*, Gilbert showed how the inclination of the magnet changes over different parts of the sphere. His prediction of increase in the magnetic dip angle with increasing latitude was confirmed in 1608 by the navigator-explorer Henry Hudson in the Barents Sea. A slow variation in declination was first reported by Henry Gellibrand of London in 1635; this is the beginning of the study of the geomagnetic secular variation. The earliest geomagnetic (declination) chart was published by the Astronomer Edmund Halley, who from 1697 to 1701 made two *magnetic survey* voyages in the Atlantic Ocean to obtain more accurate data on declination than those that had been available previously. After Halley the art of drawing magnetic charts was developed during the 18th Century, and the first magnetic dip charts covering the whole world was published in 1768 by Johann Carl Wilcke at Stockholm.

The daily variation of declination was discovered in 1722 by George Graham of London, and was confirmed in 1740 by Andreas Celsius at Upsala. Celsius and his colleague Hiorter first noted in 1741 that the aurora and the magnetic disturbance were closely related. Celsius requested Graham by letter to make observations simultaneously, and confirmed that the magnetic disturbance accompanying the aurora over Sweden was also observed in London. In 1770 Wilcke noted that the auroral rays are parallel to the direction of the dip-needle. These are the beginnings of the study of the polar disturbance that will be discussed on pages 41–58.

The concept of *potential* developed for gravity by Laplace and Lagrange was applied to electricity and magnetism by Poisson and Green. Carl Friedrich Gauss (1777-1855) opened the beginning of the modern potential theory by giving rigorous mathematical proofs to many of the theorems derived by his predecessors and adding new theorems found by himself. In 1838 Gauss applied the potential theory to the earth's magnetism and expressed the magnetic potential in a spherical harmonic series, the coefficients of which are now referred to as the *Gauss coefficients* (see page 13). Among his many great contributions the most notable are the establishment of the absolute units of magnetism and the development of a method of accurately determining the absolute magnitude of the magnetic field. The unit *Gauss* for magnetic induction is named after its founder. Gauss and Weber established in 1833 a new iron-free magnetic observatory in Göttingen, and together with von Humbolt initiated the *Göttingen Magnetic Union* which opened the gate to an international cooperation for geomagnetic researches that is so essential for complete descriptions and understanding of geomagnetic phenomena.

The first drawing of the auroral zone by Elias Loomis of Yale in 1860 and the publication in 1873 of the auroral catalog and the *isochasms* (contours of equal frequency of auroral activity) by Herman Frits of Zürich are important early contributions in the study of the aurora.

Geomagnetic phenomena are closely related to those in other fields of geophysics and the modern advancements in geomagnetism are interdependent on those in other disciplines. The understanding of the origin of the earth's main magnetic field was helped by the knowledge on the structure of the interior of the earth which had been gained mainly by seismological studies (see pages 18–20). For more complete solutions to many problems concerning the magnetism of the solid earth, further advancements in physics for high-temperature and high-pressure environments are needed.

Balfour Stewart predicted in 1882 that the daily magnetic variations are due to electric currents flowing in a conducting layer in the upper atmosphere (pages 24–40). The existence of such a conducting layer—the ionosphere—was discovered in 1902 by Heaviside and Kennelly independently. Subsequent studies of the ionosphere have supported Stewart's theory, and recent direct magnetic measurements by rockets have further confirmed the existence of the ionospheric currents. Extensive theoretical studies of the atmospheric oscillations have been made, but the nature of winds in the ionosphere is only partially understood. It is now becoming increasingly clear that the upper atmospheric winds cannot be treated separately from the dynamics of the atmosphere in the lower regions that are generally dealt with in meteorology.

Direct aeronomical measurements and observations of solar radiation by rocket techniques have greatly advanced our knowledge of the physics of the ionosphere which forms an important background for geomagnetic researches. The development of space exploration by satellites and probes has had a great impact on geophysical studies. Observations of the geomagnetic field are no longer limited to the surface of the earth or to its immediate vicinity. The geomagnetic field had never been thought to be continuously confined to a region of finite extent until a steady flow of plasma from the sun was suspected and then actually confirmed by direct measurements in interplanetary space. The form of the boundary between the geomagnetic field and the solar wind has been determined repeatedly near the equatorial plane, and the novel feature of a geomagnetic *tail* has been revealed.

The modern progresses made in plasma physics have made it possible to interpret many complex geomagnetic and related phenomena in the ionosphere, in the exosphere and in interplanetary space. The revelation of the physical laws governing the motion of a charged particle in magnetic fields has contributed to the understanding of the behavior of charged particles in the earth's magnetic field.

With increasing observational data on the geomagnetic field, electromagnetic phenomena, plasmas, and energetic particles our understanding of the physical state of the earth's environment and of various disturbances therein is rapidly advancing[1]. However the present knowledge on these subjects is far from complete; although efforts were made to describe well-established facts as completely as can be done in a limited space and to include recently developed views, many aspects of the discussions given in this Chapter will have to be improved, modified, and, in some cases, corrected in the not-too-distant future.

GEOMAGNETIC COORDINATES, MAPS, AND MATHEMATICAL DESCRIPTIONS

Coordinates

The magnetic field at any point can be represented by a vector which is described by its magnitude and direction relative to a selected coordinate system. Figure 1 illustrates the designation of seven magnetic elements (X, Y, Z, H, D, I, F) which are conventionally used in describing a field vector (magnetic force or intensity) at the the earth's surface[2]. The vector is of course completely described by a set of any three independent elements. Surface observatories commonly measure either H, D, and Z or X, Y, and Z. The recent advent of nuclear and atomic magnetic resonance instruments, which provide exceptionally accurate measurements of the total scalar field intensity F (or B, the magnetic induction, using conventional physics symbols) is, however, leading to the use of additional com-

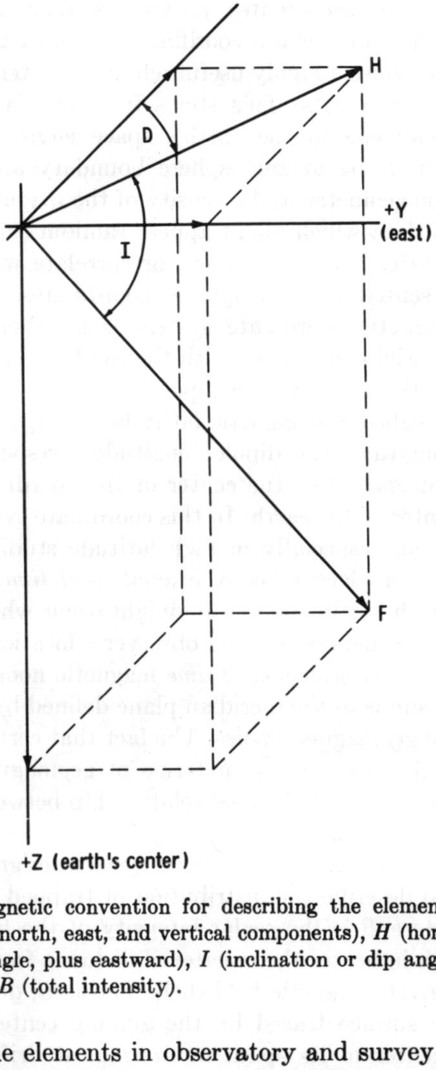

Fig. 1. Geomagnetic convention for describing the elements of a magnetic vector: *X*, *Y*, *Z* (north, east, and vertical components), *H* (horizontal intensity), *D* (declination angle, plus eastward), *I* (inclination or dip angle, plus below the horizontal), *F* or *B* (total intensity).

binations of the elements in observatory and survey measurements. In presenting space measurements the use of *B* and two angles has been most convenient and descriptive. This is largely because it provides a visual continuity of the field intensity when the field vector is transformed to coordinate systems other than the geodetic coordinates of Figure 1.

Whereas geodetic coordinates provide a reference system for describing the magnetic field, a coordinate system based on the geomagnetic field provides a highly useful reference system for describing other space measurements. This stems from the fact that a large number of phenomena in the earth's space environment between about 100 km and the magnetosphere boundary are influenced or controlled by the geometry and intensity of the magnetic field. Thus, certain types of data which might appear random or uncorrelated in geodetic coordinates become coherent or correlate well with related data when presented in geomagnetic coordinates. The most elementary geomagnetic coordinate system is a spherical coordinate system with its origin coincident with the earth's center and with its axis parallel to the geomagnetic dipole. Latitude and longitude in this system are called geomagnetic latitude (or dipole latitude) and geomagnetic longitude (or dipole longitude), respectively. Radial distances do not change as the center of the coordinate system remains at the center of the earth. In this coordinate system it has also been advantageous, especially in high latitude studies, to use *local geomagnetic time* in place of conventional *local time*. Analogous to local time in which local noon and midnight occur when the sun is in the meridian plane defined by the observer's location and the geographic poles, in *local geomagnetic time* magnetic noon and midnight occur when the sun is in the meridian plane defined by the observer's location and the geomagnetic poles. The fact that certain phenomena can be described more exactly in terms of geomagnetic time is of course an indication of the detailed relationship between the sun and geomagnetic phenomena.

A more recent geomagnetic coordinate system designed particularly for its ability to describe the distribution of trapped particles is the B, L system. Here B is the scalar intensity of the field and L is a parameter which indicates the equatorial distance from the center of the earth to a given magnetic field shell. The *shell*, designated by its L value, is the surface traced by the guiding center of a trapped particle (i.e., center of its spiral motion) as it drifts in longitude about the earth while oscillating between mirror points in the northern and southern geomagnetic hemispheres. If the drift with longitude did not exist the path of the particle's guiding center would describe a field line. The two points where a given magnetic field line intersects the earth's surface (one in the northern and one in the southern geomagnetic hemispheres) are called *conjugate points.* The relationships between B, L coordinates and the adiabatic invariants of trapped particle motion are treated in Chapter 4. An important point to note is that unlike the previously noted coordinate systems, the B, L system is designed to take into account the fact that the earth's magnetic field is not completely represented by a dipole field. In essence one way to view the B, L coordinates is to imagine that they convert the real geomagnetic field into an equivalent dipole field for the purpose of simplifying representations of the motions and distribution of trapped particles. The validity of a B, L system depends entirely on the accuracy of describing mathematically the earth's magnetic field which is the subject of the next section.

Maps and Mathematical Descriptions

A uniformly magnetized sphere or, alternatively, a perfect dipole magnet located at the center of a sphere produces a field which is described at the sphere's surface (radius $= a$) and outside the sphere (distance r from the center) by the following equations:

$$H = H_0(a/r)^3 \sin \theta \tag{1}$$

$$Z = 2H_0(a/r)^3 \cos \theta \tag{2}$$

$$B = (H^2 + Z^2)^{1/2} \tag{3}$$

where θ is the polar angle and H_0 is the horizontal intensity at the equator where $Z = 0$. The magnetic potential whose space derivatives in directions tangential and radial to the surface are H and Z, respectively, is

$$V = (M/r^2) \cos \theta, \tag{4}$$

where

$$M = H_0 a^3 \tag{5}$$

is the magnetic dipole moment. Using $H_0 = 0.31$ gauss* and given

* In geomagnetic problems that do not deal with magnetic material the magnetic permeability is unity in c.g.s. units; thus the distinction between magnetic field intensity H and magnetic induction B can be dispensed with without confusion. Hence, although the c.g.s. unit for magnetic field intensity is *oersted,* the unit for magnetic induction, namely, *gauss,* is commonly used for both quantities.

the surface locations of the dipole axis the above centered dipole equations describe the earth's main magnetic field to an average accuracy of about 90 percent. This represents a good first approximation to the true field but for many problems the 10 percent error can be serious.

It was apparent from the early maps of declination compiled primarily from ship logs of the compass direction that the earth's field was not a true dipole. Charts of declination were subsequently prepared as a navigation aid and as measurements of magnetic force accumulated charts were also drawn to show the other elements. In this century it became a common practice to redraw the charts each 5 to 10 years to correct for secular variations and to take advantage of new measurements. The fundamental problem in chart construction has always been the sparsity or complete lack of measurements over vast regions of the globe. In these areas chart construction has often been guesswork or measurements that are outdated by secular variation have had to be used. The *World Magnetic Survey* is the name given to an international endeavor currently underway to improve this situation by conducting magnetic surveys over the entire earth. Satellite measurements of the magnetic field are an important part of this effort as comprehensive coverage can be achieved in a very short time. Other measurements come primarily from surveys by airplane and ship.

Figure 2 illustrates the form of the magnetic field as shown by a chart of the total scalar intensity. The lack of dipole symmetry is apparent. It is clear also that in speaking of a magnetic pole one should discriminate between different designations of a pole. For example if poles were defined as the north and south locations where the field had maximum intensity, the north pole would be on the western edge of Hudson Bay. Two types of magnetic poles are usually referred to; these are: (1) The magnetic poles which are more properly called the dip poles because they correspond to the locations where the field is entirely vertical (i.e., $H = 0, I = 90°$), and (2) the geomagnetic poles which are the surface intercepts of the axis of the centered dipole determined by the coefficients g_1^0, g_1^1 and h_1^1 of equation (6) below. In space studies it is primarily the geomagnetic pole that becomes important as it is more related to the gross features of the field. The dip pole location is greatly influenced by local magnetism in the earth's crust.

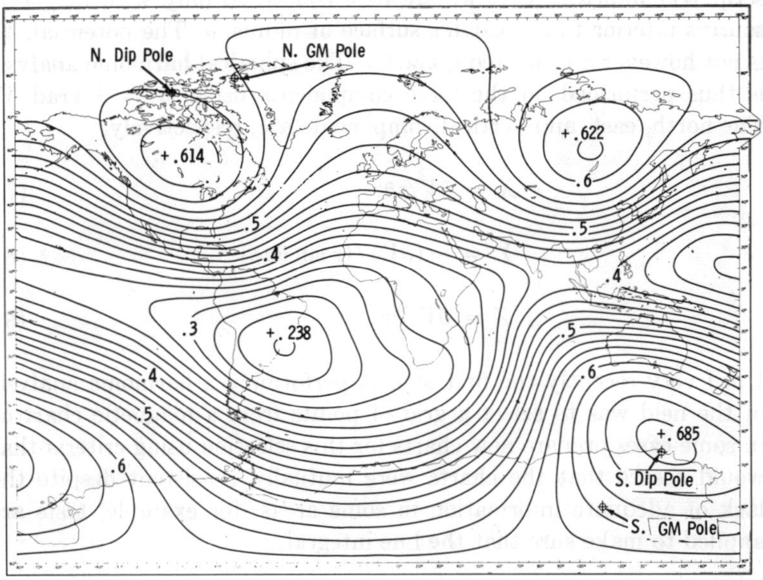

FIG. 2. Map of total field intensity, B, for 1965 with locations of the dip poles and geomagnetic poles: units are gauss (After Cain, et al., J. Geophys. Res., 1965).

Charts or maps of the geomagnetic field have limited direct use in space physics not only because of the two dimensional restriction but also because they do not facilitate data analysis. Thus a mathematical description is required. With the assumption, which appears well justified, that the amount of electric current flowing between the solid earth and space is negligible, the magnetic potential, V, satisfies Laplace's equation $\nabla^2 V = 0$ and can be expressed in spherical (polar) coordinates (r, θ, ϕ) as follows:

$$V_i = a \sum_{n=1}^{\infty} \sum_{m=0}^{n} (a/r)^{n+1} P_n^m(\cos\theta)[g_n^m \cos m\phi + h_n^m \sin m\phi] \qquad (6)$$

where θ and ϕ are now geographic colatitude and east longitude, $P_n^m(\cos\theta)$ are suitably normalized associated Legendre functions and g_n^m and h_n^m are the coefficients (called Gaussian coefficients) which one wishes to derive in obtaining the description. The subscript (i) indicates that we are here concerned only with magnetic sources interior to the earth's surface of radius, a. The potential, V, is not however a measured quantity. The spherical harmonic analysis is thus performed on the force components using $\mathbf{B} = -\text{grad } V$. The north, east, and vertical components are, respectively,

$$X = \partial V/r\partial\theta \tag{7}$$

$$Y = -(\partial V)/(r \sin\theta\partial\phi) \tag{8}$$

$$Z = \partial V/\partial r \tag{9}$$

Until very recently the first step in performing a harmonic analysis of the field was to select a grid of points from a magnetic chart or in some cases prepare new charts for this selection using criteria that would assure that the charts were mutually consistent despite the lack of adequate information in some areas; for example, tests are applied to make sure that the line integral,

$$\oint \mathbf{B} \cdot \mathbf{dl},$$

around any closed curve is zero. The selected points are then harmonically analyzed to find the set of coefficients g_n^m, h_n^m that best fit the chart data. In a number of analyses 49 coefficients have been used corresponding to terms through $m = n = 6$ although the agreement between different analyses beyond $m = n = 4$ has in general been poor, because often the higher order coefficients have been limited by the accuracy of the input data. The resulting field descriptions have represented the field with average errors in the range 1 to 3 percent.

Two primary changes in the procedure have been made recently. First, with the availability of high speed computers the measurements can be handled directly in large quantities thus eliminating the errors that appear in the process of constructing charts, and second, analysis procedures have been employed in which the total scalar magnetic field, B, is used directly as well as the field components of equations (7), (8), and (9). The direct use of B greatly increases the amount of high quality data. In particular, satellite measurements of B near the earth are highly accurate, whereas a comparably accurate determination of components or of angles requires that the orientation of the satellite magnetometer be known relative to geodetic coordinates to values such as 0.1°, but such accuracy has not been achieved. The most recent analyses taking advantages of the above innovations have reduced the errors in field description to less than 1.0 percent. The average error may, in fact, be less than 0.5 percent in representing the main field, exclusive of the large local errors given by the small scale magnetic anomalies having their origin in the earth's crust. The effects of local crustal sources disappear rapidly with altitude and thus are of little consequence in representing the field in space. This is fortunate because any attempt to represent this small scale structure would require a ridiculously large number of coefficients.

It is anticipated that, upon completion of the World Magnetic Survey satellite measurements, it will be possible to describe the earth's main magnetic field mathematically with errors of 0.1 to 0.2 percent. Such accuracies have been obtained already in the limited regions of the globe covered by the Vanguard-III satellite. Reduction of errors to less than 0.1 percent involves having highly exact information on secular field variations and external sources as discussed below.

External Sources

Equation (6) assumes that the field being analyzed originates entirely within the interior of the earth. From the existence of magnetic disturbances, diurnal variations, etc., we know, however, that there are also sources of magnetic fields in ionospheric and magnetospheric regions. Theoretically, a separation of internal and external sources can be achieved through analysis of equations (7, 8, 9) when surface harmonic terms multiplied by $(r/a)^n$ representing an external potential, V_e, are added to equation (6). Attempts to determine the existence of external sources by this technique have not met with success however, probably because the average external source contribution is less than the one percent errors in the overall description.

When we speak of field descriptions with errors of 0.1 percent or less, we must recognize that external sources become an important consideration. Similarly, if the field is to be described with proportional accuracy beyond altitudes of several thousand kilometers, the external sources must be taken into account. For this purpose, the usual techniques of including terms $(r/a)^n$ loses validity unless the external sources are located beyond the distance, r. The most desirable approach in describing the field at large distances is to have accurate independent knowledge of the location and intensity of the external sources such that they can be represented by potential terms that are added to equation (6) prior to the analysis. Of course representation of transient sources is not intended. What is desired in describing the external sources for main field studies is a description of any external source that is always present. Two types of permanent external sources that are known to exist are the *quiet day ring current* caused by the drift motion of trapped particles and the *deformation of the magnetic field* caused by the solar wind. Estimates of the combined intensity from these sources range from roughly 10 to 40 gammas near the earth. [1 gamma (γ) = 10^{-5} gauss]. This magnitude is comparable to 0.1 percent of the field near the earth's surface. At large distances it becomes an important fraction of the field.

Secular Variations

In the previous discussion we have ignored the fact that the earth's main field varies with time at a rate that is very slow compared to the transient variations seen on observatory recordings, but extremely rapid on a geologic time scale. This variation which is readily detected in terms of year to year changes at an observatory is called *secular variation*. Figure 3 illustrates the pattern of secular field variation for the total scalar intensity as determined for 1965. Lines of equal rate of change are referred to as *isopors*. Obviously the determination of secular change in vast regions that are poorly measured is somewhat subjective. It is also apparent that field measurements and descriptions rapidly become invalid in regions where the change is pronounced (e.g. inside the 100 γ/year isopor of Figure 3). Thus magnetic field charts and mathematical descriptions are labeled according to the year or period to which they apply (e.g., epoch 1945, epoch 1955, etc.). If data is available for many years, secular change may also be determined in the spherical harmonic analysis. In this case the time derivatives dg_n^m/dt, dh_n^m/dt of the Gaussian coefficients are included.

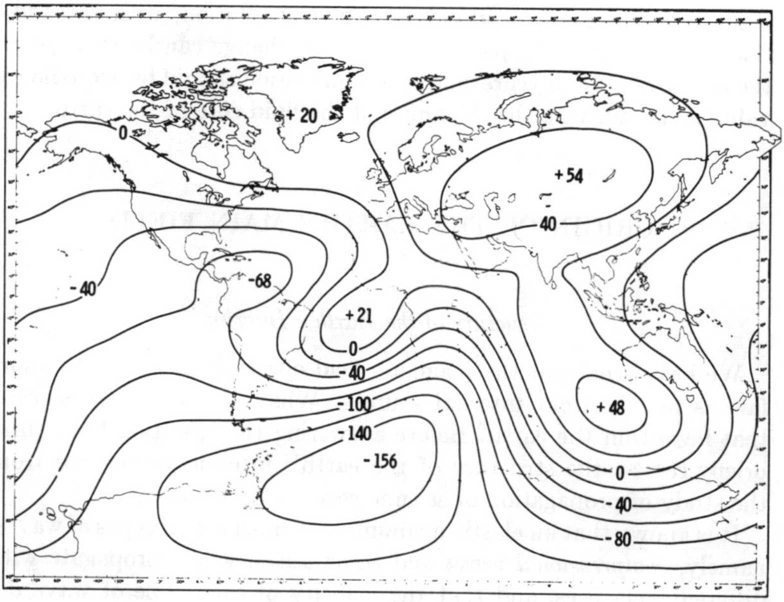

FIG. 3. Map of secular variation of the total field intensity for 1965: units are gammas/year (After Cain, et al., J. Geophys Res., 1965).

Although the errors in secular change determinations have been large, there is general agreement on the gross features. These are: (1) the dipole magnetic moment is decreasing, perhaps at a rate great enough to decrease the equatorial field by 50 γ/year, (2) the general field pattern is drifting westward at a rate of 0.2 to 0.3 degrees/year, and (3) the dipole field is shifting northward. The northward shift is revealed as a change of 21 to 27 γ/year in the term dg_2^0/dt of the harmonic analysis. This is the largest change observed in any single coefficient.

The gross rates of change as well as the more rapid changes in isolated regions of the globe make it apparent that mathematical descriptions to accuracies of 0.1 percent would need to be revised within 6 months to a year unless secular change is taken into account. This is the major source of error in using surface measurements for analysis, as global coverage is achieved only on a much longer time scale. Satellite surveys provide an obvious solution, with the additional feature that the patterns of secular change can be revealed in the course of several years, and this information should be exceedingly valuable in studying the dynamics of the fluid core of the earth.

ORIGIN OF THE EARTH'S MAIN FIELD

Structure of the Earth's Interior

We have seen that the magnetic field observed at the earth's surface is mainly from internal sources. What then are these sources that lie within the earth? Before answering this question let us first briefly review the structure of the earth's interior as inferred from the study of propagation of seismic waves within the earth.

It is known that an elastic medium can transmit two types of waves, namely, *compressional waves* and *shear waves*, which propagate with different velocities, and that the velocity of each type of wave depends on the density and the elastic property of the medium. Therefore, a careful study of the arrival of seismic waves at various points on the earth's surface can provide valuable information on the structure of the earth's interior.

The upper half of Figure 4 shows the structure of the earth schematically, and the lower half shows the distributions of the density and the velocities for compressional (P) and shear (S) waves. The earth is enclosed by the crust of rocks whose thickness is 30 to 60 km under the continents and 5 to 6 km under the oceans. The seismic wave velocities suddenly change at the bottom of the crust indicating an abrupt transition of the medium. This transition is called the *Mohorovičić discontinuity* after its discover, or simply the *Moho*. The region below this discontinuity to the depth of 2900 km is the *mantle*. At the latter depth we encounter a major discontinuity in the velocity of compressional wave and the density. Shear waves are not transmitted below that depth, indicating that the material underneath is in a fluid state. We call the region below the discontinuity at 2900 km the *core* of the earth. There is some evidence suggesting that inside the fluid core there may exist a solid inner core.

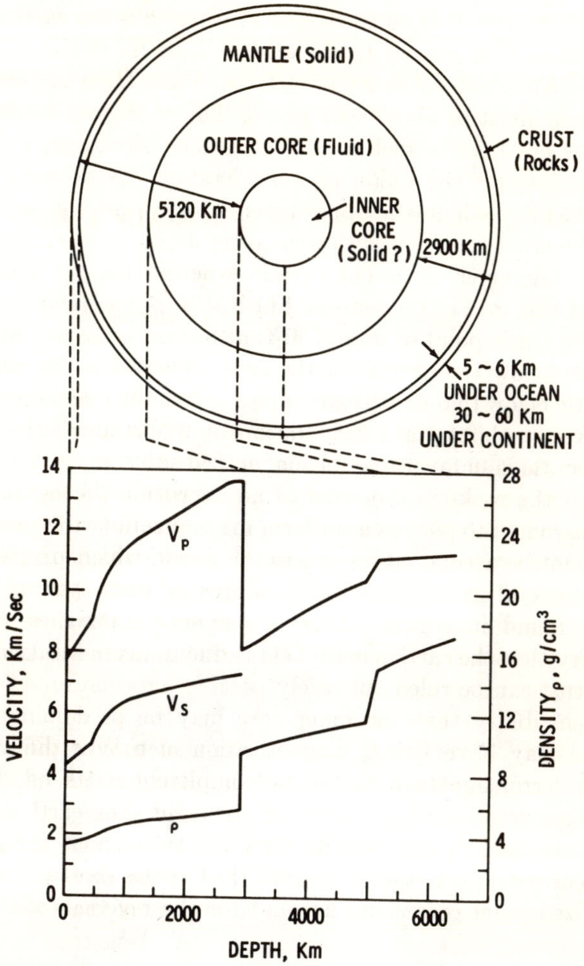

FIG. 4. The consitution of the earth's interior, and the distributions of the density and the seismic velocities for P and S waves.

Possibility of Permanent Magnetization

We now return to the question of where and what the sources of the earth's main magnetic field are. If the earth were a uniformly magnetized sphere, its specific magnetization J must be such that the magnetic moment M is equal to $\frac{4}{3}\pi a^3 J$, where a is the earth's radius. The moment M is related to the surface field H_0 at the equator by equation (5). Hence J must be $3H/4\pi$. Since H_0 is approximately 0.3 gauss, J must be about 0.07 gauss. The average material in the crust does not possess such a magnetization. Moreover, we have to take into account the following consideration. As determined from measurements made in the upper layers of the crust the temperature within the crust increases with increasing depth at the rate of about 20°C per kilometer. It is not certain whether the temperature increases at this rate in the mantle, but if it does, the temperature will exceed the Curie point of iron (750°C at normal pressure) at a depth not much below the bottom of the crust. Therefore, the material in the mantle below that depth cannot be appreciably magnetic. Furthermore, even without the consideration of temperature the material in the mantle is unfavorable for magnetization.

Then if the rocks in the crust of an average thickness of, say, 35 km are assumed to possess a uniform magnetization, the earth's main field cannot be accounted for unless the specific magnetization in the crust is several gauss. This value is greater than that for iron ore deposits found in regions of great magnetic anomalies. Thus the possibility that the earth's main field is due to magnetization of rocks in the crust can be ruled out safely.

The possibility that the inner core may be predominantly solid iron and may have strong magnetization of a type different from ordinary ferromagnetism cannot be completely ruled out. However, if the interpretation of archaeomagnetism that the earth's magnetic field has repeatedly reversed its direction throughout the geological ages is correct, it is not obvious that the hypothesis of a permanently magnetized inner core could account for the reversals of the earth's field.

Blackett's Theory

In 1947 Blackett put forward a hypothesis that a massive rotating body possesses a magnetic moment which is proportional to its angular momentum, and showed that this relation is roughly obeyed by the magnetic star 78 Virginis, the sun and the earth.

However, more recent observations of apparent reversals of stellar magnetic fields and the reversals of the earth's field mentioned above make it difficult to accept Blackett's hypothesis. Moreover, the direct measurements of the magnetic field in mines do not show a decrease with increasing depth as predicted by this theory.

Dynamo Theory of the Earth's Main Field

We saw that at the depth of 2900 km the physical properties of the material exhibits a sudden change. This discontinuity represents the boundary between the non-metallic solid material in the mantle and the metallic fluid constituents in the core. Thus it is expected that in the core both fluid motions and electric currents may exist, interacting strongly with each other. The cause for the fluid motions in the core must be mainly thermodynamical. If radioactive sources of heat are slightly concentrated in the central part of the core, the heat generated there will give rise to convective motions of the fluid in the outer core.

First let us consider a freely decaying system of currents in a spherical core without any internal fluid motions. It can be shown then that there are two different types of free modes. In one type, referred to as the *magnetic modes*, the electric currents are zonal. That is, the currents flow along circles of latitude about the axis, and the magnetic field takes the form of a dipole, a quadrupole, etc., depending on the mode; the magnetic field lines in these modes lie in the meridional planes (see Figure 5). A magnetic field of this type is called a *poloidal* field. In the other type of free modes, referred to as the *electric modes*, the configurations of the currents and the field lines in the former type are exchanged. Namely, the magnetic field lines are zonal; and the currents flow in the meridional planes (and are, of course, confined to the sphere). A magnetic field of this type is called a *toroidal* field. The current and field configurations for these free modes are illustrated in Figure 5.

The mean life of a system of an exponentially decaying free mode can readily be estimated. The mean life for the magnetic dipole mode appropriate for the earth's core is found to be of the order of 15,000 years. This is much shorter than the age of the earth, which is thought to be of the order of 10^9 years. Values of the mean life for higher modes are shorter than that for the dipole mode. Thus we can conclude that the earth's main-field cannot be a remnant of currents that might have been set up at the time when the earth was created.

If there is a fluid motion with velocity $\mathbf{v}$ in the presence of an electric

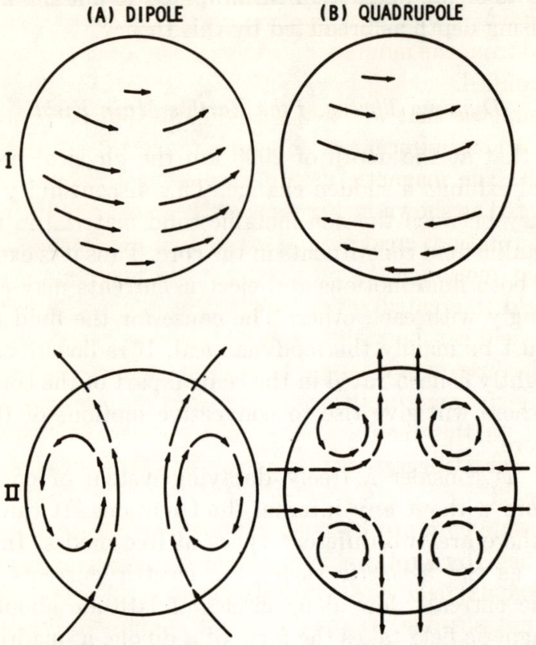

FIG. 5. Free modes of magnetic field and current in a sphere. For the magnetic mode the current is represented by full lines (I) and the poloidal magnetic field by broken lines (II) for dipole (A) and quadrupole (B) modes. For the electric mode the current is represented by broken lines (II) that are confined to the interior of the sphere and the toroidal magnetic field by full lines (I). (W. M. Elsasser, Rev. Mod. Phys. **22**, 1, 1950).

field **E** and a magnetic field **B**, the current **J** is given by

$$\mathbf{J} = \sigma(\mathbf{E} + c^{-1}\mathbf{v} \times \mathbf{B}) \tag{10}$$

where σ is the conductivity. The motion across the magnetic field induces an electric current which is represented by the second term in equation (10). The magnetic field **B** then exerts the ponderomotive force $J \times B$ on the fluid and modifies the fluid motion.

In an infinitely conducting medium (i.e., when $\sigma = \infty$) the content of the parentheses in equation (10) must be zero in order that **J** be finite; hence **E** is equal to $-c^{-1}\mathbf{v} \times \mathbf{B}$. Then it can be shown that the magnetic field lines are dragged by the fluid as if they are *frozen* into the fluid. Under such a circumstance, energy can readily be transferred from the moving fluid to the magnetic field and the latter can be amplified.

Let us suppose that we have a dipole magnetic field and that the core rotates non-uniformly with angular velocity of rotation varying with depth. The magnetic field lines passing through the core then will be twisted as shown in Figure 6, and eventually will wrap around the axis of rotation, creating a toroidal field in the core.

If a fluid motion is found whereby the magnetic energy in the toroidal field so created is fed back to the original dipole field, then the dipole field can be maintained by drawing energy from the fluid motion. Because of an analogy between such a mechanism and a self-exciting dynamo, a theory of the earth's, or a stellar, magnetic field based on this mechanism is called a dynamo theory. Such a theory was first suggested by Larmor in 1919 to explain the sun's magnetic field.

It is possible by a process such as shown in Figure 6 to create from a poloidal field, by a rotational motion, a toroidal field that is stronger than the poloidal field with which we started. However, the reciprocal process is not possible. That is, it is not possible to create a poloidal

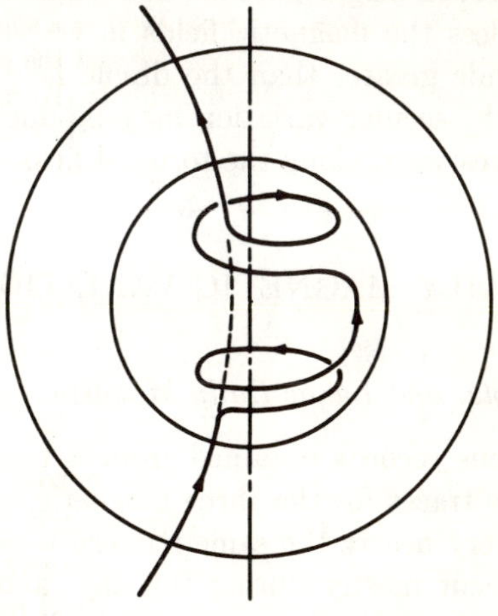

Fig. 6. A possible mechanism for creating a toroidal magnetic field from a poloidal magnetic field by a rotational fluid motion, (W. M. Elsasser, Rev. Mod. Phys. 22, 1, 1950).

field from a toroidal field by any axially symmetric rotational motions. Bullard has found a simple self-sustaining mechanism which involves only two types of motion: one is a non-uniform rotation about the axis and the other consists of two rising flows from the center to two diametrically opposite points on the equator and two inwardly sinking flows from two points on the equator such that the rising and sinking flows are perpendicular to each other on the equatorial plane.

Subsequently, several more complicated dynamo models have been proposed. To represent the actual magnetic field of the earth, a model must be highly stable. Nevertheless, the system must be such that the direction of the field is reversed when a perturbation reaches a certain level. The latter condition is required if one is to explain the archaeomagnetic evidence of reversals of the earth's field.

As was mentioned on page 17, one of the remarkable features of the secular variation is its westward drift at a rate of a few tenths of a degree per year. From such rapid movements of the isoporic foci the secular variation must be closely related to the motions in the core. The westward drift has been interpreted as an indication that the outer core is moving more slowly than the mantle.

There have been attempts made to explain the secular variation by the electromagnetic induction caused by local eddies in the core. However, the observed magnitudes of the secular variation cannot be accounted for unless the magnetic fields in the core are about two orders of magnitude greater than the dipole field. It has been suggested also that the secular variation may be due to hydromagnetic waves traveling westward along the toroidal field in the core.

DAILY MAGNETIC VARIATIONS

The Solar and Lunar Daily Magnetic Variations

When continuous records obtained from a magnetic observatory are examined, the traces for the three magnetic elements show variations in which very nearly the same pattern is repeated every day. The variations occur mostly during the day, and, if there are any during the night, they are of very small amplitude. The pattern of the variation in each element changes systematically with latitude. When the variations in the three elements observed in all parts of the world are collectively studied it becomes clear that they represent a magnetic variation of a global scope. Since the variation is controlled by the sun and is the representative variation on a magnetically quiet day, it is called the *solar quiet daily magnetic variation* and is denoted by the symbol *Sq*, with *S* for the sun and *q* for quiet. Typical magnetic records on a quiet day obtained at Huancayo, Honolulu, and Fredericksburg are shown in Figure 7.

When a long series of hourly values of the magnetic elements is available at an observatory, it is possible by a statistical method to detect another regular variation that has a period of one half of the lunar day and changes systematically with the lunar phase. This variation is called the lunar magnetic variation and is denoted by *L*. The amplitude of *L* is very much smaller than that of *Sq*, and the length of the lunar day is nearly equal to that of the solar day, the former being longer than the latter by only about 50 minutes. These two factors make it difficult to determine *L*, and a careful mathematical analysis is required to obtain this variation. Unlike *Sq*, *L* is not visible in the magnetograms except at the dip equator where *L*, as well as *Sq*, is abnormally large. This special feature at the dip equator will be discussed later. Though its periodicity is governed by the moon and though its pattern changes with the lunar phase, the main variations in *L* take place during the hours of sunlight, and *L* exhibits considerable seasonal changes, indicating that *L* is also controlled by the sun.

On pages 13–16, we described a spherical harmonic analysis of the earth's main field. Applying the same method of analysis to magnetic data from a network of observatories distributed over the world, a magnetic variation such as *Sq* or *L* can be analyzed into two parts: the part produced by sources above the surface of the earth and the part due to sources within the earth. When such an analysis is made of *Sq* and *L*, about two-thirds of the variations is found to be due to sources external to the earth and about one-third to internal sources. When the earth is in a magnetic field varying with time, electric currents are induced in the earth because the earth is a good conductor. The part of *Sq* or *L* of internal origin can be explained as the magnetic field produced by the electric current induced in the earth by the varying magnetic field of external origin.

After examining various possibilities for the mechanism to produce the daily magnetic variation Balfour Stewart concluded in 1882

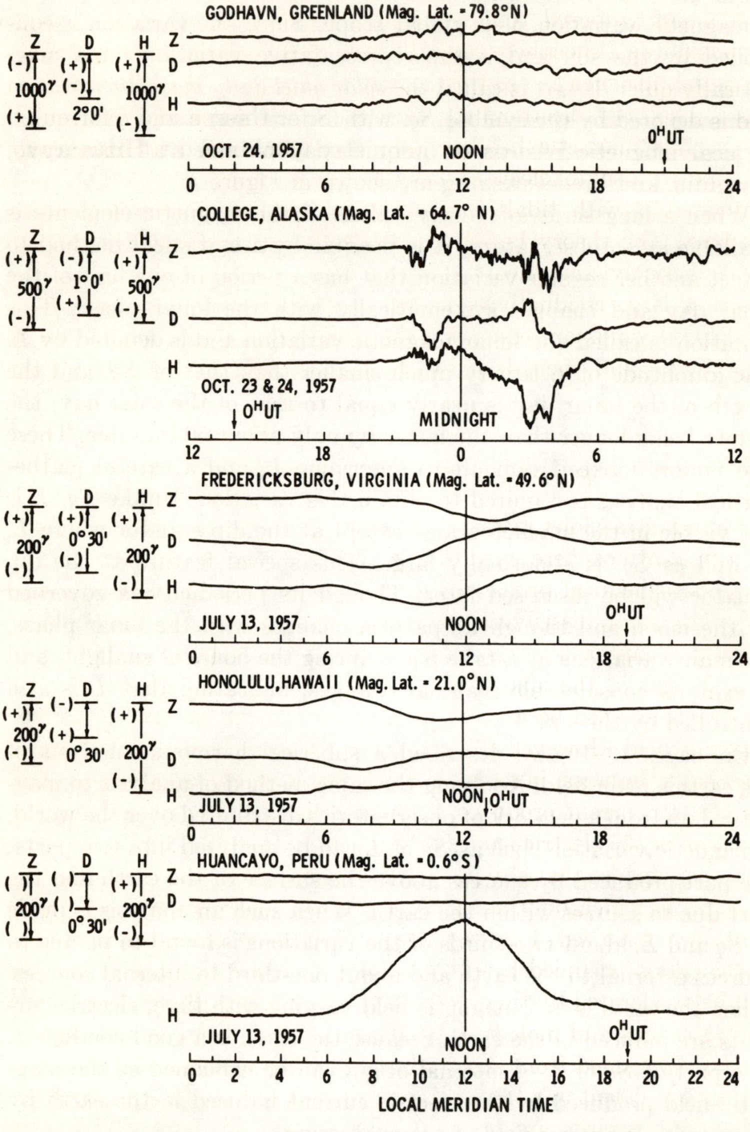

FIG. 7. Magnetograms obtained at different magnetic latitudes on an average quiet day. The scales and signs for *H*, *D*, and *Z* are indicated to the left; *H*, *D*, and *Z* are measured positively northward, eastward, and downward, respectively.

that it must be due to an electric current flowing in the upper atmosphere, and proposed that convective motions of air in the upper regions of the atmosphere caused by the solar heating are responsible for generating such an electric current. He supposed that these regions were sufficiently electrically conducting so that air motions across the earth's magnetic field would induce electric currents. He attributed L also to electric currents in the upper atmosphere and associated it with tidal action without indicating the mechanism involved. The theory based on Stewart's idea is referred to as the atmospheric dynamo theory of Sq (or L). In the atmospheric dynamo theory a large conductivity in the ionosphere and the existence of high speed winds are essential. We will discuss these two factors in some detail and then describe the electric current systems for Sq and L.

Electric conductivity in the ionosphere

X- and ultra-violet rays from the sun ionize the constituents of the the upper atmosphere, resulting in the formation of the ionosphere[3]. The photochemical processes that take place in the ionosphere and the consequent distribution of electrons and ions are discussed in detail in Chapter 3.

If the gas in the ionospheric heights is set in motion, an electric current is induced because of the presence of the earth's permanent magnetic field. However, the manner in which the current is produced is complex, because, in an ionized gas permeated by a magnetic field, the electric current produced by an electric field does not flow in the direction of the electric field.

Let us suppose that a uniform electric field of intensity E is applied to a gas in which only a small fraction of the constituent atoms or molecules is ionized. First we assume that there is no magnetic field in the region occupied by the gas. The electrons and ions in the gas will be accelerated by the electric force, but they will collide with the neutral atoms and molecules in the gas and in so doing lose part of the energy gained from the electric field. The effect of the collisions is thus equivalent to a frictional force acting on the charged particles. When due regards are given to this collisional effect the current density i can be expressed as follows:

$$i = \sigma_0 E, \tag{11}$$

where

$$\sigma_0 = Ne^2[(m_i\nu_i)^{-1} + (m_e\nu_e)^{-1}] \tag{12}$$

Here the subscript, i or e, refers to the ions or electrons, respectively. The constant σ_0 relating i to E is called the conductivity. To distinguish σ_0 from other elements of conductivity discussed later, σ_0 is often referred to as the *direct* conductivity.

When a magnetic field is present, the matter is considerably more complex than it is in the absence of the magnetic field. The reason for this is as follows: if a charge moves in the direction parallel to the magnetic field, the magnetic field exerts no force on the charge. However, if the charge moves in a plane perpendicular to the magnetic field, which we assume to be uniform in space, the magnetic field exerts a force on the charge in the direction perpendicular to both the velocity of the particle and the magnetic field.

When this force, called the *Lorentz force*, is balanced by the centrifugal force, the charge moves in a circular orbit with radius mvc/eB and with frequency eB/mc radians per second or $eB/2\pi mc$ cycles per second, where m is the particle mass, e the magnitude of the charge, v the velocity, c the velocity of light, and B the magnetic induction. This frequency, termed the *gyrofrequency* or the *cyclotron frequency*, plays an important role in the electrodynamical behavior of an ionized gas in the presence of a magnetic field. When the motion of a charge has components both parallel and perpendicular to the uniform magnetic field, the particle spirals about a line of magnetic force along a helical path, with the component of its velocity parallel to the magnetic field unaffected by the presence of the field. It is this gyration of charges about the magnetic field that introduces complexity in the relation between the electric field and the current.

Let us now suppose that an ionized gas is in both an electric field and a magnetic field. If the electric field is parallel to the magnetic field, the motions of the charges will be parallel to both the electric field and the magnetic field, and the latter has no effect on the charges. Therefore the current is proportional to the electric field, and the appropriate conductivity is σ_0.

We next consider the case when the electric field and the magnetic field are mutually orthogonal. Let us assume that the electric field is in the direction of the x-axis, and the magnetic field is in the z-axis of a right-handed orthogonal coordinate system (see Figure 8). The

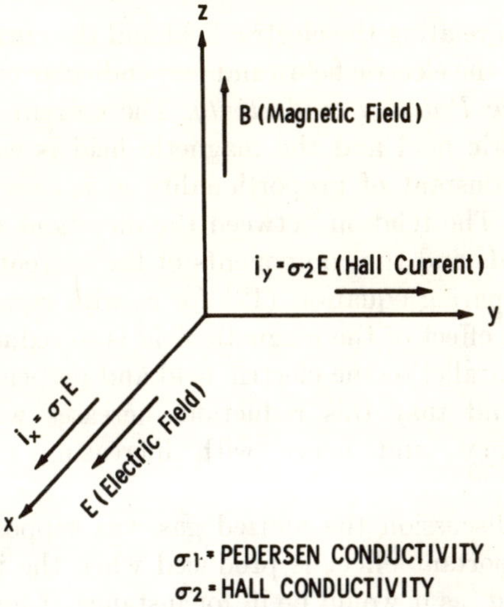

FIG. 8. Illustrating electric currents driven by a uniform electric field E perpendicular to a uniform magnetic field B.

component of motion along the x-axis produces a Lorentz force in the y-direction which, in equilibrium, must be balanced by the frictional force. Parallel to the x-axis, the frictional force is balanced by the combination of the electric force and the Lorentz force due to the motion parallel to the y-axis. These relations give the mean velocities in the x and y directions and hence enable us to derive the expressions for the components of the electric current density i_x and i_y:

$$i_x = \sigma_1 E \tag{13}$$

$$i_y = \sigma_2 E \tag{14}$$

where

$$\sigma_1 = Ne^2\{(m_i\nu_i)^{-1}[\nu_i^2/(\nu_i^2 + \omega_i^2)] + (m_e\nu_e)^{-1}[\nu_e^2/(\nu_e^2 + \omega_e^2)]\} \tag{15}$$

$$\sigma_2 = Ne^2\{-(m_i\nu_i)^{-1}[\nu_i\omega_i/(\nu_i^2 + \omega_i^2)] + (m_e\nu_e)^{-1}[\nu_e\omega_e/(\nu_e^2 + \omega_e^2)]\} \tag{16}$$

Here ω_i and ω_e are the cyclotron frequencies (in angular measure) for the ion and the electron, respectively.

The constant σ_1 relating the electric field and the current i_x flowing in the direction of the electric field (and perpendicular to the magnetic field) is called the *Pedersen conductivity.* The current perpendicular to both the electric field and the magnetic field is called the *Hall current* and the constant of proportionality σ_2 is referred to as the *Hall conductivity.* The relation between the directions of the electric and magnetic fields and the components of the current is illustrated in Figure 8. Comparing equation (12) for σ_0 with equation (15) for σ_1 we see that the effect of the magnetic field is to reduce the current in the direction parallel to the electric field and perpendicular to the magnetic field, and that this reduction increases with increasing cyclotron frequency, and hence with increasing magnetic field intensity.

In the above discussion the ionized gas was supposed to extend infinitely. An important effect is produced when the ionized region is limited in extent, as it would be if, for instance, it forms a layer of finite thickness, extending, say, parallel to the x–z plane in the configuration shown in Figure 8.

The motions of the electrons and ions contributing to the Hall current are in the same direction; in Figure 8 the electrons and ions both move in the direction of the negative y-axis. If the electrons move, on the average, faster than the ions, as they will in general, the net current is in the direction of the positive y-axis as indicated in Figure 8. Due to the difference in the mean velocities for the electrons and ions, the boundaries of the ionized layer cannot remain electrically neutral, and the two sides of the layer will be oppositely charged.

The electric field due to this charge separation, or the polarization of the medium as it is called, is such as to oppose the Hall current. In equilibrium the Hall current is thus completely inhibited. Under this circumstance, the polarization electric field, say E_p, produces the Pedersen current $\sigma_1 E_p$; and this current and the Hall current $\sigma_2 E$ are equal in magnitude and opposite in direction. Thus, in equilibrium, the polarization field is the ratio σ_2/σ_1 times the original electric field E. Now, due to the polarization electric field E_p there will be a second Hall current driven in the direction of the x-axis, which is $\sigma_2 E_p$, or $(\sigma_2^2/\sigma_1)E$. Hence in the x-direction the total current is the sum of the direct Pedersen current due to E and the Hall current due to the polarization field:

$$i_x = (\sigma_1 + \sigma_2^2/\sigma_1)E \equiv \sigma_3 E \tag{17}$$

where we introduced a new conductivity element $\sigma_3 (= \sigma_1 + \sigma_2^2/\sigma_1)$ called the *Cowling conductivity*. One can observe that if σ_2 is much larger than σ_1 the current in the direction of the electric field is much larger than it would be without the polarization. We will see later that this effect is very important in the ionosphere especially above the magnetic equator.

Let us now briefly examine how the elements of conductivity vary with height during the daylight hours. At low heights below 70 km, the direct conductivity σ_0 and the Pedersen conductivity σ_1 are nearly equal, and the Hall conductivity σ_2 is much smaller than σ_0 and σ_1. That is to say, at these heights the earth's magnetic field has little effect on the conduction of electric current. In the region from 90 km to 130 km, σ_2 is greater than σ_1, and σ_0 exceeds σ_2. In this region, therefore, the Cowling conductivity σ_3 is expected to be important. Above 160 km, σ_2 and σ_1 become much smaller than the region below this height, and σ_0 becomes extremely large.

Ionospheric wind system

The distributions of electrons and ions in the ionosphere are now fairly well known experimentally, and hence the values of the conductivity elements, of which σ_3 is the most relevant in the dynamo theory, can be estimated with a reasonably good accuracy. When these conductivity estimates are used it becomes evident that steady winds of velocities of the order of several tens to one hundred meters per second (i.e., roughly a few hundred kilometers per hour) are required to explain the observed magnetic variations. Such wind velocities are considerably higher than those observed near the ground, and thus a brief discussion of the upper atmospheric winds seems appropriate at this point.

The gravitational forces of the sun and the moon cause the atmosphere to oscillate with a period of one half solar or lunar day in much the same way as the tides in the ocean. In the atmosphere, however, the rapid variation of the density with height, and the solar heating, introduce features that have no counterpart in the ocean tides. In the latter the solar tide is much smaller than the lunar; whereas in the atmospheric tide the pressure variation as observed at ground level shows that this relation is reversed; namely, the solar semidiurnal component is much larger than the lunar semidiurnal component. Moreover the observed solar semidiurnal pressure variation at ground level is about 100 times greater than the value estimated on a basis of simple gravitational tide. To reconcile this difference it was thought in the early stage of the study of the atmospheric oscillation that the atmosphere might have a resonance period of 12 hours. Such a possibility was first suggested by Kelvin in 1882. In 1929 G. I. Taylor analyzed the propagation of sound waves from the great volcanic eruption at Krakatoa in 1882 and found that there indeed was a free oscillation of the atmosphere with a period of about 11 hours. But this resonance period is a little too short to be effective for the amplification of the semidiurnal tidal oscillation. In the free oscillation of the atmosphere investigated by Taylor the energy of the oscillation is confined mainly to the region below the tropopause at about 10 km height, where the atmospheric temperature is a minimum; at this height the velocity of a wave propagating upward becomes so small that it cannot escape through this *barrier* to higher levels; the wave may be thought to be trapped below the tropopause. There is another minimum in the temperature distribution at a height of about 80 km, i.e., in the region referred to as the *mesopause.* Then there must be a possibility of trapping waves below the mesopause. The period of free oscillation corresponding to this trapping of wave energy is found to be approximately 12 hours. At first sight this resonance oscillation may appear to be likely to account for the large semidiurnal pressure variation. However the resonance period critically depends on the distribution of temperature in the middle atmosphere, and it is difficult to believe that the atmosphere can be steadily tuned so exactly to the tidal period of 12 hours. Thus it is clear that a resonance amplification of the gravitational tide is not a likely mechanism for the large semidiurnal pressure variation.

However, through these investigations some important properties of atmospheric oscillation became known. Among others it was revealed that the velocity in the atmospheric oscillation increases with height, because when waves propagate upward their energy fluxes tend to be conserved. Since the density decreases rapidly with height the velocity in the oscillation increases with increasing height. It should be noted that this feature is not related to the amplification of tides discussed earlier, and that it is a general property of all atmospheric waves including planetary waves and internal gravity waves. For the atmospheric dynamo theory the discovery of this property was a great step forward, though the tidal theory itself was not successful in explaining the pressure variation.

After it was realized that the resonance mechanism based on a gravitational tide was unsatisfactory, the solar heating was considered as an alternative source of energy for the solar semidiurnal oscillation. Ozone in the middle atmosphere absorbs solar radiation and becomes an important heat source. When this heat source was taken into account it was found that the gross features of the pressure variations observed at ground level and in middle atmosphere can be explained without any amplification mechanism. The solar heating is, of course, diurnal in nature. It now appears that the large amplitude of the semidiurnal component, as compared to the diurnal component in the atmospheric pressure variation observed at ground level led the early workers to over-emphasize the importance of the *semidiurnal* aspect. However, as to the lunar semidiurnal pressure variation at ground level the atmospheric oscillation due to the lunar gravitational force seems to account for this truly semidiurnal phenomenon quite adequately. It might be thought that since the ionospheric conductivity varies diurnally, semidiurnal winds could still produce a diurnal magnetic variation. However the existing theories of the atmospheric dynamo indicate that the wind system, which is compatible with the observed magnetic variations and a reasonably realistic ionospheric conductivity distribution, contains a diurnal component as its major constituent.

There are several means of measuring wind velocities at high altitudes. The first two methods described below utilize radio techniques. As meteors fall into the upper atmosphere they leave trails of ionized gas along their trajectories. Since these ionized meteor trails can reflect radio waves, we can measure the drift velocities of the trails by transmitting radio waves and receiving the waves reflected by them. By this method the semidiurnal wind velocities and their variation with height in the ionospheric region from 80 to 100 km altitude have been measured, and the results are found to be in reasonable agreement with the theoretically expected values if the observations are averaged over a long interval of time. However, the observations show a very large seasonal variation that is not predicted by a simple theory. Another radio method of determining wind velocities at the ionospheric heights is to measure the movements of irregularities in the ionosphere. In this technique the velocities of the moving diffraction patterns are determined by a set of closely spaced receivers. However, the motions of the irregularities so measured are not the same as those of the neutral air.

In a more direct method of observing winds in the upper atmosphere a luminous vapor such as sodium is ejected from a rocket and the movements of the vapor are determined by examining the photographs of the vapor cloud. This method, however, can be used only in twilight. Experimental determination of the wind velocities in the ionospheric regions that are of particular interest to the study of the geomagnetic variations *Sq* and *L* is still in a rudimentary stage, and our knowledge on the wind system in these regions is very incomplete.

Ionospheric current systems for Sq and L

Large scale motions of the air in high altitudes are probably mainly horizontal. Such air motions in the presence of the earth's magnetic field induce an electric field which is perpendicular to both the air velocity and the magnetic field. We call this electric field the *dynamo electric field.* In the ionosphere where the elements of the electric conductivity are large, an electric current will be driven by the dynamo electric field. The horizontal current generated in the ionosphere in this way accumulates charges over some parts of the globe, creating an electrostatic field which, in turn, drives an additional current.

If the distribution of winds at the ionospheric heights is known, one can determine the system of currents produced by the combination of the dynamo electric field and the electrostatic field due to the polarization. In such a calculation due regards must be given to the elements of conductivity explained on pages 27–31. With an idealized wind distribution, a current system that approximately accounts for the magnetic variation *Sq* can be derived.

From the magnetic observations at a network of stations distributed over the earth one can also draw an equivalent current system that could produce the observed variations. An idealized current system so determined for *Sq* is sketched in Figure 9, in which representative current flow lines are shown. The pattern of the current remains stationary with respect to the sun, and the earth rotates under it once a day from left to right in the diagram.

Strong currents are mostly limited to the sunlit side of the earth, and to the region from the equator to mid-latitudes. In each hemisphere there is a vortex of current with its center at about 30° geomagnetic latitude and near the noon meridian. The current whirls about this center in the counterclockwise direction in the northern hemisphere and in the clockwise direction in the southern hemisphere. We call the vortex centers the foci of the current system. Though the current density at any one point is not so large, the total current flowing across the meridian plane between the northern and southern foci is about 120,000 amperes.

The current system shown in Figure 9 is a schematic model, and the actual current pattern contains many complex features. For instance, the focus in each hemisphere roughly follows a line of equal dip angle rather than the circle of a particular geographic latitude. The northern and southern foci are not exactly on the same meridian.

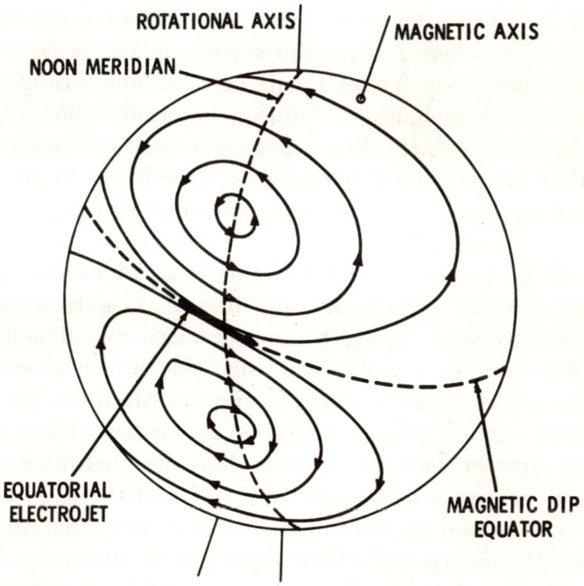

FIG. 9. *Sq* current system schematically represented. Currents flow in the direction of arrows; the thick arrow on the dip equator represents the electrojet.

Moreover, the current intensities in the summer hemisphere are greater than those in the winter hemisphere, and the current vortex in the summer hemisphere extends to the winter hemisphere crossing over the equator. In addition, the current pattern and intensity change to some degree from day to day.

It would appear that ionospheric currents causing the diurnal field variations could be detected quite simply by traversing the ionosphere with a rocket-borne magnetometer and observing the field discontinuity resulting from the reversal of the Sq field in crossing the current carrying layer. In the case of the *equatorial electrojet*, to be discussed, this proved true, and detection of the electrojet by this means was one of the earliest accomplishments of sounding rocket investigations. At middle latitude locations where the Sq currents are expected to resemble current flow in a sheet-like distribution over a wide range of latitudes the detection is not so simple and in fact it was not until 1964 that this was accomplished. The basic limitation is that the rocket instrument measures the scalar field B (vector measurements requiring prohibitively exact knowledge of the rocket orientation) whereas the Sq field is horizontal. Thus the total discontinuity in B from the vector addition is approximately twice the value $\Delta H_e \cos I$, where I is the dip angle and ΔH_e is the horizontal field of the ionospheric (i.e. external) current. The total ΔH seen at the earth's surface includes the induced (internal) field ΔH_i as well; thus $\Delta H = \Delta H_e + \Delta H_i$. From analysis of surface variations it is thought that $\Delta H_e \doteq 0.6\Delta H$. This gives a total $\Delta B \doteq 2(0.6)\Delta H \cos I$ which becomes a small quantity when the dip angle is large and ΔH is small.

Figure 10 illustrates a recent detection of an Sq current at a location where $I \doteq 70°$. The flight near 10 a.m. local time took place when ΔH due to Sq at the earth's surface was -42γ. The flight near 8 p.m. (20^h) local time took place when the surface magnetograms indicated a lack of any appreciable Sq current. The difference between B (measured) and B (computed) at 8 p.m. is believed to represent merely the error in the reference field description used for computation. The 17 gamma displacement in the case of the 10 a.m. flight is believed to be caused by ionospheric currents. The magnitude agrees very well with the expected difference of $2(0.6)\Delta H \cos I$.

Along a narrow belt of a few degrees in width over the magnetic dip equator, where the magnetic field is horizontal, the amplitude of

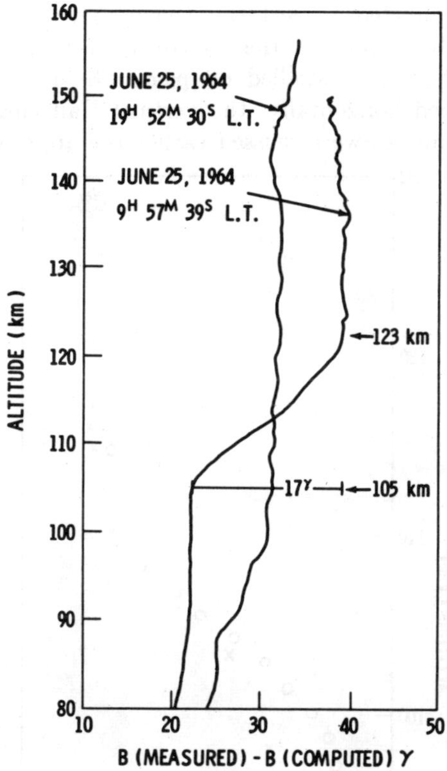

FIG. 10. B (measured) minus B (computed from a 63 coefficient field model) as a function of altitude for two rocket flights at Wallops Island, Va. near 10 and 20 hours local time (after Davis et al., J. Geomag. Geoelec., **18**, 183, 1966).

Sq in the horizontal component is abnormally large, indicating a concentrated current flow along the dip equator. By analogy with the jet stream of air flow in meteolology, this intense current is called the *equatorial electrojet.* The record from Huancayo reproduced in Figure 7 shows an example of the enhanced amplitude of *Sq* at the dip equator.

How this electrojet is created over the dip equator can be explained in the following manner. In low latitudes the current is driven mainly by the electrostatic field due to the global polarization of the dynamo region of the ionosphere, and this electric field is eastward during several hours near noon. At the dip equator the magnetic field is horizontal and directed toward the (magnetic) north. Hence the (eastward) electric field and the (northward) magnetic field constitute the configuration studied on pages 28–31. Since the ionosphere is stratified horizontally, the vertical Hall current polarizes the medium. Thus, as we discussed earlier the appropriate conduc-

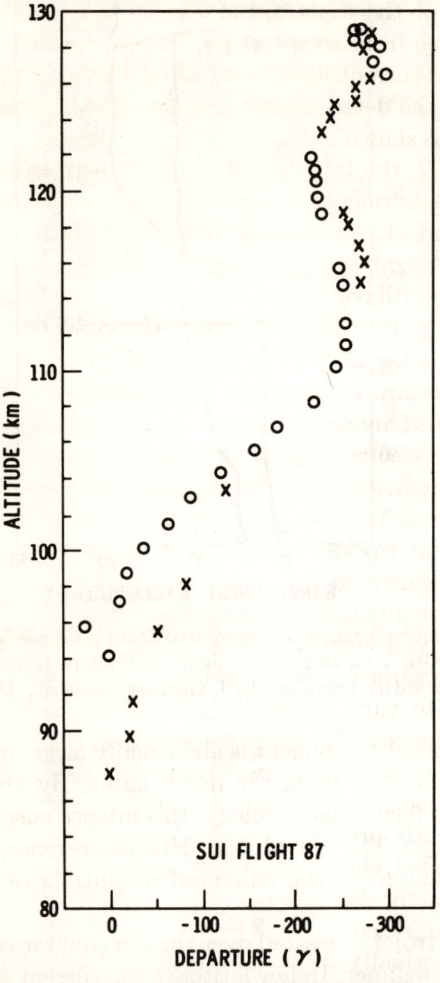

FIG. 11. Departure of the measured magnetic field from an inverse cube $(a/r)^3$ field variation with altitude for a rocket penetrating the equatorial electrojet, 09:07 LT, October 19, 1957 (After Cahill, J. Geophys. Res., **64**, 489, 1959).

tivity in the direction of the electric field (i.e. eastward) is the *Cowling conductivity* σ_3. In other words, the eastward current produced is the sum of the Pedersen current due to the original eastward electrostatic field and the Hall current due to the vertical polarization electric field.

The Cowling conductivity σ_3 is found to be extremely large within a few degrees of the dip equator and to decrease rapidly with increasing distance from the dip equator. The result is that a concentrated eastward current flows in a narrow belt along the dip equator. An example of the detection of the intense current by a rocket-borne magnetometer is shown in Figure 11. At the equator, where $\cos I = 1$, $\Delta B = 2(0.6)\Delta H$, the total effect is shown by a scalar measurement. This flight in addition to detecting the principal electrojet current between 100 and 115 km showed indications that a second maximum in the current might exist near 130 km. The second maximum may result from a multilayered electron density distribution.

An interesting effect has been observed in the equatorial electrojet region with a radar technique. It is well known in plasma physics that when there are interpenetrating streams of electrons and ions, wave-type disturbances are generated by the so-called *two-stream instability*. The electrojet, in which electrons and ions stream with great speeds, creates a favorable condition for such an instability to develop. A part of the energy of a beam of radio waves transmitted from the ground toward the electrojet is reflected back from the plasma waves generated by the instability.

So far we have discussed the ionospheric current system responsible for *Sq*. Similarly, a current system that could produce the observed lunar magnetic variation *L* can be drawn. The current system for *L* is found then to vary markedly with lunar phase as well as with season. In equinoxes the current system averaged over a lunation has four vortices, and that for new moon two major vortices, in each hemisphere. In solstitial seasons the current vortices in the summer hemisphere greatly predominate over those in the winter hemisphere. However, detailed characteristics of the current system for *L* are less certain than for *Sq*.

For the same reason as for *Sq* the amplitude of *L* over the dip equator is abnormally large. But, for some reason not well understood, the rate of amplification for *L* is larger than that for *Sq*.

The essential feature of the dynamo mechanism lies in the interaction between the winds in the neutral atmosphere and the earth's main magnetic field. It is mentioned here that motions of plasmas in the ionosphere are closely associated with the dynamics of the magnetosphere. Electrons existing above about a 90 km level and ions above about a 130 km level tend to move together with the magnetic field lines about which they gyrate; hence, motions of the field lines in the magnetosphere will drag charged particles in the ionosphere, and if adequate energy is available, through collisions between the ions and neutral particles the neutral atmosphere can also be set in motion. Conversely, due to the extremely high conductivity along the magnetic field lines, the equipotential lines in the dynamo region tend to be extended to the magnetosphere and to the opposite hemisphere along the field lines, forming equipotential surfaces. The plasma in the magnetosphere are then constrained to move parallel to the equipotential surfaces. Thus the dynamics in the ionosphere must in general be considered together with that in the magnetosphere. With respect to *Sq* the coupling between the ionosphere and the magnetosphere is weak in low latitudes, but in polar regions the dynamics in the magnetosphere may dominate over that in the ionosphere. This problem will be further discussed later (pages 50–58) in connection with polar disturbances.

Solar flare effect

Rocket measurements of solar radiation in short wavelengths have revealed that X-rays are emitted by solar flares. At the peak activity of a moderately large flare the solar emission in the entire X-ray spectrum increases to several times the normal level. These X-rays increase the ionization in the lower region of the ionosphere, causing radio fade-out. The *Sq* current is then greatly enhanced over the sunlit hemisphere, producing a magnetic change called a *solar flare effect* or a *crochet.* The existence of such an effect gives a strong support to the dynamo theory of *Sq*.

As in *Sq*, solar flare effects are augmented over the dip equator. The rate of enhancement is larger on days of large *Sq* than on days of small *Sq*. This proportionality is expected if the electrojet is driven by an eastward electric field as we discussed above; the amplitude of *Sq* is then proportional to the product of this field and the conductivity, and the latter is increased by a solar flare.

POLAR AND AURORAL DISTURBANCES

Observed Magnetic Activity: Auroral Zone

Coincident with the occurrence of visual aurora, most frequently between magnetic latitudes 63 and 72 degrees in the northern and southern hemispheres, the magnetic field variations are considerably more intense than at lower latitudes. Even during years of low solar activity magnetograms from an observatory located in the center of this zone will show some degree of disturbance during the night hours on at least 8 or 9 out of every 10 nights. The field variations at first glance may appear unsystematic and highly irregular but on further examination it is found that they follow specific patterns and that smaller scale irregularities appear to be superimposed on these patterns. On an average night a model pattern may be quite obvious whereas on some nights and even extending well into the daylight hours the disturbance will appear extremely complex and undescribable in terms of any model. However, when examined with the aid of simultaneous observations of the aurora it is found that most complex disturbances can be described in terms of the repetition of simple patterns in which a new cycle of activity begins before an existing cycle is completed.

The magnetogram shown for College, Alaska in Figure 7 would fall into the category of being a typical auroral zone magnetogram. This type of disturbance is often referred to as a *polar elementary storm* or as *magnetic bay activity.* The use of the term magnetic bay avoids possible confusion and is generally preferred. The disturbance is most readily described in terms of the variations of the horizontal component and accordingly a positive bay is a period when the magnetogram shows a $+\Delta H$ variation (e.g., near 23^h in Figure 7) and a negative bay corresponds to a $-\Delta H$ period (e.g., between 2^h and 5^h in Figure 7 at College, Alaska). The $-\Delta H$ deviations labeled A, B, C and D in Figure 12, to be discussed later, provide additional examples of what is meant by the term *negative bay.*

There is considerable observational evidence for attributing the principal parts of the $+\Delta H$ and $-\Delta H$ variations, respectively, to west to east and east to west electric currents in the ionosphere. Observations also strongly suggest, but by no means absolutely prove, that

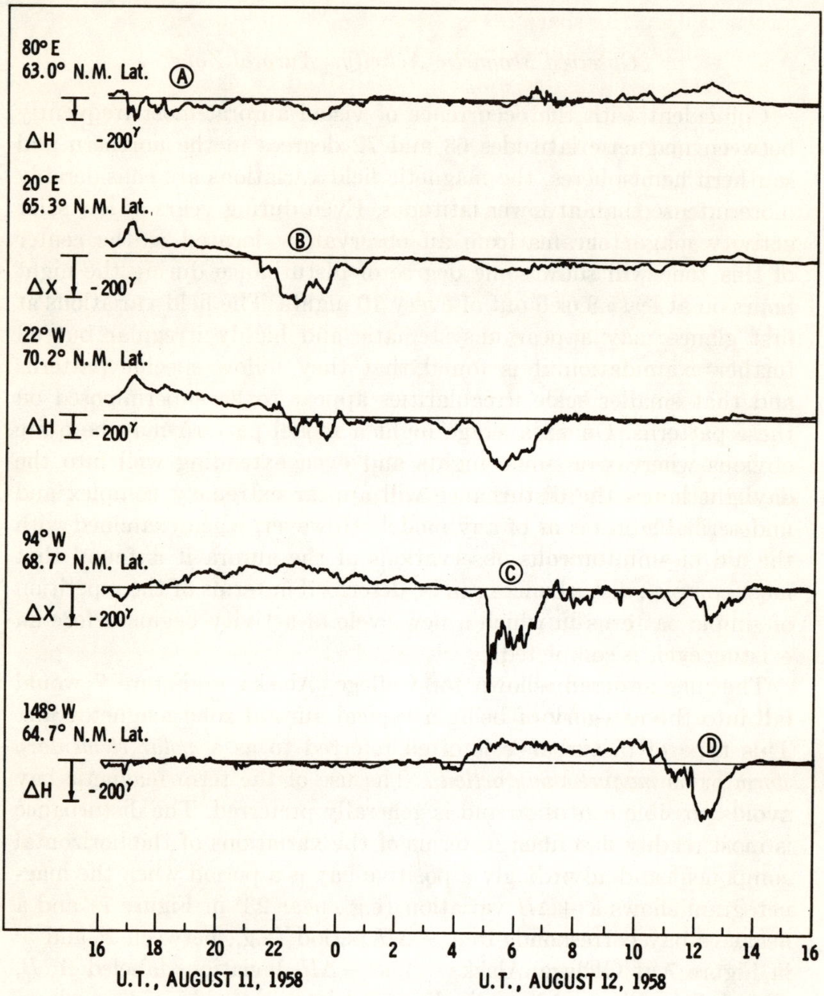

FIG. 12. Simultaneous recordings of the horizontal (H) or north (X) component from five observatories distributed in longitude along or near the auroral zone. A, B, C, and D identify principal negative bays occurring within the 24 hours shown.

the currents tend to be filamentary and are more concentrated and intense within the space occupied by visible aurora than they are in regions slightly removed from the visible aurora. Thus, in the case of horizontally elongated auroral forms such as arcs which frequently extend several thousand or more kilometers in the east–west direction, there is good reason to believe that the electric current follows the same path.

The most distinctive characteristic of auroral zone magnetic disturbances (and the feature which probably contains the principal clue to the mechanisms creating aurora and to magnetosphere dynamics) is the time sequence of the ΔH variation and its coincidence with a time sequence of distinct auroral forms and motions (see Chapter 5 for descriptions of auroral forms). The sequence of events as seen from a single station in the auroral zone is illustrated in Figure 13. Using simple logic one can attempt to translate the sequence as seen by a single station to the behavior that would be expected throughout 360 degrees in longitude or 24 hours in time. The clue to doing this is in recognizing that the reversal of current with the breakup of auroral arcs is an event. This event is observed at a single station within several hours of midnight and its occurrence is usually evident within a matter of minutes over more than 30 degrees in longitude. (Note: taking a number of nights these events are statistically grouped about *geomagnetic midnight* which will deviate within ± 2 hours of local midnight. For discussion this difference is ignored here.) Thus, if each station around 360° of longitude sees one major negative bay per night, there must be a finite number of events per 24 hours of universal time. For example, if the local time range of events is 3 hours there would be 8 per day, 4 hours gives 6 per day, 6 hours gives 4 per day, etc. Correspondingly there must be a finite number of magnetic bays per universal day, or alternatively, we can say that there must be a finite number of cycles per universal day of the type diagramed in Figure 13. An example of the validity of this logic is shown in Figure 12 which indicates that on this particular universal day there were four cycles of activity. This assumes that the gap between 80°E longitude and 148°W longitude has not caused the omission of a fifth cycle.

Figure 12 also gives some indication of the longitudinal extent of a magnetic bay. For example, each of the negative bays labelled *B* and *C*, which are bracketed by observatories both to the east and to

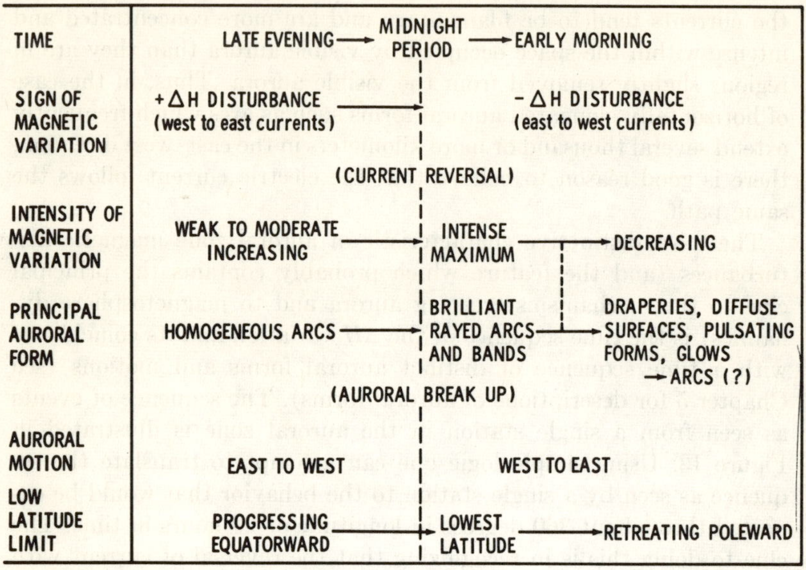

Fig. 13. Typical time sequence for one cycle of magnetic bay disturbance.

the west in this figure, appear as recognizable $-\Delta H$ variations on three of the five traces. Thus, bays B and C, respectively, extended at least 102 and 114 degrees in longitude. The intensity and sharpness of the onset of a negative bay as seen by a given station depends on the proximity of the station to the exact location where the event originated. Thus maximum intensity occurs at the observatory nearest midnight in a distribution like Figure 12. The existence of a $+\Delta H$, positive bay, on the evening side of the earth accompanying each negative bay is also evident in Figure 12. The corresponding electric currents often referred to as *auroral electrojets* are represented by the heavy black arrows in Figure 14. A diagram such as Figure 14 is necessarily an over-simplification of the true currents, especially at the time of an event in the region of current reversal.

Considering now the highly complex magnetic variations that occur during a period of intense activity, often but not always accompanying a world-wide magnetic storm, we find that the basic cycle of Figure 13 appears to remain valid if we assume that a new cycle can begin before an existing cycle is complete. Each cycle is also likely to be

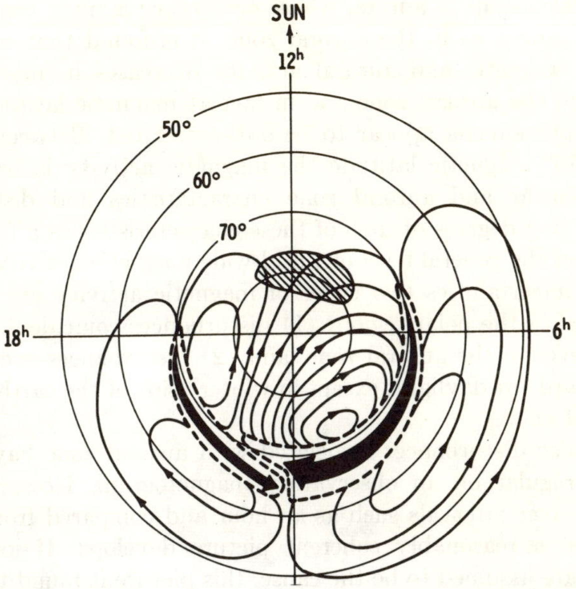

FIG. 14. Illustrative pattern of ionospheric electric currents causing magnetic bay disturbances in the auroral zone and associated currents at higher and lower altitudes. The shaded area near 80 degrees magnetic latitude locates approximately a region of enhanced daytime field fluctuations.

more intense and the low latitude limit of strong currents and visible aurora is moved further toward the equator. The magnetogram from College, Alaska during the magnetic storm shown in Figure 18 illustrates the large excursions in the field which are characteristic of storm conditions. The gross pattern is also seen to bear resemblance to that of a more average day such as shown in Figure 7. Frequently during a magnetic storm there will be several large negative bays of the type seen in Figure 18 during the post-midnight, early morning, hours.

Observed Magnetic Activity: Polar Cap

Unlike the auroral zone where the detailed correlation of magnetic activity and the behavior of aurora aids description in terms of simple patterns, the magnetic activity in polar regions is less readily classified.

In its relationship to aurora, whose occurrence is quite common but not so frequent as in the auroral zone, it is found that correlation between magnetic and auroral activity decreases in moving poleward from the auroral zone and in fact at magnetic latitudes $>80°$ the two phenomena appear to be anti-correlated. Between roughly 70° and 80° magnetic latitude the magnetic activity is transitional between polar and auroral zone characteristics and disturbances show varying degrees of each of these characteristics as a function of season and the general level of world-wide magnetic activity. Subject to these uncertainties two types of magnetic activity are generally recognized in the polar region: (1) disturbances coincident with bay disturbances in the auroral zone, and (2) disturbances occurring on the sunward (or daylight, except in winter) side of the earth centered near local noon.

The polar disturbances associated with auroral zone bays appear highly irregular on an observatory magnetogram. However, when averaged over intervals such as an hour and compared from station to station a reasonably coherent picture develops. If ionospheric currents are assumed to be the cause, this picture is found to be consistent with an electric current directed primarily across the polar cap from the night side to the day side as illustrated schematically in Figure 14. Proof of the existence of such ionospheric currents is not so substantial as in the case of auroral electrojets. However, at least one rocket-borne magnetometer has been flown into the polar cap ionosphere and a discontinuity in the field variation was seen. A more general support for the existence of these currents comes from attempts to analyze the distribution of current required to explain the magnetic variations. This process, although subject to limitations imposed by the lack of an adequate distribution of observatories, usually produces the result that the total integrated current across the polar cap is consistent with assuming that the auroral electrojet circuit is completed largely by this polar cap ionospheric current and by less intense return currents at sub-auroral latitudes such as those drawn in Figure 14. The uncertainties, however, are such that one should probably allow for an error of factor of 2 in this continuity argument. Although representations of the polar cap current, such as Figure 14, suggest that the current over the polar cap is a relatively uniform horizontal current, the large variations seen in the vertical component of the field tell us that the current must frequently be asymmetrically distributed and/or locally concentrated in completing the electrojet circuit. Alternatively, there may be mechanisms other than those directly associated with the auroral electrojets that are important in driving currents across the polar cap.

The second type of polar cap disturbance is most obvious at times when the general level of world-wide and auroral zone activity is low. During these periods there remains a region near 80 degrees magnetic latitude on the day side of the earth where fluctuations of the magnetic field are usually still observed. The central portion of this region is indicated by the shaded area in Figure 14. The magnetogram for Godhavn, Greenland in Figure 7 illustrates this type of disturbance centered near midday. Various terms such as *region of high latitude field agitation*, and *area of confusion*, have been used in referring to this activity. Studies have not revealed whether or not the fluctuations could reasonably be attributed to ionospheric currents. To do this on the basis of continuity of current would require a density of observatories that does not exist at the appropriate latitude. However it is well known that the amplitude of the fluctuations is a marked function of season with a maximum in summer and a minimum in winter.

Early Studies of Birkeland and Störmer

In 1896 Birkeland discovered that a magnetic pole has an influence on a stream of cathode-ray electrons analogous to that of a lens on light, and based on this finding he supposed that the aurora is due to electrons emitted by the sun. (The idea that the aurora is produced by high energy electrons had been proposed in 1894 by the Danish meteorologist, A. Poulsen, who, however, supposed the electrons to have originated in the atmosphere rather than in the sun.) Soon after Birkeland's experiment was published Poincaré showed mathematically that each trajectory of an electron describes a geodetic line on a conical surface of revolution with its apex at the pole and that this magnetic deflection causes no change in the energy of the electron. To test his theory of the aurora Birkeland conducted the classical experiment in which a *terrella*—a spherical electro-magnet with a thin brass crust coated with a phosphorescent material—was exposed to a stream of electrons. He observed that a narrow phosphorescent belt was formed around each of the two poles as if the polar aurora was simulated.

He also made an extensive analysis of the magnetic field changes observed in the auroral zone and recognized the occurrence of positive ($+\Delta H$) and negative ($-\Delta H$) bays which he called *polar elementary storms*. He supposed that the polar elementary storms were produced by currents flowing along the auroral zone, and that those currents consisted of moving electrons originating from the sun. To explain the zonal current Birkeland proposed the following model. Electrons streaming in interplanetary space follow the magnetic field lines leading to the auroral zone as his terrella experiment suggested. Each electron follows a spiral path of decreasing radius while descending, and upon reaching a certain altitude it recedes from the earth along a similar spiral path to interplanetary space; near the turning point the electron is supposed to have a small component of horizontal movement along the auroral zone. When a large number of electrons take such trajectories over a wide range of longitude, the vertical components of current (which is opposite to the direction of the motion of the electrons) nearly cancel one another, leaving a horizontal current along the auroral zone (eastward in the afternoon hemisphere and westward in the morning hemisphere). There will be a net intake of current near the noon meridian and a net outflow near the midnight meridian.

Birkeland's experiments and his speculative theory of the polar magnetic disturbances were followed by Störmer's extensive mathematical analysis of trajectories of a charged particle in the magnetic field of a dipole. Störmer's theory has been found to be most valuable in the study of the behavior of cosmic rays, but a certain aspect of his studies has led to the development of a powerful approximate method (guiding center approximation) of calculating trajectories of charged particles in a magnetic field. Investigations made by Birkeland and Störmer may be said to mark the starting point of the later researches of the polar disturbance.

Electric Fields for the Polar Disturbance

In Birkeland's theory the electrons, ejected from the sun, themselves constituted the currents in the upper atmosphere that are responsible for positive and negative bays. However, the collisions between the precipitating electrons and the atmospheric molecules would make it impossible for the majority of the electrons to return to high altitudes as was supposed in Birkeland's theory unless the height of the horizontal current-flow is much above the auroral heights. The currents for the magnetic disturbance need not be the currents due to the motions of the electrons from the sun. Conduction currents will do as well. The problem is then to find the electric fields driving the required currents. All the theories proposed in recent years are of this type, and they differ from each other in their mechanisms for creating the electric fields.

In the theory put forward by Alfvén an electric field was conceived as an inherent property of a neutral stream of charged particles emitted by the sun. The stream is supposed to be electrically polarized due to its motion in the solar magnetic field, the direction of which was assumed to be transverse to the direction of the streaming velocity of the particles. Alfvén's theory then concerns the behavior of charged particles in the combined effect of this polarization electric field and the earth's magnetic field. At the time the theory was proposed the solar magnetic field in interplanetary space was thought to be in a dipole configuration. However, the recent magnetic measurements in space have revealed that the magnetic fields in interplanetary space are directed either parallel or anti-parallel to the streaming direction of solar wind particles. Since the kinetic energy density in the solar wind is greater than the magnetic field energy density, the plasma flow dominates over the magnetic fields. Thus, the reality of the polarization electric field in the stream assumed in Alfvén's theory is doubtful. However, the idea of an electric field playing an essential role is of considerable significance.

The recent experimental confirmations of the steady solar wind and the magnetosphere boundary have added a new feature that must be taken into account in a theory of magnetic disturbance. A steady state problem of determining the shape of the magnetosphere boundary in a uniform flow of solar plasma has been solved by various approximate numerical methods. An ideal theory of magnetic disturbances must deal with a time-dependent problem in which the solar plasma flow is non-uniform. However, such a problem is not tractable even by approximate methods because the problem will involve a diversity of interrelated processes which themselves are not amenable to complete theoretical treatments. Thus, the problem has been divided into various parts and phases in the hope that they can be ultimately combined to give a complete picture. Since the primary energy source of the polar magnetic disturbance lies in the solar wind, the transfer of energy, with or without a direct transport of plasmas themselves, from the solar wind to the magnetosphere comprises an important aspect of the modern theories of the magnetic disturbance. This subject will be discussed below (pages 55–58).

Recognizing the difficulty mentioned above, one may start from the other end of the problem and try to construct an electric field distribution that would give rise to the observed features of the magnetic disturbance; finding a mechanism that would produce such an electric field configuration follows as a next step. In this approach the electric fields are in general treated as being static. It should be kept in mind that this is done only for the sake of simplicity and that the actual problem is one of dynamic, rather than static, nature, involving energy exchanges between electromagnetic fields and charged particles. With this in mind, however, we pursue the study of an electrostatic field configuration that would produce a steady current system for the polar magnetic disturbance of negative and positive bays. For this purpose we will first discuss the relevant conductivity elements in the ionosphere where the disturbance current flows.

Earlier in this chapter (pages 27–31) we discussed the direct (σ_0), Pedersen (σ_1), Hall (σ_2), and Cowling (σ_3) conductivities. It has been found that the auroral luminosity is proportional to the square of the maximum electron density, N^2. Using the average height-distribution of the luminosity the height of this maximum is between 100 km and 130 km. In this range of altitude σ_1 and σ_2 are of the same order of magnitude and σ_2 slightly exceeds σ_1. Because of the large magnetic inclination in the auroral zone the vertical polarization field due to the vertical Hall current is not as important as in lower latitudes (see pages 36–39) and we can take σ_1 and σ_2 approximately as the relevant conductivities when considered appropriately relative to the directions of the magnetic field B (taken to be vertical in this approximation) and the electric field E. It should be recognized that given a distribution of E the current pattern depends on the relative significance of the Pedersen current (in the direction of E) and the Hall current (in the direction transverse to both B and E). For the current-flow extending to lower latitudes the effect of the vertical polarization also must be taken into account. Conversely, if a current system is given, the configuration of E depends on the relative magnitudes of σ_1 and σ_2. Above the 100 km level the ratio σ_2/σ_1 is highly sensitive to the ratio of the frequency of collisions between the ion and the neutral molecules to the ion cyclotron frequency. Above the 130 km level σ_1 is greater than σ_2. In the past the dominance of the Hall current over the Pedersen current in the auroral electrojets has been almost invariably assumed. However if we allow for uncertainty in the atmospheric model on the basis of which the ion-neutral collision frequency is computed, and if we also recognize that the aurora itself varies greatly in altitude, then the problem appears to require careful re-assessment. The relative significance of the Pedersen and Hall currents over the polar cap has not as yet been critically examined.

So far horizontal electric fields alone have been considered and no mention has been made of electric fields along the magnetic field. The direct conductivity σ_0 along the magnetic field becomes very large above the ionosphere, and for practical purposes σ_0 can be taken to be infinite. The magnetic field lines then can be considered to be approximately equipotential lines of the electric field. Highly correlated magnetic (and other) disturbances at conjugate points give a strong support to the idea that electric fields in the magnetosphere are transported along the magnetic field lines to the northern and southern auroral zones. An example of such correlated disturbances is given in Figure 16. Correlations at conjugate areas are not limited to the major features of disturbance but often appear even in minor details. If the magnetic field lines are assumed to be equipotential lines of the electric field, and if the ionospheric current is assumed to be a Hall current (corresponding to the condition $\sigma_2 \gg \sigma_1$) that flows in the $-\mathbf{E}\times\mathbf{B}$ direction (Figure 8), the radial electric fields along the magnetic field shell necessary to produce the electrojets of Figure 14 would appear as shown in Figure 15. The assumption that the field lines are equipotential lines leads to the conclusion that the electric field configuration in the magnetosphere can be determined from knowledge of the relative significance of σ_1 and σ_2 in the ionosphere.

Although electrostatic models are useful in explaining some of the important aspects of the high latitude disturbances, there are other features that require dynamical considerations. For instance, the particle precipitation accompanying auroras and magnetic bays must

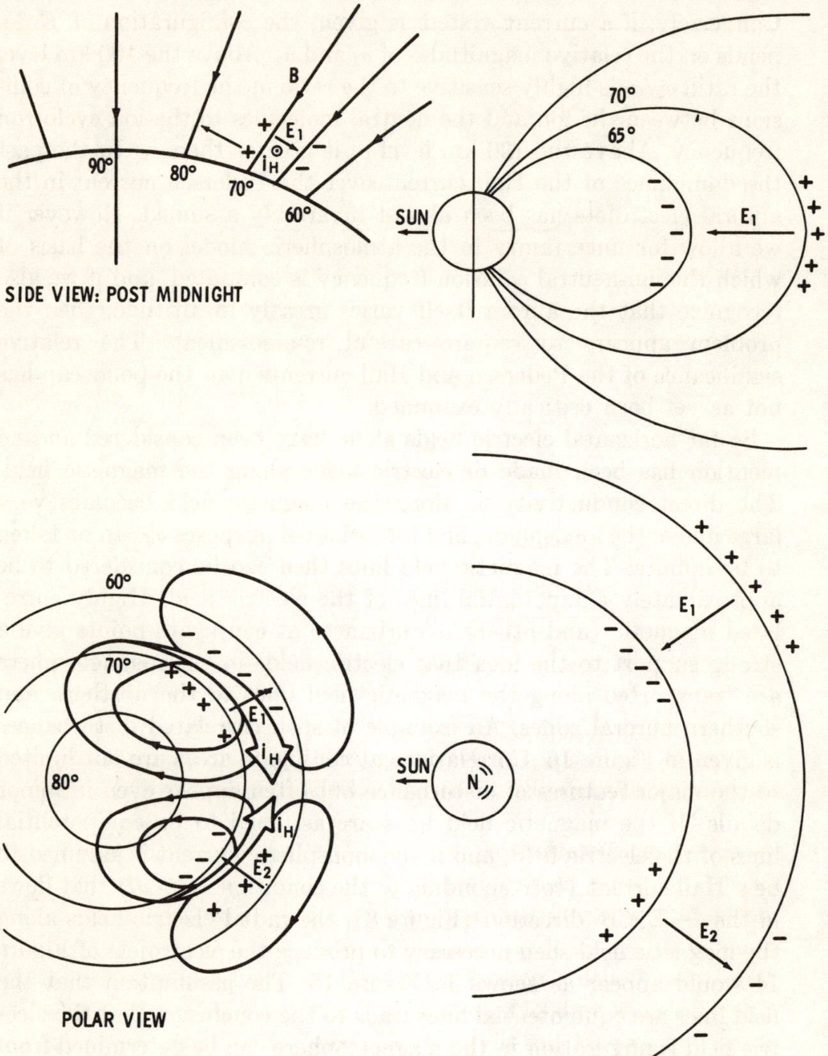

FIG. 15. Illustration of the gross features of the electric field that must exist if auroral electrojet currents are Hall currents.

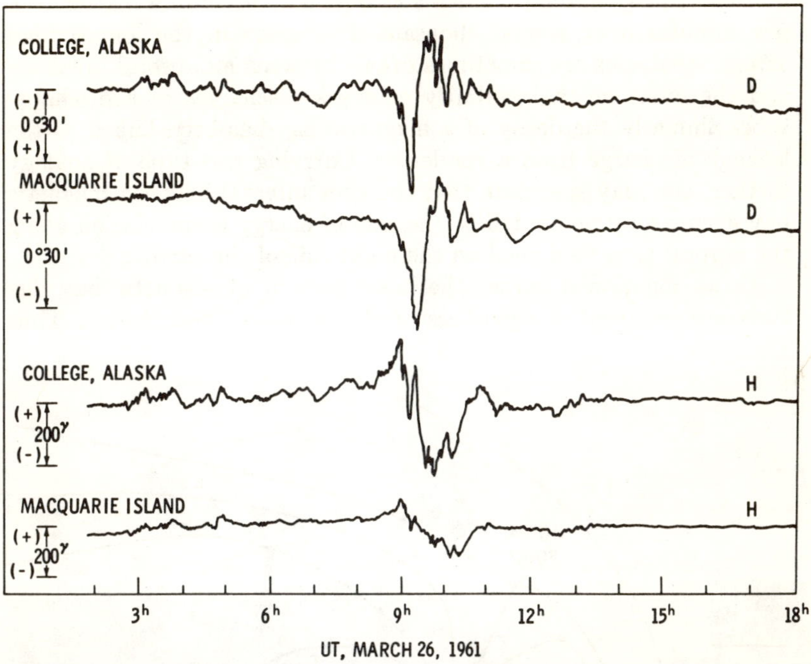

FIG. 16. Simultaneous recordings of H and D elements at two stations which are approximately conjugate in the northern and southern hemispheres (in comparing amplitudes note that the sensitivities are different at the two stations).

be due to acceleration of particles or changes in their pitch angles caused by some electromagnetic fields in the magnetosphere; the mechanism whereby the particles are replenished must also involve some dynamical process. Satellite measurements of energetic particles indicate that such a replenishment and energization of particles must occur on a time scale of hours rather than days. However, the behavior of low energy particles that are more directly related to the magnetic disturbances has not been investigated by satellite measurements.

Another dynamic feature is evident in the reversal of the electrojet current discussed above (pages 41–45). This event is not a simple stable transition from one current pattern to another. It represents a major rapid change both in the magnetic variation and in the aurora; the break-up, or transition of auroral arcs into brilliant rayed aurora, is often visually dramatic. Such a change may occur in a matter of a few minutes over several thousand kilometers in the longitudinal extent. Analogies are sometimes drawn between an auroral break-up and the onset of an instability or spark discharges in a discharge tube. Similarly the decay of a magnetic bay has been linked to the leakage of charge from a condenser. Carrying this type of analogy further, one may speculate that the time interval between magnetic bay events may be related to the rate of energy accumulation along the auroral zone field shell on the night side of the earth.

As we mentioned earlier the basic pattern of magnetic bay disturbance is repeated regardless of the degree of disturbance. Thus

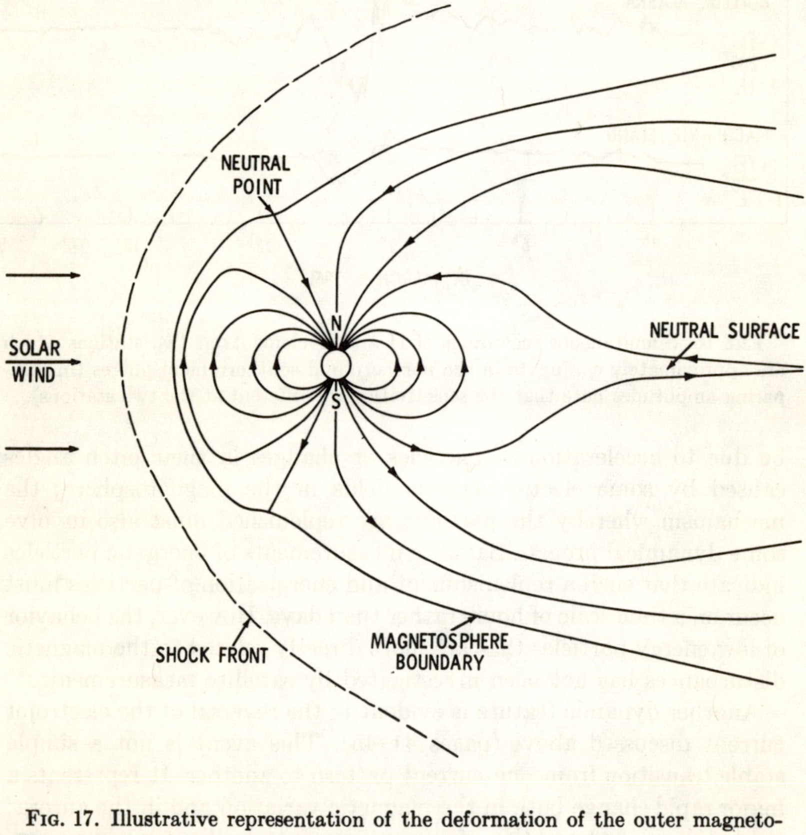

FIG. 17. Illustrative representation of the deformation of the outer magnetosphere as inferred from theory and measurement.

there is merit in seeking mechanisms for a creation of electric fields and for particle precipitation that are steadily present and continuously acting to some degree. This in turn makes it attractive to assume that the principal mechanisms are closely related to continuous interactions between the solar wind and the geomagnetic field. (See Chapter 9 for a more complete discussion of the interaction.) Figure 17 illustrates schematically the deformation of the geomagnetic field by the solar wind as determined by satellite measurements in low latitudes and by theoretical studies for high latitudes. On the antisolar side of the earth at magnetic latitudes above 75° the magnetic field lines are drawn away from the earth to form an elongated geomagnetic tail with an equatorial neutral sheet. The lines of force along the circle of magnetic latitude at 70° are shown to be closed. The division of the field lines at a latitude in the 70°–75° region is largely based on the satellite observations of the high latitude limit of trapped radiation belt particles. An additional support for this division comes from the observations of well correlated magnetic bay phenomena at conjugate points up to about 70° goemagnetic latitude. However the lack of conjugate stations at higher latitudes has not permitted tests of conjugate behavior to be made above that latitude. The apparent coincidence of the outermost magnetic shell for the trapped particles with the latitude at which the transition from the magnetic disturbance of the auroral zone type to that of the polar cap type may be given further a real physical significance.

Energy Transfer from the Solar Wind to the Magnetosphere

To explain the magnetic field variations in the auroral zone we are concerned first with mechanisms to transfer energy from the solar wind to the outermost field shells (containing radiation belt particles) on the night side of the earth, and second, with mechanisms to release this energy to produce the observed pattern of magnetic disturbance. The energy transfer from the solar wind to the magnetosphere may be effected by a number of different ways. A direct penetration of solar particles into the magnetosphere may occur in the vicinity of neutral points, lines, or surfaces which separate the domains of opposing magnetic fields. Such a process will accomplish a transfer of particles, but is likely to have little direct dynamical effect on the plasma in the magnetosphere. The behavior of plasmas at these magnetic singularities has not been as fully explored in geophysical theories as in laboratory plasma studies. However the two neutral points, one in each hemisphere on the sunward side of the earth, that appear in the theoretical models of the magnetosphere boundary (Figure 17) have been linked to the daytime magnetic activity in the area near 80° magnetic latitude, which is shown in Figure 14 by the shaded area. The exact latitude at which the magnetic field line originating at the neutral point intersects the earth's surface is subject to uncertainties, and the actual process that gives rise to the observed magnetic activity has not been specified.

There may be several different types of interactions at the interface between the solar wind and the plasma in the outer magnetosphere. The exact nature of these interactions is not as yet known but it has been thought that these may include a viscous-like component by which part of the momentum of the solar wind is transferred across the boundary to the plasma in the magnetosphere. It has been suggested that such a viscous-like interaction at the magnetosphere boundary (which we suppose for the moment to be closed on the night side of the earth) would cause the magnetospheric plasma to convect; the convection pattern is derived from an analogy with similar problems in fluid dynamics. The plasma in a thin boundary layer is dragged downstream by the *viscosity* and a return flow in the upstream direction is set up in the interior of the magnetosphere, creating two large convective cells. Since the low-energy plasma in the magnetosphere is *frozen* to the magnetic field lines and tends to move with them, the rotation of the field lines with the earth modifies the convective pattern. When the resultant flow pattern is projected onto the ionosphere along the magnetic field lines it resembles the current pattern for the polar disturbance. This similarity of the current and flow patterns has a physical significance for the following reason. The electrons in the ionosphere above about 90 km are nearly frozen to the magnetic field lines, and the current is directed opposite to the motion of the electrons; hence it follows that roughly speaking, the ionospheric current is anti-parallel to the projection of the velocity of the field lines in the magnetosphere onto the ionosphere along the field lines. Thus the convective motion in the magnetosphere described above is, in gross features, consistent with the ionospheric current pattern for the polar disturbance. When this model was proposed the magnetosphere was supposed to be of the shape of a *tear-drop*. What modifications the long tail of the magnetosphere would require in the proposed convective model is not obvious.

The field lines originating in the polar cap are dragged behind the earth to form the long magnetosphere tail by some continuous interaction between the solar wind and the boundary region of the magnetosphere. A model recently proposed supposes that near the noon meridian the polar cap field lines connect themselves to lines of force of the interplanetary magnetic field and are subsequently dragged downstream by the solar wind. The *feet* of these field lines in the ionosphere move across the polar cap toward the night side. (This corresponds to a nearly uniform flow over the polar cap toward the sun shown in Figure 14.) In the tail region the field lines continuously move toward the neutral sheet on the equatorial plane where they are gradually sucked in. This equatorward $(\mathbf{E}\times\mathbf{B})$ drift of the field lines is due to an electric field across the tail region, which also drives a current on the neutral sheet separating the opposing magnetic fields above and below it. The potential drop across the tail region corresponds to the potential across the polar cap ionosphere from the sunrise to sunset meridians; the latter potential drives the sunward Hall current over the polar cap as mentioned above. When a pair of field lines originating from the northern and southern polar regions is annihilated at the neutral sheet, they reconnect themselves and return to the day side by moving around the earth via either the morning or evening meridians; when the field line reaches near the noon meridian it again is severed and is joined to a field line of the interplanetary field, and the process continues steadily. The current flow in the ionosphere expected from this model is similar to that for weak polar disturbances. It is not certain how this model might be modified when the solar wind is intensified. However the long tail of the magnetosphere, whether open or closed at large distances, and its neutral sheet probably play an essential part in the processes involved in the high latitude disturbance phenomena.

The distortion of the trapping region of the magnetosphere due to the solar wind has been considered also as a possible cause for the creation of electric fields. The distortion would shift the trapping region of high-energy particles toward the sun, making the equatorial crossing points of energetic trapped particles closer to the earth on the night side than on the day side. The rotational motions of low-energy particles, on the other hand, are off-centered in the opposite direction from the paths of the energetic particles. These axially nonsymmetric motions of high-energy particles and low-energy plasma may upset the balance of space charge in the magnetosphere and the resultant polarization electric fields may drive currents in the ionosphere. A suggestion has been made that such currents might contribute to the polar disturbance. It is unlikely, however, that this mechanism could account for the high concentrations of the electrojet currents and their extremely large variability. There have been a few other theories of this type in which trapped particles are supposed to become polarized due to the magnetic field distortion in the magnetosphere; the electric fields so produced are then supposed either to drive ionospheric currents directly or to initiate plasma instabilities in the magnetosphere which in turn generate such currents.

MAGNETIC STORMS

General Characteristics

Intense world-wide transient variations in the earth's magnetic field are called magnetic storms. As an example, magnetograms obtained at widely different latitudes are shown in Figure 18. This storm began at 23^h50^m on November 27, 1959 with a sudden increase in the horizontal component H over most parts of the globe. Such an abrupt beginning is a common feature in magnetic storms, and is referred to as a *sudden commencement*, or SC (or *s.s.c.*, for storm sudden commencement). In Figure 18 the onset of the storm is marked with the symbol SC.

The record from Honolulu shows that after the SC, H remains above the pre-storm level for a few hours. This increase in H is another notable characteristic of magnetic storms and is called the *initial phase* of the storm. A large decrease in H follows the initial phase, indicating that the storm is now in its *main phase.* In Figure 18 the Honolulu record shows that H reached its minimum in a few hours after the beginning of its sharp decrease. A gradual return toward the normal level follows the minimum; in the Honolulu magnetogram the H trace is still below its pre-storm level at the end of the record, some fourteen hours after the H minimum. This *recovery phase* is usually several days in length.

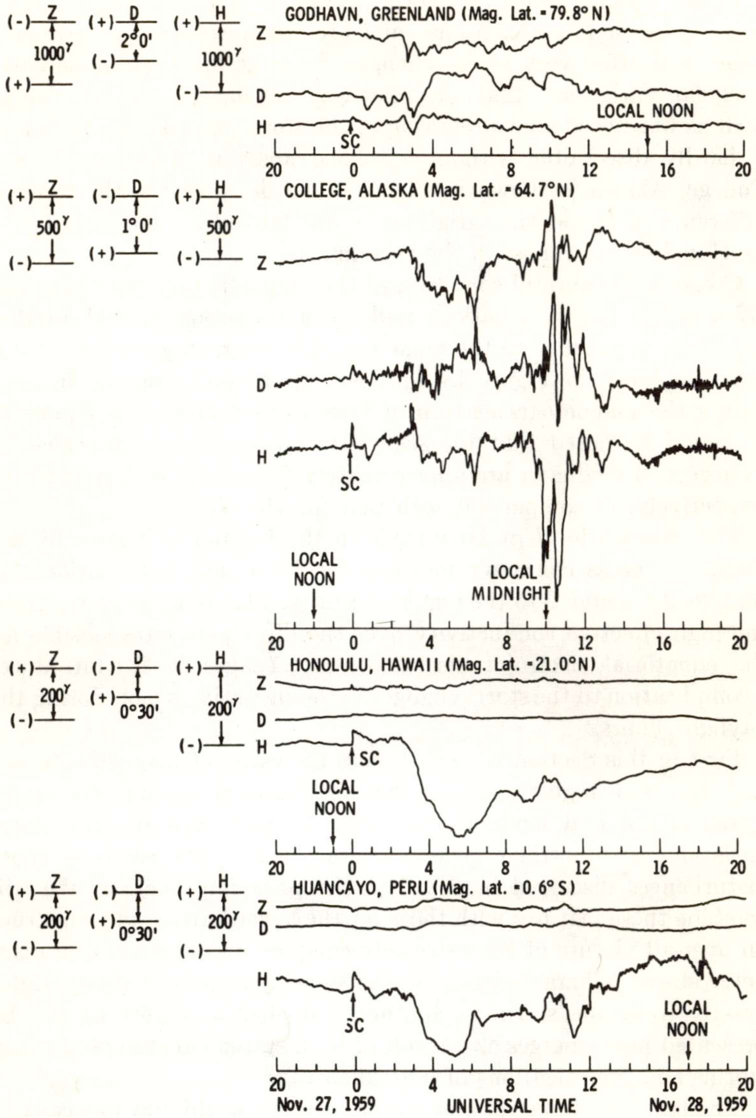

Fig. 18. Example of a magnetic storm of moderate intensity as recorded at different latitudes.

In the auroral zones the storm variations are markedly different from those in lower latitudes, and are characterized by extremely large, and often very rapid, changes. So large are these apparently irregular variations, that the relatively regular pattern of change such as was noted above referring to the storm record from Honolulu is hardly discernible without a refined analysis. The record from College, Alaska is shown in Figure 18 to demonstrate the striking difference in the storm variations in low latitudes, represented here by Honolulu, and those in the auroral zone.

Crossing the auroral zone toward the magnetic pole the characteristics of the storm variations undergo a transition to still another type that is peculiar to the polar cap. The magnetogram from Godhavn is shown in Figure 18 representing the latter region. In comparing the magnetograms from different observatories in Figure 18 it should be noted that the scales are not uniform; the scales for College and Godhavn are approximately 3- and 5.5-fold contracted, respectively, in comparison with that for Honolulu.

The changes in H at Huancayo, in the bottom of Figure 18, are similar in gross character to those for Honolulu, but considerable complexity is added to the regular features. This is an indication that the high effective conductivity over the dip equator responsible for the equatorial electrojet discussed above (pages 36–39) introduces a complication to the storm changes in the electrojet region during the daylight hours.

First in this Section we will discuss the cause of magnetic storms, and then their regular features that are observed in relatively simple forms in low and moderate latitudes. In high latitudes the storm variations are, in their general characteristics, the same as those disturbances discussed in detail in the preceding Section. We will combine these features with those for the lower latitudes to construct an over-all picture of the extremely complex phenomenon of a magnetic storm. Characteristics of the development and decay differ greatly from one storm to another, but such a picture as will be presented here emerges as a result of both statistical studies of many storms and investigations of individual cases.

Though the number of observatories in the world now exceeds one hundred, the coverage of observations, in particular, in polar regions is not adequate to determine unambiguously the distribution of currents responsible for the storm variations. Observations of storm variations in space are as yet greatly limited both in time and location, and no conclusive results have been presented by magnetic field measurements in space regarding the location of currents in the magnetosphere. Refinements and changes in our views of magnetic storms are anticipated in the future as observations become available of magnetic and electric fields and particles in the magnetosphere and its vicinity.

Solar Plasma Streams as the Cause of Magnetic Storms

We shall discuss later (pages 71–73) some of the indices that represent world-wide magnetic activity. Such activity indices and the frequency of occurrences of magnetic storms show marked variations with solar cycle, indicating that the cause of magnetic disturbances lies in the sun.

From direct observations in interplanetary space we now know that there exists a continuous flow of plasma from the sun. The observations of the solar corona provide evidence strongly suggesting that plasmas are ejected by solar flares. An intense magnetic storm is often observed on the earth following a large solar flare with a time lag of one or two days. Unlike the radiation in the optical range that is also emitted from the solar flare and reaches the earth in less than 9 minutes the ejected plasma travels interplanetary space with a speed of the order of one or two thousand kilometers per second, thus arriving at the earth's orbit, 150 million kilometers from the sun, in one to two days.

Although it is not possible with the present observational means to trace back all the terrestrial magnetic storms to the sources on the sun, there is little doubt that the primary cause of a magnetic storm is a plasma ejected by an eruption on the sun. Some of the magnetic storms recur with a period of about 27 days. This is an indication that the emission of plasma from an active region on the sun lasts for several solar rotations.

The complexity of the magnetic storm effects on the earth is the result of various interactions of the emitted solar plasma with the earth's magnetic field and its ionized atmosphere.

Sudden Commencement

When a transient plasma stream arrives at the boundary of the magnetosphere with a speed substantially greater than that of the steady solar wind, the outer region of the magnetosphere must experience a strong impact. A direct effect of the impact is a sudden compression of the magnetic field and the plasma in that region. This compression is transmitted inward as a hydromagnetic perturbation, which, arriving at the earth, is observed as a sudden increase in the magnetic field that we called the sudden commencement. This compressional effect is observed in its simplest form in low and moderate latitudes. In higher latitudes the consequence of the impact is less straightforward. We will describe below what is likely to happen in high latitudes.

There is an analogy between magnetic field lines embedded in a plasma and elastic strings. Just as transverse waves propagate along an elastic string, transverse hydromagnetic waves propagate along a magnetic field line. The existence of the latter waves was first predicted by Alfvén, and hence these waves are referred to as *Alfvén waves.* In a uniform elastic string the velocity, say c, of the wave is given by $c = (T/\rho)^{1/2}$, where T and ρ are the tension and the (linear) density of the string, respectively. For Alfvén waves the corresponding velocity, say V_A, can be expressed as $V_A = B/(4\pi\rho)^{1/2}$, where B and ρ are the magnetic induction and the plasma density, respectively.

Let us imagine an elastic string of length l stretched between two fixed points A and B so that waves reaching the endpoints are reflected back. The time it takes for a wave starting from the midpoint P between A and B to complete a path $PAPBP$ is $2l/c$. This gives the period for the lowest mode of free oscillation of the string. Periods for higher modes are obtained by dividing this period by integers 2, 3, 4, $\cdots$.

Let us suppose that a small section of the string near P is slightly displaced transversely from its equilibrium position and then immediately released. If this is done in a time short compared with l/c, that is, the time for a wave to propagate from P to A (or to B), then the perturbation given initially in the neighborhood of P will propagate toward A and B as a pair of transverse waves.

When the outer region of the magnetosphere is suddenly compressed by the impact of the solar plasma, a similar situation must happen to the magnetic field lines originating from high latitudes on the earth and crossing the equator at large distances. It is the equatorial portions of these field lines that are first distorted inward by an impulsive force. Then the distortion is transmitted along the field lines toward the earth in high latitudes as Alfvén waves. In the case of the string referred to above, the force initially applied to produce the perturbation near the midpoint was removed immediately to let the string vibrate freely. In the case of a sudden commencement first the field lines are suddenly pushed inward, and then, instead of being made free to return to their initial positions, they must eventually assume new (compressed) equilibrium positions. It is the initial strong blow by the impact that generates the Alfvén waves.

In addition, there must be a direct compressional effect propagated more or less radially inward from the boundary to the high latitude regions. However, since the impact effect is likely to be largest in the equatorial region, the Alfvén wave will have a larger amplitude than that of the direct compressional wave. Thus, we may expect to see at the time of a sudden commencement large amplitude Alfvén waves in high latitudes. The observed magnetic field changes on the earth's surface do actually show such waves. Furthermore, as will be seen below, analysis of the polarization of the Alfvén waves gives further support to such a view.

For a dipole the projections of the field lines onto the equatorial plane are all radial straight lines. But due to the steady solar wind pressure, the earth's field lines are distorted and the corresponding equatorial projections of the field lines originating from high latitudes are bent away from the sun as diagrammatically illustrated by solid lines in Figure 19.

In the final equilibrium state after the magnetosphere is compressed the lines of force must be displaced as shown in Figure 19 by the broken lines. Since the impact effect is a sudden blow rather than a gradual compression the equatorial portions of the lines of force will be set in a rotational motion as indicated in Figure 19. When these rotational perturbations propagate along the field lines to the earth, they will make the magnetic perturbation vector rotate in northern high latitudes counterclockwise before noon and clockwise in the afternoon when viewed downward; these directions are reversed in

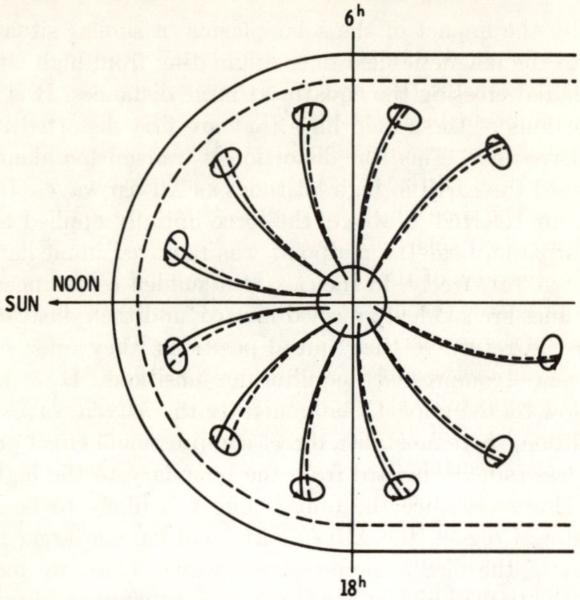

FIG. 19. Schematic drawing of the equatorial cross section of the magnetosphere boundary and the projections of magnetic field lines originating at high latitudes; solid lines for the equilibrium state before *SC* and broken lines for the new equilibrium state after *SC*. Small loops with an arrow indicate the direction of motion of the field lines.

the southern hemisphere. Such rotational characteristics are indeed observed in the magnetic records taken in high latitudes, and the sense of rotation, on the whole, agrees with the above picture.

Initial Phase

In low latitudes the increased level of H by the sudden commencement is maintained for a few hours until a large decrease of the main phase begins. At times additional increases, though not so abrupt as in the sudden commencement, may further raise the level of H.

In a relatively short time after the first contact of the front of the solar stream with the sunward surface of the magnetospheric boundary the magnetosphere must be immersed completely in the storm plasma stream, and hence the magnetosphere will now be in a more compressed state than it was prior to the arrival of the storm plasma. This is the reason for the increased level of H in the initial phase. The new equilibrium between the intensified solar wind and the earth's magnetic field is maintained until the solar plasma cloud has completely passed the earth and the solar wind pressure returns to its normal level. The time when the magnetospheric boundary is released from the increased solar wind pressure cannot be inferred from the ground observations of the magnetic field, because the increase in H is compensated by the decrease in the main phase as soon as the latter exceeds the former.

The duration of the initial (positive) phase varies from storm to storm, but it tends to be shorter in strong storms than in weak storms. The increased level of H may last for several hours or even longer without any appreciable decrease following it. However, it is difficult to establish that the increased level of H is actually maintained for an extended period of time because of the difficulties in defining the quiet level with certainty; such factors as slow variations due to the recovery from preceding storms and day-to-day variability of Sq make it difficult to separate the small increase from other effects.

Main Phase

Analyses of storm records from many observatories over the world indicate that the world-wide decrease in H in the main phase is due to a uniform magnetic field. Such a field can be produced by a ring-shaped westward current encircling the earth. The first suggestion of a current of this type was made by Störmer in 1911, who proposed that the decrease in H in the main phase is due to a flow of electrons deflected eastward around the earth on its afternoon side. Since then several theories have been put forward to explain the main phase of magnetic storms. Among others Chapman and Ferraro investigated in great detail the formation, equilibrium and stability of a ring current.

These authors were the first to treat the problem of a plasma stream approaching a dipole magnetic field, and give the essential idea on the theory of the sudden commencement and the initial phase of a magnetic storm. At the time they presented their theory the existence of a plasma around the earth extending to great distances of several tens of thousand kilometers was not known. Nor was it known then that there exists a steady plasma flow from the sun. Nevertheless, with reference to a solar stream producing a magnetic storm Chapman and Ferraro conceived the idea of a *hollow* carved out in the solar stream by the earth's magnetic field. Their theory of a ring current, however, met great difficulties, and a more satisfactory understanding of the problem had to await a more complete study of the behavior of charged particles in magnetic fields. As will be explained below, if charged particles are trapped in the earth's magnetic field they will indeed constitute a westward current.

In an earlier section (pages 27–28), we described the motion of a charge in a uniform magnetic field. An important characteristic was that the charge gyrates about the magnetic field lines with the radius of gyration mvc/qB and with the cyclotron frequency qB/mc, where q, m, v, B, and c are, respectively, the charge, the mass, the velocity of the particle, the magnitude of the magnetic induction, and the velocity of light. If the magnetic field is not uniform the radius and frequency of gyration will change along the trajectory, and the motion of the charge becomes more complex than it was in a uniform field. However, it is found that there are three invariants, or constants, of the motion that the particle conserves throughout its trajectory. Using these invariants the characteristics of the orbits of charged particles can be deduced and, in particular, their behavior in a dipole magnetic field or the actual geomagnetic field can be conveniently studied.

Since the details of this subject are discussed in Chapter 4, we only refer to some of the important properties of the particle motion that are relevant to our problem of the ring current.

A charge gyrating about a magnetic field line constitutes a small loop of current, and hence acts as if it were a small magnetic dipole. The dipole moment of this particle is found to be conserved with a great accuracy throughout its motion. This constant of motion is called the first adiabatic invariant of the particle (Chapter 4).

Let the components of the velocity parallel, and perpendicular, to the magnetic field be denoted by $v_{\|}$ and $v_{\perp}$, respectively. The magnetic moment μ is given by $\frac{1}{2}mv_{\perp}^2/B$. Thus if the particle moves into a region of strong magnetic field, the transverse velocity v must increase so as to keep μ constant. However, in the absence of acceleration the linear velocity $[= (v_{\|}^2 + v_{\perp}^2)^{1/2}]$ must remain constant. Hence, when the transverse component $v_{\perp}$ becomes equal to the linear velocity (and hence $v_{||} = 0$), $v_{\perp}$ can no longer increase; the particle then returns to the region of weaker magnetic field. Taking analogy with light reflected by a mirror, we say that the charge was reflected by a magnetic mirror, and we call the point of reflection a mirror point.

Along a field line of a magnetic dipole the magnetic field is weakest at the equator and monotonically increases away from it. Thus a charged particle could be trapped, and oscillate back and forth between a pair of mirror points. When a particle is trapped in this manner, there are additional *drift* motions that must be considered in addition to the oscillation between the mirror points.

First, the dipole field decreases with increasing distance from it by the inverse cube law. Thus, while gyrating about a magnetic field line, the particle does not follow a circular orbit exactly. Along the orbit of one gyration the magnetic field is stronger when the particle is on the portion of the orbit close to the dipole than when the particle is on the portion farther away from it. Therefore, the radius of curvature ($= mvc/qB$) is smaller in the former portion of the orbit than in the latter, and hence after the gyration the particle will have drifted longitudinally. In the case of the earth's dipole field a positive charge will thus drift westward and a negative eastward.

Secondly, since the dipole field lines are curved concave toward the earth, a particle traveling along a field line of the dipole experiences a centrifugal force directed approximately away from the dipole. Hence in a gyrating orbit the particle, irrespective of the sign of its charge, is accelerated while it is moving away from the dipole and decelerated while moving inward. Thus, again the radius of curvature will be made larger when the particle is away from the dipole than when it is closer to the dipole, and a drift motion results which, for each sign of charge, is in the same direction as that due to the non-uniformity of the magnetic field.

Finally, the earth's gravitational force acts on the particle in a way similar to the centrifugal force just discussed above, but the acceleration and deceleration due to the gravity are in the opposite direction to those due to the latter force, and hence the drift motions for charges are likewise reversed. In the case of the earth, however, the drift motions due to the gravitational force are negligible.

We have given a descriptive account of the drifting of charges in a dipole field. The manner in which a particle drifts from one field line to another is determined accurately by the second invariant of the motion that is discussed in Chapter 4. It is found on the basis of the conservation of the second invariant that the particle drifts on a shell, which, in the case of a dipole, is an axially symmetric surface generated by rotating a line of force about the axis of the dipole. If the magnetic field is not axially symmetric the shell is distorted accordingly.

Thus the charged particles trapped in the earth's magnetic field drift longitudinally, the direction of the drift being eastward for electrons and westward for positive ions. The net current is therefore westward.

As we have seen earlier a charge gyrating about a magnetic field line constitutes a small current loop which is equivalent to a small magnetic dipole. This dipole is so oriented as to reduce the magnetic field. Thus the particle has a diamagnetic effect. The total diamagnetic effect of a distribution of particles can be obtained by integrating the dipoles over the distribution. If a large number of particles of uniform density gyrate in a uniform magnetic field, the loop currents of the particles cancel each other except at discontinuous boundaries, reducing the volume current in the interior of the boundaries to zero. However, if the particle density is nonuniform, a net volume current results. The total effect of an aggregate of charged particles is the sum of the drift current and the current due to the nonuniformity in the density. Particles trapped in the earth's magnetic field thus constitute a ring current.

The discovery of energetic particles trapped in the earth's magnetic field in the regions now referred to as the radiation belts (Chapter 4) gave a strong support to the hypothesis of a ring current. However, the high-energy particles in the radiation belts were soon found to be inadequate to produce any appreciable reduction of the magnetic field. With recent measurements of a large flux of trapped protons in the energy range 100 kev to 4.5 Mev the decrease of the magnetic field at the earth's surface on the average quiet day is estimated to be 9 gammas, and the maximum reduction to be 23 gammas at the distance of 3.6 earth-radii on the equatorial plane. The magnetic field produced by this ring current belt composed of protons in the above energy range is represented in Figure 20.

Although the particle measurements so far made have not identified the particles constituting the storm time ring current, there is some evidence that the population of trapped particles in a certain energy range changes greatly during magnetic storms. It is likely that the

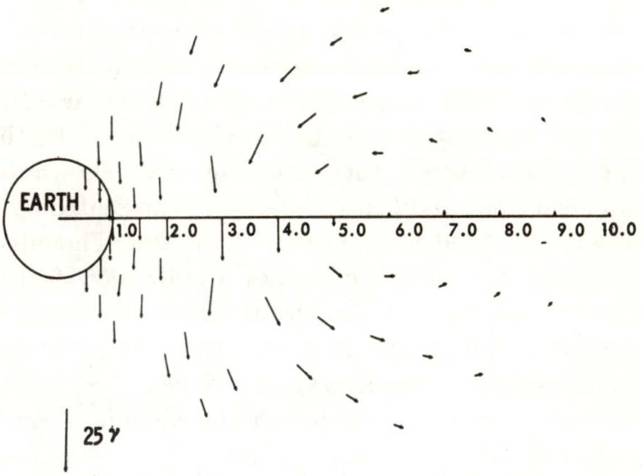

Fig. 20. Computed magnetic field distribution for a ring current consisting of the observed trapped protons in the energy range 100 kev to 4.5 Mev; radial distance is indicated in units of earth-radii (after Hoffman and Bracken, *J. Geophys. Res.*, **70**, 3541, 1965).

particles composing the ring current are mainly protons and some electrons of energies of the order of several tens kev, and that the position of the ring current is from 2 to 4 earth-radii. There may be a secondary ring current belt at a greater distance. The slow decay of the ring current is probably due to the charge exchange process between the protons in the ring current and the ambient slow neutral hydrogen in the magnetosphere. Although there have been some indications of the presence of a ring current in some of the magnetic measurements made in the magnetosphere, its precise location has not as yet been determined.

It is not known at present whether the particles forming the ring current are, by some mechanism not yet understood, injected into the inner region of the magnetosphere, or whether the low energy particles that were already existing prior to the storm are by some mechanism accelerated to higher energies to build a ring current.

Storm Variations in High Latitudes

As was discussed above (pages 41–47) the polar regions are more often disturbed magnetically than the regions in lower latitudes. We have seen that in spite of their apparent complexity and large variability these polar disturbances have a common structure with certain definite characteristics. During a magnetic storm the magnetic variations in high latitudes take, in gross character, a similar form to those disturbances not associated with world-wide storms. In the discussions of polar disturbances, the variations were shown to depend markedly on local time, and hence they were represented by a current pattern more or less fixed with respect to the sun. Though the polar disturbance of the type discussed is not a stationary feature of the geomagnetic field like *Sq*, it exists during a large fraction of time with varying intensity and duration. During a storm the polar disturbance is greatly enhanced above the average day level.

In describing the storm variation over the earth it is convenient to analyze it into the following two parts. At each epoch of a storm and for a specified geomagnetic latitude (which is a more significant variable than geographic latitude), the storm variation is averaged over longitude around the circle of that latitude. Then we determine the difference between the observed variation and this average. We denote the former by *Dst* and the latter by *DS*; these quantities refer to that epoch of the storm and to the geomagnetic latitude for which they were derived. By definition *Dst* is symmetric about the geomagnetic axis, and is a function of *storm-time* alone, where storm-time is the time measured from the beginning of the storm; the symbol *D* stands for disturbance and *st* for storm-time, and *Dst* is conventionally called the storm-time variation. The other part *DS* is a function of storm-time as well as local time, and may be called the disturbance longitudinal inequality, emphasizing that *local time* used here is only a geometrical variable, namely, the longitude with respect to the sun (hence the symbol *S*), and that the true time variable is the storm-time.

The storm variations so far discussed with reference to low and moderate latitudes, namely, the increase in *H* during the initial phase and the decrease in the same component in the main phase are all axially symmetric, and hence they belong to the *Dst* variation for these latitudes. However, a closer examination of the actual storm variations shows that there is an asymmetric part in them, indicating that *DS* is not zero. The latter variation is partially due to the asymmetry about the magnetic axis in the variation produced by the compression of the magnetosphere by the solar plasma stream and to the asymmetry, if any, in the magnetic field produced by the ring current, but it is mainly the effects of the storm disturbance in the auroral zones.

If the magnetic condition is relatively quiet at the time of the sudden commencement of a world-wide storm, an abrupt impulse is usually identifiable in the records taken in high latitudes marking the beginning of the storm. We have already indicated that the magnetic changes in high latitudes at the time of a sudden commencement are mainly due to oscillations of the field lines initiated by the impact of the solar plasma upon the magnetosphere. At times the impact is so strong that some of the high latitude field lines show many cycles of oscillation as if by resonance.

The polar part of the initial phase increase and of the subsequent main phase decrease in H is, in general, much smaller than DS and cannot be so easily identified in high latitude records as in those obtained at lower latitudes.

The DS variation in the polar regions consists of two types of changes, namely, large disturbances with time scale of the order of several hours or longer and rapid, often violent, fluctuations. In general, the former involves large areas in the polar regions, and the latter is more regional in character.

The most notable feature of the polar DS is the formation of an intense westward electrojet along the auroral zone after geomagnetic midnight. A much weaker eastward electrojet is generated before geomagnetic midnight. These concentrated currents along the auroral zone close their circuit by flowing across the polar cap and partially flowing over large areas below the auroral zone. These features are essentially the same as those illustrated in Figure 14 in the previous Section, suggesting that the DS variations at high latitudes are not fundamentally different from those occurring on average days. Detailed discussions of polar disturbances have been given earlier in this chapter (pages 41–58).

MAGNETIC ACTIVITY INDICES

It was recognized by Sabine more than 100 years ago that the annual average intensity of magnetic disturbances varied with the annual number of sunspot groups. Although it has been subsequently shown that on a daily time scale the sunspot correlation is poor, other types of solar-terrestrial correlations and more recent space observations demonstrate, when viewed in total, that magnetic activity on the earth is very closely related to solar phenomena.

To facilitate correlations with other phenomena, both terrestrial and solar, and to have simple criteria for relating changes in magnetic activity from hour to hour, day to day, month to month, etc., various indices of magnetic activity have been proposed and used. Two of the oldest are: C, the daily character figure on a scale of 0 to 2, which is a simple numerical, but qualitative, estimate of the day's activity at each observatory, and U which is a measure of the change in the average value of the horizontal component from day to day.

In the past two decades considerably more use has been made of an index called K, when given for a single observatory, and Kp, when given for a select group of observatories which we will describe below. Kp, in particular, has been used extensively (and possibly sometimes misused!) in space studies both for characterizing solar-terrestrial conditions and for correlating in detail magnetic activity with numerous types of space measurements ranging from magnetospheric neutral particle densities to cosmic rays. Thus it is important that we have some understanding of the meaning of Kp. Each observatory for each of the eight 3-hour intervals per universal day determines an index K on a quasi-logarithmic scale of 0 to 9 that has been determined for that particular observatory and standardized relative to other observatories. The quantity measured by the scale is essentially the range of the magnetic component showing the largest variation during the 3-hour interval after Sq and L daily variations, *crochet* type solar flare effects, and long term recovery effects from magnetic storms have been eliminated. The 3-hour planetary index, Kp, is obtained by averaging the K indices from twelve observatories selected by an international committee. The twelve observatories are all located between 47.7 and 62.5 degrees geomagnetic latitude but are distributed in longitude to reduce diurnal influences. The average geomagnetic latitude of the twelve observatories is 56 degrees. Thus, if we wish to attach physical meaning to the Kp index, we need to find the principal causes of magnetic disturbance near this latitude. When this is done, it is found that by far the most important disturbances are those associated with magnetic bay activity in the auroral zones. In the ionospheric current representation of Figure 14 these disturbances would be caused by the lower latitude return currents for the auroral electrojets. *Kp* is thus also an approximate indirect index of auroral activity. The fact that it is highly sensitive to magnetic storms is because of the combination of *Dst* (see pages 70–71) variations, the increased intensity of auroral electrojets, and the tendency for the auroral zone to shift to lower latitudes during storms.

For studies in which a linear scale is desired another index, the 3-hour equivalent planetary amplitude, called *ap* is used. The index *ap*, ranging from 0 to 400 in units of 2 gammas, comes entirely from converting *Kp* to a normalized linear scale and thus is a measure of the identical disturbance phenomena.

The fact that *Kp* and *ap* represent a mixture of disturbance effects associated with auroral electrojets, *Dst*, and other magnetic variations of smaller amplitude such as sudden impulses makes them highly useful for very general correlations. However, for more discriminating studies it is desirable to have indices for each type of disturbance. Toward this objective there are currently efforts to establish separate indices for *Dst* and auroral electrojet activity but these have not come into common use.

RAPID VARIATIONS

General Remarks

Besides the large scale geomagnetic variations discussed so far, there are variations with much shorter periods and smaller amplitudes, some with considerable regularity and others with irregular appearance like random noise. Abrupt, impulsive changes are called *sudden impulses,* and more continuous variations with periods roughly from ten minutes to one-tenth of a second are grouped together and generally called *geomagnetic micropulsations.*[4]

Sudden impulses are often found simultaneously over the world. In spite of their small sizes these impulses are of considerable interest because some of them represent responses of the magnetosphere to sudden changes in solar wind pressure. While micropulsations with periods of several minutes may have amplitudes as large as a few hundred gammas in the auroral zones, those of periods of about one second have much smaller amplitudes of the order of one-thousandth or one-hundredth of one gamma. Consequently, physical processes involved in generating micropulsations may differ greatly from one type to another. In this Section we will describe some of the distinct classes of sudden impulses and micropulsations, and discuss briefly how they might be produced.

Sudden Inpulses

In low latitudes the *H* trace in magnetograms shows, at times, an abrupt change in level, generally an increase of several gammas, observed simultaneously over the globe. The raised level of *H* usually returns to its normal level after one hour or so with a decrease which is in general less abrupt than the increase. A typical example is shown in Figure 21. The abruptness of these sudden changes in *H* is less pronounced, and their amplitude smaller, than in storm sudden commencements, but they are similar in general characteristics to the latter. In low latitudes the corresponding changes in *D* and *Z* are much less than in *H*. These sudden increases in *H* can be interpreted as being due to a sudden compression of the magnetosphere by an abrupt change in solar wind pressure. When the compression is very sudden and impulsive as in a storm sudden commencement, polarization characteristics of the field changes become similar to those in a sudden commencement.

Along the dip equator, where the amplitude of *Sq* is greatly enhanced by the electrojet, these sudden changes in *H* are augmented as are sudden commencements. In polar regions a disturbance of small magnitude is usually observed simultaneously with a sudden increase in lower latitudes, indicating that an abrupt change in solar wind pressure, even though only slight, can produce a small, but complex, polar disturbance.

These sudden changes in the magnetic field observed at the earth's surface have been detected in the magnetosphere, confirming the interpretation that they are indeed due to a sudden compression of the magnetosphere.

Large Amplitude Hydromagnetic Waves

In the auroral zones large amplitude regular waves are observed frequently. Their periods range from one to several minutes and their amplitudes from several gammas to a few hundred gammas. The

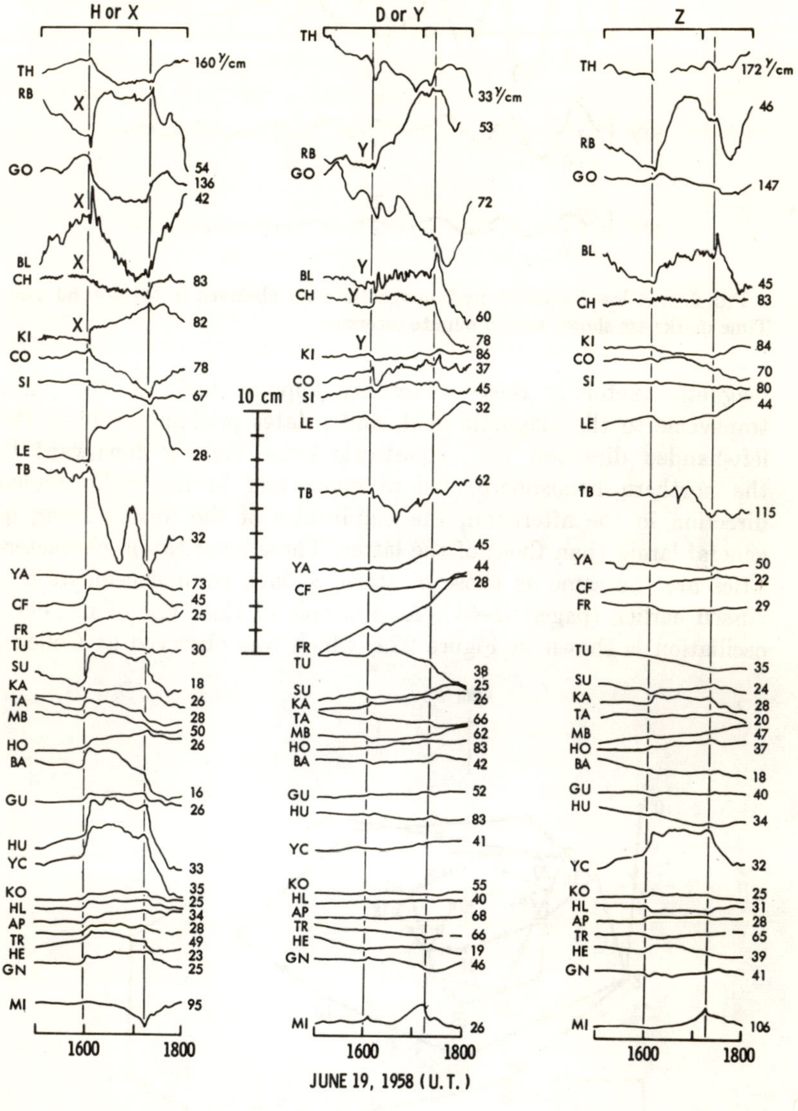

FIG. 21. Magnetogram traces showing a world-wide change in the H level, arranged in the order of decreasing latitude from top to bottom. Station abbreviations are marked to the left of the curves, and the numbers to the right give scales in gamma per cm when the contraction of length scale given in the space between the column for H or X and that for D or Y is used (after Sugiura, Davis, and Heppner).

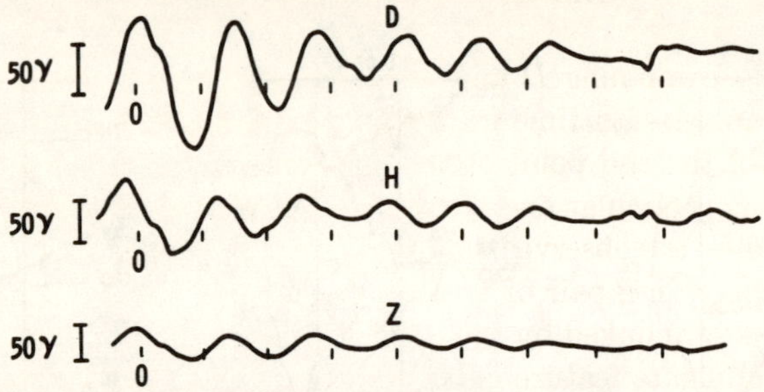

Fig. 22a. A low frequency hydromagnetic wave observed in the auroral zone. Time marks are shown with 5 minute intervals.

magnetic vector in these waves lies approximately in the plane transverse to the magnetic field, and rotates predominantly in the left-handed direction (i.e., counterclockwise looking downward in the northern hemisphere) before noon, and in the right-handed direction in the afternoon, the amplitudes of the former being in general larger than those of the latter. These polarization characteristics are the same as those of storm sudden commencements discussed earlier (pages 62–64). An example of this type of magnetic oscillation is shown in Figure 22a, which was observed at College,

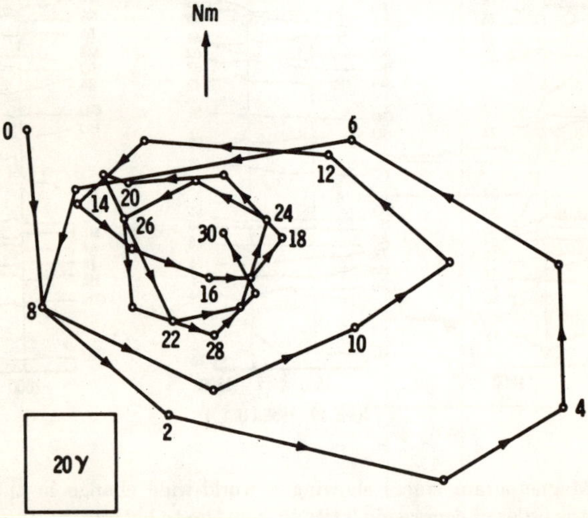

Fig. 22b. Polarization in the horizontal plane for the wave shown in Figure 22a. Time is indicated in minutes.

Alaska. By combining D and H we can examine how the magnetic vector changed with time in the horizontal plane. Figure 22b shows the locus of the end point of the magnetic vector in this plane. The polarization is circular and counterclockwise, i.e., left-handed.

Such waves are observed simultaneously at magnetically conjugate areas, namely, in a pair of areas in the northern and southern hemispheres that are linked by the magnetic field lines. An example showing this conjugate feature is presented in Figure 23; College, Alaska and Macquarie Island, Australia for which records are shown are very nearly conjugate. We can interpret this simultaneous occurrence of waves at magnetically conjugate points as the indication that these waves are hydromagnetic waves generated in the magnetosphere at a distance of several earth-radii and propagated along the magnetic field lines to the earth in the auroral zones.

A plasma in a uniform magnetic field can transmit waves of fre-

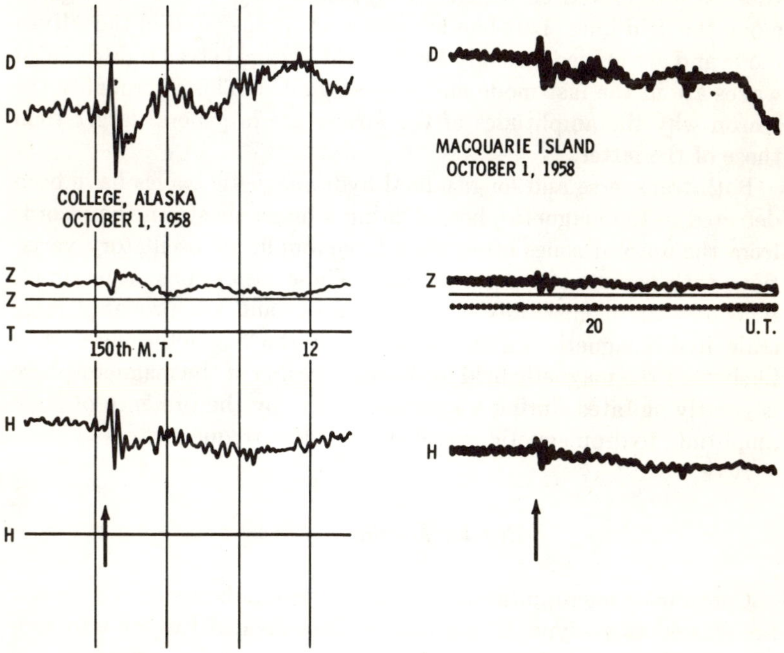

Fig. 23. A low frequency hydromagnetic wave observed simultaneously at a pair of magnetically conjugate points.

quencies below the ion cyclotron frequency in the direction parallel to the magnetic field in two modes. In both modes the magnetic (and electric) perturbation is transverse to the steady magnetic field and is circularly polarized, the sense of polarization being either left- or right-handed. Alfvén waves that we referred to earlier are those with left-handed, circular polarization. They are strongly guided along the field lines and do not propagate across the field. The other mode is called the *fast mode* because the phase velocity for this mode is greater than that for the Alfvén mode (or the slow mode), though in most regions in the magnetosphere the difference in the phase velocities for the two modes is negligible. Unlike waves in the Alfvén mode those in the fast mode can propagate in all directions. When fast waves propagate across the magnetic field, they become longitudinal (i.e., compressional) waves; the compressional waves discussed in connection with storm sudden commencements are of this type. Circularly polarized waves, such as shown in Figure 23, can thus be interpreted as transverse hydromagnetic waves propagated along the field lines. Left-handed waves are propagated in the Alfvén mode and are guided strongly by the field lines; whereas right-handed waves are in the fast mode and are not guided. This is probably the reason why the amplitudes of the former are in general larger than those of the latter.

Both transverse and longitudinal hydromagnetic waves have been detected in the magnetosphere. During a magnetic storm the records from the auroral zones often show large amplitude oscillatory variations with periods of several minutes. These oscillations occur simultaneously at magnetically conjugate areas, and are probably large scale hydromagnetic waves generated in the magnetosphere. It is likely that the magnetic field in the outer region of the magnetosphere is greatly agitated during a magnetic storm by the presence of large amplitude hydromagnetic waves of very low frequency.

Regular Micropulsations

Continuous micropulsations lasting for several hours or even longer are classed as *pc*-type. The group is often divided further into subgroups according to period, but the division is rather arbitrary and artificial. For the sake of simplicity we refer here by *pc-type micropulsations* to those with periods from several seconds to about 40

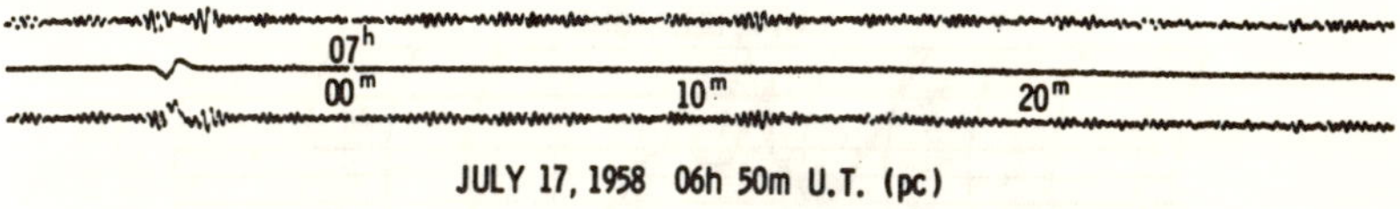

FIG. 24. An example of pc-type micropulsations observed at Onagawa, Japan (after Saito, Report of Onagawa Magnetic Observatory).

seconds. An example observed at Onagawa, Japan is shown in Figure 24.

The occurrence of *pc*'s has a diurnal variation with a maximum near noon. Their period varies from day to day and even within one day. It has been suggested that *pc*'s might be due to poloidal resonance oscillation of the magnetosphere, but their large variability in period and the lack of harmonic structure make this suggestion doubtful. They probably represent small hydromagnetic perturbation existing nearly at all times in the daytime magnetosphere, but the possibility that they are of ionospheric origin cannot be ruled out.

There is a group of regular, continuous type of micropulsations with longer preiods, 60 to 150 seconds, having similar characteristics to those described above. It is uncertain whether micropulsations of these long periods are a different phenomenon or the division is artificial.

Irregular Micropulsations

Trains of micropulsations consisting of series of damped oscillations lasting for several minutes to about one hour are often classed as *pt*-type. Their periods range roughly from 40 to 100 seconds. The occurrence of *pt*'s is closely related to polar magnetic disturbances. In particular, they are often observed in the beginning of a bay-type magnetic disturbance. Their amplitude is largest in the auroral zone and decreases with decreasing latitude. As evidenced by their association with polar disturbances, *pt*-type micropulsations are essentially nighttime phenomena. However, they propagate to large distances and can be detected even on the day side of the earth with a sensitive instrument. The sources of *pt*'s are not known, but they are probably small hydromagnetic disturbances generated in the nighttime magnetosphere, possibly in the tail of the magnetosphere.

In high latitudes more rapid, irregular micropulsations are observed during polar disturbances. Their periods range from several to 20

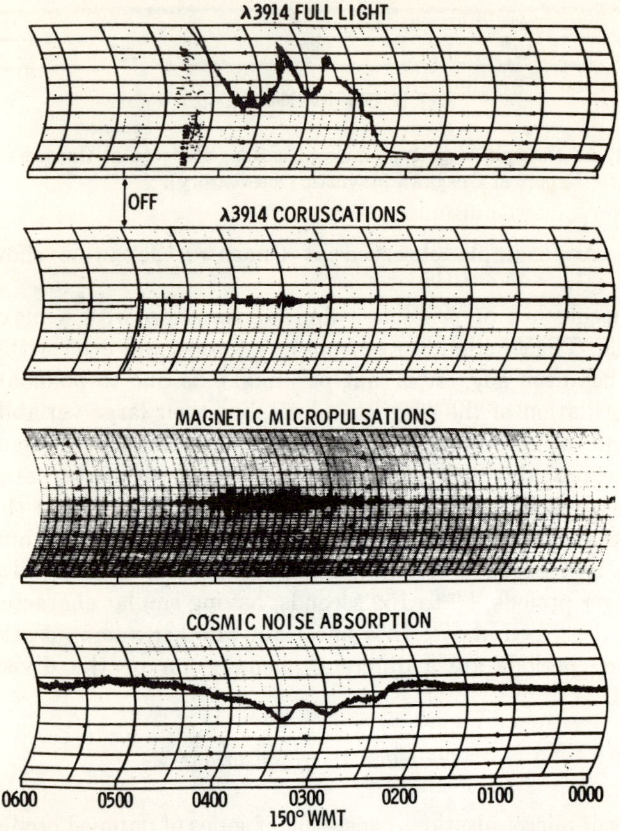

Fig. 25. Micropulsations associated with rapid changes in the luminosity in the aurora (in 3914 Å) and an increase in cosmic radio noise absorption (after Campbell and Rees, J. Geophys. Res., **66**, 41, 1961).

seconds. These micropulsations are closely related with auroras, cosmic radio noise absorption, bremsstrahlung X-rays, and magnetic disturbances. Figure 25 gives an example showing simultaneous activity in the aurora (as observed in 3914 Å wavelength), micropulsations, and cosmic noise absorption all observed at College, Alaska.

When abrupt bursts of X-rays and of cosmic noise absorption are observed, micropulsation activity also takes the form of bursts. Thus these micropulsations are directly related to precipitation of high-energy electrons. They can be interpreted as hydromagnetic waves that are greatly amplified by a plasma instability created in the magnetosphere by high-energy electron beams.

Hydromagnetic Emissions

When regular sinusoidal oscillations with periods from 0.2 to 5 seconds are recorded on a magnetic tape and are displayed on a sonagram, it is found that they have a fine structure such as shown in Figure 26. (In a sonagram the presence of wave energy is represented by a darkening on a photographic record in which wave frequency and time are taken along the vertical and horizontal axes, respectively, thus showing the frequencies of the waves present as a function of time.) The sonagram in Figure 26 indicates a series of regularly spaced emissions of rising frequency. The term *emissions* is used here, because as will be discussed below, they can be interpreted as hydromagnetic emissions in the magnetosphere, with analogy to VLF emissions which are discussed in the following Section.

The majority of these hydromagnetic emissions observed on magnetically quiet days show a rising frequency as in the example in Figure 26. During magnetic disturbances the elements in the fine structure may not be repeated regularly. While the frequency in the example shown in Figure 26 is spread over a wide band, there are cases in which emissions occur in one or more narrow bands. When

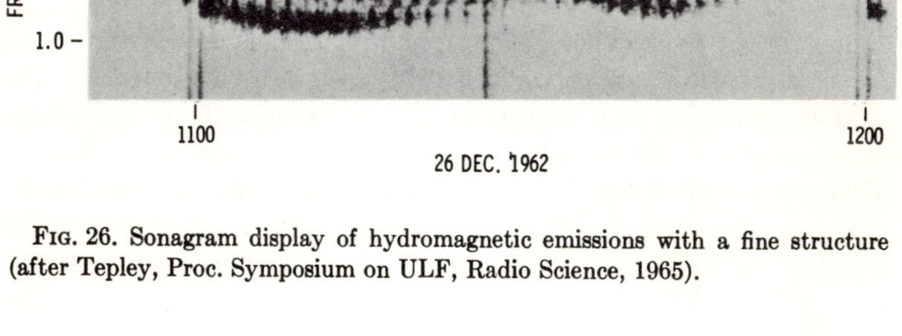

FIG. 26. Sonagram display of hydromagnetic emissions with a fine structure (after Tepley, Proc. Symposium on ULF, Radio Science, 1965).

the frequency is limited to a narrow band, the wave-form (in an ordinary amplitude-time record) is relatively simple, showing a regular sinusoidal oscillation with undulating amplitude. Such waves often resemble a pearl necklace, and hence are called *pearl-type* micropulsations.

When observations are made at a pair of magnetically conjugate points it is found that a series of emissions received in one hemisphere is displaced from the series observed in the opposite hemisphere by approximately half the interval between the repeated emissions. This suggests that the agency responsible for the emission is bouncing back and forth between the two hemispheres along the magnetic field lines. It has been proposed that beams of high-speed protons in the radiation belt produce these emissions. When the speed of the protons, say, V, exceeds the local Alfvén wave velocity, V_A, waves of frequency approximately equal to $(V_A/V)\omega_i$, where ω_i is the ion cyclotron frequency, are greatly amplified by an instability induced in the plasma. A hydromagnetic pulse generated by this instability is supposed to bounce back and forth along a magnetic field line, and the echoing waves are reinforced in energy on each pass by the instability, the energy required for the amplification of the waves being supplied by the streaming protons. Protons of energies of 200 to 500 kilovolts at the distance of about 4 earth-radii may constitute the beams that are responsible for the emissions.

Absorption of Hydromagnetic Waves in the Ionosphere

Collisions between particles are not frequent above the ionosphere, and hence are unimportant as a dissipative cause for hydromagnetic waves. However, as a wave propagates into the ionosphere the frequencies of collisions between the charged and the neutral constituents of the atmosphere at first become comparable with the wave frequency, and, in the lower regions of the ionosphere, exceed the latter. Under these circumstances, absorption of hydromagnetic waves takes place, and the wave energy is converted to the thermal energy of the medium, resulting in a heating of the atmosphere. Thus hydromagnetic waves with frequencies above one cycle per second are heavily attenuated in the ionosphere, making the latter *opaque* to these waves.

It has been suggested that hydromagnetic waves generated at the interface between solar wind and the magnetosphere propagate to the ionosphere and constitute a significant heat source, contributing to the diurnal variation of the atmospheric density. In order for this process to have an appreciable effect in the atmospheric heating, an almost continuous influx of hydromagnetic waves of large amplitude is required; however such waves have not been detected by satellite measurements. Moreover, the existence of a maximum in the distribution of Alfvén velocity at an altitude of several thousand kilometers forms a barrier to hydromagnetic waves propagating inward from the magnetosphere boundary to the earth. This is because these waves are refracted away from the earth at the region of high Alfvén velocity unless the incident angle falls within a very narrow cone about the vertical. The accessibility to the earth of hydromagnetic waves with frequencies of interest (i.e., about 1 cps) is thus rather poor. Therefore, it seems unlikely that hydromagnetic waves generated at the magnetosphere boundary would contribute significantly to the daily atmospheric heating.

However, during magnetic disturbances the heating of the atmosphere by hydromagnetic waves may be appreciable in the polar regions. Observations of the atmospheric drag of a satellite with a high inclination have revealed that in the auroral zones the heating of the atmosphere during magnetic disturbances is several times greater than that found in low latitudes during disturbances of comparable magnitude. However, the Joule heating by disturbance currents of large scale probably dominates over that of hydromagnetic waves.

WHISTLERS AND VLF EMISSIONS

Whistlers

A large fraction of energy emitted by a lightning discharge is contained in electromagnetic waves of frequencies from 300 to 30,000 cycles per second. These signals can be heard if the electromagnetic signals received by the antenna are fed into an audio-amplifier. Since the earth and the ionosphere are conductors, they form a wave guide for the audio signals from lightning discharges and transmit them to large distances. These signals are called *atmospherics*. Part of the energy leaks through the ionosphere and is guided along the magnetic field lines to the opposite hemisphere. Because of the dispersive property of wave propagation in a plasma, which will be explained below, these signals have a quasi-musical tonal quality; for this reason they are called *whistlers*.[4,5] Frequently, whistlers are reflected back and forth between the two hemispheres many times.

The remarkable guidance of whistlers along the magnetic field is due to the propagation characteristics of very low frequency electromagnetic waves in a plasma. Earlier we discussed a similar guidance of Alfvén waves for frequencies below the ion cyclotron frequency. Of the two modes we mentioned there, the propagation in the Alfvén mode becomes impossible above the ion cyclotron frequency. But the fast mode exists above this frequency, and when the wave frequency reaches a certain critical value, propagation across the magnetic field becomes impossible due to a resonance called the (*lower*) *hybrid resonance*. For a plasma consisting of electrons and one species of ions (and with a plasma frequency much higher than the ion cyclotron frequency), the hybrid resonance frequency is approximately equal to the geometric mean of the electron and ion cyclotron frequencies [i.e., $(\omega_e\omega_i)^{1/2}$]; at the latter frequencies a plasma has the well-known cyclotron resonances, and the term hybrid resonance derives from this circumstance. (To distinguish from another mixed resonance at a much higher frequency the adjective *lower* may be added to the hybrid resonance.) Between the ion cyclotron frequency and the hybrid resonance frequency, waves can propagate in all directions, but above the hybrid resonance frequency the propagation becomes highly directional and waves are strongly guided along the magnetic field lines. It is this mode of propagation that gives the guidance of whistlers along the magnetic field lines in the magnetosphere, and the mode is called the *whistler mode* even when applied to wave propagations that are not related to whistlers. The whistler mode propagation becomes forbidden above the electron cyclotron frequency.

Besides the guidance along the magnetic field lines the whistler mode has another important property, namely, that the wave velocity is strongly dependent on the frequency. Consequently, if a wave packet containing frequencies in a finite range is transmitted through a plasma, the frequency of the wave received at a distant point will vary with time. If we assume that a whistler is guided in a duct along a magnetic field line and that its constituent waves of different frequencies all propagate along the same path, the time delay, say t, measured from the instant of the lightning discharge to the time of arrival of the signal of frequency f via a one-hop path (from one hemisphere to the other) is related to the frequency by the equation;

$$t = (2c)^{-1} \int_{\text{path}} \frac{f_p f_H}{f^{1/2}(f_H - f)^{3/2}} ds \tag{18}$$

where f_p and f_H are the plasma frequency and the electron cyclotron frequency (in cycles per second), respectively, and c is the velocity of light; the integral is to be taken along the propagation path of the whistler.

If the wave frequency f is much less than the electron cyclotron frequency f_H, equation (18) can be approximated by the equation:

$$t = Df^{-1/2} \tag{19}$$

where

$$D = (2c)^{-1} \int_{\text{path}} (f_p/f_H{}^{1/2})\, ds \tag{20}$$

This constant D is called the *dispersion constant* of the whistler. With appropriate expressions for f_p and f_H, equation (20) can be rewritten as follows:

$$D = A \int_{\text{path}} (N/B)^{1/2}\, ds \tag{21}$$

where A is a constant given by $A = (e/2c)^{1/2}$, and where N and B are the electron density (per cm^3) and the absolute value of the magnetic induction (in gauss). Equation (19) shows that t increases with decreasing f; this is the reason for the descending tone of a whistler. The dispersion of a whistler can be demonstrated most clearly by a sonagram. Figure 27 shows an example of the dispersion of a whistler observed from the Vanguard III satellite at an altitude of about 2900 km at geomagnetic latitude 18°S.

The paths through which whistlers propagate above the ionosphere appear to remain the same for a considerable length of time regardless of the positions of the sources and the receivers, suggesting that at a

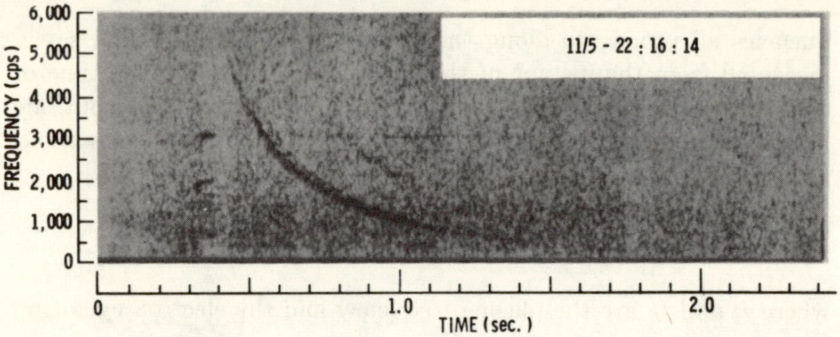

FIG. 27. Sonagram showing the dispersion of a whistler observed by a proton magnetometer coil aboard Vanguard III.

given moment there may exist a few stable ducts aligned with the magnetic field lines through which whistlers are transmitted. In such ducts, only a small amount of ionization excess compared with the surrounding regions is required to trap whistler energy. Although we mentioned that in the whistler mode, waves are guided along the magnetic field, an exact calculation of a ray-path of a whistler shows that if the electron density distribution is smooth in the magnetosphere the ray-path does not precisely coincide with a magnetic field line; however, only a small gradient in the electron density is required to guide the whistler exactly along the field line.

It is of interest to see how the amplitudes of the electric and magnetic fields in a wave propagating in a (cold) plasma in the whistler mode differ from those for a wave propagating in free space. Let the amplitudes of the electric field and the magnetic field in the latter case be denoted by E_f and B_f, respectively, and the corresponding amplitudes in the former by E_p and B_p, respectively. Then it can be shown that if the energy flux which is in the direction of the wave propagation (and hence in the direction of the stationary magnetic field for the whistler-mode propagation) is the same for both cases, the following relations hold:

$$E_p/E_f = n^{-1/2}$$

$$B_p/B_f = n^{1/2}$$

where n is the refractive index for the whistler mode, appropriate for the wave frequency. The value of n in the magnetosphere is of the order of 10 to 100, making $n^{1/2}$ of the order of 3 to 10. Hence, for the whistlers in the magnetosphere the amplitude of the electric field is reduced and that for the magnetic field enhanced in comparison with the corresponding amplitudes in free space. The above equations show that if the energy flux is kept constant the ratio of E to B varies inversely proportionately with the refractive index n.

Equation (18) indicates that when the frequency f is comparable with, but still less than, the electron cyclotron frequency f_H, f may increase with increasing t. In a whistler this corresponds to the case when the high frequencies of the whistler are comparable with the minimum electron cyclotron frequency in the distant part of the propagation path. In this instance there will be a rising tone as well as a descending tone, and these two branches will be adjoined at a frequency which may be called the frequency of minimum time delay. Whistlers of this type are in fact observed, and are called *nose whistlers* (because of their appearance in sonagrams). The frequency of minimum time delay is then called the *nose frequency*.

Equation (21) shows that apart from the numerical constant A the dispersion D is an integral of $N^{1/2}$ weighted by $B^{-1/2}$ along the path. The value of B is small in the outer portion of the propagation path with a minimum at the magnetic equator, and the path length is long for that portion in comparison with the portion in the ionosphere. Therefore, it is anticipated that the main contribution to the integral comes from the outer part of the propagation path. Such a feature makes it possible to use values of the whistler dispersion D to estimate the electron density in the magnetosphere. Nose whistlers are found to be particularly suited for this purpose, and on this basis the electron density in the equatorial region in the magnetosphere has been inferred to distances of several earth-radii. This method also provides a means of studying temporal variations in the electron density in the magnetosphere. It has been found that the electron density (as deduced from the whistler data) at distances of 2 to 7 or 8 earth-radii decreases during magnetic storms usually by an amount of the order of 20 percent; in some very severe storms much greater decreases have been observed. In addition, during magnetic storms a sharp decrease, referred to as a *knee*, may be created in the electron density at about 3 earth-radii.

Recent measurements by rockets show that whistler signals observed in and above the ionosphere are stronger by a factor of several hundred to several thousand than those observed on the ground. Such large differences between the signal intensities in the ionosphere and those on the ground cannot be explained by absorption in the low ionosphere. For instance, for 5 kc/sec the ratio of signal intensity above the ionosphere to that on the ground is typically 1.5 at night and about 40 during the day. The large observed ratios are probably due to total reflection in the low ionosphere, because whistlers propagating downward cannot penetrate this region unless their wave normals lie within a narrow cone about the vertical. This interpretation is supported by the evidence that whistlers are much more frequently observed in and above the ionosphere than at ground level.

The observations made by the Alouette satellite and also by rockets have shown that there are whistlers whose propagation paths are restricted in the region between roughly 100 km and 1000 km altitude and whose energy echoes back and forth between these levels. Earlier we mentioned that waves with frequencies between the ion cyclotron frequency and the lower hybrid frequency are not guided by the magnetic field and can propagate in all directions. The transition from the guided whistler mode to this relatively isotropic mode is essentially due to the presence of ions. The refraction of whistlers at 1000 km level can be interpreted as the result of the combined effect of the presence of hydrogen ions above this level and the presence of small horizontal gradients in the refractive index.

While whistlers are electromagnetic signals provided by nature, man-made signals at 15.5 kilocycles per second have been transmitted from Annapolis, Maryland, station NSS, and received at Cape Horn in South America. The signals propagated in the whistler mode, and confirmed the theory of whistler propagation.

VLF Emissions

Electromagnetic noises of natural origin at very low frequencies other than whistlers are generally called *VLF emissions*. They are not generated by lightning discharges, but some of them are observed in close association with whistlers. Many different types of *VLF* emissions have been observed, but the mechanisms responsible for these emissions are not as yet well established. We will only attempt here to describe a few types of *VLF* emissions and relate some of the mechanisms that have been proposed. Most of the observations and the theoretical studies have been made relatively recently, and many of the conclusions drawn have not as yet been thoroughly tested. In the following discussions it should be kept in mind that *VLF* emissions propagate in the same modes as whistlers.

One of the distinct types of *VLF* emissions consists of short bursts repeated at regular intervals typically of a few seconds. These are called *periodic VLF* emissions. They often show a dispersive characteristic similar to whistlers. The period between successive bursts varies systematically with frequency, and this variation is often found to be exactly the same as that in the whistlers associated with the emissions. The following picture appears to be consistent with the observed morphological features of these periodic emissions.

A wave packet traveling in a duct in the magnetosphere in the whistler mode initiates a *VLF* emission (by some mechanism which we do not specify for the moment), where the wave packet may be a natural whistler or a whistler-mode echo of a previous emission. If the whistler energy (or the energy in the whistler-mode echo) dominates over the emitted energy, a whistler-type dispersion will be observed when the signal reaches the earth's surface. There will be in general an enhancement of the wave energy in some narrow frequency range due to the emission. Part of the energy is reflected back into the same duct in the magnetosphere as the one in which the wave packet initially propagated. The reflected wave packet then triggers another emission in the same region of the magnetosphere where the previous emission took place, and propagates to the opposite hemisphere together with the newly emitted noise. This process may be repeated a large number of times (even more than 30 or 40 times), giving a long train of whistler-echoes with associated emissions.

Sometimes the triggering whistler may be relatively weak, but the emitted waves may be quite strong. Under these conditions a whistler-type dispersion will not be so evident as in the case previously discussed. However, the repetition period of the emissions will be exactly the same as the whistler-echo period for the same frequency. If there is a relatively strong absorption, only the emitted wave may be received at the ground. Periodic *VLF* emissions in which a whistler-type dispersion is not evident can be interpreted as examples of such events. In a complicated version of this type, one of the bursts in a train may trigger another train of emissions with a slight time delay.

A number of examples of periodic emissions of both dispersive and nondispersive types have been shown to fit the above process quite well. In both types of emissions it has been demonstrated with examples, though not as yet so numerous as to be conclusive, that the origin of a train of emissions can be traced to a whistler. However, such an association of emissions with whistlers is not really new. In the early phase of the observation of whistlers it was already noted that if a whistler occurred during an active period of emissions of the type called *risers* it would usually be followed by a riser. Many examples of whistlers followed by risers have since been presented. Thus there is a good reason for supposing that whistlers are the cause of some of the emissions. Then it can be supposed reasonably that a whistler-mode echo of an emission can likewise trigger another emission. The process described above for periodic *VLF* emissions thus appears to be well-founded. There is another set of observational facts that indicates that *VLF* waves themselves can trigger emissions. Observations at Wellington, New Zealand have shown that Morse code transmissions at 18.6 kc/sec from station NPG at Jim Creek, Washington trigger *VLF* emissions. A detailed analysis has indicated that the emissions (mostly of the riser type) were associated with 96 percent of the dashes and only 4 percent of the dots in the Morse code. The whistler-mode propagation of the Morse code signals themselves and the propagation of the emissions have been shown to fit the model for the natural *VLF* emissions remarkably well. Similar examples of artificially triggered *VLF* emissions have been recorded by the research ship *USNS Eltanin* off Palmer Peninsula; in this observation the transmitting station was at Cutler, Maine and the frequency was at 14.70 kc/sec.

As to the actual mechanism for the triggering emissions several models have been put forward. *Čerenkov radiation* has been considered as a possibility; in this case the particle velocity must be nearly equal to the component of wave velocity in the direction of the particle motion, and this condition is realized in the magnetosphere for *VLF* waves. Another idea has come from an analogy with a traveling-wave tube; in this model a stream of electrons selectively amplifies traveling waves. Doppler-shifted cyclotron radiation from electrons has also been considered as a possibility. Of these mechanisms the Čerenkov radiation and the cyclotron radiation would have to be coherent to produce the observed signal strengths. It has been suggested that triggering whistlers may momentarily trap electrons and group them in bunches, thus increasing the degree of coherence of their radiation. The efficiency of such a process and also that of the traveling-wave amplification mechanism have not been examined adequately to determine whether these mechanisms have significance in the *VLF* phenomena. The modern treatment of a *stream-plasma* system appears to be more promising than the approaches described above. In this theory the resonance between circularly polarized whistler-mode waves and gyrating electrons in a stream leads to an instability. We have mentioned already that for the whistler-mode propagation the wave frequency is always below the electron cyclotron frequency. Hence, for a cyclotron resonance to occur, the wave and the stream have to propagate in opposite directions. If the streaming electrons have a finite velocity component transverse to the magnetic field, a plasma instability just mentioned can occur in the stream-plasma system at very low frequencies. This mechanism has been proposed for the triggering of *VLF* emissions. The mechanism referred to earlier (pages 81–82) for hydromagnetic emissions involves essentially the same type of instability; there the streaming particles are protons instead of electrons.

In connection with the resonance between electrons and whistler-mode waves a possible effect of this resonance on the electrons in the radiation belts may be briefly mentioned. The particle population in the radiation belts is gradually depleted by scattering of the trapped particles by atmospheric molecules near the mirror altitudes. It has been thought that when the trapped electrons resonate with whistler waves their pitch angle distribution will be slightly changed and that an accumulation of this effect may appreciably change the rate of loss of trapped electrons. It has been suggested that this effect may be largest between the inner and outer belts and that the *slot* between these two trapping regions may indeed be the result of the *stirring* of electrons by whistlers. If whistlers have such an effect, the *VLF* emissions and even man-made *VLF* waves (of high power) will also have a similar effect on the trapped electrons.

Continuous, broad-band emissions, called *hiss*, are observed in the auroral zones. Their frequency ranges roughly from 1 to 20 kc/sec with a center-frequency at about 8 kc/sec. Observations in Antarctica show that the occurrence of hiss is most frequent during the several hours near magnetic midnight and closely related with auroral activity. Continuous, moderate hiss activity is associated with the presence of quiet homogeneous arcs and bands in the early phase of the diurnal auroral activity. During the *break-up* phase of the aurora, usually after magnetic midnight, hiss is observed in bursts coincident with auroral flare-ups; on such occasions magnetic micropulsations also show sudden bursts of activity (see page 79 for micropulsations), suggesting that all these phenomena have a common origin. However, the enhanced ionization due to the precipitation of electrons causes heavy absorption of *VLF* emissions and hence the peak activity in the latter may not coincide with that in auroral activity.

Not all of the occurrences of hiss are correlated with the aurora, and non-auroral hiss has a diurnal variation of occurrence that is similar to that of still another type of *VLF* emissions classed as *chorus* (or *dawn chorus*). This latter type consists of a multitude of discrete, often overlapping emissions giving rise to sounds resembling birds' singing at dawn. The mechanisms responsible for these and other types of *VLF* emissions are as yet not well established.

REFERENCES

1. Physics of Geomagnetic Phenomena, S. Matsushita and W. H. Campbell, Eds., Academic Press Inc., in press.
2. S. Chapman and J. Bartels, Geomagnetism, Oxford University Press, 1940.
3. Physics of the Upper Atmosphere, J. A. Ratcliffe, Ed., Academic Press, New York, 1960.
4. Natural Electromagnetic Phenomena below 30 kc/s, D. F. Bleil, Ed., Plenum Press, New York, 1964.
5. R. A. Helliwell, Whistlers and Related Phenomena, Stanford University Press, Stanford, California, 1965.

Chapter 2

THE EARTH'S ATMOSPHERE

I. Harris and N. W. Spencer

The earth's atmosphere is a large body of gas, with a density of nearly 10^{19} particles per cm^3 at the surface. As the distance from the surface increases, the density of the gas, the neutral particle concentration, decreases roughly at a rate of about a factor of e for each 15 km, to altitudes of 120 km. Above this region the general character of the atmosphere changes. The gas temperature rapidly increases by a large factor, causing a slower density decrease as a function of altitude at higher altitudes. Also, the density becomes sufficiently low so that the various species of gas separate diffusively, the density of each falling off with increasing altitude according to its mass and the temperature. This body of neutral gas, the atmosphere, absorbs incident solar energy differently for each gas as a function of wavelength of the solar radiation. Various phenomena result, including a particular temperature distribution, an ionosphere, chemical formation of various other gases from the normally existing gases, and other effects including winds, radiation and current flows.

The neutral particles govern significantly the transmission of the absorbed solar energy through the atmosphere and thus detailed knowledge of the amount of gas, the kinds present, their temperatures and other factors are very important considerations in understanding the controlling processes. This chapter is concerned with a description of these various properties of the high altitude neutral atmosphere, a discussion of their variations as caused by changes in solar input, and how they are measured.

The first experimental determination of some of these quantities, made possible by direct experimentation with rockets, was obtained nearly 20 years ago. The very short duration of the rockets' flights conducted at that time enabled only a limited amount of data concerning the many-dimensioned unknown atmosphere. During recent years, however, since the launching of the first Sputnik, a great deal of information about the neutral upper atmosphere above 200 km has been obtained from the analysis of the orbital decay of satellites. The acquisition of this new knowledge may be described as progress from a state of almost complete ignorance to a state of puzzlement.

INTERPRETATION OF SATELLITE ORBIT DECAY

Since the mean free path in the upper thermosphere is large compared to the dimensions of the satellites, the atmospheric drag can be taken from the free molecular flow model to be proportional to the square of the satellite velocity and in the opposite direction and is given in magnitude by

$$f_d = \tfrac{1}{2} C_D A \rho v^2, \tag{1}$$

where ρ is the air density, A the effective cross section area of the satellite and C_D the drag coefficient. There may be great uncertainty in the effective cross section area of a non-spherical satellite (for a sphere this is equal to one-fourth of the total area). However if one assumes that the satellite is randomly oriented along its path and its surface has no concavities, then one may take for the average presentation area one fourth of the total area.

The drag coefficient depends upon the mechanism of molecular reflection, on the ratio of the mean molecular speed and the satellite velocity and the surface temperature of the satellite. For a sphere, moving with the satellite's velocity the value is 2 when we have the special case of complete specular reflection. In most cases, however, the value lies between 2.1 and 2.3. Below 200 km the value may be less than 2. Thus the error in the drag coefficient is less than ten percent.

As an example of the procedure used to determine density from satellite orbits let us consider the case of a circular orbit above a spherical earth. The total energy of a satellite in a simpler circular orbit above a spherical earth is given by

$$E = -\tfrac{1}{2}(GMm/r) \tag{2}$$

where G is the universal gravitational constant, M the mass of the earth, m the mass of the satellite and r the radius of the orbit as measured from the center of the earth. The period of revolution is given by

$$P = 2\pi r^{3/2}(GM)^{-1/2}; \tag{3}$$

the velocity of the satellite is given by

$$v = 2\pi r/P = (GM/r)^{1/2} \tag{4}$$

The rate at which work done by the drag force is given by

$$f_d \cdot v = \tfrac{1}{2}C_D \rho A v^3, \tag{5}$$

which we equate to the rate at which the total energy of the satellite is decreasing:

$$dE/dt = -\tfrac{1}{2}C_D \rho A v^3 = \tfrac{1}{2}(GMm/r^2)(dr/dt). \tag{6}$$

From equation (3) we have

$$P^{-1}(dP/dt) = \tfrac{3}{2}r^{-1}(dr/dt).$$

Substituting in (6) and using equation (4) we finally have the simple formula

$$C_D \rho A/m = -\tfrac{1}{3}r^{-1}(dP/dt). \tag{7}$$

Taking as an example the Vanguard I satellite ($1958\beta_2$), with a mean radius of the orbit of 7247 km, and an effective cross section area of 308 cm^2 and using an air density of 10^{-16} gm/cm^3 yields a drag force of the order of 10^{-2} dynes. The perturbing forces arising from the oblateness of the earth are of the order of one dyne. This value of the drag force yields a rate of change of period due to drag of the order of .003 sec/day. If we use the density at Vanguard's perigee height the drag would be about a factor five higher. This rate of change of period is measurable, especially by its cumulative effect.

In actual practice practically all satellites have non-circular orbits. Because the atmosphere decreases exponentially with altitude most of the drag occurs at perigee. This has the effect of gradually decreasing the apogee of the satellite keeping perigee nearly the same height until the orbit is nearly circular. Various authors have developed formulas relating the change of period to the density at perigee, including corrections due to atmospheric rotation and the eccentricity of the orbit. Nevertheless, equation (7) is useful for making estimates of the density, the fractional error being of the order of the eccentricity and we take ρ to be the density at perigee. Usually several orbital revolutions of the satellite are needed to determine the period and rate of change of the period. Thus the densities as determined by satellite drag values are average values in which time variations of less than a satellite period are averaged out.

THE NEUTRAL ATMOSPHERE ABOVE THE MIXING REGION

The upper atmosphere above 200 km is believed to be in diffusive equilibrium, as contrasted to photochemical or mixing equilibrium which occur at lower altitudes. However, the level at which diffusive equilibrium may begin (the turbopause height) is not well known. Most authors have taken it to be about 120 km.[1]

The principal constituents of the atmosphere at this height are molecular nitrogen, molecular oxygen and atomic oxygen. There is some dissociation of molecular oxygen and recombination of atomic oxygen at this height but as the atmospheric densities drop off rapidly in height the time for establishing diffusive equilibrium becomes so small (of the order of minutes) that the atmosphere at 120 km can be taken as in diffusive equilibrium in so far as the major properties of the atmosphere at 200 km are effected. Thus the density of each constituent decreases with height as given by the barometric law

$$n_i(z, t) = n_i(z_0)[T(z_0)/T(z, t)] \exp - \int_{z_0}^{z} dz/H_i, \qquad (8)$$

where the scale height,

$$H_i = RT/m_i g(z). \qquad (9)$$

Here n_i is the density at height z, at time t, and T is the temperature, R the gas constant, g the acceleration due to gravity at height z, and m_i the atomic mass of the ith constituent. In the case of complete mixing equilibrium all the constituents would fall off with the same scale height determined by the mean molecular weight. There are corrections due to thermal diffusion and escape[2] to equation 8 for the lighter constituents, helium and atomic hydrogen, which are important at much higher altitudes.

$T(z_0)$, $n_i(z_0)$ are the temperature and number densities at the turbopause height, which are relatively constant in time over a period of a day.

Horizontal wind velocities of the order of 70 m/sec have been measured from sodium cloud experiments[3] up to 190 km. However, it has been adequate up to present to describe the time dependent dynamical behavior of the upper thermosphere by the use of only the thermal conductivity equations.[4] The heating of the upper atmosphere can be attributed to the absorption of extreme ultraviolet radiation and to a second not understood heat source, to be discussed later. Thus solutions of the time-dependent heat conduction equation with appropriate heat sources and sinks yield the temperature profiles (variation with height and time). Then by applying the barometric law [equation (8)] the time variations of the density of the upper atmosphere can be determined.

TIME VARIATIONS

The following classes of time variations have been established from satellite drag data for the time interval from 1958 through 1963:

(a) a diurnal variation with a density maximum at 1400 hours local time and a minimum at 0400 hours local time.

(b) a variation correlated with the solar decimeter radio flux in the 3 to 30 cm range and with an average period of 27 days.

(c) a semi-annual variation with a pronounced minimum in June–July and a smaller one in December–January.

(d) a long term variation associated with the eleven year solar cycle.

(e) a variation correlated with geomagnetic activity.

The Diurnal Variation

At the altitude of 120 km the time constant for heat loss by conduction is of the order of one day or greater. So, we do not expect any significant diurnal variation. Density determinations at about 200 km from satellite drag data have shown that the diurnal variations are less than 10% for levels of medium solar activity.

However, as we go up in altitude the variations from day to night increase rapidly with altitude. At 600 km the daytime density can be higher than the nighttime value by a factor eight.

Results of many investigations from drag studies of many satellites have shown that the density has a sharp maximum at 1400 hours local solar time with a shallow minimum at about 400 hours local time for all altitudes above 200 km. The 1400 hour local time maximum has been found to be valid for the entire phase of decreasing solar activity. Because of the tilt of the ecliptic a correct picture would seem to be of an atmospheric bulge at and symmetric around the sub solar point. But later and more detailed investigations by Jacchia and Slowey[5] have indicated that the center of the atmospheric bulge does not move much from the equator and has a pronounced elongation in the north-south direction. Thus local time is a very convenient parameter to use for the analysis of the satellite results and for theoretical model calculations. Figure 1 illustrates the diurnal behavior of the upper atmosphere for various altitudes. The increase of density at high altitudes due to this "breathing" mode of the atmosphere must be accompanied by a decrease of

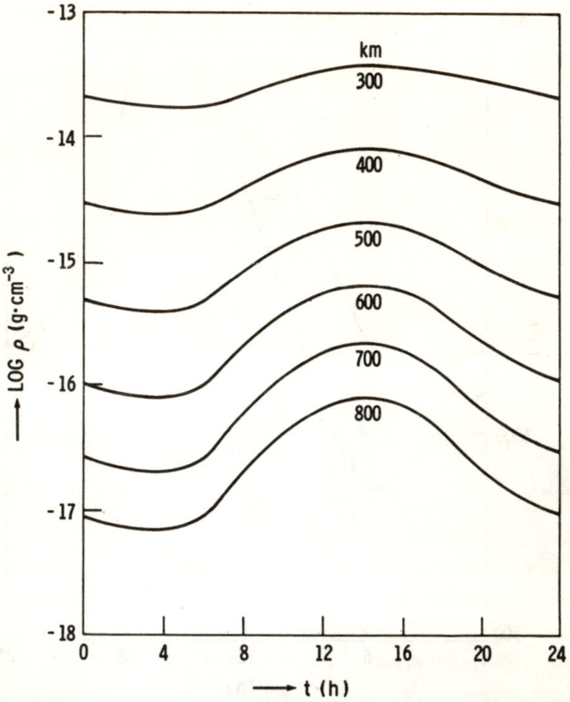

FIG. 1. The diurnal variation of the upper atmospheric density for various altitudes for a medium level of solar activity. (COSPAR International Reference Atmosphere, 1964.)

density at some lower altitude. Thus we can expect at some point, called the isopycnic level, the density to be constant.

This diurnal variation is entirely a thermal effect due to the alternate heating during the day and cooling during the night of the upper atmosphere. Tidal effects are much smaller and do not yield the correct period.

Harris and Priester[4] have integrated the one dimensional time dependent heat conduction equation, starting at 120 km. The result they have obtained is that heating due to extreme ultraviolet radiation above 120 km would yield a diurnal maximum at 1700 hours local time with a much greater ratio of daytime maximum to nighttime minimum than observed. The incoming flux was adjusted to yield

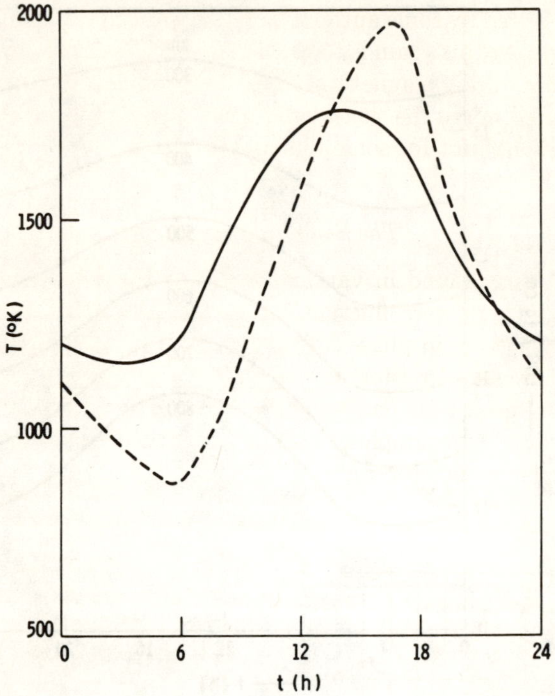

FIG. 2. Diurnal variation of the exospheric temperature calculated with extreme ultraviolet radiation as the only heat source (dotted curve) and with an additional heat source (solid curve). (After Harris and Priester, Ref. 7.)

the diurnal average densities observed by satellite drag. To obtain agreement, they introduced a second heat source which peaked at about 900 hours local time and had a flux comparable to radiation in the extreme ultraviolet range. Figure 2 illustrates their calculations of the diurnal variation of the exospheric temperature with heating due to extreme ultraviolet radiation and also the case where the second heat source is included. Some authors have suggested that correcting certain deficiencies in the theory, such as neglect of horizontal advection would then yield agreement with observations. But more precise estimates of the magnitude of the heat transfer by such a mechanism is needed. But qualitative estimates[5] of such corrections yield results much too small to alter their results. The source of the hypothetical heat source is unknown. Hydromagnetic heating has been suggested by some authors, but estimates by various people have indicated that this again is too small to effect the exospheric temperature except during magnetic storms. Nevertheless the use of a second heat source might be considered as a necessary phenomenological artifice to account for some dynamical effects.

The Solar Activity Effects

It was early found in various satellites (Sputnik II, Vanguard I, etc.) that there were fluctuations in the drag with a period of 27 days[7] which were in phase for various satellites thus indicating their global character. Priester[8] found a strong correlation with the solar decimeter flux at 20 cm, which was confirmed by Jacchia for the 10.7 cm flux. An example of such a variation is given in figure 3.

The solar decimeter radio flux cannot itself heat the upper at-

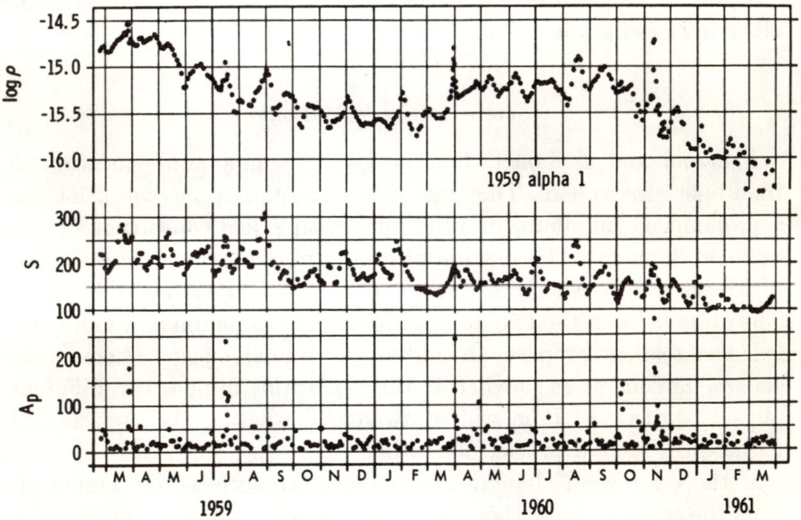

FIG. 3. Comparison of the variation of exospheric density from 1959 $\alpha 1$ and solar activity, exhibiting the 27 day periodicity. S is measured in units of 10^{-22} watts/m^2. Also the variation of magnetic activity through the A_P index is indicated (From M. Roemer, Report Number 68 of the Astronomical Observatory of Bonn University.)

mosphere but it can be taken as an indicator of the variation of the extreme solar ultraviolet radiation. From theoretical considerations we can expect that the extreme ultraviolet flux will vary in proportion to the decimeter flux. Neupert, et al.[9] have confirmed such relationships by monitoring individual spectral lines from the OSO–I satellite. However, they found that spectral lines which are emitted from the sun's disk as a whole rather than from plages alone showed a much smaller variation with the 10 cm radio flux.

Bourdeau, et al. have made a comparison of the extreme ultraviolet flux as measured from the OSO–I satellite and exospheric temperatures as determined by Jacchia and Slowey from studies of drag on Explorer IX. They obtained about a one day delay time in the response of the upper atmosphere and the fluxes measured by OSO–I. This can be due to the relaxation time of the atmosphere to re-establish thermal equilibrium and the method of reduction of the satellite drag data. Thus it is always to be preferred to use the previous day fluxes when one makes a comparison between the exospheric temperature (a deceleration of satellites) and extreme ultraviolet radiation.

Semi-Annual Variations

Paetzold and Zschöner[10] first noticed a semi-annual variation in the upper atmosphere. The main characteristics of this variation are a pronounced minimum in June–July with a less conspicuous minimum in January. The maximum occurs in September–October and in March–April, with the former somewhat larger than the latter. The differences in the two amplitudes of the maximum and minimum led Paetzold to propose an additional annual effect. There is no general agreement as to whether this separation implies two different physical processes. It is still controversial whether the semi-annual variation is related to the solar wind.

If the solar wind changes markedly in strength as one moves off from the ecliptic plane, then as the earth moves up and down through the ecliptic plane during the year a semi-annual effect might be produced by the variable solar wind. However, it seems difficult to imagine that the solar wind has the required rapid change with heliocentric latitude to produce this effect.

The semi-annual effect has been observed now over a period of

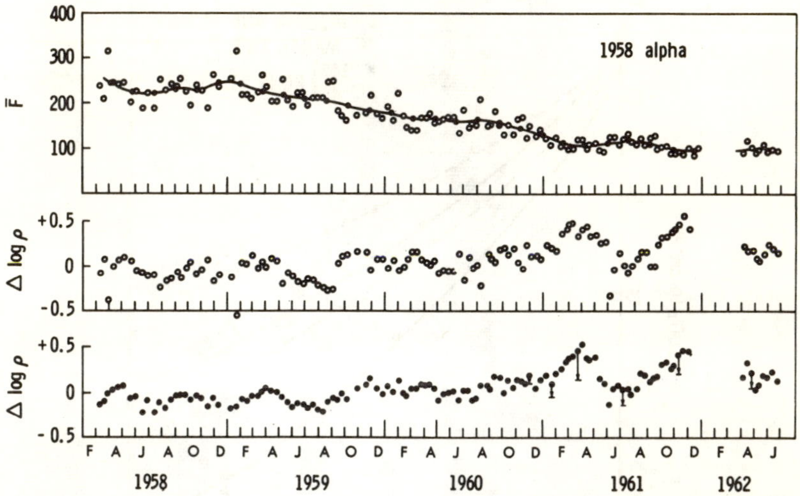

FIG. 4. Semi annual variation at 355 km altitude. The logarithm of the ratio of the observed density to a model density is plotted (Δ log ρ) so as to secure shorter period variations. The curve with the open circles are based upon taking monthly means of the 10.7 cm flux, while the dotted curve is 100 day mean value. (From M. Roemer, op. cit.)

six years in the height range from 200 to 1000 km (see figure 4). The variation seems to occur with more regularity than a similar variation in the geomagnetic indices.

Johnson has proposed that the semi-annual variation is due to convective flows from the summer pole to the winter pole. These currents would be descending at the winter pole thus transporting heat down to the cooler mesosphere from higher altitudes.

Solar Cycle Variations

The period when the first satellites were launched (1957–58) was a time of very high solar activity. Data on upper atmospheric density is now available covering the period of decreasing solar activity to the phase of very low solar activity. During this period it has been found that density at altitudes of several hundred kilometers has changed by a rather large factor of 10 to 50. The change in upper

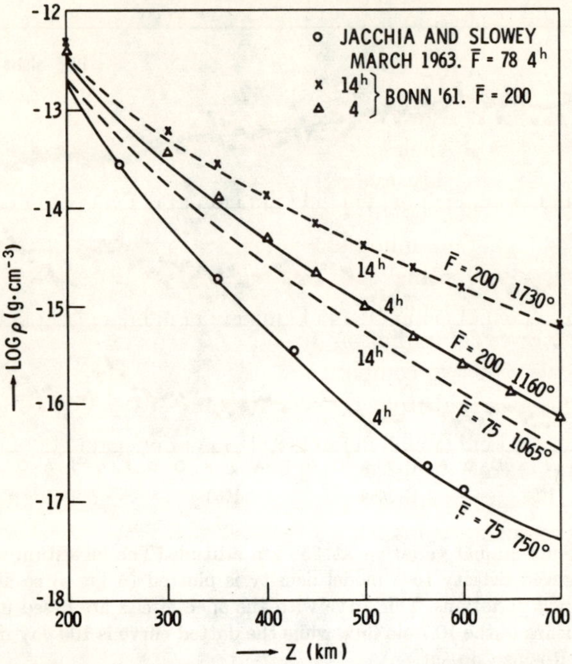

FIG. 5. The variation of the exospheric density from levels of medium solar activity and low solar activity. The curves are model curves and comparison is given with some observations. On the curves are given the exospheric temperature and the local time. (From COSPAR International Reference Atmosphere, 1964.)

atmospheric density over half a solar cycle is illustrated in Figure 5. This general decrease can also be seen in Figure 3.

As an index of solar activity it has become customary to take the monthly averages of the 10.7 cm flux to smooth out the 27-day variations. The variations of the monthly average of the 10.7 cm flux over a solar period is illustrated in Figure 5. This large decrease is attributed to the decrease of flux in the extreme ultraviolet range which should (as mentioned earlier) decrease in proportion to the 10.7 cm flux.

Jacchia[11] has found that the rate of change of the exospheric temperature per unit change in the monthly average of the 10.7 cm flux,

$$dT/d\bar{F}_{10.7}$$

is 4.2° at nighttime and 5.4° at daytime. However, they had found that the rate of change of the exospheric temperature per unit change of the unaveraged 10.7 cm flux (from the 27-day variation) is about 2.4°. Harris and Priester[12] by decreasing with the extreme-ultraviolet flux and the flux in their phenomological second heat source in proportion to the monthly averages of the 10.7 cm flux in their solutions of the time dependent heat conductive equation have obtained 4.5° for the nighttime minimum value and 7° for the daytime maximum. However, Jacchia[11] has pointed out this cannot be taken as evidence of the existence of a second independent heat source. The extreme ultra-violet radiation from the sun (as mentioned previously) consists of two components, one from plage-active regions of the sun and a second from the sun's disk as a whole. The first component would determine the 27-day variation during which the second component would not vary. However, both would decrease with solar activity yielding a much larger rate of change of exospheric temperature with the monthly averages of the 10.7 cm flux.

In their calculations (solutions of the time dependent heat conduction equation) of the variation of the upper atmosphere density variations with solar activity, Harris and Priester kept the diurnal behavior of the second hypothetical heat source constant. This led to a shift of the local time of the diurnal maximum from 14 hours local time to 1200 hours local time in their models for very low solar activity. However, satellite observations have shown that the phase of the diurnal maximum has been constant over the entire period of decreasing solar activity.

Also it has been found that the phase of the semi-annual variation is roughly constant though the amplitude decreases in proportion to the monthly averages of the 10.7 cm flux. Roemer has found that rather than a monthly average a three or five month average of the 10.7 cm flux constitutes a better measure for the long term variations of solar activity in the 11 year sunspot cycle.

The Geomagnetic Variation

In addition to the variations produced by the changing solar EUV flux there exists a variation that is well-correlated with variations in magnetic field of the earth. Though only of short duration the density may increase by an order of magnitude. An example is illustrated in

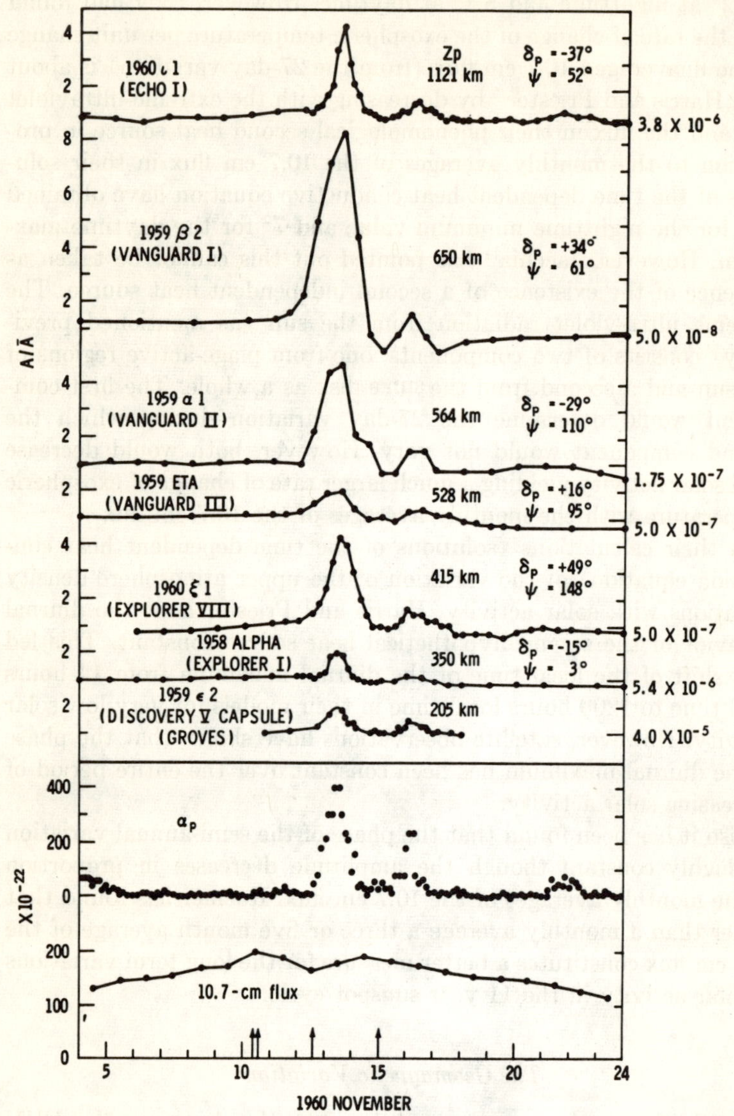
1960 ι 1
(ECHO I)
Zp
1121 km
δp = -37°
ψ = 52°
3.8 X 10⁻⁶
1959 β 2
(VANGUARD I)
650 km
δp = +34°
ψ = 61°
5.0 X 10⁻⁸
A/Ā
1959 α 1
(VANGUARD II)
564 km
δp = -29°
ψ = 110°
dP/dt
1.75 X 10⁻⁷
1959 ETA
(VANGUARD III)
528 km
δp = +16°
ψ = 95°
5.0 X 10⁻⁷
1960 ξ 1
(EXPLORER VIII)
415 km
δp = +49°
ψ = 148°
5.0 X 10⁻⁷
1958 ALPHA
(EXPLORER I)
350 km
δp = -15°
ψ = 3°
5.4 X 10⁻⁶
1959 ε 2
(DISCOVERY V CAPSULE)
(GROVES)
205 km
4.0 X 10⁻⁵
αp
X10⁻²²
10.7-cm flux
1960 NOVEMBER

Figure 6. The variation occurs world wide at about the same time with a greater amplitude in the auroral zone. However, on quiet days there is no essential differences in the temperature between auroral and equatorial zones. Jacchia has found a time delay of about five hours between the onslaught of the magnetic storm and the temperature rise of the exosphere. This could be due to the finite conduction time of the atmosphere.

The index customarily used to describe magnetic time variations is the three-hourly planetary index a_p. The daily average, $A_p = \frac{1}{8} \sum a_p$, can also be used if the values of a_p do not differ greatly from the mean values. Jacchia has found a nearly linear relationship between changes in the exospheric temperature and a_p, the temperature increase being about 1°K for every unit of a_p for medium values of a_p. However, at very low levels of solar activity where A_p is less than 20, it has been found[13] that the exospheric temperature depends much stronger on a_p and is nearly proportional to K_p, which is a logarithmic scale for measuring geomagnetic activity. This suggests that even at zero magnetic activity there still exists some heating, other than extreme ultraviolet heating.

There have been several possible theories on the heating during geomagnetic activity. Cole has suggested Joule heating by ionospheric currents; Dessler has suggested the dissipation of hydromagnetic waves. But as yet, there is no clear cut explanation.

ATMOSPHERIC MODELS

From the wealth of satellite drag data a clearer picture of the behavior of the atmosphere is now available. Also now there are

---

FIG. 6. Atmospheric drag of seven artificial satellites during the November, 1960 events, compared with the 3-hourly geomagnetic index A_P and the solar flux at 10.7 cm. The figures outside the right margin give the mean value $\bar{A}$ of the acceleration for each satellite before and after the perturbations of 12 through 17 November. The plotted points represent the ratio of the instantaneous acceleration A to this mean value $\bar{A}$. Other pertinent data—the declination of the satellite perigee δ_P and the geocentric angle ψ between the perigee and the sun—are given inside the right margin of the diagram. The arrows at the top and at the bottom of the diagram mark the instants corresponding to the appearance of 3^+ flares on the sun (From Jacchia, Ref. 10.)

becoming available many direct measurements of atmospheric properties. Most measurements of atmospheric properties are isolated in space and time and to correlate and compare the data, atmospheric models are valuable tools. Because of the large dynamic variations both diurnally and during the solar cycle any model which gives only average values cannot fulfill the requirements of a reference model.

Harris and Priester[4,12] have presented models which cover the diurnal variation and the solar cycle variation. The other dynamic variations, the geomagnetic variation, solar activity variation and the semi-annual variation can be treated as inducing small corrections to these models and empirical formulas are available for making such corrections. They are summarized in the new proposed Cospar International Reference Atmosphere[14] which presents improved models based upon all the satellite drag data available at that time. These models cover the entire 11-year period range of solar activity and the diurnal variation. Figure 7 illustrates how the diurnal variation

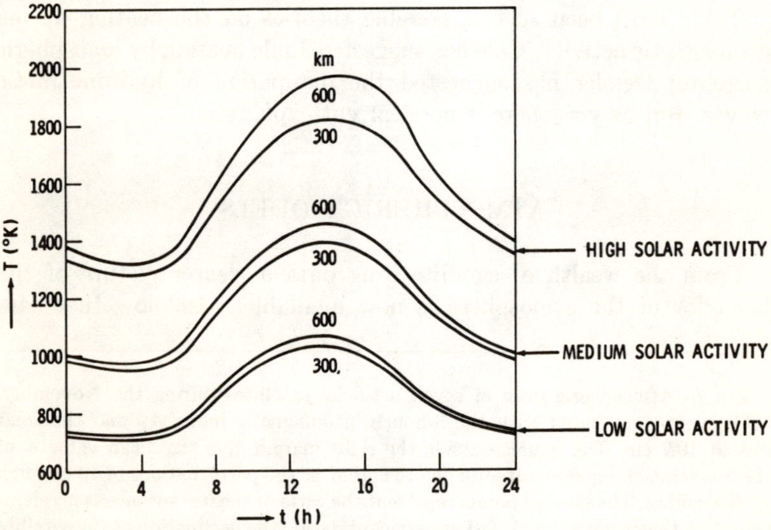

FIG. 7. Illustration of how the diurnal variation of the exospheric temperature changes from high, medium to low levels of solar activity at the altitudes of 600 and 300 km. (COSPAR International Reference Atmosphere, 1964.)

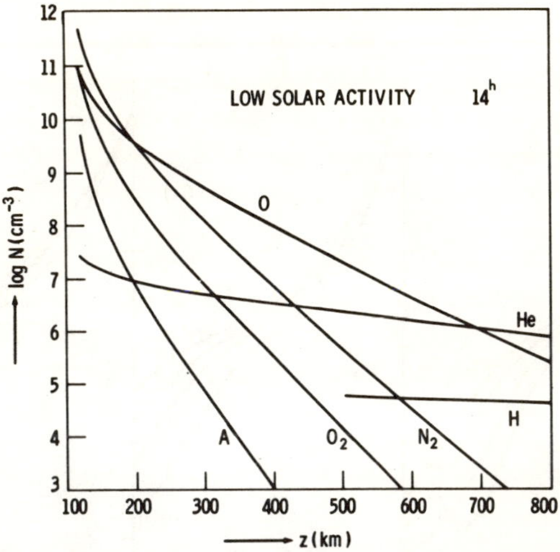

FIG. 8a. The number densities of N_2, O_2, O, A, He, and H (logarithms to base 10) versus height for 14 hours local time and low solar activity. (COSPAR International Reference Atmosphere, 1964.)

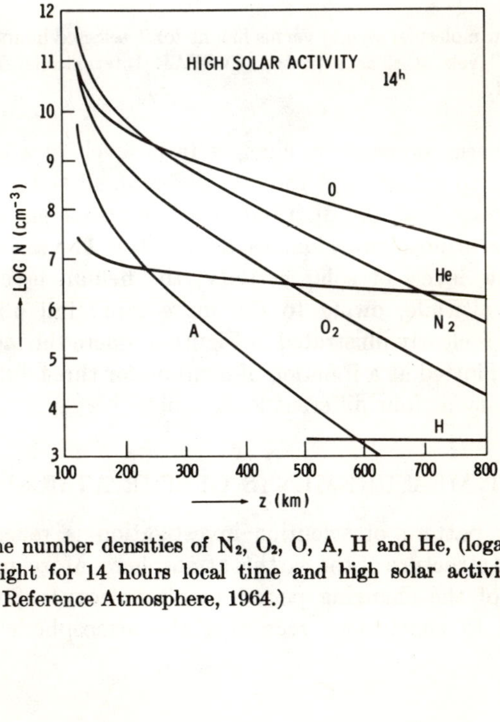

FIG. 8b. The number densities of N_2, O_2, O, A, H and He, (logarithms to base 10) versus height for 14 hours local time and high solar activity. (COSPAR International Reference Atmosphere, 1964.)

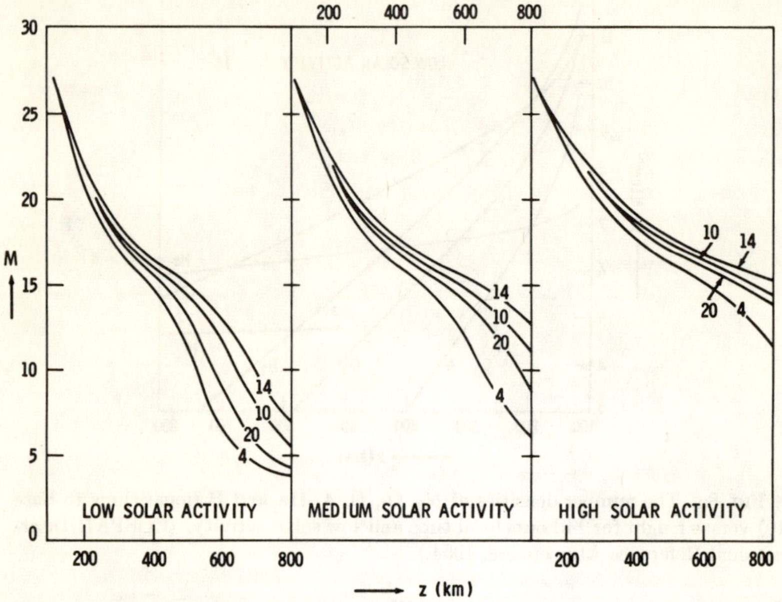

FIG. 9. Mean molecular weight versus height for 4 selected hours of local time and for three levels of solar activity. (COSPAR International Reference Atmosphere, 1964.)

in the exospheric temperature changes from levels of a high level of solar activity to a low level for two different altitudes. Figures 8a and 8b illustrate how the distribution of the principal constituents of the upper atmosphere changes during the 11-year solar cycle. Notice at low levels of solar activity, the helium layer begins at much lower altitude, owing to the more rapid fall off of atomic oxygen. This is clearly illustrated in Figure 9, where the mean molecular weight is plotted as a function of altitude for three different levels of solar activity at four different local times.

DIRECT MEASUREMENTS OF THE ATMOSPHERE

A familiar pattern of scientific investigation is revealed in the studies of the structure of the earth's atmosphere. More or less chance observation of the changing period (orbital decay) of the various satellites led to conclusions regarding the atmospheric density of neutral gas, and, subsequently, to the temperature and the composition, or nature, of the predominant gases at various altitudes. To attain a self-consistent picture, it was necessary to postulate that oxygen and helium were the chief gases present at altitudes of a few hundred kilometers.[15] As more measurements became available from additional satellite launchings, and the interpretative techniques were perfected, the determination of density from drag measurements became more precise. Thus, the theory of satellite drag and the related density of the atmosphere, was perfected through exposure to experimental data, and apparent systematic errors were revealed to be real variations of the atmosphere. These were correlated with diurnal change, solar activity variations, and other effects, as discussed in detail above. These findings led to the conclusion that more detailed and localized experimental data, as opposed to drag-derived perigee densities, were needed to improve understanding of the atmosphere. The advancing technology of atmospheric probing made experiments possible that employed direct measurement techniques, which, because of the previously relatively limited rocket capability had been applicable only to the lower atmosphere.

The early, first attempts at direct measurements of the atmospheric gases at the higher altitudes—above 120 km where, in general, a free-molecular flow state prevails—provided interesting and useful data. These gave, generally, isolated measurements of a highly complex and variable atmosphere. Quite naturally, laboratory vacuum measurement techniques were carried over to these early atmospheric vacuum measurements. Various thermionic as well as cold-cathode ionization gauges were adapted to flight use. Although the attainable laboratory vacuums were roughly equivalent to the vacuum encountered at the highest altitudes then attainable, the added difficulties of conducting the vacuum measurements under the rocket/space conditions posed real measurement difficulties. A discussion of these and other factors, and techniques which provided the basis for present-day measurements is given in the following sections.

Direct Neutral Particle Total Density Measurements

At altitudes greater than about 120 km, one may reason that the various gases of which the atmosphere is composed move about freely and thus experience negligible collisions over distances large with respect to rocket dimensions. This leads to the important conclusion that the kinetic theory of gases is applicable to the study of these gases and the measurement of their state. Thus, a rocket carrying a vacuum gauge provides a measurement circumstance, to at least a first approximation, where the vacuum gauge is immersed in a body of gas in thermal equilibrium, and where the individual gas particles are moving about essentially independently. Stated in another way, the mean-free-paths are large in comparison with the vacuum gauge and even the carrier rocket dimensions.

If the rocket were at rest, with respect to the gas, the measurement problem would be straightforward and very much like the laboratory vacuum measurement problem. However, this is not the case, for the rocket and the gauge are usually moving through the gas with a velocity that is generally significantly greater than the mean thermal velocity of the particles. In the case of typical satellites, the gauge velocity greatly exceeds the thermal velocity. A satellite travels roughly 10 km/sec while an oxygen atom at 300°K has a speed of about 0.5 km/sec. In the case of rocket-borne devices, the sensor velocity can be comparable to the thermal velocity, particularly at the apogee of the trajectory where only the horizontal component of rocket velocity remains.

As an additional complication, the rocket generally experiences a periodic motion, such as roll or tumble, or both, about its center of mass. For the satellite case, the motion is usually either pure spin, or a fixed orientation with respect to either the direction of flight (velocity vector) or inertial space.

The apparent translational velocity of the gas with respect to a randomly positioned gauge opening is

$$V_0 = V_s \cos \theta, \tag{10}$$

where V_s = satellite velocity, θ = angle between velocity vector and the gauge opening centerline.

The partial pressure of a gas in a chamber with an arbitrary orientation in space has been shown from kinetic theory to be

$$P_i = P_0(T_i/T_0)^{1/2}[\exp(-s^2) + s\sqrt{\pi}(1 + \operatorname{erf} s)], \tag{11}$$

where

$$f(s) = \exp(-s^2) + s\sqrt{\pi}(1 + \text{erf } s), \tag{12}$$

and P_i = internal pressure, P_0 = ambient pressure, T_i = internal temperature, T_0 = ambient temperature,

$$s = V_s/V_0,$$
$$V_0 = (2KT_0/m_0)^{1/2}, \tag{13}$$

where k = gas constant per molecule, 1.381×10^{-16} erg/°K, T = temperature (°K), m_0 = mass per particle, in gm.

P_i is defined as the pressure in the chamber, implying total pressure. But it can be considered as the sum of the partial pressures of the component gases; thus,

$$P_i = p(N_2) + p(N) + p(O_2) + p(O) + p(A) + p(H_2) + p(H) + p(H_e) + \cdots. \tag{14}$$

Similarly, P_0 may be considered as the sum of the external partial

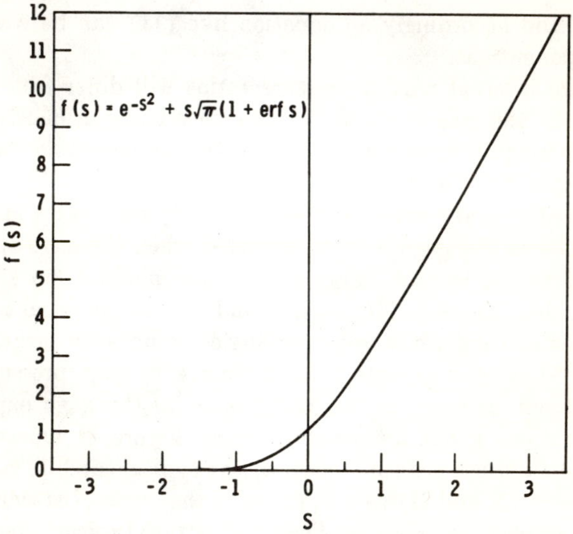

FIG. 10. Plot of function $f(s)$ vs. s.

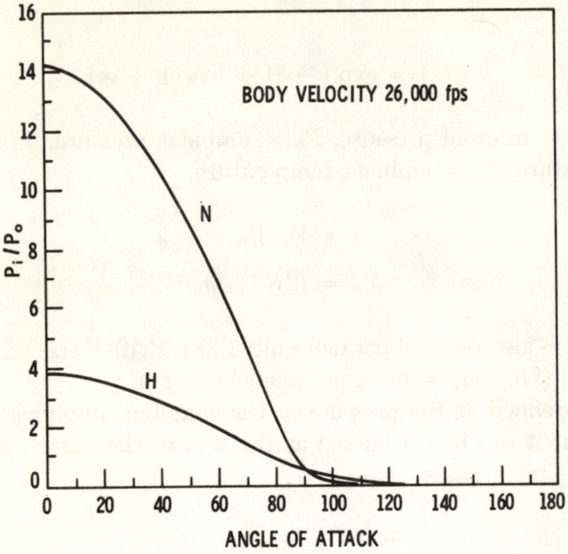

FIG. 11. Theoretical variation of the ratio of chamber pressure to ambient pressure as a function of angle of attack for the gases N and H.

pressures, and accordingly an equation like (11) can be written for each component gas.

The gage internal partial pressure ratios will differ from the external ratios, the magnitude of the difference being primarily a function of the magnitude of the body velocity component normal to the chamber opening. The $f(s)$ function shown in Fig. 10 illustrates the predominant velocity effect, and shows clearly that (a) a large pressure increase in the gage is to be expected when the gage opening is oriented into the stream (large s); (b) no modification should be observed when the $\theta = 90°$, $s = 0$; and (c) no significant number of particles can enter the gage for angles sufficiently large, s large negative. Thus, one expects a large change in gage pressure for a given ambient pressure as the body carrying the gage experiences various motions about its center of mass. Figure 11 illustrates the theoretical chamber pressure variation for typical satellite conditions for two gases (N and H) as a function of the angle of attack θ.

Two examples are presented here to illustrate typical experimental applications of this technique, which provides a basis for most direct

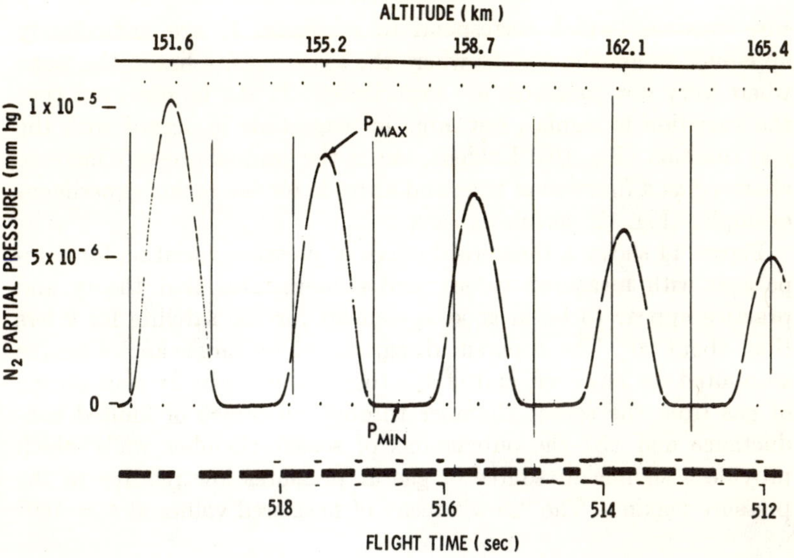

FIG. 12. Photograph of flight telemetry record showing actual variation of chamber pressure for a particular rocket flight (from Spencer, Brace, Carignan, Taeusch, and Niemann, Ref. 16).

neutral particle sampling experiments in use today. The first is illustrated by Fig. 12 which shows sensor N_2 pressure as a function of altitude[16] and flight time as observed during a typical rocket experiment. Figure 13 shows, similarly, sensor total pressure experienced in a satellite application of the experiment.[18] In the figures the pres-

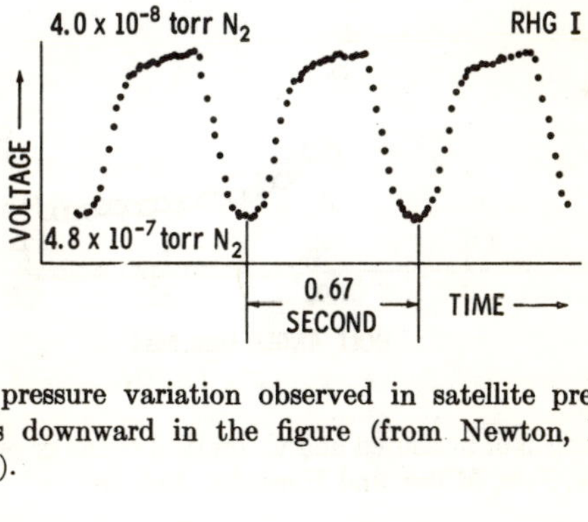

FIG. 13. Total pressure variation observed in satellite pressure gage. Note pressure increases downward in the figure (from Newton, Pelz, Miller, and Horowitz, Ref. 18).

sure maxima (which correspond to minimum θ) are immediately apparent, as are the regions where the sensor orifice lies in the wake where very low pressures are experienced. It can be observed that the variation is regular, but large in magnitude in accord with the $f(s)$ function (Fig. 10). Evident, too, is the gradual pressure increase observed as a function of time and altitude for the rocket experiment example, (Fig. 12) as one expects.

Figure 14 shows a theoretical curve of pressure variation for comparison with measured values, and demonstrates that theory and practice appear to be in good agreement for the satellite for θ less than about 90°. The apparent disagreement for larger angles can be accounted for however, and is due to (a) the relatively slow escape of gas from the sensor chamber through an orifice of limited conductance and (b) the outgassings of sensor chamber walls which provide a significant source of gas at pressures low relative to the pressure maxima. The "slow" decay of measured values at θ = 180°

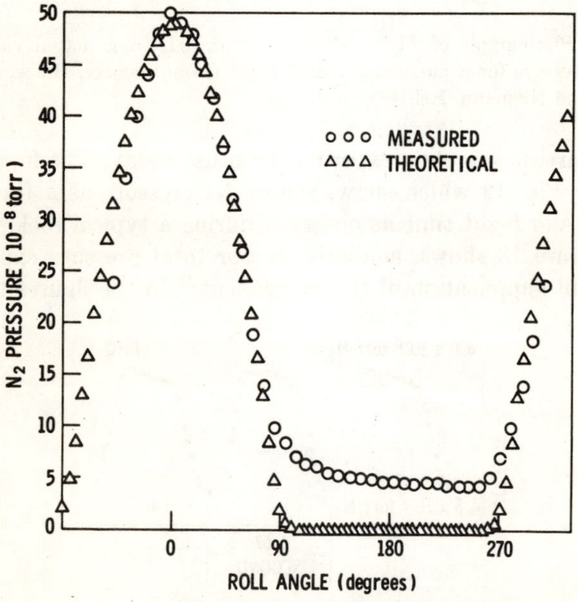

FIG. 14. Comparison of measured and theoretical chamber pressure for satellite case (from Newton, Pelz, Miller, and Horowitz, Ref. 18).

provides a direct measure of the outgassing, as constrained by the conductance of the orifice. These effects do not inhibit obtaining useful results from flight data, but on the contrary, permit a better understanding and evaluation of sensor properties.

The ambient atmospheric density is readily determined from such data by referring to equation (11), for $s \sim 2$, erf $s \rightarrow 1$, and $f(s) = 2(\sqrt{\pi}s)$.

Thus, for $\theta = 0$, and appropriate substitutions,

$$P_i = 3.55V_s(m_aT_i/2k)^{1/2}\,(P_0/T_0). \tag{15}$$

Also

$$P_i = N_akT_0, \tag{16}$$

where N_a is the number density of particles of mass a, and k = Boltzmann's constant. Finally

$$N_a = (0.58km_aT_i)^{-1/2}\,(P_i/V_s). \tag{17}$$

The peak values of pressure, corresponding to $\theta = 0$, and labelled on Fig. 11 as $P_{\max}$, can then be used in equation (17) to compute the ambient number density (can be readily modified for mass density) of the gas a. The factor T_i is taken as the sensor wall temperature, the assumption being that the thermal accommodation of the particles at the walls is perfect, as experience has shown to be adequately true. It should be noted that the technique, as represented by the equation (17) approximation, is applicable for all θ which correspond to s greater than approximately 2. This condition restricts use of the equation to sensor orientations where the sensor velocity effects predominate over thermal velocity effects.

Figure 15 shows two resulting N_2 density vs altitude profiles that were obtained through application of this technique in a rocket experiment.[16] The measurements were made using an omegatron mass spectrometer, which responded only to N_2, and thus permitted direct implementation of the theory. The Harris and Priester model values for conditions approximating those existing at the time of the experiment are plotted for comparison. The differences observed between the measured values and the model values can be attributed

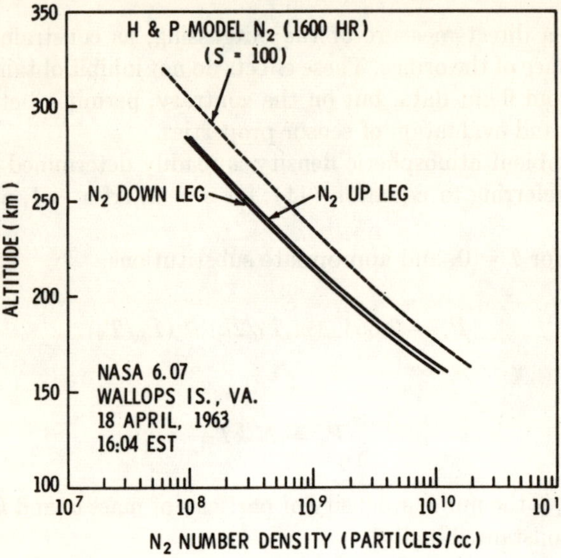

FIG. 15. Experimentally determined values of molecular nitrogen density vs. altitude for a particular rocket flight (from Spencer, Brace, Carignan, Taeusch, and Niemann, Ref. 16).

to uncertainties in gage calibration, selection of model (uncertain solar activity level) to use for comparison, other unknown errors in the measurements, and other uncertainties in the model.

Neutral Particle Temperature Measurement

A concept useful for the study of real as well as theoretical atmospheres is the "scale height", defined previously in this chapter by equation (9). The concept can be visualized as follows: assuming an isothermal, single-gas atmosphere of temperature T and mass m, the vertical distance that must be considered to realize a pressure change of a factor of e is a scale height. Considering the real atmosphere, the temperature is nearly constant above an altitude of about 200 km, and thus, since the constituent gases are separated gravitationally, a determination of the scale height of a particular gas leads directly to the temperature of the gas. As a practical matter, the scale height concept can be defined to account for typical atmospheric temperature gradients.

Referring to Figure 15, it can be seen that the N_2 profile is nearly linear over considerable altitude intervals, particularly above 200 km. Accordingly, the temperature can be readily determined, with acceptable accuracy. Thus, as shown by Taeusch,[16] using the hydrostatic equation, the gas law and equation 2 above, one obtains

$$T = -[(g/R) + (dT/dh)] \frac{\Delta P_i V z}{(d\Delta P_i/dt) - (\Delta P_i/V)(dV/dt)}, \qquad (18)$$

where $Vz = dh/dt$ (vertical velocity), V = vehicle velocity, R = gas constant, $\Delta P_i = P_{\text{max}} - P_{\text{min}}$.

The equation is directly applicable to measured values like the successive maxima shown in Figure 12, each pair of maxima enabling an independent temperature measurement. In making the computations, a value of dT/dh is assumed, and then later corrected as necessary by successive approximation. Representative temperature profiles derived from measured N_2 density profiles are shown in figure 16, in illustration of typical results obtained by this technique. The Harris and Priester model values appropriate to the state of the atmosphere at the time of the measurements are also shown for comparison. Individual flight date, time and other pertinent information are shown on the figure sequentially, matching the temperature profiles.

A brief study of these data yield a number of significant points: although the far right temperature data largely support the model values, the daytime measured values show a cooler atmosphere by at least 200°. The nighttime temperature profile when normalized for $a_p = 5$ shows an exospheric value lower by over 100 degrees (see page 105–106 for discussion of geomagnetic influence). The temperature gradients are seen to vary widely, reflecting the great variability of the thermosphere. Of particular interest are the 3rd and 5th curves from the left showing a near midnight temperature, and the temperature approximately 12 hours later. Although the exospheric values are in accord with our concept of the temperature, the nearly equivalent gradients in the 200 km region have not been explained. The daytime gradient, dependent upon the input of solar energy, the conductivity and other factors, appears generally acceptable, and comparable to the adjacent curve labeled "*C*".

However, the nighttime curve does not show the expected decrease in temperature as compared with the other nighttime data and displays a gradient which, being comparable to the daytime value, implies either greatly reduced conductivity or a substantial nighttime heat source.

For additional detail and discussion of measured data as presented above the references 16 and 17 should be consulted.

A more direct and localized measurement of the nitrogen temperature is possible through additional analysis of the "shape" of the individual pressure modulation cycles like those shown in Figure 12, and discussed above in connection with the determination of density. In this regard, when the orifice axis is normal to the velocity vector, the "ram" velocity component is zero and the chamber experiences only the thermal velocity components of the gases. For some angle slightly less than 90°, the translational and thermal velocities are comparable. Thus as the orifice passes through the transition from ram to rarefaction, the velocity distribution of the particles is, in effect, scanned. This is evident in Figure 12 as the pressure approaches or leaves P_{min}. The curvature as zero pressure is approached is a measure of the velocity distribution, and thus the temperature. To permit adequate temperature determination accuracy, the velocity-scan technique requires exceptionally good resolution of the transition portion of the curves near $\theta = 90°$, as well as a background pres-

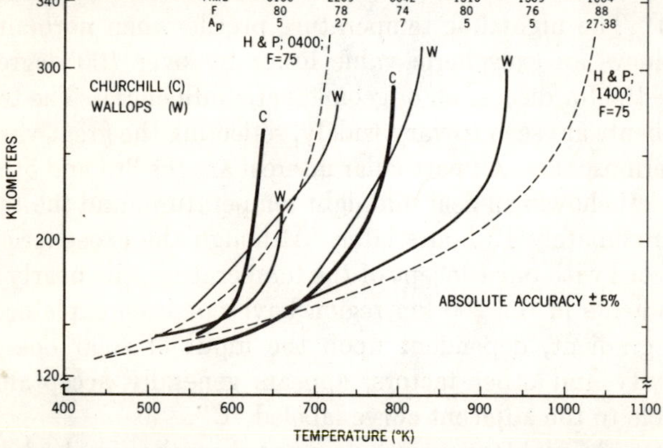

FIG. 16. Molecular nitrogen temperatures values vs. altitude derived from nitrogen density profiles.

sure ($P_{\min}$), which is usually negligible compared to the transition pressure.

To evaluate temperature by this technique, the data of Figure 12 are used. The value corresponding to $\theta_{\min}(P_{\max})$, the value at $\theta = 90°(P_{90°})$, and the minimum value (P_{bg}) are used to determine $f(s)$ as follows:

$$f(s) = (P_{\max} - P_{bg})/(P_{90°} - P_{bg}); \tag{19}$$

use of equation (2–2) then permits the calculation:

$$T_{N_2} = (m_{N_2}/2K)[(V \cos \theta)/s]^2 \tag{20}$$

In this manner two temperature values can be determined for each cycle of pressure variations such as shown in Figure 12.

Although there have been several opportunities to apply the velocity scan technique to density measurements, this technique cannot yet be considered fully developed in upper atmosphere temperature measurement. Present results still exhibit rather large scatter in the computed temperatures in contrast to those temperatures obtained using the scale height technique. As noted above, the experiment requires good precision in measurement of the angle (θ) in the neighborhood of $\theta = 90°$, and the corresponding pressure change. Accomplishment of measurements of adequate precision will, however, permit a direct measurement of the distribution of velocities among the assemblage of particles, or, if the gas is in thermal equilibrium, the gas temperature. The method therefore permits, possibly, the most direct measurement of gas temperature possible, where measurements based upon conduction are not feasible.

Quantitative Measurements of Various Species

As a practical matter, the use of ionization gages for total density measurement leads to certain difficulties because of the fact that the gages do not generally respond in an entirely known manner to the various gases of which the atmosphere is composed. Laboratory studies as well as considerations of flight data reveal that the gages respond differently to different gases and moreover, that their response is time dependent. The accuracy of the atmospheric data obtained is inevitably reduced by these effects, and also by inadequacies in calibration of the gages. The laboratory calibration capability existing today permits satisfactory results only, in general, for molecular nitrogen and helium, and possibly for molecular oxygen. Providing accurately known concentrations of atomic oxygen or nitrogen of adequate purity or even as the predominant gas in a laboratory vacuum system, is not easily accomplished. Thus calibrations for these gases cannot be considered entirely satisfactory. Even considering all these factors, however, meaningful measurements based upon attainable calibrations can still be made. For example, determinations for the limiting cases wherein calibration factors having admittedly large errors are assumed permit acceptable accuracies, considering parameter variation now known to occur. The absolute accuracy realized is believed to be about 25% as typically good, and 50% as readily attainable.

As techniques for the application of gages to atmospheric measurements improve and, consequently, as more precision in knowledge of the sensors and their calibration is demanded, the problems discussed above are of increasing concern. It becomes clear that the use of a device which responds in a not-too-well known fashion to atmospheric gases does not permit precision measurements. As a result, mass spectrometers, which respond to the various species independently, as though the other species did not exist, are being used used more and more. Although the same limitations regarding calibration exist for these devices as for ionization gages, the independent response of the spectrometer as a function of mass greatly eases the data analysis problem.

Thus present day direct measurements of the concentration of atmospheric components are in many cases best carried out by mass spectrometers, using techniques essentially identical to the technique discussed above under total density measurements.

There are also several interesting problems associated with the use of the mass spectrometer.[19] For example, atomic oxygen particles need only collide with some metallic surfaces a few times to recombine into molecular form, and thus possibly be measured (and misinterpreted) as a molecule of oxygen. As a compensating mechanism, molecular oxygen is in part dissociated into atomic form by contact with the hot filaments that are usually a necessary element of mass spectrometer ion sources. Other chemical reactions are likewise experienced, all of which tend to increase the magnitude of the error flag associated with the resulting data. Thus in designing a mass spectrometer for atmospheric measurements a number of questions must be considered. For measurements in the satellite regime above about 250 km where the atomic oxygen concentration predominates, the most accurate measurements of atomic oxygen *may* be obtained by arranging the spectrometer gas system so that all oxygen atoms can have many surface collisions. If the surfaces are prepared to enhance recombination then the atomic oxygen can be recombined and subsequently measured as molecular oxygen.

At the lower altitudes however, where atomic and molecular oxygen exist in comparable concentrations, ambiguity results. In this region the best measurements are made with an "open" ion source spectrometer where an effort is made to minimize surface collisions. Any ionizing source must, of course present some surface where recombination can take place, thus making it difficult to fully identify the source of the measured gas with ambient atmosphere.

An additional problem is establishing the temperature of the gas being measured in the instrument. In the case of a hypothetical system where a negligibly small number of surface collisions of the gases to be measured occurs, the temperature of the particles is clearly ambient. However, for practical open sources, the majority of the particles experience at least a few collisions with surfaces, and depending upon the accomodation coefficient (generally unknown), assume to a varying degree, the wall temperature. Their apparent density is thus increased as indicated by equation (11). For fully enclosed sources, full accomodation is expected and assumed meaning equivalence between particle and wall temperatures (see page 117, middle).

Although all these factors must be carefully evaluated in the analysis of flight data resulting from mass spectrometers, the most useful instrument known today which permits quantitative determination of the concentration of the various atmospheric constituents is, the mass spectrometer.

Determination of Other Atmospheric Gases

Although nitrogen, oxygen, helium, and hydrogen are the chief species in the thermosphere, from the point of view of numerical concentrations, there are also some minor constituents. In the lower thermosphere there is an easily measurable, but not significant amount of Argon. Near the mesopause, the most significant minor constituent is nitric oxide which plays an important role in the formation of the ionosphere. Its presence results from the recombination of N and O, each resulting from the dissociation of the more plentiful gases. Quantitatively, first measurements by Barth[20] indicate a total column density, of $1.7° \times 10^{14}$ molecules/cm^2 above 85 km. This value, compares, for example, with approximately 10^{20} particles for the other gases above each square centimeter at 85 km. Thus, the relative concentration of NO appears to be, at most, the order of 1 in 10^6 at 85 km, and it is therefore, very much a trace element. Barth's measurement is accomplished optically, by observing the day-glow spectrum from 1500 to 3200 A. The radiation corresponding to NO is thus observed and interpreted in terms of the amount of neutral NO.

The importance of the relatively small density of NO in the atmosphere stems from the fact that it is readily ionized by the very intense Lyman α line at 1216 A, and is thus responsible for much of the D-Region electron and ion concentration. This has been verified by direct ion concentration measurements which show a strong predominance of NO ions in the E-Region.

The NO neutral particle has never been detected by mass spectrometric means because of the lack of an instrument that would respond to the very low concentration, in a background of the much larger concentration of other molecules and atoms. In addition, the NO exists in a region that does not lend itself readily to spectrometry because of the relative high density, and the consequent short mean free path, which causes scattering with the consequent difficulties for mass spectrometers. It is probable however, that given adequate attention, an instrument could be developed, which could accomplish the measurement.

Other "trace" gases are present in the atmosphere, but are, in general, found at lower altitudes than considered in this chapter. The penetration of dissociating or ionizing radiation to the lower levels is sufficiently attenuated by the ionospheric regions so that other molecules can exist at the lower altitudes. The two principle gases are CO_2 and O_3, which are discussed in chapter 18.

Direct Measurements of Electron and Ion Properties

This chapter has been concerned primarily with the gross earth's atmosphere and thus so far has considered only neutral, or uncharged, particles. However, electrons and ions, although charged particles, are

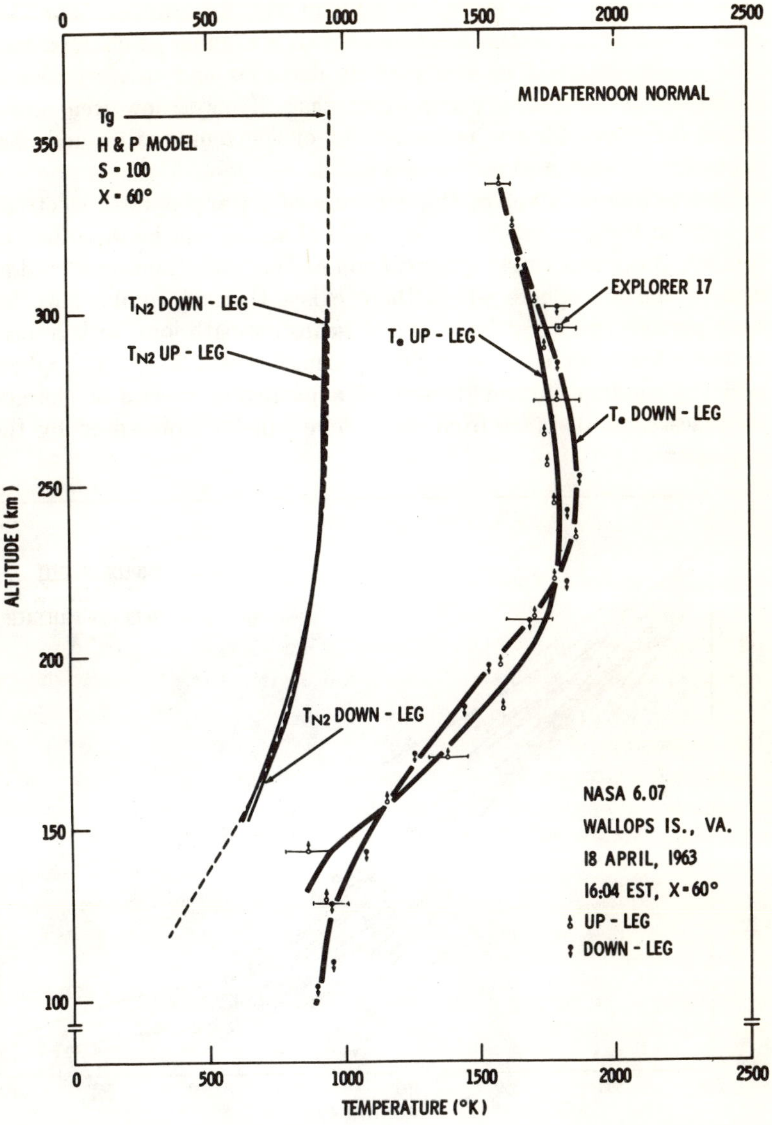

FIG. 17. Plot of simultaneously measured electron and molecular nitrogen temperatures illustration typical thermal non-equilibrium existing in the daytime *F*-region ionosphere (from Spencer, Brace, Carignan, Tauesch, and Niemann, Ref. 16).

important factors in considerations of the neutral atmosphere for they play important roles in the heating of the atmosphere. For this reason, the simultaneous measurement of a number of charged particle parameters and neutral particle densities and temperatures is becoming an increasingly important task. Thus a now well established technique for the measurement of the temperature and concentration of electrons will be discussed.

The technique involves the exposure of an appropriate electrode system to the ionosphere, an example of which can be described as follows: Assume a single spherical conducting body immersed and at rest in the ionosphere where there exists thermal equilibrium, for each particle species at least, and a mean-free-path long with respect to the sphere diameter. In this case, a current will flow to the sphere and the random current density of a particular species of charged particles some distance from the sphere can be represented by the expression

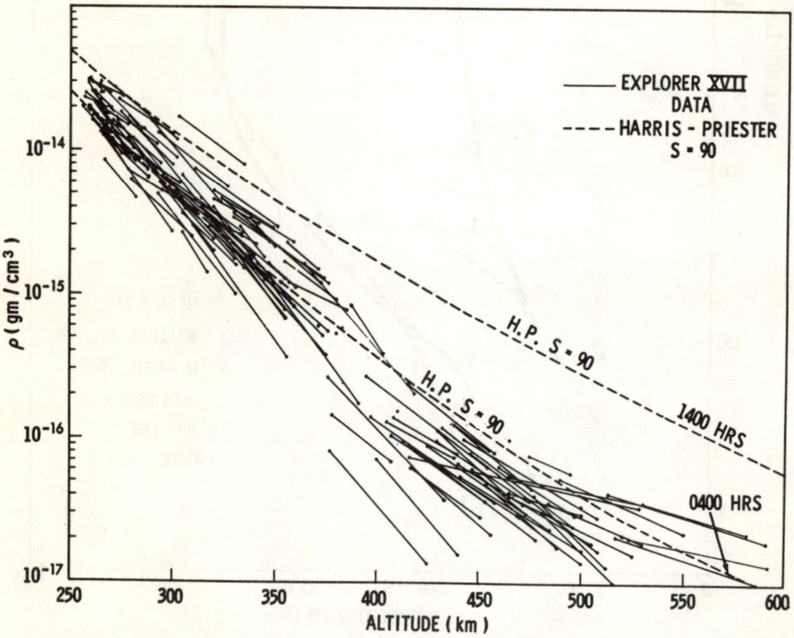

FIG. 18. Gross atmospheric total neutral particle density variation observed with Explorer XVII Satellite (from Newton, Horowitz, and Priester, Ref. 22).

$$J = Ne(kT/2\pi m)^{1/2} \tag{21}$$

where

J = random current density,
N = particle number density,
e = electron charge,
T = particle kinetic temperature,
m = particle mass,
k = Boltzmann constant.

Assuming that the predominant ionospheric ion present is O^+, the ratio of the electron to ion current density is about 170 or the square root of the mass ratio of these two components. This is the ratio of the number of electrons to the number of positive ions (number density of negative ions is considered negligible) that would cross a unit area in a unit time. This will not, however, be the case at the surface of the sphere initially because the predominance of the electron

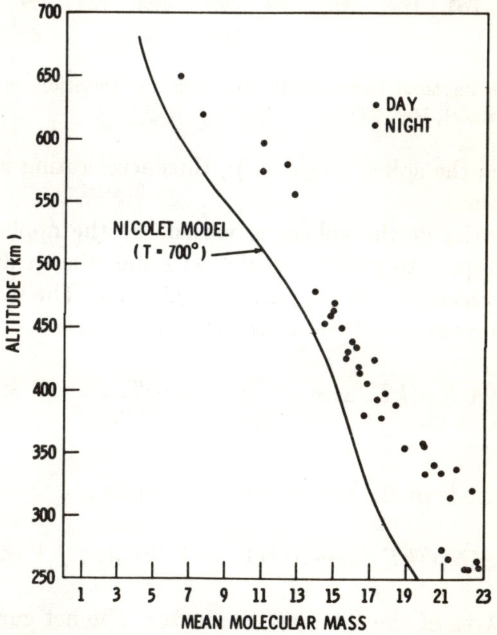

FIG. 19. Measured mean molecular weight vs. altitude compared with Nicolet model value (from Reber and Nicolet, Ref. 19).

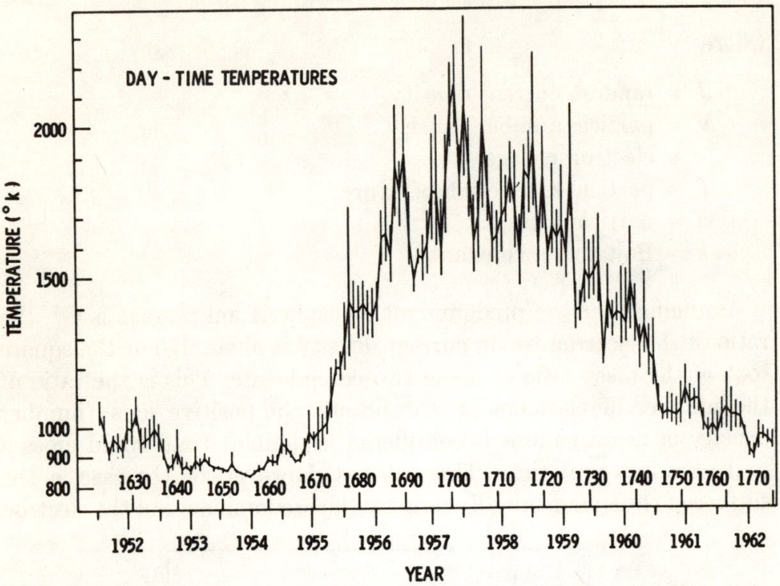

FIG. 20. Gross daytime thermopause temperature variation as reported by Nicolet (from Nicolet, Ref. 23).

flux will charge the sphere negatively, thus accelerating ions and retarding electrons.

If the potential of the sphere is varied, by the application of a voltage with respect to a rocket vehicle or some other relatively large electrode, the electron and ion currents will vary. The magnitude of the electron current that flows to the sphere is given by

$$I_e = A_s N_e e (kT_e/2\pi m_e)^{1/2} \exp(eV/kT_e), \qquad V < 0 \tag{2}$$

and the magnitude of the ion current is given by

$$I_p = A_s N_p e (kT_p/2\pi m_p)^{1/2} (1 + eV/kT_p), \qquad V < 0, \tag{3}$$

where A_s = Area of the spherical conductor. The net current to the sphere I is the sum of the electron current (2) and the ion current (3). Assuming $N_p = N_e$ and $T_p = T_e$, the net current is given by

$$I = A_s Ne(kT/2\pi)^{1/2}[(1 + eV/kT) - (m_p/m_e)^{1/2} \exp (eV/kT)], \qquad V < 0 \quad (4)$$

To compute the electron temperature one takes the logarithm of equation (2) and its derivative, which leads to

$$d \ln I_e/dV = -e/kT_e \quad (5)$$

Since k and e are known constant, T_e can be determined. Hence, each volt-ampere characteristic measured may be interpreted in terms of the electron temperature of the plasma.

Use of an experiment of this type has resulted in essentially direct measurements of the temperature of the ionospheric electrons. Knowledge of the temperature has subsequently led to an understanding of how ultra-violet energy absorbed in the ionosphere is conducted to the ionic and neutral atmosphere. For example, the He radiation at 304 Angstroms is strongly absorbed by and ionizes the atomic oxygen in the general region centered around 150 km altitude, as discussed in Chapter 3. The electrons so produced rapidly conduct their energy to other electrons and then to the neutral particles, thus heating the atmosphere. Figure 17 illustrates the difference that typically exists in electron and neutral particle temperatures during midday of the quiet sun period. The difference in temperature is a clear manifestation of the energy flowing from the solar source through the electrons to the neutral particles. Recent measurements during an eclipse confirm the phenomena, showing the rapid decay of electron and neutral particle temperature observed when the solar input was temporarily, but substantially, reduced. The direct sensitivity of electron temperature to solar energy input is thus clearly evident, or in other words, that the electron temperature is a very sensitive indicator of the intensity of the solar radiation. This statement need only be tempered by the dependence of the electron temperature upon the electron density, as pointed out by Hanson.[21]

SUMMARY OF NEUTRAL PARTICLE ATMOSPHERE

Both indirect measurements (satellite drag) and direct measurements (gauges) have demonstrated the very great dependence of the structure of the neutral particle atmosphere upon the intensity of the solar input. Absorption of energy by the earth and re-emission in the infrared heat the lower atmosphere and cause the well-known weather phenomena, as discussed in Chapter 6. The absorption of ultra-violet and corpuscular borne energy at higher altitudes, where the mean free path is long, heats, dissociates, and ionizes the constituents. Variation of the energy input results in large variations in density, temperature, and mean molecular weight, as well as an ionosphere.

Figures 18 and 20 illustrate the variation of density and temperature which occurs in the upper thermosphere. Figure 19 shows the theoretically predicted (Nicolet) mean molecular weight of the atmosphere from 250 to about 700 km, and a number of measured data points for comparison. The significant variations observed can be attributed largely to the changing conditions in the lower thermosphere, in the 100 to 200 km region, which is relatively unexplored. Measurements in this region, which is not yet readily accessible to scientific satellites, are badly needed to help test and confirm theories relating to the complex response of this region to the changing solar energy input. Experimentation devoted to the study of this altitude region is becoming a major goal in upper atmosphere experimental studies.

REFERENCES

1. I. Harris and W. Priester, On the Diurnal Variation of the Upper Atmosphere, J. Atmos. Sci., (to be published).
2. Kockarts, G. and M. Nicolet, Helium and Hydrogen in a Period of Minimum Solar Activity, Ann. Geophysique, **19**, 370 (1963).
3. A. Kochauski, Atmospheric Motions from Sodium Cloud Experiments, J. Geophys. Res. **69**, 3651 (1964).
4. I. Harris and W. Priester, Time Dependent Structure of the Upper Atmosphere, J. Atmos. Sci., **19**, 286 (1962).
5. C. G. Jacchia and J. Slowey, "The Shape and Location of the Diurnal Bulge in the Upper Atmosphere" SOA special report No. 207, Smithsonian Institution Astrophysical Observatory, Cambridge, Massachusetts.
6. G. J. F. MacDonald, The Escape of Helium from the Earth's Atmosphere, Rev. Geophys, **1**, 305 (1963).
7. L. G. Jacchia and R. E. Briggs, Smithsonian Astrophy., Obs. Spec. Rept. No. 18, 1958.
8. W. Priester, Solar Activity Effect and Diurnal Variation in the Upper Atmosphere, J. Geophys. Res. **66**, 4143 (1961).

9. W. N. Neupert, W. E. Behring, and J. C. Lindsay, The Solar Spectrum from 50 Å to 400 Å, in *Space Research, V*, (P. Muller, ed.), pp. 719–729, John Wiley, New York, 1964.
10. H. K. Paetzold,. and H. Zschoerner, The Structure of the Upper Atmosphere and Its Variations after Satellite Observations, in *Space Research, II*, (H. C. van de Hulst, C. de Jager and A. F. Moore, eds.), p. 958, John Wiley, New York, 1961.
11. L. G. Jacchia, Variation in the Earth's Upper Atmosphere as Revealed by Satellite Drag, Rev. Mod. Phys. **35,** 973 (1963).
12. I. Harris and W. Priester, Theoretical Models for the Solar-Cycle Variation of the Upper Atmosphere, J. Geophys. Res. **67,** 4585 (1962).
13. G. P. Newton, R. Horowitz, and W. Priester, Atmospheric Densities from Explorer XVII Density Gauges and a Comparison with Satellite Drag Data, J. Geophys. Res. **69,** 4690 (1964).
14. COSPAR working group IV, CIRA 1965 COSPAR International Reference Atmosphere 1965, North Holland Publishing Company, Amsterdam 1965.
15. M. Nicolet, Helium, an Important Constituent in the Lower Exosphere, J Geophys. Res. **66,** 2263 (1961).
16. N. W. Spencer, L. H. Brace, G. R. Carignan, D. R. Taeusch, and H. Niemann, Electron and Molecular Nitrogen Temperature and Density in the Thermosphere, *J. Geophys. Res.*, **70,** 2665, (1965).
17. N. W. Spencer, D. R. Taeusch, and G. R. Corignan, N_2 Temperature and Density Data for the 150 to 300 km Region and Their Implications, Annales de Geophysique, Tome 22 No. 2 Avril-Juin 1966.
18. G. P. Newton, D. T. Pelz, G. E. Miller, and R. Horowitz, Response of Modified Redhead Magnetron and Bayard-Alpert Vacuum Gauges Aboard Explorer XVII, *Transactions of the Tenth National Vacuum Symposium*, edited by George H. Bancroft, pp. 208–212, The MacMillan Co., New York, 1963.
19. C. A. Reber and M. Nicolet, Investigation of the Major Constituents of the April–May, 1963 Heterosphere by the Explorer XVII Satellite Planet. Space Sciences 1965, Vol. 13 pp. 617-646.
20. C. A. Barth, Rocket Measurement of Nitric Oxide Dayglow, J. Geophys. Res. **69,** 3301 (1964).
21. W. B. Hanson and F. S. Johnson, Electron Temperatures in the Ionosphere, Mem. Soc. Sci. Liege (Series 5), **4,** 390 (1961).
22. G. P. Newton, R. Horowitz, W. Priester, Atmospheric Density and Temperature Variations from the Explorer XVII Satellite and a Further Comparison with Satellite Drag, (Planet. Space Science 1965 Vol. 13 pp. 599-616).
23. M. Nicolet, Solar Radio Flux and Temperature of the Upper Atmosphere, J. Geophys. Res. **68,** 6121 (1963).

Chapter 3

THE IONOSPHERE

A. C. Aikin and S. J. Bauer

INTRODUCTION

Traditionally, the term *ionosphere* applies to the ionized portion of the earth's upper atmosphere above about 70 km. Cosmic rays produce electrons and ions at altitudes lower than 70 km, but such phenomena have been considered in the past as *atmospheric electricity*. The ion and electron density distribution at high altitudes and out to several earth radii distance has only recently been the subject of much study. In this context the term *magnetosphere* has been introduced for the region where the geomagnetic field has dominant control over the motion of low energy plasma *and* the fast charged particles constituting the radiation belt. The magnetosphere extends from about 160 km altitude to the outer boundary of the geomagnetic field, usually at geocentric distances greater than 10 earth radii. Whereas the magnetosphere encompasses both low and high energy charged particles, the *ionosphere*, to which this chapter will be restricted, represents only charged particles of *thermal* energies.

The existence of an ionosphere was first postulated in 1882 by

Balfour Stewart to explain the daily variations of the earth's magnetic field. Again in 1902 the suggestion for the existence of an ionosphere was made independently by Kennelly and Heaviside to explain Marconi's transatlantic radio transmissions. The actual verification of its existence, however, came much later when in 1925 Appleton and Barnett in England and shortly thereafter Breit and Tuve in the United States established the existence and altitude of the Kennelly–Heaviside layer, as the ionosphere was then called. The original method of Breit and Tuve which employs short pulses of radio energy at vertical incidence provides the basis for the present-day ionospheric soundings. From the delay time between the transmitted and reflected pulses the (virtual) height of reflection, assuming waves propagated at the velocity of light, can be determined, while transmissions at different frequencies show different heights of reflection. This is now known to be the result of an altitude variation in the number density of free electrons. E. V. Appleton, who received the 1947 Nobel Prize in Physics for his pioneering work in ionospheric research is also responsible for the expression of the refractive index of a magneto-ionic medium[1] as a function of frequency of the exploring wave, the earth's magnetic field strength, the electron density and electron neutral particle collision frequency. With the aid of this mathematical analysis and variable frequency pulse soundings it has been possible to obtain electron density versus true height profiles. A worldwide network of such ionospheric sounders has provided a large amount of data on the temporal, seasonal, and geographic variations of the ionosphere.

Reflections of radio signals occur from three ionospheric regions designated as the *D*, *E* and *F* layers. The originator of this terminology was E. V. Appleton. In a letter to J. H. Dellinger, an American pioneer in ionospheric research, written in 1943 he explains how he arrived at the nomenclature.

> "The story of how I came to give them the names D, E and F is really a very simple one. In the early work with our broadcasting wavelengths, I obtained reflections from the Kennelly-Heaviside layer, and on my diagrams I used the letter E for the electric vector of the down-coming wave. When therefore in the winter of 1925 I found that I could get reflections from a higher and completely different layer, I used the term F for electric vector of

the waves reflected from it. Then at about the same time I got occasional reflections from a very low height and so naturally used the letter D for the electric vector of the return waves. Then I suddenly realized that I must name these discrete strata and being rather fearful of assuming any finality about my measurements I felt I ought not to call them layers A, B and C since there might be undiscovered layers both below and above them. I therefore felt that the original designation for the electric vector D, E and F might be used for the layers themselves since there was considerable latitude for naming of any layers that might come to light as a result of future work. I am afraid that is all there is in the story".

The lowermost part of the ionosphere (below 70 km) has become known recently as the C-region while the uppermost part (above 300 km) is now often designated as the *topside ionosphere.*

The *origin* of the ionosphere is the ionization of atmospheric constituents by solar radiation of wavelengths shorter than 2400 Å. The solar flux in this wavelength range is very dependent on the 11 year solar cycle. A detailed discussion of this variation and the intensities at different wavelengths can be found in chapter 16 on solar radiation. Similarly, the effect of dissociation of molecular oxygen and heating of the atmosphere by solar radiation will lead to an altitude-dependent distribution of constituents and even to the formation of molecules such as nitric oxide. A description of the terrestrial atmosphere and its variation can be found in chapter 2, dealing with the neutral atmosphere.

In this chapter, the processes which determine the charge density and ionic composition at different levels of the terrestrial atmosphere will be discussed in some detail (pages 141–147). Then, a synopsis will be given of the formation of the different ionospheric regions under ideal conditions (pages 147–156). Detailed observations of the electron density distribution have shown that such conditions are rarely encountered. Instead one finds layers of a sporadic nature and anomalous latitude and seasonal distributions of electron density. Solar activity also produces marked changes in the ionosphere. These effects and their origin will be discussed (pages 160–162). Although most of our information on the ionosphere has been derived from ground-based sounders, the advent of rockets and satellites has made possible the measurement of additional parameters which lead to a better understanding of the origin of the ionosphere. The application of rocket and satellite borne experiments to the ionosphere will be the subject of the final section of this chapter.

THE PRODUCTION OF IONIZATION BY SOLAR PHOTONS

Absorption of Solar Radiation

For the case of monochromatic solar radiation the number of ion pairs produced (cm^{-3} sec^{-1}) at an altitude z by incident photons p, is

$$q_{pj} = n_j \sigma_{ij} Q_\infty \exp(-\tau), \tag{1}$$

where σ_{ij} is the ionization cross section in cm^2 of the jth constituent, n_j the number density cm^{-3} of that atmospheric constituent, and Q_∞ the number of photons cm^{-2} sec^{-1} incident on the earth's atmosphere.

The optical depth, τ, is the sum of the separate products of the absorption cross section σ_{ak} and the total number of absorbing molecules of species k between the height z and the sum. It is expressed as

$$\tau = \sum_k \sigma_{ak} \int_z^\infty n_k \sec\chi \, dz = \sum_k \sigma_{ak} n_k H_k \sec\chi, \tag{2}$$

where χ is the solar zenith angle, n_k the number of molecules of species k at height z, and H_k is the local scale height $H_k = kT/m_k g$ with k the Boltzmann constant, T the absolute temperature, g the acceleration of gravity and m_k the mass of the molecule of species k. The quantity $n_k H_k \sec\chi$ represents the total number of absorbing molecules between the height, z and the sun for a plane earth. However, for solar zenith angles greater than 85° it is necessary to take into account the curvature of the earth. This has been considered in various calculations for the sunrise effect in the ionosphere. Chapman[2] has derived a function which can be used in place of sec χ in equation (2) provided that the scale height is constant.

The product of the exponential decrease of particle density with increasing altitude and the exponential factor for the attenuation of the incident solar flux results in a maximum of the production function. The altitude at which this peak of the production function occurs is obtained by differentiating equation (1). For the case of an overhead sun, $\chi = 0°$, and a single absorbing constituent

$$\tau = 1 + \beta, \tag{3}$$

where $\beta = dH_k/dz$ is the gradient of the scale height H_k, whose value lies between 0.15 and 0.6 for the major absorbing constituents in the altitude region 85 to 200 km.

The absorption cross sections of O_2, N_2, and O are variable in the far ultraviolet and X-ray wavelength range. For instance the Lyman alpha line of hydrogen located at 1215.7 A has the following absorption cross sections $\sigma_{aO_2} = 10^{-20}$ cm^2 and $\sigma_{aN_2} = 6 \times 10^{-23}$ cm^2 so that 10^{20} oxygen molecules/vertical column are required to reduce this flux to $1/e$ of its incident value which corresponds to unit optical depth. Similarly the Lyman alpha line of helium II at 303.8 A has absorption cross sections of 2×10^{-17} cm^2 for O_2, 1×10^{-17} cm^2 for O and 5×10^{-18} cm^2 for N_2 indicating that this portion of the solar spectrum is absorbed at higher altitudes. Table I lists the altitude at which unit optical depth occurs for different wavelength intervals and prominent lines of the solar spectrum. In the case of a wavelength interval it is necessary to employ average absorption cross sections. This table gives a rough idea of the depth of penetration of different portions of the solar spectrum. Wavelengths shorter than 40 A and greater than 1026 A are absorbed below 100 km. The inter-

TABLE I

Penetration depth of some of the strong solar emission wavelength regions

Line	Wavelength, A	Altitude of unit optical depth, km
H Lyα	1215.7	73
H Lyβ	1025.6	105
C III	977.0	120
H ly cont.	911–840	105–120
He I	584.3	164
He II Lyα	303.8	130
X-rays	50	105
	4	80

mediate wavelengths are absorbed primarily in the altitude interval 100 to 160 km.

Photoionization Rates

Table II gives the first ionization potential and corresponding threshold wavelength for different atmospheric constituents. The abundance of these atoms and molecules at different altitudes coupled with the solar flux of different wavelengths available at a given altitude will determine the rate at which an ion of a given species is produced. Figure 1 illustrates the result of such an analysis.[3] Here are shown the rates of primary ion pair production in the altitudes region 90 to 350 km. It is seen that from 90 to 140 km O_2^+ and N_2^+ are the primary photoions with O^+ being of secondary importance. Above this altitude O^+ is produced in large quantities.

Below 100 km the principal neutral constituents are O_2 and N_2 so that photoionization of these molecules will produce O_2^+ and N_2^+ ions in the altitude range 70 to 100 km. A more important ion in this

TABLE II

The ionization potentials and threshold wavelengths for photoionization of atoms and molecules

	Ionization potential (eV)	Wavelength (Å)
Na	5.14	2410
Ca	6.11	2060
Mg	7.43	1670
Fe	7.83	1585
Si	8.15	1520
NO	9.25	1340
O_2	12.08	1025
H	13.59	912
O	13.61	910
N	14.54	852
N_2	15.58	795
He	24.56	504

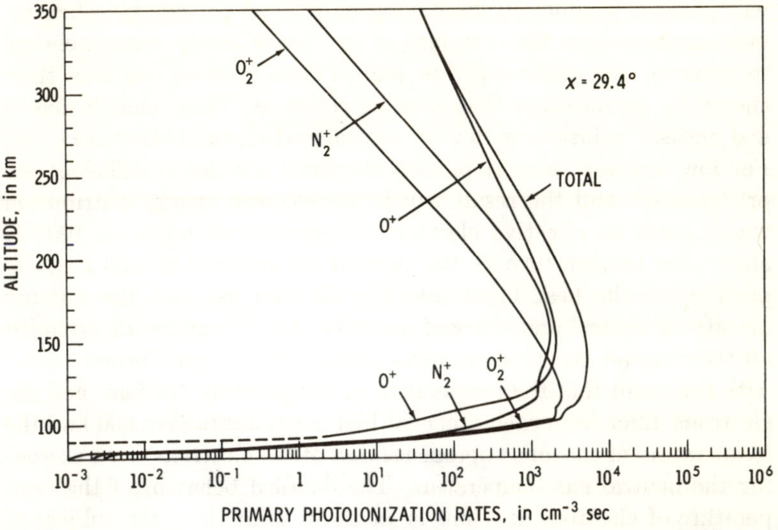

FIG. 1. Primary photoion production rates (after Hinteregger et al., Ref. 3).

altitude range is NO^+, which is produced by the photoionization of the minor constituent NO by the strong solar Lyman alpha line at 1215.7 A.

Energy Dissipation of Photoelectrons

The photoelectrons produced as a result of the absorption of photons will have a kinetic energy equal to the energy difference between the photon and the binding energy of the electron. If the electron energy lies between 20 and 70 eV, it should be able to produce one additional ion pair cm^{-3} sec^{-1} for each photoelectron. Electrons with energy greater than 70 eV will produce on the average 1 ion pair for each 34 eV of energy they possess. These additional ion pairs will contribute substantially to the total production function at short wavelengths. For instance, even if an electron is removed by a 6.2 keV photon from the K shell of nitrogen (requiring 400 eV of energy) the photoelectron will possess enough kinetic energy to produce 170 additional ion pairs. A photon whose wavelength is 304 A (41 eV) will produce photoelectrons having only 27 eV of energy when removing them from the first shell of O. This is barely enough energy to produce one additional ion pair. Photoelectrons can lose their energy by means other than ionizing collisions. These include elastic and inelastic collisions with ions, neutral particles and other electrons. For low ion and neutral particle densities, Coulomb collisions are predominant and the result will be an electron energy distribution which gives an effective electron temperature as much as 1000°K above the temperature of the neutral atmosphere at 200 km, depending on the heat input into the electron gas and the number density of neutral and charged particles. There may be an extended altitude region where departures from thermal equilibrium occur, with the result that the temperature of the neutrals, the ions, and the electrons differ from each other; at high altitudes (above 600 km) the temperature of the ions approaches that of the electrons, both exceeding the neutral gas temperature. The detailed behavior of the temperature of electrons and ions is, at the present time, the subject of intensive theoretical and experimental study.[4]

Minor Constituents

Airglow observations have shown that sodium is present as a minor constituent of the earth's atmosphere. Between 80 and 100 km the sodium concentration is of the order of a few atoms per cm^3. The wavelength corresponding to the threshold of photon energy necessary to ionize sodium is 2410 Å. Since radiation of this wavelength has unit optical depth at 40 km, a certain amount of the sodium will be ionized. Sodium ions have been detected between 80 and 110 km by a U. S. rocket-borne ion mass spectrometer.[5] This same flight also detected ions of mass 24 and 40 which may be identified as Mg^+ and Ca^+ These ions together with Fe^+ and Si^+ have also been detected in flights conducted by Russian investigators in the 100 to 120 km altitude region. The source of metallic ions such as Mg^+, Ca^+, Fe^+, and Si^+ is considered to be meteors. Although atmospheric dissipation of the energy of meteors can result in the formation of ions, (meteoric) atoms also can be ionized by solar radiation having wavelengths shorter than the thresholds listed in table II. Diffusion of such minor constituents as Ca and Mg in the major atmospheric gases can result in a layer of metalic ions centered between 100 and 110 km. The ion mass spectrometer results support this theory.

The number of atomic ions such as Na^+ and Ca^+ present in the atmosphere will be determined not only by their rate of production, but also by radiative recombination of an ion and electron. This process which is represented as

$$X^+ + e \rightarrow X + h\nu \tag{4}$$

has a rate coefficient of the order of 10^{-12} cm^3 sec^{-1} compared to molecular ions which recombine at rates of the order of 10^{-7} cm^3 sec^{-1} illustrating the relative lifetime of atomic versus molecular ions. The topic of recombination will be explored in more detail in the next section.

IONIC REACTIONS

Because of the large number of pertinent reactions whose rate coefficients are not known accurately, it is only possible to give at this time a qualitative description of the ion kinetics of the earth's atmosphere. The different reactions will be grouped according to the region of the ionosphere where they are most important.

In the D region where the main neutral atmospheric constituents are O_2 and N_2 and the ionic constituents NO^+, O_2^+ and N_2^+, the following ion-molecule reactions can occur;

$$O_2^+ + N_2 \rightarrow NO^+ + NO \tag{5}$$

$$N_2^+ + O_2 \rightarrow NO^+ + NO \tag{6}$$

and

$$N_2^+ + O_2 \rightarrow O_2^+ + N_2 \tag{7}$$

leading to NO^+ as the primary ion in the D-region with O_2^+ and N_2^+ being of secondary importance even during the times of enhanced production of N_2^+ and O_2^+, such as during solar flares and particle bombardment events.

The dissociative recombination of positive molecular ions with electrons through the processes

$$N_2^+ + e \rightarrow N + N \tag{8}$$

$$O_2^+ + e \rightarrow O + O \tag{9}$$

and

$$NO^+ + e \rightarrow N + O \tag{10}$$

is an important type of reaction for the loss of electron ion-pairs throughout the lower ionosphere.

Electrons also attach themselves to oxygen molecules to form the negative ion O_2^-. Since this ion can be photodetached by visible light, some of the negative ions formed will be destroyed by solar photons and there will be a diurnal variation of the negative ion density. This will also result in changes of the electron density in the lowest ionosphere since it is required that charge neutrality be maintained. The different rates of formation of negative ions are such that the

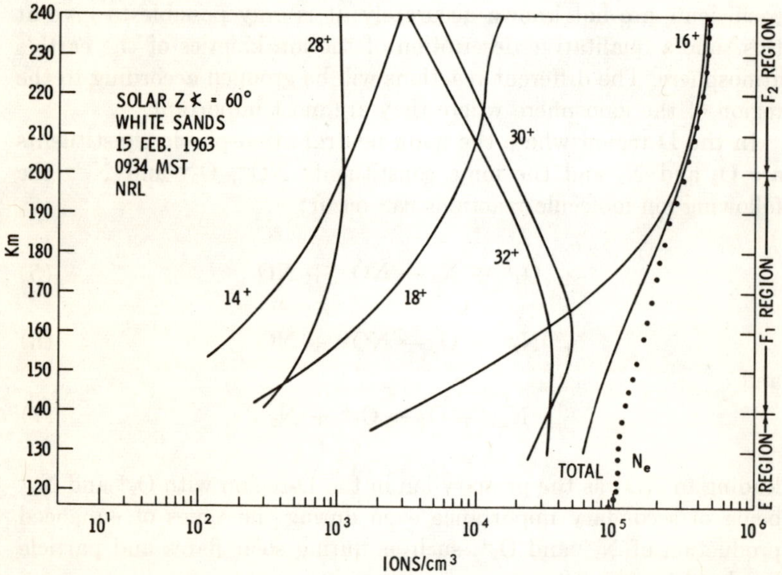

FIG. 2. Ion composition of the daytime ionosphere (after Holmes, Johnson, and Young, Ref. 6).

ratio of the negative ion density to electron density is unity at 68 km during the day and 90 km at night.

Figure 2 shows the results of a rocket borne ion mass spectrometer flown over White Sands, New Mexico, by workers of the Naval Research Laboratory.[6] This flight occurred on 15 February 1963 during the same period of the sunspot cycle as the solar ultraviolet flux measurement which was the basis for figure 1. The solar zenith angles differ by only 30° for the two flights. The dominant species of positive ions in the E region, i.e., ions below 160 km, are O_2^+ and NO^+. It is seen that N_2^+ is an ion of secondary importance even though large quantities of this ion are produced as a result of photoionization. This result implies that in addition to the dissociative recombination reactions described above other processes such as ion-atom interchange and charge transfer are acting to form NO^+. Among the most important of these reactions are

$$O^+ + N_2 \rightarrow NO^+ + N \quad (11)$$

$$O^+ + O_2 \rightarrow O_2^+ + O \quad (12)$$

The NO^+ and O_2^+ formed in this manner can then recombine with electrons through dissociative recombination. The reactions involving N_2^+ and atomic oxygen, namely

$$N_2^+ + O \rightarrow N_2 + O^+ \quad (13)$$

and

$$N_2^+ + O \rightarrow NO^+ + N \quad (14)$$

may also be of importance in determining the final ion composition.

At altitudes above 160 km corresponding to the F region the atmosphere is composed principally of atomic oxygen and the atomic oxygen ion formed by photoionization remains as the primary ion of this region. Although it can disappear through radiative recombination with an electron by the following process

$$O^+ + e \rightarrow O + h\nu$$

the slow rate of recombination allows ion-atom interchange processes to predominate.

As indicated in figure 2 atomic nitrogen ions are present in the atmosphere as a minor ionic constituent. This follows from the low concentration of neutral atomic nitrogen in the atmosphere; however the reaction

$$N_2^+ + N \rightarrow N^+ + N_2 \quad (15)$$

and the reaction (18) below can increase the N^+ concentration.

As atomic hydrogen becomes more predominant the charge exchange process

$$O^+ + H \rightleftarrows H^+ + O \quad (16)$$

must be considered. As noted earlier atomic oxygen and hydrogen have nearly identical ionization potentials so that the accidental asymmetric resonance reaction shown above serves as both a source and sink for H^+.

The helium ions produced by photoionization can react with O_2 and N_2 through the processes

$$He^+ + O_2 \rightarrow He + O^+ + O \quad (17)$$

$$He^+ + N_2 \rightarrow He + N^+ + N \quad (18)$$

Nighttime measurements of ion composition are further evidence of the important part played by ion-molecule reactions in the formation of the ionosphere. Figure 3 illustrates the results of a rocket-borne ion mass spectrometer flown at night by the Naval Research Laboratory.[6] The flight was conducted during the same part of the solar cycle as the day flight discussed previously. The decay of the ion content in the absence of sunlight is apparent by comparing figures 2 and 3. The ion species which show the greatest rate of decay are N^+, N_2^+, O^+, and O_2^+ all of which are ions produced by photoionization. During the night these ions are not produced and their number decays either through recombination with electrons or by ion-molecule reactions producing NO^+ and some O_2^+. These ions in turn recombine with electrons. At altitudes above 240 km the decay of the ionosphere is less apparent because of the absence of ion-molecule reactions producing molecular ionic species which have shorter lifetimes than atomic ions.

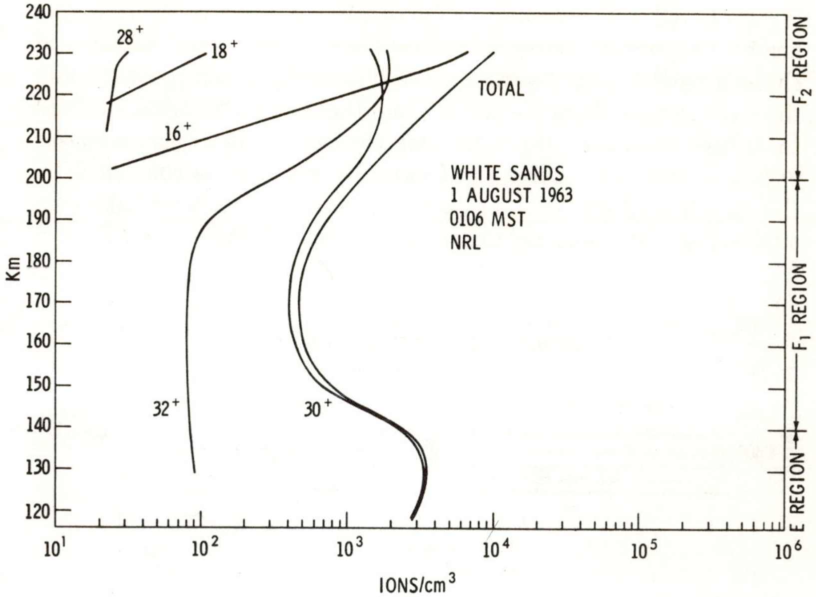

FIG. 3. Ion composition of the nighttime ionosphere (after Holmes, Johnson, and Young, Ref. 6).

MASS-TRANSPORT OF CHARGED PARTICLES

In addition to the production of ionization and its loss by chemical processes which were described in the preceding sections, *mass transport* of charged particles plays a role in establishing the observed charged particle density distribution with altitude, especially at altitudes above 300 km in the upper ionosphere. By mass transport we understand motion of charged particles due to concentration gradients, and gravity (*diffusion*) and Lorentz (e.m.) forces, also called *electrodynamic drift.*[7] The physical process responsible for most of this motion is diffusion of charged particles as the result of pressure gradients and gravity. The diffusion of charged particles in the ionosphere differs in two ways from diffusion of neutral species which leads to the exponential (hydrostatic) distribution of the neutral atmospheric constituents above about 120 km where each constituent is distributed independently of the others according to its own scale height, i.e., corresponding to Dalton's law. First, in the diffusion of charged particles, an electrostatic field is set up which prevents only the most minute charge separation and thus causes electrons and ions to diffuse essentially together. This concept is known as *ambipolar diffusion*. Secondly, diffusion of electrons and ions is controlled by the earth's magnetic field, i.e., electrons and ions are essentially constrained to move along field lines. The concept of ambipolar diffusion,

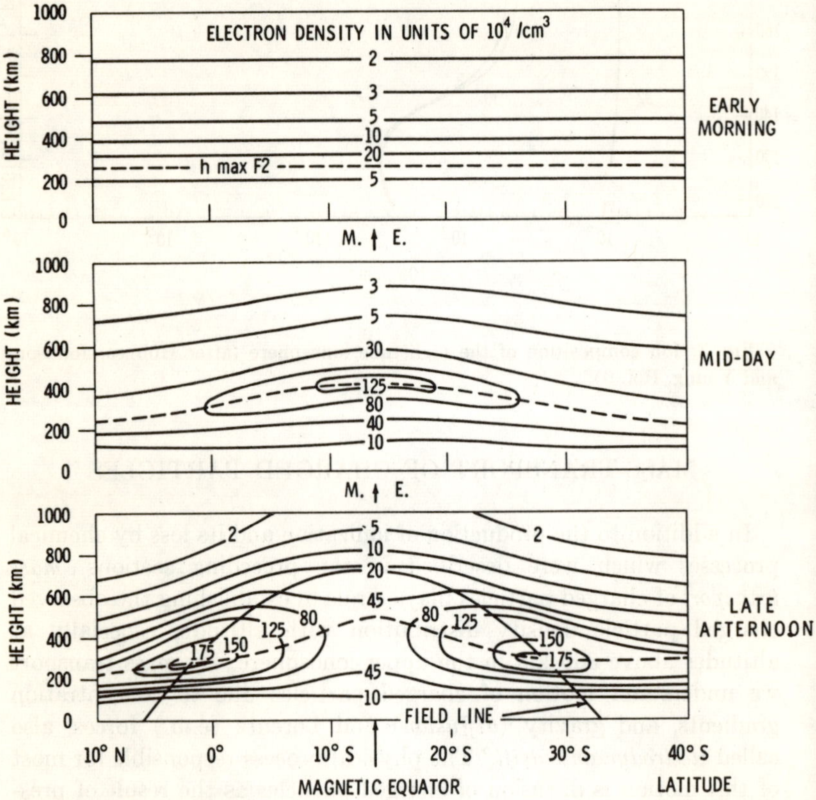

FIG. 4. Idealized representation of equatorial anomaly along 75° W meridian, based upon data by Lockwood and Nelms (topside) and J. W. Wright (bottomside).

however, meets with some difficulties if it is assumed that this diffusion occurs along field lines, and the velocity of electrons and ions is the same. In this case one does not obtain any geomagnetic control of diffusion since there is no current and, therefore, no interaction with the geomagnetic field, i.e., the charged particle distribution does not show any dependence on the geomagnetic field. This is contrary to observations which show indeed a geomagnetic control of the F region density distribution, which is known as the *equatorial anomaly* (Fig. 4). To explain this anomaly, the concept of strictly ambipolar diffusion has to be abandoned, and no a priori assumption about the equality of electron and ion vector velocity can be made. In fact, although the differences in electron and ion diffusion velocity may only be second order effects, they are required to describe the geomagnetic anomaly in the ionospheric F-region.[8] However, in most cases, excluding the region near the geomagnetic equator where field lines are horizontal, a one-dimensional approach, i.e. ambipolar diffusion along field lines, represents a reasonable approximation. In this case the vertical ambipolar diffusion velocity v_z is given by[7]

$$v_z = -D \sin^2 I[N^{-1}(\partial N/\partial z) + H'^{-1}], \tag{19}$$

where D is the diffusion coefficient which is proportional to T and inversely proportional to the density of the gas through which the charged particles diffuse; I the magnetic dip angle, and $H' = k(T_e + T_i)/m_+ g$ is the scale height of the electron–ion gas, with T_e and T_i the electron and ion temperature, respectively, k the Boltzmann constant, m_+ the mean ionic mass, and g the acceleration of gravity. It should be noted that v_z could be either upward or downward, depending upon the relative importance of density gradients and gravity.

FORMATION OF THE IONOSPHERIC LAYERS

In the previous sections a qualitative discussion has been given of the photochemistry of the terrestrial atmosphere leading to the altitude distribution of the different ionic species. In order to give a quantitative picture it is necessary to write the time-dependent continuity equation for each species of ion as well as electrons. These equations are then solved simultaneously. This is difficult particularly at lower altitudes where many reactions are involved including the formation of negative ions. In the ionosphere above 300 km the equation of continuity for the electron distribution must include the effect of mass motion. This is expressed in the form of the divergence of a flux $F = Nv$, which may be expressed by $\partial F/\partial z$ for the one-dimensional case considered here, as shown below

$$\partial N/\partial t = q - L - (\partial F/\partial z), \tag{20}$$

where q and L refer to all processes involved in the production and loss of electrons. A steady-state solution of this equation is usually valid under most circumstances except during such times as twilight. Neglecting diffusion, the electron density is controlled by chemical equilibrium, i.e., $q = L$. The electron density distribution under such conditions has been the subject of many investigations the best known and one of the earliest of which is the Chapman distribution.[2] In Chapman's derivation of the expression for the electron density, equation (1) for q is modified by the assumption of a single wavelength of ionizing radiation in an isothermal atmosphere. As shown before (pages 136–137) the maximum of ion production occurs where optical depth (τ) becomes unity. For overhead sun we can write this in the form

$$\tau = \sigma_a H n_0 \exp(-z_m/H) = 1 \tag{21}$$

Thus, the height z_m where this maximum occurs is given by

$$\exp(z_m/H) = \sigma_a n_0 H \tag{22}$$

where H is the scale height, σ_a the absorption cross section, and n_0 the density at the reference level. The maximum of production is then given by

$$q_m = (\sigma_i/\sigma_a) Q_\infty / He^1, \tag{23}$$

where σ_i is the ionization cross section and Q_∞ the photon flux outside the atmosphere. (For an arbitrary zenith angle of the sun ($\chi \leq 85°$) (22) is modified by a factor sec χ and (23) by cos χ. Thus, for oblique incidence of solar radiation, the maximum of ion production is reduced, compared to overhead sun, while the altitude where this maximum occurs is increased.) Making use of the above relations one can rewrite the production function for overhead sun in the following form

$$q = q_m \exp \{(1 - [(z - z_m)/H] - \exp -[(z - z_m)/H]\}. \quad (24)$$

Since the recombination coefficient α is assumed to be independent of altitude, the expression for equilibrium

$$q = \alpha N^2 \quad (25)$$

is solved for N giving the Chapman distribution

$$N = N_m \exp \{\tfrac{1}{2}[1 - (h/H) - \exp(-h/H)]\}, \quad (26)$$

where N_m is the electron density at the peak of the distribution, at an altitude z_m, h is a height parameter $(z - z_m) = h$, and H the scale height. This distribution can often be used to describe the *shape* of observed F-region electron density profiles. However, the electron density distribution of the F region does not result from the processes which are assumed in the derivation of the Chapman layer. The origin of the F region will be discussed in the latter part of this section.

D and E Region Formation

The *undisturbed D region* is formed by the ionizing action of Lyman Alpha radiation on nitric oxide.[9] This occurs in the altitude range 70 to 85 km and leads to electron densities of the order of 10^3 cm^{-3} at 80 km for an overhead sun. X-rays of 2 to 8 Å also can produce ionization in this altitude interval but their intensity is too low to be of importance except during times of solar disturbances (see pages 160–162). Below 70 km the atmosphere is ionized by cosmic rays, the rate of ionization being proportional to the neutral particle density. Thus, one would expect that below 70 km the electron density would increase. However, attachment of electrons to neutral molecules resulting in the formation of negative ions leads to an electron density distribution which decreases with increasing molecular oxygen density. Electron densities of the order of 10^2 cm^{-3} are found during the day at 65 km in what has become known recently as the cosmic ray layer or C-region.

The altitude region between 85 and 150 km is termed the *E* region. From the upper edge of the D region to about 105 km there is a steep gradient of electron density resulting from the absorption of X-rays whose wavelength is about 30 Å. Longer wavelength X-rays are important sources of O_2^+ and N_2^+ ions at higher altitudes. In addition, the very strong solar chromospheric lines of H Lyman beta at 1025.6 Å and, C III at 976 Å lead to O_2^+ as the primary ion produced by photoionization as shown in Figure 1. The effect of the monochromatic Lyman beta radiation is to produce a Chapman-like layer centered around 105 km with a noontime electron density of the order of 10^5 cm^{-3}. The effect of the H Lyman continuum beginning at 910 Å is to produce O^+ ions as shown in Figure 1. Although N_2^+ is an ion produced in large quantities, Figure 2 shows that it is not present in the E region. This is due to the transformation of N_2^+ into O_2^+ and NO^+, by the different reactions discussed on pages 141–144 as well as dissociative recombination of N_2^+. Dissociative recombination of NO^+ and O_2^+ are very important reactions in determining the electron density distribution in the E region.

F-Region and Topside Ionosphere

It is now well established that solar EUV radiation in the wavelength range 300–800 Å is responsible for the formation of the F region.[10] The peak of electron and ion production occurs in the vicinity of 150 km as shown in Figure 1. The so-called F_1 *region* which has been identified from ground-based soundings and which occurs in summer during the daytime has been associated with the ion production peak. In terms of the over-all F region (the upper part of which is also called the F_2 region) electron density distribution, however, it merely represents a *bump* rather than a distinct layer. The *peak* in the *electron density distribution* occurs in the vicinity of 300 km. Although this altitude varies with time of day, season, solar cycle, and latitude it always lies within the region from 200 to 400 km. The fact that the height of maximum density is not identical to the height of maximum production indicates that the formation of the F region is of a more complex nature. Current theories[7] explain this behavior by the transition from a quadratic loss law $L = \alpha N^2$ representative of recombination processes at heights of the maximum of production (F_1 layer) to a linear loss law of the form $L = N\beta_0 \times \exp(-h/H_\beta)$ at heights above 250 km (F_2 layer) according to which the electron density will increase beyond the peak of production. This height-dependent loss above, however, can still not explain the formation of the F_2 peak, for which the additional effects of diffusion are invoked. The linear loss law responsible for the increasing electron density in the lower F region results from the ion-atom interchange process between atomic oxygen ions and neutral molecular constituents (N_2, O_2) {e.g., ($\beta = K[N_2]$)} forming a molecular ion which disappears quickly by dissociative recombination. The presence of NO^+ in the F region is evidence for the occurrence of this process in the ionosphere (see pages 141–144). The equilibrium distribution arising from the height-dependent linear loss law would lead to an electron density distribution increasing with altitude even beyond the observed altitude of the F_2 peak (Figure 5) according to $N = q/\beta$. As discussed previously, diffusion of charged particles becomes important when the neutral density is sufficiently low. Thus, at higher altitudes diffusion will tend to bring the original chemical equilibrium distribution toward a diffusive equilibrium and the com-

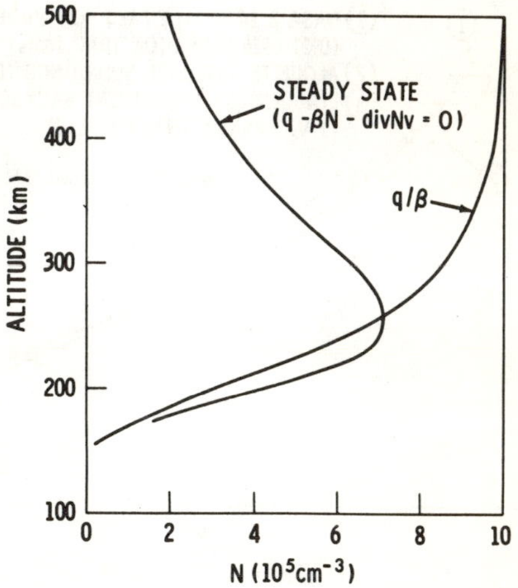

FIG. 5. Chemical equilibrium and steady-state *F*-region model (after Rishbeth and Barron, Ref. 11).

bination of these processes leads to an electron distribution having a peak where

$$\beta_m \simeq D/H^2 \tag{27}$$

and having a maximum density at the altitude where the time constants for chemical processes and plasma diffusion are about equal,

$$N_m \simeq q_m/\beta_m, \tag{28}$$

where D is the diffusion coefficient, H the scale height, q the production rate, β the loss coefficient, and the subscript m refers to the peak in electron (ion) density as shown in Figure 5. Observed electron density distributions exhibit clearly the diffusive control of the distribution above the peak; which is represented by an exponential behavior

$$N = N_0 \exp(-z/H'). \tag{29}$$

where H' is the electron-ion scale height [cf. (19)], which for thermal equilibrium is equal to twice the scale height of the neutral species ($H' = 2H$).

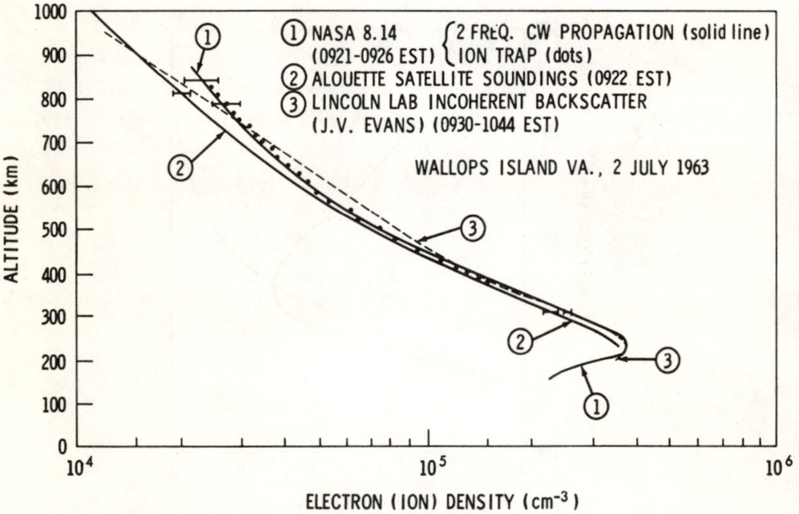

FIG. 6. Charged particle density profiles of the upper ionosphere measured simultaneously by rocket, satellite, and incoherent backscatter radar [after Bauer et al., JGR, **69**, 186 (1964)].

Rocket and satellite measurements have increased tremendously our knowledge about the ionosphere above the F_2 peak, which was inaccessible to ground observations only until the very recent development of the incoherent backscatter radar technique.[12] Currently the Canadian Topside Sounder Satellite Alouette is exploring the electron density distribution in the *topside* ionosphere on an almost routine basis and providing a multitude of data where only a few years ago there was a complete lack of observations (Figure 6). Based on these observations, as well as earlier and concurrent satellite and rocket observations it appears now that very often the electron temperature may be higher than the ion (and neutral gas temperature) as well as that the ionic composition changes with altitude in a region where a few years ago it was assumed that O^+ was the only important ion. It was realized even then, that at higher altitudes (>1000 km) H^+ (protons) will eventually become the predominant ions. This region is also called the *protonosphere*. In 1961, Nicolet[13] suggested that helium should also become an important atmospheric constituent at these high altitudes. Within less than a year of this suggestion the presence of helium ions (He^+) was identified by rocket and satellite measurements; more recently the concentration of neutral helium has also been measured by the Explorer XVII satellite. Thus, the topside ionosphere must now be considered as a ternary ion mixture of O^+, He^+, and H^+. The presence of three ions in an altitude region largely controlled by diffusion also led to a re-examination of the diffusive behavior of minor ions in an ion mixture. While previously the diffusive equilibrium distribution of an ionic species was generally considered to be governed by a scale height twice that of the corresponding neutral species (see Eq. (29), it was pointed out by Mange[14] that this concept is not justified for a minor ion in the presence of other ions. In contrast to neutral constituents the diffusive equilibrium distribution of ions is not independent of the other ionic constituents because of the presence of an electrostatic field which is set up to prevent further charge separation between electrons and ions diffusing under gravity. This electrostatic field, which depends on the mean ionic mass of the mixture and on the electron and ion temperatures is sensed by all ions in the mixture and tends to counteract the gravitational force. Although this fact had been known to astrophysicists for many years[15] it seems to have escaped the workers in the field of upper atmosphere physics until it was again considered by Dungey,[16] and more recently and independently by Mange.[14] According to this concept the equation for electrons and ions can be written as follows

$$kT_e(d/dz)\ln n_e = -m_e g - eE \tag{30}$$

$$kT_i(d/dz)\ln n_i = -m_i g + eE, \tag{31}$$

where n is the number density, m the mass, T the temperature, g the acceleration of gravity, and E the polarization field. Assuming charge equilibrium $n_e = \sum n_i$ and more than one ionic constituent we can write

$$kT_i(\partial/\partial z)\ln \sum n_i = -m_+ g + eE, \tag{32}$$

where the mean ionic mass $m_+ = \sum n_i m_i / \sum n_i$. From these equations we can eliminate the electrostatic field E and obtain the distribution of an ion constituent in diffusive equilibrium

$$n(X_k^+) = n_0(X_k^+)\exp\left(-\int_0^z \{m(X_k^+) - [T_e/(T_e + T_i)]m_+\}(g/kT_i)\,dz\right), \tag{33}$$

where $m(X^+)$ is the ionic mass, T_i and T_e are the electron and ion temperature, respectively, and $m_+ = \sum n(X_i^+)m(X_i^+)/n(X_i^+)$ is the mean ionic mass. Assuming temperature equilibrium ($T_e = T_i$) it can be seen that the ion density of mass $m(X_k^+)$ will *increase* with altitude as long as $m(X^+) < m_+/2$, i.e., as long as it is a minor ion, and only after it becomes predominant will it fall off with a scale height approaching that of twice its neutral species.

Figure 7 shows the behavior of the altitude distribution of He^+, H^+, and O^+ in the topside ionosphere assuming a given concentration of these constituents at a reference level (which may lie above 500 km) and thermal equilibrium with $T = 1200°K$. It can be seen that although at the reference level both the He^+ and H^+ concentrations are small compared to the concentration of O^+, there are altitude regions where either one can become the predominant ion. (Geopotential altitude

$$z' = \int_0^z (g/g_0)\,dz \approx R_0 z/R_0 + z,$$

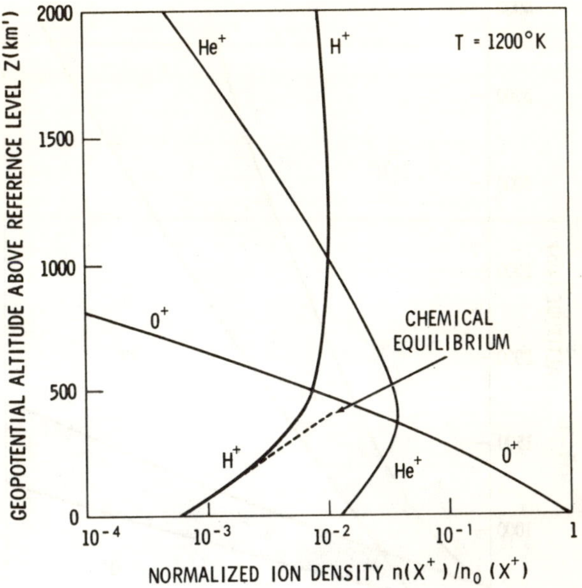

FIG. 7. Distribution of ions in diffusive equilibrium.

where R_0 is the earth's radius, is frequently used as a height parameter in upper atmosphere work, since it incorporates the altitude variation of the acceleration of gravity g. On such an altitude scale, any change in the logarithmic decrement of the density is an indication of a variation in T/m).

Helium ions are produced by photo-ionization of neutral helium and lost by charge transfer processes with molecular oxygen and nitrogen while the source and sink for protons is the charge-transfer process with oxygen. (See page 144). Thus, the relative concentration of the light ions at a suitably chosen reference level above which they can be assumed to be controlled by diffusion will behave similar to the ratio of the corresponding neutral species. While the concentrations of He and O are decreasing with decreasing atmospheric temperature, hydrogen, because of the importance of escape at high temperatures, will actually increase when the temperature is decreasing. Thus, the relative abundance of the three ionic species in the topside ionosphere behaves qualitatively as shown in Figure 8. At high atmospheric temperatures there may be a region where

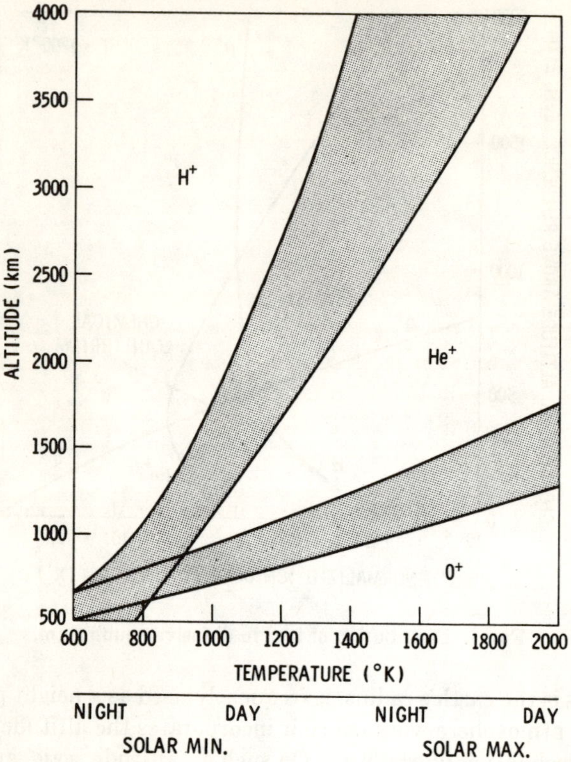

FIG. 8. Regions of predominance of O^+, He^+, and H^+ in the topside ionosphere as function of atmospheric temperature.

He^+ is the predominant ion which extends for as much as 2000 km, while at low temperatures, the relative importance of He^+ becomes insignificant, and the upper ionosphere where O^+ predominates, merges almost directly into the protonosphere.

Sporadic E

Superimposed on the normal E layer are sharp gradients of electron density and irregularities both of which cause distinctive echoes on ionosonde records. The temporal and spatial distributions of these phenomena have been studied in great detail. The results of the extensive International Geophysical Year (1957–1958) program are summarized in a monograph[17] which includes chapters dealing with the current theories of formation of sporadic E.

At temperate latitudes observations of sporadic E have been made by both rockets and ionosondes. Thin layers superimposed on the normal E layer have been observed with the aid of rockets.[18] These layers, which vary in thickness between 500 meters and 2 kilometers, have electron density gradients of between 10^5 and 10^6 electrons cm^3/km. The electron density enhancement may be a factor of 2 or more. Preferred heights of occurrence at 100, 105, 111, 117, and 120 km have led to a theory of formation based on the redistribution of electrons by wind shears in the presence of the terrestrial magnetic field. It has also been postulated that these layers may result from the increased ion pair production due to the ionization of minor constituents such as Mg and Ca. As discussed above (pages 140–141), such ions have been identified by means of rocket-borne mass spectrometers. An example of a mid-latitude sporadic E layer is shown on the E region distribution of Figure 9.

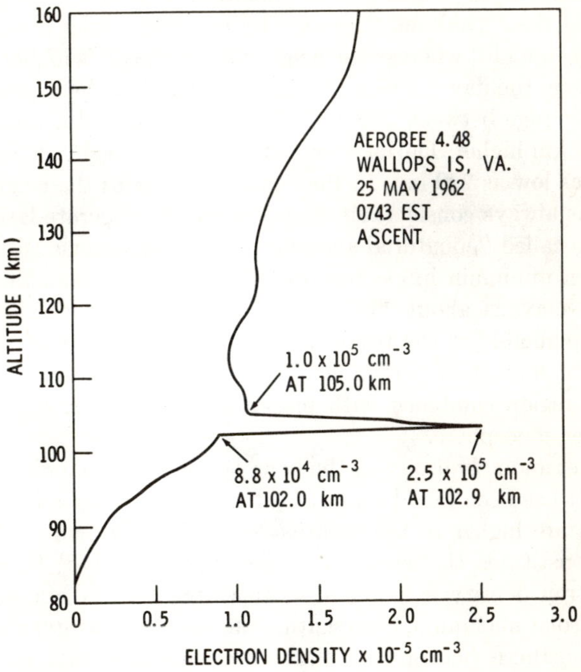

FIG. 9. Electron density distribution showing the occurrence of sporadic *E* (after L. G. Smith, see Ref. 18).

Within 5 degrees of the magnetic dip equator sporadic E is characterized by magnetic field aligned irregularities. Often disturbances are observed traveling horizontally with a speed of tens of km/sec. It has been postulated that such phenomena may be caused by a two-stream plasma instability.[17a] The character of sporadic E in the auroral zone is similar to that occurring in the equatorial belt.

Anomalies in the F Region

In contrast to the lower ionosphere which exhibits a diurnal and seasonal behavior closely linked to the position of the sun, the behavior of the F region is much more complex.[7] The diurnal variation of electron density is generally not symmetric with local noon; the maximum electron density may often be found late in the afternoon. This anomalous behavior of the F region is primarily the result of mass-transport of ionization as mentioned before. The electron density of the F peak may decrease by up to one order of magnitude from day to night, whereas the height of the F peak is higher at night than during the day. During the day the height of the F peak occurs on the average between 250 and 300 km; the nighttime value being about 50 km higher. During solar minimum the height of the F peak may be as low as 220 km. In the equatorial regions the height of the F peak is always considerably higher than at temperate latitudes.

The so-called "equatorial anomaly" mentioned previously exhibits itself as a minimum in electron density at the geomagnetic equator and two maxima about 20° off the equator. This anomaly appears to be explainable as the result of mass transport; that is, the effect of drift of ionization perpendicular to the geomagnetic field in the lower F region combined with plasma diffusion along field lines in the upper ionosphere.

Another anomaly occurs in the behavior of F region ionization with season. It has been found that at latitudes 50°N to 35°S the daytime densities are higher in the months November, December, January, which constitutes the so-called "December anomaly." In addition, the electron density is larger in local winter than in summer, especially at solar maximum, representing the so-called "winter anomaly." In the northern hemisphere, these two anomalies add to give an especially pronounced winter anomaly. Although suggestions to explain these anomalies have been made, a satisfactory physical explanation of this anomalous seasonal behavior is still lacking.

In radio soundings from the ground it has been found that the ionogram trace representing the echo from the F region may sometimes have a *spread* appearance, hence the name *spread F*. This phenomenon occurs especially in two latitude regions, one near the magnetic equator, the other at high latitudes. However, it appears that its origin is not the same for both regions, since equatorial spread F is found to be negatively correlated with geomagnetic and solar activity, whereas high latitude spread F is correlated positively with these geophysical characteristics. Furthermore, spread F has been found to be associated with magnetic field-aligned irregularities and with scintillations of signals from radio stars and earth satellites. Spread F is now also being observed with the topside sounder satellites.[19] Phenomenological explanations of various types of spread F have been given in terms of scattering of radio waves from ionization irregularities, ducting along field aligned irregularities or by refractions in large regions of reduced electron density. While the observed features of spread F can be described in this way, no theory of the physical cause of spread F is as yet available, although a connection with phenomena involving energetic particles is suspected.

The Nighttime Ionosphere

It has become apparent that the persistence of rather high ion densities throughout the night is not wholly explainable as the result of slow decay of ionization produced during the day by photo-ionization. This is especially true for the E region which still persists during the night in spite of a rather short time constant of recombination. The possibility of a nighttime ionization source has been suggested such as corpuscular radiation and/or scattering of Lyman α. However, no adequate quantitative explanation is available at the present time. The same is true for the nighttime F-region. It is often observed that the charged particle density is almost constant or even increases during the night. Since in the F region mass motions are of great importance, it has been suggested that the nighttime behavior is the result of downward diffusion of electrons and ions from the protonosphere. Although such downward fluxes of ionization during the night could maintain the observed nighttime ion concentrations, it has been questioned that the amount required can be replenished by upward fluxes during the daytime. An alternate suggestion explains the observed nighttime concentration at F region levels as the result of an upward lifting of ions by electro-dynamic forces into a region where the loss rate is smaller.[20] However, presently no fully satisfactory explanation of the behavior of the nighttime ionosphere is available.

Polar Ionosphere

Similar to the problem of the nighttime ionosphere, the existence of the winter polar ionosphere remains a challenge. It has been found that a fairly dense F layer exists in the mid-winter season, although the photoionization in the F region should be small and practically constant. This implies an additional source of ionization, either in the form of corpuscular radiation and/or transport of ionization from lower latitudes. A control of the ionization with universal time has been found, however most of the ideas about the existence and the behavior of the polar ionosphere are still the subject of much discussion.[21] There is good evidence that magnetospheric phenomena, particularly those associated with the "tail" of the magnetosphere dominate the behavior of the polar ionosphere.[22]

Disturbances

During a solar flare the X-ray flux between 2 and 8 Å may rise by a factor of as much as 10^5. Under such circumstances X-rays are the major source of D region ionization and the electron density of the layer is enhanced greatly. These events, which are of 10 to 15 minutes duration, are termed sudden ionospheric disturbances or SID, also known as the Mögel–Dellinger effect. Only a limited amount of information is available on variations of other portions of the solar spectrum during flares. However, available evidence seems to indicate that the solar flux at wavelengths greater than 100 A varies by less than a factor of 2 during solar flares. Thus, while some modification of the electron density distribution occurs above 100 km the major effect of solar flare photon radiation is in the D region.

The lower ionosphere is also affected by energetic charged particles. One example of such an effect is the ionization by cosmic rays. The most dramatic events occur in the auroral zone between 60° and 70° geomagnetic latitude and across the polar cap. Because of the large dip angle, electrons and protons impinge on the atmosphere in these areas. The altitudes at which the particles dissipate most of their

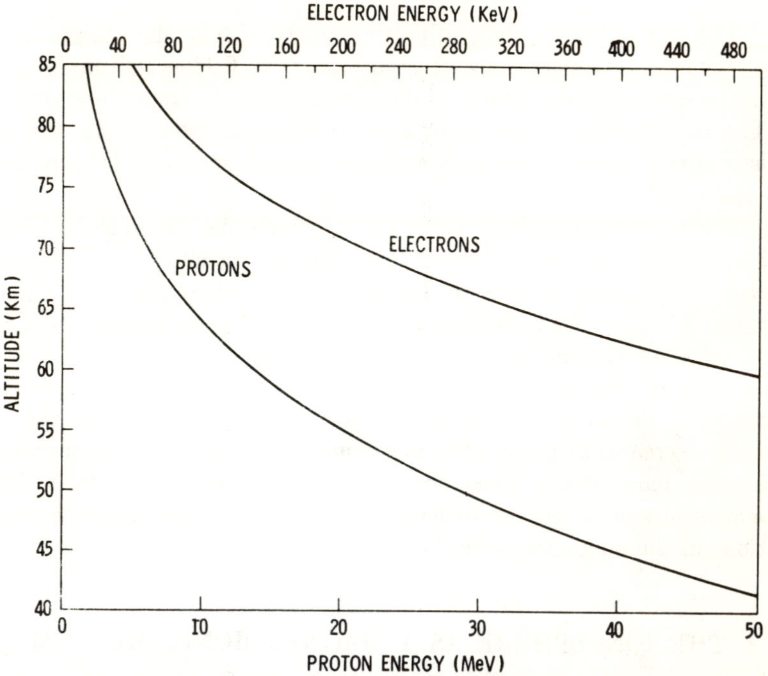

FIG. 10. Atmospheric penetration depth of energetic particles.

energy is determined by the range of the particles in air. This is of course a function of the energy of the particle. Figure 10 illustrates the depth of penetration of electrons and protons as a function of their energy.

Electrons in the energy range 10 to 200 keV are precipitated in the auroral zone. These electrons, which are the primary source of the aurora, cause large increases in D and lower E region ionization.

Protons of 10 to 100 MeV produced in the course of large solar flares arrive at the earth a few hours after the onset of the flare. The resultant increases in ionization occurring over the entire polar cap and extending south to a geomagnetic latitude of 65° are termed *Polar Cap Absorption* events or PCA since hf radio waves are heavily absorbed by this increased ionization. Some 20 hours after a large flare the polar cap absorbing region is extended to a geomagnetic latitude of 55°. This effect coincides with what is commonly termed a magnetic storm.

The interaction of charged particle fluxes of solar flare origin with the geomagnetic field causes fluctuations of the field as well as changes in the total electron content and distribution in the upper ionosphere. The morphology of the ionosphere during such geomagnetic storms is complex, since factors such as season and geomagnetic latitude are involved.[23]

During the main phase of the storm, which may begin as late as 3 hours after its commencement and last for as much as 70 hours, there is a general increase of the total electron content of the equatorial F region. This increase is accompanied by an enhancement of the electron density at the peak of the F_2 layer and by an altitude increase of the layer. In contrast, at geomagnetic latitudes greater than 47° the total electron content and maximum density at the F_2 layer decreases at times when one would normally expect an increase in such parameters. These changes in the upper ionosphere during magnetic storms appear to have their origin in hydromagnetic motions in the magnetosphere.[22]

THE IONOSPHERE AS A MAGNETOIONIC MEDIUM

In deriving the Appleton-Hartree equation it is assumed that the ionosphere is a non-iosotropic dielectric medium, which will have a volume polarization **P**. The form of Maxwell's equations describing the propagation of a wave in such a medium will be

$$\nabla \times \mathbf{H} = \dot{\mathbf{D}} \qquad \text{div } \mathbf{H} = 0 \tag{34}$$

$$\nabla \times \mathbf{E} = -\mu_o \dot{\mathbf{H}} \qquad \text{div } \mathbf{D} = 0$$

The displacement vector **D** is related to **P** by

$$\mathbf{D} = \epsilon_o \mathbf{E} + 4\pi \mathbf{P} \tag{35}$$

Because the medium is non-isotropic **P** is not necessarily in the direction of the electric field **E**.

Assume that a plane wave with components E_x, E_y, E_z and H_x, H_y, H_z is traveling with its wave-normal along the z direction (see figure 11), and that all components of the wave vary with z as exp. $(-ikz)$. Here $i = \sqrt{-1}$ and k is the propagation constant which is related

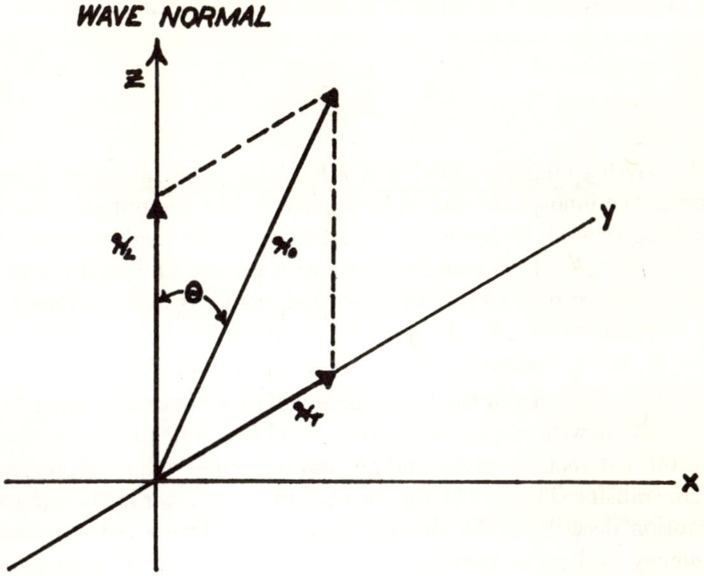

FIG. 11. Coordinate system relating the wave normal to the terrestrial dipole magnetic field vector.

to the index of refraction, n of the ionosphere by

$$n = ck/\omega \tag{36}$$

where c is the velocity of light and ω is 2π times the exploring frequency. With the aid of the curl equations it can be shown that $H_z = 0$ i. e. that the magnetic field lies entirely in the wave front and furthermore that $D_z = 0$ so that although **D** lies in the wave front **E** may have a component perpendicular to it.

Again using the curl equations together with (36) the relationship between **P**, **E** and n is derived

$$n^2 = 1 + \frac{4\pi}{\epsilon_o}\left(\frac{P_x}{E_x}\right) \qquad \text{a}$$

$$n^2 = 1 + \frac{4\pi}{\epsilon_o}\left(\frac{P_y}{E_y}\right) \qquad \text{b} \tag{37}$$

The polarization of the wave R is obtained from (37)

$$R = \frac{E_x}{E_y} = \frac{P_z}{P_y} \tag{38}$$

The earth's magnetic field will affect the polarization of a wave entering the ionosphere and in fact behaves like the optical axis of a polarizing crystal. In figure 11 the earth magnetic field vector is represented by $\mathcal{H}_o$. It is contained in the y, z plane and inclined at an angle θ with respect to the wave normal, which is in the z direction. The component of $\mathcal{H}_o$ along or longitudinal to the wave normal is $H_L = H_o \cos \theta$, transverse to the wave normal $H_T = H_o \sin \theta$.

The free electrons in the ionosphere will experience a Lorentz force when moving with respect to the terrestrial magnetic field. In addition collisions between electrons and neutral molecules will result in momentum transfer. This can be represented by a term $m\nu\dot{\mathbf{r}}$ in the equation of motion describing the electron trajectory. Here ν is the collision frequency and $\dot{\mathbf{r}}$ the velocity vector of the electron. If a sinusoidal wave with electric field $\mathbf{E} = \mathcal{E} \exp(i\omega t)$ is superimposed on the system then the Lorentz equation can be written as

$$m\ddot{\mathbf{r}} = e\mathbf{E} + e(\dot{\mathbf{r}} \times \mathcal{H}_o) - m\nu\dot{\mathbf{r}} \tag{39}$$

The polarization vector $\mathbf{P}$ is given by

$$\mathbf{P} = Ne\mathbf{r} = \mathbf{P} \exp(i\omega t) \tag{40}$$

where N is the electron concentration per unit volume. Equations (39) and (40) can be used to derive the relations between $\mathbf{E}$ and $\mathbf{P}$ i.e. the constitutive relations. These relations are then used to form a quadratic equation for the wave polarization defined by (38).

$$i\, Y_L R^2 - \frac{Y_T^2 R}{1 - X - iZ} + i\, Y_L = 0 \tag{41}$$

New quantities have been introduced which have the following definitions

$$Y_L = Y\cos\theta \qquad Y_T = Y\sin\theta$$

$$Y = \frac{\omega_H}{\omega} \qquad \omega_H = \frac{eH_o}{m_e}$$

$$Z = \nu/\omega$$

$$X = (\omega_N/\omega)^2 \qquad \omega_N = \left(\frac{Ne^2}{\epsilon_o m}\right)^{\frac{1}{2}}$$

The solution of equation (41) is

$$R = \frac{-i}{2Y_L}\left\{\frac{Y_T{}^2}{1 - XiZ} \mp \left(\frac{Y_T{}^4}{(1 - X - iZ)^2} + 4Y_L^2\right)^{\frac{1}{2}}\right\} \tag{42}$$

This result can be utilized with one of the constitutive relations and (37) to give the Appleton-Hartree equation

$$n^2 = (\mu - i\chi)^2 =$$

$$1 - \frac{X}{1 - iZ - \dfrac{Y_T{}^2}{2(1 - X - iZ)} \pm \sqrt{\dfrac{Y_T{}^4}{4(1 - X - iZ)^2} + Y_L{}^2}} \tag{43}$$

The wave for which the refractive index is given by the positive sign in (43) is defined as the ordinary wave. The negative sign refers to the extraordinary wave. This, however, is not a general rule, since it is required that n and R be continuous functions. Ratcliffe[1] has given a comprehensive treatment of the choice of sign in (43) for different values of X, Y, and Z.

The Appleton-Hartree equation in its present form does not include the effects of ions or the fact that the momentum transfer cross section for electrons in molecular nitrogen is proportional to the electron velocity i. e. ν is not a constant. The ion effects have been considered by Goubau[24] and Sen and Wyller[25] have included the energy dependence of the collision frequency in their derivation.

Equation (43) is complicated to solve. In practical circumstances one is usually able to make some approximations such as neglecting collisions, $Z = 0$, or wave propagation quasi-longitudinal, *Q.L.*, or quasi-transverse, *Q.T.*, to the magnetic field vector so that terms involving Y_T or Y_L can be neglected. Some of the more common approximations are

$$Q.L:\ n^2 = 1 - \frac{X}{1 - iZ \pm Y_L} \tag{44}$$

or if $Z = 0$

$$n^2 = 1 - \frac{X}{1 \pm Y_L}$$

The form with the minus sign is employed in describing whistlers which are ducted along the magnetic field and where $Y > 1$.

ROCKETS AND SATELLITES IN IONOSPHERIC RESEARCH

Radio Wave Diagnostic Techniques

Most of the information concerning the behavior of the ionosphere obtained over the past three decades was gathered with the aid of ground-based instruments called *ionosondes*. These are actually low frequency radars, sounding the ionosphere at vertical incidence and providing information on delay time (virtual height) of echoes versus probing frequency, f, which is a measure of the electron density $[N(cm^{-3}) = 1.24f^2\ (\mathrm{Mc})]$. In a time, $d\tau = c\,dz$ a radio wave pulse will travel in direction z a virtual distance dh' as determined by the expression

$$d\tau = V_g\, dh' \tag{45}$$

Where $V_g = c/\mu'$ is the propagation velocity of the wave packet in the ionosphere. It is related to the velocity of light, c, through the group refractive index μ'. The group refractive index can be determined by means of the expression

$$\mu' = \mu + \frac{d\mu}{df} \tag{46}$$

The quantity μ is the phase refractive index and is the real part of the refractive index of the ionosphere as determined by the Appleton-Hartree equation. The virtual height h' (f) is found by integrating (45)

$$h'(f) = \int_0^z \mu' dz \tag{47}$$

to the true height of reflection z.

An ionogram is a record of the virtual height as a function of exploring frequency. An example is shown in figure 12a where the virtual height is given as a function of exploring frequency in megacycles. Figure 12b is a tracing of the ionogram with the ordinary and extraordinary traces labeled and identification of the E and F regions. The electron density corresponding to $X = 1$ as a function of frequency is given at the top of the figure. In Figure 12c the resulting electron density profile is shown in comparison with the ordinary ray virtual height record.

If the ionosonde is placed within the ionosphere, as was accomplished with the Alouette I and II and Explorer XX satellites, then the $h'(f)$ record will be quite different. Equation (47) now has the form

$$h'(f) = \int_{\phi(N_s)}^{\phi(N)} \mu' \, dz \tag{48}$$

Ionograms of the daytime bottomside ionosphere.

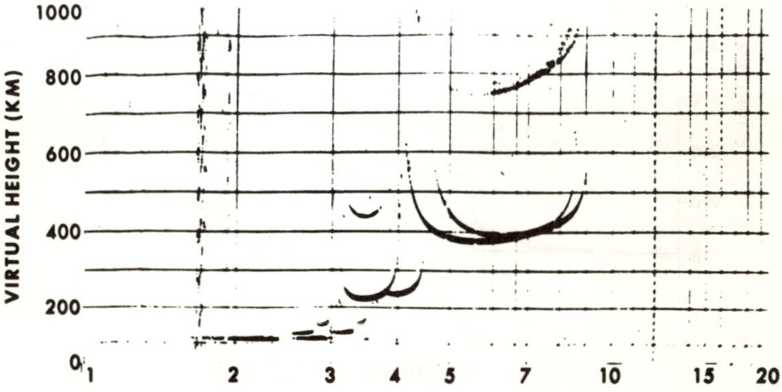

FIG. 12a. Virtual height vs exploring frequency record.

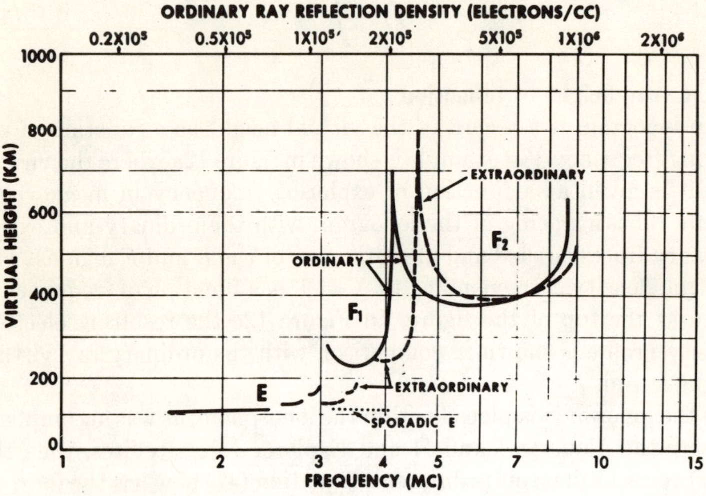

FIG. 12b. Tracing of the ionogram showing the extraordinary and ordinary wave virtual height profiles.

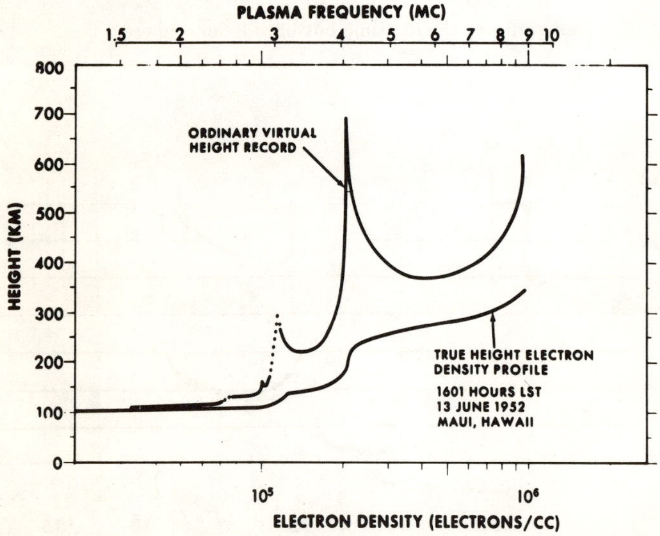

FIG. 12c. Ordinary wave virtual height tracing and resulting electron density profile.

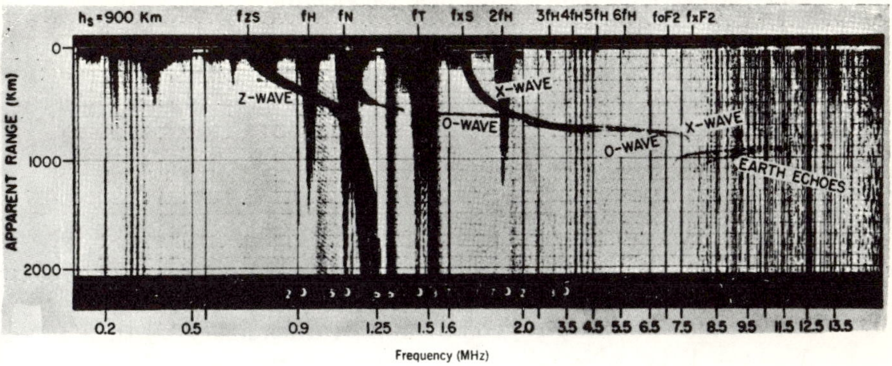

FIG. 13a. An Alouette II topside-sounder ionogram showing resonances and reflection traces.

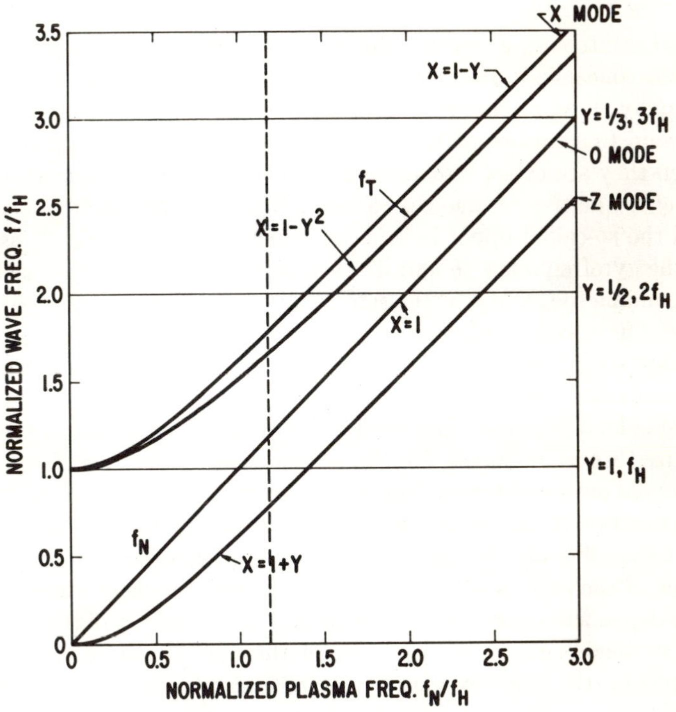

FIG. 13b. Propagation and resonance conditions in a magnetoionic medium.

where $\phi(N)$ is an assumed functional behavior of N with N_s the electron density at the satellite, which is a known quantity.

As with equation (47) an inversion of the integral equation (48) will yield a true height electron density profile, the actual analysis being performed by the use of electronic computers. Figure 13 shows an ionogram observed from a topside sounder satellite and the $N(h)$ profile derived therefrom. The ionogram exhibits the ordinary (o) and extraordinary (x) and (z) traces as well as ground reflections at frequencies well above the plasma (critical) frequency of the F_2 peak.

The reflection levels for the ordinary (o) trace correspond to the condition $X = 1$, i.e. $f = f_N$ and for the extraordinary (x) trace to $X = 1 - Y$, i.e. $f_x^2 = f_N^2 + f_x f_H$, where f_N is the plasma frequency for the o and x waves and f_H is the gyrofrequency. Another extraordinary trace the so-called z mode, is also observable in soundings from within the ionosphere, corresponding to the condition $X = 1 + Y$, i.e. $f_z^2 = f_N^2 - f_z f_H$.

Other interesting features observable with a sounder operating *within* the ionosphere are so-called *plasma "spikes"* or *local resonances.* Actually these plasma responses seen on the topside ionograms, correspond to solutions of the Appleton-Hartree equation for $n^2 = \infty$, when they are called *resonances* and for $n^2 = 0$, when they are called *cut-offs*. The true resonances correspond to the plasma frequency f_N and the so-called upper hybrid resonance $f_T = (f_N^2 + f_H^2)^{1/2}$ as well as the gyrofrequency f_H and its harmonics. Cutoffs are observed corresponding to f_x and f_z at the satellite altitude. These resonance phenomena have been explained in terms of electrostatic oscillations in the ionospheric plasma excited by the *r. f.* energy of the sounder. The sequence of reflection and resonance conditions which are observed on a topside ionogram is summarized in figure 13b. The values of the normalized wave frequency f/f_H for which some of the most frequently observed poles and zeros of the refractive index are observed are shown as a function of the normalized plasma frequency f_N/f_H. The position of the satellite determines the normalized plasma frequency since the value of the earth's magnetic field at the satellite determines f_H and f_N is dependent upon the electron density in the vicinity of the satellite. If a vertical line is drawn on the graph through the normalized plasma frequency, the sequence of reflection traces and resonances, which will be observed as the exploring frequency is increased, can be seen.

VLF (very low frequency) observations on the topside sounder satellites also exhibit other resonance phenomena associated with the *ions* in the vicinity of the spacecraft. These VLF phenomena (ion cyclotron whistlers, i.e. proton and helium whistlers and the VLF noise cutoff at the lower hybrid resonance frequency) allow the determination of *ion* composition and temperature. Thus parameters which were once considered to be measurable only by means of direct (probe) techniques can now also be obtained by radio methods[26].

In addition to radio pulse soundings of the ionosphere two other radio wave techniques are commonly used to investigate the ionosphere. The first of these makes use of a moving rocket or satellite to give a Doppler shift in frequency of a wave either transmitted from the moving vehicle and received on the ground or vice versa. One of the most accurate methods of measuring the electron density distribution in the ionosphere is the use of a rocket-borne transmitter which emits continuous waves on two harmonically related frequencies f and $6f$. The Doppler effect produced by the motion of the transmitter within the ionosphere can be expressed as a change in the phase path P of the transmitted radio wave of frequency f

$$\dot{P} = \frac{2\pi f}{c}\left[\mu\,\dot{\mathbf{r}} + \int_{o}^{R} \frac{d\mu}{dt}\,dr\right] \tag{49}$$

where c is the velocity of light in vacuo, μ is the refractive index at the rocket, which is related to the electron density N through the Appleton-Hartree equation; and $\dot{\mathbf{r}}$ is the velocity component of the rocket in the ray-direction. The integral term represents the time variation of the refractive index along the ray path. The Doppler effect can also be utilized to obtain total electron content. Either the ordinary or extraordinary propagation mode will give rise to the Doppler effect.

Another method which is in common use is the Faraday effect, or rotation of the plane of polarization of a linearly polarized radio wave in the ionosphere. This is the result of the opposite sence of rotation of the ordinary and extraordinary polarizations and thus refers to a phase change, which is proportional to $(\mu_o - \mu_x)/2$, i.e. the difference of the refractive indices for the ordinary and extraordinary modes.

Direct Measurement Techniques

Rockets and satellites have also made possible the *in situ* measurements of ionospheric parameters by means of plasma probes. These are based on the Langmuir probe used in laboratory plasmas to determine plasma density and temperature. (The plasma resonance phenomena mentioned before, actually also belong in this category, although they depend on magneto-ionic effects). The direct measurement techniques utilize probes in a number of geometrical configurations, from simple cylindrical (wire) probes to planar and sperical "traps" containing in addition to the collector a number of grids which perform a variety of functions. Basic to all these devices is the analysis method and the

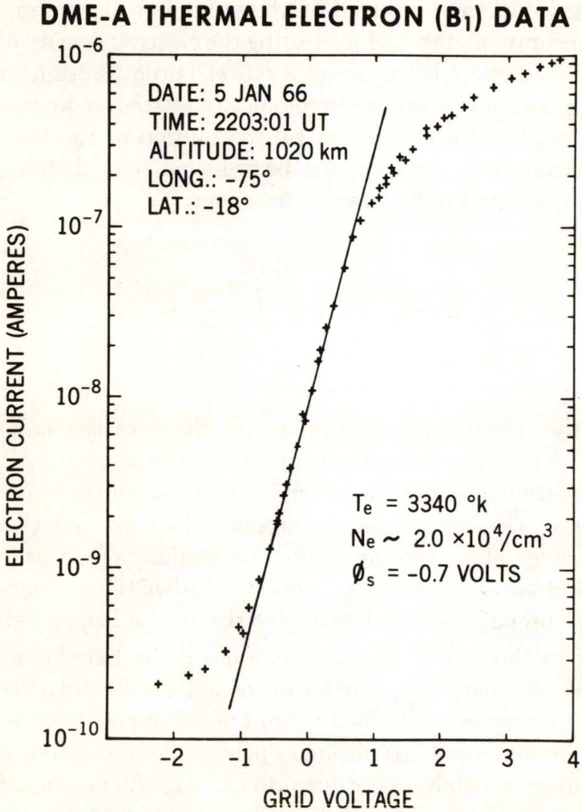

FIG. 14. I–V curve obtained with a satellite-borne planar trap.

application of Langmuir probe theory: By varying the potential of the probe V_p and observing the corresponding current I, density and temperature of the charged particles can be determined from the characteristics of the I-V curve. A typical I-V curve observed on board a satellite with a planar trap is shown in Figure 14. The potential of the probe with respect to the plasma is given by

$$V = V_p - V_o$$

where V_p is the applied probe voltage and V_o the plasma potential. The measured electron current can be expressed as a function of probe potential with respect to the plasma V. In the retarding region ($V<0$) it is given by

$$I_e = I_o \exp \frac{eV}{kT} \tag{50}$$

with

$$I_o = eNA \left(\frac{kT_e}{2\pi m_e}\right)^{\frac{1}{2}}$$

where N is the electron density, e is the charge and m_e is the mass of electron, k is Boltzmann's constant, T_e is the electron temperature and A is the probe area. The value I_o is the current at plasma potential *i.e.* the potential at which both electrons and ions are collected without the aid of an accelerating potential ($V_p = V_o$). This potential can be identified by the break in the I-V curve; below it, the particles are retarded, above it the current approaches saturation because of attractive potentials. The fact that the observed break corresponding to $V_p = V_o$ does not occur at $V_p = 0$ results from the potential, ϕ_s, a vehicle (probe) immersed in a plasma acquires and which has to be compensated for by the applied potential. Thus the vehicle potential $\phi_s = - V_o$ can be identified from the break point in terms of the voltage applied to the probe. The electron density can be determined from I_o according to

$$N = \frac{I_o}{Ae}\left(\frac{2\pi m_e}{kT_e}\right)^{\frac{1}{2}} \tag{51}$$

and the electron temperature T_e from the logarithmic decrement of the current

$$T_e = \frac{e}{k \cdot d\,(lnI)\;/dV} \tag{52}$$

A linear relationship of lnI vs V is indicative of a Maxwellian velocity distribution.

Ion characteristics can be determined in similar fashion except for some complicating factors: a) The ion current is about two orders of magnitude less than the electron current corresponding to the ratio $(m_e/m_i)^{1/2}$ and b) the ion thermal velocity is generally less or comparable to the (rocket) satellite velocity.

Because of the fact that dI_+/dV is actually a measure of kinetic energy rather than temperature, ion temperature cannot easily be determined, especially in a mixture of ions. However it is possible to use this characteristic of the I-V curve as an indicator of ion energy and thus as a crude measure of ion composition. The determination of ion temperature T_+ on the other hand presupposes a knowledge of composition. Figure 15 shows an $I_+ - V$ curve exhibiting the presence

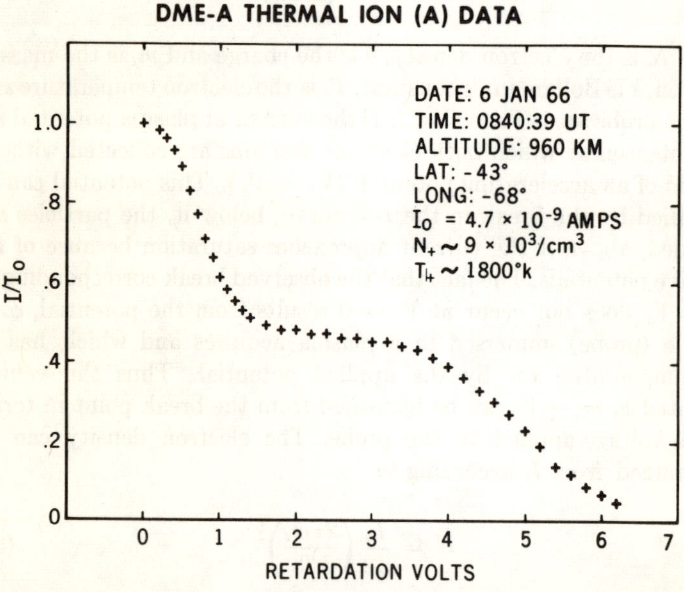

FIG. 15. Example of the effect of different ionic species on the ion portion of an I–V curve for a satellite-borne planar trap.

of two ionic constituents (H^+ and O^+).

Since the ion thermal speed is generally less than the speed of a satellite ($\backsim$8 km/sec), the ion current I_+ which is a measure of ion density, is then given by

$$I_+ = eN_+ \, Av \cos \theta, \text{ for } \theta < 45° \tag{53}$$

where $v \cos \theta$ is the satellite velocity component parallel to the probe axis. When ion thermal velocity and vehicle velocity become comparable, a much more complicated expression must be used which takes into account this velocity ratio.

Electron and ion measurements have been performed on rockets and satellites by means of probes of different geometric configuration. Often planar or spherical "traps" are used which use a number of grids to exclude unwanted currents by an application of appropriate bias voltage (e.g. photoemission currents and currents of opposite polartiy). The same trap can work in both an electron and ion "mode" by appropriate selection and programming of applied potentials[18].

Ion mass spectrometers measure ion composition with high mass resolution ($\backsim$ 1 a.m.u.). These instruments separate ions of different mass to charge ratio m/e by means of r.f. fields (*Bennett R. F. Mass Spectrometer*), by deflection in a magnetic field (*Magnetic Mass Spectrometers*). or by a combination of d.c. and a.c. fields (*Quadrupole Mass Spectrometers*). All types of ion mass spectrometers have been flown on rockets and satellites and have provided altitude and latitude behavior of atmospheric ions.

All types of instruments discussed above have been flown on both rockets and satellites. The main advantage of satellites is their ability to make local measurements as a function of time and geographic location, while rockets allow simultaneous measurements of the *vertical* structure of most parameters determining the ionosphere, such as solar ionizing radiations, neutral and ionic composition and temperature.

PLANETARY IONOSPHERES

With the presence of ionizing radiations from the sun those planets with an atmosphere will also have an ionosphere. Until recently, our knowledge of planetary ionospheres was based strictly on speculations.

The first experimental information on the properties of a planetary ionosphere became available in 1965 when the phase shifts of the telemetry carrier-wave ($\backsim$ 2000 Mc/s) of Mariner IV were measured on earth as the spacecraft passed behind the Martian atmosphere[27]: This occultation experiment provided information about the neutral atmosphere as well as the ionosphere; the two regions can easily be distinguished in the phase shift data since the lower (neutral) atmosphere's refractive index μ(Troposphere) > 1, whereas the refractive index μ (Ionosphere) < 1. According to these observations, the Martian ionosphere had a maximum electron density of $\backsim 10^5$ electrons cm^{-3} at an altitude of 120-125 km and a scale height $H' = 20$-25 km, at the time of observation (local) winter afternoon. Compared with the earth's ionosphere the scale height of the Martian ionosphere is surprisingly low. Possible explanations of the observed ionospheric characteristics have been advanced in terms of analogy with the terrestrial ionosphere. According to one interpretation[28] the maximum of ionization observed at 120 km represents an F_2 – type peak, i.e., resulting from the combined action of a linear electron loss law and ambipolar diffusion [(cf. eqn. (27) and (28)], with 0^+ as the predominant ion. Another explanation of the observations[29] considers the Martian ionosphere to be the analogue of the terrestrial E region, with molecular ions (O_2^+ or NO^+) as the predominant species. The resolution of these alternatives may not be at hand until further experimental data becomes available.

Experimental data on the ionosphere of Venus will hopefully become available from the refined occultation experiment on the Mariner Venus 1967 fly-by mission. With the availability of new experimental data we shall have the opportunity to apply our theoretical knowledge based on more than three decades of studying the terrestrial ionosphere to the understanding of planetary ionospheres.

REFERENCES

1. J. A. Ratcliffe, The Magneto-Ionic Theory and its Applications to the Ionosphere, Cambridge University Press, Cambridge, 1959.
2. S. Chapman, The Absorption and Dissociative or Ionizing Effect of Monochromatic Radiation in an Atmosphere on a Rotating Earth, Proc. Phys. Soc. **43,** 26, 483 (1931).
3. H. E. Hinteregger, L. A. Hall, and S. Schmidtke, Solar XUV Radiation and Neutral Particle Distribution in July 1963 Thermosphere, in *Space Research V* (R. Muller, ed.), pp. 282–302, John Wiley, New York, 1965.

4. W. B. Hanson, Electron Temperatures in the Upper Atmosphere, in *Space Research III* (W. Priester, ed.), John Wiley, New York, 1963; A. Dalgarno, M. B. McElroy, and R. J. Moffett, Electron Temperatures in the Ionosphere, Planet. Space Sci. **11,** 463 (1963); J. V. Evans and M. Loewenthal, Ionospheric Backscatter Observations, Planet. Space Sci. **12,** 915 (1964); N. W. Spencer, L. H. Brace, G. R. Carignan, D. R. Taeusch and H. Niemann, Electron and Molecular Nitrogen Temperature and Density in the Thermosphere, J. Geophys. Res. **70,** 2665 (1965).
5. R. S. Narcisi and A. D. Bailey, Mass-Spectrometric Measurements of Positive Ions at Altitudes from 64 to 112 km, in *Space Research V*, (R. Muller, ed.), pp. 753–754 (Abstract), John Wiley, New York, 1965.
6. J. C. Holmes, C. Y. Johnson, and J. M. Young, Ionospheric Chemistry, in *Space Research V*, (R. Muller, ed.), pp. 756–776, John Wiley, New York, 1965.
7. J. A. Ratcliffe and K. Weekes, The Ionosphere, in *Physics of the Upper Atmosphere*, J. A. Ratcliffe, Ed., Academic Press, 1960.
8. S. Chandra and R. A. Goldberg, Geomagnetic Control of Diffusion in the Upper Atmosphere, J. Geophys. Res. **69,** 3187 (1964).
9. M. Nicolet and A. C. Aikin, The Formation of the *D*-region of the Ionosphere, J. Geophys. Res. **65,** 1469 (1960).
10. H. E. Hinteregger and K. Watanabe, Photoionization Rates in the E and F Region, J. Geophys. Res. **67,** 3373 (1962).
11. H. Rishbeth and D. W. Barron, Equilibrium Electron Distribution in the Ionospheric F_2 Layer, J. Atm. Terr. Phys. **18**(2–3), 234 (1960).
12. K. L. Bowles, Measuring the Plasma Density of the Magnetosphere, Science **139,** 389 (1963).
13. M. Nicolet, Helium, an Important Constituent in the Lower Exosphere, J. Geophys. Res. **66,** 2263 (1961).
14. P. Mange, The Distribution of Minor Ions in Electrostatic Equilibrium in the High Atmosphere, J. Geophys. Res. **65,** 3833 (1960).
15. Sir Arthur Eddington, The Internal Constitution of the Stars, p. 272, Cambridge, 1926.
16. J. W. Dungey, Electrodynamics of the Outer Atmosphere, in *The Physics of the Ionosphere*, p. 229, Physical Society, London, 1955.
17. Ionospheric Sporadic E, (S. Matsushita and E. K. Smith, eds.), Pergamon Press, 1963.
17a. D. T. Farley, A Plasma Instability Resulting in Field-Aligned Irregularities in the Ionosphere, J. Geophys. Res. **68,** 6083 (1963).
18. R. E. Bourdeau, Ionospheric Research from Space Vehicles, Space Sci. Rev. **1,** 683 (1963).
19. W. Calvert and C. W. Schmid, Spread F Observations by the Alouette Topside Sounder Satellite, J. Geophys. Res. **69,** 1839 (1964).
20. W. B. Hanson and T. N. L. Patterson, The Maintenance of the Nighttime F Layer, Planet. Space Sci. **12,** 979 (1964).
21. W. R. Piggott and A. H. Shapley, Antarctic Research (Geophys. Monograph No. 7), 1962; G. E. Hill, Sudden Enhancement of F Layer Ionization in Polar Regions, J. Atmos. Sci. **20,** 492 (1963).
22. J. H. Piddington, Ionospheric and Magnetospheric Anomalies and Disturbances, Planet. Space Sci. **12,** 533 (1964); C. O. Hines, Hydromagnetic Motions in the Magnetosphere, Space Sci. Revs. **3,** 342 (1964).

23. T. Obayashi, Morphology of Storms in the Ionosphere, J. Geomagn. Geoelectr. (Japan) **16,** 1 (1964).
24. Goubau, G., "Dispersion in einem Electronen-Ionen-Gemisch das unter dem Einfluss eines äusseren Magnetfeldes steht", *Hochfrequenz* u. *Electroakust* **46,** 37, (1935).
25. Sen, H. H. and A. A. Wyller, "On the generalizations of the Appleton-Hartree magnetoionic formulas" *J. Geophys. Res.,* **67**, 3931, (1960).
26. Calvert, W., "Ionospheric Topside Sounding" *Science* **154**, 228, 1966.
27. Kliore, A., D. L. Cain, G. S. Levy, V. R. Eshleman, G. Fjeldbo, F. D. Drake, "Occultation Experiment: Results of the first direct measurements of Mars's Atmosphere and Ionosphere", *Science,* **149**, 1243–1248, 1965.
28. Johnson, F. S., "Atmosphere of Mars" *Science,* **150,** 1445–1448, 1965.
29. Chamberlain, J. W. and M. B. McElroy, "Martian Atmosphere: The Mariner Occultation Experiment", *Science,* **152**, 21–25, 1966.

Chapter 4

THE EARTH'S RADIATION BELT

Wilmot N. Hess

INTRODUCTION

In 1958, when Explorer I was launched with a geiger counter on board, it discovered a region of high count rate starting at about 1000 km altitude. This was unexpected and in fact, it was suggested that the counter might have malfunctioned. Van Allen,[1] who had conducted the experiments on Explorer I realized very soon that the measured high count rates were due to charged particles trapped in the earth's magnetic field. When Explorer III showed the same results as Explorer I a little later, it was demonstrated that the high count rate was real and confirmed that the effect was due to geomagnetically trapped particles. Störmer had worked extensively on this general subject and had even calculated orbits of trapped particles years earlier, but the actual existence of trapped particles had not been suggested in this work. Ideas about the existence of a terrestrial ring current had also essentially included the idea of trapped particles.

At the same time that Van Allen's experiments in space were

going on, experiments with trapped particles were being carried out in various laboratories. Project Sherwood is an attempt by the AEC to create a controlled thermonuclear reaction on a small scale by confining charged particles in a magnetic field. Christofilos,[2] who was working on Project Sherwood, extrapolated the laboratory idea to earth scale and suggested the possibility of trapping a large number of charged particles in the magnetic field of the earth by using a nuclear explosion to inject the particles. This idea was carried out in the Argus Experiment and demonstrated experimentally that charged particles could readily be trapped in the earth's field. It had even been suggested in planning the Argus Experiment that a natural radiation belt might exist.

Before exploring in detail the concept of a radiation belt, let us briefly outline the components of the belt that have been found experimentally.

The picture of the radiation belt that Van Allen made[3] after the flights of Explorer IV and Pioneer III and IV is shown in Figure 1. This picture of the belt shows only particles that are energetic enough to penetrate 1 gm/cm² of absorber. This double-peaked structure led to the concept of two zones of the radiation belt—an inner zone inside $2R_e$ and an outer zone outside $2R_e$. We now know that these data in Fig. 1 are made up of two components. In the inner zone, fluxes of energetic protons of $E > 30$ MeV of

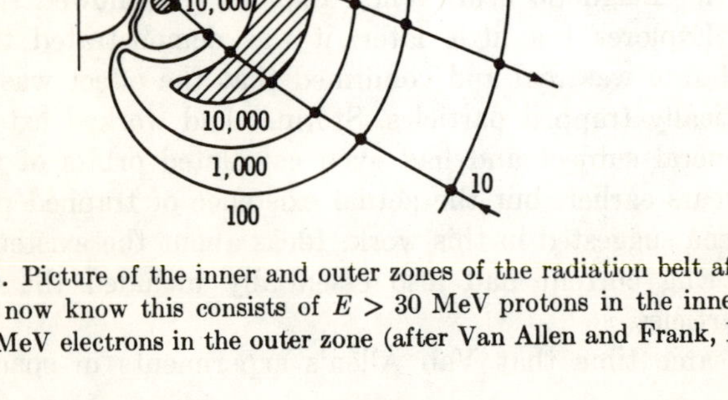

FIG. 1. Picture of the inner and outer zones of the radiation belt after Pioneer IV. We now know this consists of $E > 30$ MeV protons in the inner zone and $E > 1.5$ MeV electrons in the outer zone (after Van Allen and Frank, Ref. 3).

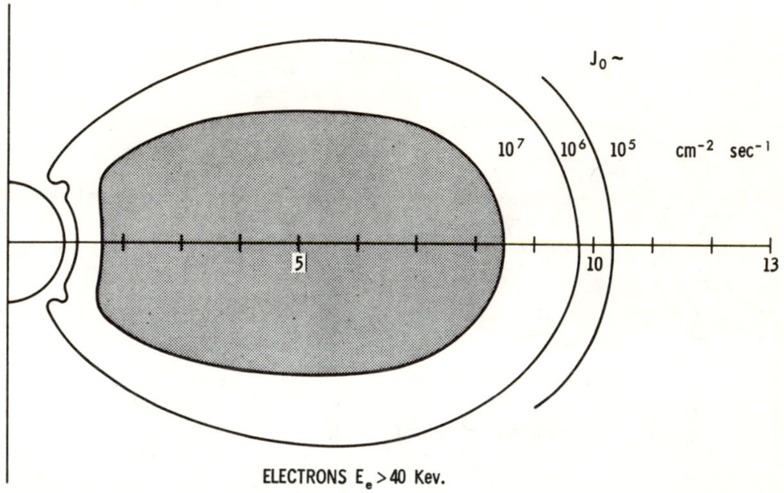

FIG. 2. Rough spatial distribution of $E > 40$ keV radiation belt electrons.

about 10^4 cm^{-2} sec^{-1} were counted while in the outer zone maximum readings were obtained from counting electrons of $E > 1.5$ MeV. The inner zone proton flux is quite stable in time, but the outer zone electron flux is extremely variable, often changing by more than an order of magnitude, especially at the times of magnetic storms.

Instruments carried on Injun and Explorer XII allowed a survey to be made of lower energy electrons[4] of $E > 40$ keV as shown in Figure 2. This particle population does not show the inner zone—outer zone two peak structure of Figure 1. The outer edge of this particle population at about $10R_e$ is quite abrupt and is controlled by the sun. The solar wind pushes in the earth's magnetic field to about $10R_e$ on the solar side. Outside this boundary (called the *magnetopause*, see chapter 9) we have the solar environment. Inside the boundary we have the somewhat compressed magnetic field of the earth. Only in this inner region is particle trapping possible.

Davis[5] has studied low energy outer zone protons on Explorer XII, Explorer XIV and Explorer XV and found the large fluxes shown in Figure 3. Most of the energy of the radiation belt particles is contained in these particles. This population is very stable in time.

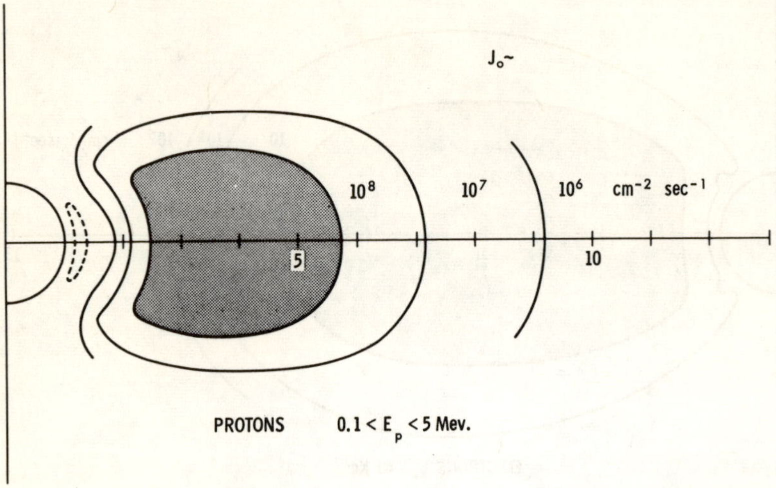

FIG. 3. Rough spatial distribution of $0.1 < E < 5$ MeV radiation belt protons.

Partial measurements have been made on other particle populations. Inner zone electrons are not yet well studied. Various groups of medium energy protons have been found, but not well sorted out yet. Recently large fluxes of low energy $E \sim 5$ keV protons have been found in the inner zone which may have interesting geophysical properties. Studies of many of these particles are going on now.

This briefly reviews the classes of particles found experimentally in the radiation belt. Now let us consider how a radiation belt can exist and with what classes of particles it might be populated. First, why is it possible to build up large fluxes of particles trapped in the geomagnetic field? To understand this we must consider the motions of a charged particle in a magnetic field.

MOTION OF PARTICLES IN A DIPOLE FIELD

The general problem of charged particle motion in a dipole field is complicated. Fortunately, in the case for radiation belt particles, an approximation can be used which considerably simplifies the

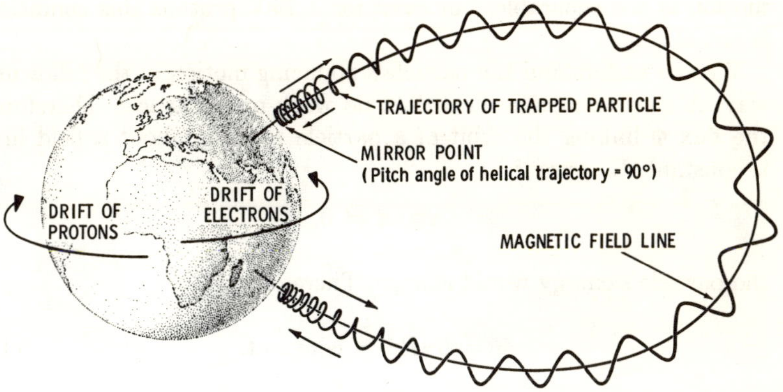

FIG. 4. Motion of charged particles in a dipole field.

situation. Alfven[6] introduced the idea of a guiding center of a particle. The particle motion here is described in terms of (1) a rapid gyration about a field line with a cyclotron period T_c and radius of gyration R_c and (2) motion of the guiding center of the particle along the line of force. The motion along the line is also periodic. The particle is reflected by the converging magnetic field near the earth and bounces back and forth in the exosphere with a bounce period T_B. There is another motion, a slow drift in longitude around the earth with a period of revolution of T_R (See Figure 4). Particles at 2000 km altitude near the equator will have the following characteristics:

	R_c, cm	T_c, sec	T_B, sec	T_R, min
50 keV Electron	5×10^3	2.5×10^{-6}	0.25	690
1 MeV Electron	3.2×10^4	7×10^{-6}	0.10	53
1 MeV Proton	1×10^6	4×10^{-3}	2.2	32
10 MeV Proton	3×10^6	4.2×10^{-3}	0.65	3.2
500 MeV Proton	2.5×10^7	6×10^{-3}	0.11	0.084

Because the three periods are so different, the particle motion is separable into these three components. If the cyclotron radius of the particle R_c becomes comparable to the diameter of the earth, the motion is not separable, but even for 1 BeV protons this condition is not attained.

We can understand the particle's bouncing motion in the following way. A static magnetic field does no work on a particle. Therefore, the flux Φ linking the orbit of a particle rotating about a field line is constant. Because if

$$d\Phi/dt \neq 0,$$

the particle's energy would change. Therefore,

$$\Phi = B\pi R_c^2 = \text{constant}. \tag{1}$$

We can write for the particle's perpendicular energy

$$E_\perp = \tfrac{1}{2}(mv_\perp^2) = \tfrac{1}{2}(m\omega^2 R_c^2) = \tfrac{1}{2}(mR_c^2)(eB/mc)^2 = e^2B^2R_c^2/2mc^2. \tag{2}$$

Substituting in (1) we get

$$\Phi = \text{constant} = 2\pi mc^2 E_\perp/e^2B. \tag{3}$$

Therefore

$$E_\perp/B = \text{constant} = \mu. \tag{4}$$

The constant μ in equation (4) is the magnetic moment due to the particle's motion around the field line. Writing $E_\perp = mv_\perp^2/2 = mv^2 \sin^2 \alpha/2$, where α is the angle between the velocity vector and the field line (called the pitch angle), we get from equation (4)

$$(\sin^2 \alpha_1)/B_1 = (\sin^2 \alpha_2)/B_2 = \text{constant}. \tag{5}$$

The particle will move into a region of increasing field B until $\sin \alpha = 1$ when it must stop. It then moves out of the high field region and repeats the process at the other end of the field line for the same value of B.

The drift in longitude of a charged particle results from a force on the particle perpendicular to the field lines and lying in the plane through the center of the earth. The magnitude and direction of this drift velocity can be obtained from the cyclotron equation

$$\mathbf{R} = (mc/eB^2)(\mathbf{v} \times \mathbf{B}) = (c/eB^2)(\mathbf{p} \times \mathbf{B}). \tag{6}$$

If a force $f_\perp$ acts perpendicular to B for a time Δt there is a change of momentum of

$$\Delta p = f_\perp \Delta t. \tag{7}$$

This results in a displacement of the guiding center of the particle of

$$\Delta\mathbf{R} = (c/eB^2)(\Delta\mathbf{p} \times \mathbf{B}). \tag{8}$$

Differentiating with respect to time gives

$$v_D = d\mathbf{R}/dt = (c/eB^2)(\mathbf{f}_\perp \times \mathbf{B}). \tag{9}$$

This is the drift velocity. It is perpendicular to both $f_\perp$ and B and, therefore, if $f_\perp$ lies in the plane through the center of the earth containing the earth's axis, the drift velocity will be azimuthal and will refer to a drift in longitude.

One force that produces a drift is due to the gradient of the earth's magnetic field

$$F = \mu\nabla B \cong 3\mu B/R. \tag{10}$$

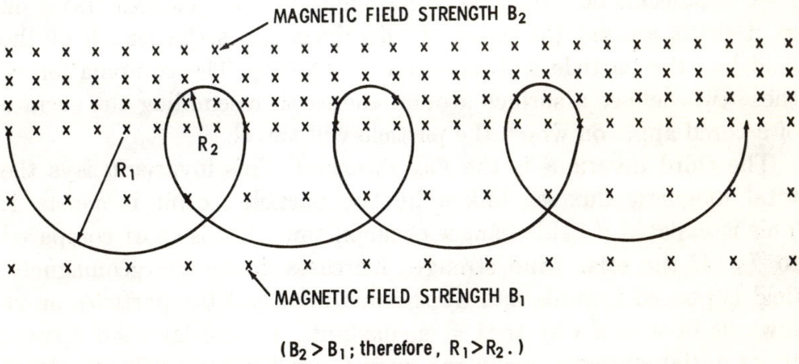

FIG. 5. Drift of a charged particle in a magnetic field with a gradient.

The sideways drift of a particle due to a magnetic field gradient is shown in Fig. 5. A second force that produces a drift is the centrifugal force on a particle due to the curvature of field lines.

$$F = mv_{\|}^2/R_l. \tag{11}$$

Adiabatic Invariants

Associated with the motion of particles in a dipole field are three constants of the motion. They are actually only *adiabatic constants* which means they are constant unless magnetic fields change rapidly. We showed the magnetic moment μ to be a constant of the motion, but it is not constant if fields change in times short compared to T_c or in distances short compared to R_c.

The second adiabatic invariant I is called the *integral invariant*

$$I = \int_{m_1}^{m_2} v\, dl = v \int_{m_1}^{m_2} (1 - B/B_m)^{1/2}\, dl, \tag{12}$$

where the integral is taken along a field line between the two mirror points. This invariant is violated if field changes occur in times short compared to T_B.

We can understand the motion of a particle in the earth's field rather well by considering only the constancy of μ and I. From μ we know that the particle mirrors at a particular value of B. From I we can decide on which set of field lines the particle will travel on as it drifts around the earth. I effectively gives the length of the field line the particle will move on as it drifts. The combination of these two defines a surface around the earth resembling the surface of a cored apple on which the particle will travel.

The third invariant is the *flux invariant.* This invariant says the total magnetic flux Φ_m linked by the particle's orbit is constant. This is violated if field changes occur in time that is short compared to T_R. If the solar wind strength increases slowly the geomagnetic field is pushed inwards and by the flux invariant the particles move inwards in such a way that Φ is constant. If the solar wind returns to its initial strength slowly the particles will return to their starting points.

PARTICLE SOURCES

There are several possible sources of the trapped radiation belt particles. We will list and consider some of the possibilities.

Cosmic Rays

A flux of about 2 particles/cm^2-sec of galactic cosmic rays reach the earth regularly. One possibility concerning the radiation belts that was suggested early was that they represented quasi-trapped particles, that is, they consisted of particles on certain special Störmer orbits that can stay near the earth for a long time and then finally wander out of the earth's field and escape. Estimates of the intensity of the radiation belt that would be produced this way show that this is a small effect. We would need a flux increase of $\times 10^4$ or more from galactic cosmic rays to achieve radiation belt fluxes. Nothing approaching this multiplication is obtained by this quasi-trapping.

Another possibility is that cosmic rays might produce trapped particles by interacting with the very thin atmosphere at very high altitude. We can estimate the importance of this source. If we take an atmospheric density of 10^5 atoms/cm^3 of oxygen corresponding to about 1000 km altitude and a cross section for producing high energy charged particles of 0.2×10^{-24} cm^2, we get a source strength S of

$$S = (2 \text{ protons/cm}^2\text{-sec})\,(10^5 \text{ atoms/cm}^3)\,(0.2 \times 10^{-24} \text{ cm}^2/\text{atom})$$

$$= 4 \times 10^{-20} \text{ particles/cm}^3\text{-sec.}$$

This also is negligible when compared to other source strengths.

Still another possibility is *splash albedo protons* made by interactions of the high-energy cosmic rays with the atmosphere. Some of the secondary protons made this way emerge upwards out of the atmosphere. We know that the particles must return to roughly their birth altitude or even lower in order to mirror at the other end of their line of force. This will mean they encounter a rather thick atmosphere and will rapidly be lost, so this source contributes little, if anything.

Solar Wind

There are about 1 to 10 protons/cm^3 in the keV energy range in the solar wind (see Chapter 8). This is a large flux of particles. It has been found on the IMP satellite that inside the shock wave upstream of the magnetopause, certain classes of energetic particles appear that are not present in the solar wind outside the shock wave. Electrons of kilovolt energy are present in fluxes of 10^{10} cm^{-2} sec^{-1} and, in addition, fluxes of 10^7 protons of energies up to 5 keV are seen. These particles may be the raw material for building parts of the outer radiation zone. Processes must be found to allow these particles to be brought into the magnetic field of the earth. There are several possibilities here; particles might be injected at the field nulls near the poles; field convection processes may bring them into the field or electric fields may also help them get in. This last source appears important and will be examined in greater detail later.

Neutrons

Neutrons are produced by cosmic ray protons colliding with oxygen and nitrogen nuclei in the earths atmosphere. Several neutrons are produced in the nuclear cascade in the atmosphere. A 5 BeV cosmic ray proton will produce about 7 neutrons in the atmosphere. About 15% of all neutrons formed in this way in the atmosphere diffuse out of the top of the atmosphere into space. This leakage flux of neutrons out of the atmosphere is about 0.2 neutrons/cm^2-sec at the equator and about 2 neutrons/cm^2-sec at the pole. The neutron flux in space close to the earth at low latitudes has been measured to be about 1.0 neutrons/cm^2-sec which is in reasonable agreement with the calculated fluxes. Few neutrons reach the earth from the sun or more distant space because the neutron is radioactive. When the neutron is bound up in a complex nucleus it is stable, but by itself it is radioactive with a mean life of 1000 sec, and decays by the reaction

$$n \rightarrow p + e + \bar{\nu}. \tag{13}$$

The anti-neutrino $\bar{\nu}$ does not interest us here, but the proton and electron are both important in forming the radiation belt.

We need to know how many neutrons of what energies decay at

different places in space to produce protons and electrons in order to estimate the neutron-decay source strength. First, we must know the neutron energy spectrum, $\Phi(E, R, \lambda)$, in space. This has been calculated[7] from a knowledge of the neutron energy spectrum inside the atmosphere. The flux decreases with increasing distance from the earth, especially for low energy neutrons. Neutrons of less than $\frac{2}{3}$ eV are trapped by the earth's gravitational field so that they essentially all decay in space near the earth. Only about 1 percent of neutrons of 1 MeV energy decay near the earth. The density of neutrons decaying dn/dV is given by

$$(dn/dV)(E, R, \lambda) = (\gamma v T_n)^{-1}\Phi(E, R, \lambda), \tag{14}$$

where v is the neutron velocity, T_n the neutron mean life, and λ is latitude.

The electrons that are produced by neutron decay will have the normal β decay spectrum. The electron's energy will be essentially unchanged by the neutron's kinetic energy. To demonstrate this, consider a neutron of 10 MeV which has a velocity of about 4×10^9 cm/sec. When an electron of about 300 keV is produced by neutron decay it has a velocity of about 2×10^{10} cm/sec. When the neutron's velocity is compounded relativistically with this velocity it changes by, at most, 10 percent. Almost all neutrons have energies of less than 10 MeV, so the effect of the neutron's motion in changing the electron's energy can be neglected. Therefore, to determine the total electron source strength S_e for the radiation belt due to neutron decay, we simply add up all neutron decay events of all energies

$$S_e(R, \lambda) = \int (dn/dV)(E, R, \lambda)\, dE = \int (\gamma v T_n)^{-1}\Phi(E, R, \lambda)\, dE. \tag{15}$$

This electron source strength S_e is about 10^{-11} cm^{-3} sec^{-1} near the earth and about 10^{-13} at 5 earth radii.

The situation is different when we consider the protons produced by neutron decay. The kinetic energy of the proton is very nearly the kinetic energy of the neutron that decayed to form the proton. The electron kinetic energy is supplied by the mass difference of the neutron and proton. The recoil energy given the proton by the electron is only about 100 volts; therefore, if we consider protons above about 10 keV, we can accurately take the proton's energy and direction of motion as that of the neutron. Because of this the decay density energy spectrum is also the proton source energy spectrum $S_p(E, R, \lambda)$ from about 10 keV up.

$$S_p(E, R, \lambda) = (dn/dV)(E, R, \lambda). \tag{16}$$

There are about 10^{-14} protons/cm³-sec of $E > 10$ MeV produced in space near the earth. The proton source near the earth at the equator has been evaluated to be

$$S_p(E, R_e, 0) = 0.8E^{-2.0}/\gamma v T_n. \tag{17}$$

Thus, we have a quantitative picture of the neutron-decay proton and electron sources. Neutrons are probably an important source of the inner radiation zone.

LOSS PROCESSES

To build up a steady-state radiation belt, we must have particle sources and also loss processes.

Protons

There are at least three ways that protons are lost from the radiation belt. In the inner belt, high-energy protons of $E > 300$ MeV are lost primarily by inelastic nuclear collisions. Protons of $E < 100$ MeV are removed most rapidly by slowing down until they reach about 100 keV. The amounts of oxygen that must be traversed, R, to stop protons of various energies are listed below:

E	R
0.1 MeV	6×10^{-5} gm/cm²
1	0.003
10	0.14
100	8.6

Below 100 keV the protons are more rapidly lost by charge exchange with slow protons by the reaction

$$p' + H \rightarrow p + H',$$

where the prime refers to the fast particle. At 50 keV the cross section for the charge exchange is $\sigma_{cH} = 2 \times 10^{-16}$ cm^2 and a charge exchange lifetime τ_{cH} is given by

$$\tau_{cH} = [\sigma_{cH} n(\mathrm{H}) v]^{-1} = [(2 \times 10^{-16})(10^4)(3 \times 10^8)]^{-1} = 1600 \text{ sec}, \tag{18}$$

where $n(\mathrm{H})$ is the atomic density of hydrogen taken here to be 10^4 atom/cm^3. This time is less than the slowing down time above about 1000 km altitude so the charge exchange process will dominate here.

In the outer belt, protons are rapidly lost by some additional process. This process probably involves interactions with electromagnetic waves which produce a breakdown of the magnetic moment invariant. We will return to this later.

Electrons

Different processes are responsible for the loss of electrons than for protons. Because electrons are lighter, they scatter more easily and are lost in the inner belt principally by coulomb scattering into the loss cone where $\alpha \approx 0$ and they enter the atmosphere and are lost. This process is more important than slowing down for electrons. This can be seen by comparing the range of the particle, R, and the amount of material necessary to scatter a particle out of the radiation belt, D. The range of a 1 MeV electron from *Feather's rule* is

$$R = 0.54E - 0.13 = 0.41 \text{ g cm}^{-2}. \tag{19}$$

The value of D can be obtained by using the scattering formula

$$\theta^2 = 7000D/E^2. \tag{20}$$

where θ is the *mean angle of scattering*, D the *path length* of NTP air traversed in cm, and E the *particle energy* in keV. Taking $\theta = 0.5$ rad as the necessary scattering angle required to lose the particle we find

$$D = 0.04 \text{ g cm}^{-2}.$$

This shows that the electron will be scattered out of the inner zone before it slows down. There is direct experimental evidence that electrons have short lifetimes in the outer zone—much shorter time than would be expected from atmospheric scattering. They are almost certainly lost here by certain types of wave-particle processes.

HIGH ENERGY PROTONS

The first experiment performed in the radiation belt that unambiguously identified the particles that were counted involved flying a stack of nuclear emulsions on an Atlas rocket. The emulsion stack was recovered and developed and the nuclear tracks read. The range and ionization of the particles were measured, the particles identified and their energies determined.

Protons of $E > 75$ MeV and electrons of $E > 12$ MeV could get through the 6 gm/cm^2 shielding into the nuclear emulsions. No electrons were found, but a large number of protons were found. The energy spectrum of protons measured by Freden and White on a later flight[8] is shown in Figure 6. The inner-belt proton flux seems quite constant in time, varying by less than a factor of 2 for several flights. This indicates that the particle lifetime is quite long.

We can get a quantitative picture of the flux and energy spectrum of the inner radiation belt protons produced by neutron β decay by considering the conservation of particles. The continuity equation in energy can be written[9]

$$dN(E)/dt = S_p(E) - L(E) + (\partial/\partial E)[J(E)], \tag{21}$$

where $N(E)$ is the *equilibrium-proton-density* energy spectrum and $S_p(E)$ is the source of protons and $L(E)$ is the loss term due to

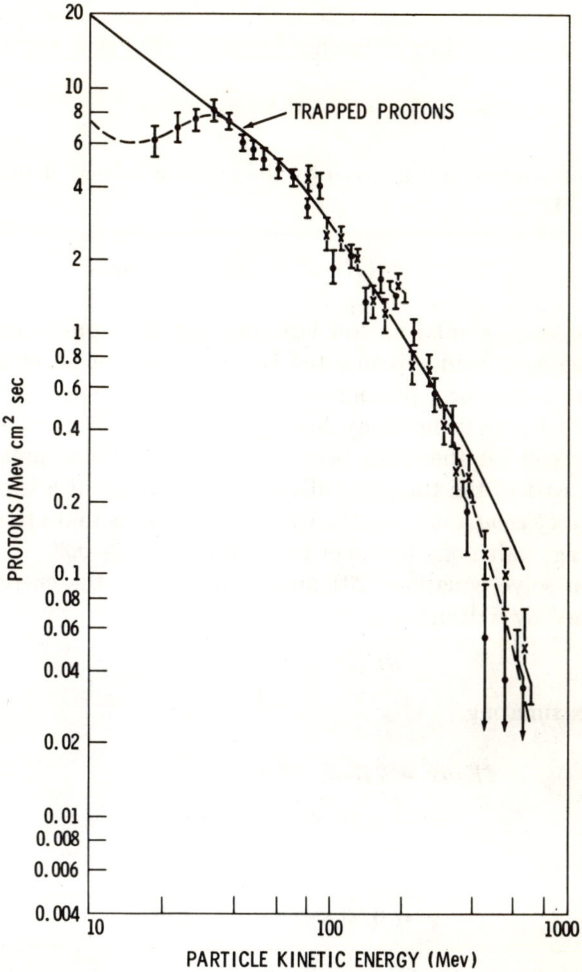

FIG. 6. Energy spectrum of inner zone protons (after Freden and White, Ref. 8).

inelastic nuclear collisions and $J(E)$ is the current in energy space

$$J(E) = N(E)(dE/dt),$$

for equilibrium

$$dN(E)/dt = 0.$$

For protons between 5 and 100 MeV we can ignore losses such as those caused by nuclear collisions. Equation (21) now becomes

$$S_p(E) = (\partial/\partial E)[N(E)(dE/dt)]. \tag{22}$$

The proton-source term is given by the neutron's decay density from equation (17).

$$S_p(E) = dn(E)/dV = \eta 0.8E^{-2.0}/\gamma v T_n. \tag{23}$$

This expression is valid for low latitudes and close proximity to the earth. The coefficient η is inserted here because not all of the neutrons that decay form protons that are trapped.[10] Some of the protons made by neutron decay have pitch angles that are so small that they will hit the earth before they mirror. These protons will not form part of the trapped radiation. The coefficient η is called an *injection coefficient* and gives the fraction of protons that are trapped. The average value of η for inner-belt protons is $\eta \cong 0.30$.

We can solve equation (23) approximately for the energy range 10–80 MeV by writing

$$dE/dt = v(dE/dx) \tag{24}$$

and approximating

$$dE/dx = 243\rho E^{-0.79} \text{ MeV/cm},$$

$$v = 1.45 \times 10^9 \times E^{0.477} \text{ cm/sec}$$

and

$$\gamma = 0.93E^{0.032},$$

where ρ is the air density in gm/cm³. Substituting in equation (23) we get the proton flux

$$\Phi(E) = vN(E) = vkE^{-n} = [(2.9 \times 10^{-16})/\rho]E^{-0.72}$$

$$\text{for } 10 < E < 80 \text{ MeV}. \tag{25}$$

Using a time-averaged density of $\rho = 2.8 \times 10^{-18}$ gm/cm³, corresponding to an atomic density of 1.0×10^5 atoms/cm³, this gives

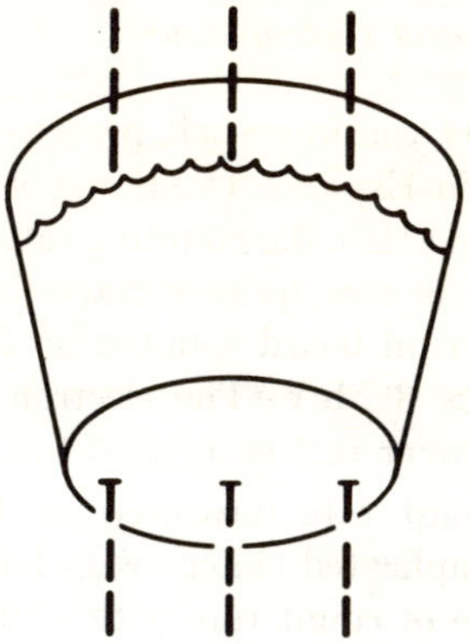

FIG. 7. Leaky bucket model of the inner zone protons. The source feeds particles into the trapping zone or bucket and losses remove them.

$\Phi(E) = 110E^{-0.72}$ protons/cm²-sec MeV. This expression for $\Phi(E)$ fits the Freden and White data well in this energy range (see Fig. 6).

The lifetime τ_p of these protons can be obtained by using the "*leaking bucket*" equation (see Fig. 7). We have for equilibrium for this problem

$$\text{Input} = \text{Output} = \text{Contents}/\tau_p, \tag{26}$$

where τ_p is the average time a particle stays trapped, which gives

$$\tau_p = \text{Contents/Input} = N(E)/S(E) = 7 \times 10^5 E^{1.3} \text{ sec}$$

for $10 < E < 80$ MeV at 1100 km. Proton lifetimes of several years occur here.

The general form of equation (21) has been solved to give the solid curve in Figure 6. From 10 to 80 MeV the curve is exactly given by equation (25). There is no arbitrary normalization involved here and there are no adjustable parameters in the theory so the comparison of experiment and calculation is direct. We know the source strength and properties, the loss processes and rate, and we know the properties of the atmosphere. It seems quite clear that neutrons produce many of the protons in Fig. 6 of $E < 30$ Mev. However, when quantitative studies are made of these high energy protons especially off-equator, it seems that neutrons cannot make all of them and that some additional source is needed.

Spatial Distribution

The spatial distribution of the inner-belt protons was first measured by the geiger counter in Explorer IV. It was not known at the time these data were obtained that the counting rate of this detector was due to protons, but it is now quite certain that this was the case. The 302 geiger counter on board counted protons of $E_p > 30$ MeV and electrons of $E_e > 3$ MeV. The electron flux above this 3 MeV energy limit in the inner belt is thought to be quite small.

Contours of constant count rate measured on Explorer IV by Van Allen show a fairly complicated change with latitude and longitude. Curves of the increase of count rate with altitude at different locations are shown in Figure 8a. The curves labeled O are data near Singapore, those labeled E are near Nigeria, Africa, the A_1 curve is in Northern South America and A_2 and A_3 are in Central South America. All of these locations are at about the same magnetic latitude, but the curves show quite different altitude behaviors. The reason for this is that the earth's surface magnetic field is different at these different locations. Replotting this data in terms of the magnetic field instead of altitude at each location (see Figure 8b) demonstrates that the curves then become essentially indistinguishable. The reason for this is easily understood in terms of the motion of charged particles in a dipole magnetic field. The particles bounce back and forth, mirroring at one particular value of the magnetic field B. The particles also drift around the earth, staying on a surface of constant integral invariant I which corresponds to staying at a constant magnetic latitude. Therefore, for observations at one latitude the count rate should vary with B as it does in Figure 8b. The altitude of the mirror points represented by the count rates in Figure 8b get as low as 400 km over South America. Below this altitude galactic cosmic rays provide most of the count rate. The lower edge of the trapped radiation belt is clearly controlled by the atmosphere and protons are lost here by slowing down. For a 50 MeV proton to be brought to rest it must transit about 2.5 gm/cm^2 of oxygen. At 400 km the atmospheric density is about 5×10^8 atoms/cm^3 or about 10^{-14} gms/cm^3. At a velocity of 10^{10} cm/sec the proton will take about 2×10^4 sec to slow down to rest. But, when considering the variations of atmospheric density along the particle orbit, especially the variation due to the drift in longitude, the lifetime is increased probably to 10^6 sec.

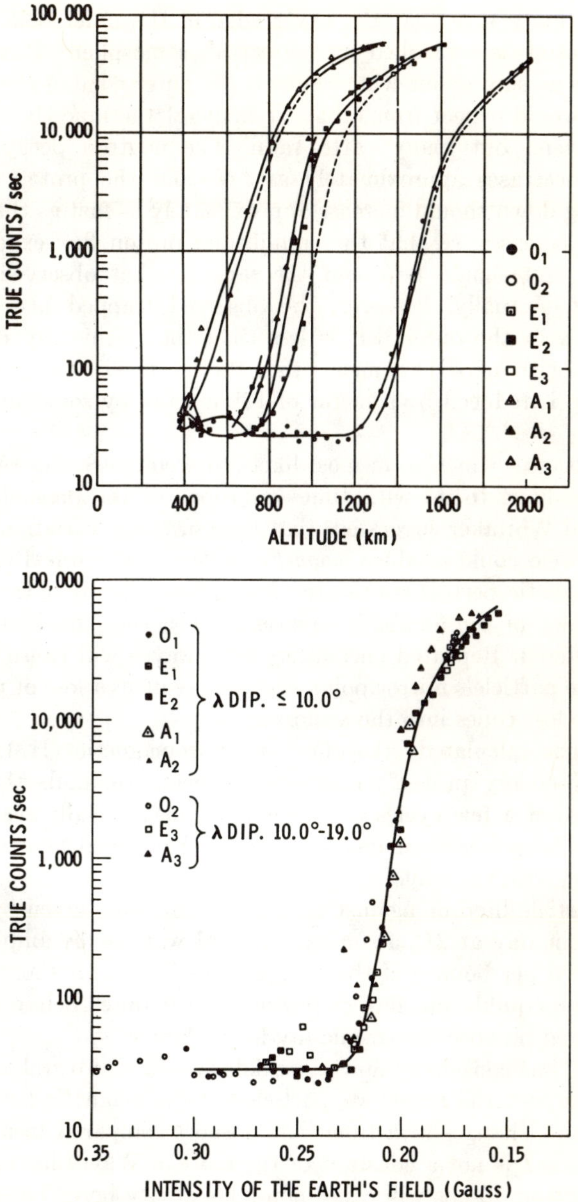

FIG. 8. (a) Counting rate vs. altitude of the geiger counter on Explorer I. (b) Counting rate vs B of the geiger counter on Explorer I (after Van Allen).

We are quite sure that the lower edge of the inner belt of high energy protons is controlled by the earth's atmosphere. But we are not nearly as sure about what controls the outer edge of the proton belt. We would expect from what we know of the inner belt that it should extend out many earth radii. The neutron decay source strength decreases approximately as $1/R^2$ but the proton lifetime for slowing down should increase approximately as fast as the source strength decreases, so that the equilibrium proton flux expected in the outer belt should be about the same as that observed in the inner belt. Actually, however, the observed trapped high energy proton flux in the outer belt is less than the cosmic ray flux and consistent with zero. This means that the lifetime of the protons in this region is reduced by a factor of 10^4 or more by some additional process.

The processes which seem most likely to accomplish this reduction in lifetime have to do with time variations of the magnetic field. Welch and Whitaker suggested[2] that time or space variations in the magnetic field could produce *magnetic scattering* of trapped particles. If the magnetic perturbations are of such character that they cause a breakdown of the adiabatic invariants then the particle's motion will be altered. Repeated encounters with waves will cause a diffusion of the particle's mirror point and will result in a loss of particles out of the loss cones into the atmosphere.

Dragt has calculated[11] the effect of hydromagnetic (HM) waves on a high-energy proton's magnetic moment. He finds that with HM waves of a few cycles per second can magnetically scatter the protons. The particle's magnetic moment is changed as a result of interacting with the wave.

The particle lifetime against HM wave scattering given by Dragt is about one day at $2R_e$ assuming one HM wave of 3γ amplitude is encountered per bounce of the proton. This lifetime is short enough so that the equilibrium flux of protons in the outer belt would not be measured because the cosmic ray flux is larger.

Parker[12] has considered how a breakdown of the integral invariant I can transport and accelerate particles. If the magnetic field at the mirror point changes in a time that is short compared to a bounce period, then I is not a constant of the motion. Waves having a frequency of 5 or 10 cps will cause non-adiabaticity here. These waves at the mirror point result in a Fermi acceleration. Some particles encounter mirror points moving towards them due to the field perturbation ΔB, and some particles find mirror points moving away from them. But, because statistically more approaching collisions take place, there is a net energy gain.

The interesting feature here is that as a particle gains energy its mirror point is systematically lowered because all the energy gain is in $E_{\parallel}$ and therefore the pitch angle decreases and the particle is lost into the atmosphere. For a $\Delta B/B$ of 10^{-3} and constantly moving mirror points, the characteristic time for a particle to diffuse down into the atmosphere is about 3×10^6 sec. It is not known whether these hydromagnetic wave conditions are really met, so we do not know if this lifetime is reasonable or not. Either of these theories may explain the absence of outer belt protons but neither has yet been tested.

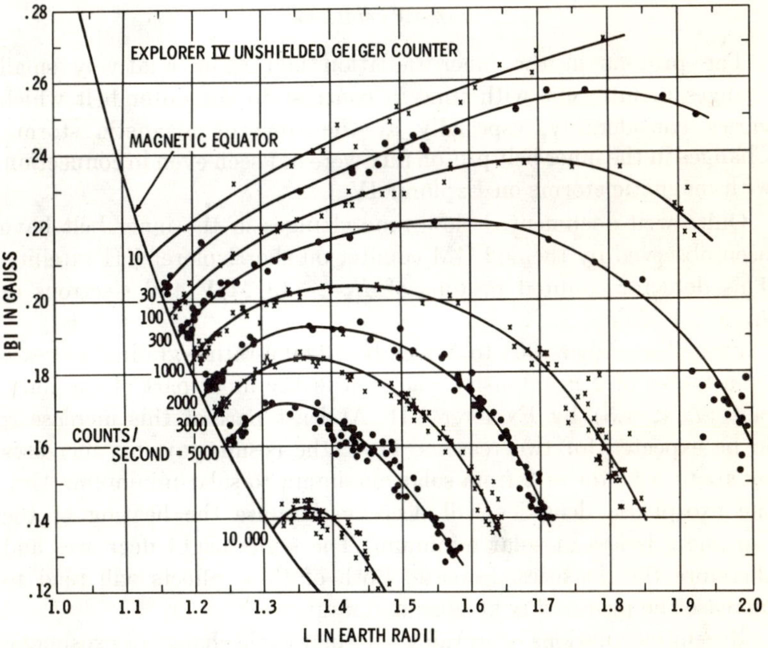

FIG. 9. Explorer IV geiger counter count rates in B–L space (McIlwain, Ref. 13).

In order to organize satellite data for different longitudes into some easily manageable form, McIlwain[13] developed the B–L coordinate system. B is the scalar magnetic field and L is a distance which, in a dipole field, is the distance to the equator for a particular field line. It is defined in terms of the Integral invariant I in such a way that the real earth's field is used. L is very nearly constant along a field line, so it can replace the commonly used equatorial radius R_0, but use real values of the earth's field. The Explorer IV data, when plotted in terms of BL coordinates, can be combined for different geographic latitudes, longitudes, and altitudes into the simple form shown in Fig. 9. Reading down a line of constant L here corresponds to going out from the earth along a field line. The B–L coordinate system is commonly used in plotting radiation belt fluxes.

Time Variations

The protons in the inner radiation belt show relatively small changes in intensity with time in contrast to the outer belt which varies considerably, especially at the time of magnetic storms. Changes in the inner belt proton flux were not seen even in connection with magnetic storms on Explorer IV.

Quite well documented slow time changes in the inner belt have been observed by the 302 GM counter on the Explorer VII satellite. This detector counted protons of $E_p > 18$ MeV and electrons of $E_e > 1.1$ MeV.

From November 1959 to November 1960 a rather gradual increase by a factor of 2 in intensity was seen in the inner part of the inner belt ($L < 1.5$) by Explorer VII. At least part of this increase is to be expected for two reasons. First, the cosmic ray flux increases by about a factor of 2 from solar maximum to solar minimum. Also, the exospheric densities will decrease because the heating of the exosphere is less at solar minimum. The scale height decreases and therefore the densities decrease. Both of these effects will tend to increase the proton flux at solar minimum.

Recent calculations concerning the solar cycle change of exospheric density show that at several hundred kilometers altitude the density varies by nearly a factor of 100 during the solar cycle (see Chapter 2). Using equation (21) again we can study time variations in the high

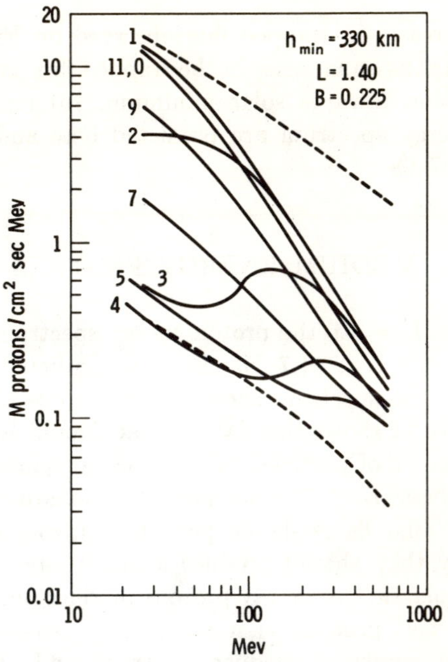

FIG. 10. Solar cycle time variations in inner zone protons for $B = 0.225$ and $L = 1.40$ (Blanchard and Hess, Ref. 14).

energy protons. We now do not assume equilibrium and use

$$dN/dt \neq 0. \qquad (27)$$

Integrating equation (21) in time starting with $N(E) = 0$ we build up an oscillating population of protons[14] as shown in Figure 10 for one point in BL space. The most important process in producing this changing flux is the changing atmospheric density. This density enters into both the nuclear collision loss L and the slowing down term

$$(\partial/\partial E)[N(E)(dE/dt)].$$

For protons of $10 < E < 500$ MeV whose minimum mirror point altitude in the South Atlantic is between 200 km and 700 km the solar cycle changes are significant. This calculation shows that the factor of 2 increases in trapped flux observed on Explorer VII can be well explained as an increase in the trapped flux as the atmosphere cools down doing towards solar minimum. Interesting changes in the proton energy spectrum are predicted here and should be observable in 1967–68.

MEDIUM ENERGY PROTONS

Out to about $L = 1.6$ the proton energy spectrum is as shown in Figure 6. But for $L > 1.7$ Naugle and Kniffen found[15] that for $E_p < 30$ MeV the spectrum showed a sharp rise above this curve in Fig. 6. Figure 11 shows this data of Naugle and Kniffen.

What new source of particles can produce medium-energy trapped protons only above $L = 1.7$? Solar protons bombarding the polar cap at the time of solar flares should produce neutrons and when these neutrons decay they should produce a new source of protons. We are sure that galactic cosmic ray protons produce neutrons that then produce inner belt protons so it seems very reasonable that solar protons should similarly produce neutrons and thereby produce trapped protons.

The trapped protons made from solar protons will be different from the other trapped protons in two ways. First, their spacial distribution will be different. Solar protons arrive at the earth only in the polar regions. They are of low enough energy so that the earth's field prevents them from getting further from the poles than about 55° magnetic latitude. Because of this the inner part of the inner belt is geometrically shadowed from this source,[10] so that none of these neutrons can produce trapped protons near the inside of the inner belt. Thus trapped protons will not be produced for $L < 1.65$. The second feature of this source that is distinctive is that the neutrons produced this way have lower average energies than for the galactic cosmic ray source. The solar proton energy spectrum usually extends up past 100 MeV, but most of the protons are of much lower energy. Therefore, the polar cap neutron energy spectrum will not contain many particles of $E > 50$ MeV and the peak of the neutron spectrum should be at about 1 MeV.

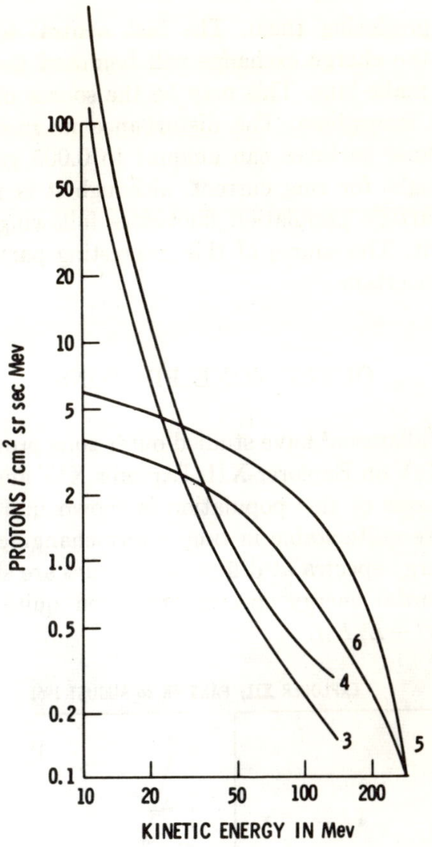

FIG. 11. Proton energy spectra measured by Naugle and Kniffen (Ref. 15) on the NERV probe. Curve 3, $L = 1.79$, 1600 km altitude. Curve 4, $L = 1.72$, 1884 km. Curve 5, $L = 1.54$, 1600 km. Curve 6, $L = 1.47$, 1400 km.

Although this polar cap neutron source seems to have the proper characteristics to explain the Naugle and Kniffen protons, when it is examined quantitatively it does not work and some other source is needed. A large flux of low energy ($E \backsim 5$ ke V) protons may have been found in the inner belt. The flux if it is real is variable but sometimes $\Phi_p > 10^{10}$ protons/cm²-sec. These protons must have a lifetime against charge exchange of $\tau \sim 3000$ sec so there must be a very strong source producing them. The fast neutral hydrogen atoms resulting from the charge exchange will bombard the upper atmosphere and will make ions. This may be the source of the nighttime F layer of the ionosphere. The disturbance magnetic field at the earth due to these particles can amount to 0.005 gauss. This may be the long sought for ring current, although it is not certain yet whether this particle population fluctuates following magnetic disturbances or not. The source of this interesting particle population is completely uncertain.

OUTER ZONE PROTONS

Davis and Williamson[5] have studied outer zone protons of energies $0.1 < E < 5$ MeV on Explorer XII, Explorer XIV and Explorer XV. Approximate fluxes of this population is shown in Figure 3. These proton fluxes are quite stable in time rarely changing by more than 20 percent. Energy spectra at different L values are shown in Figure 12. The differential energy spectra are given quite accurately by $N(E) = k \exp(-E/E_0)$.

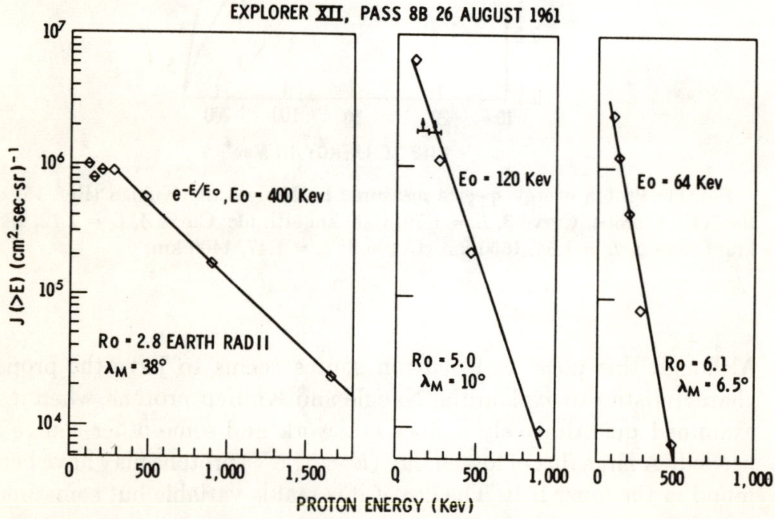

FIG. 12. Energy spectra at different L values of outer zone protons (after Davis, Ref. 5).

These energy spectra change in an orderly way with L. The changes in energy suggest that the protons are drifting in L and changing energy as they drift. In order to drift in L the third adiabatic invariant must be violated. Let us assume that the first two invariants are conserved and observe how particles will behave as they drift in L. From equation (4) we obtain

$$\mu = (E \sin^2 \alpha_0)/B_0 = EL^3 \sin^2 \alpha_0/0.312 = \text{constant} \quad (28)$$

and from equation (12)

$$I = \oint v \cos \alpha \, dl = E^{1/2} L F(\alpha_0) = \text{constant}. \quad (29)$$

By using these two relations we can see how a particle's energy and equatorial pitch angle, α_0, will change as the particle drifts in L. For a particle moving inwards in L the pitch angle, α_0, increases slowly, that is, the particle tends to move closer to the equator. A proton having $\alpha_0 \approx 45°$ at $L = 10$ will have $\alpha_0 = 55°$ at $L = 2$. Also the particle's energy increases as it drifts inwards. For particles of $I = 0$ from equation (28) the energy transforms as $E \propto L^{-3}$ so that a 1 MeV proton at $L = 6$ becomes 27 MeV at $L = 2$. It can be easily shown that the e-folding energy E_0 transforms the same way as a single particle's energy. Experimental data on E_0 from Davis and the theoretical energy transformation from equations (28) and (29) agree very well. This provides strong evidence that the L drift process really does take place. By applying Liouville's Theorem it appears that the drift is from large L inwards.

What kind of physical process can produce the L drift of protons? The most reasonable process seems to be compressions and expansions of the terrestrial magnetic field caused by changed in the solar wind strength. Repeated compressions of the field occurring in times short compared to the drift period and therefore violating the third invariant will result in a diffusion of trapped particles. This occurs because the magnetosphere is not symmetrical and protons on the back side of the earth react differently to the compression from protons on the sun side of the earth.

In this process the radial drift velocity v_d is a strong function of L giving approximately

$$v_d \propto L^8.$$

The protons would move rapidly inwards from the magnetopause at $10R_e$, possibly taking days to get to $L = 7$, weeks to $L = 4$, and years to $L = 2$. Experimentally some time variations in the proton flux are seen for $L > 6$ but for lower L values the fluxes are quite stable. This situation would be expected if the L drift was rapid at large L's. As the protons drift radially inward, charge exchange on neutral hydrogen will tend to remove the protons of $E < 100$ keV. By $L \sim 4$ these low energy protons should be gone and a low energy cut-off should appear in the energy spectrum. At $L \simeq 3$ this cut-off energy would be transformed up to $E_c \sim 250$ keV. There is some evidence in the experimental data that such a cut-off exists, but it is not certain.

This model of L drift of protons inwards from the magnetopause explains the observed features of the Davis protons quite well. What does this model predict in the way of higher energy protons in the outer belt?

Applying the energy and angle transformations to Davis protons at $L = 5$ and using the flux variations with L found applying the Liouville Theorem we find that at the equator we expect roughly 200 protons/cm^2-sec-ster at $L = 2.3$ of $E > 40$ MeV and $\alpha_0 = 90°$. McIlwain[16] measured a flux of about 500 of $40 < E < 110$ MeV at this location on Explorer XV. This measured proton group is clearly separate from the inner zone protons. Both of these measured proton populations are narrow in angular distribution as would be expected from the transformation. It seems to be reasonable to assume that these two measurements on outer belt protons both involve the same proton population convected in from the magnetopause by an L drift process related to magnetic time variations.

ELECTRONS

We have a much less complete understanding of the electrons in the radiation belt than we have of the protons.[4,17] We have considerable information on electron fluxes, energy spectra, and time variations, but until recently we did not know what the electron lifetimes were and therefore could not judge what sources could supply the particles. Study of the artificial radiation belts has given us direct measurements of electron lifetimes. Let us consider these events and then return to the natural belt.

Artificial Radiation Belts

Seven artificial radiation belts have been produced by the explosion of high altitude nuclear bombs since 1958. These artificial belts resulted from the release of energetic charged particles, mostly electrons from the β decay of the fission fragments from the nuclear explosions. These seven explosions are:

Explosion	Locale	Time	Yield	Altitude
Argus I	South Atlantic	1958	1 kt	300 miles
Argus II	South Atlantic	1958	1 kt	300 miles
Argus III	South Atlantic	1958	1 kt	300 miles
Starfish	Johnston Island, Pacific Ocean	July 9, 1962	1.4 Mt	400 km
USSR	Siberia	Oct. 22, 1962	Several hundred kt	?
USSR	Siberia	Oct. 28, 1962	?	?
USSR	Siberia	Nov. 1, 1962	?	?

The Argus explosions[2] of 1958 were carried out in order to study the trapping of energetic particles by the earth's magnetic field. After each of the Argus explosions, trapped particles were observed by Van Allen from data taken by the Explorer IV satellite.[2]

The electrons from the Argus explosions spread out to form a shell around the earth at $L \sim 2$. This shell was quite thin—only about 100 km thick near its base. Measurements on the shell showed that it did not drift in L by a measurable amount during its life of a few weeks.

The Starfish explosion of July 9, 1962, was of higher yield than Argus and produced not only a more intense artificial belt but a considerably more extensive belt.[18]

Four satellites carried charged particle detectors that studied the early history of the Starfish belt. These were Injun I, Telstar, TRAAC and Ariel I. A quite good composite picture of the artificial belt has emerged from these and later measurements.[18] A maximum flux of about 10^9 electrons/cm^2-sec resulted from the explosion at about $L = 1.3$ and large fluxes occurred up to $L = 4$ and beyond. The electron energy spectrum changed with distance from the explosion—lower average energy electrons were found at larger distances. This is probably due to upward motion of the bomb debris and adiabatic cooling of the electrons.

Studies of the decay of the large transient electron population injected by Starfish have provided electron lifetime data that were not available before the explosion. For $L < 1.5$ the electron decay was observed by Injun and Alouette and Telstar. In this region the decay is quite slow. Particle lifetimes are measured in years and the decay is consistent with what is expected from coulomb scattering of the electrons in the atmosphere. By studying the mirror point distribution of the electrons and following the particle diffusion in space and loss of energy from coulomb collisions until they leak out the loss cone and are lost in the atmosphere, one can construct theoretical decay curves.[18] These agree well with the measured decay curves for this low L region. For $L > 1.7$ very different decay curves were observed on Telstar[19] (see Figure 13). At $L = 1.7$ the mean life τ is many months but at $L = 2.5$, $\tau \sim 1$ week. This rapid decay cannot be due to the atmosphere. It probably is due to some type

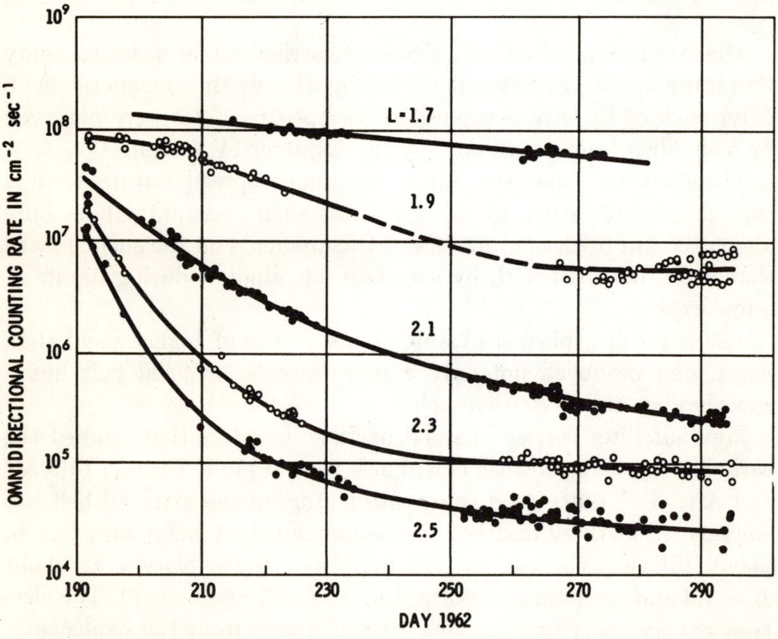

FIG. 13. Decay curves of Starfish electrons measured on the Telstar satellite (Brown and Gabbe, Ref. 19).

of magnetic disturbance. Dungey has suggested that a resonant interaction of certain electromagnetic waves, with the trapped electrons, resulting in changes of the pitch angle of the electrons, may be responsible for the electron loss. The equation of motion of an electron moving in the geomagnetic field (taken in the z direction) and interacting with a magnetic wave is

$$(m/e)(dv_{||}/dt) = v_{\perp}/c(b_y \sin \Omega t + b_x \cos \Omega t), \tag{30}$$

where $\Omega/2\pi$ is the particles' cyclotron frequency and b_x and b_y are the components of the waves' magnetic field given by

$$b_y = b_y \cos (ky + \alpha)$$

$$b_x = b_x \cos (ky + \delta). \tag{31}$$

The change in $v_{||}$ per cyclotron period is

$$(m/e)\,\delta v_{||} = (v_{\perp}/c) \int_0^{2\pi/\Omega} [b_y \cos (ky + \alpha) \sin \Omega t + b_x \cos (ky + \delta) \cos \Omega t]\, dt. \tag{32}$$

In order for the integral to be nonzero, k and Ω must be related by

$$kv = \Omega. \tag{33}$$

This is a resonance condition that requires that the wave frequency is Doppler shifted up to the particle cyclotron frequency by the motion of the electron or equivalently that the electron travel on wavelength λ in a cyclotron period.

Dungey showed this frequency resonance could be achieved by circularly polarized waves that travel along field lines called whistlers (see Chapter 1). By integrating equation (32) he found the change in the electrons' pitch angle due to interacting with one wave is

$$\delta\alpha = \delta v_{||}/v \sim b/B_0. \tag{34}$$

Considering the random walk in α due to interacting with many waves the total change in pitch angle is

$$\Delta\alpha = [M(\delta\alpha)^2]^{1/2}, \tag{35}$$

where M = number of waves. When $\Delta\alpha \approx 1$ the particle is scattered into the loss cone. Taking $M = NT$, where N is the number of waves/day, we get the electrons lifetime as

$$T = [N(\delta\alpha)^2]^{-1} = B_0^2/Nb^2 \tag{36}$$

which Dungey evaluated to be about

$$T = 5 \text{ days} \quad \text{at} \quad L = 2.5.$$

The requirements in this theory on whistler occurrence and amplitude seem reasonable and the L dependence of the process is about as observed; however, the angular distribution of the electrons that would result from whistler scattering alone is not what is observed and something besides whistler is needed.

The three Soviet explosions of 1962 made artificial belts somewhat less intense than Starfish. These belts were in the region $L = 1.7$ to 3.5. The decay of these belts were quite similar to Starfish indicating the loss process is a normal feature of the magnetosphere and not related to unusual magnetic conditions.

A few minutes after Starfish, synchrotron radiation from the trapped electron was observed in Peru.[18] This is the only effect of the artificial radiation belts that is. observed on the ground for long periods. *Synchrotron radiation* is the electromagnetic radiation given off when an electric charge is accelerated in a circular path. It was first observed as light emitted from a synchrotron electron accelerator. If the charged particles have $V \ll c$, then the radiation is emitted only at the cyclotron frequency and is called *cyclotron radiation.*

The total power radiated by a particle by synchrotron radiation is

$$P = \tfrac{2}{3}(e^2/c^3)(dV/dt)^2. \tag{37}$$

An expression that is more useful, in comparison with experiments is the power spectrum radiated in the electron's orbital plane for a relativistic electron:

$$P(f) = 4.1 \times 10^{-30} B\gamma F(f) \text{ watts/cps-ster}, \qquad (38)$$

where B is the magnetic field strength in gauss, γ is the relativistic energy factor E/m_0c^2, and $F(f)$ is a function of the frequency f. Evaluating $P(f)$ at 50 Mc for $B = 0.16$ gauss and $E = 2$ MeV gives $P(f) = 4 \times 10^{-31}$ watts/cps-ster. Integrating $P(f)$ over the electron spatial distribution and fission energy spectrum for Starfish, and integrating over antenna patterns gave calculated signal strengths in very good agreement with those measured shortly after Starfish.

Attempts were made to observe synchrotron radiation from the natural Van Allen belt prior to Starfish, but it could not be measured because of the background of other natural radio noises. After the Starfish explosion, synchrotron radiation was observed by several radio observatories. The newly trapped electrons from Starfish produced more synchrotron noise than the natural belt electrons because there were more of them and they were of higher energy.

From synchrotron radiation we know that Jupiter has a radiation belt. Radio noise emissions from Jupiter in decimeter wavelengths show polarization and a source that extends outside the disc of the planet as expected from synchrotron noise from a radiation belt. Fluxes up to 10^8 electrons/cm^2-sec of $5 < E < 75$ MeV are needed to generate the observed Jupiter emissions. This is a much more substantial electron component than is present in the earth's radiation belt.

From studying artificial radiation belts we see that the natural electron belt can be divided into two zones. In the inner zone of $L < 1.7$ the electrons have long lifetimes. For $L > 1.7$ the electrons have much shorter lifetimes and some different processes must operate. What do we know about the natural electrons in these two regions?

Natural Outer Zone Electrons

Electrons in the outer radiation zone show large time variations. The flux of $E > 1.5$ MeV electrons in the outer zone in Figure 1

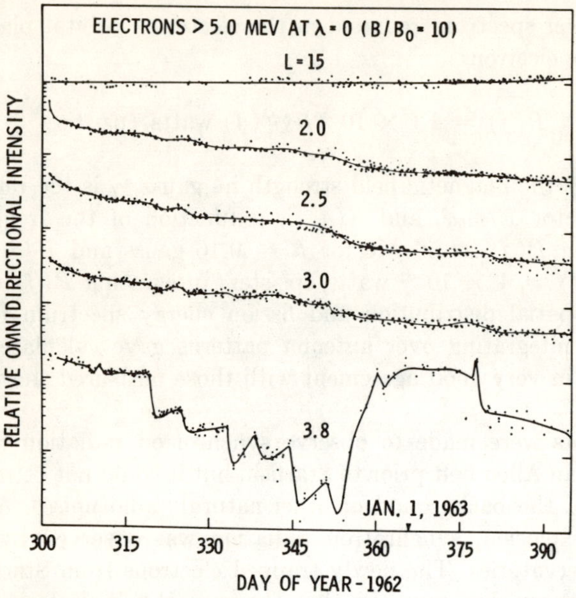

FIG. 14. Time variations of electrons near the equator as seen on Explorer XV (McIlwain, Ref. 16).

can vary up and down in time by factors of 10 or more especially at the time of magnetic disturbances. Lower energy electrons also show time variations but usually not as large as this. Figure 14 shows time histories[16] of electrons of $E > 5$ MeV measured on Explorer XV. The population in this case contains many artificial electrons. Less than two months previous to this, there had been three high altitude Soviet nuclear explosions. Figure 14 shows that at large L's there are stepwise losses of electrons especially at times of magnetic storms. Also in Figure 14 on December 18 the electron fluxes increased suddenly by a factor of more than 10. It is not believed that many $E > 5$ MeV electrons exist naturally here so this December 18 flux increase is considered to be evidence of acceleration of electrons already in the geomagnetic field. It is obvious that magnetic disturbances play a considerable part in the dynamics of the natural outer zone electrons.

What are the requirements that can be put on the source of these outer zone electrons? If a typical flux is $\Phi \sim 10^7$ elec/cm^2-sec of

$E \sim 100$ KeV and if the lifetime of an average electron is 5×10^5 seconds, then the source must supply

$$E/t = \{[10^7/(3 \times 10^{10})](\text{elec/cm}^3)\}[1.6 \times 10^{-7}(\text{ergs/electron})] \times [(4\pi/3)(5 \times 10^9)^3 \text{ cm}^3](5 \times 10^5 \text{ sec})^{-1} = 5 \times 10^{13}(\text{ergs/sec}). \quad (39)$$

The solar wind striking the magnetopause carries about 10^{19} ergs/sec which is more than adequate. Neutrons are inadequate by three orders of magnitude. It seems obvious that the energy for the outer zone electrons comes from the solar wind. But the electrons may or may not come from the solar wind. They may be electrons that are already in the magnetic field and are being accelerated by magnetic disturbances produced by the solar wind. L drift may be important for these electrons as well as for outer zone protons but other mechanisms must be involved too.

At low altitudes large fluxes of electron precipitating into the atmosphere are frequently observed. An "average" precipitating flux in the outer zone is about 10^5 elec/cm²-sec.[4] We can compare the average worldwide precipitated flux of electrons

$$P = 10^5[\pi(7 \times 10^8 \text{ cm})^2/2] = 7 \times 10^{22}(\text{electrons/sec}) \quad (40)$$

with the particle source of the trapped electrons

$$S = \{[10^7/(3 \times 10^{10})](\text{elec/cm}^3)\}[(4\pi/3)(5 \times 10^9)^3 \text{ cm}^3] \times (5 \times 10^5 \text{ sec})^{-1} = 3 \times 10^{20}(\text{elec/sec}). \quad (41)$$

These numbers are crude but we see here that the loss is larger than the source. This implies that not all of the electrons flowing through the outer zone actually become residents of it long enough to contribute to the measured fluxes. O'Brien[21] has described the outer zone electrons in terms of a "*splash catcher*" model (see Fig. 15). where some small fraction of the particles moving through the region became trapped. The observation that trapped fluxes are larger after an intense period of precipitation agrees with this model. This model was developed by comparison with the leaky bucket model of the inner zone.

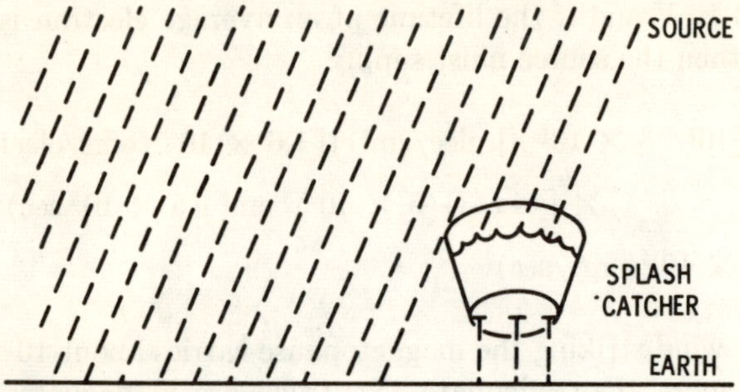

Fig. 15. Splash catcher model of the outer radiation belt. Most particles from the source directly hit the earth's atmosphere and only a few are trapped in the bucket (after O'Brien, Ref. 21).

Natural Inner Zone Electrons

Experimentally we know a good deal less about the natural electrons in the inner belt than about those in the outer belt. Inner zone protons are counted with high efficiency by most detectors and penetrate quite considerable thickness of shielding. This fact has caused confusion in several early measurements of inner zone electrons. Recently artificial belt electrons have hidden the natural electrons.

Measurements of the inner zone natural electron spectrum with magnetic spectrometers have given conflicting results. It is uncertain whether electrons of $E < 1$ MeV are present or not.[9] It does seem that there are lots of low energy electrons of $E < 0.20$ MeV. We expect that electrons will be produced in the inner zone from neutron decay. Their energies should extend up to 0.78 MeV and the maximum in the spectrum should be at about 0.5 MeV. If there are many electrons of $E > 1$ MeV they cannot be from neutron decay. Also the equilibrium electron spectrum from neutron decay should not have many $E < 0.2$ MeV electrons.

There are several measurements of the electron flux at low altitudes in the inner zone. At about 1000 km, $\Phi_e \sim 10^6$ electrons/cm²-sec of $E > 40$ keV. We can estimate the flux in the center of the inner zone by assuming that the electron flux is inversely proportioned to the atmosphere density. This gives $\Phi \sim 10^7$ electrons/cm²-sec.

The inner-belt electron flux expected from neutron decay is of the same order of magnitude as the observed flux.[9] The electron lifetime for coulomb scattering for a 300 keV electron at 2000 km is about 10^8 sec. Using this lifetime and the calculated neutron decay electron source strength of 3×10^{-12} electrons/cm^3 sec we get an equilibrium electron flux of

$$\Phi_e = (3 \times 10^{-12})(3 \times 10^{-10})(10^8) = 10^7 (\text{elec/cm}^2\text{-sec}) \tag{42}$$

In comparing the observations on electrons with calculations based on the neutron-decay theory, we find that the calculated and observed fluxes are similar but that the spectra are not in good agreement. It is quite certain that neutron-decay electrons produce a considerable fraction of the trapped inner zone electrons. It is also quite certain that other processes must operate as well in order to generate the 100 keV electrons and $E > 1$ MeV electrons if they exist.

Low altitude precipitating electrons are observed in the inner zone as well as in the outer zone. They seem to have a quite soft energy spectrum with most of the electrons of $E < 200$ keV. The precipitated flux is larger at periods of magnetic disturbance and the higher energy electrons in the inner zone do not seem to be disturbed at these times. It would appear that a second mechanism, other than neutron decay, involving magnetic disturbances produces these precipitating electrons and probably also the low energy trapped electrons.

OTHER PARTICLES

If the sun were the source of particles in the inner radiation zone we would expect to find not only protons, but other heavier particles such as deuterons, tritons and He^3 and He^4 nuclei. The sun contains about 15 percent He nuclei. Solar cosmic rays are known to contain ~5 percent He nuclei. The lifetime of a He^4 nucleous in the inner belt would be about five times less that that of a proton of the same energy because the rate of slowing down is faster for $Z = 2$. On this basis we would expect about 1 percent He^4 in the inner belt. However, experimentally not even one $Z = 2$ track has been found in nuclear emulsions so far.[9] An upper limit of the α flux is 0.1 ± 0.05 percent of the proton flux in the energy interval 125 to 185 MeV. This quite clearly shows that the sun contributes few, if any, of the heavy particles in the inner belt.

A few deuterons and tritons (roughly 0.5 percent each) were found in the emulsion experiments. These particle fluxes can be explained as being the result of nuclear collisions of trapped protons with O and N nuclei in the very thin atmosphere present at radiation belt altitudes. No heavy particles that cannot be accounted for by the neutron-decay source have been observed in the inner belt. Recently Van Allen has found a small population of low energy α particles in the outer radiation zone. This is to be expected if radial diffusion of particles inward from the magnetopause is the source of the outer belt protons. This diffusion should bring in α particles too.

REFERENCES

1. J. A. Van Allen, Dynamics Composition and Origin of the Geomagnetically-Trapped Corpuscular Radiation, in *Space Science* (Le Galley) pp. 226–274, John Wiley, New York, 1963.
2. Symposium on Scientific Effects of Artificially Introduced Radiations at High Altitudes, J. Geophys. Res. **64,** 865 (1959).
3. J. A. Van Allen and L. A. Frank, Radiation Around the Earth to a Radial Distance of 107,400 km, Nature **183,** 430 (1959).
4. B. J. O'Brien, Review of Studies of Trapped Radiation with Satellite-Borne Apparatus, Space Sci. Rev. **1,** 415 (1962).
5. L. R. Davis and J. M. Williamson, Low Energy Trapped Protons, in *Space Research III*, (W. Priester, ed.), pp. 365–375, John Wiley, New York, 1962.
6. H. Alfven, Cosmical Electrodynamics, Oxford University Press, London, 1950.
7. W. N. Hess, E. H. Canfield, and R. E. Lingenfelter, Cosmic Ray Neutron Demography, J. Geophys. Res. **66,** 665 (1961).
8. S. C. Freden and R. S. White, Trapped Proton and Cosmic-Ray Albedo Neutron Fluxes, J. Geophys. Res. **67,** 25 (1962).
9. W. N. Hess, Energetic Particles in the Inner Van Allen Belt, Space Sci. Rev. **1,** 278 (1962).
10. A. M. Lenchek and S. F. Singer, Geomagnetically Trapped Protons From Cosmic-Ray Albedo Neutrons, J. Geophys. Res. **67,** 1263 (1962).
11. A. J. Dragt, Effect of Hydromagnetic Waves on the Lifetime of Van Allen Radiation Protons, J. Geophys. Res. **66,** 1641 (1961).
12. E. N. Parker, Effect of Hydromagnetic Waves in a Dipole Field on the Longitudinal Invariant, J. Geophys. Res. **66,** 693 (1961).
13. C. E. McIlwain, Coordinates for Mapping the Distribution of Magnetically Trapped Particles, J. Geophys. Res. **66,** 3681 (1961).

14. R. C. Blanchard and W. N. Hess, Solar Cycle Changes in Inner-Zone Protons, J. Geophys. Res. **69,** 3927 (1964).
15. J. E. Naugle and D. A. Kniffen, The Flux and Energy Spectrum of the Protons in the Inner Van Allen Belt, Phys. Rev. Letters **7,** 3 (1961).
16. C. E. McIlwain, The Radiation Belts Natural and Artificial, Science **142,** 355 (1963).
17. T. A. Farley, The Growth of our Knowledge of the Earth's Outer Radiation Belt, Revs. Geophys. **1,** 3 (1963).
18. Collected Papers on the Artificial Radiation Belt from the July 9, 1962 Nuclear Detonation, W. N. Hess, ed., J. Geophys. Res. **68,** 605 (1963).
19. W. L. Brown, J. D. Gabbe, and W. Rosenzweig, Results of the Telstar Radiation Experiments, Bell System Technical Journal **42,** 1505 (1963).
20. B. J. O'Brien, et al., High Latitude Geophysical Studies with Satellite Injun 3 (in 4 parts), J. Geophys. Res. **69,** (January, 1964).
21. B. J. O'Brien, The Trapped Radiation Zones, in *Space Physics*, pp. 505–572, John Wiley, New York, 1964.

Chapter 5

THE AURORA

T. N. Davis

INTRODUCTION

The aurora is a beautiful and complex phenomenon resulting from a series of interrelated physical processes occurring in the whole of the region between the earth and the sun. Often thought of as existing only in conjunction with magnetic storms, the aurora is, in fact, a continuously occurring part of the earth environment.

We begin our discussion with the *visible* aurora; it is that portion of the phenomenon which can be perceived by the human eye and which has received the most study. The visible aurora is due primarily to the interaction of energetic particles with the oxygen and nitrogen gases of the upper atmosphere. The important emissions in the visible region are the discrete green (5577 Å) and red (6300–6364 Å) lines of atomic oxygen and the molecular bands of N_2^+ in the blue and N_2 in the red regions of the spectrum. These emissions occur primarily at altitudes of 100 km or above and are usually confined at any one time to relatively narrow regions so that isolated auroral forms result.

The *optical* aurora includes the visible aurora and a large number of emissions in the spectral region extending from the infrared to the far ultraviolet; the combined intensity of non-visible part of the optical aurora somewhat exceeds that of the visible part. Another broad category is the *radio* aurora in which we may group radio-frequency emissions from the auora and structures detected by radar means. Associated with the aurora are a number of aurorally related phenomena (or observed effects) which include absorption of cosmic radio noise, sporadic E ionization, radio-star and satellite scintillations, electric current in the auroral ionosphere, earth currents, infrasonic waves and perhaps audible-range sound waves. Each of these types of auroral phenomena shares a common origin either directly or indirectly through the interaction of energetic particles with the earth's atmosphere. Of the various types of auroral phenomena, the available observational information deals mainly with the optical, and more specifically, the visual aurora.

MORPHOLOGY OF THE VISUAL AURORA

Classification of Auroral Forms

Form Types

A significant and entirely unexplained aspect of aurora morphology is the fact that most auroras occur as discrete forms in an otherwise relatively black sky. Rather elaborate schemes have been devised to classify the various types of auroral forms, however one need consider only three truly distinctive types. One common type is the *band* which is characterized by its horizontal thinness (0.2 to perhaps 10 km) relative to its horizontal length (hundreds to thousands of kilometers). The band extends from its *lower border* upwards along the direction of the local magnetic field to heights ranging from several tens to several hundreds of kilometers; the lower border of a normal green aurora lies at an altitude of near 100 km. Bands usually show some curvature along their length and may be highly contorted; those which show only slight curvature are called *arcs*. A second auroral type is the *patch* or *surface* which is a more-or-less isolated region of luminosity having no particular shape. Patches may be completely isolated but often are interconnected. The third and more uncommonly observed auroral form type is the *veil*; it is an extensive and usually uniform luminosity covering a major portion of the visible sky.

Several self-explanatory terms are used in the description of the internal structure of auroral forms: these terms are homogeneous, rayed, striated and diffuse. In general, the rayed and striated forms appear to be thinner in horizontal width and more extended in height than the homogeneous and diffuse forms.

Forms By Color

Auroras are classified by color as well. Auroral forms appear colorless if they are so weak as to be below the color threshold of the eye; if of sufficient brightness, the normal aurora appears green due to [OI] 5577 Å. *Type A* aurora is one with a red upper portion or one which is entirely red. The red color, due to [OI] 6300 Å and 6364 Å emission, dominates the green color of the [OI] 5577 Å emission only at high altitude; thus the Type A aurora is a high altitude phenomenon (300–400 km). *Type B* auroras are those which exhibit red lower borders arising from enhancement of N_2 First Positive and O_2 First Negative bands. Type A aurora usually occurs only during times of high global magnetic activity while Type B aurora usually occurs only during the most active phases of the normal auroral display.

TABLE I

IBC	I	II	III	IV
Visual equiv. brightness	Milky way	Thin moonlit cirrus clouds	Moonlit cumulus clouds	Full monlight
Intensity [OI] 5577 Å, (kilorayleighs)	1	10	100	1000
Intensity N_2^+, First Neg. 3914 Å, kR	1	10	100	1000
Energy deposition, ergs cm^{-2} sec^{-1}	3	30	300	3000

The brightness of auroras varies from below the visual threshold to a brightness which produces an illumination on the ground equivalent to full moonlight. Forms are rated in brightness according to the International Brightness Coefficient (IBC) as indicated in Table I.

Measures of the Occurrence of Visual Auroras

Occurrence is the most commonly-used index of visual auroral activity. Occurrence is the percentage hours or nights on which auroral forms are detected, either within a limited region of the observer's sky or in his whole sky. The occurrence index only takes into account whether or not aurora is present and ignores the extent and brightness of auroral forms. Therefore, it is difficult to give the index any quantitative meaning or to relate it meaningfully to another measurement, as for example, one of particle flux incident on the atmosphere. A slight improvement results in the use of auroral *incidence*, which is similar to occurrence and measures the number of auroral forms visible in the overhead portion of the sky within a given time interval. Either visual observations or all-sky camera data are used to determine the occurrence and incidence of auroras.

The likelihood of seeing aurora (that is, its occurrence) depends

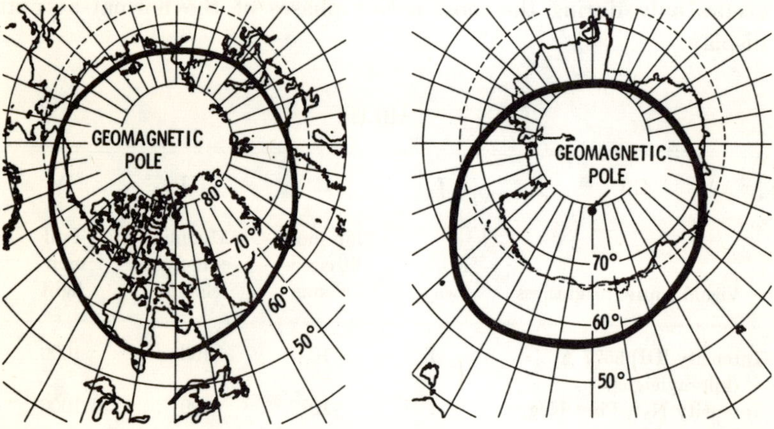

FIG. 1. Location of the northern and southern auroral zones.

upon several factors other than the masking effects of sunlight or cloudiness. Roughly in order of their importance these are; latitude, state of short-term solar activity (geomagnetic disturbance), local time, long-term solar activity (time within the 11-year solar cycle), and season. Due to the overall complexity of the auroral occurrence, it is necessary to consider these various factors separately.

Latitudinal and Diurnal Dependence of Auroral Occurrence

Auroral occurrence is greatest at the two *auroral zones* (see Fig. 1) which, at solar maximum, are located approximately at 68° north and south geomagnetic latitudes. The positions of the auroral zones are given more accurately by specifying them as the earth intersections of L-shell 6.4 (see Chapter 4). The diurnal variation at the auroral zones is such that there is a single peak in occurrence near local geomagnetic midnight and a minimum near local geomagnetic noon.* Aurora is observed at the auroral zone during most dark hours and on essentially 100 percent of the days (24-hour periods) in the year. Equatorward of the auroral zones, the occurrence falls sharply with decreasing geomagnetic latitude but maintains the same form of diurnal variation. The statistical occurrence of aurora poleward of the auroral zone declines somewhat more slowly with increasing latitude and also exhibits a more complex form of diurnal variation which is not yet completely documented. At geomagnetic latitude 75°, the midnight peak in the overhead occurrence is still observed, and in addition, two lesser maximums appear near geomagnetic times 16 and 04 hours. Farther poleward, near geomagnetic latitude 80°, the midnight maximum in overhead aurora disappears leaving only the morning and evening maximums. Near the pole, the morning and evening maximums merge into a single peak (which is observed at Thule, Greenland, near the geomagnetic pole, at 03 hours local geographic time)[1] (see Fig. 2).

---

* The geomagnetic time of a station is reckoned from the meridian plane containing the north and south geomagnetic poles and the antipodal point to the meridian plane containing the north and south geomagnetic poles and the station.

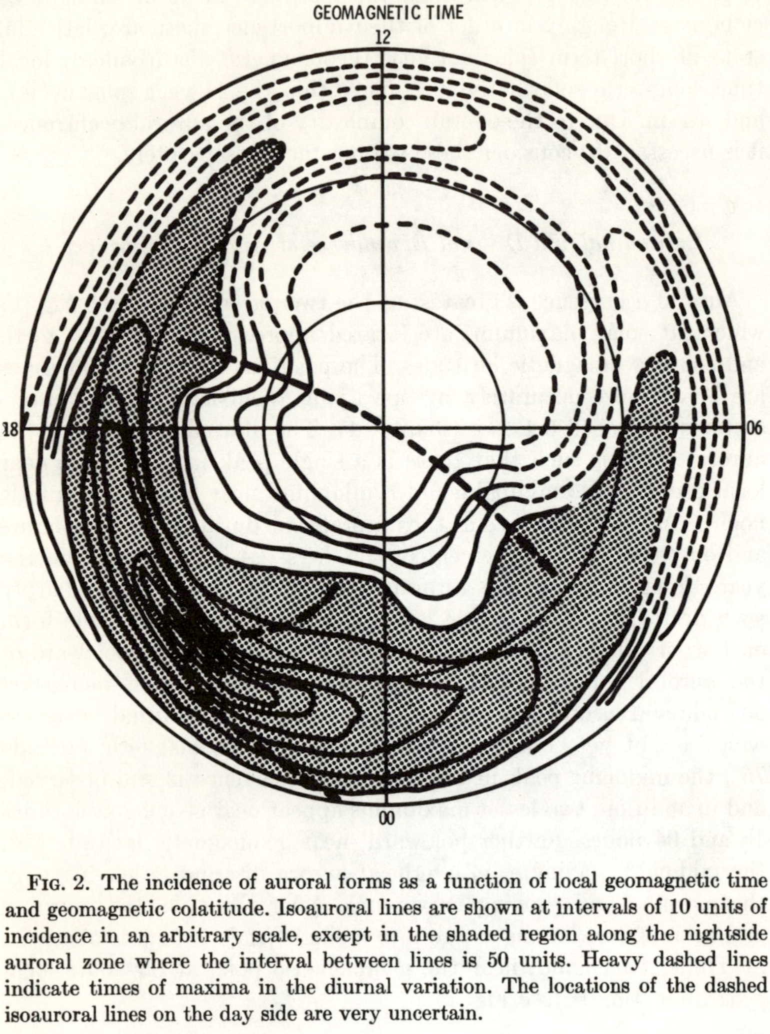

FIG. 2. The incidence of auroral forms as a function of local geomagnetic time and geomagnetic colatitude. Isoauroral lines are shown at intervals of 10 units of incidence in an arbitrary scale, except in the shaded region along the nightside auroral zone where the interval between lines is 50 units. Heavy dashed lines indicate times of maxima in the diurnal variation. The locations of the dashed isoauroral lines on the day side are very uncertain.

Progression of Auroral Form Types

Observed at the auroral zone, auroral form types show a rather general progression in local geomagnetic time. The entire progression is not observed at most auroral zone locations due to the limitations caused by daylight; it can be seen best near the intersection of Longitude 120° and the Southern auroral zone during mid-winter. During the evening hours, diffuse or well-defined homogeneous arcs and bands are the forms most commonly observed. As the night progresses, the bands dominate over the arcs and the tendency is more toward rayed forms than homogeneous ones. Near geomagnetic midnight the so-called *breakup* may occur, in which the forms increase in intensity and may rapidly change shape and position. Following the initial portion of the breakup period, irregular patches appear often with pulsating luminosity. Later in the morning these tend to reform into arcs and bands. Interconnected patches and diffuse bands are the dominant form types on the day side of the auroral zone. Towards

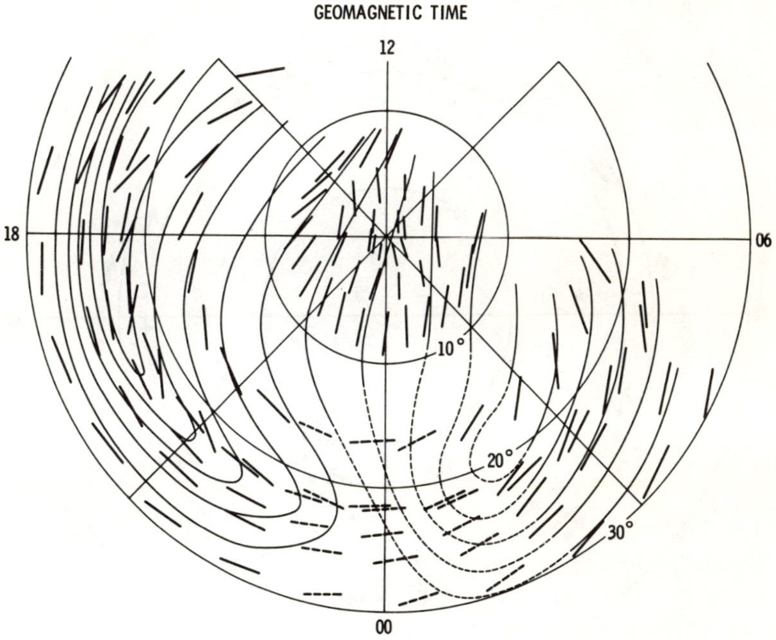

FIG. 3. The average alignment of auroral forms over the polar cap. Short heavy bars represent measured alignments of extended auroral arcs. The near-midnight dashed portion of the smooth pattern deduced from the measurements takes into account the tendency of less-extended forms occurring near midnight to be meridionally aligned.

higher latitude, the tendency is toward rayed arcs and bands throughout the 24-hour day.

Auroral Form Alignments and Motions

A definite local time progression in the horizontal alignment of arcs and bands appear at all latitudes, as shown in Fig. 3. The alignment is roughly east–west at and outside the auroral zones; however, if short arcs and bands are considered together with the greatly ex-

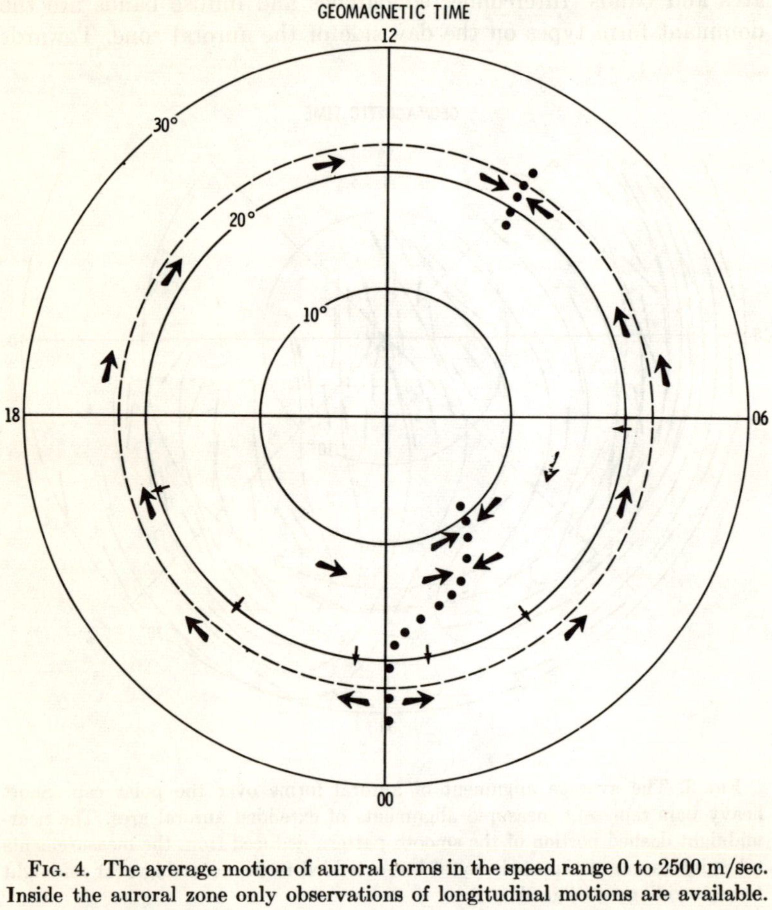

FIG. 4. The average motion of auroral forms in the speed range 0 to 2500 m/sec. Inside the auroral zone only observations of longitudinal motions are available.

tended forms, a north-south alignment appears in the hours near magnetic midnight. Above geomagnetic latitude 80° the orientation varies continuously in the sky so as to be directed somewhat east of the sun-earth line. Horizontal auroral motions show a clear pattern of diurnal behavior as well, Fig. 4. The motions referred to here are horizontal drifts of large irregularities of auroral forms or of the entire form. The apparent speeds range up to 2,500 m/sec. On the night side of the earth, both poleward and equatorward meridional motions occur at the auroral zones, but the equatorward motions predominate. Eastward longitudinal motions are most prevalent along the auroral zones from local geomagnetic midnight until 10 hours. A reversal in the longitudinal motion then occurs and westward motion continues until midnight when the eastward motion resumes. The 10 hour and midnight reversals in longitudinal motion are quite abrupt on any given night. Little information on auroral motions inside the auroral zones exists due to the sparsity of both aurora and observing stations in those regions. However, it is known that near the midnight meridian the longitudinal motion just inside the auroral zones is opposite to that in the auroral zones.

Seasonal and Yearly Variations

The study of seasonal effects in the aurora is complicated by the large annual variation in darkness at high latitude where the aurora is most prevalent. Consequently not much is known about high-latitude seasonal variations. At low latitude, however, strong spring and fall maxima appear in the occurrence of aurora. As yet it is not possible to distinguish whether these maxima result from the earth's being at maximum heliographic latitudes in early March and early September or from it being at the equinoctial positions in late March and September.

Both the geographic distribution and frequency of strong auroral displays show a marked correlation with the 11-year variation in solar activity. Minimum auroral activity apparently coincides with minimum solar activity in the 11-year cycle, however the maximum auroral activity seems to lag the maximum solar activity by approximately two years. The likelihood of observing aurora at low latitude is greatly enhanced in those years of maximum activity within the 11-year cycle. Large displays at the auroral zones tend to be more frequent at this time and there may be an equatorwards expansion of the statistically-determined auroral zones in the maximum years. Deep inside the auroral zones, near the geomagnetic poles, there is evidence suggesting that the occurrence of visual aurora actually decreases in the years of maximum activity; it is clear, however, that this polar decrease does not arise simply from a general equatorward shift in the aurora.

Conjugacy in the Aurora

Study of the degree of conjugacy in the auora has been hampered by the lack of conjugate station pairs in the auroral zones, and almost none of the few existing station pairs are exactly conjugated by the geomagnetic field. Further, it is possible to obtain visual observation only near the equinoxes, at which times there is darkness at both ends of the geomagnetic field lines linking the auroral zones. Consequently, very little data are available and few analyses are in the literature. By using data from Campbell Island near the southern auroral zone and from Farewell, Alaska near the northern auroral zone (stations conjugate within approximately 100 km), it was found that the general behavior of the aurora over these stations was identical, that auroral breakups occurred simultaneously at the two stations, and that the intensity of the auroras over the two stations varied together.[2] It was not possible to determine whether or not exactly conjugate counterparts of individual auroral forms were occurring.

Magnetic Activity at the Auroral Zone

In the preceding sections are described those aspects of auroral morphology which are amenable to both statistical and detailed study and which appear to be strongly influenced by the geometry of the geomagnetic field and the geometrical relationship between the geomagnetic field and the solar wind. An equally important aspect of auroral morphology is its irregular temporal (dynamical) behavior. We now take up this subject in order to examine the observational material and to try to see the relative importance and inter-relations of the dynamical and geometrical aspects of auroral morphology.

In discussing this topic it is necessary first to consider a related subject, geomagnetic activity. Geomagnetic activity is covered more

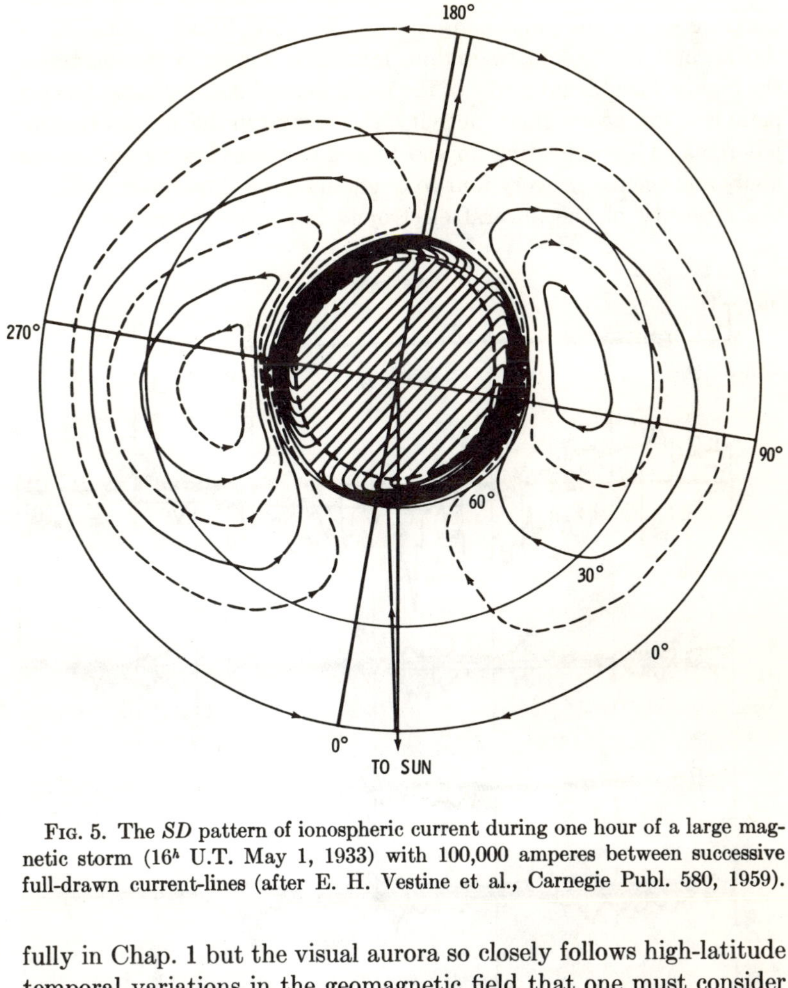

FIG. 5. The *SD* pattern of ionospheric current during one hour of a large magnetic storm (16^h U.T. May 1, 1933) with 100,000 amperes between successive full-drawn current-lines (after E. H. Vestine et al., Carnegie Publ. 580, 1959).

fully in Chap. 1 but the visual aurora so closely follows high-latitude temporal variations in the geomagnetic field that one must consider both together. The magnetic variations of primary concern are those due to intense electric currents in the auroral zone ionosphere. While highly variable in intensity and configuration, the polar ionospheric current system essentially maintains the configuration illustrated in Fig. 5. The essential features of this configuration are 1) the intense eastward current (electrojet) along the evening side of the auroral zone, 2) the more intense westward electrojet along the morning side of the auroral zone, 3) the diffuse extensions of the auroral electrojets over the polar cap, and 4) the more diffuse extensions of the auroral electrojets to middle latitudes. Figure 6 demonstrates the highly variable intensity of the polar ionospheric currents. During periods of low geomagnetic activity, the current in the auroral electrojets remains low excepting for short-lived increases lasting one or two hours and occurring every four to ten hours. As is illustrated in Fig. 6, the intensity of the auroral electrojets is markedly greater during

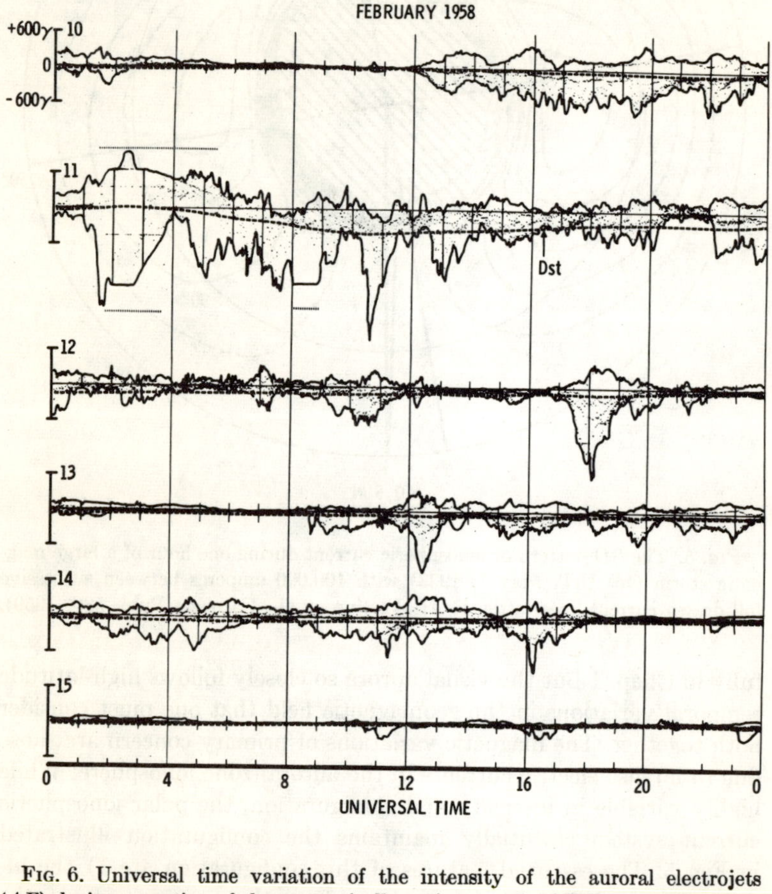

FIG. 6. Universal time variation of the intensity of the auroral electrojets (AE) during a portion of the magnetically-active month of February 1958 (after Davis and Sugiura, in press).

periods of magnetic storm. Even then short-lived and intense bursts of activity appear every few hours; each lasts from one to three hours. There appears to be a tendency for the bursts to occur more frequently during storm conditions than during times of magnetic quiet. When these bursts occur, magnetic observatories located under the eastward electrojet observe positive bays in the horizontal component of the magnetic field and those under the westward electrojet observe negative bays. In general, the westward electrojet is more intense than the eastward electrojet, hence the negative bays usually have greater amplitude than the positive bays.

Examination of the magnetic records taken at an auroral zone observatory reveals much similarity in the appearance of one day's record with another. The usual behavior of the horizontal component H trace is to show little activity during the day and to exhibit a positive bay in the evening hours. Then near local geomagnetic midnight there is usually a fairly abrupt reversal in sign of the disturbance and a negative bay in H develops. Much of this regular variation exhibits the strong controlling effect of the sun-earth geometry. On the other hand, the very abruptness in the onset of the negative bay suggests a temporal or dynamical control. Also, there is some variation in the time of onset of the positive and negative bays; this variation seems to be controlled primarily by dynamical factors. These effects of dynamical control are observed also by noting the variation in amplitude of the magnetic bays and seeing the development of multiple bay-like excursions on some nights. During especially stormy periods the auroral zone magnetogram may have very erratic traces. If the onset of a large bay occurs at one auroral zone observatory, a nearly simultaneous onset normally is observed many degrees to the east and west of that observatory. The sign of the bay recorded at an observatory normally is related to the station's position relative to the geomagnetic midnight meridian; if the observatory is west of the meridian the bay is usually positive, while the bay is usually negative if the observatory is to the east. The relative amplitude of the observed disturbance is normally greater the closer the observatory is to the midnight meridian. Thus, while the onset time and overall amplitude of a widespread polar magnetic disturbance appears to be controlled by temporal parameters, the geometrical parameters determine the relative amplitude and sign of the disturbance at any location.

Temporal Variations in the Aurora and Relationships Between Aurora and Magnetic Activity

The strong positive correlation between auroral activity and global magnetic activity has been known for many years. It is especially pronounced at sub-auroral latitudes where the likelihood of auroras occurring is greatly enhanced during periods of magnetic storm. The relationship is so strong there that it is possible to predict the equatorwards boundaries of auroral displays by knowing only the value of the *global magnetic index Kp*.

At the auroral zones, the connection between auroral activity and magnetic activity is evidenced in many ways. As the intensity of the auroral electrojet increases, the polewards and equatorwards boundaries of display tend to separate and the display brightens. This relationship is so close that Fig. 6 satisfactorily serves as an indicator of overall auroral activity in the auroral zone even though it was compiled strictly from magnetic information. If the Dst component (see Chapter 1) of magnetic activity becomes especially large, both the polewards and equatorwards boundaries of the display may shift equatorward for periods of several hours. Within the polar cap region, deep inside the auroral zone, a curious relationship exists. There too a strong correlation exists between auroral occurrence and local or global magnetic activity (as measured by K or Kp), but it is a negative correlation. That is, there tends to be less aurora the greater the magnetic activity.

There are a number of examples of detailed relationships existing between auroras and ionospheric currents at the auroral zones. Several investigations indicate strong electrojet current exists in or near at least the equatorwards edge of the auroral display. Heppner found examples in which the drift of an auroral arc over College, Alaska was associated with a change in sign of the Z-component of the magnetic field disturbance at that station.[3] This observation indicated that a line current or sheet current was centered in or near the auroral arc. Figure 7 shows examples which indicate that there is a tendency for the horizontal disturbance vector to lie at right angles to the horizontal projection of overhead auroral forms. This relationship implies that the auroral forms and ionospheric currents tend to be parallel at the auroral zone.

The normal progression of auroral forms and the various other

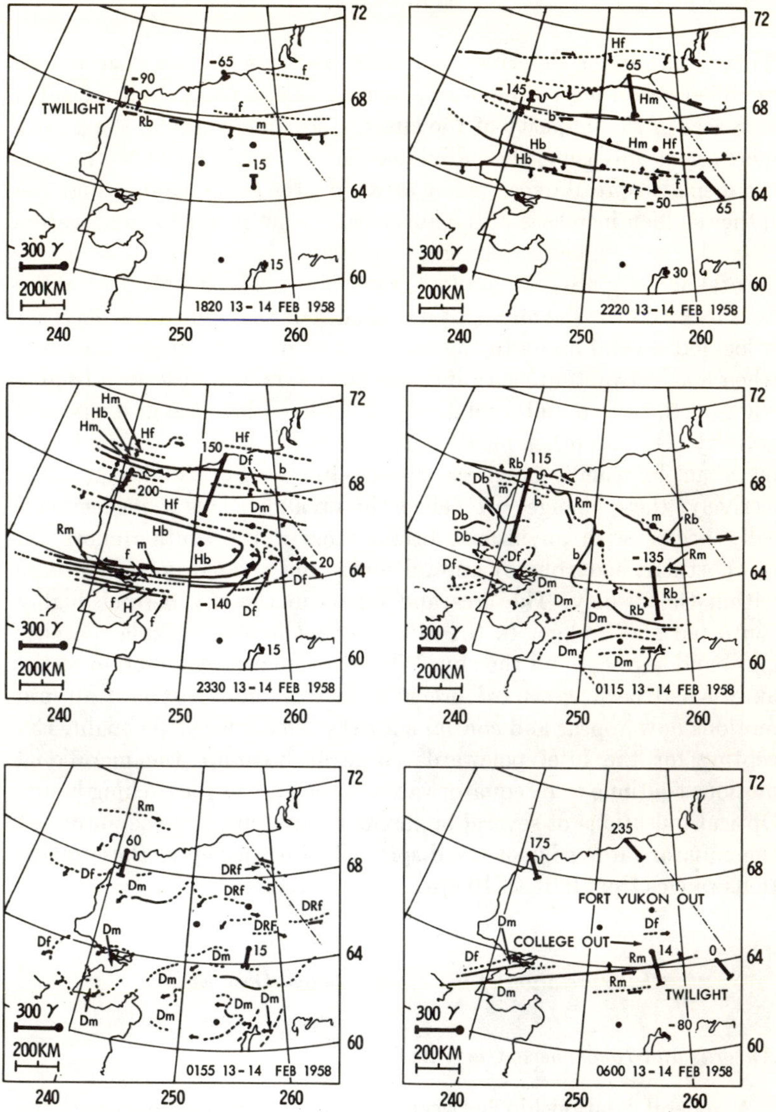

FIG. 7. Synoptic maps of aurora over Alaska with horizontal magnetic disturbance vectors (square-tipped bars) and numerical values of vertical disturbance (at bases of disturbance vectors). Directions of auroral motions are shown by short arrows (after Davis and Kimball, Geophysical Institute Rept. UAG–RI20, 1962).

diurnal variations described in earlier sections of this chapter are modified according to the degree of magnetic activity. When sizable increases in the intensity of the auroral electrojets occurs, the aurora increases in intensity, the individual forms become more active, and the displays spread over greater areas.[4] In the same manner that the rather sudden increases in the aurora electrojet intensity are greatest near the geomagnetic midnight meridian, the related increases in auroral activity are most profound and obvious to the observer when he is located near the midnight meridian. For example, if the observer is located several hours to the east or west of the midnight meridian when an electrojet intensity increase occurs, he will observe a brightening in the auora and a relatively short-lived increase in its general activity. On the other hand, if he is located near to the midnight meridian, he usually observes violent alteration in the display. Collectively, these changes are called the breakup. This sequence typically begins with an intense brightening of the southernmost (in the northern hemisphere) auroral forms and a rapid poleward surge within the display. The arcs and bands in the sky become highly contorted and broken. In the minutes that follow, irregular patches and band segments fill the sky. While the longitudinal motion of the aurora had been westward previous to this time, eastward auroral motions now appear and continue for the remainder of the night. Excepting for the brief polewards surge at breakup, the meridional motions continue to be equatorwards as they are in the evening hours. Optical pulsations of several apparent types appear in the aurora at the equatorwards edge of the display. The frequency of these pulsations ranges from 0.01 to 10 cps.

Interpretation of the Morphological Observations

Auroras and Ionospheric Currents

A general relationship between auroras and geomagnetic activity has been recognized for many decades. It is only within the past ten years that data has accumulated to indicate that the relationship between the aurora and the auroral zone ionospheric current-systems may be quite close in detail. Certainly now it is obvious that the two

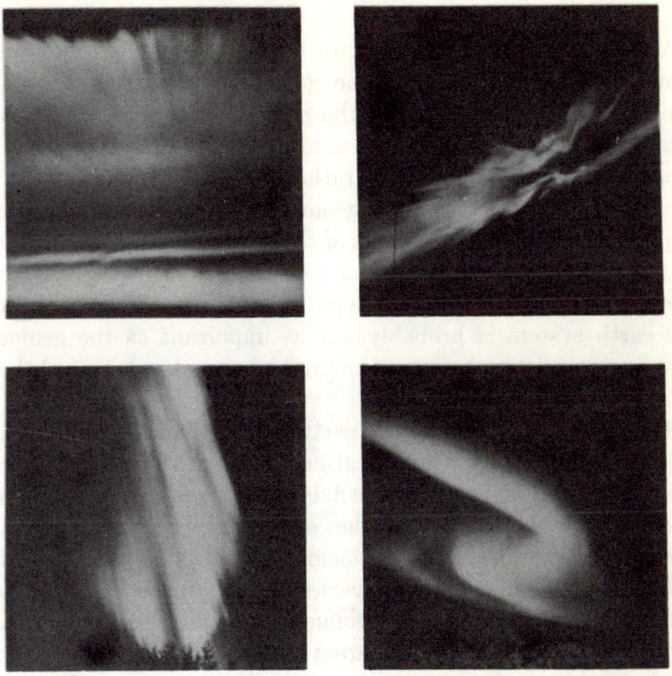

PLATE 1

Auroral forms photographed at College, Alaska and from the Arctic ice pack north of Alaska by V. P. Hessler.

phenomena must have a common origin, or at least be controlled mainly by the same parameters.

Geometrical Factors Controlling the Aurora and the Auroral Electrojet Currents

Auroral occurrence, alignments, form progressions and motions show a strong dependence upon geomagnetic latitude as does the intensity and configuration of the polar ionospheric current system. Also, the daily variations in the aurora and the auroral electrojets are dependent upon geomagnetic rather than geographic time. There is a strong tendency towards geomagnetic conjugacy. Therefore, it is quite clear that many aspects of the auroral phenomena are related to and controlled by the geometry of the geomagnetic field and the geometry of the sun–earth system. In actuality, the geometry of the sun–earth system is probably not as important as the geometrical relationship existing between the earth's magnetosphere and the solar wind flowing past it.

The 24-hour rotation of the earth within the solar wind is a first-order geometrical variation creating the diurnal variations observed at any location. Since the dipole axis of the geomagnetic field does not coincide with the earth's rotation axis, the earth's rotation causes a precessional motion of the geomagnetic field. This precession of the geomagnetic field effectively varies the approach direction of the solar wind relative to the geomagnetic equatorial plane. Such a periodic variation in the apparent direction of arrival of the solar wind should provide a second order effect in auroral diurnal variation and probably creates a diurnal variation in the degree of auroral conjugacy. Similarly, as the earth moves in its orbit around the sun, there is variation in the angle between the earth rotation axis and the sun-earth line. This variation probably contributes to seasonal variations in the aurora since it affects the incidence of direct sunlight onto the polar atmospheres and it also modifies the apparent approach direction of the solar wind. The latter effect probably influences the precipitation into the atmosphere of the energetic particles producing the aurora.

Detailed information as to why the geometry of the geomagnetic field and the geometrical relationship between this field and the onrushing solar wind produce the auroral variations is not yet available.

However, the fact that the geometrical relationships are important is, in itself, a valuable clue to an understanding of the aurora.

Temporal Factors Controlling the Aurora and the Auroral Electrojet Currents

The auroral phenomena are basically dynamical in the sense that they are active and continuously changing. Much of this change is manifested within the constraints of the geometrical controls described above. However, the observations indicate profound temporal changes, often almost explosive in nature, which are clearly independent of local time control.

A significant aspect of the *universal time* variations is that they do not appear to be entirely random in their occurrence. During periods of general low-level activity, the variations appear in the form of low-amplitude enhancements lasting near one hour and spaced four to ten hours apart. At times of high activity (during magnetic storms) the enhancements often are much larger and may last up to three hours. The onset can be quite abrupt; there also is some indication that the enhancements then tend to occur more frequently during the storm periods than in the periods of relative quiet. These universal time variations may reflect changes in the energy distribution or energy content of the solar wind moving past the earth. Another alternative is the possible existence of some instability near the outer edge of the magnetosphere that allows relatively sudden changes in the rate at which particles or energy are transferred from the solar wind to the magnetosphere.

RELATED GEOPHYSICAL PHENOMENA

Earth Currents

Horizontal voltage differences of the order of 0.1 to 1.0 volts/km appear at the earth's surface during periods of polar magnetic disturbances. Apparently induced by the changing magnetic fields of ionospheric currents, these potentials and the resulting earth currents are related sufficiently well to auroral occurrence to serve as warning or indicators of auroral activity. No quantitative relationships have been demonstrated; the problem is especially difficult because of insufficient knowledge of earth conductivity and ionospheric current distributions.

Radio Aurora

Radio aurora refers to aurorally-associated ionization which gives rise to certain types of radio reflections at very-high and ultra-high frequencies (30–3000 Mcps). The radio aurora is studied normally by radar techniques and is found to occur in the same general regions as the optical aurora. A significant but imperfect correlation exists between the occurrence of the optical and the radio aurora and either can occur in a region of space without the occurrence of the other.

Radar echoes from auroral ionization are interpreted as being the result of weak scattering from irregularities in the ionization and as resulting from critical reflection; the later cause is the more likely at low frequencies. The reflection or scattering mechanism occurs in the narrow altitude range 90 to 120 km, the same altitude range in which auroral lower borders normally occur. Spatial correlations between the optical and radio aurora are difficult to determine because the latter are highly aspect sensitive. That is, radio signals are returned preferentially when the line of sight is perpendicular to the magnetic field at the reflection point. Thus, at an observing location slightly polewards of the auroral zone, most of the radio aurora will be seen in the direction of the pole whereas most of the optical aurora will occur on the equatorward side of the station. The aspect sensitivity of the radio aurora observations is interpreted generally as indicating that the reflecting or scattering ionization is in the form of strongly field-aligned bundles. These reflecting regions show an apparent horizontal motion at the auroral zone which is essentially the same as that observed there in the visual aurora: westward in evening and eastward in morning at speeds of several hundred to several thousand meters per second. The diurnal variation of the radio aurora at the auroral zone is bimodal (as contrasted with the single maximum in the visual aurora occurrence) with the maximums occurring in the late afternoon, nighttime or early morning hours.

Radio-Noise Emission

Sporadic VHF and UHF emissions apparently associated with auroras have been reported but their nature and origin are obscure. Broad-band VLF emissions (hiss) are routinely observed at high latitude and one type called *auroral hiss* is closely associated with the occurrence of visual aurora. Intense auroral hiss sometimes is observed without the concurrent appearance of aurora, but ground and satellite observations indicate that at least part of the phenomena is due to the precipitated electrons causing the visual aurora.

Variations in Ionospheric Layers

During magnetic disturbance the height of the F_2 layer increases and its density decreases; also there is a tendency for the F-region ionization to be at higher altitude at the auroral zone than at sub-auroral latitudes. Oblique-incidence sounding of the auroral ionosphere reveal anomalous F-region echoes which may be associated with auroral ionization.

Sporadic E-layer ionization (Es) occurs at auroral zone and higher latitudes with a diurnal variation not unlike that of visual aurora occurrence. Study of the detailed correlation between Es and visual aurora indicates that the critical frequency of Es may double when a visual aurora moves into the zenith. Although there appears to be a good correlation between auroral Es and visual aurora, it is not obvious that the Es ionization is directly related to the ionization associated with auroral production.

Phase and amplitude scintillations of radio signals from point sources outside the atmosphere (radio stars and satellites) are observed under conditions of auroral activity. It is found that when the path of the radio wave passes through an auroral form the scintillation is increased.

Auroral Absorption

Several types of radio-wave absorption are recognized in the high-latitude ionosphere. These are: 1) Sudden cosmic noise absorption immediately following a solar flare. This type of absorption is due to intense ionization in the D-region by solar X-rays of 3–8 A. 2) Sudden commencement absorption which occurs in the auroral zone at the times of magnetic storm sudden commencements; it is attributed to bremsstrahlung X-rays due to energetic electrons entering the atmosphere. 3) Widespread polar cap absorption due to penetration into the polar cap ionosphere of 10–100 MeV solar protons emitted following major solar flares, and 4) Auroral absorption.

Auroral absorption is so-named because of its apparent association with the visual aurora in space and time. It occurs in the altitude range 60–90 km, somewhat lower than the visual aurora. A recent study of auroral absorption using the riometer technique indicates two main types.[5] One type of absorption occurs in the six hours preceeding local magnetic midnight and correlates well with intensity fluctuations of the 5577 Å auroral emission. This type of absorption is limited to the luminous regions of the sky. The second type is observed in the post-midnight hours, it does not correlate with the intensity of 5577 Å emission and appears not to be limited to luminous regions of the sky. This second type of absorption generally is much stronger than the first. It has been suggested that the primary particles responsible for the first type of absorption are electrons in the 10–20 keV range with a flux of 10^7 to 10^8 cm^{-2} sec^{-1} and that the second type is associated with a hardening of the primary particle spectrum to include a flux of 10^6 to 10^7 electrons cm^{-2} sec^{-1} in the energy range 30–100 keV. Studies of the statistical distribution in latitude of auroral absorption show that the zone of maximum absorption lies several degrees of latitude equatorward of the zone of maximum auroral occurrence.

X-Rays at Balloon Altitudes

X-rays in the energy range 20 keV to several hundred keV are observed by balloon-borne instruments flying at altitudes near 30 km. These x-rays are attributed to bremsstrahlung from electrons with energies mainly above 30 keV. A strong correlation exists between x-ray bursts and aurorally associated absorption of extraterrestrial radio waves, although the correlation between bremsstrahlung x-rays and visual aurora may not be as good. Balloon observations at sub-auroral latitudes, during periods when visible auroras are observed, show strong correlation between x-rays and auroras, especially for active, rayed auroral structures. The correlation at the auroral zone is reported to be much poorer, although examples are available in which the auroral luminosity and the x-ray flux follow closely together.

Sonic and Infrasonic Waves

Infrasonic waves with periods ranging from 10–110 seconds are observed at auroral and middle latitudes. These appear only on nights of moderate or strong auroral activity and emanate from the local magnetic midnight position of the auroral zone, the position where intense, active auroras usually prevail. These waves are suggested to originate from periodic heating of the ionosphere by primary auroral particles.

Audible sounds near the upper limit of human perception are reported by many untrained observers and several trained ones. Since propagation of sound waves in this frequency range to the earth's surface is thought to be impossible, the *auroral sounds* if real, are probably produced by some mechanism near the ground.

AURORAL EMISSION AND EXCITATION

Forbidden Atomic Lines[6]

Transitions between the ground state configurations of singly and doubly-ionized oxygen and nitrogen atoms are forbidden by electric-dipole radiation but have finite probabilities for electric-quadrupole and magnetic dipole radiation. These transitions (Fig. 8), especially those of [OI] and [NI], yield prominent features in the optical aurora. In fact, the $^1D \rightarrow {}^1S$ transition of [OI] at 5577 Å gives the strongest line in the visible aurora, and the $^2D^0 \rightarrow {}^2P^0$ doublet transition at 10,400 Å is equally as strong in the infrared portion of the spectrum.

The terms in the ground configurations of [OI], [OII], [NI] and [NII] have excitation energies of 5 eV or less and therefore are more easily excited by collisions with slow electrons than are the other configurations, which have much higher energies.

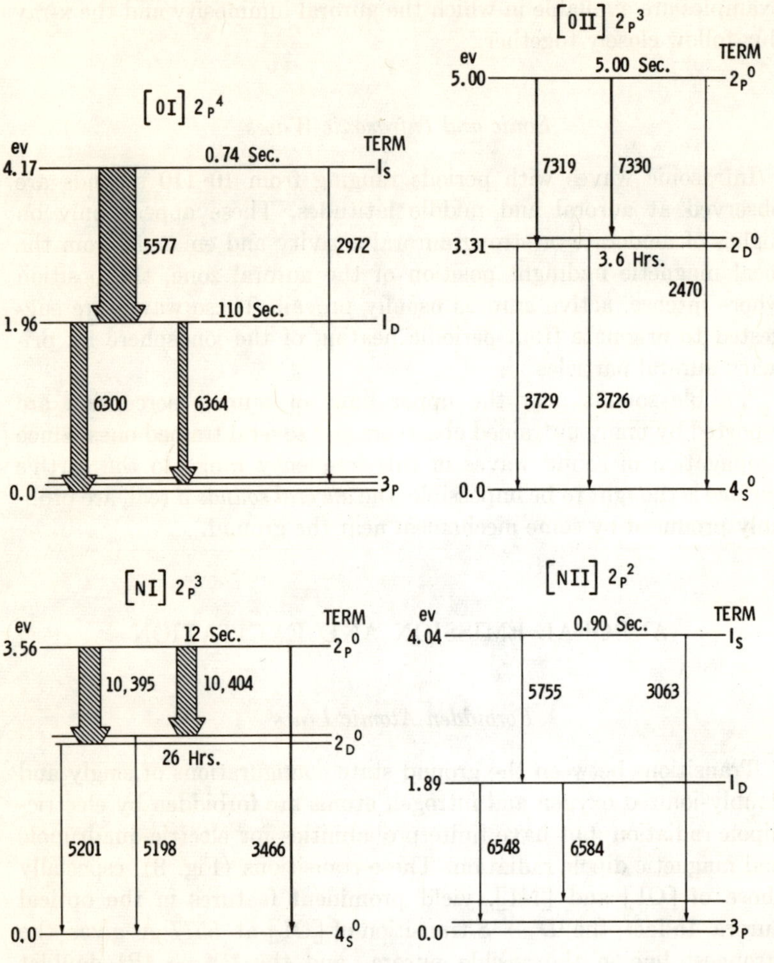

FIG. 8. Energy level diagram showing oxygen and nitrogen forbidden transitions in the aurora. An indication of the relative intensities of the emissions is given by the widths of the lines drawn to indicate each transition.

Permitted Atomic Lines

Approximately a dozen permitted multiplet lines of singly- and doubly-ionized oxygen and nitrogen are definitely identified in the aurora and others are uncertain. The excitation energies of the electronic configurations involved in these transitions lie in the range 9 to 30 eV, well above those of the forbidden lines of oxygen and nitrogen (cf. Figures 8 and 9). The combined intensity of the OI lines is approximately equivalent to that of the NII lines; it is equal to about one-quarter of the intensity of the [OI] 5577 Å green line and approximately five times the combined intensities of OII or NI lines in normal auroras. Thus, none of the permitted lines of oxygen

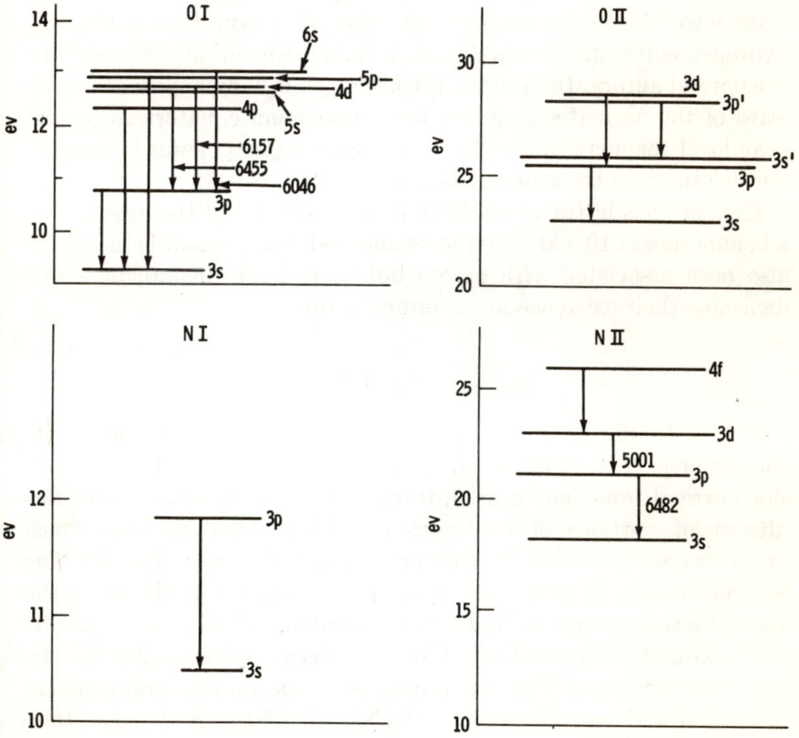

FIG. 9. Energy level diagram showing oxygen and nitrogen permitted transitions in the aurora.

and nitrogen are especially strong features of the aurora, although NII multiplets near 5002 Å and 5680 Å are readily identified on many spectra.

The Balmer lines of hydrogen, H_α (6563 Å), H_β (4861 Å), and H_γ (4340 Å) are seen in auroral spectra; the upper levels of these transitions have excitation energies between 12 and 13 eV. The intensity of the hydrogen lines is much less than that of [OI] 5577 Å in the normal aurora, and is quite variable relative to that emission. The hydrogen emissions are of special interest for several reasons. Over a decade ago the first concrete evidence that fast protons were entering the atmosphere was found in the profile of H_α. The H_α line, Doppler broadened and shifted toward shorter wavelengths, indicated that protons with a mean velocity of about 500 km/sec penetrate into the earth's atmosphere. Also, it is now known that the hydrogen emissions show a diurnal behavior somewhat different from the normal aurora. In evening, the hydrogen usually appears equatorward of the normal aurora, reaches a maximum equatorward extent near local magnetic midnight, and then recedes polewards into the normal auora in the morning hours.

Helium lines in the auora have been reported and the existence of a helium line at 10,830 Å is well established. The sodium D lines have also been associated with aurora but there is no unambiguous evidence for their existence in an auroral form.

Molecular Band Systems

The molecular band systems of O_2, O_2^+, N_2, and N_2^+ contribute the major portion of the intensity in the auora. The molecular emissions extend from the infrared, through the visible, and into the far-ultraviolet portions of the spectrum. Those band systems which are definitely identified in the aurora are indicated in Fig. 10. The Second Positive System of N_2 has been identified in the blue and the First Positive System of N_2 has been found in the red. Enhancements of N_2 First Positive and O_2^+ First Negative are responsible for the red lower borders of Type B auroras. An important auroral emission in the blue and near ultraviolet is the N_2^+ First Negative band system, which is thought to result from simultaneous ionization-emission of N_2 by particles striking the atmosphere. Thus, its presence or absence in an upper-atmosphere spectrum is used frequently as an in-

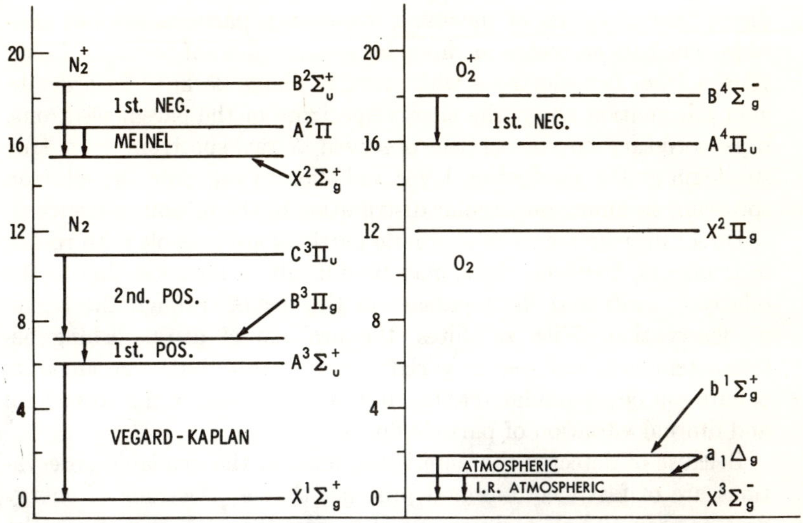

FIG. 10. Energy level diagram showing electronic states and band systems of N_2, N_2^+, O_2, and O_2^+ observed in the aurora.

dicator of whether or not aurora is present. The upper state of this system has an excitation energy above 18 eV; the lowest excitation energies of the nitrogen and oxygen band systems are those of O_2, as is seen in Fig. 10.

ENTRY OF ENERGETIC PARTICLES INTO THE ATMOSPHERE

Methods of Observation

The precipitation of charged particles into the atmosphere has been studied over the past few years primarily with Geiger–Mueller tubes and scintillation counters carried on balloons, rockets and satellites. As yet, only a rough picture of the precipitation has emerged because of limitations of the particle detectors and limitations brought about by the trajectories and small number of instrument vehicles used.

Balloon altitudes are limited to below 40 km, at which heights direct measurements of incoming low-energy particles are not possible. The balloon technique involves measurement of bremsstrahlung X-rays from fast electrons with energies above 25 keV. In order to gain information about the energy spectrum of the parent electrons, it is necessary to correct the observed X-ray spectrum to a true spectrum at the production layer and then to calculate the electron spectrum assuming an angular distribution of the incoming electrons.

Direct measurements of energetic particles are possible with rocket instruments, however, these may be difficult to interpret due to the relatively swift and short passage of the rocket through the region of observation. With satellites, the problem of distinguishing between temporal and spatial variations still exists, but huge amounts of data can be accumulated to facilitate investigation of the latitudinal and diurnal variation of particle fluxes.

Because of detector characteristics, most of the available observations are of fast electrons and protons; that is, electrons of energy greater than 40 keV and protons of greater than 100 keV. Recently, measurements have included electron energies to less than 100 eV and protons down to several tens of keV.

Observations of Precipitated Particles

Only very rough determinations of energy spectra are available and these indicate spectra which vary greatly with time and position. An early rocket flight into a faint auroral glow indicated an electron energy flux near 20 ergs/sec cm^2 while a second into a bright auroral arc was as high as 2000 ergs/sec cm^2.[7] Detectors on the first flight indicated that there was approximately one proton incident for every 1000 electrons, and detectors on the second indicated relatively few protons. The integral number energy spectrum of electrons in the energy range 3 to 30 keV as observed on the first flight was $N(E) = 2.5 \times 10^9 \exp(-E/5)$, while the second flight indicated that most of the light in the aurora was produced by nearly monoenergetic electrons of energy 6 keV with peak electron fluxes of 5×10^{10} electrons (E is the particle energy in keV). The measured energy spectrum of protons on the first flight was equal to $2.5 \times 10^6 \exp(-E/30)$.

Electron energy spectra, measured to date, fall into two broad categories which appear to be related to the low energy limits of the measuring instruments. When the results are expressed in the form

$n(E) = C \exp(-E/E_0)$, values of E_0 between 2 and 8 keV are found if the lower energy limit of the instruments are 10 keV or less. Less steep spectra, that is, E_0 values of 20 to 40 keV, are deduced from observations with instruments having low energy cutoffs greater than 10 keV.

The fact that particle precipitation is highly variable in both space and time is illustrated by the comparison in Fig. 11 which

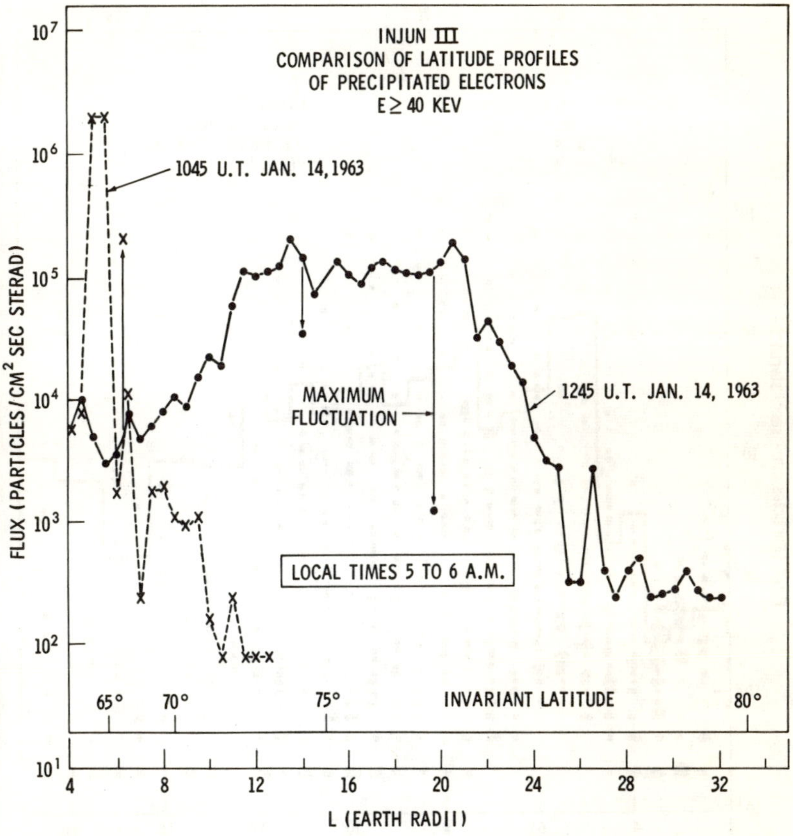

FIG. 11. Comparison of the latitude profile of precipitation for two successive passes at about the same local time of the Injun III satellite illustrating the temporal variability in the precipitation [after B. J. O'Brien, J. Geophys. Res., **69**, 29 (1964)].

shows two passes of the Injun III satellite moving at an altitude of approximately 600 km.[8] The variability of the electron flux is demonstrated further by the Injun III data assembled in Fig. 12. It is indicated there that the electron flux in the energy range $E \geq 40$ keV can vary by a factor of 10^6 at the auroral zones. Electron flux measure-

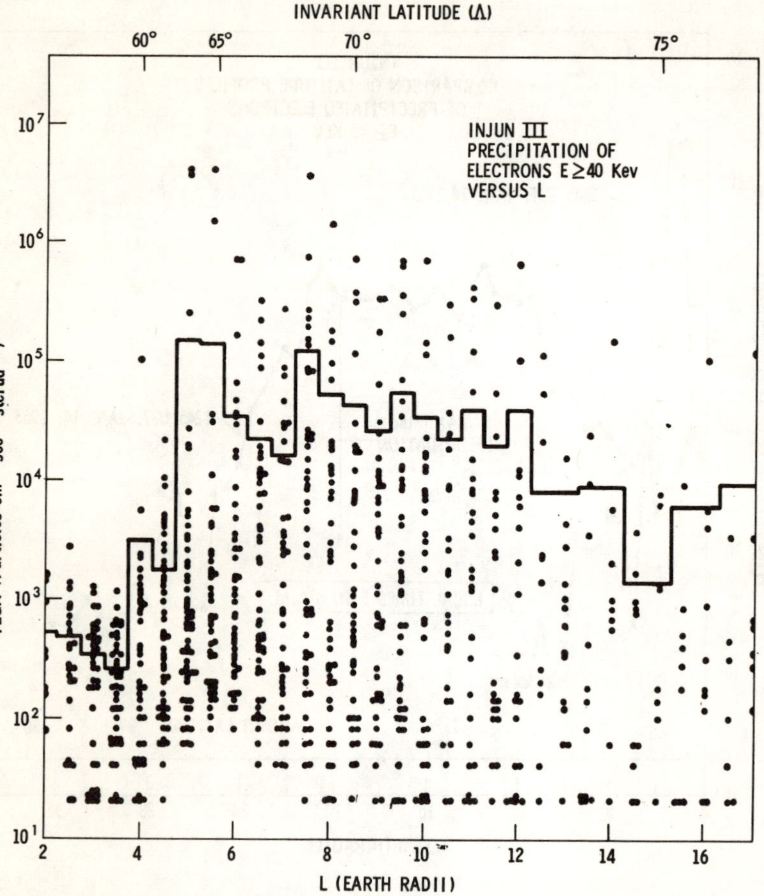

FIG. 12. Samples of precipitated fluxes over North America in January 1963 measured by the Injun III satellite [After B. J. O'Brien, J. Geophys. Res., **69**, 29 (1964)].

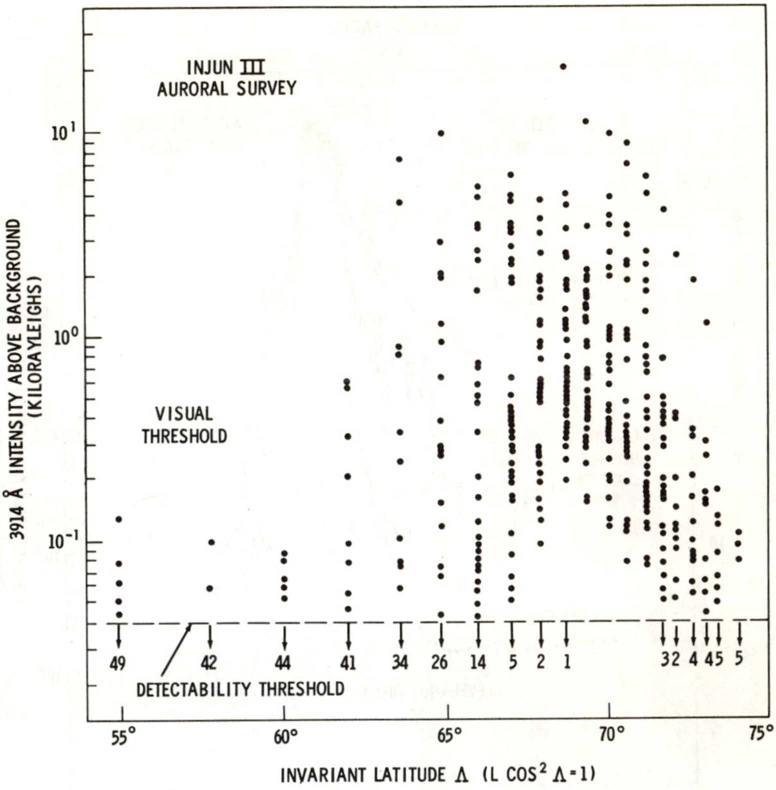

FIG. 13. Intensity of N_2^+ 3914 A auroral light measured by the Injun III satellite during approximately fifty passes early in 1963 [after B. J. O'Brien and H. Taylor, J. Geophys. Res., **69,** 51 (1964)].

ments from satellites indicate that the flux is statistically greatest near the auroral zones and that there always is some precipitation at auroral zone latitudes. Similarly, measurements of the N_2^+ First Negative Band at 3914 Å made from Injun III indicate that the 3914 Å auroral emission is statistically greatest near the auroral zone and that there is always some emission at this latitude: see Fig. 13.

The relations between precipitated electrons and auroral luminosity as well as between precipitated electrons and trapped electrons is

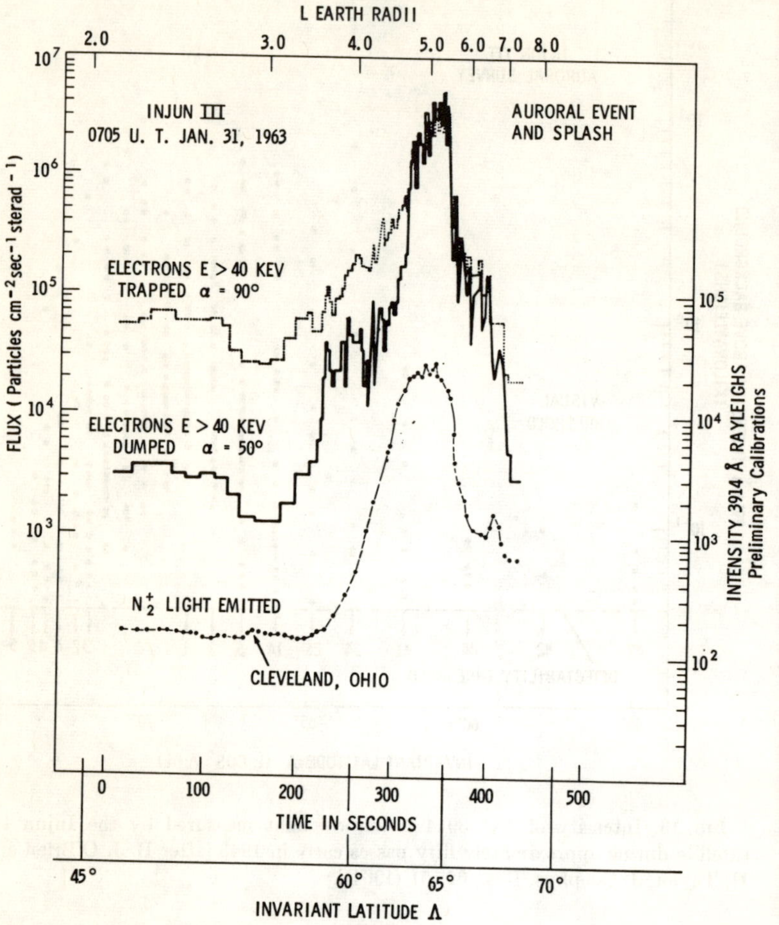

FIG. 14. Measurements of trapped and precipitating particles and N_2^+ 3914 A auroral emission from a pass of the Injun III satellite showing the simultaneous increases of trapped and precipitating particle fluxes [after B. J. O'Brien, J. Geophys. Res., **69,** 16 (1964)].

demonstrated by the Injun III pass of January 31, 1963, Fig. 14. This diagram illustrates a general result from the Injun III measurements that the intensity of trapped electrons of energy $E > 40$ keV mirroring near the satellite altitude increases with increasing intensity of precipitation such that the distribution of electron pitch angles tends to be isotropic over the hemisphere above the satellite. A similar result for electrons of energy near 10 keV also has been obtained. This observation, together with the Injun III observation that the upward flux of particles never exceeded the downward flux, suggests that the mechanism leading to electron precipitation takes place high above the atmosphere.

Quite recently some very interesting results have been obtained from low altitude polar-orbiting satellites carrying instruments capable of detecting low-energy electron and proton fluxes.[9] Typical meridional passes of these satellites across the auroral zone showed that low-energy electrons ($E \sim 100$ eV) were precipitated in relatively wide zones which contained narrower zones of precipitation by higher energy ($E = 10$–30 keV) electrons. In some cases, the narrow zones of higher energy electron precipitation appeared to correspond to positions of auroral forms. Large variations in the angular distribution of precipitating electrons were found, and it was observed that the distribution of electrons precipitating into the northern auroral zone was harder (containing relatively more high-energy particles) on the day-side as compared to the night-side of the earth. Simultaneous measurements of low-energy protons indicated that the proton flux was spatially less variable than the electron flux and usually contained within it.

The flux of precipitated electrons in all energy ranges has been found to increase markedly with increasing magnetic activity as indicated by the 3-hour planetary index Kp. There is also indication that the region of maximum precipitation of high energy ($E \geq 40$ keV) electrons moves equatorward during periods of strong magnetic disturbance.

THE INTERACTION OF ENERGETIC PARTICLES WITH THE ATMOSPHERE

The entry of an energetic electron or proton into the atmosphere initiates a complex series of processes. The most important processes contributing both to the dissipation of kinetic energy carried by incoming fast particles and to the production of auroral luminosity are direct collisional excitation and ionization of atmospheric gases by electrons;

$$e + X \rightarrow e + X^* \tag{1}$$

$$e + X \rightarrow 2e + X^{+*} \tag{2}$$

and by protons or neutral hydrogen atoms,

$$H + X \rightarrow H^* + X^* \tag{3}$$

$$H + X \rightarrow H^* + X^{+*} + e, \tag{4}$$

where the asterisk signifies and excited atom or molecule, and X is an atmospheric molecule.

Through studies of the intensity of the principal emission features of the auroral spectrum and measurements of particle fluxes above the atmosphere, it is known that energetic electrons rather than protons or neutral hydrogen produce most of the ionization and excitation in normal aurora.

On the average, about 35 eV are required to produce an ion pair by the process in Equation (2). Hence a *primary* electron with initial energy of several tens or hundreds keV carries enough to produce many ion pairs. These are created by multiple inelastic collisions of the primary particle and by further collisions of *secondary* electrons, both *first-generation* secondaries and later-generation secondaries. The principal excitation of the N_2^+ auroral emissions is believed to result from simultaneous ionization and excitation of N_2 by electrons [the process indicated by Equation (2)], with primary and secondary electrons perhaps being equally important. It is generally assumed that approximately one quantum of N_2^+ First Negative Group radiation at 3914 Å is emitted per 50 ion pairs formed. Using various assumed energy and angular distributions for incident streams of primary electrons, 3914 Å luminosity profiles have been computed; some examples are given in Fig. 15.[10] These computations are in satisfactory agreement with observed luminosity variations when measured electron energy distributions are employed.

Direct collisions also are thought to be responsible for the excitation of the N_2 bands in the aurora. Their excitation may be important in removing most of the energy from electrons in the 10–15 eV range and thereby reducing excitation of emissions such as the forbidden green line of [OI] 5577 Å which requires 4.17 eV.

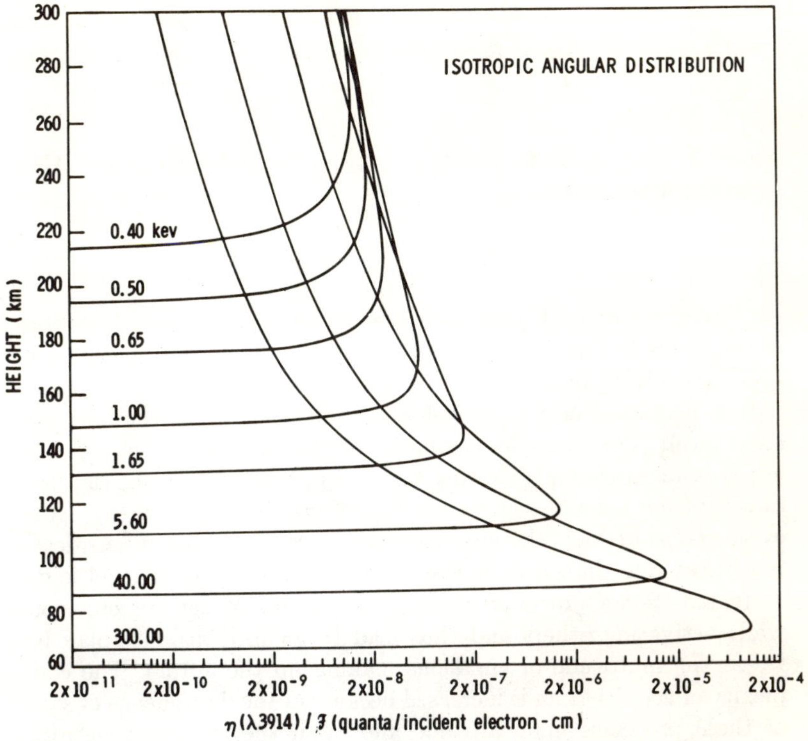

FIG. 15. Excitation rate of N_2^+ 3914 Å radiation by isotropic streams (0°–80°) of mono-energetic electrons [after M. H. Rees, Plan. and Space Sci., **11,** 1214 (1963)].

There is still uncertainty as to the relative importance of various possible mechanisms giving excitation to the forbidden lines of atomic oxygen and nitrogen, however it appears likely that the most important is collisional excitation by electrons: in the case of the [OI] lines,

$$e + \mathrm{O}\ (^3P) \rightarrow e + \mathrm{O}\ (^1D) \tag{5}$$

and

$$e + \mathrm{O}\ (^3P) \rightarrow e + \mathrm{O}\ (^1S). \tag{6}$$

Another possibility is collisional dissociation involving fast particles,

$$O_2 + X \rightarrow O + O^* + X \qquad (7)$$

$$O_2 + X \rightarrow O + O^{+*} + e + X, \qquad (8)$$

where X represents an electron, proton, or hydrogen atom. Also, radiative dissociation

$$O_2 + h\nu \rightarrow O^* + O^* \qquad (9)$$

can leave one or both product atoms in an excited state. Reactions of the types in Equations (6)–(9) apply equally to the excitation of [NI] and [NII] lines.

It is necessary also to consider mechanisms which deactivate excited levels. For example, excitation of the oxygen red lines $[OI]_{21}$ arises from cascading following the $[OI]_{32}$ green line transition and directly from reactions such as in Equations (5), (7), and (9). However, except at high altitude, the ratio of red to green intensities is much less than unity, so it is clear that the [OI] (^1D) level is deactivated. Some processes cause deactivation of certain emissions while activating others and thus lead to an intricate interplay between the multitude of emissions making up the aurora. The complexity of the situation is increased because of the dependence of some of these processes upon altitude and upon the relative concentration of the constituents involved. Deactivation by slow electrons

$$e + X^* \rightarrow e + X \qquad (10)$$

is important for the long-lived $[NI]_{21}$ and $[OII]_{21}$ transitions and may be important in deactivation of the $[OI]_{21}$ red lines in bright auroras where the electron concentration is high. Another process which may be effective in deactivating the red lines while exciting the O_2 Atmospheric system is molecular deactivation. An example is

$$O\ (^1D) + O_2\ (X\ ^1\Sigma_g{}^+, \nu = 0) \rightarrow O\ (^3P) + O_2\ (b\ ^1\Sigma_g{}^+, \nu = 2). \qquad (11)$$

Conversely, an important process which may act to excite [OI] and

[NI] transitions while reducing emission from the N_2^+ and O_2^+ systems is dissociative recombination

$$XY^+ + e \rightarrow X^* + Y^*, \tag{12}$$

where

$$XY^+ = N_2^+ \quad \text{or} \quad O_2^+.$$

As mentioned above, energetic protons entering the atmosphere are slowed down by ionizing collisions; the resultant secondary electrons may undergo the same interactions as secondaries produced by primary electrons. After the primary protons undergo a sufficient number of collisions to reduce their energy to about 100 keV, charge-transfer collisions

$$H^+ + X \rightarrow H^* + X^{+*} \tag{13}$$

become important and yield excited neutral hydrogen. The excited hydrogen atom may undergo ionizing collisions and thereby cycle between H and H^+ many times before losing all its energy. According to Chamberlain, a fast proton yields total H_α, H_β, and Ly_α emissions of 60, 16, and 460 quanta, respectively. In an aurora produced solely by primary protons, the ratio of intensity of N_2^+ 1. N.G. 3914 Å to H_β is dependent upon the initial energy of the protons and is calculated to be in the range 1–3 whereas in a normal aurora the ratio is observed to be 10 or much greater. This result alone indicates that electrons rather than protons must be the primary source of the normal aurora.

THEORIES OF THE AURORA

It has been generally recognized for some time that the aurora and geomagnetic activity are closely related and originate through the interaction of the geomagnetic field with plasma flowing outward from the sun. No theory has yet evolved which satisfactorily explains even a major portion of the observed features of the aurora and geomagnetic activity. This failure arises in part from incomplete observation of the aurora and magnetic activity at the earth's surface and partly from lack of knowledge of the physical parameters governing particle motion in the magnetosphere and in the solar wind.

One of the most important theories of magnetic storms is that of Chapman and Ferraro and which has been extended by Martyn in an attempt to explain auroras and related phenomena.[11] The essential feature of the Chapman–Ferraro theory is an electrically neutral stream of charged particles which approaches the earth until currents induced in the conducting boundary of the stream grow sufficiently large to prevent further penetration of the stream into the geomagnetic field. The geomagnetic field carves out a hollow in the neutral stream; Chapman and Ferraro supposed that the hollow extended to infinity behind the earth but Martyn's extension of the model assumed closure of the stream at some distance behind the earth. Surface charges on the hollow (Fig. 16) were thought to jump the gap

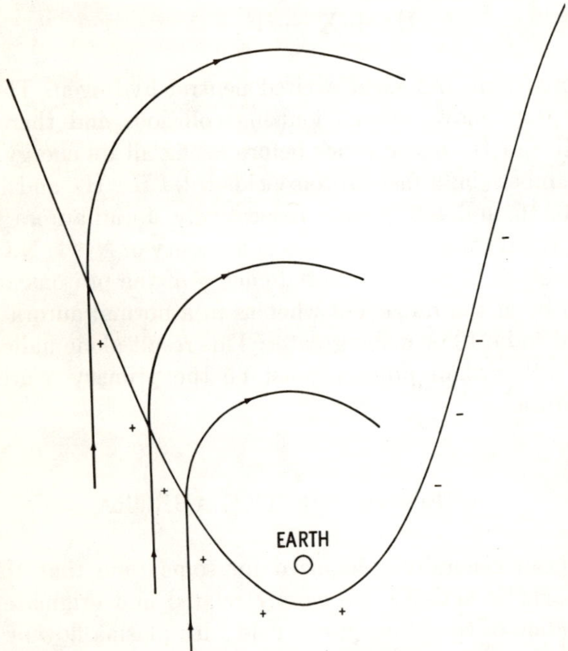

FIG. 16. Illustration of the hollow in the neutral solar stream produced by the geomagnetic field as proposed by Chapman and Ferraro, and the subsequent development of ring currents.

across the hollow and contribute to the growth of a equatorial ring current which would explain the main phase of the magnetic storm. Some of the charged particles were thought to be accelerated near the stream boundary and to travel down the geomagnetic field lines into the polar regions where they would give rise to the aurora and the intense ionospheric currents in the auroral atmosphere. While not providing much explanation of auroral phenomena, this model does fit the main features of a magnetic disturbance and predicts the magnetosphere boundary. The boundary on the sunward side of the earth is now observationally well established; the initial concept of Chapman and Ferraro is still retained and forms the basis of improved calculations of the boundary shape (see Chapter 9).

In 1939, Alfvén put forth a model in which a general solar geomagnetic field was assumed. Such a field would tend to polarize the neutral stream of charged particles passing through it so that the stream coming near the earth would act under the influence of a constant electric field as well as the geomagnetic field.[11] Positive and negative particles in the stream drift westward and eastward, respectively, as they enter the inhomogeneous geomagnetic field. If the constant electric field is properly oriented, these particles drift so as to increase their energy because of the presence of the electric field. As in the Chapman–Ferraro model, there is a forbidden region sur-

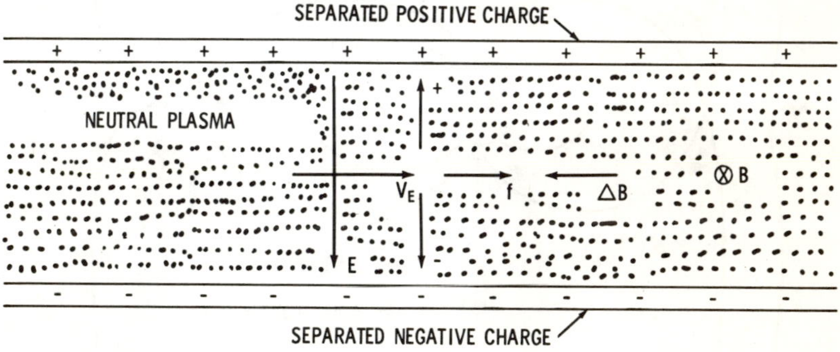

FIG. 17. A perspective drawing showing the paths of ions and electrons as they invade the magnetosphere according to the theory of Alfvén [redrawn after T. G. Cowling, Terr. Mag. and Atmos. Elect., **47,** 209 (1942)].

rounding the earth, and in the Alfvén model, positive space charge is built up on the day side of this region and negative space charge on the night side. These space charges are neutralized in the Alfvén model (Fig. 17) by discharge into the polar atmosphere of the charged particles which become auroral primaries. The original Alfvén theory was strongly criticized on several grounds. These criticisms included the suggestion that electrostatic fields growing as charges of opposite sign tended to separate would prevent more than negligible separation of charge, that the solar stream itself would be dispersed before arriving at the earth, and that the model required a general solar magnetic field of which there was no evidence at the time.

More recently Dungey proposed a model which he suggested could yield the main features of polar ionospheric currents and auroras if there were a general interplanetary magnetic field directed southward.[12] The result of such a general field is to produce a magnetic null line around the earth, approximately in the earth's equatorial plane. Two null points on the null line are shown in Fig. 18; the diagram also shows the magnetic field configuration and the direction of plasma flow in that configuration. In Dungey's model, plasma flow-

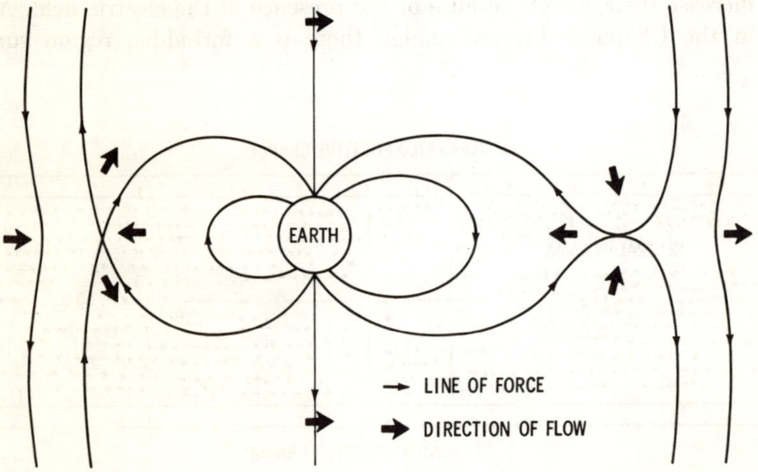

FIG. 18. Interplanetary plasma flow in a plane containing neutral points according to the theory of Dungey [after J. W. Dungey, Phys. Rev. Letters, **6**, 47 (1961)].

ing with velocity **u** through the interplanetary magnetic field creates an electric field $\mathbf{E} = -\mathbf{u} \times \mathbf{H}/c$. Assuming infinite conductivity so that magnetic field lines become equipotentials, the field **E** maps onto the polar ionosphere in such a way that the polar SD current configuration would result. While one may question the required existence of a southward interplanetary magnetic field, it is interesting that this model does yield an inward flow of plasma on the night side of the earth, and we do know that it is on the night side that the strongest magnetic and auroral activity occurs.

The discovery of the Van Allen belts during the IGY had a profound effect on auroral theory in the following few years. New interest was awakened and, despite the fact that the trapped radiation was found to be located relatively deep within the magnetic field, it was looked upon as a particle reservoir from which to draw auroral primaries. The problem then became one of finding mecha-

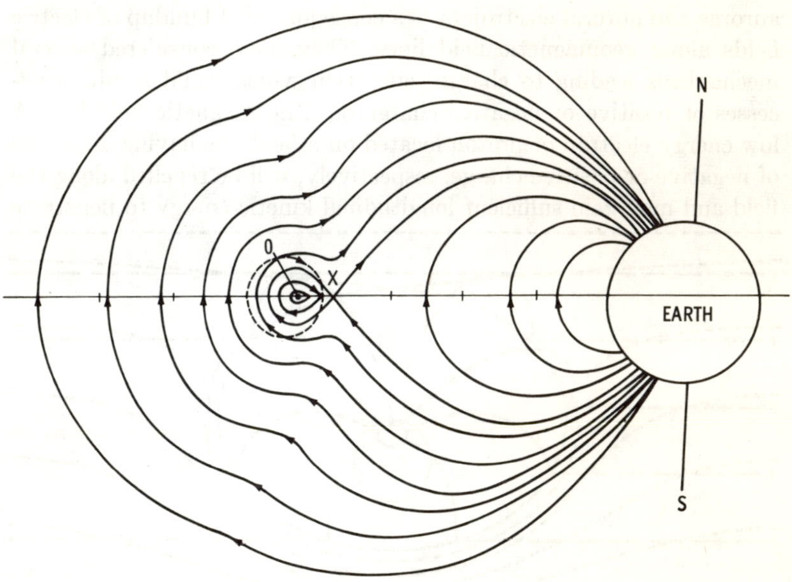

FIG. 19. Illustration of the neutral line theory of Akasofu and Chapman showing the distortion of the geomagnetic field by a ring current so as to create X- and O-type neutral lines. [After S. I. Akasofu and S. Chapman, Phil. Trans. Roy. Soc. (London) **A253**, 359 (1961)].

nisms whereby the low energy particles in the trapped radiation could be accelerated and ejected into the auroral zones.

Akasofu and Chapman proposed a model (Fig. 19) whereby the trapped radiation could be accelerated in very narrow regions near X-type neutral lines in the geomagnetic field.[13] Their mechanism requires that ring currents near 6 earth radii be sufficiently intense and confined to produce reversals of the geomagnetic field, with each generated X-type neutral line being responsible for producing an auroral arc. The model has the feature that it explains the thinness of auroral forms and gives the observed multiplicity of auroral forms if a number of neutral lines can be formed. However satellite observations indicate that it is unlikely that the required magnetic field reversals are occurring regularly in the magnetosphere, and it is now thought that the main ring currents lie much nearer the earth than 6 earth radii.

Chamberlain[14] and Kern[15] have studied the possibility of producing auroras and auroral electrojets as a consequence of buildup of electric fields along geomagnetic field lines. They have considered several mechanisms leading to charge separation which could result in excesses of positive or negative charge on a geomagnetic field line. A low energy electron or proton located on a field line having an excess of negative or positive charge, respectively, will be repelled along the field and may gain sufficient longitudinal kinetic energy to penetrate

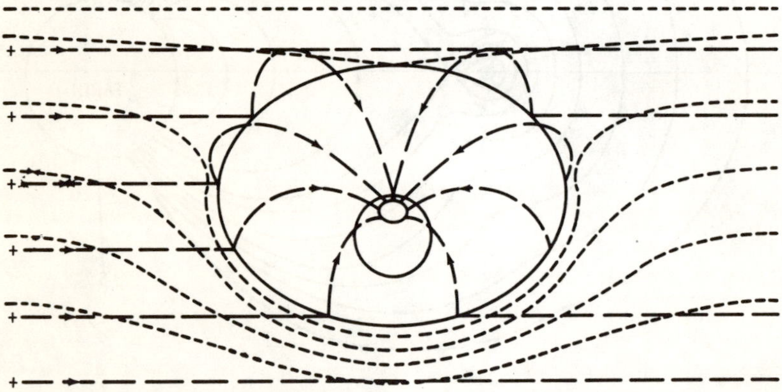

FIG. 20. The charge separation theory of Kern. A longitudinal magnetic field gradient (B) drives positive and negative charges in opposite directions giving rise to a polarization field (E) which in turn causes particle drift V_E [after J. W. Kern, J. Geophys. Res., **67**, 2651 (1962)].

into the atmosphere. Chamberlain suggested that charge separation could result from energetic particle distributions which are irregular in longitude since positive and negative charges tend to drift longitudinally in opposite directions through the action of the inhomogeneous geomagnetic field (∇B directed radially inward). Kern, in particular, also examined the effect of longitudinal geomagnetic field gradients in producing charge separation (Fig. 20) and related these to the generation of sheet-type auroras, e.g., bandlike structures. In essence the mechanisms considered by Chamberlain and Kern involve means of transferring transverse kinetic energy of particle distributions into, first, potential energy and then into longitudinal kinetic energy. Quite reasonable explanations of the observed auroral configurations, motions, and growth rates have evolved from these studies. More recently Chamberlain has suggested a mechanism of accelerating trapped particles with parallel (to **B**) electric fields induced by instability in the magnetosphere plasma.[16] The theory requires an outward density gradient in a magnetospheric plasma which has a mean ion energy much greater than the mean electron energy. Under these conditions an instability is generated in such a way that some of the electrons and ions in the plasma are continuously accelerated along the geomagnetic field and eventually ejected into the atmosphere.

The theories devised by Akasofu and Chapman, Kern, and Chamberlain all assume a two-step process: relatively low energy particles are introduced into the magnetosphere and then they are accelerated sufficiently to be injected into the atmosphere. A somewhat different approach was taken by Axford and Hines who suggested that a viscous-like interaction between the outer portion of the magnetosphere and the solar wind might drive plasma in the outer part of the magnetosphere back into the magnetosphere tail where, due to the constraint of the geomagnetic field, the plasma would be convected inward toward the earth (Fig. 21). Both previously trapped particles and those convected from the solar wind into the magnetosphere tail are energized as they are pushed deeper into the geomagnetic field. In the Axford-Hines model the magnetosphere is considered to be infinitely conducting so that motions and the resulting electric field $\mathbf{E} = -\mathbf{V} \times \mathbf{B}$ are mapped through the field onto the ionosphere where the resulting patterns resemble the observed pattern of auroral motion and the polar DS current-system configuration.[17]

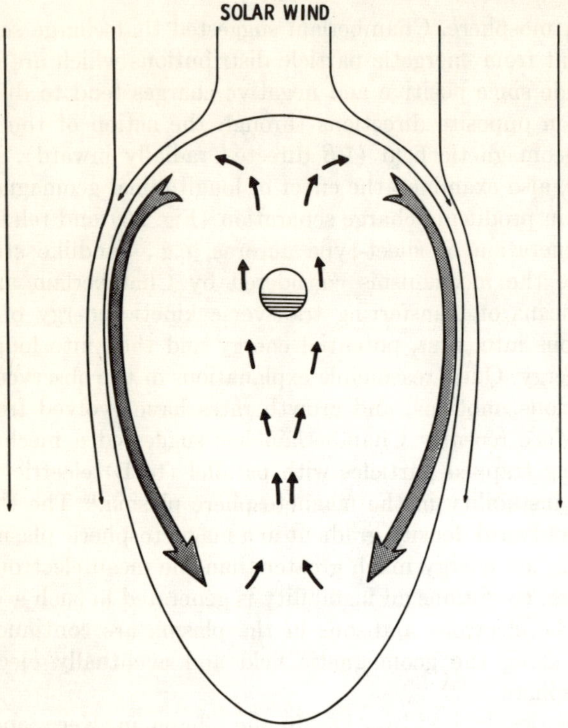

FIG. 21. The motion impressed on the magnetosphere by viscous-like interaction with the solar wind according to the theory of Axford and Hines. [After W. I. Axford and C. O. Hines, Can. J. Phys., **39,** 1438 (1961)].

Fejer also has considered convection in the magnetosphere and its relation to polar ionospheric current systems.[18] He has suggested three possible convection mechanisms other than the viscous interaction between the solar wind and the magnetosphere: namely, temporal asymmetries in the distribution of geomagnetically trapped particles, electric polarization fields of tidal (dynamo) origin in the region of particle trapping, and the co-rotation of the magnetosphere with the earth.

Since the introduction of his first model in 1939, Alfvén and several of his co-workers have continued to investigate the part which electric fields can play in auroral production. From this work has come the suggestion that the low-density magnetized plasma of the solar wind can penetrate into the magnetosphere with the aid of the electric field associated with the moving magnetized plasma. Charge separation associated with the penetration of the plasma is the suggested source of particle acceleration.

Recently Hones has calculated particle motion due to magnetic gradient and line-curvature drifts, the co-rotational electric field and an assumed electric field across the night side of the magnetosphere.[19] From this work he concludes that it is possible to explain the main features of the high latitude auroral phenomena with thermalized solar wind plasma entering the front of the magnetosphere rather than the back, as proposed by the convection theories.

The recent observations of O'Brien cited above (pages 233–237) indicate that previously trapped particles are not the sources of auroral primaries; instead it now appears that the energetic primaries may be accelerated simultaneously with their injection into the magnetosphere from the plasma of the solar wind. This turn of affairs poses problems to some of the above auroral theories and tends to favor those theoretical models which specifically include mechanisms of injecting particles into the magnetosphere.

REFERENCES

1. T. N. Davis, The Morphology of the Auroral Displays of 1957–1958, J. Geophys. Res. **67,** 59 (1962).
2. R. N. DeWitt, The Occurrence of Aurora in Geomagnetically Conjugate Areas, J. Geophys. Res. **67,** 1347 (1962).
3. J. P. Heppner, A Study of the Relationships Between the Aurora Borealis and the Geomagnetic Disturbances Caused by Electric Currents in the Ionosphere, Thesis, California Institute of Technology, Pasadena, 1954.
4. S.-I. Akasofu, The Development of the Auroral Substorm, Planet. Space Sci. **12,** 273 (1964).
5. Z. A. Ansari, The Spatial and Temporal Vibrations in High Latitude Cosmic Noise Absorption and Their Relation to Luminous Aurora. Thesis, University of Alaska, College, 1963.
6. J. W. Chamberlain, *Physics of the Aurora and Airglow*, Academic Press, New York, 704 pp, 1961. (A general reference for much of the material in the Sections on *Auroral Emission and Excitation* and *The Interaction of Energetic Particles with the Atmosphere*.)
7. C. E. McIlwain, Direct Measurement of Particles Producing Visible Auroras, J. Geophys. Res. **65,** 2727 (1960).
8. B. J. O'Brien, High-latitude Geophysical Studies with the Satellite Injun III, J. Geophys. Res. **69,** 13 (1964). (Also, see related articles in the same issue.)

9. R. G. Johnson, J. E. Evans, R. D. Sharp, and J. B. Reagan, Satellite Measurements on Precipitated Electrons with Energies Greater than 80 ev, Trans. Am. Geophys. Union **45,** 41 (1964) (Abstract).
10. M. H. Rees, Auroral Ionization and Excitation by Incident Energetic Electrons, Planet. Space Sci. **11,** 1209 (1963).
11. S. K. Mitra, The Upper Atmosphere, The Asiatic Society, Calcutta, 713 pp., 1952.
12. J. W. Dungey, Interplanetary Magnetic Field and the Auroral Zones, Phys. Rev. Letters **6,** 47 (1961).
13. S.-I. Akasofu, and S. Chapman, A Neutral Line Discharge Theory of the Aurora Polaris, Phil. Trans. Roy. Soc. (London) **A253,** 359 (1961).
14. J. W. Chamberlain, Theory of Auroral Bombardment, Astrophys. J. **134,** 401 (1961).
15. J. W. Kern, a Charge Separation Mechanism for the Production of Polar Auroras and Electrojets, J. Geophys. Res. **67,** 2649 (1962).
16. J. W. Chamberlain, Plasma Instability as a Mechanism for Auroral Bombardment, J. Geophys. Res. **68,** 5667 (1963).
17. W. I. Axford, and C. O. Hines, A Unifying Theory of High-Latitude Geophysical Phenomena and Geomagnetic Storms, Canad. J. Phys. **39,** 1443 (1961).
18. J. A. Fejer, Theory of Auroral Electrojets, J. Geophys. Res. **68,** 2147 (1963).
19. E. W. Hones, Jr., A Theoretical Interpretation of High-Latitude Geophysical Phenomena, (in press).

Chapter 6

METEOROLOGY FROM SPACE

W. Nordberg and S. I. Rasool

ROCKETS AND SPACECRAFT AS TOOLS FOR METEOROLOGICAL OBSERVATIONS

Space-borne platforms are used for meteorological observations primarily for two reasons. First, they enable the measurement of atmospheric characteristics at high altitudes not accessible by any other means, and second, their use is more efficient and economical than that of conventional techniques in performing continuous synoptic measurements in the lower regions of the atmosphere and at the surface of the earth covering the entire globe. The first of these reasons brought about the development of a variety of meteorological sounding rockets which carry sensors up to the ionosphere, while the second produced the meteorological satellites orbiting beyond the region of interest, but observing the earth with radiation sensors and imaging devices. Because of their indiscriminate continuity of measurement, meteorological satellites have generated a number of challenging problems in the field of data technology requiring first, the storage of vast amounts of data in the spacecraft,

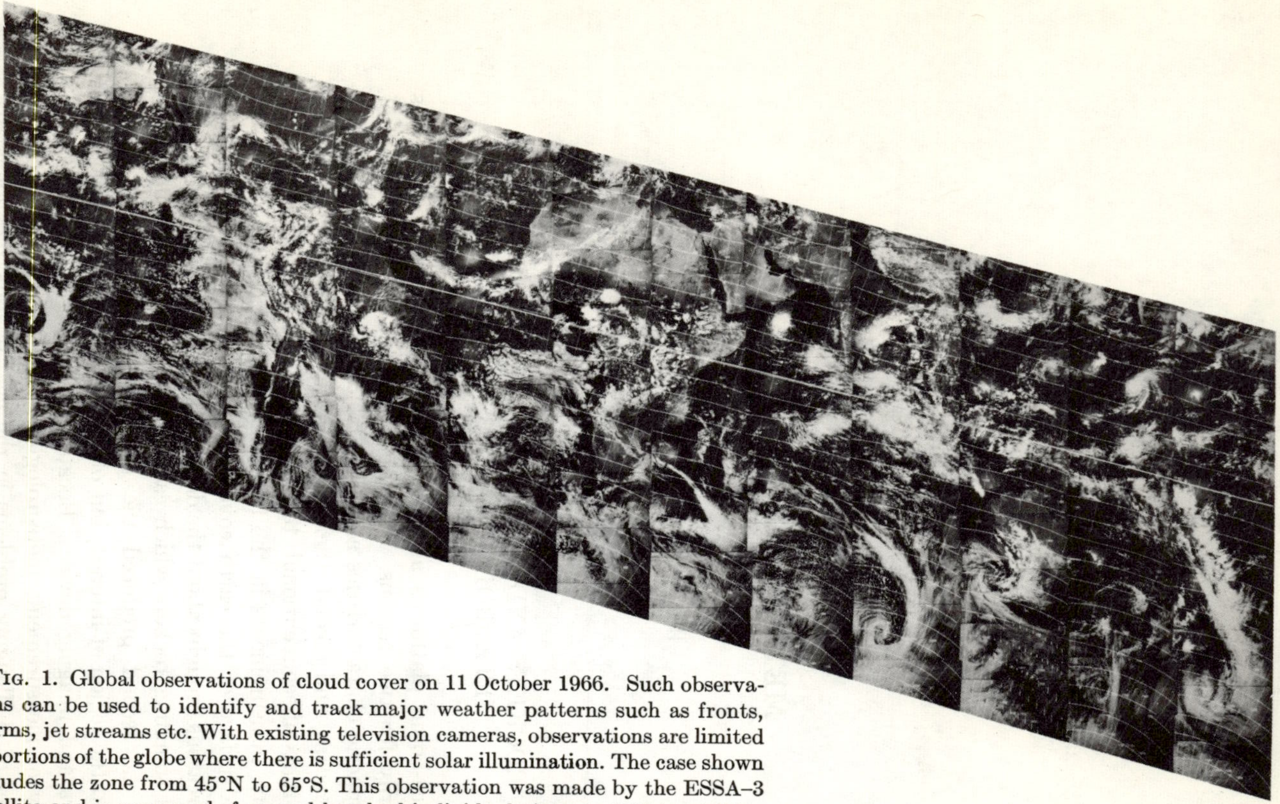

FIG. 1. Global observations of cloud cover on 11 October 1966. Such observations can be used to identify and track major weather patterns such as fronts, storms, jet streams etc. With existing television cameras, observations are limited to portions of the globe where there is sufficient solar illumination. The case shown includes the zone from 45°N to 65°S. This observation was made by the ESSA–3 satellite and is composed of several hundred individual pictures. ESSA satellites are operational and provide such cloud cover observations now routinely once every 24 hours.

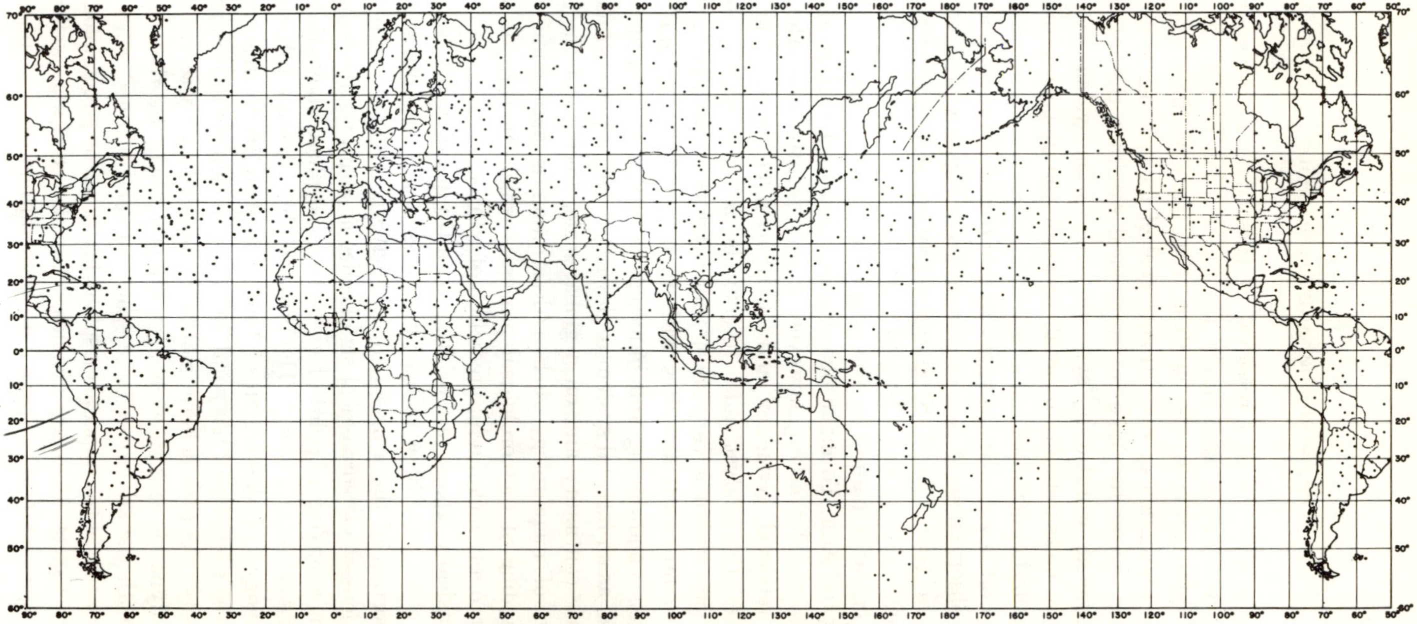

FIG. 2. Map of locations at which conventional, surface meteorological observations are being made routinely. Compare with satellite observations of cloud cover shown in Fig. 1. Each dot represents the location of an observation site.

then the acquisition and retrieval of these data by ground-based receiving complexes, and finally the distribution of the efficiently reduced information to the analyst. Execution of these steps will bring the realization of the ultimate goal of meteorological observations from space: to make meteorological satellites the nucleus of an operational, world-wide, synoptic observation system providing continuous measurements over all areas of the globe regardless of their population densities.

This point is well demonstrated by comparing Figures 1 and 2: the satellite observations of cloud cover shown in Figure 1 cover an entire global zone (45° N to 65° S) giving nearly equal weight to all areas while conventional, ground-based observations shown in Figure 2 are concentrated mainly in the populated areas of Europe, North America and Asia. To make observations from meteorological satellites operationally useful, however, it is imperative that the steps of data retrieval, reduction and distribution be executed in "real time", i.e., the analyst must receive the relevant content of the observations within a very short time after they have been taken. Thus, meteorological satellites with their associated apparatus for acquisition, conditioning and immediate analysis of data are being developed as an integral system to describe weather phenomena with systematic continuity and on a global scale. Because of the global character of meteorology such a description is paramount to an understanding of meteorological processes which, in turn, is a prerequisite to any meaningful prediction of weather.

Paradoxically, spacecraft orbiting at altitudes of many hundreds of kilometers have been primarily successful in observing the very lowest portion of the atmosphere and have not, so far, contributed observations of meteorological significance at altitudes above 30 km. From a meteorological point of view the atmosphere above this level is still of significant interest. The primary objective in establishing a "meteorology" of this region of the atmosphere, is to observe and analyze the spatial and temporal variations of temperature, pressure and density, the atmospheric motions generated by these variations and the dynamic energy exchanges brought about by the motions.

Also, with the advent of large spacecraft and manned spacecraft, the exploration of the upper atmosphere has taken on a technologically significant aspect. In many instances, launch vehicles as well as reentering spacecraft are expected to pass through critical phases of their flight in the 30–80 km region. In order to design appropriately for these conditions, engineers are keenly interested in obtaining the best possible knowledge of the anticipated flight environment. Accurate predictions of density and wind are of greatest importance for these purposes. Because of the altitude limit of about 30 km for radiosonde balloons and because of the inability of satellites at this time to "see" the atmosphere above 30 km, meteorological observations in this region must be carried out by large numbers of relatively simple and economical rocket-sondes. An increasing number of soundings are taking place systematically at various key locations over the entire world.

METEOROLOGY OF THE UPPER ATMOSPHERE

The variation of temperature with altitude is the most outstanding single parameter to reflect the physics of a particular segment of the atmosphere. Indeed, it is customary to categorize the atmosphere into various segments characterized by alternatingly increasing and decreasing temperature with altitude. Nearly isothermal layers separate these segments at the transition boundaries (Figure 3). The region nearest to the ground where temperature decreases rapidly with altitude is called the troposphere. Near 10 km there is a pronounced reversal in the temperature gradient. This transition layer is called the tropopause. The abruptness of this reversal as well as its altitude depend strongly on latitude and season. In certain cases multiple tropopause layers may appear. The region of increasing temperature with altitude, generally between 12 and 50 km, is referred to as the stratosphere. Then follows another region of generally decreasing temperatures between 50 and 80 km which is called the mesosphere. Again the altitudes of the transition regions near 50 km (stratopause) and 80 km (mesopause) as well as the magnitude of the negative temperature gradient (lapse rate) and its uniformity depend strongly on latitude and season. Above the mesopause temperatures increase extremely rapidly with increasing altitude. This region is called the thermosphere. Thorough mixing of all major atmospheric gases prevails up to altitudes of about 80 km. This mixing is mainly due to the rapid decrease of temperature with altitude in the troposphere and mesosphere. Because of this homo-

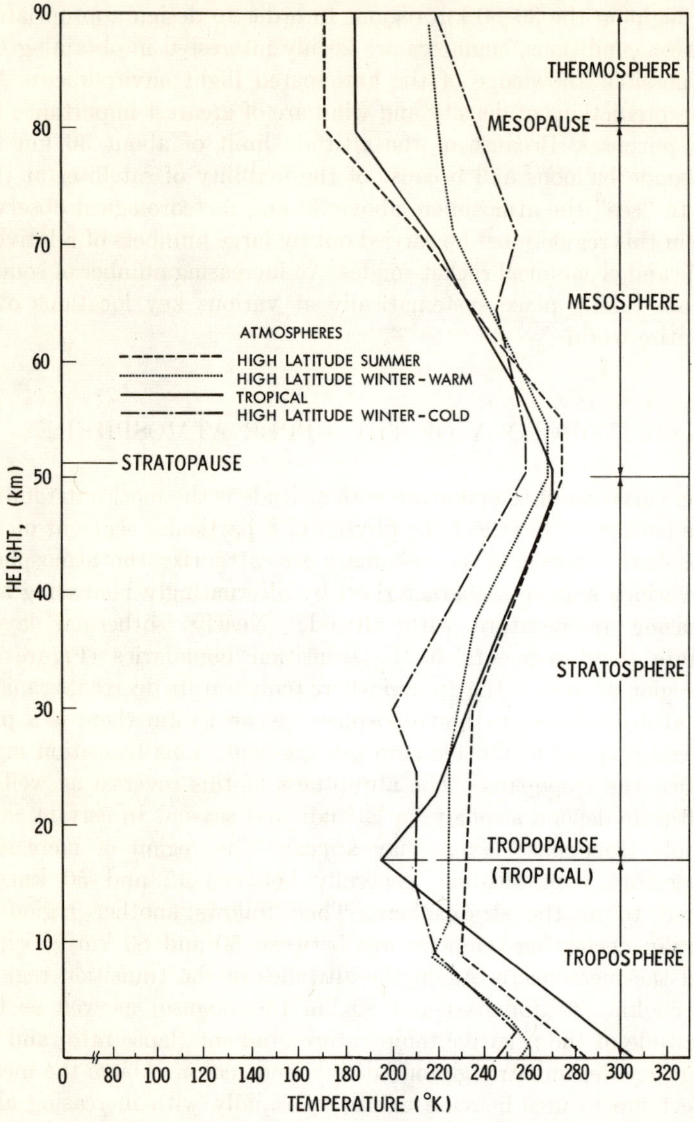

FIG. 3. Typical upper atmosphere temperature variations with height, latitude and season. Typical profiles were derived from a large number of rocket observations performed primarily over North America since 1950. A "U. S. Standard Atmosphere" has been based on these profiles.[1] Considerable variations from these typical profiles occur from day to day at all altitudes.

geneity in chemical composition the troposphere, stratosphere and mesosphere together are sometimes referred to as the "homosphere". The thermosphere and higher regions are also called "heterosphere" because, there, the heavier gases separate from the lighter ones by diffusion.

Many of the basic physical processes in the atmosphere are reflected in the temperature profile. The temperature maxima occur at levels of maximum absorption of solar energy: at the earth's surface where about 50 percent of the visible solar radiation impinging on the earth is absorbed, at about 50 km above the surface where ozone absorbs much of the solar ultraviolet radiation between 2000 and 3000 A (about 2 percent of the total solar energy) and at the top of the thermosphere where the far ultraviolet solar radiation is absorbed, mainly by oxygen. The actual temperature at each altitude level, below 80 km, reflects the balance between heating by absorbed radiation and cooling by emission of infrared radiation from the triatomic atmosphere constituents such as carbon dioxide, water vapor and ozone. Near the 12 km and 80 km levels very little absorption of solar radiation occurs, necessitating very low temperatures in order to produce equilibrium between solar heating and cooling by infrared emission which is proportional to the fourth power of the absolute temperature. At the levels of maximum absorption of solar radiation however, much higher temperatures are reached in order to produce sufficient cooling by infrared emission to maintain the balance. Thus, the atmospheric temperature profile is a convenient indicator of the type and amount of radiative energy exchange at each altitude level.

Furthermore the vertical temperature gradient also governs the stability of the atmosphere by determining the thermodynamic properties which control the buoyancy forces responsible for convection (the upward and downward motion of air parcels). For example, any air parcel which by chance is initially moving upward will expand and thus change its density. If this density change which takes place adiabatically is such that the air parcel still remains lighter than the surrounding air the parcel will continue to rise demonstrating instability. If the density change of the arbitrarily rising and adiabatically expanding air parcel causes it to be of equal weight or heavier than the surrounding air, the parcel will remain at its initial level, demonstrating stability.

Therefore, instability occurs when the temperature of the surrounding air decreases more rapidly with altitude than the temperature of the adiabatically expanding air parcel causing it to remain warmer; stability occurs when the air parcel remains colder than the surrounding air. Thus, that temperature-height gradient which matches the temperature decrease of an adiabatically expanding volume of air moving from a lower to a higher level marks the limit between stability and instability between the two levels. In dry air this adiabatic temperature decrease with height, called the adiabatic lapse rate, amounts to approximately 10°C per kilometer. In moist air, the adiabatic lapse rate may be as small as 6°C per kilometer. In the troposphere and mesosphere, the lapse rate often approaches instability which results in thorough mixing of the atmosphere in these regions. In the stratosphere where the lapse rate is negative (temperature increases with height) maximum stability results. Thus, the vertical temperature gradient is intimately related to vertical motions. These motions, in turn, bring about the convective exchange of energy and, in the troposphere, bring about such "weather" phenomena as the upward transport and condensation of water vapor.

One of the primary objectives of meteorological science is the pursuit of this complex cycle of exchange of potential and kinetic energy. Potential energy, available, to drive wind systems, at each height level, is generated by the variation of temperature with latitude and longitude due to variations in the balance of the absorption and re-emission of radiation and due to the inhomogeneous transfer of latent and sensible heat. The variations in temperature structure lead to pressure variations over the entire globe at each altitude level. A variety of wind patterns result from the pressure variations converting the radiatively stored potential energy into kinetic energy through dynamic processes. Since these winds do not necessarily flow exactly along latitudinal circles, but, are generally perturbed in a wavelike fashion they convert energy very effectively: They transport heat by advection (horizontal winds) and convection (vertical winds), cause phase changes such as the condensation of water vapor and transport angular momentum (meridional winds) which creates potential energy and closes the cycle. The horizontal and vertical motions resulting from this energy exchange on a global scale form what might be considered the backbone of meteorology, namely, the patterns of the general circulation. On a smaller scale, these circulation patterns are responsible for the generation and movements of cyclones and high pressure systems, the formation and dissipation of frontal zones, etc. In short, they are weather. Obviously, the vertical temperature profile at one given location is not sufficient to determine the global circulation patterns. But these patterns can be derived from the geographic variation of the temperature profiles and of the atmospheric pressure measured near the surface of the earth.

The driving force for the horizontal wind at any altitude level is provided by horizontal pressure gradients. The resulting wind vector, **w** is derived from the consideration that the force of friction and the force generated by the rotation of the earth (Coriolis force) balance exactly the force generated by the pressure gradients. Neglecting friction and considering only horizontal motions, the results of this derivation can be expressed as:

$$w_y = (2\rho\omega \sin \phi)^{-1}(\partial p/\partial x) \qquad -w_x = (2\rho\omega \sin \phi)^{-1}(\partial p/\partial y) \qquad (1)$$

where w_y and w_x are the components of the horizontal wind blowing to the north and east, respectively, p, ρ, and ω denote pressure, density and angular velocity of the earth's rotation, respectively. ϕ measures the latitude ($\phi = -\pi/2$ at the South Pole and $\phi = +\pi/2$ at the North Pole).

Equations (1), which are called the geostrophic wind equations, are a sufficiently good approximation for most cases outside the tropics.[2] It is obvious that neglecting friction in the tropics where the Coriolis force becomes negligible leaves no force to balance the pressure gradient force, causing the geostrophic wind equations to break down. Thus, for tropical regions, friction must be included in the equation which changes the resulting relationship between wind direction and the pressure gradients, but does not materially alter the concept that the winds are driven by horizontal pressure gradients.

In the troposphere, the above considerations have governed the thinking of meteorologists for many decades. But, because of the long recognized convective stability of the stratosphere, it was thought that in that region, processes were sufficiently simple and undisturbed so as not to justify a meteorology of the upper atmosphere. Most of the early rocket soundings exploring altitudes above 30 km were conducted at the White Sands Missile Range, New Mexico, and it was not until the onset of the International Geophysical

Year in 1957 that measurements at various geographic sites became available at these altitude regions. Results from these measurements as well as from very high altitude balloon soundings developed during the 1950's proved that meteorology does play an important role in the upper atmosphere. It was found that the temperature maximum at 50 km, induced by absorption of solar radiation, varies greatly with season and latitude. A very warm stratosphere is found during times and at locations of strong solar illumination (high latitude summer, Fig. 3) and a very cold stratosphere is found in winter at high latitudes. The fact that the solar radiation in the stratosphere is absorbed by a volume of air rather than by a solid surface has the consequence that the heating depends much more on the total time of illumination than on the solar zenith angle. A lower zenith angle simply moves the level of maximum heating to a higher altitude. This results in the somewhat paradoxical situation that the high latitude summer stratosphere which is under continuous solar illumination shows a higher temperature than the tropical stratosphere. In addition, the tropical tropopause and lower stratosphere are extremely cold, usually colder than during the high latitude winter, probably because of the infrared transmission properties of the large amounts of water vapor transported to higher altitudes in the tropics.

In the upper mesosphere, temperatures are about 100°K higher over the winter pole than over the summer pole. For example, very low temperatures at 80 km (below 140°K) have been observed in August at Pt. Barrow, Alaska (71°N), while temperatures of 230°K were measured at the same altitude in February 1963 in Northern Canada. Temperatures near the equator have been shown to undergo the smallest seasonal variation, as expected, for there is a minimum variation of solar illumination at the low latitudes. The reason for the unusually cold and warm high latitude mesosphere in summer and winter respectively is not yet fully explained. Infrared cooling rates cannot be as readily calculated for these altitudes as they have been for the tropopause level.[4] The major infrared emitter at these altitudes is carbon dioxide but absorption coefficients at the low pressures are not known well enough to permit calculations of cooling rates with sufficient accuracy. Calculations made several years ago[5] based on the best possible estimates of cooling by carbon dioxide and ozone did not predict the observed temperature extremes and variations. Chemical processes such as heating by recombination of atomic oxygen brought down from higher altitudes and dynamic processes such as transport of heat and angular momentum by meridional motion and adiabatic heating and cooling by vertical motion must be considered to explain the temperature distribution in the mesosphere.[6,7]

Up to 80 km large scale pressure patterns governing the general circulation can be readily derived from the observed geographic variations in the temperature profiles. The very low surface and stratosphere temperatures at high latitudes in winter cause a pronounced shrinking of the atmosphere in the vertical producing an intense low pressure system in the stratosphere and mesosphere centered over the general vicinity of the winter pole. Conversely, the high latitude summer atmosphere is greatly expanded upwards, resulting in high pressure centered over the summer pole throughout the stratosphere and mesosphere (Fig. 4a.) Although this high pressure system is of lesser intensity it is less variable and much more symmetrical with respect to the pole than the winter low pressure system. This symmetry is achieved when the pressure pattern is primarily a result of the solar heating which is symmetrical to the pole and if the atmosphere is sufficiently uniform so that infrared cooling also takes place uniformly.

A coarse picture of the circulation up to 80 km resulting from these pressure patterns can be derived from Eq. (1) and has been verified by wind observations. The winter low pressure system causes a strong cyclonic motion (west winds) around the winter pole, called the polar vortex. The intensity of this motion is considerably weaker in the tropics but the vortex has been observed at latitudes as low as 7° from the equator. Anticyclonic motion (easterly winds) of lesser intensity than the winter vortex occurs around the high pressure system of the summer hemisphere (Fig. 4b). Although the existence of these very general circulation patterns is now fairly well known in a climatological sense, there are still far too few observations to practice meteorology on a day-to-day basis above 30 km. Recent rocket soundings carried out over limited regions, mainly North America, and during limited time periods, have demonstrated that significant deviations from these purely climatological patterns exist. These deviations are manifested by disturbances on the theoretically circumpolar pressure patterns causing the winds to depart from their purely zonal character. The resulting meridional wind components

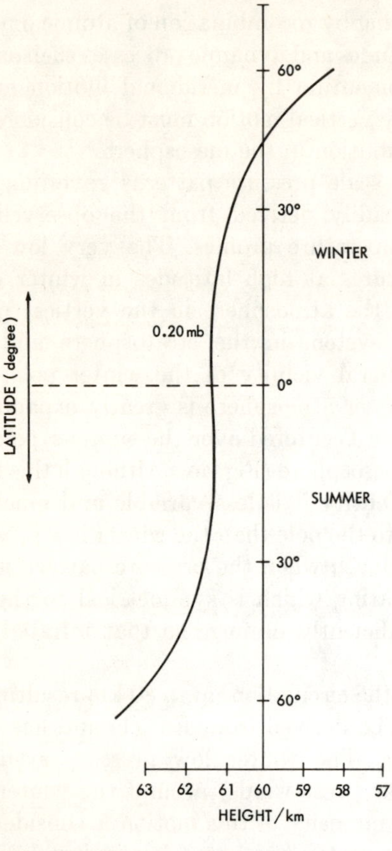

FIG. 4a. Typical variation of the height of the 0.20 millibar pressure level with latitude and season derived from rocket sondes. The 0.20 millibar pressure level is about 6 km higher near the summer pole than at the winter pole. This indicates low pressure centered over the winter pole and high pressure over the summer pole. Within 20 degrees on each side of the equator, the pressure level lies approximately midway between the summer and winter pole heights.

are a major factor in transporting thermal and kinetic energy across latitudinal circles, thus upsetting the radiationally established equilibrium. Typical examples of such events are found mostly in the winter hemisphere and especially in late winter. At such times the circular pressure pattern around the pole becomes strongly elongated and assumes elliptical shape with the center often considerably re-

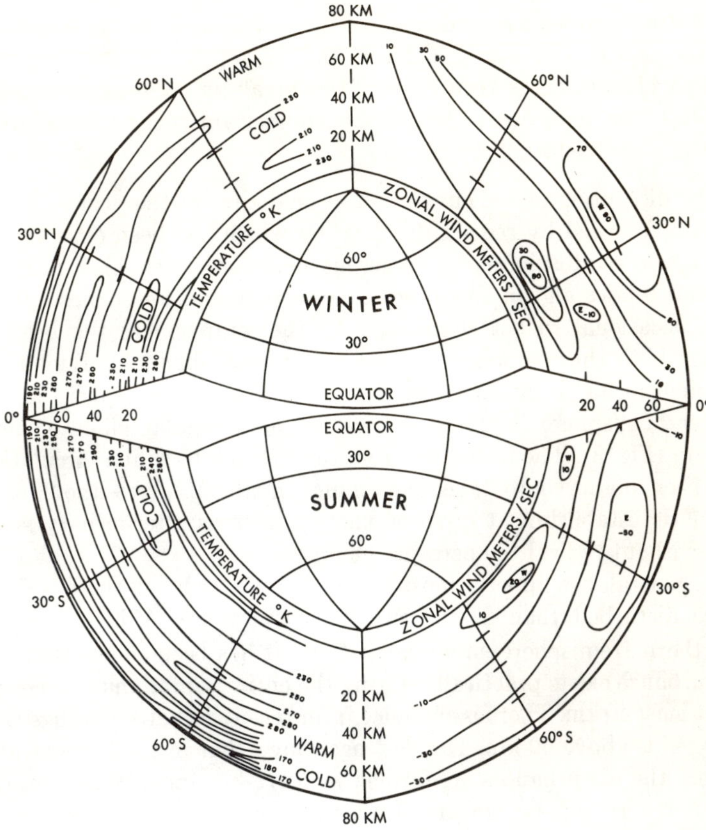

FIG. 4b. Typical distribution of temperature and zonal wind with height, season and latitude derived mainly from rocket soundings near 30°N, 40°N, 60°N, and 70°N over North America, 30°S over Australia, 13°N over the Pacific Ocean, 7°S over the Atlantic Ocean and 65°N over Sweden. W indicates winds from the west and E from the east.

moved from the pole, causing northerly flow along some meridians and southerly along others. Approximately once every year, usually in late January or in February (in the northern hemisphere) the polar vortex breaks up, temporarily, into a number of low pressure cells scattered around the winter hemisphere. Rocket soundings in Northern Canada in January 1958 showed that the disruption of the vortex during such a temporary break-up took place throughout the entire stratosphere and reached up to 70 km in the mesosphere. At the end of the winter (March and September in northern and southern hemispheres, respectively) this break-up becomes permanent and lasts for many weeks before the summer anticyclonic circulation is firmly established.

The disruption of the circulation at that time of the year in many cases is followed by remarkable heating which has been observed in the lower stratosphere on numerous occasions with high altitude balloon sondes and in the upper stratosphere and lower mesosphere with occasional rocket soundings. In the stratosphere the heating during both the temporary or permanent break-ups may amount to as much as 45°K within a few days. This warm stratosphere in the winter hemisphere has been recognized as a regular climatological feature (Fig. 3). The transition from the summer to winter circulation has been observed to take place much more rapidly without such dramatic establishment of secondary systems as intermediary steps and without radical temperature changes.

In general, the upper atmosphere in the winter hemisphere is much more disturbed than in the summer hemisphere. A high pressure Northern Hemisphere circumpolar flow. It has been shown that this disturbance exists practically during the entire winter and reaches up to at least 40 km. There is evidence from recent rocket soundings that it extends above 60 km. Teweles[8] has shown that in the lower stratosphere the disturbances are formed as wave-like indentations on the circular, circumpolar isobars. Under certain conditions these waves amplify, causing the break-up into separate cyclones and anticyclones which, though larger in scale, propagate similarly to storm systems in the troposphere. It could well be that the disturbances find their origin in the low latitude regions of the winter hemisphere since the contrast in heating between low latitudes and high latitudes is much greater in the winter hemisphere than in the summer hemisphere. Conditions in the stratosphere and mesosphere at certain times may permit transfer of energy from low to high latitudes in winter, which results in these major disturbances.

Thus, it is evident that the huge heat engine which constitutes the meteorology of our lower atmosphere is still at work at the higher altitudes. In the mesosphere the scale of motions and their annual cycle is naturally somewhat different than in the troposphere, but the meteorological considerations of converting and transporting energy by zonal and meridional motions and by convection are still very much in evidence. The only meteorological phenomenon absent at the high altitudes is the evaporation of water and precipitation by formation of clouds, because upward transport of water vapor is generally unlikely above the tropopause due to the very great vertical stability of the stratosphere. Nevertheless, part of the role of water vapor as an infrared radiator may be taken over by ozone in the stratosphere and mesosphere. Regarding the absorption and storage of solar energy, ozone serves a similar function as the earth's surface. Even clouds have been observed in the mesosphere; these are noctilucent clouds occurring at high latitudes in summer, at the mesopause where temperatures are extremely low. They are probably formed by the condensation of the small traces of water vapor or other volatiles present at these altitudes with dust particles serving as condensation nuclei. The meteorological significance of these clouds is not so much due to the condensation energy involved or their influence on radiative properties, but rather due to the interest in the mechanism responsible for transporting dust particles and condensing vapors to this selected level. Interest is also due to the concern about the relationship between the formation of the clouds and the very low temperatures at these altitudes.

Thus, all facets of meteorology do indeed exist up to 80 km. Up to 60 km many thousands of wind soundings so far have been made. Up to 80 km in the order of 100 temperature and wind soundings were conducted during the past decade. Between 90 and 200 km very few temperature soundings have been made. Most soundings in that region measured winds during twilight by the sodium vapor trail method. Analyses of these sodium trails have indicated that between 80 and 120 km there is a region of extreme wind shear where very strong winds from all directions rapidly alternate with relatively calm regions. Vertical wind speed gradients of 40–50 m/sec/km are not uncommon in this region. These winds do not follow the seasonal and latitudinal pattern found at lower altitudes. Above 120 km greater uniformity seems to return, but samples at these altitudes are too few to derive any definite circulation patterns.

At about 80 km there seems to be a sharp and remarkable boundary separating the circulation below 80 km from the circulation in the regions where ionization of the atmosphere and dissociation of oxygen sets in. This boundary seems to suggest that the physical causes which sustain the motions of the atmosphere are quite different in the two regions. Such an abrupt change in the wind structure may very well be due to the increasing importance of tidal motions and may possibly be due to the fact the atmospheric structure as a whole undergoes basic and major changes at the 80 km level. Some evidence of wave motions with periods of several hours propagating vertically through this region have been observed. There is also strong evidence of periodic variations of wind speeds and direction from indirect observations of the drift of meteor trails. The periodic variations are very strong and have the typical characteristics of tides. Periods of 24 hours, 12 hours and 8 hours have been observed. These short-term variations are so strong that they obscure much of any seasonal or day-to-day variations of large scale meteorological features known to exist below 80 km.[9] Thus a meteorology of the region above 80 km is not yet ready to be developed.

Perhaps the most challenging question in upper air meteorology is to what degree there is interaction between various altitude regions of the atmosphere. The pressure difference between any two given altitudes is determined by the entire temperature profile between these altitudes; thus, for example, the magnitude of the temperature maximum at the stratopause has an influence on the pressure levels as high up as 80 km. There is also the possibility that energy could be propagated over long vertical distances linking physical events at very high altitudes to the meteorology of the lower atmosphere. Of particular interest here are solar-terrestrial relationships as they might apply to the lower atmosphere. At high altitudes (200–300 km) direct solar influences have been established to the extent that temperature, pressure and density of that region follow very closely the short-term variations in the short wave solar ultraviolet radiation absorbed at these altitudes. Since this radiation does not penetrate to lower levels and since the longer wave radiation absorbed at these lower levels is rather independent of the active regions on the sun, the lower atmosphere will not directly reflect this solar activity. If the solar controlled events occurring in the thermosphere could be propagated to the tropopause dynamically through the mesosphere and stratosphere a link between solar activity and weather could be established. Although some connections between the troposphere, and thermosphere have been demonstrated statistically, no physical evidence for any connection has been found as yet. In fact, it is more likely that all the statistical relationships can be explained on the basis of events taking place near the surface propagating to higher altitudes. For example, it is easier to conceive of a physical mechanism which explains the winter-time high pressure disturbance found to exist in the stratosphere over the North Pacific as being caused by the peculiar distribution of the surface temperatures of water and land and by the cooling rates resulting from this distribution, rather than to speculate that this high pressure regime exerts an influence on the weather at the lower levels. In fact, no such influence has been found.

SENSORS FOR METEOROLOGICAL OBSERVATIONS FROM SPACE

Meteorological Soundings

The purpose of all meteorological soundings is to perform measurements to fully describe and understand the physical state of the atmosphere and to carry out a sufficient number of observations, both in space and time, in order to make predictions regarding the evolution of the atmospheric state, at least within the region where the observations have been carried out. The atmospheric parameters whose variability with latitude, longitude, altitude and time must be measured are: pressure, density, temperature, wind and composition. They are sometimes referred to as atmospheric structure parameters. From a technological point of view three different atmospheric regions can be distinguished.

Up to 30 *km*

In the troposphere and lower stratosphere (Fig. 3) observations are made routinely by meteorological agencies all over the world with radiosonde balloons carrying inexpensive sensors to measure temperature, pressure and humidity. Wind speed and direction are determined by tracking the balloon-borne radio transmitter with a direction-finding ground receiver. It is assumed that the balloon drifts perfectly with the wind. The only compositional parameter of major meteorological importance in the lower atmosphere is the amount of water vapor since considerable heat can be released or consumed by phase changes of water. Standard radiosondes measure water vapor content up to about 8 km by means of a hygroscopic material whose electrical resistance is a function of moisture. The temperature measurement is accomplished by means of a thermistor, whose electrical resistance changes with temperature. The thermistor element is shielded against solar radiation and it is assumed to be in thermal equilibrium with the surrounding air. The balloon-sonde also measures pressure by means of an aneroid barometer. Pressure, temperature, and humidity measurements are telemetered to the ground. The temperature and pressure measurements suffice to obtain a complete description of the variation of not only temperature and pressure but also of density with altitude. This is accomplished by means of a very fundamental and widely used relationship governing the state of the atmosphere and derived from the ideal gas and hydrostatic laws. It relates the "static" structure parameters, pressure (p), temperature (T), density (ρ), and molecular weight (M) to altitude (h):

$$-dp/dh = p(Mg/RT) = \rho g, \tag{2}$$

where R is the universal gas constant. In the case of the balloon radiosonde, M is assumed to be known and constant with altitude; g, the gravitational acceleration, is a well-known function of altitude, and T and p are measured as function of time; therefore, by integrating Equation (2), p, ρ, and T can be expressed as functions of altitude. With h known as a function of time and the azimuth and elevation angles determined by the tracking receiver, the balloon trajectory and its time derivative can be computed. This gives the required wind velocity and direction as a function of the altitude.

Up to about 60 *km*

The same principles used on radiosonde balloons to measure temperature and winds have been adapted to small and simple meteorological rockets capable of performing these measurements throughout the stratosphere extending through the lower boundary of the mesosphere, i.e., up to 60 km. These rockets generally carry "payloads" of less than 12 cm in diameter, weighing about 1–5 kg. They are relatively inexpensive and the entire launch operation can be performed with a minimum of complexity, generally requiring only a simple, easily transportable launching tube. Their simplicity has made them useful in carrying out synoptic soundings as part of a Meteorological Rocket Network performing temperature and wind soundings at least once per week from about half a dozen launch sites in North America. The network is gradually expanding with increasing participation of sites in other parts of the world.[10]

A variety of rockets have been introduced to carry the simple sensors and in general, they can be classified into two categories: a) the "Loki" type system is a rapidly burning rocket motor which launches an instrumented projectile (dart) carrying the sensors to the desired altitude on a ballistic trajectory after the rocket motor has dropped off; b) a system based on the "Arcas" rocket has a slowly burning motor from which the sensors (payload) separate when the entire rocket reaches apogee. The sensors perform their measurements while descending by parachute. The metallized parachute drifts with the horizontal wind as it descends and is tracked by radar. The thermistor temperature sensor is suspended from the parachute and the temperature is telemetered to the same type of receiver as required for the balloon radiosonde. These wind and temperature sensing techniques are generally limited to altitudes below 60 km. Descent rates of the parachute are usually too rapid at higher altitudes to permit satisfactory drifting of the parachute with the wind. The temperature sensor becomes ineffective above 60 km because the size of the thermistor sensor (a few microns) becomes comparable to the mean free path between collisions of air molecules, resulting in insufficient exchange of thermal energy between the air and the sensor. Thermistors must be as small as possible to minimize thermal mass, thus providing a short time constant which permits the rapid adjustment of the sensor to the temperature of the medium. There are other types of sensors used on small meteorological rockets but most of them are subjected to similar altitude limitations.

The direct measurement of pressure or density is much more complex from on-board a rocket than from a balloon. This is because the rapidly moving rocket probe, usually at several times the speed of sound, perturbs the surrounding atmosphere aerodynamically to such an extent that the behavior of the vehicle (its speed and angle of attack) and the configuration of the sensor must be known and/or controlled to relate a pressure measurement in the vicinity of the sensor to the ambient atmosphere. For this reason such techniques have been limited to larger sounding rockets except for one especially successful technique for deriving atmospheric density by measuring the drag on a lightweight, inflatable, metallized, falling sphere. The sphere can be tracked by radar after ejection from the rocket. This method has been adapted to small sounding rockets but techniques for measuring pressure directly with gauges mounted on the rocket still are of such complexity that they require larger and more expensive rockets.

Up to 200 km

To carry sensors through the mesosphere and into the lower thermosphere rocket vehicles of generally 10 times greater weight (and cost) are used. They possess sufficient thrust to carry relatively large payloads (30–40 kg, 15–20 cm diameters) to altitudes of 100–200 km. Their cost permits only "semi-synoptic" programs at a number of sites with launchings at periodic intervals. In the United States, two stage rockets, using the "Nike" booster as a first stage, are most common. The "Cajun" rocket, which has been most commonly used as a second stage since the International Geophysical Year (IGY), has now been succeeded by the more powerful "Apache."

In this altitude region, pressure and density are more susceptible to measurement by *in situ* sensors than temperature. The simplest method in concept is the measurement of density by means of a falling sphere. This technique was originally developed at the University of Michigan in 1952, and a large number of successful flights have been carried out since then.[11]

The drag force (D) exerted on a perfect sphere dropped from a rocket at high altitude is measured and ambient density (ρ) is derived by means of the relationship:

$$mg - D = m\ddot{h} = mg - \tfrac{1}{2}\rho\dot{h}^2 C_D A, \tag{3}$$

where h is the altitude of the sphere, A its cross sectional area and C_D its aerodynamic drag coefficient which is a function of Mach number which depends on $\dot{h}$ and h and must be determined empirically; m is the mass of the sphere and $\dot{h}$ indicates the time derivative of h; g is the acceleration of gravity and depends on h.

Falling spheres have been used in many different combinations; as rigid shells or as inflatable balloons; with built-in accelerometers (active sphere), to measure $\ddot{h}$ directly, or with high precision radar tracking (passive sphere) to determine the drag acceleration from the second derivative of the altitude vs. time function given by radar track. Differentiation of the tracking data or integration of the measured accelerations, in case of the active sphere, leads to the velocity, $\dot{h}$, required to solve Equation (3) for ρ.

Spheres are ejected from the rocket at altitudes of about 60 km and travel along ballistic trajectories up to about 140 km. Acceleration is determined both along the upward and downward leg of the trajectory. Winds can be measured with inflatable spheres at altitudes below 50 km from the horizontal drift determined by radar. Density measurements with large (1 to 2 meters) passive inflatable spheres have been obtained up to 120 km.

A variety of techniques exist to perform rocket-borne *in situ* pressure measurements. These techniques are always based on the premise that ambient conditions of pressure, density or temperature can be derived from a variety of direct pressure measurements carried out on board the vehicle by applying aerodynamic theories which relate the measured pressure to the ambient parameters.

In general, the pressure measurement is performed at a minimum of two aerodynamically different locations on the rocket nose or along the surface of a pitot-static tube carried at the tip of the rocket.[12] Pressure sensors are usually placed in chambers which are exposed to the air flow by means of orifices in the skin of the rocket or the pitot tube. In one chamber the stagnation pressure (P_i) at the tip of the probe is measured, and in the second chamber, a measurement of static pressure (P_s) along the wall of the same probe is made several calibers to the rear of the stagnation point. Aerodynamic tests of this configuration show that the pressure measured at the latter point is equal to ambient pressure. Additional chambers may be provided for redundancy. As a pressure sensor each chamber consists of a radioactive ionization source and a multi-range electrometer which measures the pressure sensitive ion current. The current, calibrated as a function of chamber pressure and susceptible to measurement in the altitude range of about 40–120 km, is telemetered to the ground.

Ambient density ρ is derived from the interpretation of the basic pitot static tube equation (Rayleigh Equation) and from the equation of state:

$$\rho \propto P_i/V^2. \tag{4}$$

This proportionality holds essentially over a range of Mach numbers between 3.5 and 7.5; V being the tangential velocity of the rocket which must be determined by accurate tracking.

Equation (4), relating impact pressure to ambient density, and the measurement of ambient pressure directly at the side of the pitot tube, are valid only in the region of continuous flow where the mean free path between air molecule collisions is small compared to the dimensions of the sensor. At higher altitudes, in the free molecular flow region, where the mean free path becomes larger than the sensor dimensions, these methods break down. The transition from continuous to free molecular flow occurs near the upper boundary of the mesosphere at about 90 km. Above that altitude equation (4) is replaced by the following expression, which relates the ambient density ρ to the impact pressure P_i in the region of free molecular flow:

$$\rho = P_i/KV \cos \alpha, \tag{5}$$

where α is the angle of attack of the rocket and K is a function of the temperature in the chamber and of the molecular mass. With present techniques this type of density measurement can be performed up to 120 km.

Since the measurement of temperature, which can be performed directly in the stratosphere by means of thermistors, becomes questionable at altitudes above 60 km, temperature in the mesosphere must be measured by more complex, indirect methods. The method most commonly used is the "Grenade" technique.

Charges of high explosives, each weighing up to 1.5 kg, are ejected and exploded at altitudes up to 90 km at regular intervals during the ascent of the rocket. Average temperatures and winds of the atmosphere between two explosions are determined by measuring exactly the time of explosion of each grenade, the time of arrival of each sound wave at various ground-based microphones, and the exact position of each grenade explosion. Thus, the speed of sound in the layer between two explosions is measured. Temperatures can be derived since the speed of sound is proportional to the square root of the temperature and molecular weight which is assumed to be known. Wind speed and direction can be derived from the horizontal drift of the sound in the layer.[13]

The highest altitude from which sound returns can be consistently received with present explosive charges and existing sound ranging techniques is about 90 km. Temperatures and winds can be measured simultaneously which is of great importance in describing the dynamics of the atmosphere.

Complete profiles of density (ρ), pressure (p) and temperature (T) can be derived from each one of the three experiments described above, although the primarily determined parameter is density in the sphere experiment, pressure in the pitot tube experiment and temperature in the grenade experiment. The three parameters are related to each other by Equation (2).

The measurement of winds at high altitudes by tracking the drift of trails from smoke or other visible substances released by rockets has been pursued for many years. This method requires continuous tracking and triangulating of the trail in order to establish its space-time coordinates the derivative of which will yield the wind vector for the air mass in which the trail is suspended. To obtain sufficient accuracy, the base lines for this triangulation must be in the order of 50 or more kilometers if optical tracking is used. Photographs of such trails show very interesting detail even of the smaller scale wind shears and turbulence at various altitude levels (Fig. 5). Most commonly, sodium vapor is injected into the atmosphere between 70–200 km during twilight conditions. Interaction of sunlight with the sodium vapor (resonance radiation) causes the sodium trails to be luminous, and with the proper ejection technique the sodium remains visible for periods up to 30 minutes at middle latitudes. The trail moves instantly with the atmospheric wind field and its track is recorded by an array of cinetheodolite cameras. This experiment functions only in the upper mesosphere and in the thermosphere because the

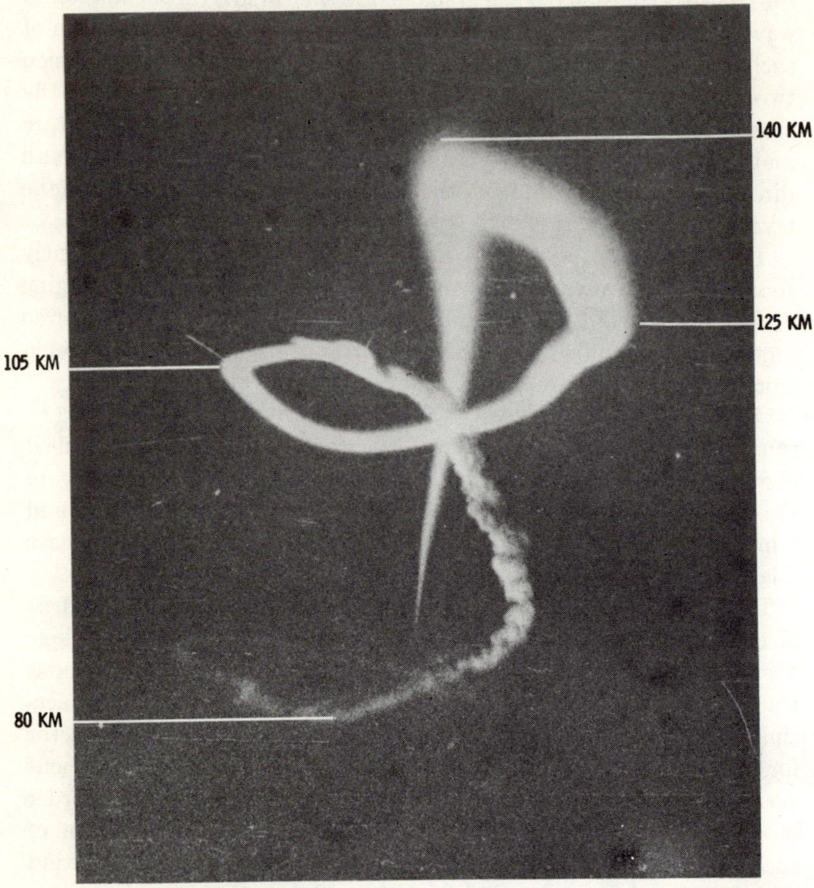

FIG. 5. Photograph of luminescent trail of sodium vapor released by a Nike-Cajun sounding rocket over Wallops Island, Virginia, between the altitudes of 80 and 140 km. Motions and distortions of the trail describe wind and turbulence patterns in that region of the atmosphere. (Reproduced from "On the Question of Turbulence in the Upper Atmosphere," Owen Cote, Geophysics Corporation of America, Technical Report 62–12–N, Bedford, Massachusetts.)

resonance glow cannot be induced below 60–70 km owing to the absorption of incident sunlight by natural sodium in the atmosphere. It has produced valuable information on the tremendous wind shears in the transition region between the mesosphere and thermosphere.

Metallized chaff, tracked by radar, has been used to measure winds below 75 km. Above that level it is very difficult to measure wind by methods of dispersing solid materials such as chaff because the fall velocity of the chaff approaches free fall very rapidly with increasing altitude. Also, the chaff, ejected from the rocket near apogee, disperses rather rapidly (in many cases within about 20 km), as it falls through the atmosphere.

Satellite Sensors

Satellite-borne instruments for the observation of meteorological phenomena are based entirely on the sensing of electromagnetic radiation received from the earth's surface or from the lower atmosphere.[14] At the present time these instruments are: television cameras, radiometers (photometers) and spectrometers. Long wave radiation (infrared, millimeter and centimeter waves) is emitted into space by the earth and atmosphere, and its intensity is primarily a function of the temperatures of the emitting surfaces or gases but is also influenced by such factors as the morphology of a surface or the concentration and distribution of emitting gases along the optical path. Short wave (near ultraviolet, visible and near infrared) solar radiation is scattered back into space by the earth and atmosphere; its intensity compared to the incident radiation, depends primarily on the nature and orientation of the scattering surfaces and to some extent on the size and concentration of the atmospheric scattering constituents. All measurements depend strongly on the geometry between the position of the satellite sensor, the area viewed on the earth, the local vertical and, in the case of solar radiation, the position of the sun.

The dependence of the radiation received by the satellite sensors on such atmospheric characteristics as temperature, composition and particle size, and on the temperature and nature of the surfaces beneath the atmosphere is used to derive meteorologically significant information. Mathematical relations between these parameters and the sensed radiation are extremely complex, and cannot be stated explicity for all cases. For example, it is not exactly known at this time how the backscattered solar radiation measured at the satellite depends on the size and distribution of the scattering particles and on the scattering geometry or how each atmospheric parameter contributes precisely to the long wave emitted radiation. The interpretation of radiation measurements from present-day meteorological satellites is therefore still crude and requires a number of simplifying assumptions. Nevertheless, even such very crude simplifications as the assumptions of isotropic scattering and pure blackbody emission by all surfaces and gases has led to the global mapping of cloud cover and cloud top heights in day and nightime, the global distribution of the balance between emitted terrestrial and absorbed solar energy, the determination of water vapor content in the upper troposphere, and the global mapping of stratospheric temperature patterns. To accomplish this meteorological satellites have carried radiometers operating, respectively, in the 4 and 11 micron atmospheric "windows", the broad band range of terrestrial (7–30 microns) and solar (0.2–5 microns) radiation, the 6.3 micron water vapor absorption band, and in the 15 micron carbon dioxide band.

The choice of these intervals, called channels, and their widths depend on the atmospheric parameter which is to be derived from measurement and on the sensitivity of the radiometer. On TIROS the 11 micron window channel was 5 microns wide while the water vapor and carbon dioxide channels were less than one micron wide. Relatively high resolutions of intensity, about 100 levels or better, were typical for TIROS radiometers with spatial resolutions of 30 to 50 kilometers on the earth's surface. In the NIMBUS I satellite the spatial resolution of the 4 micron window radiometer was about 6 km with a spectral band width of about 0.8 microns. Despite the very high relative intensity resolution of radiometric measurements, the absolute accuracies of radiation intensities measured by TIROS radiometers over long periods of time (several months) were still unsatisfactory because the calibration of the sensors did not remain stable for such long periods in space. Continuous inflight calibration which was accomplished on NIMBUS is therefore very important.

More accurate atmospheric derivations can be made from spectrometer measurements which provide not only intensity resolutions similar to radiometric measurements but, also, very high spectral resolutions over a wide range of emitted and reflected radiation. Spectrometers with which the vertical structure of atmospheric parameters can be measured are more complex than radiometers. They have not yet been flown on meteorological satellites, but their use is contemplated on future spacecraft with spectral resolutions of one or two orders of magnitude better than those stated for radiometers. However, the spatial resolution will probably be not better than 100–200 km on the earth's surface.

Imaging with television cameras gives the best possible spatial resolution (fractions of one kilometer on the earth's surface), but permits only poor relative intensity resolution (about 10 levels) with practically no capability of absolute intensity measurements. Also, at this time, TV imaging can be used only for scattered solar radiation in the limited spectral region from 0.5 to 0.8 microns and is not suited for observations of emitted radiation which would especially be desirable for nighttime mapping of cloud cover. The same high spatial resolution, but better photometric accuracy can be achieved with the use of scanning photometers in the visible part of the spectrum. Such photometers produce "images" by sequentially scanning along a line and by repeating each subsequent scan contiguous to the previous one. Such scanning photometers always use scanning mirrors or similar rotating parts and are mechanically somewhat more complicated than TV cameras. They possess, however, the best potential as nighttime mapping devices operating in the 10–11 micron atmospheric "window". Flown on Nimbus satellites, spatial resolutions of about 5 km seem quite feasible within the next few years.

Aside from the physical significance of the measurements, there are two considerations of technological nature which are of eminent importance especially in an operational meteorological satellite. The spacecraft's orbit and attitude must be chosen to provide optimum data coverage and efficient transmission of the data accumulated during the orbit. These criteria, as well as automatic processing and conditioning of the information within a few hours, must be assured to enable interpretation by the user while the data are still valid. Meteorological satellites to date are not yet capable of fulfilling these considerations completely.

Early meteorological satellites, such as TIROS, were stabilized in space by spinning in the range 9–12 rpm. When the spin decayed in the earth magnetic field to about 9 rpm, a pair of small peripheral rockets were fired on command from the ground, spinning the satellite up to 12 rpm again. The TIROS orbit was nominally circular, 740 km above the earth with a period of about 100 minutes. Its motion was "direct" (in the same sense as the earth's rotation) and its inclination to the equatorial plane ranged from 40° for earlier TIROS to about 58° for more recent launches. A total of 64 pictures taken at 30 sec intervals were stored in the TIROS spacecraft between interrogations. The 64 picture sequence covered an area of approximately 13 by 120 degrees of great circle arc. Thus, television coverage from TIROS was not only limited to the daylight portion of the orbit, but also restricted by the number of photographs which could be stored in each orbit. Coverage by radiation measurements, which were obtained and stored continually over the orbit was limited by the inclination of the orbit to the global zone from about 58°N to 58°S.

The "TIROS Operational Satellite" (TOS) system became fully operational during 1966. In this system, five spacecraft, named ESSA have been launched by mid-1967. They are spin-stabilized in the same manner as TIROS and utilize the same basic spacecraft design. Some aspects are, however, greatly improved over TIROS. The orbit is more nearly polar and the altitude is increased to about 1300 km. The spacecraft spin axis is perpendicular to the orbital plane instead of within the orbital plane as on TIROS. Thus, it rolls along its orbital path in a cartwheel fashion with cameras viewing out of the cartwheel's rim. These improvements extend coverage to the entire sunlit portion of the earth.

Many motions of satellite, orbit, earth, and sun combine to create major problems in data reduction and frustrations in programming the satellite. Due to the earth's equatorial bulge, the orbital nodes of a 58° orbit regress to the west at the rate of about 4.6°/day relative to the earth-sun line. This motion, in absence of other forces, would cause the camera axis of a solely spin-stabilized TIROS to look first continuously at the sunlit, then continuously at the dark portions of the earth, alternating with a period of about 40 days. Fortunately, a torque is exerted on the spin axis by the interaction between the magnetic fields of the spacecraft and the earth. This torque, in some measure, counteracts the changes of attitude caused by the precession of the orbit and by creating controlled electric currents in the spacecraft, it is possible to influence its magnetic field. Thus, it is possible in some measure to steer the spin axis in space in order to effect the most favorable geometry for picture taking and to prevent the direct rays of the sun from shining on the scanning radiometer sensors except for occasional short periods of time. In addition, the attitude of the spacecraft relative to the sun has an important bearing on the spacecraft's power and thermal budget because of the solar illumination of solar cells and heat absorbing surfaces, respectively.The primary method for displaying cloud observation for operational use is by a "nephanalysis" chart showing graphical representations of areas of similar cloud amount and type. Initially, the nephanalyses were prepared manually by a team of metorologists at each data acquisition station utilizing perspective latitude-longitude grids produced in advance by a simple, onsite digital computer-plotter system. The grids were overlayed on the TV photos within minutes following a satellite interrogation. The nephanalyses were transmitted via fascimile from the acquisition station to the users. Because of the complicated scan geometry of the radiation sensors on TIROS, radiation measurements were processed in a very complex fashion by large digital computers. Such data could not therefore, be used for immediate operation.

Present operational meteorological satellites, operated in the U.S.A. by the Environmental Sciences Service Administration, channel all operationally useful measurements immediately to an automatic computing and analysis center from where the geographically referenced and perspectively rectified information is distributed worldwide. A limited amount of directly interpretable data, such as cloud pictures, are transmitted from the satellite directly by an "Automatic Picture Transmission" (APT) system to the meteorologist in the field who receives these data via a simple, inexpensive ground station. "High Resolution Infrared" observations of nighttime cloud cover to be obtained from these operational spacecraft in the near future will be displayed in the form of pictures produced by a photo facsimile process on the ground within minutes of the readout of the radiation data either stored in the spacecraft or transmitted directly via the APT system. This process will be patterned after similar nighttime cloud observations made experimentally from NIMBUS satellites.

More advanced meteorological satellites such as NIMBUS incorporate active attitude control so that the sensor axes are always oriented toward the earth and solar cells are oriented toward the sun. Also, as in the TOS, NIMBUS orbits are nearly polar (about eight degrees inclination of the orbital plane with the earth's axis) and provide complete coverage at the same local time every day and at each location. The latter requires that the precession of the orbit follows exactly the motion of the right ascension of the sun so that, if at launch the orbital ascending node occurs at local noon, the ascending node of every subsequent orbit throughout the lifetime of the satellite will also occur at local noon. The satellite then is in "sun-synchronous" orbit and passes over almost every location on earth at local noon and midnight.

Meteorological satellites orbiting at relatively low altitudes (1000 km) have one major disadvantage, even if they provide full global, and regular coverage such as the "sun-synchronous" TOS and NIMBUS. The spacecraft generally passes over the same region only once every 12 hours. This is the case especially in tropical and low latitude regions. Such frequency of passes will generally suffice for large scale measurements of atmospheric parameters sure as temperature, moisture etc. However, observations of cloud patterns, which can be carried out so well with satellites, should ideally be made continuously especially over areas where short lived mesoscale systems (tropical storms, tornados, thermally driven convective cells etc.) dominate the weather. For that purpose the satellite must be in a "geo-synchronous" orbit, i.e. the angular velocity of the spacecraft around the equator must be equal to that of the earth around its axis; namely one revolution in 24 hours. Thus, relative to the earth's surface, the spacecraft will remain stationary over whichever point along the equator it is inserted into that orbit. A geo-synchronous orbit requires a distance of about 36,000 km from the earth's surface. From that distance a major portion of the globe can be viewed continuously and permanently. Cloud observations from a geosynchronous orbit were first carried out with the ATS-1 satellite, launched in December 1966 over the Pacific ocean stationary at approximately 150°W. The view from ATS-1 encompasses the entire Pacific Ocean and cloud systems with sufficient detail and are resolved between the east coast of Asia and the west coasts of the Americas from the tropics to moderately high latitudes in both Hemispheres (Fig. 6). The excessive distance from the earth poses some problems in the choice of sensors for a spacecraft; the energy gathering power of the sensors must be enormous to resolve a small

FIG. 6. Cloud observations obtained with a scanning phototelescope on the ATS–1 satellite in geosynchronous orbit. Reflected sunlight from the Earth and clouds is sensed by a phototube scanning from horizon to horizon along straight lines from West to East. After each scan the direction of viewing is tilted southward by the width of one scan line (about 2 km). Thus, after about 20 minutes the entire disc of the earth, visible from the satellite, is imaged. Repeating the scanning and tilting sequence, pictures of cloud formations over the entire disc are obtained in 20 minute intervals. This permits the analysis of the growth and movement of meteorological systems almost continuously during daytime. Such observations are extremely important in the study of energy transfer in the tropics and of severe storms. They are possible on this scale only from geosynchronous orbits. This view ranges from about 60°N (top) to about 50°S (bottom) and from Indonesia (left) to Central America (right). Lower California is visible in the upper right and West Coast of South America in the lower right. (Photograph made available through the courtesy of Prof. V. E. Suomi, University of Wisconsin, who was responsible for this experiment on ATS–1).

area element on the earth's surface. By far the largest amount of energy for remote sensing is available in the visible part of the spectrum. Thus, at this time, only sensors operating in that part of the spectrum provide the spatial resolution necessary for the observation of mesoscale meteorological systems from geosynchronous orbits. The ideal operational meteorological satellite system of the future should consist of a set of spacecraft in two different orbits: A near earth sunsynchronous orbit to provide regular quantitative observations of a variety of meteorological parameters every twelve hours; and a geosynchronous orbit to afford continuous observations of cloud formations. Tracking of certain cloud formations at relatively short intervals (½ hour to 1 hour) will provide the only measurement of winds (at the cloud top level) which are feasible by remote sensing from satellites at this time.

CONTRIBUTION OF SATELLITE EXPERIMENTS OF METEOROLOGY

Observations made from the TIROS and NIMBUS series of satellites have already made a significant impact on operational and physical meteorology. To date, more than a million TV photographs of cloud cover and other features of the earth have been taken by TIROS, Nimbus and TOS satellites. Continuous measurements of the energy emitted and received by the earth and the atmosphere for periods longer than one year are now available for use in basic meteorological research. The Nimbus satellite has demonstrated a capability of obtaining contiguous cloud photography, including cloud heights, of almost the entire earth every 12 hours as well as of mapping temperatures, moisture and circulation patterns in the atmosphere. Finally, the ATS–1 satellite at geosynchronous altitudes has provided continuous cloud observations, every day from sunrise to sunset over a major portion of the world.

Cloud Cover

The view of the earth from space shows that more than 50 percent of the globe is usually covered by clouds. The formation of clouds over a given region, however, is governed by local meteorological conditions. Low pressure systems are accompanied by intense and characteristic cloud cover, while high pressure systems are often free of

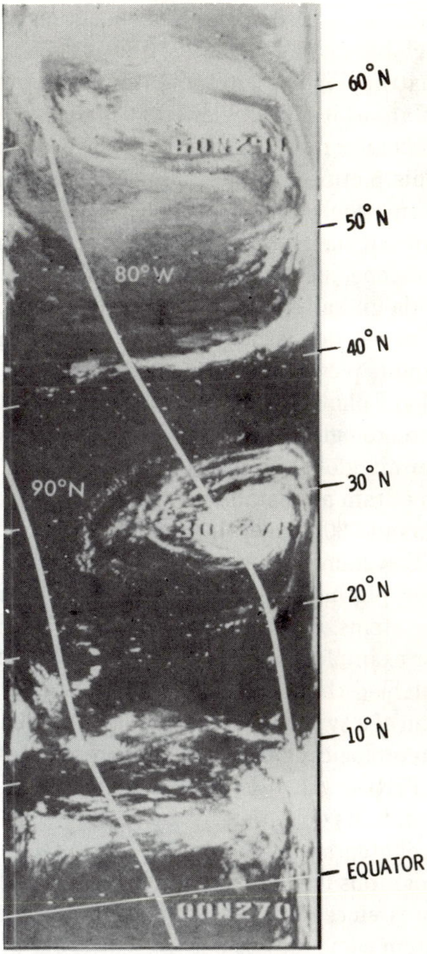

FIG. 7. Nighttime cloud observations from Nimbus I over the western North Atlantic on 9 September 1964. Clouds associated with storms and frontal systems stand out clearly in the satellite picture. The large cloud spiral near 30°N is associated with hurricane Dora. The cloud band near 40°N signifies a cold front generated by a low pressure system indicated by the cloud spiral near 60°N. The cloud mass near 10°N is indicative of strong convective activity in this tropical zone.

clouds. A continuous surveillance of the cloud cover around the globe, therefore, gives information about the location of large scale weather systems. A global map of cloud distribution resembles a weather map plotted by nature. Figures 1, 6 and 7 are such examples. Figure 7 shows a portion of an orbital strip viewed at night by the High Resolution Infrared Radiometer of Nimbus 1 over the eastern seaboard of North America. This picture exhibits the major features of a weather map. The cloud formations indicate the locations of the large scale pressure systems. Due to their particular cloud formations, the Intertropical Convergence Zone, just north of the equator, a major hurricane (Dora) east of Florida and a large frontal system east of Hudson Bay can be clearly recognized in the picture. This type of information obtained from the satellite pictures is extremely valuable for operational meteororology. Cloud photographs of distant parts of the globe are in the hands of meteorological centers within two hours of the time they were taken to permit global analyses. On the other hand, local weather stations can obtain an instantaneous view of weather patterns within a radius of about 2000 km of their station from cloud pictures received via the APT system during each overpass of the satellite. To date, these photographs have been used to detect, identify and track storms and frontal systems. In this regard the early TIROS have already made history. For example, on April 10, 1960 analysis of photographs from TIROS I enabled the Australian Meteorological Service to give the exact location of a typhoon 800 miles east of Brisbane in a region where surface meteorological observations were virtually nonexistent. In July 1961, TIROS III discovered a hurricane in the Caribbean Sea two days prior to its detection by routine meteorological observations. TIROS III photographed 18 tropical storms in all stages of development. Photographs from TIROS have been used for routine forecasting operations since 1960, six years before an operational weather satellite system came into being.

Pictures from meteorological satellites also provide valuable information in cloud-free areas. Photographs taken by TIROS II clearly showed the break-up of ice packs in the St. Lawrence River, thus demonstrating that weather satellites could also be helpful in navigation by monitoring ice movements. Since then, observations of ice formations with infrared sensors during the polar night have played an important part in the prediction of navigation routes in high latitude waters as well as contributing to oceanographic and glaciological research.

Aside from pure operational aspects, cloud photography from satellites has found its way into the areas of fundamental meteorological research. The satellite pictures have shown highly organized cloud patterns, revealing the dynamic mechanisms that may be acting to produce these formations. The most important patterns observed in this regard are large scale vortices related to storms and smaller scale cellular-type clouds related to local disturbances. The dynamics leading to these latter types has been extensively investigated by both theory and laboratory experiments. However, the laboratory experiments are carried out under very limiting conditions, and results are often constrained by questionable assumptions. The cloud systems photographed by satellites (Fig. 8) have none of these constraints; they are immensely useful in supplementing the results from laboratory experiments.[15] Photographs of vortices on various scales ranging from frontal clouds and tropical storms (planetary scale, Figs. 1 and 7) to smaller eddies (mesoscale, Fig. 8) have been especially useful because they permit the instantaneous viewing of the *entire* phenomenon extending over many hundreds of kilometers. Thus, cloud photography from meteorological satellites has been very successful in testing the validity of theories and laboratory experiments attempting to link dynamic processes with the formation of certain cloud types and patterns. More importantly, features of the general as well as local circulations may also be inferred from cloud patterns. Such inferences include tracing of the jet stream,[16] determination of convective heat transport, revelations of vertical wind shears, pinpointing areas of condensation and, therefore, release of latent heat, and many others.

Intensities of back-scattered solar radiation range from about 5 percent of the total incoming solar radiation over oceans to about 30 percent over deserts to nearly 95 percent over some snow, ice and cloud surfaces. Thus, over most surfaces, the reflectance from clouds is usually much larger than that from the clear atmosphere. However, in some exceptional cases such as over large areas of snow and ice, ambiguities result. Also, the reflectance of clouds depends strongly on their thickness, and very thin clouds may reflect less than 30 percent of the incoming energy. In such cases, additional information such as persistence, time variability of the measurements and local conditions must be taken into account to resolve the ambiguities. Since it is desirable to keep the scattered radiation from the clear

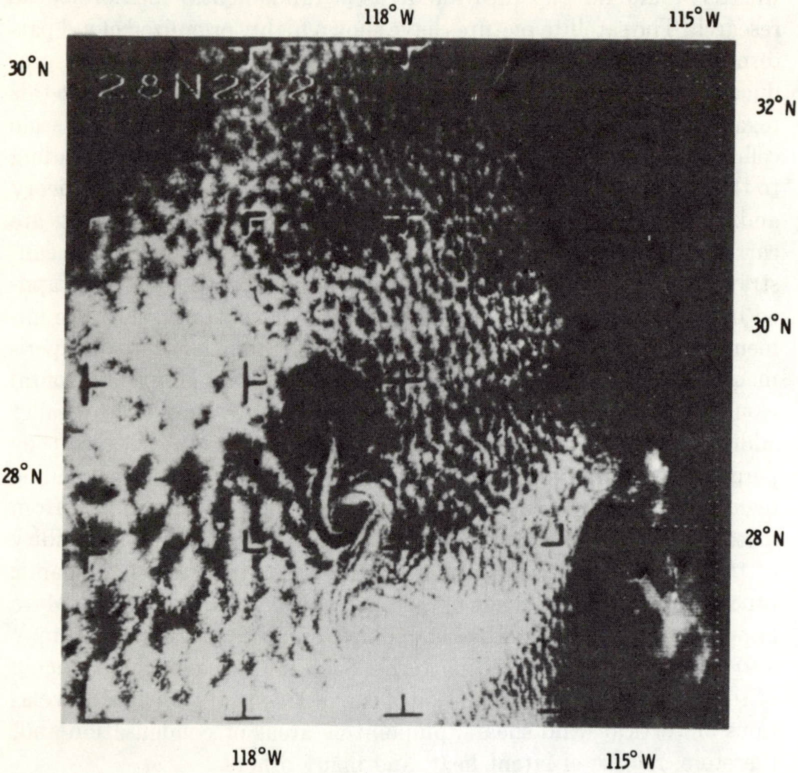

NIMBUS I (14 SEPT. 1964 1927 U. T.)

FIG. 8. Vortices in the air flow formed in the lee of Guadalupe Island in the Pacific Ocean are reflected by the cloud formations photographed with high resolution television cameras on NIMBUS I. The characteristic dimensions of the vortex pattern and of the cellular cloud structure surrounding the vortices permits the derivation of parameters relating to the hydrodynamic properties of the atmosphere. Such parameters are usually difficult to obtain from theory or from laboratory experiments.

atmosphere to a minimum, cloud cover observations are best made in a relatively narrow region in the red or near infrared of the solar spectrum.

Energy Balance

The energy available for meteorological processes can ultimately be traced to the global variations of heat stored in the atmosphere. The surplus or deficit in the amounts of radiative energy received from the sun and lost to outer space by the earth-atmosphere system plays an important role in the generation of these heat gradients. The global distribution of this radiative energy balance can be derived from satellite measurements of the total absorbed solar radiation and of the total emitted radiation. Such measurements have been made with a number of satellites including Explorer VII launched in 1959, TIROS IV, VII, Nimbus II and COSMOS 122 of the USSR.

The total amount of solar radiation absorbed by the earth-atmosphere is derived from satellite measurements of reflected radiation and from the assumption that one knows the total radiation received from the sun at the top of the atmosphere. From this assumption, the quantity W, which would be scattered back into space if the earth and atmosphere were a perfect 100 percent efficient isotropic reflector, is completed. The satellite radiometers are calibrated in terms of this quantity before flight. The backscattered energy at wavelengths between 0.2 and 4.0 microns from a given portion of the earth in a given direction is thus measured relative to W. Making the highly oversimplifying assumption that radiation is backscattered isotropically, one can derive the amount of radiation backscattered in *all* directions by this given area. This amount, expressed in percent of W, is referred to as the reflectance of the observed area. The absorbed energy can be calculated since that amount of the solar energy which is not scattered back into space must remain in the earth and atmosphere.

Two relatively broad spectral intervals have been used on the TIROS and Nimbus to derive the total emitted long wave radiation. One channel operated between 8 and 12 microns, the other between 7 and 30 microns. About 60 percent of the total emitted radiation is contained in the former channel, about 80 percent in the latter. Also, as in the case of solar radiation, a given spot of the earth is seen by the satellite only from one direction at a given time. Since, again, the emitted radiation varies in a complex manner with angle and with wavelength one must apply certain physical reasoning to extrapolate the total flux in all directions from the satellite measurements which were made in a limited portion of the spectrum and in one singular direction. These extrapolations are arrived at, in part theoretically, in part empirically.

Attempts to derive estimates of the global heat budget from satellite measurements of emitted and reflected radiation have been made and have shown that on a large scale the satellite results confirm previous theoretical estimates of the global distribution of total emitted long-wave radiation. The TIROS results clearly show a minimum of outgoing radiation just north of the "meteorological" equator and a maximum in the 20°–30° latitudinal belts (Fig. 9). The minimum is due to the extensive cloud cover in the equatorial zone, while the maxima in the two subtropical belts reflect the clear skies and warm temperatures in these regions. Beyond the subtropical belts, the total outgoing radiation decreases rapidly with latitude, especially toward the winter pole.

The important parameter in the study of the dynamics of the atmosphere is, however, the net atmospheric energy balance and not the outgoing radiation alone. The net energy balance is made up of the difference between the incoming solar radiation, mostly in the visible, and the outgoing terrestrial radiation in the infrared. It is well known that the latitudinal variation of the energy balance shows an excess of incoming solar radiation over outgoing radiation near the equator and a deficiency at the poles. It is this variation of of the energy balance with latitude that drives the tropospheric heat engine. Using the satellite data to measure the energy balance of the entire earth-atmosphere system one may derive the variations in the latitudinal averages of the net energy balance in the atmosphere alone which is necessary to understand the general circulation of the atmosphere.

TIROS IV and VII have given fairly reliable estimates of the variation of the albedo of the earth for all parts of the globe between about 60°N and 60°S for periods of over two years. The main result obtained from the analysis of these albedo measurements is that the latitudinal variation in the albedo shows a relative maximum at the equator and a minimum in the two subtropical belts, with the albedo increasing with the latitude up to a value of about 50 percent at 60° latitude (Fig. 9). The latitudinal gradient of albedo is found to be steeper than that given by the theoretical estimates. However, absolute measurements of the total albedo of the earth derived so far from TIROS and Nimbus measurements leave much to be desired. In all cases the total amount of reflected radiation measured by the satellites was much too low (by about 40 percent) to balance the measured as well as the theoretically computed outgoing long wave radiation. Since such an inbalance cannot obviously exist on the time scale observed by the satellites, the albedos shown in Fig. 9 were arbitrarily increased to achieve balance. It is suspected that the discrepancy stems from the models used in deriving the albedo, i.e. the total, omnidirectional short wave radiation flux, from the highly directional satellite measurements.

The incoming radiation calculated from these albedo measurements, combined with the measured outgoing radiation then gives the latitudinal and longitudinal variation of the energy balance of the earth-atmosphere system.

Cloud Heights

The radiation intensities measured in the 8–12 micron channel on TIROS are used not only to derive the total emitted long wave radiation but also to derive the temperatures of surfaces "seen" by the satellite. Clouds are opaque at wavelengths longer than 3 microns

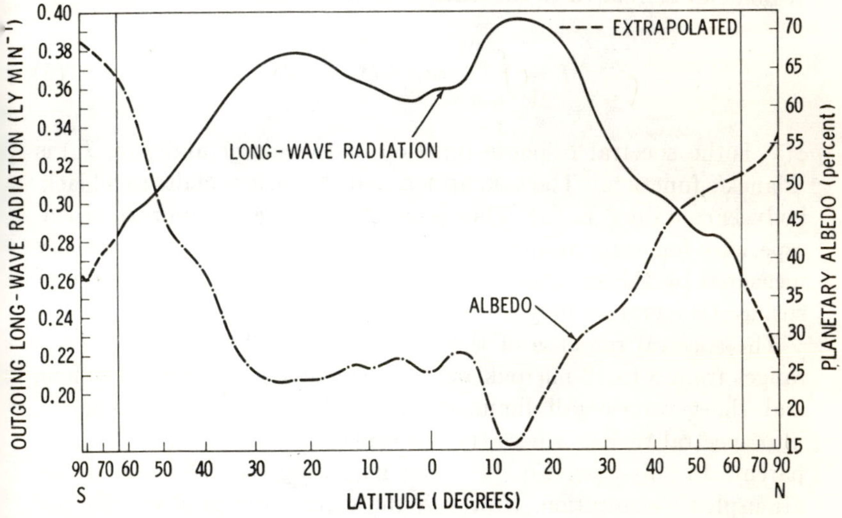

Fig. 9. Average outgoing longwave radiation determined from 8–12 micron and albedo determined from 0.55–0.75 micron channel of TIROS VII for December, January, February 1963/64 as a function of latitude. (Reproduced from "A Radiation Climatology in the Visible and Infrared from the TIROS Meteorological Satellites," Bandeen, Halev, and Strange, NASA Technical Note D2534, 1965.)

and trap the far infrared radiation from below. Therefore, in the presence of, clouds the emission to space occurs only from the top of the clouds where the temperatures are generally lower than at the surface. Consequently, the outgoing radiation is more intense over clear areas than over cloudy regions. This fact is used to quantitatively derive heights of cloud tops from radiation measurements. As will be shown the radiation intensity in this channel is a direct measure of the cloud surface (top), temperature. A unique relationship can be established between temperature and height either on the basis of climatology such as the profiles shown in Fig. 3 or from actual balloon sonde temperature measurements provided that it can be assumed that clouds do no penetrate above the temperature minimum at the tropopause.

Ideally, the atmosphere transmits radiation from the earth's surface or from cloud tops to the satellite almost unattenuated in the region between 10.5 and 11.5 microns. The intensity of radiation I, which would be measured by the satellite in the 10.5–11.5 micron region can be related to the surface temperature T_s by

$$I = \int_{\lambda=10.5}^{\lambda=11.5} \phi(\lambda) B(\lambda, T_s)\, d\lambda. \tag{6}$$

$\phi(\lambda)$ is the spectral response function of the sensor and $B(\lambda, T_s)$ is Planck's function.[17] The assumption that the surface emits as a blackbody is contained in (6). This assumption is a good approximation, especially for water and cloud surfaces. Thus, the radiation intensity measured by the satellite in this spectral "window" is a function of surface temperature only.

The spectral response of this channel of the TIROS radiometers ranges from 8 to 12 microns and thus extends both beyond the long and short wavelength limits of the 10.5–11.5 micron "window", into spectral regions where the atmosphere is not completely transparent. In this case, Eq. (6) is no longer rigidly valid due to the atmospheric absorption. However, the surface temperature can still be determined from the measured radiation by making the appropriate corrections for atmospheric absorption. These corrections can be obtained if approximate estimates of the vertical temperature profile and the distribution of water vapor and ozone above the radiating surface are known. In general, climatological "standard" models for these quantities will suffice as the corrections are small. The corrections are even less important in the much narrower spectral window of 10.6 to 11.6 microns which was used on Nimbus II and will be used for this purpose in the future as well as in the 4 micron window used in the Nimbus High Resolution Infrared Radiometer. Both of these windows are considerably "cleaner" i.e. free of atmospheric absorption, than the 8-12 micron interval used earlier. Radiometric temperature measurements in these windows on TIROS and in the 4 micron window on NIMBUS have produced large scale, "three dimensional" analyses of cloud heights. Cloud height differences of a few hundred meters can be detected and mapped by this method. The nighttime cloud patterns show in Fig. 7 were mapped in this fashion in the 4 micron window. Here the satellite radiometer can resolve about 100 steps of temperature (corresponding to shades of grey in the picture) from 200 to 310°K. Such cloud height differences could not be detected by TV cloud photography. Thus, radiation measured in the atmospheric windows has become an excellent tool to map cloud cover and height on a global scale at night since this method is independent of illumination by sunlight. However, this mapping of cloud cover is effective only if the clouds are large enough to fill the field of view of the radiometer uniformly. In the case of the present 10.6 to 11.6 micron channel on Nimbus this field of view corresponds to about 50 km on the surface of the earth. For the 4 micron radiometer on Nimbus the instantaneous field of view is much smaller, corresponding to about 8 km. In the future 10.6 to 11.6 micron radiometers with a field of view corresponding to about 5 km will be flown.

The Nimbus satellite, with its High Resolution Infrared Radiation sensor, has produced global cloud cover and height analyses. Radiation intensities are expressed in the shades of grey as in Fig. 7 where bright shades correspond to low intensities (cold temperatures) and dark shades to high intensities (warm temperatures). Thus, the highest clouds are indicated by the brightest shades of grey. Since clouds represent an indication of convective activity, with the most active regions (tropics) producing the highest clouds, detailed analyses of cloud top heights on a scale given by the satellite observations provide a new insight into the dynamics of weather systems.

Water Vapor and Stratospheric Temperature

From measurements in certain, sufficiently narrow spectral intervals, one can infer either the temperature or the amount of constituents, such as water vapor or carbon dioxide, distributed with height in the atmosphere. The spectral region must be chosen such that the absorption coefficient k which depends on pressure (p) temperature (T) and on the wavelength (λ) for a given constituent is very large; e.g., near 6.3 microns for water vapor, near 15 microns for CO_2.

The intensity of radiation, measured at the satellite, for an absorbing spectral interval $\Delta\lambda$ is:

$$I_{\Delta\lambda} = \int_{\Delta\lambda} \phi(\lambda) I_s \tau_s \, d\lambda + \int_{\text{surface}}^{\text{top}} \psi(h) \, dh, \tag{7}$$

where

$$\psi(h) = \int_{\Delta\lambda} \phi(\lambda) B(\lambda, T) [\partial\tau(h, \lambda)/\partial h] d\lambda$$

and

$$\tau(p, T, \rho, \lambda) = \exp\left[- \int_0^h k(p, T, \lambda) \rho \, dh \right]$$

is the transmission from the level h to the top of the atmosphere, ρ is the number density of the absorbing gas and the subscript s refers to the underlying radiating surface. Scattering has been neglected and it is also assumed that the satellite is viewing in the vertical. The quantity $\psi(h)$ represents the contribution of different altitude levels to the measured outgoing radiation.

If a temperature profile and the surface pressure is assumed and if $\rho(h)$ is known, as is the case for CO_2, ψ can be determined as a function of altitude. The weighting functions, $\psi(h)$, for the 15 micron CO_2 channel of TIROS VII are shown in Figure 10 for four typical temperature profiles. Figure 10 shows that absorption by CO_2 in the 15 micron region is sufficiently strong so that contributions from an underlying radiating surface [first term in Equation (7)] can be

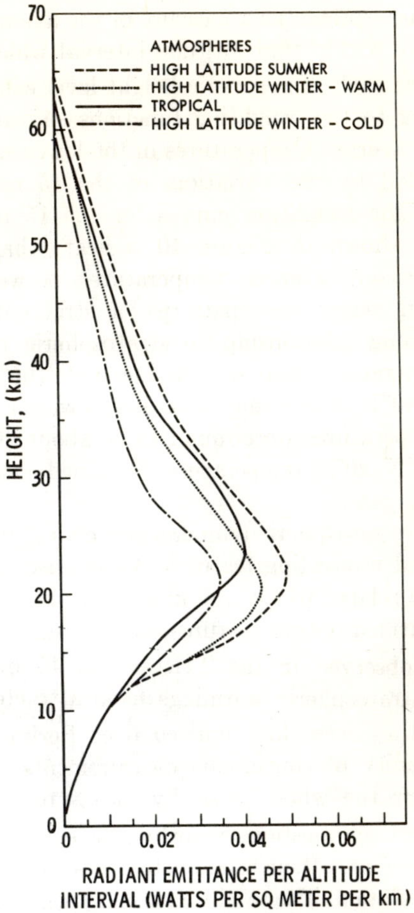

FIG. 10 Radiation ψ within the spectral range of the 15 micron channel of TIROS VII emitted to the satellite in each altitude interval as a function of altitude, latitude and season. Curves are based on the four typical temperature profiles shown in Figure 3.

neglected under all conditions if that surface lies at altitudes below about 10 km where the curves of Figure 10 indicate small values of ψ. Most radiative surfaces (cloud tops) can be assumed to be below about 10 km. Thus, the radiation measurements in the 15 micron channel may be interpreted as a measure of the average temperature of the atmosphere within that altitude interval which is "seen" by the satellite where each atmospheric height level contributes to the average according to the weighting functions shown in Figure 10. Variations in the average temperatures in the lower stratosphere can be readily detected by the variations of the 15 micron radiation measurements. The weighting curves for the 15 micron channel on TIROS VII shown in Figure 10 indicate that this channel will generally measure average temperatures between 15 and 35 km. On a global scale, such patterns of stratospheric temperatures bear a strong relationship to stratospheric circulation. In general, the temperatures near the summer pole (65° latitude) were of the order of 240°K, decreasing towards the winter pole where the stratospheric temperatures were found to be about 200°K. In equatorial regions (20°N–20°S) temperatures remained constant at 230°K at all times of the year.

Analyses of the stratospheric temperatures on a regional scale have also shown several interesting features. An intense warm area over the North Pacific related to the well-known Aleutian anticyclone has been clearly identified. Other features, such as sudden stratospheric warmings were observed in the TIROS VII 15 micron measurements.[18] These stratospheric warmings have attracted considerable attention since they were first noticed over Berlin in 1950. Now, with the availability of continuous measurements of stratospheric temperatures over the whole globe by means of satellites, it will certainly resolve many questions regarding the extent and magnitude of their occurrences and their relation to the sudden changes in the stratospheric circulation. The satellite measurements are of particular interest since they enable the detailed study and comparison of these phenomena in both hemispheres.

In the case of water vapor, where the distribution of the vapor is not known the process can be reversed if $T(h)$ is known from some other observation. ψ vs h curves can then be drawn for one "standard" distribution with height but for different amounts of H_2O and the amount which matches the 6.3 micron channel measurements determines the proper integrated value for

$$\int \rho \, dh.$$

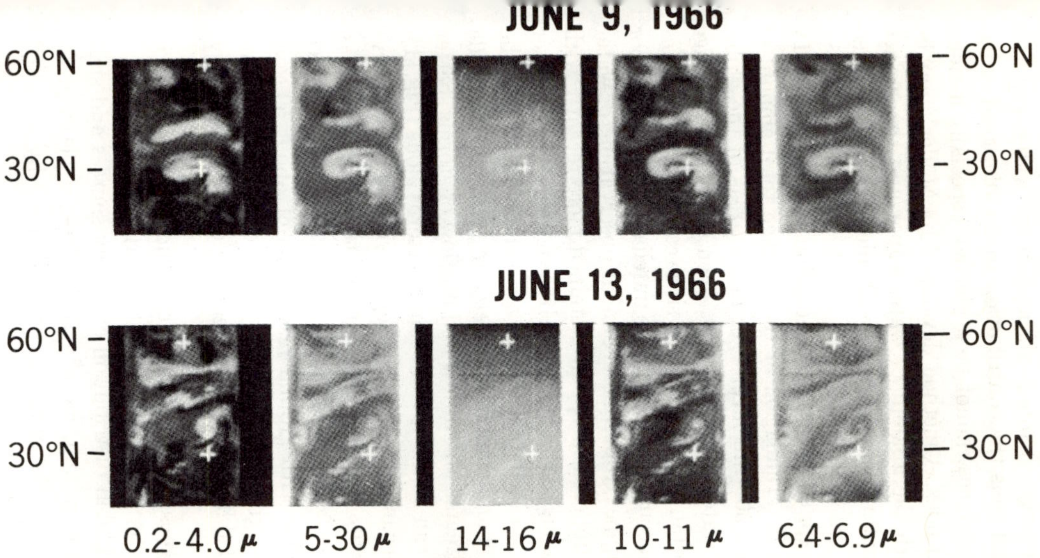

FIG. 11. Hurricane Alma in two stages of its development observed by Nimbus II in five spectral bands. In the upper picture, the storm is fully developed and its shape is essentially the same in all five spectral bands. Comparison of the emitted (10–11 microns) and reflected (0.2–4.0 microns) radiation shows that the tail of the spiral consists mainly of high but thin clouds while lower, thick clouds are concentrated near the eye. In the Water Vapor Channel note the sharp and long sector of dry air in the southwest section of the spiral indicating downward motion. The cloud tops in the spiral are generally very high, penetrating even into the Carbon Dioxide Channel; digitally inferred temperatures are 223°K and 232°K in the Water Vapor and Window Channels respectively. An entirely different picture prevails 4 days later (lower part of Fig. 11) when the storm is in its decaying stage. The brightest image occurs in reflected radiation with no noticeable spiral structure while in the Window Channel there is an indication of a wound up spiral but the clouds are relatively dark indicating low altitude. No trace of the storm can be seen in the CO_2 Channel. It has disappeared from the upper troposphere. Most noteworthy, however, the Water Vapor channel shows the most extensive and pronounced sign of a mature or decaying circulation system. The 14–16 micron channel indicates stratospheric temperatures generally increasing with increasing latitude. (Reproduced from: Bulletin of the American Meteorological Society, Nov. 1966).

Only the total integrated water vapor content can be determined by this method and the distribution of water vapor vs. height as well as of temperature vs. height must be known or assumed. In the case of the 6.3 micron channel, absorption by water vapor is small above 10 km and thus the contribution to the observed radiation from underlying surfaces cannot be neglected, especially if cloud tops reach to high altitudes. A method developed by Möller[19] may be used to determine both the amount of water vapor and the surface temperature from simultaneous 6.3 and 8–12 micron channel measurements.

This method was applied to the observations from TIROS IV which resulted in the determination of the average total amount of water vapor above the 500 mb level over many areas of the globe. The measurement of water vapor amount, even only as an average quantity is most important at these heights since conventional meteorological radiosondes become generally quite unreliable for water vapor measurements in the upper troposphere. Similar observations of water vapor content in the upper troposphere obtained more recently with Nimbus II[20] have demonstrated that the observed patterns of water vapor or thin cirrus clouds at these levels are very indicative of the atmospheric circulation and can be interpreted in terms of large scale vertical motion, jet streams and inflow and outflow of storms better than these inferences can be made from cloud patterns alone (Fig.11). Such inferences are very important since they provide the best observations of dynamic events from satellites at this time.

Thus, the TIROS and Nimbus spacecraft, which so far, have carried five-channel radiometers with ground resolutions of about 50 × 50 km and single channel high resolution radiometers of 8 × 8 km ground resolution have provided highly useful meteorological information by:

1. Mapping the large scale distribution of cloud patterns, both in day and night, and determining the overall cloud top heights.
2. Determining the global distribution of energy sources and sinks in the earth-atmosphere system, namely the difference between the solar energy absorbed and the infrared energy emitted by the earth and the atmosphere.
3. Determining the global distribution of mean temperatures in the stratosphere during a full seasonal cycle, and the distribution of water vapor in the upper troposphere.

The meteorological experiments carried out by satellites have so far been limited to broad band radiometry, providing information only on cloud cover, upper troposphere moisture fields, the gross features of the thermal structure of the atmosphere and selected features of the circulation inferred from the cloud, moisture and temperature patterns. The next step in this field is to increase the spectral resolution of the sensors, i.e., to make spectrometric measurements. The transmission of long wave radiation in the earth's atmosphere varies considerably with wavelength and altitude. If, therefore, the satellite measures the energy radiated by the earth in small spectral intervals over the range of an entire absorption band, one can obtain an estimate of the vertical distribution of temperature. This parameter is most significant in meteorological research. Besides the 15 micron infrared absorption band of carbon dioxide, there is a strong oxygen absorption band at 5 mm in the microwave region of the spectrum which could also be used for the same purpose. Spectroscopic measurements will undoubtedly be made from meteorological satellites in the future. It has also been suggested that microwave radiation ((λ = 0.3 cm) be used to measure atmospheric temperture and water vapor, percipitation, and earth surface characteristics, even in the presence of clouds. This is because some clouds are transparent to longer wavelengths in some microwave channels and there is no absorption by other atmospheric constituents. The problem, however, remains that the emissivities of the ground and ocean are not very well known for these wavelengths and surfaces are far from blackbodies in that portion of the spectrum.

Another atmospheric parameter which can be usefully observed from the satellite is the distribution of ozone in the stratosphere. Because ozone is a strong absorber of the ultraviolet radiation, the measurements by a satellite of either direct or backscattered sunlight after its passage through the atmospheric ozone can be used for the deduction of the ozone distribution in the atmosphere.

The measurement of "sferics", the radio frequency emissions by electrical discharges from cumulonimbus clouds, is still another domain in which satellite observations could contribute significantly. The detection and mapping of sferics may indicate areas of strong vertical motion related to thunderstorm development and such phenomena as strong winds, heavy rainfall and turbulence.

It must be mentioned that the most important meteorological parameters, namely the surface pressure and the direction and velocity of the wind near the ground, cannot be accurately measured by instruments on-board a satellite at this time. The way in which the satellite can provide this information is by acting as a data collector and relay. Pressure and wind of the lower troposphere would be measured by a series of strategically placed sensors on the earth's surface and in balloons, and the information from these detectors would be transmitted to the over-passing satellite which could locate and track the balloons and relay the sensor data to the analyst at a central weather station.

All these considerations will be part of future meteorological satellite programs, but their implementation is certainly a matter of formidable technological effort and will take at least the remainder of this decade to be completed.

REFERENCES

1. U. S. Standard Atmosphere 1962, Government Printing Office, Washington, D.C.
2. S. L. Hess, Introduction to Theoretical Meteorology, Henry Holt and Company, New York, 1959.
3. W. Nordberg, L. Katchen, J. Theon and W. S. Smith, Journal of the Atmospheric Sciences, Vol. 22, No. 6 pp 611–622, November 1965.
4. S. Manabe and F. Moeller, The Radiative Equilibrium and Heat Balance of the Atmosphere, Monthly Weather Review, 89, No. 12, 503 (December 1961).
5. R. J. Murgatroyd, and R. M. Goody, Quarterly Journal of the Royal Meteorological Society, **84**, No. 361. (July 1958).
6. W. W. Kellogg, Chemical Heating Above the Polar Mesopause in Winter, Journal of Meteorology **18** (3), 373 (June 1961).
7. R. E. Newell, The Circulation of the Upper Atmosphere, Scientific American, **210**, No. 3, 62 (March 1964).
8. S. Teweles, Spectral Aspects of the Stratospheric Circulation During IGY, MIT, Report No. 8 January 1963.
9. A. Kochanski, Monthly Weather Review, Vol. 94, No. 4, pp 199–212 April 1966.
10. J. E. Masterson, W. Nordberg, and S. Teweles, Meteorological Observations Above 30 Kilometers, NASA SP–49, 1964.
11. L. M. Jones, J. W. Peterson, E. J. Schafer and H. F. Schulte, Journal of Geophysical Research, Vol. 64, pp 2331, 1959.
12. J. E. Ainsworth, D. F. Fox, and H. E. La Gow, Journal of Geophysical Research, Vol. 66, No. 10., October 1961 pp 3191–3212.
13. W. G. Stroud, W. Nordberg, W. R. Bandeen, F. L. Bartman and P. Titus, Journal of Geophysical Research, Vol. 65, pp 2307–2323, August 1960.
14. W. K. Widger, Meteorological Satellites, Holt, Rinehart and Winston Inc., New York, 1966.

15. K. P. Chopra and L. F. Hubert, Journal of the Atmospheric Sciences, Vol. 6, pp. 652–657, 1965.
16. L. F. Whitney, A. Timchalk, and T. I. Gray, Monthly Weather Review, Vol. 94, pp. 127–138, 1966.
17. W. Nordberg, Science, Vol. 150, No. 3696, pp. 559–572, 29 October 1965.
18. W. Nordberg, W. R. Bandeen, G. Warnecke and V. Kunde, Space Research V, North Holland Publishing Co. Amsterdam, 1965, pp. 783–809.
19. F. Moller, Einige Vorlaufige Auswertungen der Strahlungsmessungen von TIROS II, Arch. f. Met. Geophys. und Biokl., Ser. B, pp. 78–94, 1962.
20. W. Nordberg, A. W. McCulloch, L. I. Foshee and W. R. Bandeen, Bulletin of the American Meteorological Society, Vol. 47, pp. 857–872, Nov. 1966.

Chapter 7

THE SHAPE OF THE EARTH*

W. M. Kaula

INTRODUCTION

The shape of the earth is commonly taken to mean not the form of the solid surface of the earth—the ups and downs of the mountains and valleys, etc.—but rather the form of an equipotential surface called the *geoid*. An *equipotential surface* is one which is everywhere perpendicular to a pull or an acceleration: the geoid is that equipotential surface which most nearly coincides with the mean sea level of the earth. The form of the geoid is determined by the combination of the accelerations due to the attraction of the earth's matter and the centrifugal force arising from the earth's rotation. The shape of the earth so defined is thus just a manner of expressing the earth's gravitational field, which is of interest to space science because it affects the orbits of close satellites, and because it is an indicator of conditions in the earth's interior, the most accessible example of planetary interiors in general.

* Publication No. 385, Institute of Geophysics and Planetary Physics, University of California, Los Angeles.

To draw physical inferences from the earth's gravitational field, we need a theoretical model with which to compare the actual earth. The obvious model is a perfectly fluid body which rotates at the same rate ω as the earth and which has the same radial variation in density. This fluid model would be distorted by centrifugal acceleration; at a point a distance p from the rotation axis (see Figure 1), this acceleration would be $\omega^2 p$. This acceleration, similar to an acceleration caused by gravitational attraction, can be expressed as the gradient of a scalar potential W:

$$W = \frac{\omega^2}{2} p^2 = \frac{\omega^2}{2} r^2 \cos^2 \psi \tag{1}$$

where r is the radial coordinate and ψ is the geocentric latitude. If we combine this *rotational potential* W with the *gravitational potential* V, we get a total *potential of gravity* U which applies to any point rotating with the earth:

$$U = V + W, \tag{2}$$

where

$$V = k \int_{\mathrm{vol}} \frac{\rho}{r'} \, dv, \tag{3}$$

in which k is the gravitational constant, 6.67×10^{-8} cm^3 gm^{-1} sec^{-2}, ρ is density, dv is an element of volume and r' is the distance between the point of density ρ and the point at which the potential V is calculated.

The effect of the centrifugal acceleration $\partial W/\partial p$ we would expect to cause a bulging outward from a perfect sphere which would be a maximum at the equator and zero at the poles. This bulging would be symmetric about the equator and hence of even degree, like W, so that for the radius r at latitude φ of a surface of constant potential U with equatorial radius a we should be able to approximate r by an expression of the form:

$$r \approx a(1 - f \sin^2 \psi) \tag{4}$$

where f is a function of a called the *flattening*. It can be shown that a

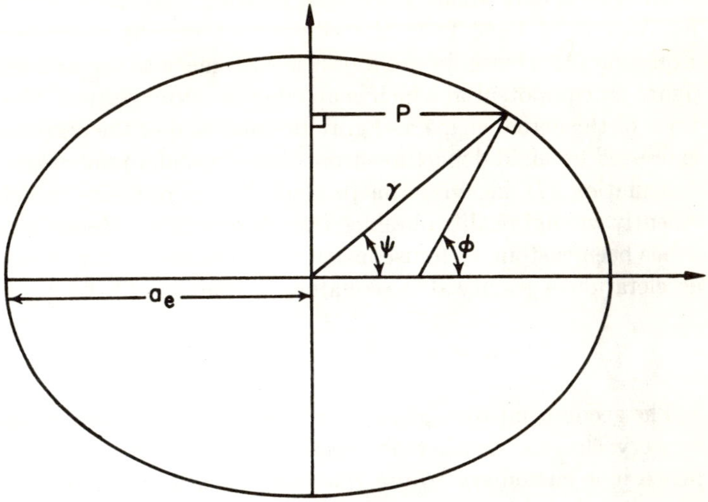

FIG. 1. Meridian Ellipse.

surface of constant potential—called an equipotential—in a rotating fluid is a surface of constant density ρ. Using this fact, we can derive an expression for f as a function of the radial distribution of density ρ, the rotation rate ω, and the equatorial radius a. At the surface, setting a_e for a:[2]

$$f = \frac{5m}{2} \Bigg/ \left[1 + \left\{\frac{5}{2} - \frac{15C}{4Ma_e^2}\right\}^2\right] + O(f^2) \tag{5}$$

in which the density ρ appears through the total mass M and the moment of inertia C about the rotation axis,

$$C = \int_{\text{vol}} \rho p^2 dv, \tag{6}$$

and m is the ratio of the centrifugal acceleration $\omega^2 a_e$ to the total acceleration of gravity γ_e at the equator:

$$m = \omega^2 a_e / \gamma_e \approx \omega^2 a_e / [kM/a_e^2] \tag{7}$$

Since m is only about 1/288, the rotating fluid model differs by a few meters at most from the ellipsoid of revolution expressed by Equation (4). Hence, in geodesy it is customary to use as a reference figure an equipotential which is an ellipsoid of revolution. The variations of the actual shape and gravitational field of the earth are then expressed as small departures from this ellipsoidal reference.

Equation (7) includes the product kM, a quantity which until recently was not readily obtained from observations. Hence in geodesy it has been customary to use instead as a fundamental parameter the acceleration of gravity at the equator, γ_e, related to kM by:[1,4]

$$kM = a_e^2\gamma_e[1 - f + 3m/2 + O(f^2)] \tag{8}$$

The geodetic latitude φ in Figure 1, the normal to the ellipsoid, will be very close to another observable, the direction of gravity, and hence it is customary to use the geodetic latitude φ instead of the geocentric latitude ψ. Then the formula for the acceleration of gravity at geodetic latitude φ on the ellipsoid which is required for comparison with measurements of the acceleration of gravity is:[1]

$$\gamma = \gamma_e[1 + \beta_2 \sin^2\varphi + \beta_4 \sin^2 2\varphi + O(f^3)], \tag{9}$$

in which β_4 is $O(f^2)$ and

$$\beta_2 = 5m/2 - f + O(f^2) \tag{10}$$

For computation of the orbits of satellites, which are not attached to the earth and hence unaffected by the centrifugal acceleration, we require only the gravitational potential V, to which we need to compare the potential V_0 of the ellipsoidal reference, conventionally expressed as:

$$\begin{aligned} V_0 = (kM)/r)\,[1 &- J_2(a_e/r)^2P_{20}(\sin\psi) \\ &- J_4(a_e/r)^4P_{40}(\sin\psi) - O(f^3)\,] \end{aligned} \tag{11}$$

where $P_{20}(\sin\psi)$ and $P_{40}(\sin\psi)$ are the Legendre Polynomials:

$$P_{20}(\sin\psi) = (3\sin^2\psi - 1)/2 \tag{12}$$

$$P_{40}(\sin \psi) = (35 \sin^4 \psi - 30 \sin^2 \psi + 3)/8 \tag{13}$$

J_2 is related to the difference of the moments of inertia C and A about the polar and equatorial axes, respectively[4]:

$$J_2 = (C - A)/Ma_e^2 = (2f - m)/3 + O(f^2), \tag{14}$$

and J_4 is $O(f^2)$.

The more interesting parts of the earth's gravitational field are the variations from the ellipsoidal model. In conventional geodesy, the quantity most directly measured is the acceleration of gravity g, which is customarily corrected for elevation to refer to the geoid:

$$g = g_h - (\partial g/\partial r)h, \tag{15}$$

where g_h is observed at elevation h above the geoid. The theoretical gravity on the ellipsoid is subtracted from the observed gravity g on the geoid to obtain the *gravity anomaly* Δg:

$$\Delta g = g - \gamma \tag{16}$$

The location of the geoid is not directly observable. However, if the anomaly were known over all the geoid, the location of the geoid referred to the ellipsoid as expressed by its height N could be obtained by an integral transform known as Stokes' theorem, using a function $S(\theta)$:[1]

$$N(\varphi_N, \lambda_N) = (r/4\pi g) \int_{\text{sphere}} S(\theta)\, \Delta g(\varphi, \lambda) d\sigma, \tag{17}$$

where mean values can be used for the radius r and the gravity g. In Stokes' theorem, N and Δg are considered as small linear departures, and the integration can be treated as over a sphere, rather than an ellipsoid, with negligible error. λ is longitude, and θ is the arc distance from (φ, λ) to (φ_N, λ_N).

Equation (17) is a special application of a general theorem in potential theory which states that the shape of a closed equipotential bounding all masses can be deduced if we know the acceleration over all the surface. The difficulty in applying Stokes' theorem is, of course, that the anomaly Δg is known only over part of the earth's surface.

Stokes derived a closed expression for the function $S(\theta)$, but it is complicated. Easier to understand, and more convenient in satellite orbit applications for which the gravitational field must be extrapolated to altitude and accelerations integrated to obtain changes in position, is the representation of the potential in terms of functions orthogonal over the sphere, known as *spherical harmonics*:[4]

$$V = (kM/r)[1 + \sum_{l=2}^{\infty} (a_e/r)^l \sum_{m=0}^{l} P_{lm}(\sin\varphi). \\ \{C_{lm} \cos m\lambda + S_{lm} \sin m\lambda\}] \tag{18}$$

where $P_{lm}(\sin\varphi)$ is the associated Legendre polynomial:

$$P_{lm}(\sin\varphi) = \frac{\cos^m \varphi}{2^l} \sum_{t=0}^{(l-m)/2} \frac{(-1)^t (2l-2t)!}{t!(l-t)!(l-m-2t)!} \sin^{l-m-2t}\varphi, \tag{19}$$

and the coefficients C_{lm}, S_{lm} are given constants which constitute a *spectral* representation of the gravitational field alternative to the *spatial* representation in terms of $\Delta g(\varphi, \lambda)$ or $N(\varphi, \lambda)$.

By orthogonal we mean that the integral over the sphere of the product of any two spherical harmonics $P_{lm}(\sin\varphi)\{\sin \text{ or } \cos\}m\lambda$ and $P_{jk}(\sin\varphi)\{\cos \text{ or } \sin\}k\lambda$, $lm \neq jk$, is zero. Hence for variations over a spherical surface, spherical harmonics are analogous to Fourier series in time. Note that an harmonic of indices l,m is a polynomial of degree m in longitude λ and of degree $(l\text{-}m)$ in latitude φ. Hence a particular spherical harmonic is best remembered as a variation over the surface of a sphere which has $(l\text{-}m)$ zeroes in the 180° along a meridian and m zeroes in 180° along a parallel. An harmonic with $m=0$ will represent solely north-south variation, while one with $m=l$ will represent predominantly east-west variation. The half-wave length of variations represented by harmonics of degree l will be $180°/l$.

In geodesy, the difference between the total potential V in (18) and the potential of the reference ellipsoid V_0 in (11) is known as the disturbing function T:

$$T = V - V_0 \tag{20}$$

Since by definition the geoid is an *equipotential* on which the potential is equal to the reference value V_0 at the ellipsoidal surface, for the displacement N we can write:

(21)
$$N = -(V - V_0)/(\partial V_0/\partial r) \approx T/g \tag{21}$$

The gravity anomaly Δg defined by Equation (16) differs from the radial gradient of T because in accordance with Equation (15), measurements can be referred to the geoid, a physical reality, but not the reference ellipsoid, a mathematical fiction, which differs by the displacement N:

$$\Delta g = -\partial T/\partial r + N\partial g/\partial r \approx -\partial T/\partial r - 2T/r \tag{22}$$

In sperical harmonics, the Stokes' function in Equation (17) becomes[1]:

$$S(\theta) = \sum_{l=2}^{\infty} \frac{2l+1}{l-1} P_{l0}(\cos\theta) \tag{23}$$

In the dynamics of close satellite orbits, the *disturbing function* is symbolized by R and is usually defined differently from Equation (20), as:

$$R = V - kM/r \tag{24}$$

METHODS OF OBSERVATION

Gravimetry

The most direct means of measuring the earth's gravity is by *gravimetry*, which measures the acceleration, g. The accurate measurement of the absolute acceleration is a delicate problem, and is carried out only in a few laboratories, either by timing the fall of an object or the period of a pendulum. Most of the measures of gravity used in geodesy are relative measurements, which reduce calibration difficulties by measuring the difference in gravity from place-to-place, either through the change in period of a pendulum or through the change in tension on a spring required to keep a constant mass at a null point. This system of differential measurements is ultimately connected to the absolute measurements through a network of reference stations.

Modern gravimetric measurements generally have relative errors less than $\pm$ 0.001 cm sec^{-2}, compared to the total acceleration g of 980 cm sec^{-2} and the magnitude of the anomalies Δg defined by Equation (16) of $\pm$ 0.035 cm sec^{-2}. The principal defect of the system of gravimetry is its irregular distribution; and, although observations can be made at sea, there are still wide stretches of ocean without gravimetry. Hence the determination of the coefficients C_{lm}, S_{lm} in Equation (18) by harmonic analysis and the determination of the geoid heights N by Stokes' theorem, Equation (17), is very much affected by the statistical methods applied either explicitly or implicitly.[1,2,3]

Astro-Geodesy

Geodetic survey yields differences in position which can be expressed in coordinates (φ, λ, h)—latitude, longitude, and height—referred to the ellipsoid, known as *geodetic position.* Another type of latitude and longitude is *astronomic position,* which actually is the direction of the acceleration of gravity measured relative to the fixed stars. In general, the astronomic position will differ from the geodetic position, because the geoid will have a slope with respect to the reference ellipsoid. Integrating this slope over the ellipsoid gives the change in geoid height N.

The *astro-geodetic* system is also important in determining the scale of the earth's figure, as expressed by its equatorial radius a_e. Connected systems of triangulation have now been extended over most of the major continents. However, the astro-geodetic method still has the principal defect of determining the gravity field only on land, and hence is of value principally as a check on the other methods.[3]

Satellite Orbits

The departure of the earth's gravitational field from a central field, as expressed by the disturbing function R of Equation (24), causes the orbit to depart from a simple Keplerian ellipse. The dominant term in the disturbing function R is the second degree zonal term with coefficient $-J_2$, or C_{20}, caused by the flattening. Using this disturbing function in the Newtonian equations of motion,

$$\ddot{\mathbf{r}} = \nabla(kM/r + R), \tag{25}$$

the most important departures from a fixed Keplerian ellipse are the consequences of the torque exerted by the equatorial bulge which J_2 represents on the orbiting satellite, and of the different radial variation—$1/r^3$ in Equation (18), instead of $1/r$. As for any torque about an axis at an angle to that of the principal revolution, the result is a precession of the orbital plane, expressed as a regression of the *node*, the point where the orbit crosses the equator referred to an inertially fixed point—according to the formula:[4]

$$d\Omega/dt = -\sqrt{kM}\,[3J_2a_e^2/2(1 - e^2)^2a^{7/2}]\cos i + O(J_2^2) \tag{26}$$

The other principal effect of the equatorial bulge is a steady motion of the *perigee*—the angle from the equator to the point of closest approach to the earth—according to the formula:[4]

$$d\omega/dt = -\sqrt{kM}\,[3J_2a_e^2/4(1 - e^2)^2a^{7/2}]\,(5\cos^2 i\text{-}1) + O(J_2^2) \tag{27}$$

The elements of the Keplerian orbit which appear in Equations (26) and (27) are shown in Figures 2 and 3. At any instant, the six quantities expressing the satellite position and velocity referred to the earth's equator and equinox, $\{x, y, z, \dot{x}, \dot{y}, \dot{z}\}$, can be transformed to six osculating Keplerian elements $\{a, e, i, M, \omega, \Omega\}$, which can be regarded as an alternative coordinate system. This representation is particularly appropriate for geodetic use of satellites, because the orbits which are most useful have only moderate periodic variations from a Keplerian ellipse with the slow rates of motion given by Equations (26) and (27); at any time, they will vary from such a precessing ellipse by less than 10 kilometers.

To use this moving Keplerian ellipse geodetically, the spherical harmonic expression of the gravitational field by Equation (18) to orbital coordinates, a purely geometrical transformation:[4]

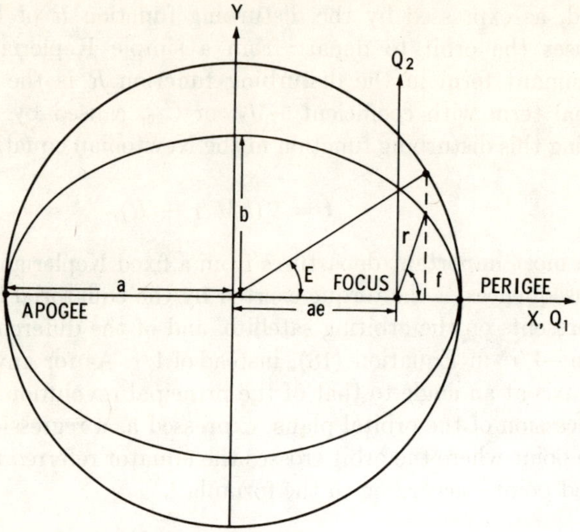

FIG. 2. Orbital Ellipse.

$$V = kM\left[\frac{1}{r} + \frac{1}{a}\sum_{l=2}^{\infty}\left(\frac{a_e}{a}\right)^l \sum_{m=0}^{l}\sum_{p=0}^{l} F_{lmp}(i) \sum_{q=-\infty}^{\infty} G_{lpq}(e)\right.$$

$$\times\left(\begin{Bmatrix} C_{lm} \\ -S_{lm}\end{Bmatrix}_{l-m\ \text{odd}}^{l-m\ \text{even}} \cos\{(l-2p)\omega + (l-2p+q)M + m(\Omega-\theta)\}\right.$$

$$\left.\left. + \begin{Bmatrix} S_{lm} \\ C_{lm}\end{Bmatrix}_{l-m\ \text{odd}}^{l-m\ \text{even}} \sin\{(l-2p)\omega + (l-2p+q)M + m(\Omega-\theta)\}\right)\right] \tag{28}$$

where θ is the Greenwich sidereal time—the angle between the equinox and the Greenwich Meridian—and $F_{lmp}(i)$ and $G_{lpq}(e)$ are polynomials in the sine and cosine of the inclination and the eccentricity respectively. The inclination function $F_{lmp}(i)$ arises from the rotation of the coordinate system to which the potential V is referred from the equatorial plane to the orbital plane, while the eccentricity function arises from the transformation of V from the circular co-

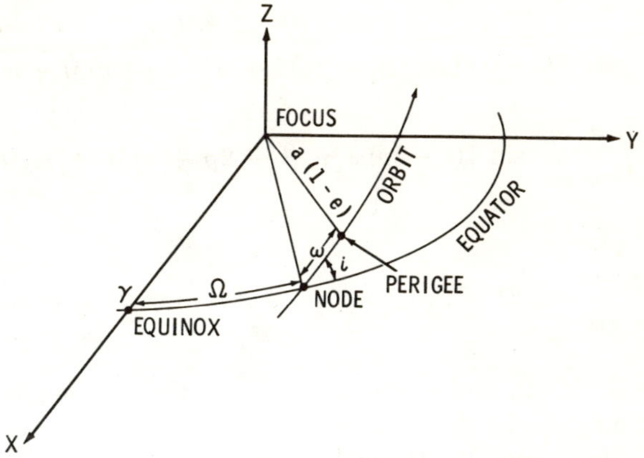

FIG. 3. Orbital Orientation.

ordinates r,f (see Figure 2) to the Keplerian elements a, e, M. In the unperturbed case, a and e are constant, while M changes secularly:

$$\dot{M} = n = [kMa^{-3}]^{\frac{1}{2}}, \tag{29}$$

known as Kepler's law, (M on the right of Equation (29) is the earth's mass).

The corresponding transformation of the equations of motion (25) from rectangular coordinates to Keplerian elements results in the Lagrangian planetary equations (4), (5), (6), which express the rates-of-change of the Keplerian elements ($\dot{a}$, $\dot{e}$, di/dt, $\dot{M}$, $\dot{\Omega}$, $\dot{\omega}$) in terms of the elements themselves and derivativates of the disturbing function R with respect to them. For example[4,6]:

$$d\Omega/dt = [na^2 (1 - e^2)^{\frac{1}{2}} \sin i]^{-1} (\partial R/\partial i) \tag{30}$$

If a, e, and i are assumed constant on the right side of Equation (30) and ω, Ω, and M to be secularly changing, then with R defined by Equations (24) and (28) the equations can be integrated. For example, for the perturbation of the node by a particular term V_{lmpq} of the potential in Equation (28) there is obtained:[4]

$$\Delta\Omega_{lmpq} = \frac{kMa_e^l(\partial F_{lmp}/\partial i)G_{\partial pq}}{na^{l+3}(l-e^2)^{\frac{1}{2}}\sin i\{(l-2p)\dot{\omega}+(l-2p+q)\dot{M}+m(\dot{\Omega}-\dot{\theta})\}}$$

$$\times\left[\begin{Bmatrix} C_{lm} \\ -S_{lm}\end{Bmatrix}^{l-m \text{ even}}_{l-m \text{ odd}} \sin\{(l-2p)\omega+(l-2p+q)M+m(\Omega-\theta)\}\right) \quad (31)$$

$$\left. -\begin{Bmatrix} S_{lm} \\ C_{lm}\end{Bmatrix}^{l-m \text{ even}}_{l-m \text{ odd}} \cos\{(l-2p)\omega+(l-2p+q)M+m(\Omega-\theta)\}\right]$$

The rates $\dot{\omega}$, $\dot{M}$, $\dot{\Omega}$, and $\dot{\theta}$ which appear in the denominator differ greatly in magnitude: $\dot{\omega}$ and $\dot{\Omega}$ are on the order of 0.01 cycles/day; $\dot{\theta}$, 1 cycle/day; and $\dot{M}$, 10 cycles/day. Hence the principal perturbation due to a particular variation V_{lm} in the gravitational field will depend on the combination (p, q) of the other two subscripts. In particular, $(l - 2p + q)$ must be zero, to eliminate the high rate $\dot{M}$. Furthermore, terms with m zero, i.e., the zonal harmonics—will have a greater effect because of the absence of the rate $\dot{\theta}$, which is the mathematization of the "averaging out" effect of the earth's rotation.

The zonal harmonics symmetric about the equator—l even, m zero—will have a zero rate $(l-2p)\dot{\omega}+(l-2p+q)\dot{M}+m(\dot{\Omega}-\dot{\theta})$, and hence have a secular, rather than periodic effect, similar to that of the leading term C_{20} given by Equations (26) and (27). The zonal harmonics skew-symmetric about the equator—l odd, m zero—have a leading effect with a period equal to one rotation of perigee, ω. The term of this type first detected was a perturbation of the eccentricity by C_{30}, notorious as the "pear-shaped" term:[4]

$$\Delta e = [3C_{30}(1 - 5\sin^2 i/4)/2na^6(1-e^2)^2\dot{\omega}]\sin\omega \quad (32)$$

Of the perturbations by tesseral harmonics, that which has probably received the most attention is by C_{22}, S_{22}, which corresponds to the equatorial ellipticity of the geoid[4]:

$$\Delta M_{22} = \frac{9\sin^2 i}{4na^5(1-e^2)^{3/2}(\dot{\Omega}-\dot{\theta})}[-C_{22}\sin 2(\Omega-\theta) + S_{22}\cos 2(\Omega-\theta)] \quad (33)$$

a semi-daily oscillation in the mean anomaly.

Nowadays several other terms are included in orbit calculation and determination, and some specialized analyses include spherical harmonics as high as the 15th degree.

Interesting exceptions to the rule that terms with the mean anomaly M present are small sometimes occur because of the phenomenon of resonance:

$$(l - 2p)\dot{\omega} + (l - 2p + q)\dot{M} + m(\dot{\Omega} - \dot{\theta}) \approx 0 \qquad (34)$$

In these cases, the important effect arises through that for da/dt, which depends on $\partial R/\partial M$. This term is significant because of the dependence of the mean motion n on the semi-major axis a, as expressed by Equation (29). Hence an acceleration results:

$$\ddot{M} = (\partial n/\partial a)\dot{a} = (3/a^2)\ (\partial R/\partial M) \qquad (35)$$

Physically, this effect constitutes a change in the energy, which depends only on the semi-major axis of the orbit.

Integrating Equation (35) twice after the manner of Equation (31) to obtain perturbation ΔM of the mean anomaly would result in the small rate of Equation (34) appearing squared in the denominator, and hence in a considerable amplification. Sometimes however, this manner of integration is not legitimate because the rate (34) is so close to zero than the angle never goes through a complete cycle; instead it liberates about an equilibrium point, analogous to a pendulum. This condition is true for the 24-hour communication satellites, for which:

$$\dot{\omega} + \dot{M} + \dot{\Omega} - \dot{\theta} \approx 0 \qquad (36)$$

Equation (34) can reduce to (36) for any term where $(l - 2p)$ can equal m, i.e., where $(l - m)$ is even. The communication satellites launched since 1963 have been very sensitive measures of C_{22}, S_{22} and C_{33}, S_{33}.

Other cases of near resonance frequently detected in near satellite orbits satisfy the condition:

$$\dot{\omega} + \dot{M} + m(\dot{\Omega} - \dot{\theta}) \approx 0 \qquad (37)$$

for $12 \leq m \leq 15$. Oscillations of 100-200 meters in periods of more than two days occur. These oscillations must be due to harmonics such as $C_{13,13}$, $S_{13,13}$.

The linear theory described above is inadequate in the case of perturbations by the oblateness C_{20}, since it is about 1,000 times as big as the other terms. Hence if numerical integration is not used, the analytic theory must be developed to include $(C_{20})^2$ terms. As usually occurs when a problem is extended to non-linear terms, it becomes more complicated, and there are a considerable variety of possible solutions.

The greatest inadequacy in the mathematical description of close satellite orbits is the expression of drag effects, due to the incomplete understanding of the variations of atmospheric density in response to varying energy and particle fluxes from the sun. As a consequence, these perturbations are usually approximated in a rather empirical manner. Also the amount of drag is an important consideration in the selection of orbits to be used geodetically. Usually satellite orbits with perigee altitudes around 800 to 1500 km are chosen because they are high enough to escape the worst drag effects, but low enough to be still sensitive to the variations of the gravitational field.

In common with the other types of measurements, the determination of the gravitational field from satellite orbits suffers from incomplete and non-uniformly distributed observations. The principal types of tracking used in geodetic studies are radio Doppler and telescope cameras. The Doppler system which has had greatest application to geodesy is the U.S. Navy "Transit" system. The Transit system utilizes a crystal-oscillator controlled beacon in the satellite, which transmits on multiple frequencies in the 100-megacycle range to enable elimination of frequency-dependent ionospheric refraction. The most advanced camera tracking system is the Smithsonian Institution Astrophysical Observatory Baker-Nunn camera, which has an $f/1$ lens with 50 cm aperture.

Both types of tracking are incomplete in the sense that at any instant they measure, respectively, only one or two components of the six required to express the satellite position and velocity. They are also incomplete in that, for economic reasons, tracking stations are limited in number and irregularly spaced in location, so that coverage of an orbit closer than 1500 km is rather non-uniform. This limitation is most severe for camera tracking, which requires darkness and clear weather. Hence, even though the daily, semi-daily, etc. perturbations of the orbit are of higher frequency than the significant drag effects, determination of the gravitational coefficients can be appreciably distorted. Further compounding the difficulty are errors in station position sometimes having apparent effects on the observations comparable to those of the real perturbations of the orbit by the gravitational variations. Hence the station coordinates must also be considered as correctible parameters.

As a consequence of the large number of parameters to be corrected, the inadequate accounting for drag effects, and the irregular distribution of observations, any attempt to determine variations in the gravitational field (other than zonal harmonics) from satellites requires many observations of a variety of orbits. Hence large computers and rather elaborate programs must be applied, and progress has been made more by empirical "computer experimentation" than by rigorous mathematical deduction.[4,5]

In addition to close satellites, the moon has also been used to obtain information about the gravitational field of the earth. Today it is most useful in determining the product kM from its mean motion, through Kepler's third law:

$$kM(1 + \mu) = n^2a^3(1 + \beta)^3 \tag{38}$$

In Equation (38), μ is the ratio of the moon's mass to the earth's —1/81.30—and β is a small correction term of about 10^{-3} for the effect of the sun on the mean motion.

The moon enters in one more way to obtain information about the earth's figure through its attraction on the earth's bulge. This attraction causes the earth's axis of rotation to precess. The rate of precession also depends on the *dynamical flattening:*

$$H = (C - A)/C \tag{39}$$

where C and A are the moments of inertia, and hence is an accurate means of determining the ratio H.[7,8,9]

NUMERICAL RESULTS

Model Parameters

As described above (pages 000-000), the parameters of the ellipsoidal reference model are the flattening, f; the rate of rotation, ω; the semi-major axis a_e; and either the equatorial gravity γ_e or the product kM of the gravitational constant and the mass. New numerical values for these quantities have recently been adopted by the International Astronomical Union, and are given in Table I.

The most accurately known parameter is the mean rate of rotation of the earth with respect to inertial space, ω, determined by observation of the stars with photographic zenith tubes. Also accurately known is the ratio H [Equation (38)] from the precessional constant.

TABLE I

Reference ellipsoidal parameters[3,4,9]

Parameter	Symbol	IAU adopted value
Rotation rate	ω	$0.7292115085 \times 10^{-4}$ sec^{-1}
Dynamical flattening	H	1/305.51
Oblateness coefficient	J_2	$0.00108270 \pm .00000003$
Flattening	f	$1/298.25 \pm .01$
Grav. const. × mass	kM	$3.98603 \pm .00001 \times 10^{14}\ m^3/\text{sec}^2$
Semi-major axis	a_e	$6378160 \pm 10\ m$
Equatorial gravity	γ_e	$9.780316 \pm .000015\ m/\text{sec}$
Centrif. acc./eq. grav.	m	0.0034678
$\text{Sin}^2\ \varphi$ coeff. in γ	β_2	0.0053025
$\text{Sin}^2\ 2\varphi$ coeff. in γ	β_4	−0.0000059
P_{40} coeff. in V_0	J_4	−0.00000234

The best value of the flattening f is now obtained (see Equation (14)) from the gravitational coefficient J_2, which is determined from the rates-of-motion of several satellite orbit nodes and perigees in Equation (26) and (27). The value adopted for the product kM is a compromise between results obtained from the moon's motion (higher) and from close satellite motion and terrestrial data (lower). The semi-major axis a_e given in Table I is determined from astro-geodetic data; it has been confirmed closely by using the value of kM with a gravity γ_e found from gravimetry in Equation (8), and by the mean radial component of shifts in position of tracking stations obtained from close satellite orbits. The quantities m, γ_e, β_2, β_4, and J_4 are defined in Equations (7), (8), (9), and (11) respectively.[3,6]

Geoidal Variations

The dominant means today for determining the broad variations—more than 3,000 km in half-wavelength—of the gravity field are the orbits of close satellites, while finer detail is better obtained by terrestrial gravimetry where it is available. The most accurately determined variations are the odd-degree zonal harmonics, because the long periodic variations of argument ω, as in Equation (32), are large and accurately determinable, and because no other effects cause perturbations of the same period. The even-degree zonal harmonics are not so accurately known because of the variety of other perturbations which appear to be secular. Values are given in Table II.[10,11,12]

The determination of the tesseral harmonics, which depends on the smaller daily, semi-daily, etc. oscillations in the orbit, is improving rapidly due to better orbits and better tracking. Since 1965, the coefficients C_{lm}, S_{lm} obtained by different investigators[13,14,15,16] agree rather well for $(l + m) \leq 9$, despite the different techniques employed. Most of the larger discrepancies are for higher degree—7th or more—harmonics, and result from insufficient orbital variety to resolve ambiguities as to which terms are causing perturbations of a particular frequency. Values of normalized coefficients:

$$\{\bar{C}_{lm}, \bar{S}_{lm}\} = [(l+m)!/\{(l-m)!\,(2l+1)\,(2-\delta_{0m})\}] \times \{C_{lm}, S_{lm}\} \qquad (40)$$

obtained from a combination of satellite data with gravimetry are given in Table II.[17] The normalized functions all have a mean square of unity over the sphere, and hence the values of $\{\bar{C}_{lm}, \bar{S}_{lm}\}$ vary much less with the index m, and the root-mean-square contribution of a coefficient to the geoid may be obtained by multiplying it by the radius of the earth.

Figure 4 is the geoid corresponding to Table II.

TABLE II

Spherical harmonic coefficients of the earth's gravitational field

l	m	$\bar{C}_{lm} \times 10^6$	$\bar{S}_{lm} \times 10^6$	l	m	$\bar{C}_{lm} \times 10^6$	$\bar{S}_{lm} \times 10^6$
2	0	−484.17		5	0	0.04	
2	2	2.42	−1.36	5	1	0.00	−0.02
3	0	0.97		5	2	0.44	−0.28
3	1	1.79	0.18	5	3	−0.31	0.03
3	2	0.78	−0.75	5	4	0.02	0.11
3	3	0.57	1.42	5	5	0.10	−0.49
4	0	0.54		6	0	−0.18	
4	1	−0.56	−0.46	6	1	−0.10	0.10
4	2	0.30	0.60	6	2	−0.01	−0.28
4	3	0.92	−0.19	6	3	0.09	0.04
4	4	−0.06	0.32	6	4	−0.19	−0.43
				6	5	−0.17	−0.60
				6	6	−0.05	−0.60
				7	0	0.11	
				7	1	0.06	0.07
				7	2	0.29	0.08

GEOPHYSICAL IMPLICATIONS

The scientific interest of the geoidal variations expressed by Table II and Figure 4 is that they necessarily entail density irregularities in the earth, which in turn entail stress differences which require some mechanism of support. The logical reference model for geophysical purposes is not an arbitrary ellipsoid of revolution, but rather the rotating fluid specified by Equation (5).

From Equations (14) and (39):

$$C/Ma_e^2 = J_2/H \tag{41}$$

The values of J_2 and H in Table I give 0.330776 for C/Ma_e^2. Using this quantity with the ω, A_e, and kM from Table I in Equation (2)

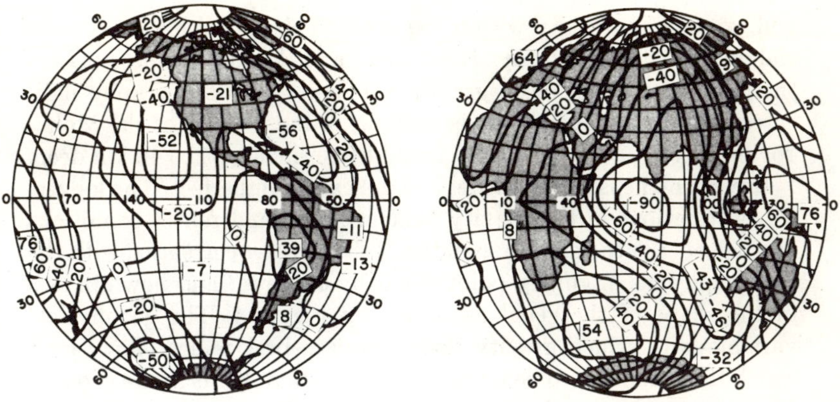

FIG. 4. Geoid heights in meters referred to an ellipsoid of flattening 1/298.25. Calculated by Stokes' formula from 300 n. mi. mean anomalies based on a combination of satellite data and terrestrial gravimetry.[17]

obtains 1/300.0 for the flattening f. This flattening f together with the quantity m from Table I in Equation (14) obtains 0.001071 for the hydrostatic J_2. Subtracting this figure from the observed J_2 of 0.0010827 in Table I and applying the normalization factor of $5^{-\frac{1}{2}}$ from Equation (40) obtains a $\Delta\bar{C}_{20}$ of about $- 5 \times 10^{-6}$. This value is appreciably larger than any of the other variations in Table II. If Figure 4 were redrawn to refer to an ellipsoid of 1/300.0, it would show mainly a progression of positive to negative geoid heights from equator to pole, and the dominant feature in the east-west direction would be a maximum near New Guinea. However, the fact that J_2 is associated with the rotation of the earth suggests a special explanation. Substituting the flattening of 1/298.25 in Equation (5) and solving for the rotation ω instead yields a rate 3 in 10^3 higher—equivalent to the rotation 1.5×10^7 years ago at the present rate of decrease due to tidal friction. Hence this discrepancy can be considered an estimate of the decay time for the response of the earth as a whole to stress.[8]

The dominant characteristic of the irregularities in the earth's gravity field, first noted more than 100 years ago, is that they are much gentler than would be expected if the topography were a load superimposed on a homogeneous earth; i.e., *isostasy* prevails: surface excesses and deficiencies of matter are *compensated* by deficiencies and excesses in density within the earth. This model has been most elaborately developed by Heiskanen and Vening-Meinesz.[1] Furthermore, the variation of gravity anomalies with topography indicates that this compensation is shallow, in the vicinity of the Mohorovicic discontinuity, or *Moho*, found by seismology 5 to 40 km deep.

However, the isostatic model is significantly imperfect. The longer-wave variations in the gravity field, which would reflect conditions much deeper in the mantle than the Moho, are not correlated with the topography, as originally pointed out by Jeffreys,[2] and now obvious in the shape of the geoid as shown by Figure 4. The geophysical variable with which gravity does correlate best is heat flow, with which there appears to be a negative correlation: highs in the geoid of Figure 4 are associated with lows in the heat flow, and geoid lows are associated with thermal highs. The heat flow also thus shows little correlation with topography, and hence must have a *compensation* more remarkable that isostasy, because the crustal rocks of the continents are estimated to contain much more radioactive heat sources than those of the oceans.

Negative correlation between gravity and heat flow would exist simply because hotter rocks are expanded and colder rocks are contracted. The density irregularities implied by the gravity may also be due to variations in chemical composition or in crystal structure, the latter of which would in turn be connected to heat flow through the thermal history of the material. Regardless of the source of the density irregularities, as previously stated they must be supported in some way. We thus must be concerned with the *rheology of the mantle:* the relationship between stress and the strain, or rate-of-strain, in the material. The simplest rheological assumption is elasticity: i.e., stress is proportional to strain. A model of the mantle elastic throughout indicates that the $\Delta\bar{C}_{20}$ of 5×10^{-6} implies stresses up to about 150×10^6 dynes/cm^2, and the sum of the remaining variations stresses up to 100×10^6 dynes/cm^2; if the upper mantle can be regarded as a fluid, then this *cushion* reduces the stresses required in the lower mantle to less than 60×10^6 dynes/cm^2.[19] These stresses are smaller than the strength of rocks at surface conditions; but extrapolating from laboratory experiments with rocks to the high temperatures and pressures of the mantle, it seems dubious that the mantle could support such stresses over geological periods of time without appreciably yielding. It is possible that the mantle is in a state of creep—matter in slow motion, but with stresses supported elastically; or repeated fracture and yielding, with elastic support of stresses re-established after yielding; or in a state of viscous flow—matter in steady motion, but stresses proportionate to strain rate, rather than to strain.[18,19] The hypothesis most often suggested to explain various evidences of large scale motions have been convection models with *viscous rheology*.[21,22,23] Aside from probably being drastic oversimplifications (like the elastic models), viscous models entail theoretical difficulties. The most definite difficulty is a zone of inhomogeneity in the mantle 400 to 1000 km deep now well established by seismology, indicating either a composition change or a phase transition. In either case, the zone would probably be an impenetrable barrier to any system of convection.

Breaking up the mantle into two systems, one say 50-500 km deep and the other 900-2900 km deep, there are two main questions: (1) is there a finite yield stress below which flow does not take place? and (2) what is the appropriate value of the viscosity: the ratio of stress to strain rate? The fact that earthquakes occur only at depths less than 700 km suggests in answer to question (1), "yes" for the upper mantle, but "no" for the lower mantle. For an answer to question (2), for the upper mantle, the principal evidence is the observed rates of uplift of areas from which ice or water loads are known to have been removed within the past few thousand years: they yield 10^{21} to 10^{22} dyne sec/cm^2, which in turn would result in rates of motion on the order of centimeters per year for convective systems, sufficient to satisfy some paleomagnetic and geologic evidence of displacement. However, these viscosities are dubious because the crust is also involved in the uplift, so it may be an elastic rebound as well as a viscous flow response; because comparable motion is not taking place in other areas of comparable disequilibrium; and because seismic evidence indicates considerable variation with depth of rheological properties in the upper mantle. For an answer to question (2) for the lower mantle, there is the previously calculated discrepancy $\Delta\bar{C}_{20}$ in the oblateness and the associated slowing down of the earth's rotation, which yields 10^{26} dyne sec/cm^2. Such a high viscosity would result in an extremely stiff and slow moving system.[1,20,21]

In summary, it appears that the mantle must be considered as comprising two main parts, with the upper envolving much more rapidly than the lower part, but that it is still unsure as to how much of each part is reflected in the long-wave variations of the geoid. Improvement should be expected with better determination of the shorter-wave variations of the geoid; better knowledge of the variations in heat flow; and better theoretical understanding of the rheology of the mantle.[1,2,8,18,19,20,21,22,23]

THE SHAPE OF THE MOON

The shape of the moon's topographic surface is determinable photogrammetrically using photos taken from observatories on the earth. However, because of the relatively small angle for stereoscopic viewing provided by the $\pm 7°$ optical libration resulting from the eccentricity and inclination of the earth-moon orbit coupled with the great distance of the moon, there are probably errors of a few hundred meters in the overall shape of the moon.

Until recently, information about the moon's gravity field was even more limited. It was confined to the mass, from its effect on the location of the earth with respect to the earth-moon barycenter, the center of mass of the earth-moon system; and to the second degree harmonics from 1) the effect of the moon on the motion of its own node and perigee; and 2) the inclination of the moon's equator and its physical libration about its rotation axis, resulting from the torques exercised by the earth. The combination of the lunar motion and the equatorial inclination yielded a ratio C/Ma_e^2 (see Equation (41)) of 0.643: much closer to a hollow sphere (2/3) than to a homogeneous body (2/5), which is difficult to reconcile with a physically reasonable picture of the moon's interior.

The advent of lunar satellites in 1966 produced considerably more information. The use of a satellite to determine the moon's gravity field is similar to that previously described for an earth satellite, except that 1) the harmonic coefficients C_{lm}, S_{lm} should be appreciably larger, because the stress differences resulting from a given magnitude density irregularity should be smaller in the less massive body; 2) the amplitudes of the perturbations should be larger, because of the slower rate of rotation $\dot{\theta}$ of the moon and slower rate of node and perigee motion $\dot{\Omega}$, $\dot{\omega}$ of the lunar satellite; 3) the satellite can be tracked almost continuously from the earth.

The results obtained thus far are all from orbits of one inclination,

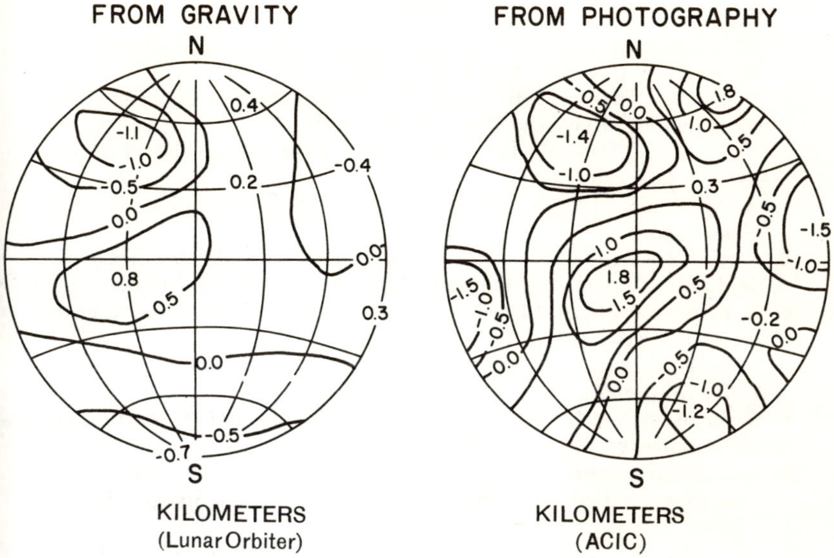

FIG. 5. Comparison of lunar topography obtained from gravity field by lunar satellite with that obtained from photography from earth.

and hence the analysis is considerably hindered by the ambiguity problem. The data are still insufficient to determine a new J_2 and thence a new C/Ma_e^2, but some interesting results have been obtained for some tesseral harmonics, from which the principal conclusions to be drawn are:

1. The variations of the lunar gravitational field are smaller than predicted from the earth's gravitational field under the equal stress-difference implication hypothesis.

2. The lunar gravity variations have a much higher correlation with the topography than is the case for the earth, but an appreciably smaller amplitude than if the lunar topography were a surface load rigidly supported by a homogeneous sphere—implying some sort of isostatic compensation. Figure 5 shows the lunar gravity field transformed into the equivalent variations of the surface topography, together with the topographic variations obtained by photography.[24,25]

REFERENCES

1. W. A. Heiskanen and F. A. Vening-Meinesz, The Earth and its Gravity Field, McGraw-Hill Book Co., New York, 1958.
2. H. Jeffreys, The Earth: Its Origin, History and Physical Constitution, 4th ed., Cambridge University Press, London, 1959.
3. W. M. Kaula, Determination of the Earth's Gravitational Field, Rev. Geophys., **1**, 507 (1963).
4. W. M. Kaula, Theory of Satellite Geodesy, Blaisdell Publ. Co., Waltham, Mass., 1966.
5. A. H. Cook, Contribution of Observations of Satellites to the Determination of the Earth's Gravitational Potential, Space Sci. Rev., **2**, 355 (1963).
6. D. Brouwer and G. M. Clemence, Methods of Celestial Mechanics, Academic Press, Inc., New York, 1961.
7. I. Fischer, Parallax of the Moon in Terms of a World Geodetic System, Astron. J., **67**, 373 (1962).
8. W. H. Munk and G. J. F. MacDonald, The Rotation of the Earth, Cambridge University Press, London, 1960.
9. W. L. Sjogren and D. W. Trask, Results on Physical Constants and Related Data from the Radio Tracking of Mariner (Venus) and Ranger III—VII missions, J. Spacecraft, **2**, 687, (1965).
10. D. G. King-Hele, and G. E. Cook, The Even Zonal Harmonics in the Earth's Gravitational Potential, Geophys. J., **10**, 17, (1965).
11. D. G. King-Hele and G. E. Cook, The Odd Zonal Harmonics in the Earth's Gravitational Potential, Planet. Space Sci., **13**, 1213, (1965).
12. Y. Kozai, New Determination of Zonal Harmonic Coefficients in the Earth's Gravitational Potential, Publ. Astron. Soc. Japan, **16**, 262 (1964).
13. R. J. Anderle, Geodetic Parameter Set NWL-5E-6 Based on Doppler Satellite Observations, NWL Rept. 1978, 32 pp., also in Use of Artificial Satellites for Geodesy, edited by G. Veis, in press, Athens, 1966.
14. E. M. Gaposhkin, A Dynamical Solution for the Tesseral Harmonics of the Geopotential and for Station Coordinates (abstract), Trans. Am. Geophys. Union, **47**, 47, (1966).
15. W. H. Guier, and R. R. Newton, The Earth's Gravity Field as Deduced from the Doppler Tracking of Five Satellites, J. Geophys. Res., **70**, 4613, (1965).
16. W. M. Kaula, Tesseral Harmonics of the Earth's Gravitational Field from Camera Tracking of Satellites, J. Geophys. Res., **71**, 4377, (1966).
17. W. M. Kaula, Tests and Combination of Satellite Determinations of the Gravity Field with Gravimetry, J. Geophys. Res., **71**, 5303, (1966).
18. R. B. Gordon, Diffusion Creep in the Earth's Mantle, J. Geophys. Res. **70**, 2413, (1965).
19. W. M. Kaula, Elastic Models of the Mantle Corresponding to Variations in the External Gravity Field, J. Geophys. Res., **68**, 4967, (1963).
20. G. J. F. MacDonald, The Deep Structure of Continents, Revs. Geophys., **1**, 587, (1963).

21. R. K. McConnell, Jr., Isostatic Adjustment in a Layered Earth, J. Geophys. Res., **70,** 5171, (1965).
22. D. P. McKenzie, The Viscosity of the Lower Mantle, J. Geophys. Res., **71,** 3995, (1966).
23. D. C. Tozer, Heat Transfer and Convection Currents, Phil. Trans. Roy. Soc., **258 A,** 252, (1965).
24. W. H. Michael, R. H. Tolson, and J. P. Gapcynski, Lunar Orbiter: Tracking Data Indicate Properties of Moon's Gravitational Field, Science, **153,** 1102, (1966).
25. W. M. Kaula, Recent Developments in Determination of the Lunar Gravitational Field from Satellite Orbits, Phil. Trans. Roy. Soc., **A,** in press, (1967).

Part II

SPACE

INTRODUCTION

A casual concept of space is emptiness and darkness. But interplanetary space is really not like that. It is filled with starlight, cosmic rays, plasma, magnetic fields, micrometeorites, comets and recently, artificial satellites.

One could say the terrestrial environment stops at the magnetopause—the end of the earth's magnetic field. Outside this we are in the solar environment. The boundary region has many interesting properties— collisionless shocks, turbulence, particle acceleration processes, to name a few. Somewhat outside the magnetopause we come to the undisturbed interplanetary environment. This region is dominated by the sun. In it we find the outflowing of the solar atmosphere—the solar wind—and also the magnetic field of the sun which is swept outwards by the solar wind. The sun also occasionally emits high energy protons, misnamed solar cosmic rays. Real "cosmic" rays diffuse into the solar system from outside. Dust also drifts into the solar system from outside.

In this section we want to consider the contents of space roughly from the edge of the earth's magnetic field outwards.

Chapter 8

THE INTERPLANETARY MEDIUM

Norman F. Ness

INTRODUCTION

In addition to the planetary, cometary and meteoritic objects and debris which are in various types of heliocentric orbit, interplanetary space is now known to be filled with a tenuous magnetized plasma of solar origin. This plasma is not stationary in space but rather is moving radially away from the Sun at a very high velocity, approximately 400 km/sec at 1 AU. At the present time the distance to which this interplanetary gas extends is unknown, but theoretical extrapolations indicate that the scale of the interplanetary "gas bubble" created by this plasma flow may be as large as 100 AU in diameter. At this distance the interplanetary medium merges with the interstellar medium at a boundary which may at times be very sharp and at other times very diffuse, the type of boundary depending upon the energy of the test particle used to define it. Because of the vast distances involved and the relatively small effects observable on the surface of the earth, investigation of these interplanetary phenomena has proceeded at a restricted rate until quite recently. The advent of the satellite or space probe "portable laboratory bench" upon which physical experiments could be placed now permits direct in-situ measurements of the properties of the interplanetary medium.

That the Sun emits ionized gases at high velocities was originally suggested on the basis of observations of very high temperatures in the solar corona. Detailed study of the precise mechanism for the acceleration of coronal gases into interplanetary space is among the most active areas of current research in solar physics. This is particularly true with respect to a study of the physics of solar flares, which are spatially rather minute on the scale of the solar disk. However, these intense eruptive disturbances are followed in time sequence by a variety of significant disturbances of the terrestrial environment. These disturbances cover a broad electromagnetic spectrum and show time durations from seconds to days.

The study of the interplanetary medium historically was begun by investigation of characteristic disturbances of the geomagnetic field following intervals of solar activity (see Chapter 1). Such disturbances or magnetic storms were in general individual and specific transient events which could be identified as being associated with specific activity on the Sun. In addition to the classic sudden commencement type magnetic storm, there exists a recurrent type of geomagnetic disturbance referred to as an M-region event by Bartels[1] in 1937. It was postulated that these were associated with persistent particle streams emitted from specific long lived regions on the Sun since they occurred at time intervals of approximately 27 days. This interval corresponds to the synodic rotation period of low latitude regions on the Sun. The development of auroral displays following solar activity has also been noted as direct evidence for the *transient* emission of high energy particles from the Sun and subsequent propagation to the Earth and interaction with the Earth's atmosphere.

The detailed study of the problem of the temporary expansion of the solar corona into interplanetary space was begun many years ago by Chapman and Ferraro[2] in an attempt to explain the geomagnetic storm phenomenon. Working with the hypothesis that the Sun emitted an electrically neutral but ionized gas or plasma, as first proposed by Lindemann in 1919, Chapman and Ferraro in the early 1930's developed a theory of geomagnetic storms which explained many aspects of the overall storm phenomena. Rather gross estimates of the plasma flux from the Sun at this time were possible, although the inherent difficulty in the overall physical model was one in which the basic mechanisms were not well understood. These yielded plasma velocities of 1000–3000 km/sec.

The physical characteristics of the interplanetary medium have also been deduced in the past indirectly by observations of scattered electromagnetic radiation by the neutral and charged components of the interplanetary medium. Propagation effects on stellar radio sources have been observed when occulted by the solar atmosphere. A rapid increase in the electron concentration near the Sun has been deduced by this and by observing scattering of white light from the solar corona. The density estimate at a distance of approximately 22 solar radii indicates on the order of 1000 electrons/cc. Observations of the polarized component of the zodiacal light which is due to free-electron scattering at great distances from the Sun permit only upper limits on densities to be determined. That the zodiacal light includes a component scattered by free electrons in interplanetary space is an important fact in itself, as well as the results which show a maximum density of approximately 300 electrons/cm^3 at approximately 1 AU.

The most important area studied recently relating to the characteristics of the interplanetary medium have been the investigations associated with Type I cometary tails (see Chapter 12). These tails are not the classical dust tails considered in the past, which are attributed to solar radiation pressure blowing material away from the comet, but rather are composed of heavy ions, principally CO^+, N_2^+, CO_2^+, and CH^+. Such ion tails are observed to be always directed generally away from the Sun with their mean axis departing slightly from a radial line from the Sun through the head of the comet. Analysis of the structure and time variations of these tails indicates accelerations of material directed away from the Sun which far exceed any possible gravitational or mechanical force, or mechanism involving light pressure. Early in the 1950's Biermann[3] suggested that the observed acceleration of ionized molecules in the Type I cometary tails was due to a *continual* flux of solar plasma. Computations by Biermann indicated that fluxes on the order of 10^{11} to 10^{13} particles/cm^2/sec were required in order that the observed accelerations be due to momentum transfer from the solar gas to the cometary tail molecules as a result of simple collisions. We now know that some form of hydromagnetic fluid-like interaction of the solar plasma and the comet must be responsible for the Type I tails rather than single particle collisions. The velocities estimated for the solar plasma flux obtained from both coronal temperatures and geomagnetic storm observations led to particle densities of approximately 10^3 to 10^5 particles/cm^3. These very high flux values have not been substantiated from satellite measurements. Historically the use of comets as space probes to investigate the interplanetary medium was principally responsible for the concept of a continual flux of plasma from the surface of the Sun.

SOLAR WIND

In a series of papers beginning in 1957 on the investigation of the expansion of the solar corona into interplanetary space, Parker[4] developed the magnetohydrodynamic theory of the "solar wind," as the solar plasma flux was termed by him. The use of the descriptive phrase "wind" was primarily to emphasize the difference between two competing contemporary models which led to appreciably different velocities of the plasma flux in interplanetary space. An essentially static solution had been considered for many years to represent the steady state interplanetary medium into which transient "tongues" of plasma would be emitted following solar activity and thereby lead to magnetic storms. Parker showed that the only reasonable model of the interplanetary medium utilizing the available information on coronal temperatures and densities was of necessity hydrodynamic and most importantly supersonic. The theory predicted that the expansion of the solar corona was such that for temperatures between 1×10^6 to 3×10^6 °K, velocities of 400 to 1000 km/sec would be observed at the orbit of the Earth.

Since then the investigation of the interplanetary medium by artificial satellites and space probes has dramatically confirmed the model of the solar wind as developed by Parker. A tabular summary of the results of satellite and space probe measurements of the interplanetary plasma is presented in Table I.

The first direct measurements of the interplanetary plasma were those conducted by the Russians in 1959 on the moon probe Lunik I. These exploratory results showed a positive ion flux of approximately 2×10^8/cm^2/sec of an ion species possessing an energy per unit charge greater than 15 eV. Additional plasma measurements have been performed on the Mariner II space probe and the Explorer X and IMP I earth satellites. These more recent measurements have

TABLE I
(Interplanetary plasma
(Range/Average)

Spacecraft	Time	Ion flux (per cm²/sec)	Equivalent proton energy (eV)	Velocity (km/sec)	Density (per cm³)
Luna I	Jan. 1959	$\sim 2 \times 10^8$	>15	>60	<30
Luna II	Sept. 1959	$\sim 4 \times 10^8$	>15	>60	<30
Explorer X	March 1961	3×10^8	300–800	250–400/300	7–20
Mariner II	Aug.–Dec. 1962			400–700/505	0.2–70
IMP–1	Nov. 1963 to Feb. 1964	10^8–10^9 3×10^8	300–1000	250–440/378	2–10
Vela 2	July 1964–Present	—	—	280–750	—
OGO–1	Sept. 1964–Present		NOT YET REPORTED		
IMP–2	Oct.–Dec., 1964		NOT YET REPORTED		
Vela 3	July 1965–Present		NOT YET REPORTED		
Pioneer 6	Dec. 1965–Present		NOT YET REPORTED		
Luna–10	April–May 1966		NOT YET REPORTED		
OGO–3	June 1966–Present		NOT YET REPORTED		
Explorer 33	July 1966–Present		NOT YET REPORTED		
Pioneer 7	Aug. 1966–Present		NOT YET REPORTED		

demonstrated a continual flux of positive ions from the Sun consisting principally of hydrogen. The Mariner II results show approximately 5% helium nuclei. At 1 AU the direction of flow is observed to be principally within a few degrees of the Sun, and in the case of satellites gravitationally anchored to the Earth, the plasma flux is aberrated approximately 5° west of the Sun. This aberration is due to the heliocentric orbital velocity of the Earth at 30 km/sec when observing a radial plasma flux of approximately 400 km/sec. The early measurements by the Pioneer V magnetometer in 1960 indicated that the interplanetary plasma was magnetized to a few gammas ($1\ \gamma = 10^{-5}$ oerstad). The IMP I magnetic field experiment in 1963 has accurately measured the interplanetary magnetic field and its topological characteristics. In the remainder of this chapter a discussion of the present physical model and quantitative parameters describing the interplanetary medium and the solar origin of the interplanetary magnetic field will be given.

THEORETICAL MODEL OF THE INTERPLANETARY MEDIUM

In the brief theoretical treatment which follows, attention will be given to presenting a simplified but none the less quantitatively accurate description of the physical characteristics of the interplanetary medium. One of the principal problems related to a discussion of the interplanetary medium is its microscopic physical properties. Beyond several solar radii it can be considered to be a collisionless plasma because the mean free paths of the individual particles are very large compared to the scale of the phenomena being studied. In particular, if one assumes the kinetic temperature of the corona to be approximately 2×10^6 °K, and assuming a density of $10^8/\text{cm}^3$, then the mean free path is approximately 1000 km. A striking difference however occurs away from the surface of the Sun. At distances of approximately 1 AU if it is assumed that the kinetic temperature is 10^3 °K, and a density of $10/\text{cm}^3$ typical values of the mean free path are of the order of 10^5 km.

A precise measure of temperature and indeed the concept of temperature in such a rarefied plasma is difficult. The usual approach is to consider the observed plasma as a particle beam which shows a velocity distribution about the mean. The velocity distribution is assumed to be isotropic. Then by measurement of the RMS deviation in the magnitude of the directed velocity it is possible to deduce an RMS thermal energy/particle, and this energy determines the equivalent temperatures of the interplanetary medium. Using this approach values of 10^5–10^6 °K have been obtained from the Mariner II and Explorer X satellites. Employing these temperatures and with measured values of the density of the interplanetary medium of approximately $10/\text{cm}^3$ the mean free path becomes *comparable* with an astronomical unit. For the present discussion we shall assume the mean free paths to be very large so that we are discussing a collisionless plasma. The mechanism by which the various particles interact and propagate disturbances is related to the presence of the interplanetary magnetic field. Particles moving relative to one another generate electric and magnetic fields which communicate the presence of the particles and distort the interplanetary field.

With the large mean free path, the appropriate scale length is related to the presence of the magnetic field and is on the order of the proton cyclotron or Larmor radius. For particles in the solar plasma and a measured interplanetary magnetic field strength of approximately 5 gammas, the Larmor radius ranges between 500 to 2000 km.

In addition to being considered as a collisionless plasma, the solar wind is also a very highly conducting medium. The conductivity is proportional to the mean free path and on the scale of interest the solar wind can be described as an approximately infinitely conducting medium. To be more precise, the important factor is the characteristic magnetic Reynolds number of the plasma flow which measures the ratio of magnetic forces to inertial forces. In general this number is high both near the Sun and at the orbit of the Earth because of the very large scale distances involved. Combining the two facts that the interplanetary medium is collisionless and highly conducting with a very high magnetic Reynolds number justifies the description of the interplanetary medium as a continuum fluid flow. The basic equations describing the motion of the continuum fluid involve a coupling of the Maxwell's equations governing the electromagnetic phenomena with the hydrogynamic equations of the motion of a fluid. Maxwell's equations can be simplified since the displacement current is negligible compared to conduction currents for a good conductor and hence can be summarized as follows.

$$\nabla \times \mathbf{E} = -\partial \mathbf{H}/\partial t \qquad \nabla \cdot \mathbf{H} = 0$$

$$\nabla \times \mathbf{H} = 4\pi \mathbf{J} \qquad \nabla \cdot \mathbf{J} = 0 \tag{1}$$

In addition, Ohm's law, including the electrical fields caused by the motion of the medium in a magnetic field is given by

$$\mathbf{J} = \sigma[\mathbf{E} + \mathbf{V} \times \mathbf{H}] \tag{2}$$

In order that finite electrical currents exist in the case of an infinitely conducting medium, it is clear that the electrical fields induced are exactly opposite to the existing electrical field ($\mathbf{E} = -\mathbf{V} \times \mathbf{H}$). It should be noted that the motion of charged particles in a magnetic field is not isotropic so that the use of a vector description is not completely correct. The electrical conductivity, permittivity and permeability should really be described as tensor quantities. The hydrodynamic equations can be written in terms of the plasma density, and pressure, by neglecting viscous and gravitational forces

$$(\partial\rho/\partial t) + \nabla \cdot \rho\mathbf{V} = 0 \tag{3}$$

$$\rho(dV/dt) = \rho(\partial V/\partial t) + \rho V \cdot \nabla V = \mathbf{J} \times \mathbf{H} - \nabla P. \tag{4}$$

These basic equations can be combined to deduce certain aspects of the flow of the interplanetary plasma in free space. We have already discussed the kinetic properties of the solar plasma. Combining the sets of equations [(1), (2), (3), and (4)] couples the motion of the fluid and the variations of the magnetic field and yields the following equation for the motion of the magnetic fields.

$$-\nabla \times \mathbf{E} = +\partial\mathbf{H}/\partial t = +(4\pi\sigma)^{-1}\nabla^2\mathbf{H} + \nabla \times \mathbf{V} \times \mathbf{H}. \tag{5}$$

Since the conductivity of the medium is high the first term in the equation describing the diffusion of the magnetic field can be ignored leaving the following equation:

$$\partial\mathbf{H}/\partial t = \nabla \times \mathbf{V} \times \mathbf{H}. \tag{6}$$

Integration of the equation over a closed surface in space and use of Stokes theorem for the right-hand side yields the following equation.

$$d\Phi/dt = \frac{d}{dt}\int \mathbf{H} \cdot d\mathbf{a} = \int (\partial\mathbf{H}/\partial t) \cdot d\mathbf{a} - \int (\mathbf{V} \times \mathbf{H}) \cdot d\mathbf{l} = 0. \tag{7}$$

Interpretation of this equation is that the change in magnetic flux associated with the time dependent part of the field must be constant if the same specimen of plasma is observed. This preservation of total flux in a contour enclosing a fixed sample of plasma, regardless of its motion, introduces the concept of "frozen-in" magnetic fields which move in unison with the plasma. This yields a magnetic field topology which is completely determined by the motion of the fluid.

In the preceding paragraphs we have shown that the motion of the fluid and the deformation of the magnetic lines of force are intimately coupled. Which of these two physical phenomena is the motivating and dominating force is not indicated thus far. One intuitively expects that if the plasma kinetic energy density of motion, as measured by $\frac{1}{2}nmV^2$, exceeds the magnetic energy density measured by $H^2/8\pi$ then fluid motion will govern the magnetic field topology rather than vice versa. It is possible to justify this statement by consideration of the conservation of momentum in the plasma, as derived from the previous equations.

$$\begin{aligned}\text{Energy density ratio} &= \text{inertial stress/magnetic stress}\\ &= \tfrac{1}{2}nmV^2/(H^2/8\pi). \qquad (8)\end{aligned}$$

The ratio of the two forces involved is numerically related to the supersonic Mach number at which the plasma flows. The Alfvén magnetohydrodynamic velocity for transverse wave propagation in a magnetized plasma is given by

$$V_a = H/(4\pi nm)^{1/2}. \qquad (9)$$

Rewriting the expression for the ratio of energy densities as below

$$\tfrac{1}{2}nmV^2/(H^2/8\pi) = V^2/V_a^2 = (M_A)^2, \qquad (10)$$

in the plasma and the magnetic field we see that the ratio is directly proportional to the square of the Alfvén Mach number of the fluid flow (M_A). In the case of the interplanetary medium, substitution of the observed values at a distance of 1 AU yields Mach numbers corresponding generally from 5 to 20. Thus the ratio of the energy density in the plasma flow is much higher than the magnetic field energy density.

Near the Sun, however, the ratio is not the same. Various parameters such as the plasma velocity, density and the magnetic field strength depend upon distance from the Sun, and hence the ratio of these parameters will vary. Indeed, the situation may well be that the magnetic field at the solar corona is a dominating factor over the plasma flow and that it is only at several solar radii that the plasma energy density exceeds the magnetic field energy density. In such a case the solar atmosphere would be described as corotating with the solar corona, and this could have an important effect upon the structure of the interplanetary medium. If such corotation existed it would imply that that magnetic field controls the flow of plasma out to an appreciable distance beyond the photosphere, at which distance these fields are observed with terrestrial instruments. At the present time the important question of whether or not the solar atmosphere corotates to an appreciable distance ($\sim$5 solar radii) has not been accurately established. However, the measurements from satellites indicate that corotation must not be a large scale phenomenon since if it were the direction of the plasma flow would tend to occur east of the Sun. Since those plasma measurements obtained to date which yield adequate directional information do not show on average any flux of plasma east of the Sun, it must be concluded that maximum corotation occurs out to only a few solar radii.

In order to deduce the magnetic field topology and the solar plasma stream characteristics, we must investigate the geometry of the plasma flow in a fixed coordinate system. The Sun rotates with a sidereal period of 2.94×10^6 radians per second at its equator. At a distance of 1 AU the moment arm associated with a fixed meridian plane in the Sun moves at approximately 400 km/sec. Consider the ejection of plasma radially from a specific point on the equator of the Sun at a given instant of time t. The locus of the particles which are continuously emitted from this source will trace out in space a spiral structure which has a characteristic Archimedean structure. This can be formally derived by inspection of the following equations. Let $d\omega = \Omega\, dt$ be the angle through which the Sun rotates in the time interval dt. In a differential time interval dt the plasma moves a radial distance $dr = v\, dt$ while the Sun rotates through an angle $d\omega$. At a distance r the rotation distance is $r\, d\omega$. The stream angle, Ψ which is the angle made by the locus of the particles emitted by the same source with a local radius vector is given by

$$\text{tangent}\ \psi = r\, d\omega/dr = r\, d\omega/V\, dt = r\Omega/V. \tag{11}$$

Integration of

$$d\omega/dr = \Omega/V$$

assuming V constant yields

$$\omega(r) = (\Omega/V)(r - r_0) + \omega_0 \tag{12}$$

which is the equation of an Archimedean spiral.

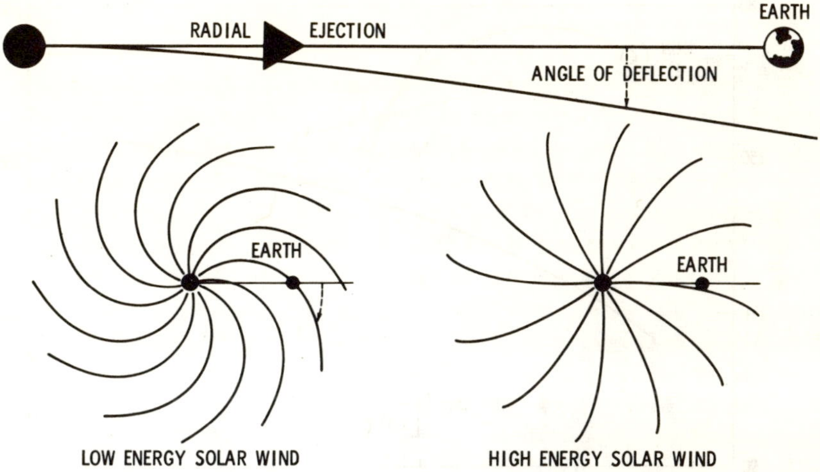

FIG. 1. Theoretical twisting of the interplanetary magnetic field lines of force caused by the rotation of the Sun and the radial flux of solar plasma wind as viewed from the North Ecliptic pole. The topology of the field lines, as well as the traces of the plasma streams, are Archimedean spirals.

The locus of the particles emitted from a fixed source on the Sun as it rotates is also a trace of a magnetic line of force since the motion of the plasma and field is so intimately coupled. Thus, a magnetic line of force originating at the solar photosphere is carried into interplanetary space along these stream lines. It is important to note that along the stream lines the direction of the plasma flow is still always radial from the Sun.

Figure 1 shows the projection on the plane of the ecliptic of an Archimedean spiral structure for two extreme limits of the velocity of the solar wind. A very "tightly wound" Archimedean spiral is observed when the velocity of the solar wind is low. The direction of the field lines approach radial lines when the velocity is high. This twisting of the interplanetary magnetic field lines in interplanetary space is a very important aspect of the structure of the medium.

The fact that an Archimedean spiral structure exists was theoretically deduced many years ago in the case of the study of transient emission of particles by the Sun. In this case particulate streams of solar corpuscular flux would have been observed to be twisted in a spiral configuration although they would not continuously fill up

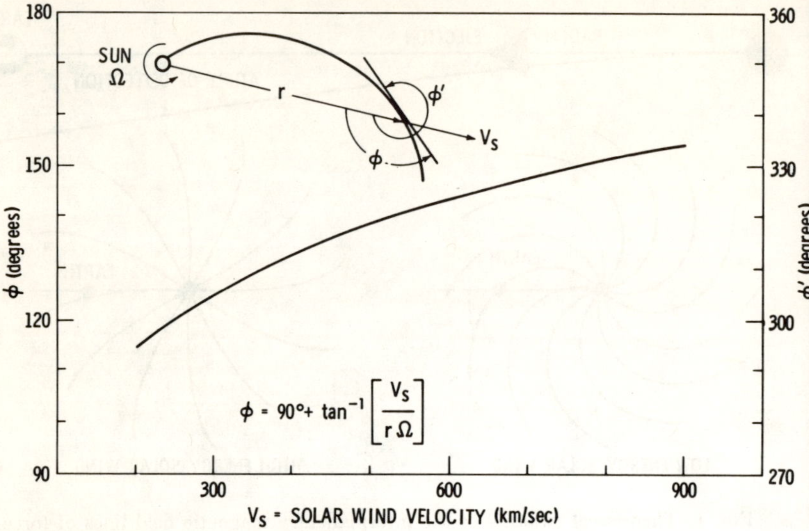

FIG. 2. Theoretical direction of the interplanetary magnetic field at 1 AU assuming a steady state expansion of the solar corona at an energy corresponding to a velocity $= V_S$ (from Ness et al., Ref. 7a).

the interplanetary gas bubble. Note that no directions of the interplanetary field lines are shown in Figure 1. The formulation deriving the direction of the interplanetary field angle at 1 AU has been computed for a range of velocity between 300 and 900 km/sec and is shown in Figure 2. In this case the "streaming angle Φ is defined as:

$$\Phi = 180° - \psi.$$

Since the Sun continuously emits plasma interplanetary space contains an approximately Archimedian spiral structure magnetic field which rotates with a synodic period of 27 days. The streaming angle ψ is approximately 45° at 1 AU and is a direct result of the approximate equivalence of the tangential velocity of the spiral structure at 1 AU with a radial plasma velocity. Parker has shown that the solar wind velocity is essentially independent of the radial distance from the sun, although it is very sensitive to the coronal temperature, when distances greater than 0.2 of an astronomical unit are considered. In order that the total mass flux be conserved it is therefore necessary that the plasma density fall off as $1/r^2$. It is possible to derive the magnetic field strengths observable as a function of distance from the Sun by knowing the radial component (B_0) of the magnetic fields on the surface of the Sun. The following two formulas express the radial, tangential components and total field components of the interplanetary field.

$$B_r(r) = B_0(r_s/r)^2 \tag{13}$$

$$B_\omega(r) = B_0(\Omega r/V_s)(r_s/r)^2 \tag{14}$$

$$B = B_0(r_s/r)^2[1 + (r\Omega/V_s)^2]^{1/2}. \tag{15}$$

PLASMA MEASUREMENTS

Following Gringauz' exploratory experiments the second direct measurement of the interplanetary plasma was made by a Faraday cup on the Explorer X satellite. However, the trajectory was such that the satellite remained always in the transition region surrounding the earth so that the plasma cannot be considered to be representative of the undisturbed interplanetary medium. The interaction of the solar plasma with the earth's magnetic field had modified the properties considerably (see Chapter 9). The high conductivity of the plasmas involved is illustrated by the fact that the solar wind and geomagnetic field do not mix or diffuse into each other appreciably. When the magnetic field showed a sizeable and rather steady field (indicative of the distorted geomagnetic field) the plasma cup response was essentially zero. When the field changed to low and disordered values (very likely of interplanetary origin) then the plasma cup indicated the presence of a significant plasma flux. Theory indicates the plasma should diffuse into the geomagnetic field only a few meters in the time it moves past the magnetosphere.

An electrostatic particle analyzer was flown on Mariner 2[5] to study the solar wind. This provided the first information on the energy spectrum of the solar wind far away from the earth so that free-stream undisturbed flow was observed. Figure 3 shows several energy spectra observed during the four month lifetime of the space probe. The instrument measures the flux of particles from the sun of a certain charge/mass ratio for several different velocities. It could not uniquely distinguish protons from alpha particles. The second peak in spectra

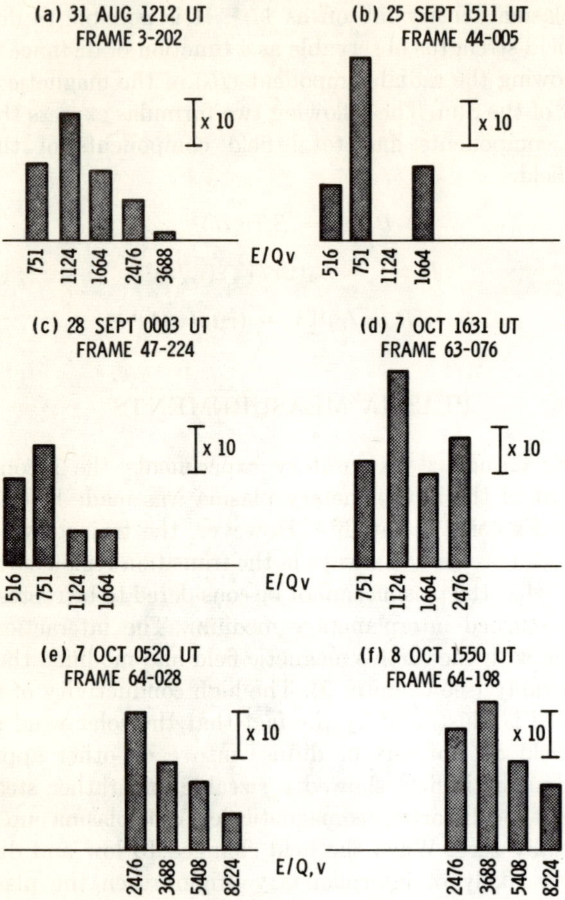

FIG. 3. Typical energy spectra of the solar wind as measured by the Mariner II curved plate electrostatic analyzer. The second peak with approximately twice the energy per unit charge (E/Q) is suggested to be due to the presence of helium nucleii in addition to hydrogen nucleii, the solar wind protons (from Snyder and Neugebauer, Ref. 5).

b and d of Figure 3 are almost certainly due to alpha particles with the same average velocity as the protons and therefore with four times their energy. Approximately 5 percent alpha particles are indicated.

The solar wind velocity measured on Mariner II showed important time variations. Figure 4 shows the velocity varying from 400 to 700

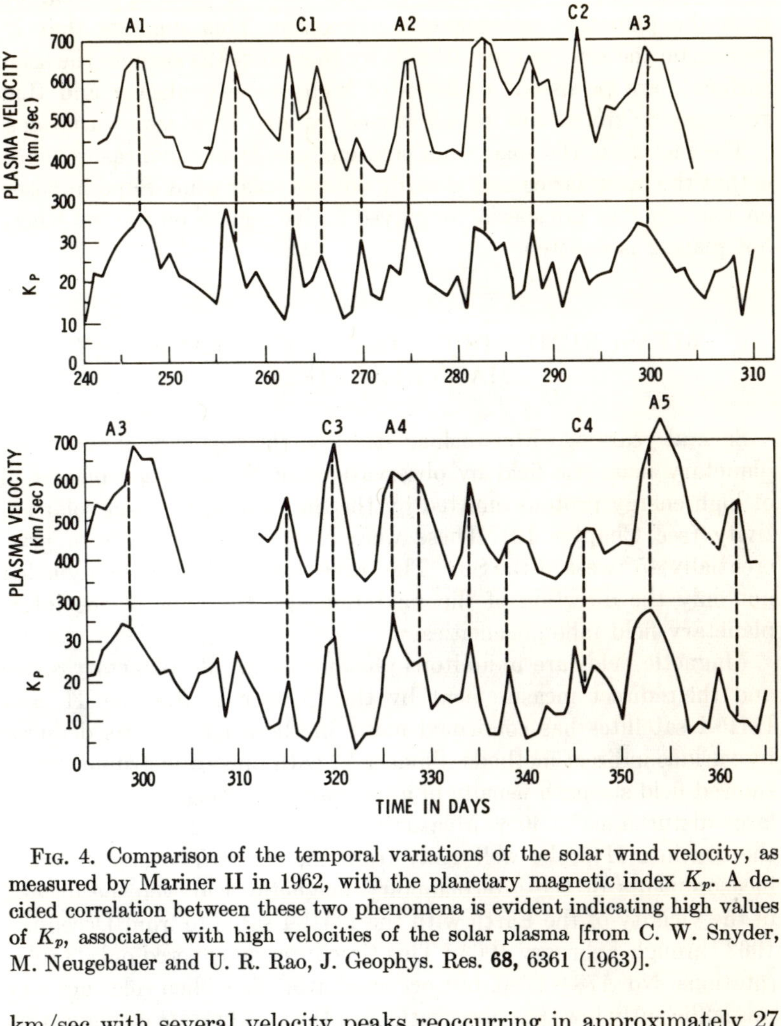

FIG. 4. Comparison of the temporal variations of the solar wind velocity, as measured by Mariner II in 1962, with the planetary magnetic index K_p. A decided correlation between these two phenomena is evident indicating high values of K_p, associated with high velocities of the solar plasma [from C. W. Snyder, M. Neugebauer and U. R. Rao, J. Geophys. Res. **68,** 6361 (1963)].

km/sec with several velocity peaks reoccurring in approximately 27 day periodicity for example A_1, A_2, A_3, A_4, A_5. This indicates a long lived stream of high velocity plasma emitted from one region on the sun that sweeps past the earth for five solar rotations. It has been known for a long time that there are geomagnetic storms observed at the earth's surface that recur with a 27 day period. This indicates a persistent solar surface feature called a *M region*. In Figure 4 the time variations of solar wind velocity show a striking correlation with the planetary geomagnetic index *Kp*. This suggests that M regions on the sun (not identifiable by themselves as such on the solar surface), are persistent emitters of high velocity plasma and that geomagnetic storms are indeed caused by solar wind time variations.

The picture of the solar wind that emerges from the measurements is that the hot solar corona steadily emits a solar wind. Superimposed on a steady flow are special long lived active regions on the sun where hot plasma is emitted.

MEASUREMENTS OF THE INTERPLANETARY MAGNETIC FIELD

Several years ago McCracken[6] deduced the topology of the interplanetary magnetic field by observations of the direction of arrival of high energy protons emitted by the Sun during times of solar activity (see Chapter 10). These were observed to come from preferentially 45° west of the Sun. The travel time and dispersion yielded not only the direction of the field but also the scale of the interplanetary field inhomogeneities.

Magnetic fields are ubiquitous phenomena in interplanetary space and their direct measurement by the Pioneer V, Mariner II, and IMP–I satellites has confirmed many of the original ideas deduced by various indirect methods. Pioneer V, with limited instrumentation showed field strength usually of a few gammas rising at the time of a large disturbance to 40 γ. Measurements were made for 4 months on the Mariner II probe well away from the earth. The first accurate measurements of the interplanetary magnetic field were performed in the vicinity of the Earth with the IMP–I satellite from December 1963 through February 1964.[7] This time interval covered only 3 solar rotations, No. 1784–1786, but occurred at a time when solar activity was sufficiently low that observations of the steady state configuration of the interplanetary magnetic field topology were possible. Measurements in the undisturbed interplanetary medium by IMP–I were possible only when the satellite was far removed from the effect of the interaction of the solar plasma with the Earth's magnetic field, as shown in Figure 5 (see Chapter 9).

These results dramatically confirm the spiral model as suggested

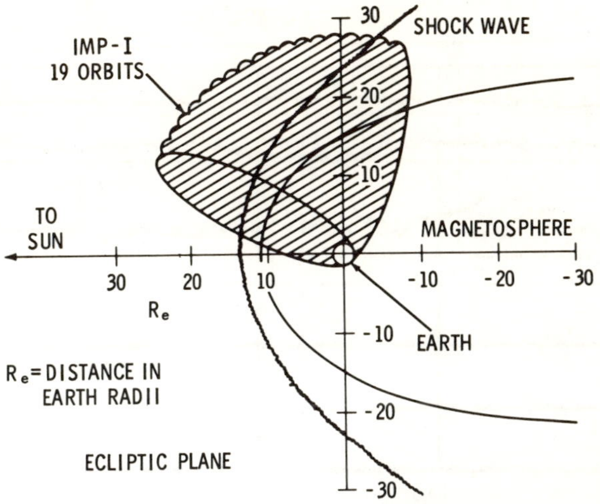

FIG. 5. Illustration of orbits 1–19 of the IMP–1 satellite as projected on the ecliptic plane showing the precession of the orbit in solar ecliptic coordinates. Sensible measurements of the interplanetary medium are possible only when the satellite is outside of the interaction region of the solar plasma with the geomagnetic field, indicated by the shock wave and magnetosphere boundary.

by Parker that the rotating Sun coupled with the continual radial ejection of ionized gases twists the interplanetary magnetic field into a classical Archimedean spiral. A sample of these measurements is shown in Figure 6. The direction of the interplanetary magnetic field is shown in a solar ecliptic coordinate system. This geocentric reference system consists of an X-axis pointed from the Earth to the Sun, the Y-axis in the plane of the ecliptic and the right-handed orthogonal Z-axis being directed northward to the pole of the ecliptic. The angle θ is the latitude of the direction of the field and Φ the longitude with the convention that the direction (0°, 0°) points to the Sun while the direction (0°, 180°) points away from the Sun. Included also are estimates of the mean square deviation of the fields over time intervals of 5.46 minutes. These are defined for each component as

$$\delta X = N^{-1} \sum_{i=1}^{N} (X_i - \bar{X})^2 \qquad N \leq 12 \qquad \bar{X} = N^{-1} \sum_{i=1}^{N} X_i. \quad (16)$$

Twelve or less discrete samples of the interplanetary magnetic field

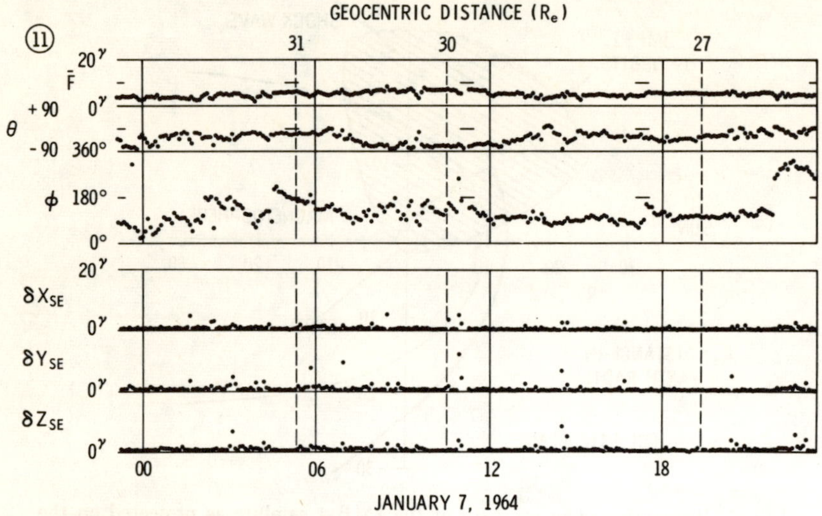

FIG. 6. Sample of the interplanetary magnetic field as obtained from the IMP–1 satellite on Orbit No. 11 in early January, 1964. The angles θ and ϕ measure the latitude and longitude of the field vector in a solar ecliptic coordinate system. Each sample point represents a 5.46 minute average of the field and the δX_{se}, δY_{se}, δZ_{se} represent the RMS deviation over that time interval. Note the constancy of magnitude, ~5 gammas and that the variations are mainly directional with a change from a positively directed field ($\phi \sim 135°$) to a negatively directed field ($\phi \sim 315°$) at 21:30 UT (from Ness et al., Ref. 7a).

determined at 20.48 second intervals are represented by each point illustrated. It is important to note in this figure that the field is directed near the ecliptic plane and approximately at the streaming angle $\psi \sim 45°$ or $\Phi = 135°$ associated with the field sense directed away from the Sun. However, at approximately 2300 on January 7, 1964, observations of the direction of the field indicate an abrupt change to a field pointed directly back along the streaming angle, $\Phi = 310°$, towards the Sun. Changes in the direction of the interplanetary field such as this have been observed many times during the three solar rotations studied. It suggests that the interplanetary field, and indeed the entire interplanetary medium, is structured into a set of sectors[8] in which the predominate sense of the interplanetary field is either positive or negative along the streaming angle depending upon in which sector the measurements are made.

A histogram of the distribution of the direction of the interplanetary

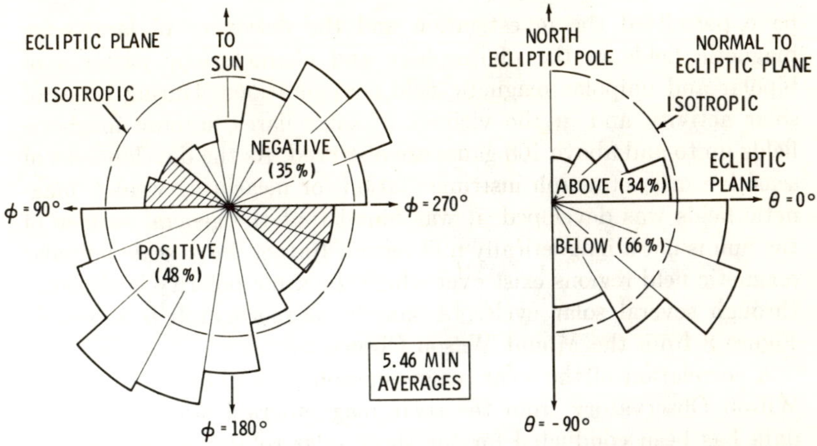

FIG. 7. Directional histogram of the interplanetary magnetic field as measured by IMP–1 during solar rotations 1784–1786 from Nov. 27, 1963 to Feb. 19, 1964. The field is observed to reflect the Archimedean spiral topology illustrated ideally in Figure 1. Although a slight but net southward component is observed. A total of approximately 13,000 sample points at 5.46 minute intervals was employed in constructing these diagrams (from Ness and Wilcox, Ref. 7b).

magnetic field as measured by IMP–I is shown in Figure 7. It is seen that there is a preferential distribution of the magnetic field direction centered about the two positions $\Phi = 135°$ and $315°$ expected from the measurements of the solar plasma velocity as shown in Table I and using the theoretical model illustrated in Figure 2. The field illustrates a strong preference to be near the ecliptic plane but with an indication of a slight but significant southward directed component. The average magnetic field strength is 5 gammas and principally between 3 and 6 gammas. At the present time analysis of the characteristics of this very recently obtained data are being conducted to deduce in more detail the over-all structure of the interplanetary medium during these three solar rotations.

It is possible to consider correlating the interplanetary magnetic field observed at 1 AU by IMP–I with that observed on the surface of the Sun. Measurements of the magnetic field of the Sun were begun early in the 1950's by Babcock[9] with his development of the solar magnetograph. This instrumentation measured the field strength along the line of sight by detecting the longitudinal Zeeman effect on an Fe line emitted in the photosphere. These solar magnetographs have permitted the investigation and the discovery of large-scale magnetic fields in the photosphere and characteristic patterns of bipolar and unipolar magnetic field configurations. During times of solar activity and in the vicinity of solar flares, intense magnetic fields up to and above 100 gauss are observed. As the development of sensitive magnetograph instrumentation for measuring weaker magnetic fields was developed, it was found that the general surface of the Sun is not a magnetically null region. Rather, large scale unipolar magnetic field regions exist over which the same polarity is observed through several solar cycles. A sample magnetograph is shown in Figure 8 from the Mount Wilson Observatory.

A correlation of the solar magnetograph data obtained at Mount Wilson Observatory from the solar magnetograph and the IMP–I data has been conducted for the three solar rotations covered. The procedure in the analysis was to assign a polarity to the magnetic field in interplanetary space which indicated whether or not the line of flux appeared to be coming away from the sun or directed towards the sun along the streaming angle. The polarity or direction over time

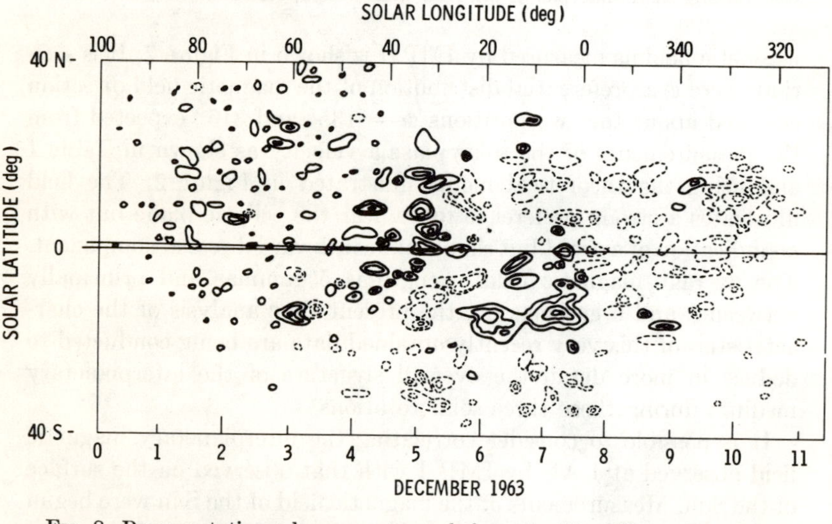

FIG. 8. Representative solar magnetograph from the Mt. Wilson Observatory showing contours of equal line of sight intensities of the photospheric magnetic fields and polarities at CMP (Central Meridian Passage) of date shown. Large scale unipolar field regions are observed. The contour intervals correspond to 2, 5, 12, and 25 gauss.

intervals of 12 hours were constructed with the requirement that a valid polarity showed coherency of 3 of the 4 subintervals into which the data was divided. Two time series, at an increment of 12 hours, were obtained represented by $+1$, -1, or 0 in the case of missing data.

An approximate auto-correlation of the interplanetary magnetic field observed at 1 AU shows that there is a decided peak at 27 days lag suggesting that indeed the Archimedean spiral pattern was being reproduced during successive solar cycles. A cross correlation between the direction of the field at 1 AU and the direction of the photospheric field was also computed. This approximate correlation function was normalized by the total number of possible comparisons as shown in the following formula

$$\begin{matrix}\text{Auto-}\\ \text{Cross-}\end{matrix}\text{correlation } (\tau) = \left[\frac{\sum S - \sum D}{\sum S + \sum D}\right], \tag{17}$$

where $S = S(\tau) = +1$ if the two samples of the time series being considered show the same value, i.e., either both $+1$ or -1. $S(\tau) = 0$ if the values are not the same. The quantity $D = D(\tau)$ is defined as the converse of $S(\tau)$, being $+1$ if the two samples of the time series are different and 0 if they are the same. The operation of computing the correlation function consists then of advancing one time series in time by τ relative to the other time series and then multiplying the corresponding terms. The normalization by the total number of terms contributing to the net value is performed to eliminate any spurious results which might be associated with the spacing of the gaps in the data. The basis of modern statistical communication theory is the mean-squared error criteria and such correlation techniques are the mathematical foundation for this work. The results of the cross correlation are shown in Figure 9, where it is seen that a coherent peak occurs at a time delay at 4.5 ± 0.5 days. This propagation indicates an average velocity of approximately 385 ± 45 km/sec for the solar plasma to transport solar magnetic fields from the surface of the sun to a distance of 1 AU where they were observed by the IMP–I satellite.

This value is substantially lower than the average of 505 km/sec obtained for the average plasma velocity from Mariner II measurements. The amount of solar activity during the flight of Mariner II

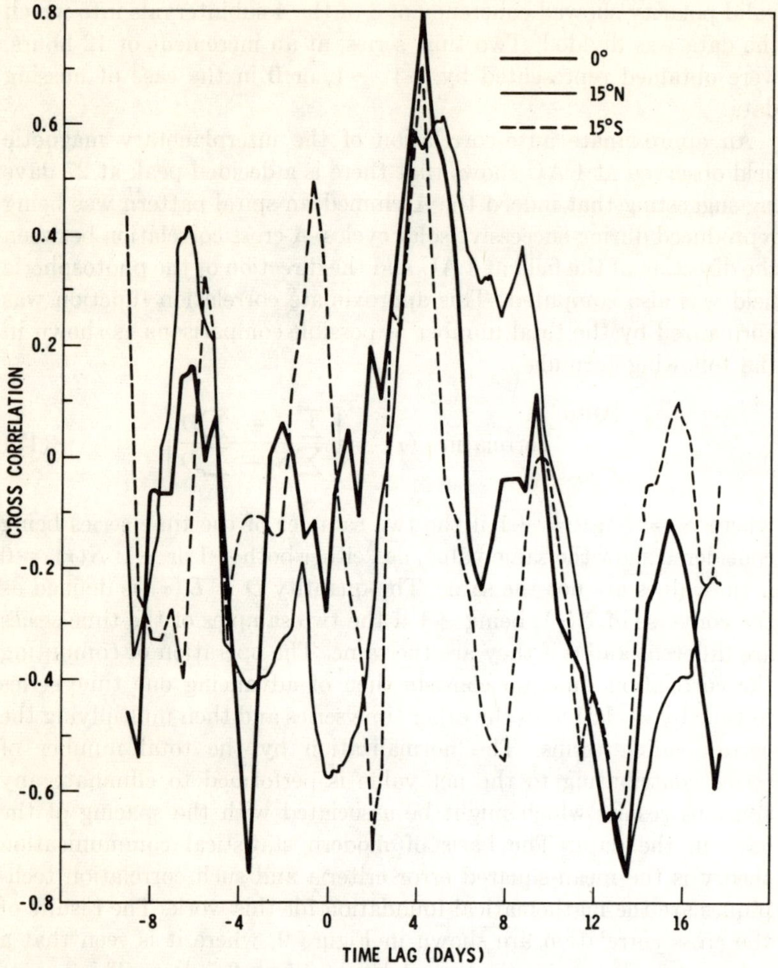

FIG. 9. Cross correlation between directions of the interplanetary magnetic field as measured at 1 a.u. near the Earth by IMP–1 and at the solar photospheric surface. Note the various latitude regions (15N, 0, 15S) used on the solar disk and the high and coherent correlation at a time lag of 4.5 ± 0.5 days (from Ness and Wilcox, Ref. 7b).

was much higher than when IMP–1 data were obtained so this difference may be a real effect of a variable plasma velocity associated with changes in solar activity. Indeed, if one inspects the Mariner data as shown in Figure 4 from the Mariner II plasma probe it is noted that at times of solar quiet (low planetary magnetic index Kp) the average velocity is close to 350 km/sec. The electrostatic analyzer measurements on the IMP–I satellite[10] have yielded an average plasma velocity of approximately 378 km/sec, a value in good agreement with the value estimated from the magnetic field correlations. This substantial agreement with the velocity obtained by cross correlating magnetic field directions observed at 1 AU with that observed on the surface of the sun lends strong evidence to the hypothesis that the interplanetary magnetic field is of solar origin.

From a superposed epoch analysis of the IMP 1 data it has been possible to deduce additional features concerning the characteristics of the interplanetary medium.[11] A quasi-stationary co-rotating structure in the interplanetary magnetic field has been observed during three solar rotations when interplanetary measurements were available. Figure 10 presents the graphical summary of results obtained by considering the directions of the interplanetary magnetic field as either directed away or towards the sun along the spiral angle. The direction of positive and negative has been defined in Figure 7. It is seen that the interplanetary field is directed predominantly away from the sun for 2/7 of a rotation, toward the sun for 2/7, away from the sun for 2/7 and finally toward the sun for 1/7 rotation.

As the structure rotates past the earth once every 27 days, it influences geomagnetic activity and cosmic ray intensity in a characteristic fashion. A recurrent stream of protons of several MEV energy was observed to be almost entirely contained within one of the four sectors. The structure has been observed not only in the interplanetary magnetic field but also in the solar wind plasma. In the leading portion of the sectors the magnitude of the interplanetary magnetic field, the solar wind velocity and the index of geomagnetic activity, Kp, tend to be larger than in the trailing portion.

Recently the spacecraft Pioneer 6 has provided very detailed information about the interplanetary magnetic field, the solar plasma and the propagation of cosmic rays in the interplanetary medium.[12] A sample of the data obtained from the magnetometer experiment for a short interval of time is shown in Figure 11. It is seen that the

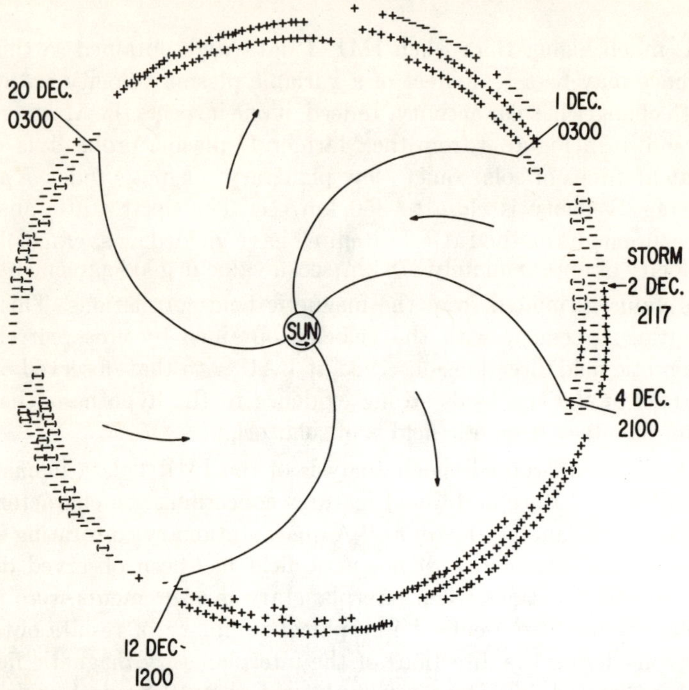

FIG. 10. The + and − signs along the circumference of the figure indicate the direction of the measured interplanetary magnetic field during the successive three hour intervals. Parenthesis indicate a time during which the field direction has moved beyond the "allowed regions" shown in Figure 7 for a few hours in a smooth and continuous manner. The inner portion of the figure is a schematic representation of the sector structure of the interplanetary magnetic field that is suggested from these observations. The deviations about the average streaming angle that are actually present are not shown (From Wilcox and Ness, reference 11).

magnitude of the field changes slowly although there are abrupt changes. Following a large solar flare event, the particle flux into interplanetary space permits a comparison of the local measurement of the interplanetary magnetic field as obtained from the magnetometer as well as an integrated measurement of the large scale magnetic field structure as deduced from the cosmic ray anisotropy. Direct comparison of the observed solar cosmic ray anisotropy with the field direction shows a high level of correlation. This indicates that the interplanetary medium is composed of a micro-structure of intertwined magnetic filaments which are rooted to the surface of the

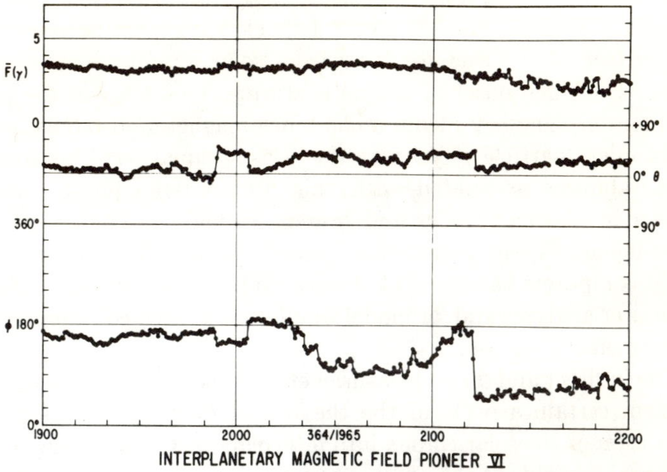

FIG. 11. Measurements of the interplanetary magnetic field on Pioneer 6 on December 30, 1965. Note the abrupt discontinuity in both magnitude and direction, which are interpretated to be filamentary bundles of flux tubes in the interplanetary magnetic field structure. At 2112 a large change in the direction of the magnetic field is detected without a noticeable change in magnitude. (Ness et al., reference 12).

sun. Finally the same spacecraft has provided information concerning the apparent anisotropy of solar plasma flux as influenced by the presence of the imbedded interplanetary magnetic field. Conservation of the magnetic moment as the solar wind adiabatically expands into interplanetary space leads to a temperature anisotropy determined by the magnetic field direction in the interplanetary plasma. The Vela satellites[12] have also yielded results on this temperature anisotropy as well as presented evidence for the non-radial flow of plasma from the sun. Deviations of up to 15° have been reported.

SUMMARY

At the present time investigations of the propagation of disturbances in the interplanetary medium are progressing both theoretically and experimentally. The suggestion that a collisionless magnetohydrodynamic shock wave existed in nature was first made by Gold in 1949 to explain the very sudden onset time of magnetic disturbances observed in geomagnetic records. He suggested that since the interplanetary medium was not a void a shock wave must propagate from the surface of the Sun resulting from solar activity to produce the observed sudden onset of magnetic storms. The study of the physics of the interplanetary medium combines magnetohydrodynamics and collisionless particle physics in the Vlasov equation. The analytical intractabilities associated with this differential equation are well known in plasma physics and present formidable challenges to the most ingenious and persevering research worker. The advent of high speed computers has stimulated research into the possibility of digital computer analyses and/or model simulation of such collisionless shock phenomena.

The indirect and direct measurements of the interplanetary medium confirm certain aspects of the theoretical models of the supersonic expansion of the solar corona into interplanetary space. The existence of magnetic fields at the photospheric surface of the Sun and the high magnetic plasma energy density values result in the stretching of the lines of force into interplanetary space which are then twisted in the plane of the ecliptic by the rotation of the Sun. The Archimedean spiral field lines provide preferential guidance of energetic particles originating in solar flares so that west limb events have a higher probability of leading to terrestrial disturbances when compared with east limb events. Among the future research problems facing space physicists today are more detailed descriptions of the interplanetary medium than those presently available and inspection of the various transient phenomena to delineate the specific characteristics of each event. As understanding of the interplanetary medium improves we can more effectively compensate for the effects of the disturbances propagating through the medium to Earth and thereby deduce the characteristics of the sources of these phenomena. Whether or not it will be possible to obtain meaningful measurements of interstellar properties by deduction from these measurements made within the interplanetary gas bubble has yet to be decided.

REFERENCES

1. S. Chapman and J. Bartels, Geomagnetism, Clarendon, Oxford, 1940.
2. S. Chapman, Solar Plasma, Geomagnetism and Aurora in Geophysics, in The Earth's Environment, Les Houches 1962 Lectures, edited by C. DeWitt et al., pp. 371–502, Gordon and Breach Scientific Publishers, New York, 1963.

3. L. Biermann, Kometenschweife und solare Korpuscular strahlung, Z. Astrophys. **29,** 274 (1951).
4. E. N. Parker, Interplanetary Dynamical Processes, Interscience Publishers, New York, 1963.
5. C. Snyder and M. Neugebauer, Interplanetary Solar Wind Measurements by Mariner II, Space Research IV, edited by P. Muller, p. 89, 1964.
6. K. G. McCracken, The Cosmic-Ray Flare Effect, 1, 2, and 3, J. Geophys. Res. **67,** 423 (1962).
7. N. F. Ness, C. S. Scearce and J. B. Seek, Initial Results of the IMP–I Magnetic Field Experiment, J. Geophys. Res. **69,** 3531 (1964); N. F. Ness and J. M. Wilcox, Solar Origin of the Interplanetary Magnetic Field, Phys. Rev. Letters **13,** 461 (1964).
8. H. S. Ahluwahlia and A. J. Dessler, Diurnal Variation of Cosmic Radiation Intensity Produced by a Solar Wind, Planet. Space Sci. **9,** 195 (1962).
9. H. W. Babcock, The Solar Magnetograph, *Astrophys. J.* **118,** 387 (1953).
10. John H. Wolfe, Richard W. Silva and Marilyn A. Myers, Observations of the Solar Wind during the flight of IMP–1, J. Geophys. Res. **71,** 1319 (1966).
11. John M. Wilcox and Norman F. Ness, A Quasi-stationary corotating structure in the Interplanetary Medium. J. Geophys. Res. **70,** 5793 (1965).
12. See Pioneer 6 Symposium, July 1966 issue J. Geophys. Res. 3287–3336 (1966).
13. A. J. Hundhausen *et al.* J. Geophys. Res. **72,** 87 (1967).

Chapter 9

THE BOUNDARY OF THE MAGNETOSPHERE

W. N. Hess and G. D. Mead

INTRODUCTION

It has been demonstrated in the previous chapter that the interplanetary medium in the vicinity of the earth is not just empty space, but instead is filled with a highly tenuous plasma which is being continuously blown radially out from the sun at speeds averaging 300–500 km/sec. This plasma consists primarily of ionized hydrogen (protons and electrons), and is electrically neutral—that is, enough electrons are present to balance out the total charge. The density at the earth is of the order of 1 ion/cm^3. Imbedded in this so-called "solar wind" is a magnetic field whose strength is of the order of 5 gammas (5×10^{-5} gauss) during quiet solar periods, but increases to many times this value during periods of high solar activity.

It has long been realized that plasmas and magnetic fields tend to confine one another. In an experimental machine such as a stellerator, for example, a strong magnetic field can compress and confine a hot dense plasma within a small region of space, without the need for confining walls. In like fashion, if a streaming plasma encounters a magnetic object such as a magnetized sphere, the plasma will confine the magnetic field to a limited region about the object. The object, in turn, will tend to exclude the plasma, creating a hole or cavity. The size of the cavity is determined by the energy density of the streaming plasma and the degree of magnetization of the object.

In addition, if the streaming velocity of the plasma is sufficiently high as to be highly supersonic in the magnetohydrodynamic sense—that is, if the velocity is much higher than the Alfvèn velocity in that medium—a detached shock wave may be produced in the region ahead of the cavity boundary. This process is analogous to the formation of a detached shock wave in front of an aerodynamic object traveling at hypersonic speeds (Mach numbers greater than 5) through the atmosphere.

In 1931 Chapman and Ferraro first predicted the confinement of the earth's magnetic field inside an elongated cavity during magnetic storms. The continual presence of such a cavity has been experimentally verified by many satellite observations, including those made by Explorers 10, 12, and 14, and the IMP and OGO satellites. The region inside the cavity is called the *magnetosphere*, and the boundary is termed the *magnetopause*. In addition, IMP 1 has verified the presence of a detached *shock wave*. The region between the magnetopause and the shock wave is usually referred to as the *magnetosheath* or *transition region*. Outside this magnetosheath, i.e., beyond the shock wave, conditions are characteristic of the interplanetary medium, and the presence of the magnetized earth has little or no effect.

The dimensions of the cavity depend, of course, on the intensity of the solar wind, although the dependence is rather weak. That is, large changes in the solar wind intensity produce comparatively small changes in the size of the cavity, as we shall see below. The distance to the magnetopause in the solar direction is typically around 10 earth radii, although distances less than $8R_e$ and more than $13R_e$ have occasionally been observed. The shock wave is located several earth radii beyond this. At 90 degrees to the solar direction, both the magnetopause and shock wave are observed to flare out to distances about 30–50 percent greater than the subsolar distances. In the region away from the sun, an extended tail is formed, similar in many respects to the tails of comets. The magnetopause has been observed at a distance of 80 R_E in the antisolar direction, well past the moon's distance of 60 R_E. Most likely, the tail extends much further than the moon, perhaps hundreds or even thousands of earth radii long.

In a semi-popular article, Cahill[1] has summarized the results of magnetic measurements obtained in the magnetosphere through 1964. Frank and Van Allen[2] have reviewed boundary phenomena as observed by both magnetic and particle probes.

HISTORICAL BACKGROUND

The concept that a plasma jet emanating from the sun could confine the earth's field inside a cavity was first suggested by Chapman and Ferraro in 1931. They were searching for an explanation for the magnetic storms that are frequently recorded by magnetic observatories. Since 1850 it had been well known that there exists a strong correlation between geomagnetic storm activity and the number of sunspots. An eleven-year variation in geomagnetic activity follows very closely the 11-year sunspot cycle, and short periods of unusually high visible solar activity are usually associated with large changes in the earth's magnetic field. This geomagnetic activity is often observed about a day or two following a large visible flare on the surface of the sun. In addition, there is a strong tendency for magnetic activity to recur in 27-day intervals, the synodic period of the sun's rotation. These observations strongly suggest that an active region on the sun's surface is in some way responsible for enhanced geomagnetic activity. The observed delay times indicate that the disturbance is transmitted via particles rather than electromagnetic waves.

A magnetic storm usually follows a rather typical pattern in low-latitude regions. The changes are most clearly seen in the horizontal component of the earth's field (see Chapter 1, Figure 18). Following a period of normal magnetic activity, a sudden increase of as much as 50 to 100 gammas is observed. The rise time for this increase, or sudden commencement, is of the order of 5 minutes, and is observed almost simultaneously all over the world. The increased field persists for a period of several hours, and is referred to as the initial phase of the storm. This is usually followed by a much larger world-wide decrease in the horizontal component, sometimes as much as 500 gammas or more. This main phase is accompanied by large fluctuations. Then follows a gradual recovery of the field intensity to the average pre-storm value, which may last several days.

Chapman and Ferraro proposed that the sudden increase in the horizontal field component coincided with the arrival at the earth of a jet of plasma emitted from the sun. As the plasma cloud approached the earth, the charged particles in the forward surface would be deflected by the magnetic field. Since particles of opposite charge would be deflected in opposite directions, an electric current would be set up at the boundary, separating the plasma from the magnetic field. The direction of this current would be such as to increase the horizontal component of the field near the earth, and to cancel out the dipole field within the plasma. In effect, because the conductivity of the plasma is essentially infinite, it would behave as a purely diamagnetic medium, thus preventing any penetration of the magnetic field.

As the plasma wave sweeps past the earth, the magnetic field would create a cavity in the plasma. This cavity, and the associated electric currents at the surface of the cavity, would persist until all the plasma has moved beyond the earth, after which the earth's magnetic field would return to its normal value.

Such a theory provides a reasonable explanation for the initial phase of a magnetic storm, where the horizontal component is increased over its average value. The boundary currents, however, cannot explain the large main phase decrease which usually follows. At present this decrease is generally considered to be caused by currents associated with the drift around the earth of particles trapped in the earth's magnetic field.

Most of the basic aspects of the Chapman–Ferraro theory are still generally accepted today. The most important modification is that we now know from space probe and satellite data that plasma is being continually emitted from the sun, instead of just during storm times. This means that the magnetosphere cavity is always present. Increased magnetic activity is thus associated with increases in the solar wind velocity, which compresses the cavity still further. In addition, Chapman and Ferraro did not predict the detached shock wave which is now known to exist.

In a review article, Chapman has summarized many of the earlier concepts of the interaction between the solar plasma and the earth's field, and has given examples of some of the earlier calculations, as well as the more recent ones, based on various models.[3]

OBSERVATIONS OF THE MAGNETOPAUSE

The first measurements of the outer regions of the magnetosphere were made in 1958 by Sonett and coworkers with instruments carried in the space probe *Pioneer I*.

Although the magnetometer could only measure the component of the magnetic field perpendicular to the spin axis of the spacecraft, this measured field was found to be reasonably close to an undistorted dipole field between 4 and 7 earth radii. Between 12 and 14 earth radii in the solar direction, the field exhibited large fluctuations, and was several times stronger than the predicted dipole field. At about 13.5 earth radii, the magnitude of the field decreased considerably. At the time, this was interpreted as possibly being due to the crossing of the magnetosphere boundary, although this supposed boundary was considerably more distant than the theoretical predictions of that day, based on image dipole calculations (see pp. 382–383). We now believe that what was probably observed was the crossing of the outer shock wave in front of the magnetosphere boundary, with the turbulent region just prior to this being the transition zone.

Two years later Pioneer V, carrying similar instruments, traversed the outer magnetosphere about half-way between the noon and dusk meridians, i.e., at about 3 p.m. local time (local time is a convenient measure of the sun-earth-probe azimuthal angle as projected on the earth's equator). The results were similar to those obtained on the earlier flight: a smoothly decreasing field out to 8 earth radii, a fluctuating field stronger than a dipole field alone beyond 10 earth radii, and a decrease to a weak, presumably interplanetary field somewhere between 15 and $20R_e$.

The first definite observation of the magnetospheric boundary was made with Explorer 10, launched on March 25, 1961, into a highly elliptical orbit with an apogee of 47 earth radii, in approximately the 9 p.m. direction, i.e., away from the sun (see Fig. 1). On this flight both the strength and direction of the geomagnetic field was measured by a magnetometer provided by Heppner et al.[4] The field beyond 8 earth radii was distorted, as though the lines of force were being pulled away from the earth, and the strength of the field was higher than predicted. Between distances of $22R_e$ and apogee, the satellite apparently crossed the boundary (or vice versa) on at least six occa-

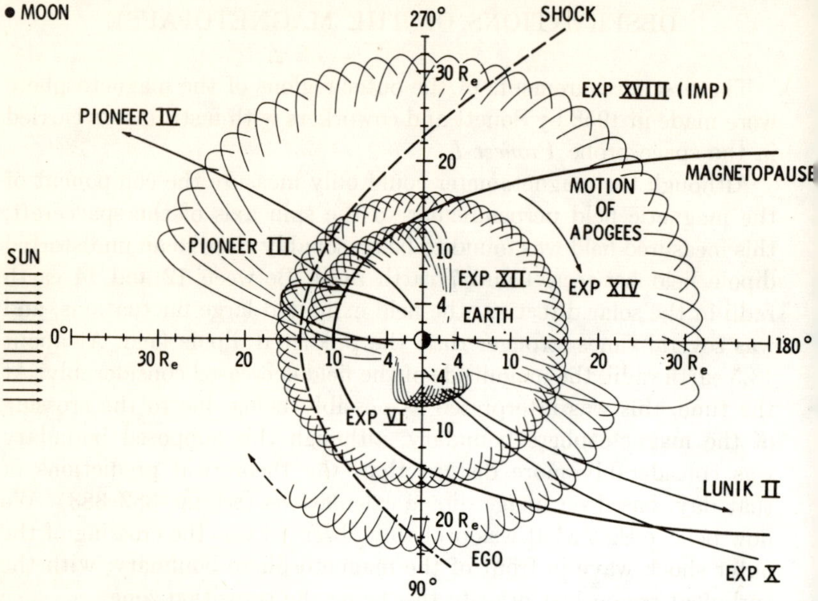

FIG. 1. Orbits of various eccentric satellites and space probes projected into the equatorial plane.

sions. This conclusion was reached after comparing the magnetometer results with those from the plasma probe experiment of Bridge and coworkers at MIT.[5] While inside the magnetosphere, the magnitude of the field was comparatively strong (20–30 gammas, where the dipole field alone would be less than 4 gammas), and there was usually no detectable plasma. Outside the boundary the field changed direction and became weaker (10–15 γ) and more disordered, and plasma was always observed. Because the boundary was apparently crossed a number of times, the satellite must have been moving approximately parallel to the boundary during this period. If the magnetopause was symmetric about the sun-earth line, the observed diameter of the magnetosphere cavity was somewhat larger than current theories had indicated, i.e., about 40–45R_e, instead of around 35R_e.

In August of 1961 Explorer 12 was launched into a highly elliptical orbit with a geocentric apogee of 13.1R_e. It was launched initially in the solar direction (see Fig. 1). However, as the earth moves around

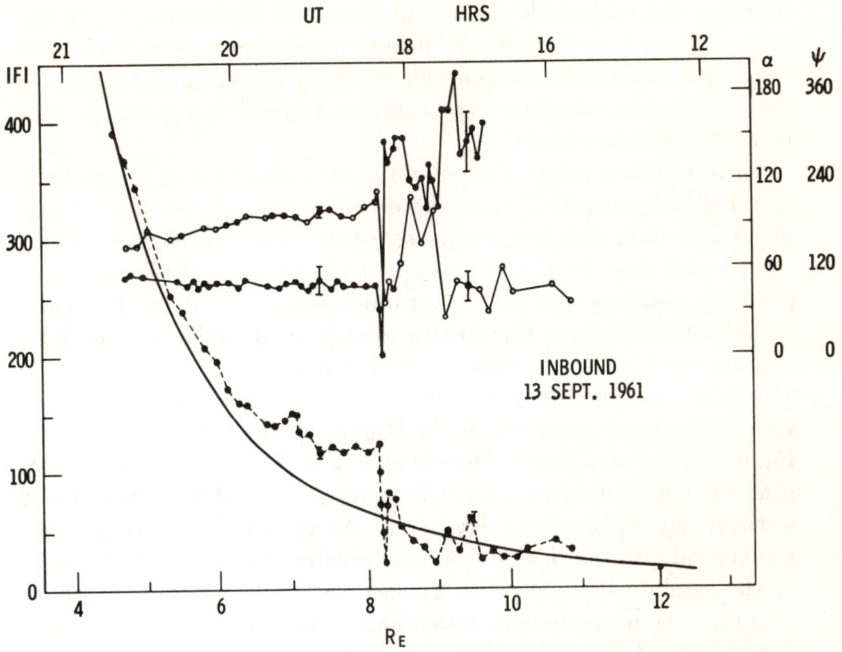

FIG. 2. Magnetic field data from Explorer 12 showing an abrupt change in the field at 8.2 R_e. This change, the magnetopause, is the outer limit of the geomagnetic field (after Cahill, Ref. 6).

the sun and the orbit of the satellite remains fixed in inertial space, the apogee measured with respect to the sun appears to move westward (i.e., clockwise as observed from above the north pole) at approximately one degree per day (360°/year). Explorer 12 transmitted for about 3 months, and therefore the region out to 13 earth radii between noon and 6 a.m. local time was rather fully explored. A magnetometer on board provided by Cahill measured both magnitude and direction of this field.[6]

A typical pass during the early part of the satellite lifetime, when it was near the noon meridian, is shown in Fig. 2. The magnitude of the field is plotted as a function of radial distance, along with the theoretical dipole field. The direction of the field as measured by the spacecraft coordinates α and ψ (angles relating the field direction to the spin axis of the satellite) is also plotted. Out to 8 earth radii the direction and magnitude of the field show little fluctuation, but in the region of 6 to 8 earth radii, the magnitude is almost twice the dipole field. At $8.2R_e$, the field magnitude suddenly decreases and the direction of the field simultaneously changes. Beyond this point the field is much more disordered.

These results are typical of all of the early Explorer 12 passes, both inbound and outbound. Since the orbital period of the satellite was about one day, two magnetopause crossings of this type were observed each day. In almost every case the position of the boundary could be uniquely specified by examination of the data. The most reliable indicator was the sudden change in direction of the field, accompanied by an increase in turbulence outside the boundary. The field magnitude usually, but not always, decreased, but never to zero. The closest approach of the boundary was 8.2 earth radii (on the pass illustrated here). More typically it was about 10 earth radii near the noon meridian, although there was considerable movement of the magnetopause from day to day. Away from the noon meridian the typical crossing distance became greater. During the latter part of the satellite lifetime, when apogee was near the dawn meridian (6 a.m.), the magnetopause was seldom observed, an indication that it was beyond 13 earth radii.

The thickness of the magnetopause appears to be quite small, of the order of 100 km or less, or about 1/600th of the overall magnetosphere dimensions. This is based on detailed observations of the magnetometer record during the time of magnetopause crossings. However, such distances are difficult to determine, as it is usually impossible to separate time variations from spatial variations, i.e., movement of the satellite past the boundary from movement of the boundary past the satellite.

We can get some idea of the expected boundary layer thickness by calculating the proton cyclotron radius in the field just inside the boundary. The protons could come into the field this depth and produce a boundary current by being bent in the opposite direction from the incoming electrons. The cyclotron radius of a 1 keV proton in a 70 γ field is 65 km. A more detailed analysis of this problem by Beard[7] indicates that twice this proton cyclotron radius is an upper limit, and that the actual thickness might be much less than this.

Explorer 14, launched in October, 1962, also carried a magnetometer experiment provided by Cahill.[1] This satellite had an apogee of

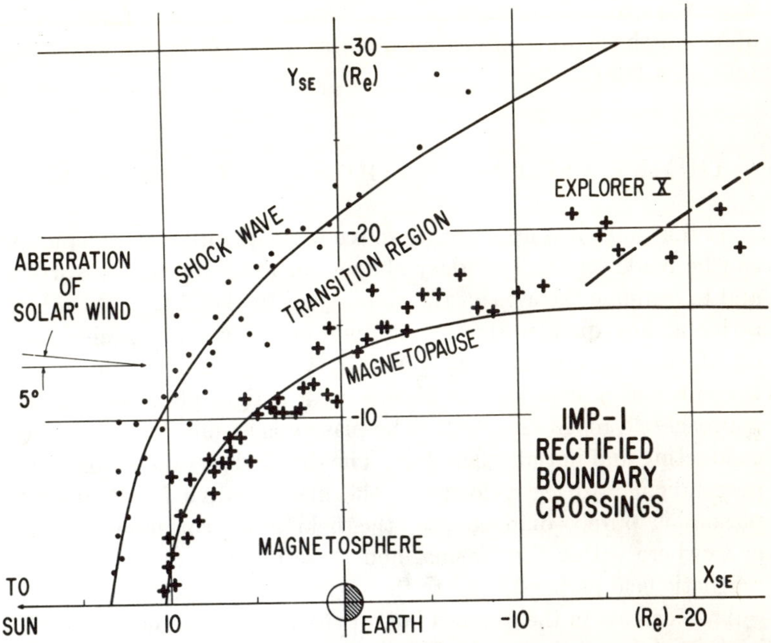

FIG. 3. The location of the magnetopause and bow shock as determined by the magnetometer on IMP-I. Shown for comparison are the theoretically expected locations (after Ness, Ref. 20).

16 earth radii, which was initially on the 7:00 a.m. meridian, but moved back towards the midnight meridian as the earth proceeded along its solar orbit. Since the magnetopause was detected only occasionally during the first month of flight, it was probably beyond $16R_e$ for most of this time.

The most accurate mapping to date of the magnetopause boundary has been made by the IMP-I satellite (Explorer 18), launched in November, 1963, with an initial apogee of $31R_e$ in the solar direction.[8] Both the magnetometer and plasma probes on IMP have obtained data on boundary crossings. Figure 3 shows the magnetopause transitions as observed by the magnetometer, together with the theoretical results on the boundary position (see pp. 382–385). It is seen that the agreement is quite good between noon and 6 a.m., but that the observed boundary flares out on the night side somewhat more than the calculations predict. Also shown in this figure are the observed shock wave transitions. These data will be discussed more fully in a later section.

THEORY AND CALCULATIONS ON THE BOUNDARY

At this point it is appropriate to see what quantitative calculations can be made on the boundary shape and resulting field distortions, and to examine theories on the stability of the boundary. In order to arrive at any quantitative results, one must set up a specific model on which to base a calculation. One very simple model is to consider the effect of a semi-infinite plasma with a rigid plane surface as it approaches a magnetic dipole. The plasma is assumed to be perfectly conducting and initially field-free. The dipole direction is taken to be perpendicular to the velocity of the plasma wave front. Since the plasma is purely diamagnetic, the field must continue to vanish everywhere within the plasma, and there must be no component of magnetic field perpendicular to the surface at any point.

The currents in the forward face of the plasma will modify the field in the vicinity of the dipole. However, it is not necessary to calculate the currents themselves in order to determine their effect. Chapman and Ferraro showed that an image dipole, equal in strength to the original dipole, placed in the plasma at the mirror image of the original dipole, would produce a magnetic field in the non-plasma region identical to that actually produced by the induced surface currents. One can easily show by symmetry, for example, that the total field produced by two parallel dipoles of equal strength has no component perpendicular to a plane located half-way between the dipoles. Therefore, no field lines cross this plane, and the dipole field lines are completely separated from one another.

This simple model now enables one to calculate the induced field produced by the approach of an infinitely-conducting plane. To be specific, let us consider the field B_d of the earth's equivalent dipole. In the magnetic equatorial plane,

$$B_d = B_0/r^3, \tag{1}$$

where B_0 is the magnitude of the field on the earth's surface at the equator, 0.31 gauss, and r the geocentric distance in earth radii. Suppose now that the plasma front has advanced to a distance of 10 earth radii. The induced field near the earth is equivalent to that of a dipole of equal strength located 20 earth radii away. Close to the earth this is approximately $0.31/(20)^3$ gauss, or about 4 gammas. Now suppose the plane advances to $5R_e$; the image dipole is at $10R_e$ and the induced field rises to 31 gammas. From such a calculation, Chapman and Ferraro concluded that for typical storms, the plasma sheet must approach to within about 5 earth radii, in order to produce the observed field increase at the earth. Note that the field immediately adjacent to the plasma sheet is doubled in the equatorial regions. Near the plane on a line joining the two dipoles, the total field (earth dipole plus image dipole) is 62 gammas in the first instance, and 496 gammas in the second instance.

So far the model has assumed that the advancing plasma front will remain a rigid plane. The magnetic field, however, will exert its maximum resistance to the approach of the plasma where the field is strongest, i.e., where the front surface is closest to the dipole. This will cause the surface to become curved, and the regions further away will flow around the dipole.

We may now inquire as to the size and shape of the cavity when equilibrium is finally reached. This has turned out to be a very difficult problem to solve, because of the three-dimensional character of the cavity. However, a few simple arguments enables one to calculate the approximate size in terms of solar wind parameters.

The fundamental approach has been to assume that the equilibrium surface represents a balance between the external pressure exerted by the solar wind and the internal pressure exerted by the geomagnetic field. In the simplest analysis, the solar wind particles are considered to be reflected from the boundary just as gas molecules exert pressure by being reflected from a rigid wall. The magnitude of the pressure is then twice the component of momentum perpendicular to the wall times the number of particles per square centimeter per second which hit the wall. If the particles are moving perpendicular to the boundary with velocity v, the pressure is

$$p = (2mv)(nv) = 2mnv^2, \tag{2}$$

where m is the mass of the ions (the electrons exert almost no pressure because of their small mass) and n the density of ions in the solar wind.

From plasma physics, one can show that if a plasma and a magnetic field are in equilibrium, the effective pressure of the magnetic field is equal to its energy density (dimensionally dynes/cm^2 = ergs/cm^3 = gauss^2):

$$p = B^2/8\pi \tag{3}$$

in c.g.s. units. The pressure in the equatorial plane can be obtained from Equation (1), remembering that the earth's dipole field will be approximately doubled just inside the boundary:

$$p = (2B_d)^2/8\pi = B_0^2/2\pi r^6. \tag{4}$$

Equating magnetic and particle pressure at this point,

$$2mnv^2 = B_0^2/2\pi r^6 \tag{5}$$

or

$$r_b = [B_0^2/4\pi mnv^2]^{1/6}. \tag{6}$$

Substituting typical values of 400 km/sec and 10 particles/cm^3 into Equation (6), we obtain a distance of $9.8R_e$, if the solar wind consists purely of ionized hydrogen.

As one attempts to proceed further in obtaining a solution to the shape of the boundary, the mathematics becomes much more complex. It is possible to write down a general equation for pressure balance at each point on the surface, but this becomes a non-linear, second-order, partial differential equation which is very difficult to solve except in the planes of symmetry, i.e., the equatorial plane and the noon–midnight meridian plane. Beard[7] has reviewed this whole problem, outlining the approach taken by various groups in attempting to find a solution. Because of the complexity of the various approaches, no attempt will be made here to review the details. One self-consistent solution to the shape of the boundary, as obtained by Mead and Beard,[9] is shown in Figure 4. Note the characteristic dimple, or null point, in the noon meridian surface near the pole.

It is important to remember that the shock wave and interplanetary field have been disregarded in these calculations, as well as the effect

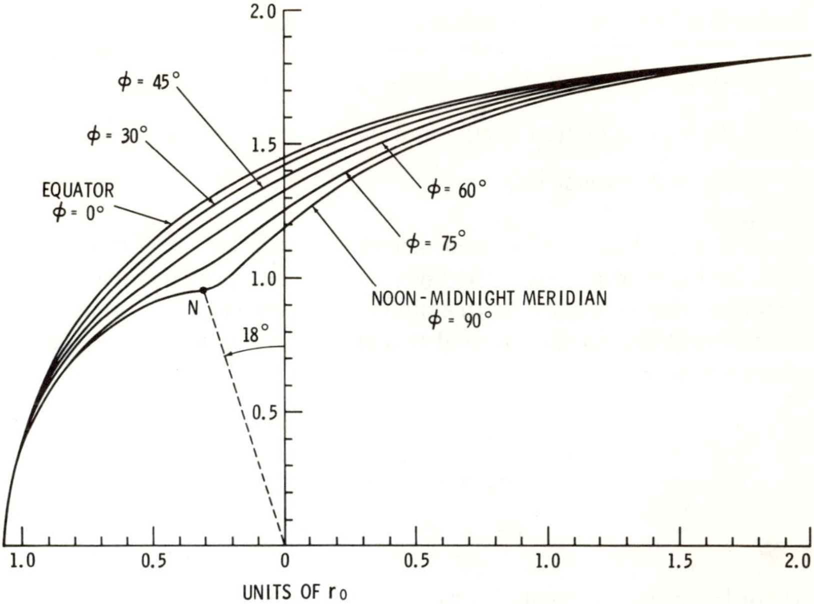

FIG. 4. Theoretical shapes of the intersection of the magnetopause with planes parallel to the earth–sun line. The surface is essentially cylindrical in the distant anti-solar regions.

of ring currents or trapped particles inside the magnetosphere. The solar wind is also assumed to be perpendicular to the dipole axis, whereas in the summer or winter this angle can deviate by as much as 35° from the perpendicular. Recent calculations of the boundary shape that consider flow of a fluid around an obstacle, rather than single-particle reflection from the boundary, give remarkably similar results.

Once the boundary surface is obtained, it is possible to calculate the currents that must flow at the magnetopause to make the field zero outside. These currents produce a field inside which will change the original field. This distorted field is most easily described by an expansion in spherical harmonics, with the coefficients in the expansion determined by making a least-squares fit to the calculated distorted field. It turns out that with only two coefficients one can make a rather good fit to the distorted field.[10] With a boundary at 10 earth radii in the solar direction, the three components of the field (including the earth's dipole terms) in gammas are given by

$$B_r = -62{,}000(\cos\theta/r^3) + 25\cos\theta - 2.1r\sin 2\theta\cos\phi, \quad (7)$$

$$B_\theta = -31{,}000(\sin\theta/r^3) - 25\sin\theta - 2.1r\cos 2\theta\cos\phi, \quad (8)$$

$$B_\phi = 2.1r\cos\theta\sin\phi, \quad (9)$$

where r, θ, and ϕ are the usual spherical coordinates, r is in earth radii, and ϕ is measured with respect to the midnight meridian. In cartesian coordinates the earth's dipole field becomes more complex, but the field due to the external boundary currents alone simplifies and is given by

$$B_x = 2.1z \quad (10)$$

$$B_y = 0 \quad (11)$$

$$B_z = 25 + 2.1x \quad (12)$$

where the x axis is pointed toward the sun, the z axis towards the north star, and x and z are in earth radii. With a boundary at a distance other than 10 earth radii, the coefficients 25 and 2.1 must be replaced by $25{,}000/r_b^3$ and $21{,}000/r_b^4$ gammas, respectively, where r_b is the boundary distance.

A comparison of the predictions of this model of the distorted field with data from a number of Explorer 12 passes has been made by Mead and Cahill.[11] They find that the measured field agrees quite well with the model on the day side of the magnetosphere, where the compression of the magnetic field by the solar wind is the greatest. Angular distortions in the field direction by as much as 60° are both predicted and observed. However, as described in a later section ("The Tail of the Magnetosphere"), field lines in the dark hemisphere have been observed to stretch out in the antisolar direction, forming an elongated tail. This stretching of field lines is not predicted by the above theoretical model, and is most likely due to the presence of plasma currents inside the magnetosphere. The Chapman-Ferraro model specifically excludes the presence of such currents. In order to simulate the effect of tail formation, Williams and Mead[12] have added to this model a magnetic field due to a postulated current sheet in the equatorial plane on the dark side of the earth. The resulting magnetic

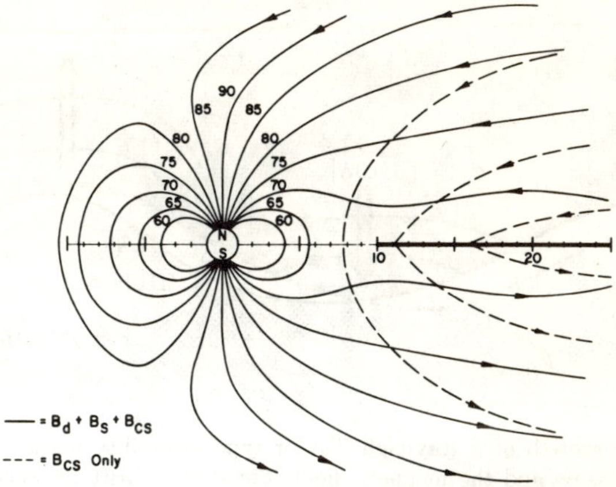

FIG. 5. Theoretical shapes of field lines in the magnetosphere noon-midnight meridian plane. Field contributions due to the earth's dipole (B_d), magnetopause surface currents (B_s), and a current sheet in the tail (B_{cs}) are included. Dashed lines show the field lines calculated from the current sheet only.

field lines as calculated in the noon–midnight meridian plane are shown in Figure 5.

We have mentioned earlier that energetic particles appear to drift closer to the earth on the night side, as observed, for instance, by Frank and Van Allen on Explorers 12 and 14.[2] The theory of particle drifts predicts that particles which mirror near the equator will remain on contours of constant magnetic field. Since the field is more compressed on the day side than the night side, particles would drift closer to the earth on the night side. To calculate how large a shift this is, Equation (8) may be used to equate the field magnitude in the solar direction to that in the anti-solar direction:

$$(31{,}000/r_d^3) + 25 + 2.1r_d = (31{,}000/r_n^3) + 25 - 2.1r_n, \quad (13)$$

where r_d and r_n are the day and night distances. Setting $r_d = 10R_e$ gives $r_n = 7.7R_e$, indicating a day-night difference of more than two earth radii. Contours of constant count rate as measured by Frank and Van Allen seem to follow rather closely contours of constant magnetic field as calculated in this fashion.

Another consideration about the magnetopause is its stability.

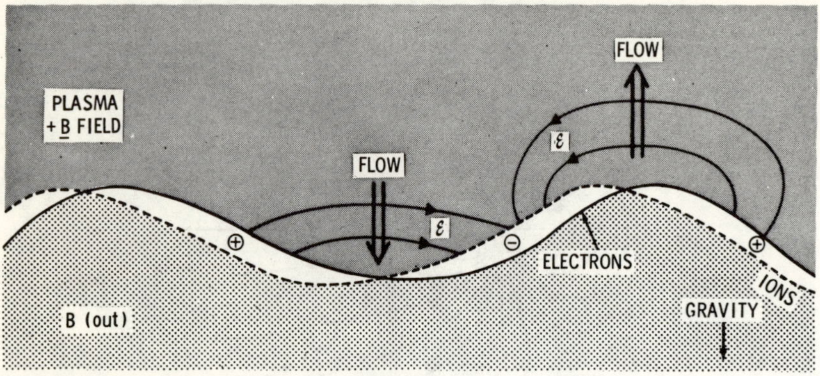

FIG. 6. The growth of a Rayleigh–Taylor type instability in the interface in between the plasma and the magnetic field. The $g \times B$ drift produces a charge separation, which produces an electric field, which produces a plasma flow, amplifying the initial disturbances.

Will the boundary oscillate around the equilibrium position we have described for it? Will it be transparent or opaque to particles? Will it let them enter the magnetosphere from the transition zone? There are many possible kinds of instabilities. When a heavy fluid is suspended above a light fluid in a gravitational field the interface will develop Rayleigh–Taylor instabilities. A perturbation of the boundary will grow and cusps will develop, and the heavy fluid will flow down through the cusps. In Figure 6 is shown what can happen to the boundary between a plasma and a magnetic field in a gravitational field.[13] As shown in Chapter 4, equation (9), the particles in the plasma will drift parallel to the boundary due to the gravitational force. Electrons drift right and protons left. This produces a charge separation electric field E as shown. The particles in the plasma will now drift up or down due to the electric field E. The $E \times B$ drift here makes the perturbation grow. If the direction of g had been reversed, the perturbation would have decayed and the boundary would be stable. Gravitational forces at the magnetopause are negligible, but the curvature of the boundary will produce a situation where Rayleigh–Taylor instabilities can occur. Due to the curvature there will be a field line curvature drift and a ∇B drift [see Chapter 4, equations (10) and (11)] that takes the place of the $g \times B$ drift.

If the center of curvature lies in the plasma the interface is unstable. If the center of curvature is on the field side the interface is stable. Looking at Figure 4, which shows the shape of the interface, it is obvious that most of the interface is stable. However, the field line curvature near the field null points at high latitudes is such that this region is unstable. Particles may be able to leak into the magnetosphere from the magnetosheath here. However, if the magnetosphere is open instead of closed, that is, if field lines from the interplanetary region connect with those from the earth, then the high latitude null points don't exist.

A second type of boundary instability is the one that produces waves on water when wind blows over it. This is called Helmholz instability. Analysis shows that the sides of the magnetosphere are Helmholz stable, due to the high velocity of the solar wind. But Parker has shown theoretically that the boundary is generally unstable when the solar wind strikes it at an angle. It seems likely that the boundary will be unstable in various ways such that some particles will be able to leak through. The solar wind proton flux shows a sharp decrease at the magnetopause, so probably the boundary is quite opaque to protons. But experimentally it seems the boundary is quite transparent to electrons of $E_e \sim 100$ eV. The retarding potential analyzer on IMP–I measuring $E > 100$ eV electrons recorded no flux change at the magnetopause; the flux was the same inside as out. This would imply that these electrons can flow through the boundary relatively freely, due probably to some instability. This is still an unsettled subject.

THE BOW SHOCK

A supersonic bullet produces a shock wave as it moves in air. If the speed is sufficiently high, the shock wave becomes detached and lies upstream of the bullet. The earth and its magnetosphere present an obstacle to the supersonic flow of the solar wind and as a result produce a shock wave in the solar wind upstream of the magnetopause.

The sound speed in the interplanetary medium is $v_s \approx 50$ km/sec. The solar wind velocity is about 400 km/sec, so that the flow has a Mach number $M = v/v_s \approx 8$. Another speed to consider for our case is the Alfvèn speed

$$v_A = B/(4\pi\rho)^{1/2} \approx 50 \text{ km/sec};$$

Thus the Alfvèn number is $M_A \approx 8$ also. Since both M and M_A indicate supersonic flow, one might expect that a shock wave would exist upstream of the magnetopause. But the analogy with the bullet fails. The shock wave upstream of the bullet is produced by collisions between air atoms. The collision mean free path in the interplanetary medium near the earth is roughly 1 AU. This distance is so much larger than the size of the magnetosphere that there will be no collisions. A shock is a rapid transition between two different states of matter. If collisions cannot produce the rapid transition here, what can? The magnetic field clearly takes over the job of making the plasma act like a fluid and communicating information between particles. All the particles on one magnetic field line are constrained to move together. As the field line is blown at the magnetosphere by the solar wind, it will be deflected and try to move around the sides of the magnetosphere. All the particles on the field line will take part in the flow, and a fluid-like motion will result. Clearly particles don't act independently in the solar wind. However, single particle calculations can give a reasonable description of the shape of the magnetosphere. But single particle action could not make a shock wave in a collisionless medium; cooperative processes involving the magnetic field are involved. Very probably, instabilities lead to growing plasma waves which scatter the particles and produce the shock. The cyclotron radius is one characteristic distance in which things can change for this situation; for a 1 keV proton in the interplanetary magnetic field (5 gammas) at 1 AU, the cyclotron radius is about 1000 km. Holzer showed that the shock was only about 50 km thick, so that the cyclotron radius and Alfvèn waves do not seem to be important in forming the shock. Tidman[14] has suggested that ion plasma waves grow at the shock due to an instability. In this case, the characteristic distance is $\approx 10\, v/\omega_p \approx 3$ km (ω_p is the proton plasma frequency), which seems reasonable. However, this matter of the origin of the bow shock must be considered unsettled.

The possibility that a bow shock would exist upstream of the earth was suggested (by analogy to supersonic aerodynamics) shortly after the magnetopause was discovered. But this was conjectural, and it was uncertain whether the shock existed or not until its discovery by the IMP–I satellite. This was launched in November, 1963

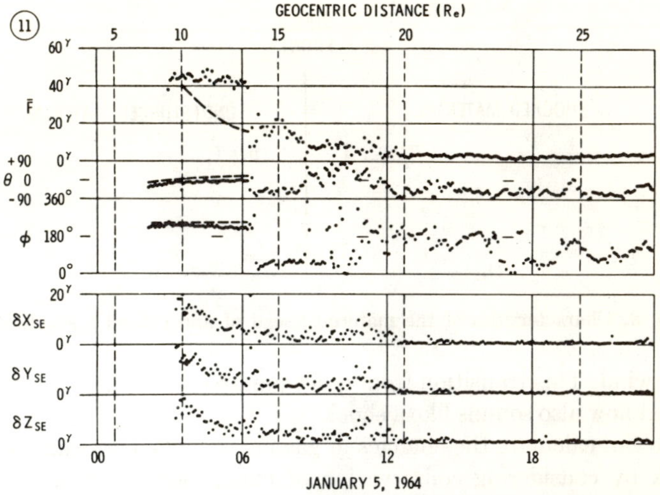

FIG. 7. Magnetic field data from orbit 11 of Explorer 18 (IMP–I). The magnetopause is at 13.6 R_e. The second transition at 20 R_e to an ordered field outside is the location of the bow shock wave (after Ness, Ref. 8).

into a highly eccentric orbit going out to 31 earth radii, with instruments on board designed to explore the outer magnetosphere and interplanetary regions. Ness[8] found that a radial survey with the IMP I magnetometer showed two discontinuities in the magnetic field. Proceeding outwards on orbit 11 (see Figure 7), at $13.6R_e$ the magnetopause was encountered, as indicated by the sharp change in magnitude and direction of the field. Then at $20R_e$ a second change occurred. The field suddenly became steady at about 5 γ, and of relatively constant direction. The variance [defined in Chapter 8, equation (16)], which is a measure of the time variability of the magnetic field, suddenly became small. These properties seem like those one would expect for a shock. Upstream of the shock the flow is quiet and steady, but behind the shock it is disturbed and turbulent. At the same place that the magnetometer found the second discontinuity, a transition was detected by the plasma probes in IMP. The solar wind outside the shock flows radially away from the sun and has a narrow spread of energies. Inside the shock location as indicated by the magnetometer, the plasma flow became much more spread in direction—the plasma probe showed a roughly isotropic distribution near the stagnation point. Also there was a much larger spread in proton energies inside the shock than in the free stream

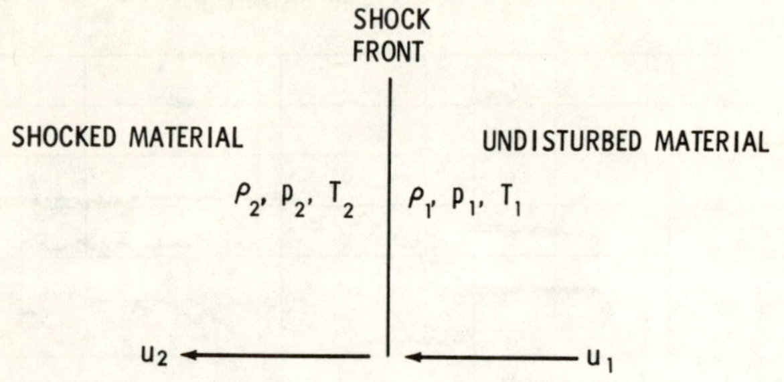

FIG. 8. Characteristics of the material ahead of and behind a shock front.

solar wind. This transition from quiet unidirectional flow to turbulent heated flow also sounds like a shock.

We can calculate the changes in gas parameters across an ordinary shock by considering conservation of mass, momentum and energy flow through the shock. Going to a frame of reference in which the shock is at rest (as in Figure 8) with temperature T, pressure p, density ρ, and velocity u as shown, continuity of mass flow gives

$$\rho_1 u_1 = \rho_2 u_2 = \mu, \tag{14}$$

where μ = mass flow in gm/cm²/sec. Change of momentum per cm² per second across the front gives

$$p_2 - p_1 = \mu(u_1 - u_2), \tag{15}$$

which can be rewritten

$$p_1 + \rho_1 u_1^2 = p_2 + \rho_2 u_2^2. \tag{16}$$

The work done per gram in crossing the front increases the energy by

$$(p_1 u_1/\mu) - (p_2 u_2/\mu) = \tfrac{1}{2}u_2^2 + \epsilon_2 - (\tfrac{1}{2}u_1^2 + \epsilon_1), \tag{17}$$

where ϵ is the internal energy per gram. From this we get

$$\tfrac{1}{2}u_1^2 + \epsilon_1 + (p_1/\rho_1) = \tfrac{1}{2}u_2^2 + \epsilon_2 + (p_2/\rho_2). \tag{18}$$

Equations (14), (16), and (18) are called the Rankine–Hugoniot equations. One more relation is needed to solve for changes across the shock. For a perfect gas the internal energy is given by

$$\epsilon = p/(\gamma - 1)\rho, \tag{19}$$

where γ is the ratio of specific heats and for a monatomic gas is given by

$$\gamma = 1 + (2/f) = \tfrac{5}{3} \qquad \text{for } f = 3 \text{ degrees of freedom.}$$

Solving the Rankine–Hugoniot equations plus equation (19) for a strong shock where $p_2/p_1 \gg 1$ we get

$$\rho_2/\rho_1 = (\gamma + 1)/(\gamma - 1) = u_1/u_2, \tag{20}$$

$$p_2/p_1 = 2\gamma M^2/(\gamma + 1), \tag{21}$$

where M = Mach number, and

$$T_2/T_1 = (p_2/p_1)[(\gamma - 1)/(\gamma + 1)]. \tag{22}$$

For a shock where $M \approx 8$, we get $p_2/p_1 \approx 80$; this clearly is a strong shock. These *jump conditions* across the shock are for an ordinary shock in air. Considering instead a shock in a magnetized plasma they can be generalized[15] by using

$$p^* = p + (B^2/8\pi) \tag{23}$$

$$\epsilon^* = \epsilon + (B^2/8\pi\rho). \tag{24}$$

For the case of a shock traveling perpendicular to B, the same *jump conditions* are obtained as before. But now we must use the starred quantities and $M \to M_A$ and also

$$\gamma^* = \beta\gamma + (1 - \beta)2, \tag{25}$$

where $\beta = \epsilon/\epsilon^*$, and we also get a new jump condition

$$B_1u_1 = B_2u_2. \tag{26}$$

For the case of propagation parallel to B we get $B_1 = B_2$.

Everything we have said about jump conditions so far holds only for the collision-dominated case, even for the magnetized plasma.

But for the case of a collisionless shock wave we cannot be really sure that these jump conditions are valid. If there is no process by which entropy can be increased across the shock, then there may not really be a definite new state behind the shock, but just wave trains which may eventually decay away and return the plasma to the original state. However, it appears likely that various processes like plasma instabilities may produce a change in entropy and a definite new state behind the shock. In this case, it should be all right to use the jump conditions (20) through (22) even for the collisionless case. Assuming this to be true and taking $\gamma^* = 5/3$ and $M_A = 8$ we get

$$u_1/u_2 = \rho_2/\rho_1 = 4 \tag{27}$$

$$1 < (B_2/B_1) < 4 \tag{28}$$

$$p_2/p_1 = 80 \tag{29}$$

$$T_2/T_1 = 20. \tag{30}$$

The location of the bow shock as determined by the IMP magnetometer is shown in Figure 3. The solid curve through the data points is the shape calculated for an ordinary collision-dominated shock wave for $M = 8.7$ and $\gamma = 2$.[16] The experimentally-determined shock shape seems to agree with this well. The shock stand-off distance Δ in aerodynamic flow is given by

$$\Delta/D = 1.1(\rho_1/\rho_2) = 1.1[(\gamma - 1)/(\gamma + 1)] \approx 0.25 \tag{31}$$

where $D \approx 13R_e$ is the effective radius of the front of the magnetosphere. Using $\gamma = 5/3$, $\Delta = 3.3R_e$, which seems about right.

It seems that using aerodynamic shock theory here gives about the right answers. That does not mean that the theory is really applicable, but only that the magnetic terms may be negligible and that there probably are processes that change the entropy across the shock. We know very little about the structure of the bow shock. The shock thickness measured by a magnetometer is $\sim$ 10 to 50 km. Tidman's theory[14] gives about this thickness for the magnetic jump, but suggests that the density jump may take only $\sim$ 1 km. We need more detailed data on the shock structure to resolve this situation.

THE MAGNETOSHEATH

The region behind the bow shock, called the magnetosheath or transition zone, shows energies and types of particles not seen in the solar wind. We have calculated what some of the characteristics of this region should be by using the jump conditions. From measurements in this region we can see how well these predictions are met.

An electrostatic curved-plate analyzer was flown on IMP-I, IMP-II, and OGO-I by Wolfe.[17] It showed a sharp change in the plasma properties at the shock. Observations in the magnetosheath consistently showed a large increase in the disorder of the plasma as compared to interplanetary space. While the convective flow velocity decreases by less than a factor of two, the plasma temperature, as measured by the spread in the energy spectrum, was observed to increase almost an order of magnitude.

Similar results have been reported from the Vela satellites flown by the Los Alamos group.[18] These satellites are in circular orbits of

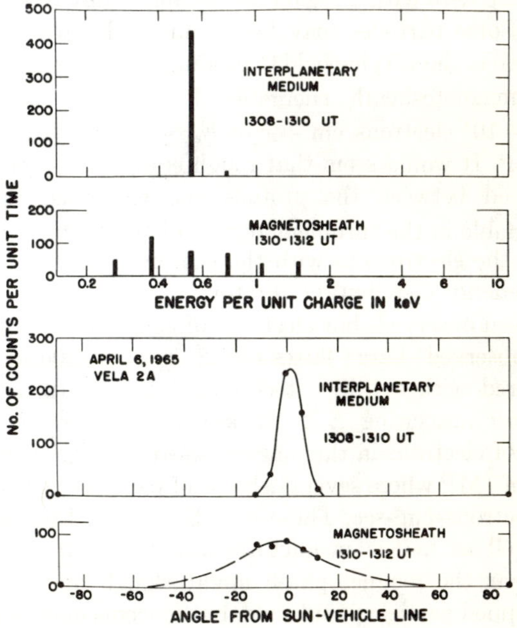

FIG. 9. Observations of the change in energy spectrum and angular distribution of the solar wind as the Vela 2 satellite traversed the bow shock. The plasma is partially thermalized and isotropized in the magnetosheath. (After Gosling et al., Reference 18.)

approximately 17 earth radii. While in the solar direction, therefore, they are in interplanetary space. Each time they move around to the side, they pass through first the shock wave, and then the magnetopause. The characteristics of the plasma upstream and downstream of the shock transition is shown in Fig. 9. In the interplanetary medium, most of the plasma is seen in only one energy channel and is limited in direction to a cone of about 10° width. In the magnetosheath the energy spectrum is much broader and the cone is very much wider. The protons have been somewhat thermalized. They have been both lowered and raised in energy. Theoretically, at the stagnation point of flow around an obstacle, the directed beam energy is all converted to thermal energy so there should be a Maxwellian energy distribution with $\frac{3}{2}kT = E$(solar wind). Trying to apply this to the bow shock doesn't work too well. The energy spectrum is really not Maxwellian. Wolfe has fit the proton energy spectrum with a Maxwellian of $T = 100$ eV *plus* a high energy tail of protons. There definitely are more high energy protons than a Maxwellian contains. We are dealing here with a region containing highly disordered magnetic fields. Some particles may be Fermi-accelerated by bouncing back and forth in the varying field to produce the high energy particles.

In the magnetosheath energetic electrons are also observed. Fluxes of $\sim 10^8$ electrons/cm^2-sec of $E_e \sim 100$ eV have been measured on IMP. It would seem that rough equipartition of energy had been achieved between the protons and electrons. Equipartition seems reasonable in the turbulent regime of the magnetosheath.

But with the electrons as with the protons there is a high energy tail in the energy distribution. Protons of energies up to about 10 keV have been observed, but electrons of considerably higher energy have been observed. Large fluxes of $0.2 < E_e < 500$ keV electrons were measured with a CdS detector on Explorer 12. On IMP–I a geiger counter measuring $E > 40$ keV electrons showed transient large fluxes of electrons in the magnetosheath.[19] Figure 10 shows one radial pass of IMP where several islands of particles were observed of $J \sim 10^6$ electrons/cm^2-sec. These islands are found outside the bow shock as well as in the magnetosheath. Anderson suggested that processes near the magnetopause might slough off these particles from the trapped particle population, but it seems more reasonable to attribute theirs to Fermi acceleration processes in the magnetosheath. Islands of these particles are also found inside the magnetosphere on the night side in regions where permanently trapped particles don't exist. They may have been transported in through the magnetopause from the magnetosheath to the tail.

On the early Pioneer IV flight it was observed that large fluctuating populations of particles existed from $10R_e$ to $15R_e$. These are not a permanent feature of this region because they were not seen on Pioneer III. These particles probably are electrons of $E \sim 1$ MeV.

From these various observations of high energy electrons it seems certain that there are non-thermal processes operating in the transition region. It is not surprising that stochastic acceleration processes can take place in this turbulent zone. Fermi acceleration or acceleration due to fluctuating electric fields might take place.

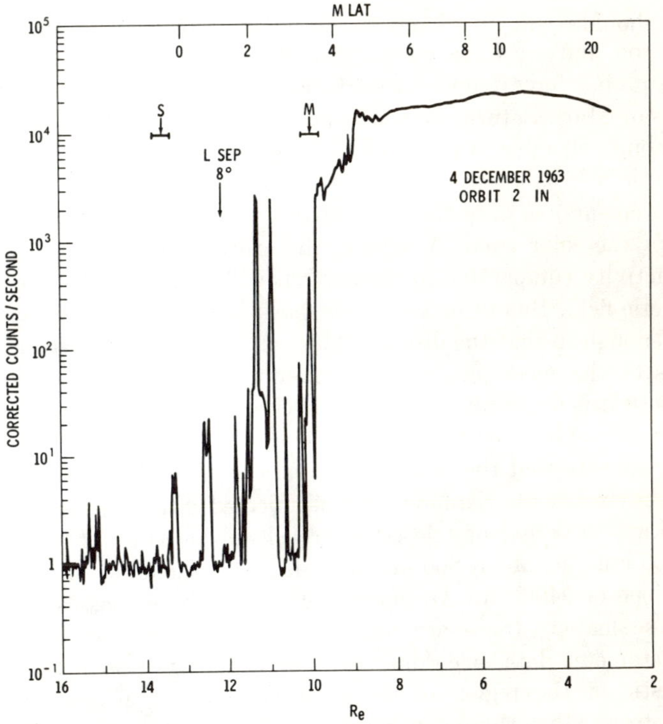

FIG. 10. Flux of $E > 40$ keV electrons detected on IMP. The radiation belt is present inside the magnetopause at $10R_e$, and outside this occasional "islands" of particles are observed (after Anderson et al., Ref. 19).

Measurements made on the sides of the magnetosphere show that the flow of the plasma in the magnetosheath becomes more directional than near the stagnation point. The particles tend to stream along the sides of the magnetopause. There seems to be some asymmetry from the dawn to the dusk side of the magnetopause. The Vela satellite detected many more energetic particles on the dawn side.

The Earth certainly has a bow shock and magnetosheath filled with energetic particles. What about other planets and the moon? It seems certain that Jupiter will have a bow shock. It quite definitely has a substantial magnetic field and a trapped radiation belt (see Chapter 4). If the Jupiter magnetic field is ~10 gauss, as is currently thought, then the Jupiter magnetopause should be at $\sim 50 R_{\text{Jup}}$. This will be a substantial obstacle to the solar wind flow and clearly a shock should result. It is likely that some other planets have magnetic fields too, but we have no direct evidence on this point. Mariner space probes have found no evidence for magnetic fields on either Venus or Mars. Saturn, even if it had no magnetic field, would be a big enough obstacle to probably make a bow shock wave. The moon magnetic field is known to be quite small ($<100\ \gamma$ from the Lunik measurements) so only the body of the moon will be an obstacle to flow of the solar wind. Assuming that the moon has an electrical conductivity comparable to the earth it will clearly distort the solar magnetic field. But to make a bow shock wave an obstacle must be large enough so that the discontinuity can have time to develop. The radius of the moon is roughly the same as the proton cyclotron radius so it is uncertain whether a shock can develop.

Recent results from the Explorer 35 satellite, launched on July 19, 1967, indicate that the moon exerts very little influence on its magnetic environment. Explorer 35 was injected into orbit around the moon with a periselene (closest approach to the center of the moon) of 2500 km, i.e., about 800 km altitude above the lunar surface, and aposelene of 9400 km. Although once each orbit it passes into the moon's shadow, there are almost no observable changes in the magnetometer data, according to Ness. The data are simply characteristic of the region of space (Interplanetary, magnetosheath, magnetotail) that the moon happens to be in. This means that the moon has no internal magnetic field and has very low conductivity. A shock wave, if it exists at all, is very poorly formed.

THE TAIL OF THE MAGNETOSPHERE

The early models of the magnetosphere, based on the Chapman–Ferraro theory, were unable to say very much about the back side of the magnetosphere, or tail region. If one assumes specular reflection of a zero-temperature, field-free solar wind from the magnetopause surface (neglecting, for the moment, the shock wave), the plasma can exert no pressure perpendicular to its direction of motion. Since the internal magnetic field never completely vanishes, the pressure balance condition predicts that the magnetopause will asymptotically become parallel to the solar wind in the distant anti-solar regions, but never curve inward. Thus the tail can never close, but extends to infinity with a roughly cylindrical cross section.

The concept of an infinitely long tail, however, was not very appealing. It was thought, therefore, that the random motion of particles in the solar wind would produce some perpendicular pressure, and that this pressure would close off the tail at some reasonable distance.

Explorer 10, with an apogee of 47 earth radii, crossed the boundary a number of times near apogee at about 2300 hours local time. There was no indication that the magnetopause was beginning to close off at this distance; thus, it appeared likely that the tail extended at least as far as the moon's distance, 60 earth radii. Results obtained by Cahill on Explorer 14 (apogee 16 earth radii) showed a marked tendency for the field to point away from the earth in the southern hemisphere, as though the field lines from the south pole region were being somehow drawn back into the tail.

The IMP satellites explored the tail region in much greater detail. A summary of IMP-I magnetopause and shock wave transitions was shown in Fig. 3, indicating that the magnetosphere is about 40 earth radii in diameter at a distance of 20 earth radii back of the earth, assuming circular symmetry about the sun-earth line. The boundary appears to be roughly parallel to the solar wind direction at this point, and shows no sign of beginning to close. Explorer 33, reaching a distance of 80 earth radii, found that the tail was still about 40 R_E wide, and not closing off. [22]

On IMP-I, Ness found that the field showed a strong tendency to be aligned parallel to the sun-earth line in the region beyond 10 R_E in the dark hemisphere.[20] This is illustrated in Fig. 11, showing typical field vectors obtained by IMP-I throughout the tail region as projected onto the solar ecliptic plane. In this region, the component of field perpendicular to the ecliptic was generally much smaller than the parallel component. Data are only shown for that portion of each pass when the satellite was well below the ecliptic plane, where it spends most of its time. Presumably the field is roughly symmetric between the northern and southern hemispheres (depending, however, on the time of year), so that the field would be pointed back toward the sun if one could get sufficiently above the ecliptic plane. Note that the field magnitude remains strong all the way out to apogee, rarely falling much below 15 or 20 gammas. The dipole field is one or two gammas at 30 earth radii.

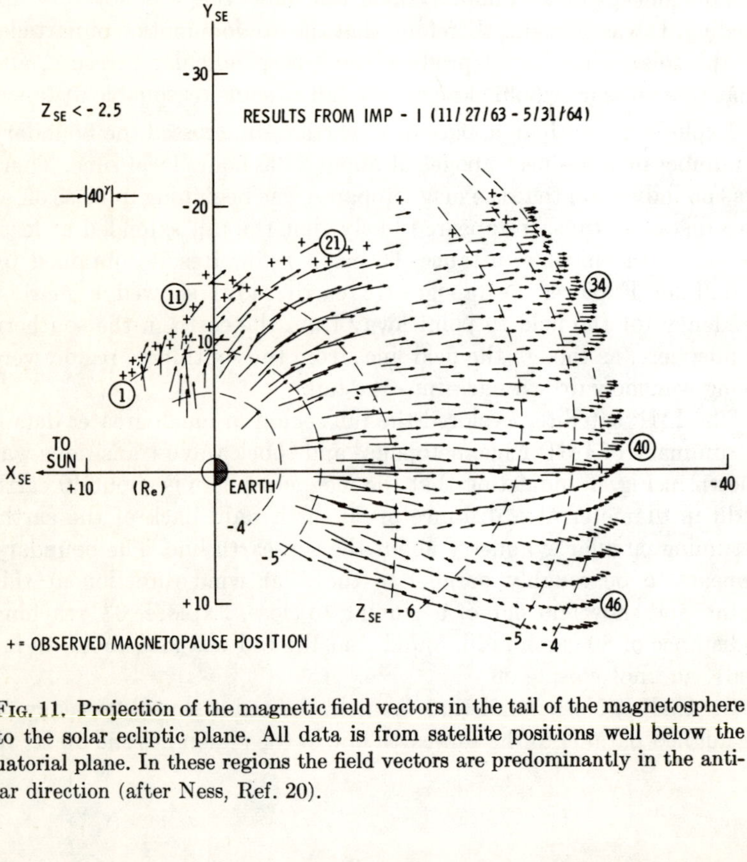

FIG. 11. Projection of the magnetic field vectors in the tail of the magnetosphere onto the solar ecliptic plane. All data is from satellite positions well below the equatorial plane. In these regions the field vectors are predominantly in the antisolar direction (after Ness, Ref. 20).

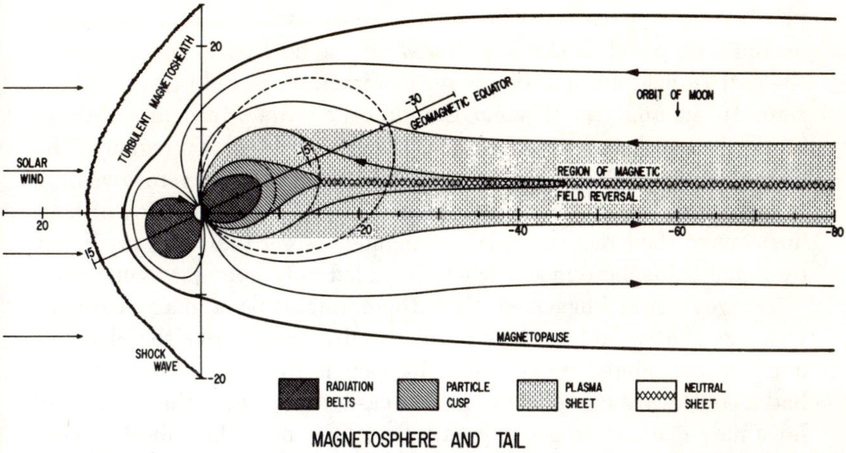

FIG. 12. Model of the magnetosphere in the noon-midnight meridian plane, as conceptualized by Ness, based on IMP data. The situation is shown for the summertime, when the north dipole axis is tilted towards the sun.

Another unusual feature emerging from this IMP data is the tendency of the field to undergo a rather sudden reversal of direction when near the equator, from pointing in the anti-solar direction in the southern hemisphere to the solar direction in the northern hemisphere. This is often accompanied by a reduction in field magnitude at the point of field reversal. Ness has termed this the "neutral sheet" region, or region of magnetic field reversal. The existence of a small but finite perpendicular field component, however, shows that the sheet is not completely "neutral." Most likely, some field line closure exists throughout the region. The experimental data obtained from several IMP satellites in the tail region is summarized in Fig. 12. This shows the field configuration in the noon-midnight meridian plane which one would expect during the summer months, when the north pole of the dipole is tilted toward the sun. It is evident from this figure that when the solar wind is not perpendicualr to the dipole, the earth's field is symmetric about the geomagnetic equator out to distances of 10 to 15 earth radii; beyond this distance the direction of the solar wind controls the plane of symmetry.

The tendency of the field to be aligned with the solar wind direction is seen along the sides of the magnetosphere as a warping or twisting of the field lines out of their meridian plane. If the earth's field were purely dipolar, each field line would lie entirely in a plane. The dipole declination, which is the angle between the horizontal component of the earth's field and the dipole north direction, would be everywhere zero. In an analysis of some Explorer 12 data, Mead and Cahill[11] found that at distances of 8 to 12 earth radii in the dawn region of the magnetosphere, this angle deviated substantially from zero, reaching as high as 60° near the magnetopause. The observed twisting of field lines out of their meridian plane was much larger than that predicted by a magnetosphere model which included only boundary currents.

Dungey[21] first suggested that there might be a magnetosphere topology that would produce a tail structure rather like the observed one. He considered what would happen if the solar magnetic field had a southward component. In this case it is possible that solar field lines may connect to geomagnetic field lines near the subsolar point in an X geometry magnetic neutral point. The connection process called Sweet's mechanism involves dissipation at the null point. This process will result in solar field lines which are joined to the terrestrial field in the polar regions and will result in particles from the sun having direct access to the polar regions of the earth. The connected solar field lines will be blown back over the terrestrial pole to the tail by the solar wind and will disconnect from the terrestrial field in the tail in a second X geometry null point. This field arrangement is shown in Chapter 5, Figure 18. The process is in dynamic equilibrium with new lines continuously connecting to the geomagnetic field at the front and disconnecting at the back. Petschek has shown how this process can proceed rapidly by involving standing Alfvèn waves near the null points. The magnetic neutral sheet in the ecliptic plane is a natural part of this model. Another interesting part of this theory is the existence of an electric field across the polar cap due to the field lines being swept over the polar cap. The voltage V across the polar cap will be given by

$$V/l \sim (\mathbf{v}/c) \times \mathbf{B},$$

where l is the distance across the cap and v is a typical ionospheric wind velocity of $\sim$0.1 km/sec. The voltage here can be estimated to be roughly 30 kilovolts. Because field lines are nearly equipotentials this source voltage will appear East to West across the neutral plane in the geomagnetic tail. This field can accelerate particles up to $\sim$10 keV. It has been suggested that this may be the source of the auroral energetic particles but it is uncertain how the particles can get from the tail all the way into the geomagnetic field to the low altitudes required to make an aurora. This voltage may also drive the *SD* current system as a Hall current (see Chapter 1). This model of the magnetosphere involving field lines connected to the sun can be called an open model to distinguish it from the closed model of Beard and Mead where all field lines are closed and there is no connection to the solar field. Axford and Hines suggested a closed model magnetosphere where viscous forces on the sides of the magnetopause would be responsible for drawing out the geomagnetic field in back into a tail. They also suggested that convective cells of plasmas might be set in motion in the tail which would move particles into the field near the equator and outward near the magnetopause.

The question of the length of the tail is not decided. Dessler has suggested that it might be 20-50 AU long, being kept inflated by the pressure of hydromagnetic waves. Dungey has estimated a length of about $1000R_e$, based on how long a solar field line will remain attached to the geomagnetic field. Mariner IV was very nearly eclipsed by the earth on its voyage to Mars, at a distance of about $3000R_e$. At this time no signficant disturbances were observed in the interplanetary medium, indicating that the tail is not observable at this distance. Explorer 33 observations show, however, that the tail must be substantially longer than $80R_e$. At this distance the magnitude of the field at the point of field reversal was observed to be about 1 gamma. If we use this value as typical throughout the "neutral sheet," then by conserving magnetic flux from the bottom of the tail to the top one can calculate a tail length of about $300R_e$.

SUMMARY

The existence of a sharp boundary to the geomagnetic field, and thus the region within which particles may be stably trapped, is seen to be a permanent feature of the earth's space environment. This boundary, or magnetopause, has been explored in all but the distant anti-solar regions and the regions above the earth's magnetic poles. The observed shape is in reasonable agreement with theoretical calculations based on the assumption of a pressure balance between the impinging solar wind and the geomagnetic field.

The existence of a bow shock wave several earth radii upstream from the magnetopause in the solar direction is also seen to be a permanent feature. Thus, as one proceeds outward from the earth towards the sun, two distinct transitions occur. The first is normally seen as a sudden change in the field direction, a decrease in field magnitude, a disordering of the field, both in direction and magnitude, the termination of energetic trapped particles, and the appearance of low-energy plasma. The second transition is marked by a reduction of field turbulence or variance and a change in the nature of the plasma, which now enters the plasma probes only from the solar direction. Beyond this point conditions are characteristic of the interplanetary medium, with the field averaging about 5 gammas in magnitude and tending to lie in the plane of the ecliptic along the streaming angle characterized by the Archimedes spiral (see Chapter 8).

Inside the magnetosphere, in the tail region beyond 8 or 10 earth radii, the field has a strong tendency to point in the solar or antisolar direction. A region of field reversal is seen to separate these two regions from one another. The magnitude of the field in the tail is many times that of the dipole field alone, indicating that field lines from the north and south polar regions are being drawn back into the tail, or that additional field is being created by large plasma currents in this region. The diameter of the geomagnetic tail is seen to be at least 40 earth radii, and no data so far have given any indication of its closing out to $80R_e$. Ness has compared the geomagnetic tail to those of comets, which have sometimes been observed to stretch out to a length of 1 AU or more. In any case, the magnetosphere must extend at least as far as the orbit of the moon, and is most likely at least several hundred earth radii long.

The energetic particle trapping region, however, extends only out to the magnetopause in the solar direction, and even less on the dark side. The day–night asymmetries in the trapping region are reasonably explained by the theory of particle drifts in a magnetic field. Generally speaking, energetic particles drift around the earth so as to remain roughly in regions of constant magnetic field. Since the solar wind compresses the field more on the day side than on the night side, this means that particles must move in closer on the night side in order to stay where the field magnitude is comparable. Blobs of energetic particles are often seen in the more distant regions on the night side, but these are not believed to be in stable trapped orbits.

BIBLIOGRAPHY

1. Laurence J. Cahill, Jr., The Magnetosphere, Scientific American **212,** 58 (March, 1965).
2. L. A. Frank and J. A. Van Allen, A Survey of Magnetospheric Boundary Phenomena, in Research in Geophysics, Vol. 1, H. Odishaw, Ed., p. 161, M.I.T. Press, Cambridge, Massachusetts, 1964.
3. Sydney Chapman, Solar Plasma, Geomagnetism and Aurora, in Geophysics, the Earth's Environment, de Witt et al., Ed., p. 373, Gordon and Breach, New York, 1963.
4. J. P. Heppner, N. F. Ness, C. S. Scearce, and T. L. Skillman, Explorer 10 Magnetic Field Measurements, J. Geophys. Res. **68,** 1 (1963).
5. A. Bonetti, H. S. Bridge, A. J. Lazarus, B. Rossi and F. Scherb, Explorer 10 Plasma Measurements, J. Geophys. Res. **68,** 4017 (1963).
6. L. J. Cahill and P. G. Amazeen, The Boundary of the Geomagnetic Field, J. Geophys. Res. **68,** 1835 (1963).
7. D. B. Beard, The Solar Wind Geomagnetic Field Boundary, Rev. Geophys. **2,** 335 (1964).
8. N. F. Ness, C. S. Scearce, and J. B. Seek, Initial Results of the IMP–1 Magnetic Field Experiment, J. Geophys. Res. **69**, 3531 (1964).
9. G. D. Mead and D. B. Beard, Shape of the Geomagnetic Field Solar Wind Boundary, J. Geophys. Res. **69**, 1169 (1964).
10. G. D. Mead, Deformation of the Geomagnetic Field by the Solar Wind, J. Geophys. Res. **69,** 1181 (1964).
11. G. D. Mead and L. J. Cahill, Jr., Explorer 12 Measurements of the Distortion of the Geomagnetic Field by the Solar Wind, J. Geophys. Res. **72,** 2737 (1967).
12. D. J. Williams and G. D. Mead, Nightside Magnetosphere Configuration as obtained from Trapped Electrons at 1100 km, J. Geophys. Res. **70,** 3017 (1965).
13. C. L. Longmire, Elementary Plasma Physics, John Wiley and Sons, Inc., New York, 1963.
14. D. A. Tidman, The Earth's Bow Shock Wave, J. Geophys. Res. **72,** 1799 (1967).
15. F. De Hoffman and E. Teller, Magnetohydrodynamic Shocks, Phys. Rev. **80**, 692 (1950).
16. J. R. Spreiter and W. P. Jones, On the Effect of a Weak Interplanetary Magnetic Field on the Interaction Between the Solar Wind and the Geomagnetic Field, J. Geophys. Res. **68,** 3555 (1963).
17. J. H. Wolfe, R. W. Silva, and M. A. Myers, Observations of the Solar Wind during the Flight of IMP–1, J. Geophys. Res. **71,** 1319 (1966).
18. J. T. Gosling, J. R. Asbridge, S. J. Bame, and I. B. Strong, Vela 2 Measurements of the Magnetopause and Bow Shock Positions, J. Geophys. Res. **72,** 101–112, 1967.
19. K. A. Anderson, H. K. Harris and R. J. Paoli, Energetic Electron Fluxes in and Beyond the Earth's Outer Magnetosphere, J. Geophys. Res. **70,** 1039 (1965).
20. N. F. Ness, The Earth's Magnetic Tail, J. Geophys. Res. **70,** 2989 (1965).

21. J. W. Dungey, The Structure of the Exosphere or Adventures in Velocity Space, in Geophysics, the Earth's Environment, C. DeWitt et al., Ed., p. 505, Gordon and Breach, New York, 1963.
22. N. F. Ness, K. W. Behannon, S. C. Cantarano, and C. S. Scearce, Observations of the Earth's Magnetic Tail and Neutral Sheet at 510,000 Kilometers by Explorer 33, J. Geophys. Res. **72**, 927 (1967).

Chapter 10

COSMIC RAYS

R. A. Palmeira and G. F. Pieper

Cosmic Rays are energetic nuclear particles found in interplanetary space, and arriving at the top of the earth's atmosphere coming from all directions. They were originally discovered because the ionization they produce in the atmosphere at the surface of the earth caused the discharge of even the best insulated electroscopes. This fact has been known since 1900, but its explanation was first ascribed to the ionization in the air produced by radioactive materials in the vicinity of the electroscope and around the laboratory in general. In order to verify this situation, Hess[1] in 1912 placed an electroscope on a balloon, only to discover that it detected more ionization at high altitude than it did on the ground. He then hypothesized that a radiation from outer space was the cause, and this radiation later became known as cosmic rays or cosmic radiation.

After this discovery by Hess and some clarification of the nature of the cosmic rays by Bothe and Kohlhörster, Clay, Compton, and others, cosmic rays served primarily as a source of very high energy particles in the advancement of the physics of fundamental particles. At present, cosmic rays are studied primarily for the astrophysical implications of their sources and acceleration processes and to probe the electromagnetic conditions of interplanetary space. It is only at extremely high energies, in excess of 100 BeV, where particles cannot be economically produced in accelerators in the laboratory, that the cosmic rays are used as a beam of high energy particles to study the details of nuclear interactions.

The purpose of this chapter is to summarize some of our present knowledge about cosmic rays.

GALACTIC COSMIC RAYS

At the time of their discovery, nothing was known about the origin of the cosmic rays and therefore their name, implying that they were coming from the cosmos, was justifiable. Today we know that on some occasions, the sun emits high energy particles which are somewhat inappropriately called *solar* cosmic rays. These will be dealt with in their section. In this section we will be concerned with those particles not originating in the sun, which will be termed *galactic* cosmic rays. We will have occasion to mention the possibility that some of the very high energy particles are of extra-galactic origin.

We will define the composition of galactic cosmic rays in terms of their charge and energy spectra.

The primary components of the galactic cosmic rays are protons and alpha particles, with a small remainder consisting of stripped nuclei of higher atomic number. (Electrons and gamma rays are discussed below.) The latter species are customarily divided into four categories as follows:

light nuclei (L)	$3 \leqslant Z \leqslant 5$	Li, Be, B
medium nuclei (M)	$6 \leqslant Z \leqslant 9$	C, N, O, F
heavy nuclei (H)	$10 \leqslant Z \leqslant 19$	Ne to K
very heavy nuclei (VH)	$20 \leqslant Z \leqslant 28$	Ca to Ni

Waddington[2] has compiled the available information on the charge spectrum of the galactic cosmic rays. Part of his findings is shown in Table I.

Included in the table are the directional intensities of the various categories above a total energy threshold of 2.4 BeV per nucleon and

TABLE I

	Flux above E = 2.4 BeV per nucleon (p/cm²-sec-ster)	Abundance by number (%)	Abundance by mass (%)	Ratio of cosmic ray abundance to cosmic abundance	
				Suess and Urey	Cameron
Hydrogen	1510 ± 150	94	74	1	1
Helium	89 ± 3	5.5	17.4	0.77	0.39
L nuclei	2.0 ± 0.2	0.5	8.6	very large	very large
M nuclei	5.6 ± 0.2			4.7	2.5
H nuclei	1.88 ± 0.3			4.5	9
VH nuclei	0.69 ± 0.16			26.5	63.3

their relative abundance by particle number and by mass. Also included are the ratios of the cosmic ray abundance to the cosmic abundances of the elemental groups as determined by Suess and Urey, and by Cameron. The ratios show an enhancement of the cosmic radiation relative to the cosmic abundance, especially in the VH nuclei. The ratio is also very large for the L nuclei because they are present only to a very small extent in the cosmic abundance.

The presence of electrons in the cosmic ray flux above the atmosphere has only recently been definitely established.[3,4] In the energy range 0.1 to 1.3 BeV their vertical flux is given approximately by $(3.5\text{–}11) \times 10^{-3}$ electrons cm²-sec-ster. One important question concerning the existence of electrons above the atmosphere is whether those electrons are truly primaries in the sense that they are produced and accelerated at the same time as the protons and heavier nuclei, or secondaries, generated by the primary component in its journey through space. One process by which primary protons would produce secondary electrons in the interstellar material is the inelastic collision between a high energy proton from the primary cosmic ray beam and a proton of the interstellar space. From this collision neutral and charged π mesons are produced. These charged mesons decay into μ mesons and neutrinos. As a result a number of electrons and positrons will be produced in each high energy p–p collision.

Hayakawa and Okuda,[5] and Jones[6] have calculated the energy spectrum of these electrons and the positron to electron ratio. The result of these calculations shows that the positron to electron ratio should vary from about 2 at an energy of 100 MeV to 1 at much higher energies. On the other hand, on the basis of the supernova theory of origin of cosmic rays (see pages 395–397) developed by Ginzburg, we expect to find the electrons much more abundant than the positrons.

The first experimental result available on the positron to electron ratio in the primary cosmic radiation was the one obtained by De Shong et al.[7] These authors found for this ratio 0.45 at an energy between 50 and 100 MeV, and 0.19 at an energy between 300 and 1000 MeV. The primary electron component contains therefore an excess of negative electrons which cannot be attributed to the proton–proton collision hypothesis. A similar substantial excess of negative electrons was found in a later experiment of Hartman, Hildebrand and Meyer.[35]

In the past years the radio astronomers have established the presence of higher energy electrons in some localized regions of space, by virtue of their radio emission when accelerated in the presence of magnetic fields. This is the so-called *magnetic bremsstrahlung* or *synchrotron radiation* since it was first seen when emitted by particles accelerated in the synchrotron. When an electron with energy $E \gg mc^2$ moves in a plane perpendicular to a magnetic field H, it emits radiation with a continuous spectrum of frequency, having a maximum intensity at a frequency ν_m given by:

$$\nu_m \approx 6H_{(\text{gauss})}E^2_{(\text{MeV})} \quad \text{Mcps.} \tag{1}$$

For a field of 10^{-5} gauss and an electron energy of 1 BeV $= 10^3$ MeV this gives:

$$\nu_m = 60 \text{ Mcps.}$$

On the other hand in this same field an electron of energy 10 BeV $= 10^4$ MeV will emit synchrotron radiation with a maximum intensity at the frequency:

$$\nu_m = 6000 \text{ Mcps.}$$

It is in this frequency interval that the radio waves from space are mostly studied. If the magnetic field in the emitting region is of the order of 10^{-5} gauss, these waves then come from electrons in the energy range (1–10) BeV.

It can be shown that if in some region of space, electrons with an energy spectrum $N(E)\,dE = KE^{-\gamma}\,dE$ emit synchrotron radiation, the spectral intensity observed at the earth is given by

$$I(\nu)\,d\nu = K'\nu^{-\alpha}\,d\nu, \tag{2}$$

where

$$\alpha = (\gamma - 1)/2.$$

Therefore, measurements of the dependence of the spectral intensity on the frequency of the received radiation can yield information on the energy spectrum of the emitting electrons.

These electrons indeed deserve the classification of cosmic electrons, since they are found in regions of space outside the solar system. However, they do not reach the earth, being confined and losing their energy in the regions from which we detect their synchrotron radiation. For this reason, whether or not we should consider them as part of the cosmic rays, depends on the definition we adopt for the term *cosmic rays.*

Experiments are presently being carried out to try to detect γ-rays in the cosmic ray flux above the atmosphere, but so far without conclusive results.[8, 36] The detection of γ-rays will have great importance because, being unaffected by magnetic fields and therefore traveling in straight lines, they will give information about the location of their sources.

The energy spectrum of the protons above a total energy of a few BeV can be represented by the formula:

$$N(>E) = CE^{-\gamma} \tag{3}$$

Where $N\,(>E)$ is the flux of protons with total energy (kinetic plus rest energy) greater than E. This formula gives a good fit to a great number of observations in a wide range of energies, with the exponent γ varying slightly from 1.4 at E between 5×10^9 and 5×10^{11} eV, to 1.5 at E between 10^{13} and 10^{15} eV, and finally being close to 2 at the highest energies observed of 10^{16} to 10^{19} eV (Figure 1).

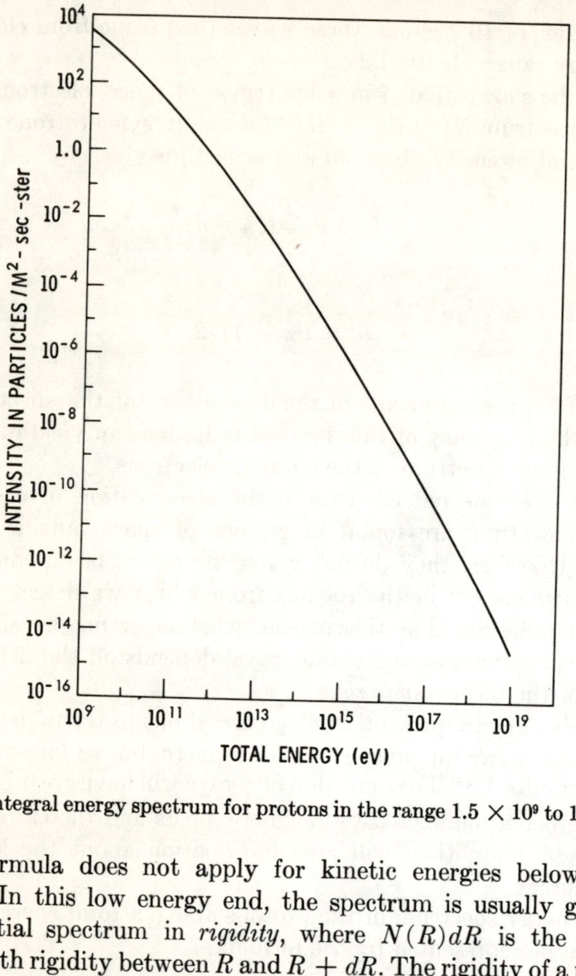

FIG. 1. Integral energy spectrum for protons in the range 1.5×10^9 to 10^{19} eV.

This formula does not apply for kinetic energies below about 0.5 BeV. In this low energy end, the spectrum is usually given as a differential spectrum in *rigidity*, where $N(R)dR$ is the flux of protons with rigidity between R and $R + dR$. The rigidity of a particle of momentum p and charge Ze is given by $R = pc/Ze$, where c is the velocity of light.*

---

*In the CGS system of units the rigidity is measured in gauss-cm, since it can be shown that in this system the rigidity R is equal to the product of a magnetic field intensity H (measured in gauss) times the radius of curvature ρ (measured in cm) that the particle of rigidity R will have when moving in a plane perpendicular to the magnetic field. In the MKS system of units the rigidity is measured in volts and is numerically equal to the particle momentum p (measured in eV/c) divided by the charge number Z.

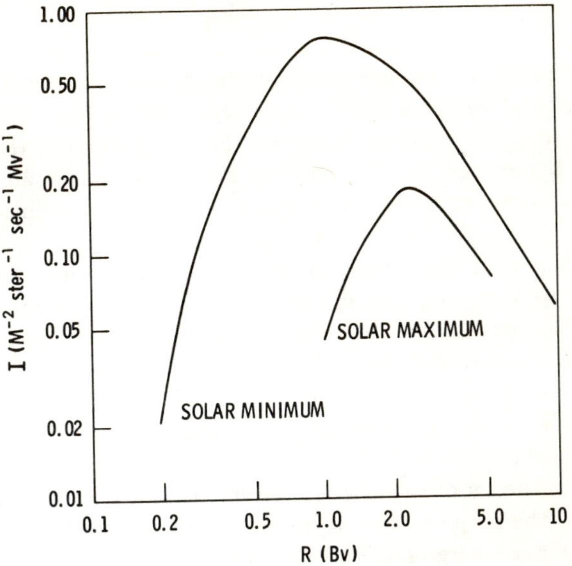

FIG. 2. Differential rigidity spectrum for protons during solar minimum and solar maximum.

The principal feature of the proton spectrum at these low energies is its variability following the 11-year period of the solar cycle. During the period of maximum solar activity the proton intensity is much lower than during the period of minimum activity. Figure 2 shows two representative rigidity spectra using data from many experiments and taken close to a maximum and a minimum of the solar activity. The difference between these two spectra is striking.

The energy spectrum of the other nuclei with charges greater than 1 has been investigated in detail only in the range of low energies. The experimental evidence is that between a total energy per nucleon of 1.5×10^9 and 10^{10} eV the spectrum of heavy nuclei is similar to the one for protons as given in (3) with E now representing the total energy per nucleon, and γ close to 1.5.

At a total energy per nucleon below about 1.5×10^9 eV only the alpha particle spectrum has been observed in detail. This spectrum shows the same variability with the solar cycle as does the proton spectrum in this energy range. However, the exact relationship between the proton and alpha particle 11-year solar modulation is still the subject of current experimental and theoretical investigations.

ORIGIN OF COSMIC RAYS

At the present time the origin of cosmic rays is one of the most important and complex problems confronting the cosmic ray physicist. Not only the solution, but even the formulation of the problem involves knowledge of solar physics and astrophysics and has implications concerning the problems of origin and evolution of the universe.

Any theory of the origin of the cosmic rays must specify the source or sources, i.e., the regions of space responsible for the injection of the particles and the mechanism of their acceleration to the high energies observed. In addition, this theory would have to explain all the important characteristics of the cosmic rays as seen on earth, in particular, the isotropy, the composition, and the shape of the energy spectrum.

An early theory of the origin of cosmic rays attempted to consider the sun as the major source of injection and acceleration of particles of cosmic ray energies. This view was in part supported by the observation, following some large solar flares, of cosmic ray intensity increases on the earth which could undoubtedly be ascribed to an extra flux of particles with energies up to a few tens of BeV coming from the direction of the sun. It was then clear that the sun is a source of cosmic rays, and the question was whether it could provide the bulk of the cosmic ray flux as seen on earth. This theory would have given the cosmic radiation the character of a local phenomenon, restricted to the vicinity of the solar system without much connection with the rest of the universe.

The theory had great difficulties, however. First, to explain the observed cosmic ray isotropy one would have to postulate the existence of very strong, large scale magnetic fields in the solar system. Second, the existence of particles with energies $\gg 10^{10}$ eV posed great difficulties since no particles with those high energies were observed to be generated by the sun. Finally, more recent observations on both the charge and energy spectra of the cosmic ray particles coming directly from the sun show appreciable deviations between these properties and those pertinent to the bulk of the cosmic ray particles. All these facts contributed to the abandonment of the theory that considered the sun as the major source of cosmic rays, although there is no doubt that it sporadically emits particles with energies up to 10–50 BeV which are detected on the earth.

The next step in trying to explain the origin of cosmic rays was to consider that all the stars in our galaxy emit cosmic ray particles, the sun being just a typical emitter, and that these particles escape from the region of emission, eventually filling the whole galactic volume. If we assume however, that all the $\sim 10^{11}$ stars in our galaxy emit cosmic rays with the same power output as the sun (10^{21}–10^{22} ergs/sec), then the total power output from all the stars in our galaxy in the form of cosmic rays would be 10^{32}–10^{33} ergs/sec. This is short by many orders of magnitude of the power of 10^{39}–10^{40} erg/sec which is the estimated rate of energy loss by cosmic rays in our galaxy. Even considering that some special types of stars such as magnetic stars, flare stars and others could be more powerful cosmic ray emitters than the sun, the problem of providing the necessary power output still remains acute.

This problem was solved in part by the first detailed and quantitative theory of the origin of cosmic rays, due to Fermi.[9] In this theory the particles gain energy by a process of gradual acceleration in interstellar space, instead of being produced with full energy by some or all of the stars of our galaxy. These stars then serve only as true injectors of particles of comparatively low energy which will be further accelerated in the interstellar space. In order to achieve this acceleration, Fermi considered the following specific model:

The whole galactic space is pervaded by a weak magnetic field. The field although more or less regular over large regions of space can be stronger and very irregular in some localized regions such as, for instance, inside the ionized hydrogen gas clouds. If, during its motion through space, a cosmic ray particle encounters such a localized region of stronger magnetic field, its direction of motion will be changed appreciably and we can consider such a process as a scattering of the particle by the magnetized cloud. Moving more or less freely between clouds and being scattered by them when coming to a close approach, the motion of the particle can be considered as a random walk with a mean free path λ of the order of the distance between clouds. We can then use the diffusion equation to describe the distribution of particles in space. These hydrogen clouds however are not stationary; they are moving through space with some average velocity V. When colliding with such a moving scattering center, the particle will gain some energy if the collision is head-on, or it will lose some energy if it collides with the scattering center when overtaking it. Since the collisions of the first type will be more numerous than those of the second, the particle will on the average gain energy in its motion through space.

This situation of diffuse motion with a net gain of energy through collisions with moving scattering centers is very similar to the situation that occurs when we mix two gases of different molecular weight and different temperatures. The lighter molecules of the gas originally at lower temperature will gain energy by collisions with the hotter and heavier gas until thermal equilibrum is reached between the two gases when they attain the same temperature. We can then consider the gain in energy of the cosmic rays when colliding with the moving gas clouds, as the tendency to thermal equilibrium between the cosmic ray "gas" and the moving cloud "gas".

It can be shown that in this model the average energy gain during one collision is $\Delta E = \alpha E$, where E is the total energy of the particle and $\alpha \approx V^2/c^2$.

In this formula, V is the cloud velocity and c is the velocity of light. We can then calculate the average energy of a particle as a function of time, with the following result:

$$E(t) = mc^2 \exp (\alpha t/\tau), \tag{4}$$

where m is the mass of the particle and τ is the average time between scattering collisions.

Let us now assume a loss mechanism by which the cosmic rays are effectively removed from the accelerating region and let T be the mean lifetime against this removal mechanism. It can then be shown that the age distribution of the cosmic rays, i.e., the number of particles with age between t and $t + dt$ is given by

$$P(t)\, dt = (N_0/T) \exp (-\, t/T)\, dt, \tag{5}$$

where N_0 is the total number of particles present.

By eliminating t between (4) and (5), we arrive at an expression giving the number of cosmic rays with energy between E and

$E + dE$; that is to say, we get the differential energy spectrum of the primary cosmic rays. This expression is

$$N(E)\, dE = \text{const. } E^{-\gamma'}\, dE, \tag{6}$$

where

$$\gamma' = 1 + (\tau c^2/TV^2).$$

We see that in this way we obtain a differential energy spectrum that is a power law in energy with an exponent equal to $-\gamma'$. The fact that this statistical process of acceleration gives an energy spectrum in the form of a power law speaks in favor of such a mechanism for building up cosmic ray energies.

We have seen earlier that in the energy range 10^{10}–10^{15} eV the observed integral primary spectrum can be fitted by the formula

$$N\ (> E) = \text{const. } E^{-\gamma}, \tag{7}$$

with $\gamma = 1.5$ in this region.

The differential spectrum is then of the form

$$N(E)\, dE = \text{const. } E^{-(\gamma+1)} dE. \tag{8}$$

Comparing the theoretical and observed spectra, we find $\gamma' = \gamma + 1$. Therefore,

$$k \equiv \tau c^2/TV^2 = 1.5 \tag{9}$$

and the question is whether we can satisfy this equation using reasonable values of the parameters τ, T and V. Astronomical evidence suggests the following values:

$$V = 30 \text{ km/sec} = 3 \times 10^6 \text{ cm/sec};$$

$$V/c = 10^{-4};$$

$$\lambda = 3 \times 10^{20} \text{ cm};$$

$$\tau = \lambda/c = (3 \times 10^{20})/(3 \times 10^{10}) = 10^{10} \text{ sec}.$$

In the early version of this theory, Fermi assumed that the mechanism responsible for the removal of particles from the cosmic ray "gas" was nuclear interaction with the protons of the interstellar medium. Using estimated values of the proton–proton collision cross-section at these energies, and the density of the interstellar medium, we arrive at a value

$$T = 2 \times 10^{15} \text{ sec}$$

and thus

$$k = (10^{10} \times 10^{8})/(2 \times 10^{15}) = 5 \times 10^{2},$$

which is of the order of 300 times the observed value.

A more serious drawback of the theory however, is that the exponent of the predicted energy spectrum will depend on the type of particle being accelerated. This is so because this exponent depends on T, the mean time for removal of the particle through nuclear interaction with protons of the interstellar medium, and this time will be shorter for heavier nuclei on account of the larger collision cross section. Such a dependence of γ' on the nuclear species of the primary cosmic rays is not observed and some modification must be made to bring the theory into agreement with the observations.

An appropriate modification was made by Morrison, Olbert and Rossi[10] who considered as a loss mechanism the diffusion out of the galaxy in place of the collision with protons of the interstellar medium. If we consider the average time for escape from the galaxy to be shorter than the average time before a nuclear collision of the heaviest nucleus found in the cosmic ray primaries, we can then have T and therefore γ' independent of the type of particle being accelerated.

The difficulty however still remains of fitting the observed exponent of the energy spectrum with the one calculated from the theory. Also the problem of removing the particles from the galaxy in a relatively short time is a serious one.

The root of these difficulties lies in the fact that the statistical process of acceleration taking place in the interstellar medium is a very inefficient one. The particle gains very little energy in each collision with the gas clouds, and it has to travel a long distance, spending therefore a long time, before suffering the next collision.

If the statistical acceleration were to take place, however, in localized regions of space where τ could be smaller and V could be larger than the values used above, we would have a much more efficient accelerating mechanism.

A theory of cosmic ray origin invoking such an accelerating mechanism in restricted regions of our galaxy has been developed in great detail, mainly by Ginzburg,[11] Shklovsky,[12] and Hayakawa.[13] In this form, the theory identifies these restricted regions of space with the gaseous envelope surrounding a star that has undergone a catastrophic explosion of the supernova type. These envelopes are made of material ejected by the star at the time of the explosion and they expand into space with great speed. The best example of such an envelope is the nebula known as the Crab Nebula, which is the remnant of a supernova explosion that took place more than 900 years ago.

These expanding envelopes formed of material ejected by the exploding star seem to provide a suitable region for the statistical acceleration of cosmic rays. In the first place, strong magnetic fields have been detected and there are undoubtedly regions of turbulence which can then act as scattering centers for the cosmic ray particles. Secondly, the strong emission of radio waves from the Crab Nebula and other strong radio sources in the sky has been postulated to originate as synchrotron radiation of high energy electrons in such magnetic fields. If the processes taking place in such envelopes can accelerate electrons, which then radiate energy as synchrotron radiation in the strong magnetic fields, they can also conceivably accelerate protons and heavier nuclei to cosmic ray energies. It should be pointed out that although the most abundant production of cosmic rays takes place in the early decades following the supernova explosion, there are reasons to believe that in the Crab Nebula in particular cosmic ray acceleration is still taking place.

These turbulent gaseous envelopes resulting from supernova explosions are more efficient regions of cosmic ray acceleration mainly because of two factors. One is the higher value of the velocity V of the turbulent regions which act as scattering centers as compared to the velocity of the clouds of the interstellar medium. The other factor is the smaller scale of the region, with the result that the mean time between collisions can be made much shorter than the corresponding time for the interstellar acceleration. Therefore, the quantity

$$k = \tau c^2/TV^2$$

can be made much smaller and thus fit the experimental value.

Two further points can be made in favor of such a selection of the accelerating region. The first is that due to the smaller scale of the region, the mean time T for loss of the particles out of the accelerating region can be made smaller than the mean time for loss due to nuclear collision of even the heaviest nuclei. As a result, the exponent $\gamma' = 1 + k$ in the differential energy spectrum can be made independent of the mass of the particle being accelerated. The second point is that the larger abundance of the heavy elements in the cosmic ray beam as compared with the "cosmic" abundance, can be explained in a very natural way. The particles being accelerated in the supernova envelopes were released by the star at the time of the explosion. This supernova type of explosion usually occurs at a very late stage of the star evolution, when, through nuclear fusion of progressively heavier elements, all elements up to iron can be formed inside the star. All these nuclear species will be accelerated, and in this way we can build a cosmic ray beam richer in heavy elements than the cosmic abundance.

At this point, we must make clear that this so-called cosmic abundance refers to samples of material taken for the most part from within the solar system, whereas the cosmic rays that arrive at the earth may represent a sampling from other regions of the universe. The fact that they differ in some respects then seems natural, and none can really claim the title of true cosmic abundance.

In synthesis then, this modified theory of statistical acceleration of cosmic rays ascribes specific regions of our galaxy, the supernova envelopes, as the locale of this accelerating mechanism. The cosmic ray particles, after gaining their energies in these regions, escape into the interstellar space where their motion is further modified by existing magnetic fields (with possibly some further acceleration) in such a way that they lose all recollection of their place of injection and acceleration. As a result an isotropic distribution of cosmic rays throughout the whole galaxy (including the halo) is obtained. The weak interstellar magnetic fields then act only to stir the cosmic ray beam in order to produce the observed isotropy, with the energy that goes into the cosmic ray particles coming, ultimately, from the nuclear energy released in the supernova explosion.

There seems to be one difficulty with this theory, and this is the necessary frequency of supernova outbursts in order to give the observed cosmic ray density in our galaxy. The frequency used by Ginzburg and Shklovsky of one outburst every 30–50 years in our galaxy seems to be a little higher than has been inferred. One can also possibly consider, as Ginzburg does, the less energetic but more frequent nova outbursts as another source for cosmic ray injection and acceleration. It is not completely clear whether this theory in its present form can account for all the observed facts concerning the primary cosmic rays detected on earth, but on the other hand there is no doubt that it is a plausible theory that does not contradict any known and established property.

The only possible exceptionable property is the presence in the cosmic rays of very high energy particles, (10^{19}–10^{20} eV). The radius of curvature of particles of these energies in the interstellar magnetic fields is of the order of magnitude of galactic dimensions, and it is difficult to see how they can be trapped in our galaxy for any appreciable time. If the very high energy particles are indeed protons it is then difficult to escape the conclusion that they must have entered our galaxy from the intergalactic space already possessing these very high energies. They were perhaps accelerated in other galaxies by similar processes, escaped into intergalactic space after reaching these high energies, and eventually penetrated our galaxy where they were finally detected.

In conclusion, following Morrison,[14] it now seems that we can divide the cosmic ray spectrum into three energy regions as far as the origin of these particles is concerned. The first region comprises the particles with energy up to 10–100 BeV: they could have been produced in the atmospheres of the sun and other sun-like or special type stars. The second region is the one containing particles from 10^2 to 10^9 BeV: they were in all probability accelerated to these energies by the Fermi statistical acceleration mechanism in the envelopes of supernova. Finally the last region comprises those very energetic particles, with energy above 10^9–10^{10} BeV which conceivably could have come from the intergalactic space, accelerated to these energies by similar processes taking place in other galaxies from which they have escaped.

The fact that we do not detect any discontinuity in the shape of the energy spectrum of the primary cosmic rays, as different origins for different energy intervals could produce, may be only the result of the inaccuracy of our measurements.

TIME VARIATIONS

One of the characteristics of the cosmic ray flux is its constancy in time. Analyses of the contents of C^{14} found in deep sea core samples and produced as a result of cosmic ray bombardment, have disclosed that the cosmic ray flux has not varied by more than 10–20 percent over the last 30,000 years. Measurements of the contents of many cosmic-ray produced radioactive isotopes in the Aroos meteorite have shown that the average cosmic ray intensity over the past 2.5 million years has not changed by more than a factor of 2 (Geiss[33]). More recently, Hintenberger et al[34] by measuring the isotopic constitution of potassium in iron meteorites have concluded that the cosmic ray intensity has been constant within $\pm 50\%$ during the last 10^9 years. It should be emphasized that these measurements refer to the average cosmic ray intensity over the time spans mentioned above, and at the location of the earth or of the volume occupied by the solar system. They do not exclude the possibility that the intensity in the vicinity of the solar system has stayed relatively constant, while the intensity somewhere else in our galaxy has suffered large variations. Notwithstanding, this measured averaged constant intensity in the neighborhood of the solar system is an important characteristic of the cosmic ray flux, and local and transient deviations from this constancy that do exist should be considered important as well, principally because of the information they bear on some of the problems of astrophysics and geophysics.

We can divide these transient cosmic ray time variations into 3 broad categories:

a) of meteorological origin
b) of solar origin
c) of galactic origin

In the following we will deal briefly with each one of these categories.

The cosmic radiation detected at the surface of the earth is not of the same composition as that incident on the top of the atmosphere. The protons and heavier nuclei that compose the so-called primary radiation, upon impinging on our atmosphere, suffer nuclear collisions with the atoms of oxygen and nitrogen and in these collisions protons and neutrons, as well as charged and neutral pions are produced. These active nuclear particles penetrate deeper into the atmosphere and suffer further collisions which produce more particles, until the energy of the incident primary particle (proton or heavy nucleus) is shared among many lower energy protons and neutrons which are detected at sea level. This is the so-called nucleonic or N component of the secondary cosmic radiation.

Some of the charged π mesons undergo decay into a μ meson and a neutrino. The μ mesons so produced interact weakly with matter, thus they survive further traversal of the atmosphere and reach sea level where they are by far the most abundant of the secondary components. For historical reasons the μ mesons component is called the hard component of the cosmic rays.

Neutral π mesons are produced in the energetic nuclear interactions at the top of the atmosphere and decay quickly into two γ rays. These high energy γ rays interact with the electric field surrounding the atomic nucleus, and as a result the γ ray disappears and one negative and one positive electron are formed (pair production process). The high energy electrons in turn are deflected by the electrical field of the atomic nucleus with the emission of a high energy γ ray (bremsstrahlung process).

These pair production and bremsstrahlung processes continue down through the atmosphere, originating the so called *electromagnetic cascade*, until the energies of the electrons and γ rays are too low for these processes to take place. The electrons and γ rays at the level of observation deep into the atmosphere form the soft component of the secondary cosmic radiation. Figure 3 shows the schematic development of the secondary cosmic radiation in the atmosphere (Simpson et al.[15]) and Figure 4 shows the variation with height of the various secondary component (Rossi[16]).

With this general understanding of the generation of the secondary component in the atmosphere, we can now give a qualitative picture of the meteorological effects in the cosmic radiation as measured at sea level. The first factor to affect the cosmic ray intensity is the barometric pressure. An increase in the barometric pressure at the point of observation means an increase in the total mass of air above this place, consequently causing a decrease in the observed

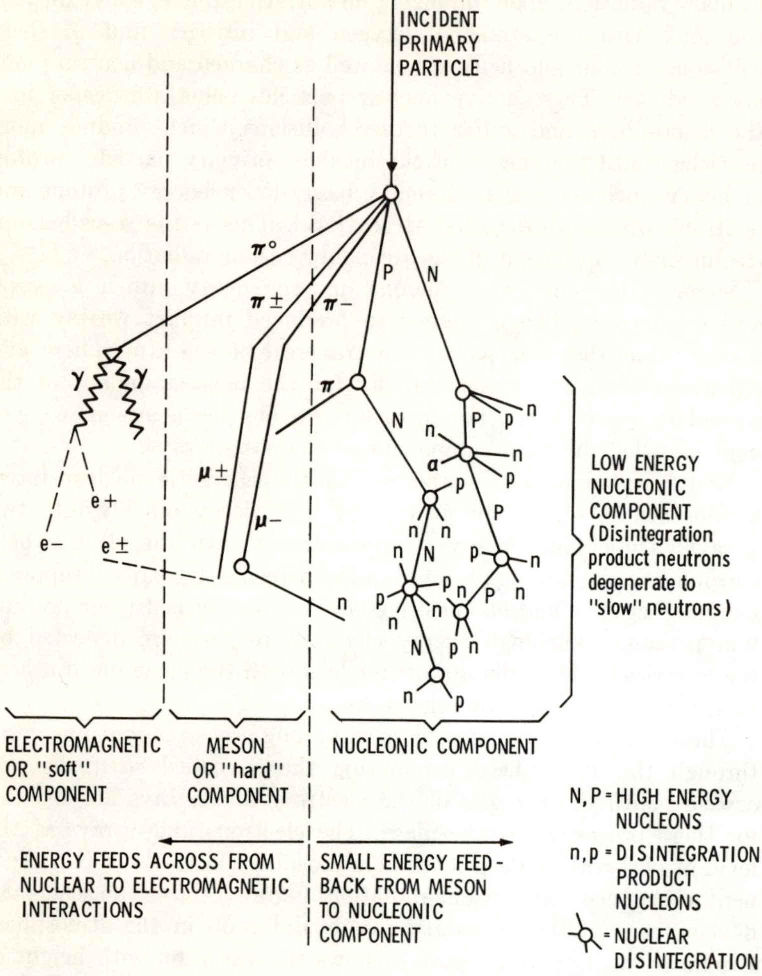

FIG. 3. Schematic representation of the typical development of the secondary cosmic radiations within the atmosphere arising from an incident primary particle (Simpson et al., Ref. 15).

intensity due to this increased absorber that the radiation has to traverse.

Another meteorological factor that affects the cosmic ray intensity is the temperature distribution in the atmosphere. With an

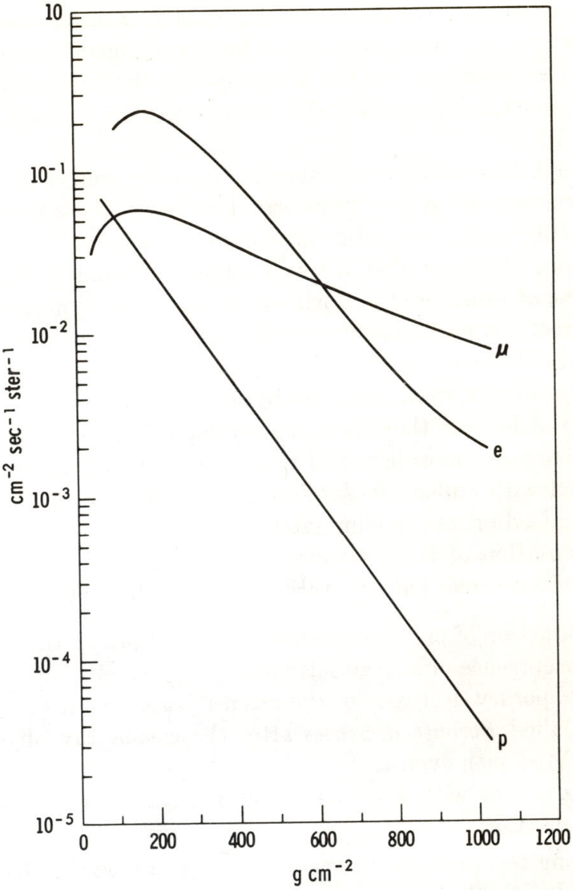

FIG. 4. Vertical intensities of various cosmic-ray components at 50° geomagnetic latitude, as functions of atmospheric depth. Curve e: positive and negative electrons of energy greater than 10 MeV. Curve μ: μ-mesons of all energies. Curve p: protons of kinetic energy greater than 400 MeV. The last curve represents a crude estimate (Rossi, Ref. 16).

increase in the temperature the atmosphere expands and as a result the mean level where the mesons are produced by the nuclear interactions increases. This makes the path length of a meson at the observation point longer and therefore enhances its probability of decaying before reaching the detector. As a consequence, we observe a decrease in the measured intensity. The temperature variation affects only the μ meson or hard component, since it is the only one composed of unstable particles (for the time scales involved between nuclear interactions the neutron can be considered a stable particle.)

Although of relative importance in the early studies of cosmic ray time variations when they were first discovered, these meteorological or atmospherically produced variations are now only of secondary importance. Their origins are well understood and in the study of variations of other origins, their effects are first eliminated by suitable corrections using experimentally determined coefficients (different for the different secondary components) therefore leaving in the data only the variation produced by other causes.

Finally it is clear that these meteorological variations are present only inside our atmosphere and therefore, in cosmic ray experiments conducted with balloon or satellite borne detectors these variations are minimized or totally eliminated.

The variation of the intensity of cosmic rays of solar origin can be subdivided somewhat arbitrarily into the following sub-categories:

i) Production of particles of cosmic ray energies by the sun, following the occurrence of a large solar flare.

ii) Temporary decreases of the intensity associated with magnetic storms, called *Forbush decreases* after the cosmic ray physicist who first reported such events.

iii) Variations with a period of one solar day; the so-called *solar diurnal variation.*

iv) Long term change in the cosmic ray intensity, which is anticorrelated with the increase and decrease of the solar activity that occurs with a period of approximately 11 years, and is thus called *the* 11-*year variation.*

We will now deal briefly with each of the above types of variations which are in one way or another under solar control.

i) On some occasions following a large flare occurring in the solar atmosphere, particles are accelerated to an energy up to 10–50 Bev, and are detected on earth through their secondary effect in our atmosphere as a sudden burst or increase in intensity of cosmic rays, which can range from a few percent up to a few hundred percent for a typical detector. Recently, balloon and satellite borne instruments have detected lower energy particles of solar origin above our atmosphere with a frequency of occurrence much higher than that of the higher energy events detected previously.

ii) The first systematic monitoring of the cosmic ray intensity at several places on the earth started in the decade of 1930 when A. Compton, S. Forbush and collaborators of the Carnegie Institution of Washington installed a network of cosmic ray detectors at such widely separated places as U.S.A., Greenland, Peru, and New Zealand. As a result of such continuous monitoring of the cosmic ray intensity, it was discovered that on many occasions, at the approximate time of occurrence of a magnetic storm on earth, the cosmic ray intensity decreased by a few to 10 percent in a few hours, and recovered to the pre-decrease level in a time of the order of a few days. These decreases came to be known as *Forbush decreases* or *F-events*. Later, with the use of detectors sensitive to the lower energy part of the primary cosmic ray spectrum, larger decreases up to 20 percent have been observed. More recently, such decreases have also been observed with instruments on satellites and space probes, at some great distance from the earth, indicating that the phenomenon is not a local one connected in some way with the earth or its magnetic field, but involves a substantial portion of the volume occupied by the solar system.

Although the exact details are not yet well known, the general mechanism responsible for these decreases is generally believed to be the following:

At the time of occurrence of a flare in the solar atmosphere, the sun may emit a cloud or stream of ionized matter, mostly hydrogen, and this cloud or stream travelling with a typical velocity of 1000 to 2000 km/sec arrives at the earth in a time of the order of 1 to 2 days. This cloud, due to its high electrical conductivity, carries with it the magnetic fields that were present in the vicinity of the sunspot area where the flare originated. When the cloud then envelopes the earth, the effect of the magnetic field is to deflect away some of the cosmic rays that otherwise would have arrived at the earth, and therefore to produce a decrease in the measured intensity. The same cloud by its effect on the geomagnetic field produces the magnetic storms seen in the magnetograms.

iii) It has been observed that the cosmic ray intensity as measured on the earth, undergoes periodic fluctuations with a period of one solar day. These fluctuations are small, highly variable and therefore difficult to observe. Many experiments have been set up and many papers have been written on the results and its explanations, but only recently, with the development of detectors of very high counting rates, and with the proper understanding of the role played by the geomagnetic field in deflecting the cosmic ray particles, has it been possible to interpret correctly the observations although the origin of these fluctuations is not yet well understood.

The observations are consistent with an anisotropy of a few tenths of a percent, with the direction of maximum intensity located at approximately 85° to the east of the earth-sun line, although this direction as well as the amplitude suffers short and long term changes. This anisotropy is then seen by a detector situated on the spinning earth as a periodic fluctuation with a period of one solar day.

The origin of the anisotropy is believed to be due to the interaction between the cosmic ray particles and the solar magnetic field that is continuously being stretched out in space by the continuous expansion of the solar corona (the so-called solar wind). Such interaction might cause the cosmic ray "gas" to co-rotate with the sun. The co-rotation velocity at the orbit of the earth is approximately 435 km/sec, which is much greater than the average orbital velocity of the earth around the sun (30 km/sec). Under these conditions the cosmic ray "gas" is continuously overtaking the earth, and this would produce a directional anisotropy with the direction of maximum intensity lying at 90° to the east of the earth-sun line.

iv) It has been observed, by the continuous monitoring of the cosmic ray intensity over a great number of years, that the average intensity varies in more or less a regular way, following the general trend of the 11-year cycle of solar activity. At the maximum of the solar cycle as measured for example by the sunspot numbers, the cosmic ray intensity reaches a minimum, and starts increasing thereafter to reach a maximum at the time of minimum solar activity. The amplitude of this variation for a typical cosmic ray detector such as the neutron monitor, is of the order 20–30 percent, but is very highly energy dependent, the low energy particles being much more strongly affected. Balloon flights carrying cosmic ray detectors and made during different phases of the solar cycle have established that particles with kinetic energy of a few hundred MeV are strongly reduced in number and sometimes are almost completely absent during the maximum of the solar activity.

The explanation for this anti-correlation between solar activity and cosmic ray intensity is now believed to be due to the continuously expanding solar corona: the solar wind mentioned above (Parker[32]). The galactic cosmic rays when penetrating the solar system space are deflected by irregularities existing in the interplanetary magnetic field. These irregularities that act as scattering centers for the cosmic ray particles are not stationary. They move radially out from the sun, being carried away by the solar wind. Under equilibrium conditions the inward diffusion of the cosmic ray particles is balanced by the outward convection due to motion of the scattering center, the net result being a cosmic ray intensity at the orbit of the earth which is depressed in relation to the galactic intensity outside the solar system. The parameters that enter in the expression for the depressed intensity vary with the 11-year solar cycle of activity and in this manner a related 11-year variation is introduced on the cosmic ray intensity at the orbit of the earth.

The story of continuous cosmic ray measurements is not very old, covering only less than three solar cycles. It is thus natural that our experimental observations on this effect are very scanty; in particular, balloon observations of the lower energy part of the primary cosmic ray spectrum have been conducted only for the second time during a minimum of solar activity. When more data accumulate during the next solar cycles and with improved technology born out of the space age, we hope to gain a more through understanding of this important type of solar cosmic ray modulation.

If the cosmic ray intensity is not isotropic within the volume of our galaxy, then, as a result of the rotation of the earth, an earth-bound detector will see a periodic fluctuation of the measured intensity with a period of one sidereal day. Many experiments have been designed to try to detect such variation with a period of one sidereal day. It is clear, that if such anisotropy exists, it should be confined to the high energy particles, because at lower energies, the galactic magnetic fields will act as an agent to destroy the original directions of motions of the particles, and it is only at higher energies, where the trajectory deflections by the magnetic fields are small, that the particles might be able to retain some of their original direction of motion.

Unfortunately, at these high energies, the intensity of the cosmic rays is much less than at lower energies and therefore, in order to obtain statistically meaningful results one has to use detectors with a very large sensitive area, and run the experiments for a long time. Due to these difficulties it has not yet been possible to detect with certainty such sidereal variations, and the conflicting results so far reported only serve to put an upper limit on the magnitude of this fluctuation. For energies up to 10^{15} eV this upper limit has been put at 0.1 percent, and for energies up to 10^{17} eV, at 10 percent.

Although this sidereal time variation has not been completely established, its importance in the problem of cosmic ray origin and acceleration cannot be minimized, because its existence will point to specific regions of our galaxy as the seat of production of cosmic ray particles, or at least to some preferential directions of acceleration.

SOLAR COSMIC RAYS

In 1942, Forbush[17] discovered that high energy particles originating at the sun can strike the earth. Although the designation is not precise, such particles have generally come to be called *solar cosmic rays.* In the years since their discovery, the techniques for detecting solar cosmic ray events have improved markedly, resulting in an increased number of observations of such events. The original widespread notion that these events are very rare has now been superseded by the view that solar cosmic ray particles are probably present near the earth 10 to 20 percent of the time, that they exceed the normal galactic cosmic ray intensity for 1 to 2 percent of the time, and that they may occasionally reach intensities very much larger ($\gtrsim 10^4$) than the galactic cosmic ray intensity.

Experimental results from the spacecraft Explorer X, Mariner II, and Explorer XVIII have confirmed the existence of a steady emission of particles from the sun, *the solar wind.* Typically, in the vicinity of earth, the energies of the solar wind particles are in the range of 100 eV to a few keV. So far as is known at this time, the production of particles of energy greater than ~ 1 MeV, i.e., solar cosmic rays as distinct from the solar wind, does not take place continuously, but rather only in conjunction with a solar flare. The size of a solar flare is rated in terms of its area on the solar surface and its brightness using a parameter called *importance,** whose range is 1^- to 3^+. Large (in terms of integrated intensity) cosmic ray events generally occur in association with large solar flares, those of importance 3 or 3^+. The solar wind and solar cosmic rays probably differ significantly in their modes of production. The energy for solar cosmic rays probably comes from the collapse of strong magnetic fields in the region of their production at the surface of the sun.

Malitson and Webber have reported a study[18] of the 30 largest solar cosmic ray events of the nearly 50 recorded between 1958 and 1961. They noted that only 16 different active centers (flare producing regions) produced all 30 major events, and furthermore, from a summation of the integrated intensities, that 4 active centers produced over 90 percent of all the solar cosmic rays observed at earth in the 6-year period. The very limited statistics of the frequency of occurrence of the very largest events—3 events in the 6 years—taken with earlier less definitive data, combine to suggest that on the average about once in 18 months an active region appears on the sun that will produce one or more major cosmic ray events. The appearance of such active regions is *not* strongly correlated with the maximum in the 11-year cycle of solar activity. Rather, it appears that the frequency during periods of increasing and, even more so, decreasing solar activity may be greater than one per 18 months, while near solar maximum and solar minimum it may be less than one per 18 months. When all recorded cosmic ray events, small as well as large, are examined, the correlation with the solar activity cycle is somewhat better, as expected, although even here there are some significant departures and again the statistics are not very good.

Solar cosmic rays originate in solar flares in the sun's *chromosphere.* Thus far every solar cosmic ray event can be correlated at least moderately well with a solar flare, including one case in which the flare took place about 20° in longitude behind the sun's west limb. On the other hand, only a relatively small fraction (0.2) of the observed flares of importance 2 or larger, produce cosmic rays that

*Recently a new classification of solar flare importance has been adopted by the International Astronomical Union. In this new classification the flare importance is given by two figures. The first one ranges from S (for sub-flares) to 1 through 4 depending on the area on the solar disk occupied by the flare. The second figure can assume the values F (faint), N (normal), and B (brilliant).

can be detected on or near earth. Malitson and Webber note that of the 30 major solar cosmic ray events in the period 1956–1961, 22 were from flares in the sun's western hemisphere and 8 from the eastern. Further analysis indicates that the probability of observing cosmic rays at the earth is greatest when the flare is in the sun's northwestern quadrant. The east–west asymmetry can be understood in terms of the propagation of solar cosmic rays through space (see below); the north–south asymmetry is not understood.

In a somewhat different analysis, Guss[19] has observed a longitudinal persistence in the occurrence of solar flares that produce solar cosmic rays. In particular, he finds that flares from a single 10° interval of heliographic longitude caused most of the large solar particle events in the last solar cycle and contributed about 75 percent of the integrated intensity of particles of energy greater than 30 MeV. This indicates the existence of a center for the formation of active regions that has persisted for more than 73 solar rotations.

The various observations of solar cosmic rays have led to theoretical models[20–22] for a solar cosmic ray event in which the plasma ejected by the flare carries the magnetic field of the sun spot with it. The lines of force are stretched irregularly outward in a more

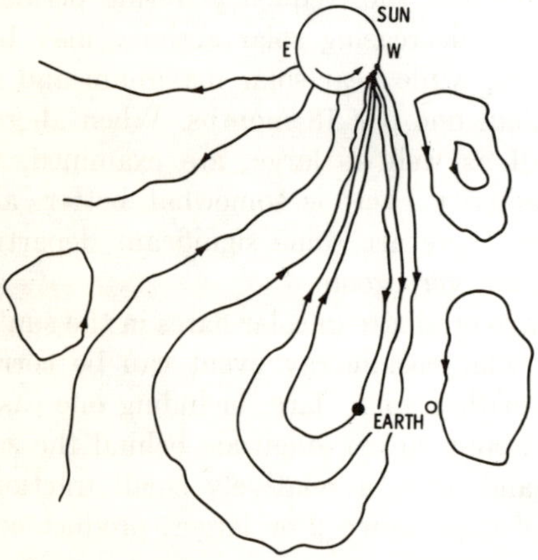

FIG. 5. A schematic representation of a tongue of magnetic lines of force drawn out from the sun. The solid circle represents a possible position of the earth within the storage region; the open circle represents a possible position outside the storage region.

or less radial fashion, with the sun's rotation causing them to curve westward in the plane of the ecliptic in an Archimedes' spiral, in the manner of a rotating garden sprinkler. Such a magnetic field pattern presumably exists in interplanetary space all the time because of the solar wind. The effect is enhanced by a flare, and in Gold's model[20] a *magnetic bottle* is formed which inhibits the escape of low energy solar cosmic rays from the inside and the entrance of galactic cosmic rays from the outside. When the earth is included within the magnetic bottle, the diminution of the flux of galactic cosmic rays at the earth is called a Forbush Decrease as described above, pp. 402–403. In times of considerable solar activity, the arrival near the earth of solar cosmic rays from a second flare in the same solar region as a recent predecessor is usually quicker than the arrival of the rays from the first flare. The field lines apparently remain continuous all the way back to the sun for a time and thus form a magnetic duct through which the later particles can readily propagate. (See also the Chapter on Interplanetary Medium for further details.)

It has also been observed by Bryant et al.[23] that solar cosmic rays can be found in space 27 days and 54 days after their initial observation. This implies the existence of the magnetic bottle and the storage of solar cosmic rays in it for long times, i.e., at least for two rotations of the sun.

In cases in which a flare occurs on the eastern half of the sun, the stretched-out field lines are located so that the magnetic container does not include the earth. In such circumstances, the more-or-less direct path to the vicinity of earth enjoyed by solar cosmic rays from west limb flares does not exist, and cosmic rays from eastern solar flares can reach the earth only by diffusing out of the magnetic bottle. Such diffusion probably takes place in close proximity to the sun where the particles make their way around to lines on the western hemisphere going directly to earth, rather than at 1 AU from the sun. It has already been noted that solar cosmic rays are observed more frequently from flares in the sun's western hemisphere than in the eastern. When they are observed, solar cosmic rays from eastern flares are generally reduced in intensity, compared to those from western flares, and slower to get to earth.

The first hint of a forthcoming solar cosmic ray event is usually the observation of a solar flare, a sudden large increase in the radiation intensity in a restricted portion of the sun. Observations are usually made in the light of the hydrogen Hα line* and the maximum brightness is generally attained in a few minutes. During and after the visible flare one or more of a whole series of events may occur at or near the earth.

Within minutes after the start of a flare, its ultraviolet and X-radiation can cause a *sudden ionospheric disturbance* on the sunlit hemisphere of the earth, that is, an abrupt large increase in absorption in the D- or lower E-region of the ionosphere. The disturbance may manifest itself through short wave fadeouts, sudden increases in cosmic noise absorption, VLF phase anomalies and signal enhancements, etc.

Flares are frequently accompanied by *solar radio noise storms*, bursts of electromagnetic noise over a wide range of frequencies, 10 to 10,000 Mcps. Five different classes of radio bursts have been distinguished, according to the variation of intensity and frequency with time. One of the 5 classes, Type IV, is strongly correlated with the production of solar cosmic rays in the flare and their subsequent arrival at the earth. Type IV emission extends over a broad band of frequencies up to 500 Mcps, is very intense and lasts for a number of hours, with a gradual decay in intensity while the center of emission moves outward from the solar disk. Type IV noise is thought to be due to synchrotron radiation from electrons accelerated along with the cosmic ray nuclei in the flare, but trapped in the expanding flare region by the local magnetic field.

On rare occasions, a neutron monitor on the ground may record an abrupt increase in counting rate, perhaps 15 to 30 minutes after the time of maximum brightness of the solar flare. This effect is caused by the incidence on the top of the atmosphere of very high energy solar cosmic rays, $\gtrsim 1$ BeV. Typically the time delay involved corresponds to only a few times the interval required for straight-line travel from sun to earth.

The signal strength of extraterrestrial radio waves is measured by an instrument called a *riometer*. The name of the device, an acronym for *relative ionospheric opacity meter*, refers to the dependence of the measured signal strength on the free electron density in the ionosphere. During a solar cosmic ray event, the ionization

* The Hα line is the radiation emitted by atomic hydrogen when it undergoes atomic transition between the energy levels characterized by quantum numbers 3 and 2. The wavelength is 6563 Å and this is the first line of the Balmer series.

of the upper atmosphere caused by energetic charged particles produces an attenuation of the normal radio noise which is a function of the intensity and energy spectrum of the incoming particles. The particles primarily responsible were originally thought to be protons in the energy range 20–200 MeV; more recently it has become clear that protons in the energy range 5–20 MeV can appreciably enhance the absorption and may indeed be the most important ones in a majority of cases. Because of the earth's magnetic field such particles can enter only near the poles, hence the term *polar cap absorption.* At high latitudes, riometers now provide continuous monitoring of the intensity of protons in this energy range and thus of their time history in the vicinity of the earth. In a typical PCA event, the arrival of the low energy flare particles is recorded within a few hours after the maximum brightness of the flare. Later, a further increase in riometer absorption may occur simultaneously with the sudden commencement of a geomagnetic storm; the absorption will decay away with the same kind of time dependence as other storm effects, typically over a period of two to four days.

A day or two after a solar flare, a *geomagnetic storm* may occur, caused by the arrival at the earth of the magnetic bottle containing the solar plasma. The plasma contains predominantly very large numbers of low energy (keV) particles, but may also contain considerable fluxes of protons with energy up to $\sim$20 MeV. During a geomagnetic storm, small changes occur in the components of the earth's magnetic field. Other important effects also occur: Forbush decreases in the galactic cosmic radiation, drastic changes in the outer Van Allen radiation belt, aurorae, and ionospheric changes that affect radio communications. Recently it has been possible to correlate measured intensities of energetic storm particles (1–15 MeV) with the phases of a geomagnetic storm.[24]

During the main phase of the storm, normal geomagnetic cut-offs can be lowered by considerable amounts. In this way, particles of a given energy can be found at lower latitude than the normal geomagnetic cut-off would allow. For example, particles of $\sim$10 Mev are normally observable in a magnetic quiet period only at magnetic latitudes above $\sim$ 61° at 1000 km altitude; during the main phase of a storm, they can be found as near the equator as magnetic latitudes of 53° to 55°. Particles of other energies have their cut-offs similarly affected.

The general world wide lowering of geomagnetic thresholds in the main phase of a geomagnetic storm had already been observed in the galactic cosmic radiation.

The characteristics of solar cosmic ray events differ considerably from one event to another with regard to their *charge spectrum* and *energy spectrum*. For example, many events contain very many low energy particles and few high energy particles; in occasional rare cases the situation is to a considerable extent reversed.

It has become clear from many measurements that the energy spectrum of the solar cosmic rays does not remain constant during an event. The highest energy particles may range up to $\gtrsim 1$ BeV and are seen soonest after the flare, with delay times corresponding to only a few times their direct sun-earth propagation times. Lower energy particles then appear in increasing numbers while the higher energy ones decrease in number and disappear. Outstanding among measurements of this effect are the Explorer XII observations of the September 28, 1961, event by Bryant, Cline, Desai, and McDonald.[25] Figure 6 shows "snapshots" of the solar proton energy spectrum recorded at seven minute intervals early in the event and clearly demonstrates the time dependence of the spectral shape. (At the time Explorer XII was outside the magnetosphere on the sunlit side of the earth.) The changing behavior shown in Figure 6 generally continues at a slower pace throughout the event; i.e., energy spectra taken successively over many hours show a continuous steepening. This will be clear from Figure 7, taken from a later

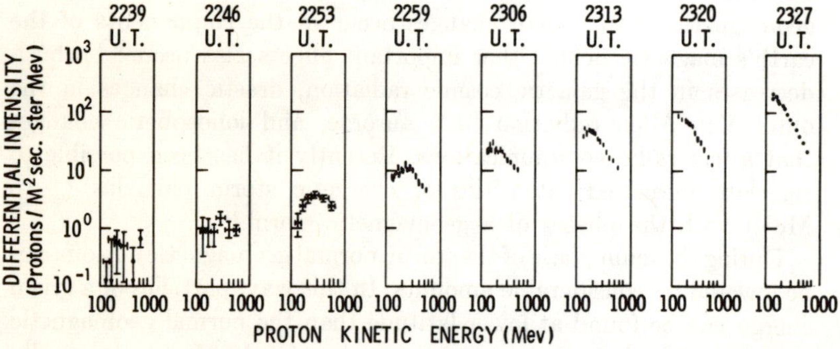

FIG. 6. The rise of the Sept. 28, 1961, solar proton event; differential energy spectra of the particle increase over galactic background are shown. These "snapshots" of the solar proton spectrum were recorded about seven minutes apart and show that the spectral shape changed rapidly early in the event. The flare began at 2202 UT and the maximum of the X-ray emission took place at 2216 UT (measured on earth) (Bryant et al., Ref. 25).

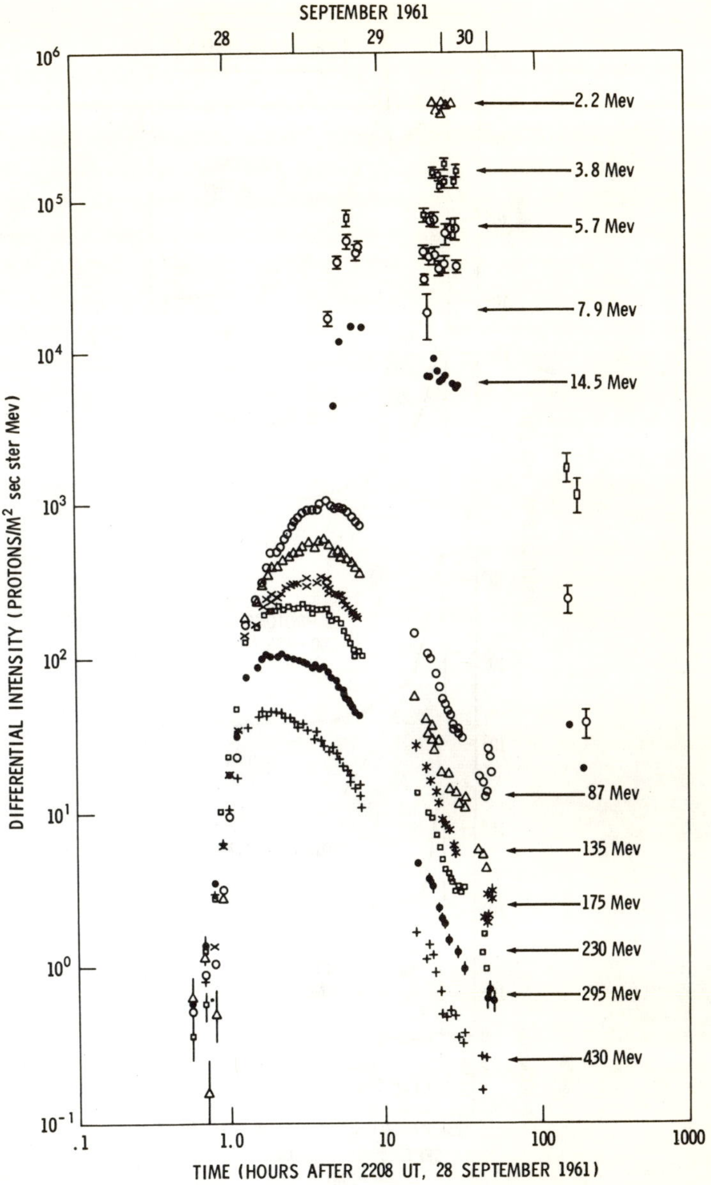

FIG. 7. The differential intensities of solar protons during the Sept. 28, 1961, event plotted against time after the X-ray burst at the sun. The data are interrupted when satellite passed through the magnetosphere and when the delayed increase occurred on Sept. 30, 1961 (Bryant et al., Ref. 28).

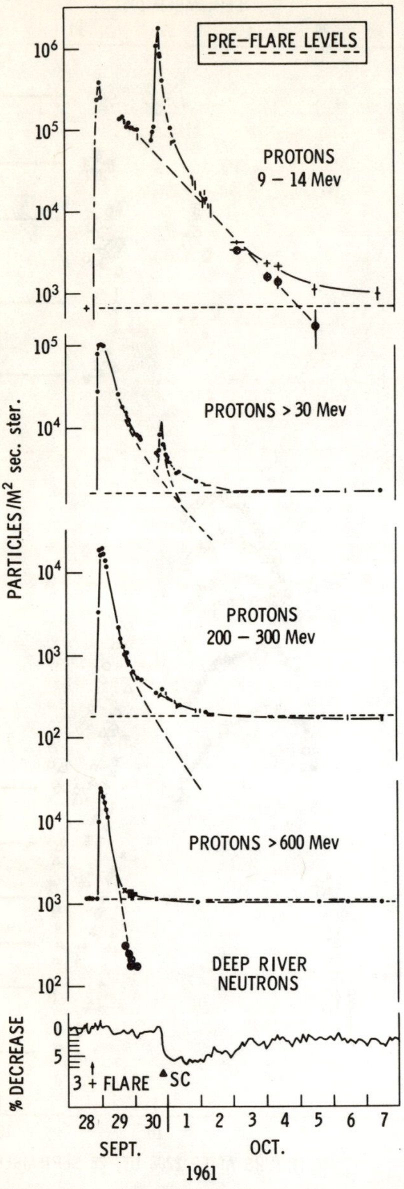

FIG. 8. Representative proton intensities between Sept. 28 and Oct. 7, 1961; the decay of the solar proton event and the arrival of the energetic storm particles late on Sept. 30 are shown. The Deep River neutron monitor record is shown for comparison (Bryant et al., Ref. 25).

paper of Bryant et al.,[28] which shows the intensity vs time profile of several energy components following the September 28, 1961, event. The number of low energy particles may show a sharp increase with the arrival of the solar plasma. Figure 8 also from Bryant et al.,[25] demonstrates this effect. In that particular case, the September 30, 1961, geomagnetic storm, a similar increase was seen essentially simultaneously in the polar region by low energy proton detectors on the high-latitude near-earth satellite, Injun I.[24,26] The combined Explorer XII–Injun I observations indicate that the upper energy limit for protons contained within the magnetic bottle on this occasion was $\sim$40 MeV. The maximum intensity of particles of energy above 3 MeV during the geomagnetic storm increase was more than ten times that observed during the preceding solar proton event.[28]

Although there is no strong theoretical reason to expect the differential energy spectrum of solar cosmic rays to follow a power-law in energy (as there is in the case of galactic cosmic rays), such a law is frequently used and is approximately valid over limited energy ranges. When the energy range is taken above 30 MeV and the integral intensity of these particles is at the maximum of the event, then μ in

$$N(E)\,dE = KE^{-\mu}\,dE \tag{10}$$

varies between 3.5 and 5.0 (E in MeV). Before the maximum of the event, μ may be less than 3.5 to 5.0, and after the maximum it may be larger. That is, the spectrum generally steepens throughout the event as we have seen.

A different approach from the power-law spectrum has been used by Freier and Webber.[27] They find that at times longer than the sun-to-earth propagation times for solar protons, an exponential number-rigidity spectrum pertains. This spectrum has the form

$$J(>P) = J_0 \exp(-P/P_0). \tag{11}$$

Both J_0 and P_0, a characteristic rigidity, vary in a systematic manner during the course of a cosmic ray event; however, the exponential nature of the relationship is retained. In the analysis of 16 solar proton events, values of P_0 between 45 and 300 MV, were found.

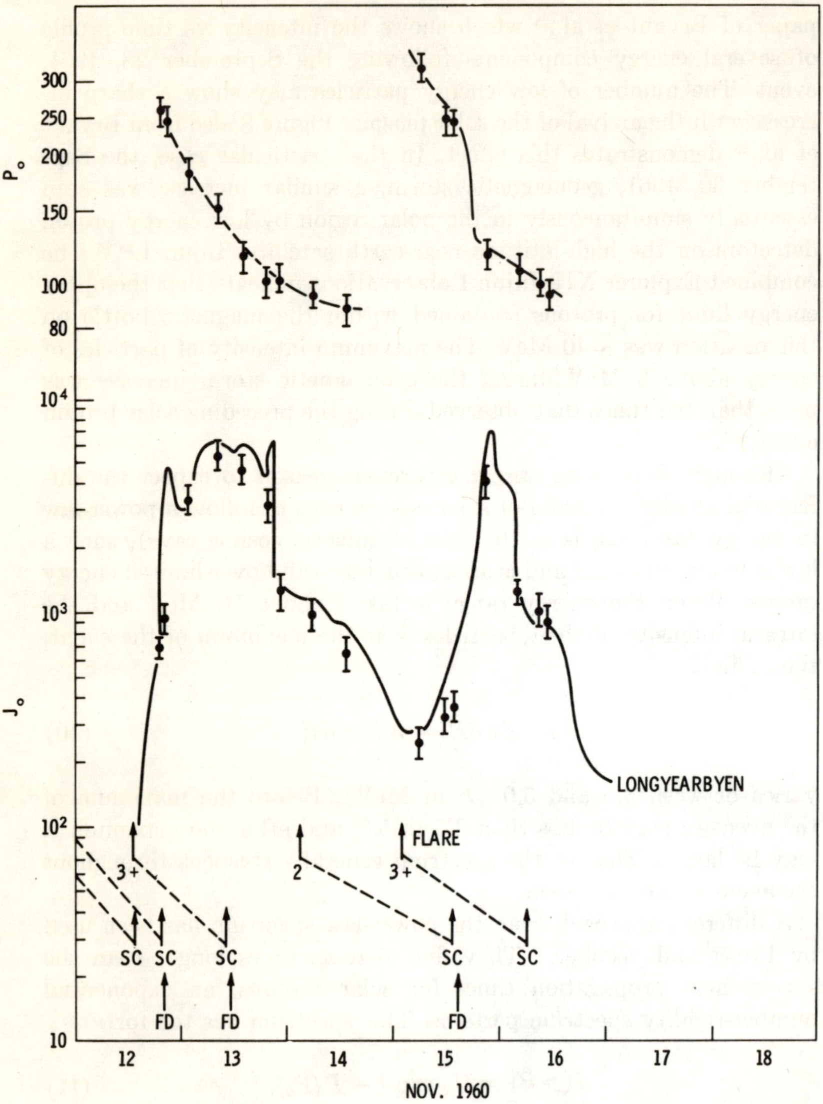

FIG. 9. The time variation of P_0 (MV) and J_0 (protons/cm²-sec-ster) is shown for the Nov. 12 and 15, 1960 flares. The measured values of P_0 are shown connected with a continuous curve. The measured values of J_0 are plotted, but the continuous curve is drawn using riometer absorption from Longyearbyen station. Time of major flares, sudden commencements and Forbush decreases are indicated (Freier and Webber, Ref. 27).

Figure 9 shows the time variation of J_0 and P_0 during the November 1960 events.

In their studies of solar proton events observed by Explorers XII and XIV, Bryant, Cline, Desai, and McDonald[28] have been able to determine the source energy spectrum of the solar protons at the sun. This has resulted from their observation that the curve of intensity vs distance travelled by the particles has the same shape for all energy groups. (Distance travelled = particle velocity $\times$ time from the x-ray burst in the flare to the moment of observation.) This result is shown in Figure 10, in which the intensity vs time data of Figure 7 are converted to intensity vs distance and scaled in intensity to give the best fit to a common curve. The relative intensity between any two energy components of the event is thus essentially constant over a range of from 2 to more than 100 astronomical units. An extrapolation back to zero distance seems perfectly reasonable: the relative intensities of the various components at zero distance give, by definition, the source specturm.

The scale factors used to prepare Figure 10 give the relative intensities. In the case of the September 28, 1961 event, the source spectrum is well represented by a power law in energy like Eq. (10) with $\mu = 1.7$. Bryant et al.[28] found in addition that the October 23, 1962, event required a μ of 2.3, while the November 10, 1961, event could not be represented by a single exponent in the power law. Here the low energy end had $\mu = 1.5$ while the high energy particles required $\mu = 3.5$. The further pursuit of this approach to solar cosmic ray energy spectra should prove very fruitful both with regard to the understanding of events at the time of their creation at the sun, and concerning the characteristics of the interplanetary medium.

The presence of α particles and heavy nuclei has been measured in several cosmic ray events. It is of interest to determine the energy spectrum of each type of nucleus and the relative intensity. The energy spectra can be represented by power-laws in energy like Eq. (10) with exponents in the range $\mu = 2$ to 6. Freier and Webber[27] found that the α particle spectrum from nine flares could be well represented by exponentials in rigidity of the same form as Eq. (11) given above for protons, and in fact *with the same* P_0 for α-particles as for simultaneously measured protons. This observation, taken with measurements of synchrotron radiation from electrons during

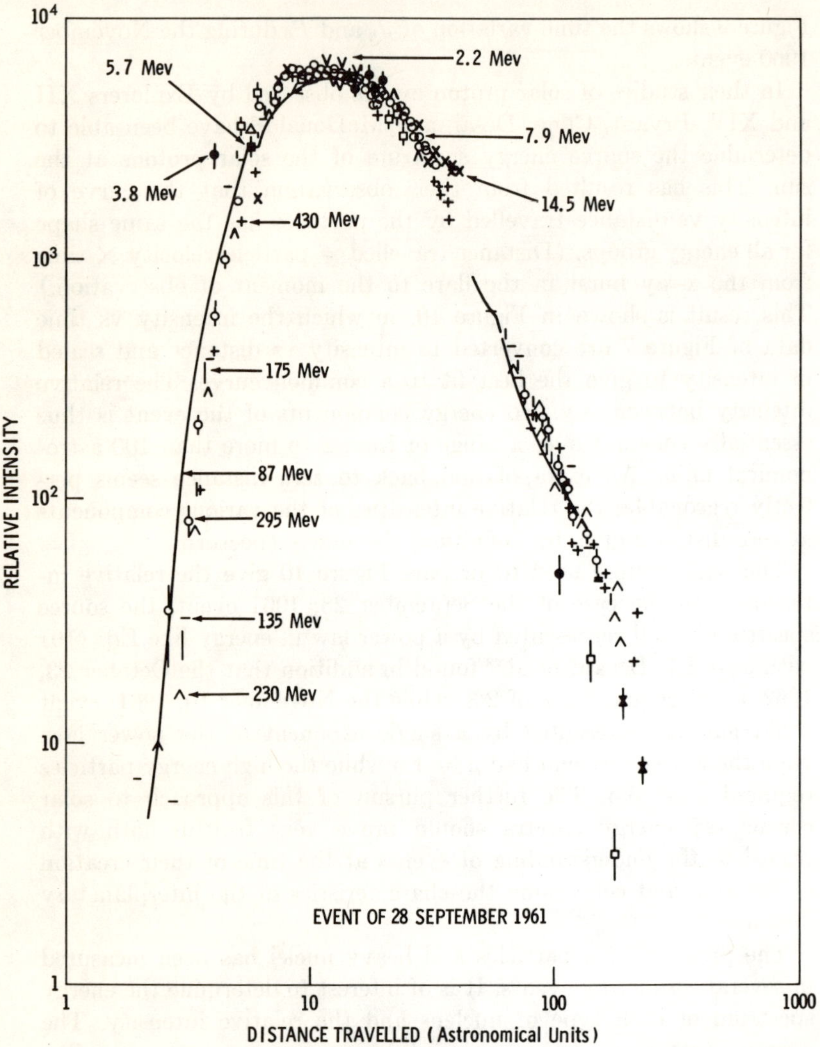

FIG. 10. The intensity vs. time plots of Fig. 7 converted to relative intensity vs. distance plots. The distance is computed for each energy component by taking the product of the corresponding particle velocity and time from the event; the intensities are scaled to give the best fit to a common propagation curve. This fit occurs over a dynamic range in energy of a few hundred and a velocity range of 14 and over a time duration of several days (Bryant et al., Ref. 28).

the initial flash part of the flares, leads to the inference that electrons, protons, and heavier nuclei are all accelerated to the same exponential rigidity spectrum in solar flares. This, in turn would place restrictions on the types of mechanisms that can occur in solar flares, for example, a time varying magnetic field would be a suitable acceleration mechanism, but a static electric field would not. However, before the matter is fully understood or even well explored, exceptions occur: Rocket flights carrying emulsion into the November, 1960, events by Biswas, Fichtel and collaborators[29] produced cases in which the values of P_0 for protons and α-particles were *not* the same, although the exponential rigidity spectrum did fit the experimental data for both protons and alphas quite well, at least above 10 MeV. Thus, it is now clear that none of the simple spectral forms, power law or exponential in either energy or rigidity, is consistent with the available data for protons and α-particles. It is also clear that there is much to be learned about the acceleration and propagation of solar cosmic rays before this situation will be satisfactorily resolved.

TABLE II

Relative abundances of nuclei normalized to a base of 1.0 for oxygen (Ref. 29)

Element	Solar cosmic rays	Sun	Universal abundances	Galactic cosmic rays
$_2$He	107 ± 14	?	150	48
$_3$Li		≪0.001	≪0.001	0.3
$_4$B–$_5$B	<0.02	≪0.001	≪0.001	0.8
$_6$C	0.59 ± 0.07	0.6	0.3	1.8
$_7$N	0.19 ± 0.04	0.1	0.2	≤0.8
$_8$O	1.0	1.0	1.0	1.0
$_9$F	<0.03	≪0.001	≪0.001	≤0.1
$_{10}$Ne	0.13 ± 0.02	?	0.40	0.30
$_{11}$Na		0.002	0.001	0.19
$_{12}$Mg	0.043 ± 0.011	0.027	0.042	0.32
$_{13}$Al		0.002	0.002	0.06
$_{14}$Si	0.033 ± 0.011	0.035	0.046	0.12
$_{15}$P–$_{21}$Sc	0.057 ± 0.017	0.032‖	0.027	0.13
$_{22}$Ti–$_{28}$Ni	≤0.02	0.006	0.030	0.28

The emulsion measurements of Biswas, Fichtel, Guss and Waddington[29–31] have also shown that for nuclei of $Z \geq 2$, the composition of the solar cosmic rays reflects the composition of the sun as determined from spectroscopic studies. This result is shown in Table II. This conclusion is strengthened by the facts that (1) these multiply-charged nuclei have had the same relative abundances in each measurement, even when the p-to-α ratios have varied, and (2) that α-particle and medium nuclei (C, N, O) have had the same energy per nucleon spectrum in the two cases of simultaneous observation, while the p and α-spectra were quite different.

CONCLUSIONS

In this chapter we have attempted to cover some of the properties and characteristics of the cosmic rays as known at present. The selection is by no means complete and some of the areas of current cosmic ray research have not been mentioned, as for example that of extensive air showers. We have also not dealt with the experimental problem of detecting and measuring the cosmic ray intensity both at the surface of the earth and on balloons, rockets and satellites. We believe, however, that this chapter is sufficient to give a general idea of what is known today about cosmic rays, and of some of the current areas of active research. In particular, the subject of the production and propagation of low energy solar cosmic rays and the solar modulation of low energy galactic cosmic rays are at present under active investigation with more and more sophisticated detectors carried on balloons and satellites. It is in these areas that we may expect rapid progress both on the experimental and on the theoretical side. Some of the conclusions presented here might have to be modified in the light of the knowledge to be gained from these investigations.

REFERENCES

1. V. F. Hess, Measurement of the Earth's Penetrating Radiation on Seven Balloon Ascensions, Physik. Z. **13,** 1084 (1912).
2. C. J. Waddington, The Composition of the Primary Cosmic Radiation, in Progress in Nuclear Physics, Vol. 8, p. 1, Pergamon Press, New York, 1960.
3. J. A. Earl, Cloud-Chamber Observations of Primary Cosmic-Ray Electrons, Phys. Rev. Letters **6,** 125 (1961).

4. P. Meyer and R. Vogt, Electrons in the Primary Cosmic Radiation, Phys. Rev. Letters **6,** 193 (1961).
5. S. Hayakawa and H. Okuda, Electronic Components in Galactic Cosmic Rays, Progr. Theoret. Phys. (Kyoto) **28,** 517 (1962).
6. F. C. Jones, Energy Spectrum of Galactic Electrons Produced by Cosmic Rays, J. Geophys. Res. **68,** 4399 (1963).
7. J. A. DeShong Jr., R. H. Hildebrand and P. Meyer, Ratio of Electrons to Positrons in the Primary Cosmic Radiation, Phys. Rev. Letters **12,** 3 (1964).
8. W. L. Kraushaar and G. W. Clark, Search for Primary Cosmic Gamma Rays with the Satellite Explorer XI, Phys. Rev. Letters **8,** 106 (1962).
9. E. Fermi, On the Origin of Cosmic Radiation, Phys. Rev. **75,** 1169 (1949).
10. P. Morrison, S. Olbert and B. Rossi, The Origin of Cosmic Rays, Phys. Rev. **94,** 440 (1954).
11. V. L. Ginzburg, The Origin of Cosmic Radiation, in Progress in Elementary Particle and Cosmic Ray Physics, Vol. 4, pp. 337–421, North-Holland Publishing Co., Amsterdam, 1958.
12. I. S. Shklovsky, Cosmic Radio Waves, Harvard University Press, Cambridge, 1960.
13. S. Hayakawa, K. Ito and Y. Terashima, Origin of Cosmic Rays, Progr. Theoret. Phys. (Kyoto), Suppl. No. 6, 1 (1958).
14. P. Morrison, The Origin of Cosmic Rays, in Encyclopedia of Physics: Cosmic Rays I, Vol. 46/I, p. 1, Springer-Verlag, Berlin, 1961.
15. J. A. Simpson, W. Fonger and S. B. Treiman, Cosmic Radiation Intensity—Time Variations and Their Origin. I. Neutron Intensity Variation Method and Meteorological Factors, Phys. Rev. **90,** 934 (1953).
16. B. Rossi, High Energy Particles, Prentice-Hall, Inc., Englewood Cliffs, 1952.
17. S. E. Forbush, Three Unusual Cosmic Ray Increases Possibly Due to Charged Particles from the Sun, Phys. Rev. **70,** 771 (1946).
18. F. B. McDonald, Ed., Solar Proton Manual, NASA Tech. Rept. R–169, Dec. 1963.
19. D. E. Guss, Distribution in Heliographic Longitude of Flares which Produce Energetic Solar Particles, Phys. Rev. Letters **13,** 363 (1964).
20. T. Gold, Plasma and Magnetic Fields in the Solar System, J. Geophys. Res. **64,** 1665 (1959).
21. E. N. Parker, Sudden Expansion of the Corona Following a Large Solar Flare and the Attendant Magnetic Field and Cosmic-Ray Effects, Astrophys. J. **133,** 1014 (1961).
22. K. G. McCracken, The Cosmic-Ray Flare Effect, III. Deductions Regarding the Interplanetary Magnetic Field, J. Geophys. Res. **67,** 447 (1962).
23. D. A. Bryant, T. L. Cline, U. D. Desai and F. B. McDonald, New Evidence for Long-Lived Solar Streams in Interplanetary Space, Phys. Rev. Letters **11,** 144 (1963).
24. G. F. Pieper, A. J. Zmuda, C. O. Bostrom and B. J. O'Brien, Solar Protons and Magnetic Storms in July, 1961, J. Geophys. Res. **67,** 4959 (1962).
25. D. A. Bryant, T. L. Cline, U. D. Desai, and F. B. McDonald, Explorer 12 Observations of Solar Cosmic Rays and Energetic Storm Particles after the Solar Flare of September 28, 1961, J. Geophys. Res. **67,** 4983 (1962).

26. J. A. Van Allen, L. A. Frank, G. F. Pieper, W. A. Whelpley, B. J. O'Brien, and W. G. V. Rosser, Explorer 12 Symposium, Goddard Space Flight Center, January 18, 1962.
27. P. S. Freier and W. R. Webber, Exponential Rigidity Spectrums for Solar-Flare Cosmic Rays, J. Geophys. Res. **68,** 1605 (1963).
28. D. A. Bryant, T. L. Cline, U. D. Desai, and F. B. McDonald, Studies of Solar Protons with Explorers XII and XIV, Astrophys. J. **141,** 428 (1965).
29. S. Biswas and C. E. Fichtel, Nuclear Composition and Rigidity Spectra of Solar Cosmic Rays, Astrophys. J. **139,** 941 (1964).
30. S. Biswas, C. E. Fichtel, and D. E. Guss, Study of the Hydrogen, Helium, and Heavy Nuclei in the November 12, 1960 Solar Cosmic-Ray Event, Phys. Rev. **128,** 2756 (1962).
31. S. Biswas, C. E. Fichtel, D. E. Guss, and C. J. Waddington, Hydrogen, Helium, and Heavy Nuclei from the Solar Event on November 15, 1960, J. Geophys. Res. **68,** 3109 (1963).
32. E. N. Parker, Interplanetary Dynamical Processes, p. 196, Interscience Publishers, New York (1963).
33. J. Geiss, Proc. Int. Conf. Cosmic Rays, Jaipur, *3*, 434 (1963).
34. H. Hintenberger, H. Voshage and H. Sarkar, Proc. Int. Conf. Cosmic Rays, London, **1**, 602 (1965).
35. R. C. Hartman, R. H. Hildebrand, and P. Meyer, J. Geophysic, Res. **70,** 2713 (1965).
36. M. Oda, Proc. Int. Conf. Cosmic Rays, London **1**, 68 (1965).

Chapter 11

INTERPLANETARY DUST PARTICLES

C. W. McCracken and W. M. Alexander

INTRODUCTION

The various components of the solar system can be divided into the following groups: the sun, the planets and their satellites, the asteroids or minor planets, the comets, the interplanetary particulate aggregates, the interplanetary gas, and a radiation field. The dimensions of the interplanetary particulate aggregates conceivably range from several tens of kilometers (characteristic of small asteroids) down to a few angstroms (characteristic of aggregates of a few atoms).

Small interplanetary particulate aggregates which have dimensions ranging from a few centimeters down to a few tenths of a micron are the subject of this chapter and can be called *interplanetary dust particles* or simply *dust particles*. The use of the term *dust particles* is not intended to focus attention on siliceous particles alone. Interplanetary dust particles may be composed of metals, minerals, ices, and mixtures of these materials.

The distribution of orbits, the spatial density, the flux, the distribution by mass or size, the mass density, the composition and structure, the shape, and the light-scattering properties of the interplanetary dust particles constitute physical properties or characteristics which appear in comprehensive studies of these dust particles.

The observational techniques employed in studying the interplanetary dust particles include the following, arranged roughly in the order of increasing particle size for which a given technique applies: photometric studies of sky brightness using balloons; collection of atmospheric aerosols using balloons and high-altitude aircraft; collection of dust particles from above the mesopause (80 km altitude) by means of rockets and manned spacecraft; studies of light scattered by dust particles in the upper atmosphere using lasers; photometric studies of the zodiacal light and the solar F-corona; direct measurements obtained with spacecraft; collection of dust particles at ground-level from polar ice fields and from deep-sea sediments; observation of meteors using radar, visual, and photographic methods; and studies of the frequency of meteorite falls.

The dimensions of dust particles for which observations are presently possible and feasible range from approximately 75 Angstroms (characteristic of the smallest dust particles collected by high-altitude aircraft) to a meter or so for the meteorites. Each observational technique covers only a small portion of the size range of the interplanetary dust particles. Some ranges of particle size can be investigated by two or more observational techniques. Simultaneous radar, visual, and photographic meteor observations constitute one such example. Other ranges of particle size are not readily accessible with any existing technique.

The observable parameters are generally different for the different observational techniques. This makes intercomparison of the results difficult and subject to some uncertainty of interpretation, since assumptions must be made in order to convert between parameters. A fundamental parameter that is closely related to the observable parameters for several techniques should be chosen whenever possible. The mass of a dust particle is one reasonable choice; the flux of dust particles of a given mass is another. The flux constitutes an observable parameter common to several of the observational techniques, but the flux of dust particles of a given mass is not generally observable.

The major objective of this chapter is a discussion of some of the astronomical aspects of interplanetary dust particles and their relation to the more limited subject of dust particles in the vicinity of the earth. A connection undoubtedly exists between these two. The exact nature of this connection is not yet well known for the small dust particles which produce the zodiacal light and the solar F-corona and which are measured through the use of spacecraft.

Selected data obtained from studies of meteors and meteorites, from studies of the zodiacal light, and from direct measurements made with spacecraft are used here to define a flux distribution that is expressed as the flux of dust particles of a given mass and larger. Only the cumulative flux distribution for average conditions in a given region of space can be considered here. Once established with some degree of certainty, such a cumulative flux distribution can be used as a basis for future analyses of the time-dependent variations in the flux.

Some of the available data apply only for the vicinity of the earth, but considerations of the distributions of orbits permit the data to be applied also to interplanetary space. The subject of distributions of orbits (to which radar and photographic observations of meteors have contributed so much) is a recurrent theme in discussions of interplanetary dust particles. Also of importance are the changes in orbit distributions with time. For this reason, let us first consider the concept of the Poynting–Robertson effect.

EQUATIONS OF MOTION AND THE POYNTING–ROBERTSON EFFECT

The equation of motion for an interplanetary dust particle of mass m moving under the gravitational influence of the sun of mass $M_\odot$ may be written (for $m \ll M_\odot$)

$$m(d^2\mathbf{r}/dt^2) = -(GM_\odot m/r^3)\mathbf{r} \tag{1}$$

where $\mathbf{r}$ is the position vector of the dust particle and G is the gravitational constant.

The use of polar coordinates r, θ in the orbital plane of the dust particle permits the separation of the equation of motion into the radial and angular scalar equations of motion,

$$\ddot{r} - r\dot{\theta}^2 = -\mu/r^2, \quad \text{(radial)} \tag{2}$$

$$r\ddot{\theta} + 2\dot{r}\dot{\theta} = 0, \quad \text{(angular)} \tag{3}$$

where $\mu = GM_{\odot}$.

The angular equation is equivalent to

$$\frac{1}{r}\frac{d}{dt}(r^2\dot{\theta}) = 0 \tag{4}$$

which can be integrated to yield

$$r^2\dot{\theta} = h_0 \tag{5}$$

where h_0 is a constant and is the angular momentum per unit mass for the dust particle. Also, one can get from (2) and (3) the integral of energy

$$v^2 = \mu[(2/r) - (1/a)] \tag{6}$$

in which v is the orbital speed of a dust particle at a distance r from the sun and a is the semi-major axis of the orbit. The quantity $1/a$ is positive, zero, or negative if the orbit is elliptic, parabolic, or hyperbolic, respectively.

The actual motion of a small interplanetary dust particle is not nearly so simple as the foregoing equations would imply. First, there are the perturbing gravitational forces exerted on the dust particle by the planets (particularly Jupiter). Second, the force exerted by sunlight falling on a dust particle becomes comparable to the gravitational force exerted by the sun for dust particles of micron-size. The subject of the effects of gravitational perturbations cannot be treated here, but the subject of radiation pressure can hardly be avoided for dust particles of the sizes under discussion.

Sunlight falling on a dust particle produces a force directed away from the sun and thereby reduces the field of force in which the dust particle is moving. The dust particle receives the sunlight from a source which is moving relative to the dust particle, so the incident sunlight is shifted in frequency by the Doppler effect and is shifted in direction by the effect of aberration. Whatever radiation is absorbed by the dust particle is then re-radiated in a frame of reference that moves with the dust particle. The absorption and subsequent re-emission of radiation by the dust particle produces a small tangential drag force which opposes the orbital motion of the dust particle. This tangential drag force removes kinetic energy and angular momentum from the dust particle and causes it to spiral toward the sun.

Poynting realized in 1920 the existence of the tangential drag force, and Robertson gave the problem a proper relativistic formulation in 1937. The action of sunlight in reducing the central force and in producing a non-conservative drag force is called the *Poynting–Robertson effect.* The important consequence of the Poynting–Robertson effect is that small dust particles have very limited lifetimes in the solar system. The lifetimes amount, in some cases, to only a few centuries.

Some auxillary quantities are needed for use in modifying the equations of motion given previously in order to include the Poynting–Robertson effect. The luminosity of the sun is

$$L_\odot = 3.86 \times 10^{33} \text{ (ergs/sec)},$$

so the flux of sunlight at a distance r from the sun is

$$\sigma(r) = L_\odot / 4\pi r^2.$$

At the distance of the earth from the sun, $r_\oplus = 1.495 \times 10^{13}$ cm, the flux is

$$\sigma_\oplus = 1.374 \times 10^6 \text{ (erg/cm}^2\text{ sec)}$$

which is just the solar constant. The energy density at a distance r from the sun is

$$u(r) = c^{-1}\sigma(r)$$

where $c = 3 \times 10^{10}$ cm/sec is the speed of light in a vacuum. Energy density is introduced, because the radiation pressure acting on a perfectly absorbing surface is equal to the energy density of the incident radiation.

Most treatments of the Poynting–Robertson effect involve the assumption that a dust particle is a perfectly absorbing (black) sphere and that the cross section for absorption is just the cross-sectional area of the sphere. Further refinement will not be attempted here, but the reader is cautioned that classical optics does not apply for dust particles having dimensions comparable to the wavelength of light. The theory of the scattering of electromagnetic waves by spheres was worked out by Mie in 1908. The interested reader may consult the book by van de Hulst[1] for an introduction to the subject of the Mie theory. The Mie theory will appear later in the discussion of the zodiacal light.

The cross-sectional area of a spherical particle is

$$A = \pi s^2 = 3m/4\rho s, \tag{7}$$

where s is the particle radius in cm and ρ is the mass density in gm/cm^3. This geometric cross section is a fair approximation to the absorption cross section for a perfectly absorbing particle as long as the dimensions of the particle are large compared to the wavelength of the incident radiation. The force exerted by the incident radiation is directed away from the sun and has a magnitude

$$F_r = u(r)A$$

or

$$F_r = (L_\odot/16\pi c r^2)(3m/\rho s). \tag{8}$$

The force of gravity is directed toward the sun and has a magnitude

$$F_g = \mu m/r^2. \tag{9}$$

The magnitude of the ratio of the radiation force to the gravitational force is

$$\beta = F_r/F_g = (3L_\odot/16\pi c\mu)(1/\rho s). \tag{10}$$

Substitution of numerical values yields

$$\beta = 5.78 \times 10^{-5}(1/\rho s)\,(\text{gm/cm}^2).$$

This expression has been used to compute the masses and sizes of dust particles of various mass densities for the selected values of β given in Table I. The size of a dust particle of a given mass density for which $\beta = 1$ as shown in Table I is commonly referred to as the

TABLE I

Relation between β and the sizes and masses of dust particles for various mass densities

β	$\rho = 1.0$ gm/cm^3		$\rho = 2.5$ gm/cm^3		$\rho = 7.8$ gm/cm^3	
	s (cm)	m (gm)	s (cm)	m (gm)	s (cm)	m (gm)
0.0001	5.8×10^{-01}	8.1×10^{-01}	2.3×10^{-01}	1.3×10^{-01}	7.4×10^{-02}	1.3×10^{-02}
0.001	5.8×10^{-02}	8.1×10^{-04}	2.3×10^{-02}	1.3×10^{-04}	7.4×10^{-03}	1.3×10^{-05}
0.003	1.9×10^{-02}	3.0×10^{-05}	7.7×10^{-03}	4.8×10^{-06}	2.5×10^{-03}	5.0×10^{-07}
0.006	9.7×10^{-03}	3.8×10^{-06}	3.9×10^{-03}	6.0×10^{-07}	1.2×10^{-03}	6.2×10^{-08}
0.01	5.8×10^{-03}	8.1×10^{-07}	2.3×10^{-03}	1.3×10^{-07}	7.4×10^{-04}	1.3×10^{-08}
0.03	1.9×10^{-03}	3.0×10^{-08}	7.7×10^{-04}	4.8×10^{-09}	2.5×10^{-04}	5.0×10^{-10}
0.06	9.7×10^{-04}	3.8×10^{-09}	3.9×10^{-04}	6.0×10^{-10}	1.2×10^{-04}	6.2×10^{-11}
0.1	5.8×10^{-04}	8.1×10^{-10}	2.3×10^{-04}	1.3×10^{-10}	7.4×10^{-05}	1.3×10^{-11}
0.2	2.9×10^{-04}	1.0×10^{-10}	1.2×10^{-04}	1.6×10^{-11}	3.7×10^{-05}	1.7×10^{-12}
0.4	1.4×10^{-04}	1.3×10^{-11}	5.8×10^{-05}	2.0×10^{-12}	1.9×10^{-05}	2.1×10^{-13}
0.6	9.7×10^{-05}	3.8×10^{-12}	3.9×10^{-05}	6.0×10^{-13}	1.2×10^{-05}	6.2×10^{-14}
0.8	7.2×10^{-05}	1.6×10^{-12}	2.9×10^{-05}	2.5×10^{-13}	9.3×10^{-06}	2.6×10^{-14}
1.0	5.8×10^{-05}	8.1×10^{-13}	2.3×10^{-05}	1.3×10^{-13}	7.4×10^{-06}	1.3×10^{-14}

Note: All numbers were computed using classical optics and perfectly absorbing spheres. The parameter β is the magnitude of the ratio of the force of radiation to the force of gravity exerted by the sun.

classical radiation pressure limit. A popular but misleading statement is that dust particles for which $\beta = 1$ (in Table I) and *all* smaller dust particles will be blown out of the solar system. This statement requires some qualification. Indeed, a dust particle for which $\beta \geq 1$ will be blown out of the solar system, since $\beta > 1$ means that the force of radiation exceeds the force of gravity exerted by the sun. In addition, a dust particle having $\beta < 1$ can at least begin its escape from the solar system if its total energy of motion is non-negative. Notice that the dust particles for which $\beta < 0.01$ in Table I have sizes measured in microns or tenths of microns. These dimensions are comparable to the wavelength of light (~0.55 micron for sunlight), so the assumption that classical optics can be used in computing the values of β breaks down. The Mie theory is available for computing the light-scattering properties of homogeneous spheres of any size, and the value of β is one of the properties which can be computed. The values $\beta \geq 1$ generally occur when the dimensions of a dust particle become comparable to the wavelength of the incident light, but β does not necessarily remain greater than unity as the size of the dust particle is further decreased. Some dielectric particles smaller than the wavelength of light have $\beta < 1$, so they will not be blown out of the solar system. The values of β given in Table I will serve the purpose of illustrating the domain of particle size which we wish to treat here. If an accurate value of β is required for a small spherical dust particle of a given size, composition, and index of refraction, then the Mie theory should be used in computing the appropriate value of β.

Use

$$\mu_{\text{eff}} = \mu(1 - \beta) \tag{11}$$

to represent the combined radiation and gravitational fields of the sun. The radial equation of motion becomes

$$\ddot{r} - r\dot{\theta}^2 = -\mu_{\text{eff}}/r^2 = -\mu(1 - \beta)/r^2. \tag{12}$$

A parameter α defined by

$$\alpha = (3L_{\odot}/16\pi c^2)(1/\rho s) \tag{13}$$

and which becomes, upon substituting numerical values,

$$\alpha = 2.56 \times 10^{11}(1/\rho s), \qquad \text{(c.g.s. units)}$$

has conventionally been used in treatments of the Poynting-Robertson effect. The parameters α and β are related by

$$\alpha = \mu\beta/c. \tag{14}$$

The equations of motion, modified to include the Poynting–Robertson effect, are

$$\ddot{r} - r\dot{\theta}^2 = -(\mu_{\text{eff}}/r^2) - (2\alpha\dot{r}/r^2), \qquad \text{(radial)} \tag{15}$$

and

$$\frac{1}{r}\frac{d}{dt}(r^2\dot{\theta}) = -(\alpha\dot{\theta}/r), \quad \text{(angular)} \tag{16}$$

where $\mu_{\text{eff}} = \mu - \alpha c$. The equations can also be written in the form

$$\ddot{r} - r\dot{\theta}^2 = -(\mu/r^2) + (\mu\beta/r^2)[1 - (\dot{r}/c)] - (\mu\beta/r^2)(\dot{r}/c) \tag{17}$$

and

$$\frac{1}{r}\frac{d}{dt}(r^2\dot{\theta}) = -(\mu\beta/r^2)(r\dot{\theta}/c), \tag{18}$$

in which the parameter α has been replaced by the dimensionless parameter β.

The terms

$$-(\mu\beta/r^2)(\dot{r}/c) \quad \text{and} \quad -(\mu\beta/r^2)(r\dot{\theta}/c),$$

which have been added to the left-hand sides of the equations of motion are the radial and transverse components, respectively, of the tangential drag force. The frequency ν of the incident radiation is Doppler shifted to a new frequency ν' where, to first order,

$$\nu' = \nu[1 - (\dot{r}/c)], \tag{19}$$

because of the radial velocity $\dot{r}/c$ of the dust particle. Thus, the energy density observed at the dust particle is also Doppler shifted and gives rise to the term

$$(\mu\beta/r^2)[1 - (\dot{r}/c)],$$

which represents the force per unit mass exerted on the dust particle by the incident radiation. The effect of the aberration of sunlight through a small angle

$$\phi \sim \frac{r\dot{\theta}}{c}.$$

can be neglected in a first approximation.

The angular equation of motion can be integrated to yield

$$r^2\dot{\theta} = h = h_0 - \alpha\theta, \tag{20}$$

where h_0 is an initial value for the angular momentum per unit mass and h is the value after the angular coordinate of the dust particle has increased by the angle θ. Neither angular momentum nor energy now constitutes a constant of motion. The angular momentum decreases as θ (or time t) increases, so the dust particle spirals *toward* the sun. The dust particle does not necessarily spiral *into* the sun, for partial vaporization of the particle or changes in the optical properties may result in the value of β increasing sufficiently for the dust particle to escape from the solar system.

The secular rates of change of the semi-major axis a and the eccentricity e of the orbit of a dust particle, as given by Wyatt and Whipple,[2] are

$$da/dt = -\frac{\alpha(2 + 3e^2)}{a(1 - e^2)^{3/2}} \tag{21}$$

and

$$de/dt = -\frac{5\alpha e}{2a^2(1 - e^2)^{1/2}} \tag{22}$$

showing that both the semi-major axis and the eccentricity decrease with time and that the orbit becomes more circular as a dust particle spirals toward the sun.

Robertson considered circular orbits ($e = 0$) for which the equation for the decrease in the semi-major axis reduces to

$$da/dt = -2\alpha/a \tag{23}$$

or

$$dt = -(1/2\alpha)a\,da. \tag{24}$$

Integration yields

$$t = a^2/4\alpha = 7.0 \times 10^6\,\rho s a^2 \tag{25}$$

for t in years and a in A.U. (astronomical unit) as the time required for a dust particle initially in a circular orbit of radius a to spiral into the sun.

Substitution of the values

$$\rho = 1 \text{ gm/cm}^3, \quad a = 1(\text{A.U.}), \quad \text{and} \quad s = 1 \text{ cm}$$

into equation (25) leads to a lifetime of 7×10^6 years. Reducing the particle radius to $s = 10$ microns and keeping the same values for the other two parameters leads to a lifetime of only 7000 years. The Poynting-Robertson effect cannot be neglected when one considers the lifetimes and dynamical behaviour of interplanetary dust particles of the sizes which produce the zodiacal light and which are being measured through the use of instruments mounted on spacecraft.

Wyatt and Whipple[2] extended the work of Robertson to include non-circular orbits ($e > 0$) and gave tables of computed quantities that can be used in calculating the times of fall for dust particles having orbits of various eccentricities. The time of fall is less for an elliptic orbit than for a circular orbit having the same semi-major axis by a factor that depends on the eccentricity of the orbit. The factor varies from $\sim$2 to $\sim$200 for some of the major meteoroid streams listed by Wyatt and Whipple.

METEORS

Interplanetary dust particles having masses between the approximate limits 10^{-5} gm and 10 gm form *meteors* during their high-speed plunge into the atmosphere of the earth. Such dust particles are called *meteoroids* and have speeds ranging from 11 km/sec to 72 km/sec. The lower limit of 11 km/sec corresponds to the speed of escape from the earth and would be acquired by a meteoroid falling to the earth from rest at a large geocentric distance. The upper limit of 72 km/sec applies for a meteoroid travelling in a retrograde parabolic orbit about the sun and meeting the earth head-on. The average speed, which is set by the distribution of the orbits of the meteoroids, is generally taken to lie between 25 km/sec and 40 km/sec.

A meteor-producing meteoroid is completely destroyed through ablation, vaporization, sputtering, and fragmentation as it enters the earth's atmosphere. The kinetic energy of the meteoroid is converted into light, ionization, and heat through collisions of the evaporated meteoric atoms and fragments with atmospheric atoms and molecules. The light constitutes the visual or photographic meteor, depending on the method of observation. The column of ionization constitutes the radar meteor, which can be observed by virtue of its ability to reflect transmitted radio waves back to a receiver. Meteors can be observed visually or photographically only during the night. An advantage of radar is that observations can be made during the hours of sunlight as well as during the night.

The intensity of the light from a meteor is commonly expressed in terms of magnitude, which is just a logarithmic measure of the intensity. A decrease in the intensity by a factor $(100)^{1/5} \approx 2.5$ corresponds to an increase of one magnitude. The ionization in a radar meteor is interpreted in terms of the line density of electrons, and radar magnitude is used as a logarithmic measure of the electron line density. The observed apparent magnitude is then converted into absolute magnitude by making corrections for the distance to the meteor and for atmospheric effects.

Radar, visual, and photographic magnitudes are roughly equivalent, at least for the purposes of a general discussion of the observational results. The ranges of applicability of the three methods of observations are different, but there is considerable overlapping of the ranges. Naked-eye observations include meteors having visual magnitudes of approximately +5 and brighter. The use of binoculars or a wide-angle telescope permits visual observations for fainter meteors and extends the range of visual observations to a visual magnitude of approximately +10. Photographic observations made with the Baker Super–Schmidt meteor cameras extend to a visual magnitude of +3 or +4. Radar equipment used in the Harvard Radio Meteor Program extends the observational range to a visual magnitude of approximately +12, while radar observations made by Gallagher and Eshleman extended to an estimated visual magnitude of +15. Meteor observations cover a total range of about 25 magnitudes or 10 decades of particle mass.

The *radiant* of a meteor is the point at which the backward extension of the path of the meteor appears to intersect the celestial sphere. Visual observations led to the separation of meteors into two groups, the *shower meteors* and the *sporadic meteors*, on the basis of the distribution of the radiants. The meteors of a given meteor shower have a common radiant, since they arise from meteoroids travelling in groups or streams and having similar orbits in space. In addition, a meteor shower occurs annually whenever the earth intersects the orbit of a given meteoroid stream. The radiants of the sporadic meteors are more randomly distributed over the celestial sphere than are the radiants of the shower meteors. Photographic and radar observations of meteors have revealed more associations among the radiants of faint meteors than could be detected by visual observations; hence, the concept of sporadic meteors has lost some of its original significance.

The combined results of radar, visual, and photographic observations of meteors show that the enhancement of flux during a meteor shower occurs mainly within the range of visual observations. Shower meteors so faint that they can be observed only by radar generally are identified through statistical groupings in the radiants rather than through an enhancement of the flux. Radar observations for a visual magnitude of +15 made by Gallagher and Eshleman showed no correlation between the observed enhancements of flux and the major meteor showers. Subsequent work by Hawkins and by Nilsson has revealed statistical groupings of radiants which suggest that many of the faint sporadic meteors are dispersed members of known meteoroid streams. The rate of dispersion is higher for the smaller meteoroids because of the Poynting–Robertson effect, so one might expect that the association of faint meteors with a given meteoroid stream would be less pronounced than for the brighter meteors.

The speed and direction of motion of a meteoroid can be established if the meteor is observed simultaneously from two or more stations spaced a few tens of kilometers apart. Such observations permit computation of the heliocentric orbit of the meteoroid. The Super–Schmidt meteor cameras have been used extensively in this application. Radar equipment is regularly operated in a similar capacity.

Correlations between the orbits of the meteor streams and the orbits of comets have led to the identification of parent comets for most of the major meteor streams. The remainder of the meteor streams probably were formed also by comets which escaped being observed, which have disintegrated, or which have subsequently been perturbed into different orbits.

A fundamental problem in meteoric astronomy is the determination of the mass of a meteoroid which produces a meteor of a given magnitude. The first theories of meteors were formulated by Öpik in 1922 and by Lindemann and Dobson in 1923. These early theories were based on the assumption that the meteor-producing meteoroids are dense, solid particles like the meteorites. Subsequent photographic and radar observations have shown that the concept of a fluffy *dust-ball* meteoroid helps to explain the deviations of the observations from the theoretically expected results. Such deviations include fragmentation and flaring in meteors and the appearance of meteors at higher altitudes than expected on the basis of the theory. McCroskey found that the photographic meteoroids typically had crushing strengths of approximately 10^4 dynes/cm^2 or about 1/50 of an atmosphere pressure. This value for the crushing strength is roughly comparable to the crushing strength of freshly fallen snow.

Assuming that the intensity of light and the line density of electrons produced in a meteor are proportional to the mass of the meteoroid for a given speed permits the absolute visual magnitude M_v and the particle mass m (measured in grams) to be related by an expression of the form

$$M_v = K - 2.5 \log m. \tag{26}$$

The absolute visual magnitude is defined as the apparent visual magnitude for a meteor observed at the zenith and at an altitude of 100 km. The factor K contains all the uncertainties in the mass-to-magnitude relation. Estimates of the mass of a meteoroid which will produce (for a given speed) a meteor of zero absolute visual magnitude have ranged from the value of 30 gm given by Hawkins and Upton down to the value of 0.04 gm given by Levin. Not all of the values that have been given are for solid meteoroids; some of the values apply to the friable dustballs that are so characteristic of the photographic meteors. Also, different values for the average speed have been used by the different investigators.

Whipple[3] has re-evaluated the mass-to-magnitude relation, making use of some new data. Values for the luminous efficiency derived for three meteoroids of which the composition was known from available spectra and a value for the luminous efficiency obtained by McCroskey for an artificial iron meteoroid were used. A value of 1 gm was obtained for the mass of a meteoroid that would produce (for a speed of 30 km/sec) a meteor of zero visual magnitude. The mass-to-magnitude relation thereby reduces to the simple form

$$M_v = -2.5 \log m$$

which is particularly convenient, because it matches the zero of the visual magnitude scale with the zero of the scale for the logarithm of the particle mass.

The flux of meteoroids of a given mass can be derived by plotting the flux of meteors as a function of visual magnitude and then using the foregoing relation to relate visual magnitude and particle mass.

More extensive discussions of the astronomical aspects of meteors and the technological aspects of observing meteors can be found in the books by McKinley,[4] Öpik,[5] Lovell,[6] and Watson.[7]

METEORITES

Meteoroids that are sufficiently large to survive their high-speed passage through the atmosphere of the earth give rise to *meteorites.* Meteorites are partially ablated and are often broken into several fragments. The *meteorite finds* (meteorites found but not observed to fall) outnumber the *meteorite falls* (meteorites recovered after having been observed to fall), but allowances must be made for the effects of observational selection before any significance can be attached to the ratio of meteorite finds to falls. Once identified, a meteorite constitutes a sample of extra-terrestrial material which can be analyzed in the laboratory.

Meteorites generally are classified as irons, stones, or stony irons on the basis of their composition and structure. The stone meteorites outnumber the iron meteorites by about 10 to 1 for the meteorite falls; a greater percentage of the meteorite finds are irons, because the iron finds are easier to distinguish from terrestrial rocks than are the stone finds. The iron meteorites typically have a high content of nickel and iron, while the stone meteorites contain mostly silicates. The admixture of metals and minerals in the stony iron meteorites indicates that a similar origin applies to all classes of meteorites and, hence, that different origins are not necessarily required for the different classes of meteorites.

An asteroidal origin for the meteorites is suggested by their structure and composition. The asteroids probably originated in a catastrophic collision of two small planets. The meteorites may be the products of this collision, or they may have originated in subsequent collisions among the asteroidal fragments. Little, if any, evidence exists for the grouping of meteorite-producing meteoroids into streams, nor is there evidence that the frequency of meteorite falls increases during the major meteor showers. A cometary origin for the meteorites is regarded as rather unlikely.

The Pribram meteorite, which was observed photographically at two stations when it fell in Czechoslovakia, is the only meteorite for which the orbit is known with any degree of certainty. The orbit had a small inclination (10° 25′) and a moderate eccentricity (0.68). This orbit and the most probable orbits for other meteorite falls (as deduced by Wood through a statistical analysis of the times of day at which the meteorites fell) are similar to the orbits of the asteroids. Most of the asteroids have orbits of low inclination and small eccentricity. A bias toward higher eccentricities can be expected for the meteorites, since an asteroidal fragment having a circular orbit in the asteroid zone cannot intersect the orbit of the earth to fall as a meteorite. The large time scales required for the planet Mars to perturb bodies in the asteroid zone into earth-crossing orbits have led Öpik[8] to suggest that meteorites are released from defunct comet nuclei.

Studies of the frequency of meteorite falls have been made by Brown, Hawkins, and Millard in order to obtain estimates of the flux of meteoroids of a given mass for the vicinity of the earth. An excellent critique of these results, considered in connection with the size distribution and frequency of craters on the moon, has been presented by Baldwin.[9] The Prairie Network currently encompasses several of the Plains States for the purposes of photographing meteorite falls so that the computed trajectories can be used for the purposes of recovering the fallen meteorites. This program (directed by McCroskey) should provide valuable data on the frequency of meteorite falls and the orbits of meteoroids in addition to providing meteorites for laboratory analyses.

MICROMETEORITES

Interplanetary dust particles having radii of approximately 4 microns and less are decelerated without an appreciable loss of mass as they enter the earth's atmosphere. The large area-to-mass ratio for the dust particles permits them to radiate the thermal energy gained in collisions with atmospheric molecules at a sufficiently high rate to prevent the surface of the dust particle from reaching the melting point.

The decelerated dust particles were called *micrometeorites* by

Whipple, who emphasized in 1949 the importance of this non-destructive mechanism by which the earth accretes interplanetary dust particles. The deceleration of dust particles of micron-size occurs at altitudes in the neighborhood of 100 km, leaving the dust particles to settle downward through the atmosphere according to Stokes' law. The motion of a dust particle is, of course, modified by turbulence and convective transport in the atmosphere. The concentration of the dust particles increases as their speeds decrease, particularly at the levels of the temperature inversions in the atmosphere. The relatively stable temperature inversion at 80 km altitude represents a particularly good region in which to obtain collections of accreted and decelerated dust particles through the use of recoverable rockets as was done by Soberman and Hemenway[10] and Farlow, et al. ([11]). The latter group conducted an experiment during the 1965 Leonid meteor shower. The noctilucent clouds, which supposedly consist of ice crystals or ice-covered dust particles, occur in the temperature inversion layer at 80 km altitude. Particles have been collected from this region (when noctilucent clouds were visible as well as when no clouds were visible) through the use of recoverable rockets flown in Sweden by Hemenway, Soberman, and Witt.[19]

The micrometeorites, together with the solidified droplets and the fragments lost through ablation and fragmentation of large meteoroids and the residue left at the ends of meteor trails, constitute the extraterrestrial aerosols present in the earth's atmosphere. Atmospheric aerosols have been collected at ground-level and extracted from ice fields and deep-sea sediments; they have been collected with rockets, balloons, and high-altitude aircraft; and they have been studied through their property of scattering light, both sunlight and the light of laser beams. Identification of the aerosols as being of extraterrestrial origin is one of the outstanding problems. The collections of dust particles obtained through the use of rockets and the dust particles studied by making measurements of the sky brightness at balloon altitudes or by the use of lasers should be relatively free from terrestrial contaminants.

A question remains, however, as to how representative a sample of the interplanetary dust particles is obtained (or can be identified) in the collected samples. Volatile components certainly will have disappeared by the time the dust particles reach ground-level. Some of the ground-based collection techniques involve magnetic separation of the collected particles, so that all the non-magnetic particles are lost from the sample. Perhaps the most fundamental question is that of whether the collected particles are true micrometeorites or are the products of ablation and fragmentation of larger meteoroids or are mixture such as this, plus the probable concentration of the particles as their speed decreases, particularly at the temperature inversions, make it very difficult to determine the cumulative flux of *interplanetary* dust particles from these types of measurements. For example, one group of rocket collection experiments by Berg, et al ([13]), demonstrate that the particles were collected at altitudes between 80 and 100 km.

No attempt will be made here to include the results from studies of micrometeorites in the compilation of a cumulative flux distribution. The importance of these results lies more in the studies of composition, shape, and structure of the micrometeorites than in some computed value for the flux that involves numerous assumptions about the transport of small dust particles in the upper atmosphere. The values derived for the flux are in satisfactory agreement with the flux measured just above the atmosphere of the earth through the use of spacecraft.

A concise survey of the older observational results obtained by means of the various dust particle collection schemes has been presented by Schmidt.[14] More recent results formed the subject of a conference, the proceedings of which have now been published.[15]

THE ZODIACAL LIGHT AND THE SOLAR CORONA

The *zodiacal light* is a band of faint, nebulous light extending along the ecliptic (or through the zodiac, from which the name is derived). The surface brightness of the zodiacal light decreases with increasing angle of elongation from the sun and with increasing ecliptic latitude. This variation in surface brightness gives the zodiacal light the appearance of a broad cone of faint light that is situated on the morning and evening horizons and points away from the sun. The *gegenschein* is a localized intensification of the zodiacal light in the direction of the antisolar point. Cassini suggested, nearly three centuries ago, that dust particles in interplanetary space reflected sunlight to produce the zodiacal light. This explanation has been widely accepted since that time. There was, however, a period in the 1950's during which there was a trend toward interpretations by which interplanetary electrons were considered to make a significant contribution to the zodiacal light.

The *solar corona* is observable out to about one degree from the center of the solar disk during a total solar eclipse, thereby opening a second range of angular distances from the sun to occasional ground-based observations. The luminous energy from the solar corona may be separated into three components. The *L-corona* (line corona) arises from line emission by highly ionized atoms (primarily metals such as Fe and Ca) in the solar atmosphere and is of no interest here. The remainder of the coronal light has a continuous spectrum like that of the sun, except that the depths of the Fraunhofer absorption lines are no so great as in the solar spectrum. This similarity between the coronal spectrum and the solar spectrum shows that the coronal continuum is scattered sunlight. The two possible scattering mechanisms are Thompson scattering by free electrons in the solar atmosphere and Mie scattering by interplanetary dust particles. The light arising from the scattering of sunlight by free electrons is called the K-*corona,* while that arising from the scattering by dust particles is called the F-*corona* (Fraunhofer corona).

Grotrian suggested, in 1934, that the same heliocentric cloud of interplanetary dust particles produced both the zodiacal light and the F-corona. Observational results show this to be true. The two phenomena are not necessarily produced by the same size of dust particle; this question merits further attention sometime in the future. One difficulty with Grotrian's suggestion was that the increase in the surface brightness of the reflected sunlight with decreasing angle of elongation from the sun required that the spatial density of dust particles increase with decreasing distance from the sun. This was clearly impossible for small elongation angles, because dust particles (even metallic particles) would be vaporized within a few tenths of an astronomical unit from the sun. This difficulty was circumvented when Allen in 1946 and van de Hulst in 1947 almost simultaneously made the suggestion that diffraction rather than reflection by the dust particles should dominate for small angles of elongation from the sun. The increase in the surface brightness of the F-corona with decreasing elongation angle was thus attributed to the shape of the

Frauhofer diffraction pattern for the dust particles. This permitted the dust particles responsible for producing the F-corona at small elongation angles to be situated at considerably larger distances from the sun than would be required if only reflection were considered.

Two methods have been used to separate the continuum of the solar corona into the two components produced through scattering of sunlight by electrons and by dust particles. The first method involves measurements of the depths of the Fraunhofer lines relative to those in the solar spectrum, while the second method involves measurements of the polarization of the scattered sunlight.

The thermal velocities of the electrons in the high-temperature solar corona lead to Doppler shifts which are sufficiently large to smooth out the Fraunhofer lines in the sunlight scattered by the electrons. On the other hand, the sunlight scattered by the dust particles will have the same spectrum as the sun, provided only that the dust particles do not possess the property of selective absorption. The amount by which the Fraunhofer lines are filled-in for the spectrum of the sunlight scattered by both the electrons and the dust particles indicates the relative contributions of the electrons and the dust particles to the scattered light.

The second method for separating the contributions of electrons and dust particles to the scattered sunlight involves the reasonable assumption that the sunlight scattered by the dust particles is not polarized at small elongation angles from the sun. Then the polarized component of the corona can be attributed to the scattering by the electrons and corresponds to the K-corona. The remainder of the scattered continuum is attributed to the scattering by the dust particles and constitutes the F-corona. The sunlight scattered by the dust particles actually is a false corona, since the dust particles are in interplanetary space and are orbiting the sun rather than being in the atmosphere of the sun.

Observations of the zodiacal light show that it also is polarized. The low intensity of the zodiacal light makes difficult the determination of the spectrum, so the polarization method used for separating the K-corona from the F-corona came to be used also for separating the corresponding components of the zodiacal light. This method of separating the two components of the zodiacal light necessitated making the assumption that the dust particles produce *no* polarization in the zodiacal light, or that the polarization was produced solely by the electrons in interplanetary space. The polarization of the sunlight scattered by the interplanetary electrons or the coronal electrons is a calculable quantity once the spatial distribution of the electrons is known, or vice versa.

Whipple and Gossner, in 1949, used the foregoing assumption in deriving an upper limit of 1000 electrons/cm^3 for the electron density at 1 A.U. from the sun. It was emphasized that the value was an upper limit and that dust particles might well produce the observed polarization of the zodiacal light. Photoelectric measurements of the polarization of the zodiacal light, together with the assumption that *all* the polarization was produced by interplanetary electrons, were used by Siedentopf et al.,[16] in deriving a value of 600 electrons/cm^3 for the electron density at 1 A.U. from the sun.

Measurements made with *spacecraft* show fluxes of positive ions (protons) of a few times 10^8 particles/cm^2/sec with speeds of a few hundred km/sec. Such values for the flux give (for a quasi-neutral plasma) electron densities of approximately 10 electrons/cm^3 at the earth's distance from the sun.

Beggs et al.,[17] have reported some recent results that were obtained by using narrow-band photoelectric photometry to study the profiles of the Fraunhofer lines in the spectrum of the zodiacal light. A value of 16 electrons/cm^3 was derived for the electron density at 1 A.U. from the sun. The photoelectric observations of the zodiacal light made by Weinberg[18] do not require any electrons to explain either the polarization or the surface brightness of the zodiacal light.

The estimates of the electron density in interplanetary space at 1 A.U. from the sun have ranged from the value of 600 electrons/cm^3 down to the current best estimates of 1 to 10 electrons/cm^3 that are indicated by both the spacecraft measurements and the results from recent studies of the zodiacal light. These low values for the electron density require that the polarization of the zodiacal light be attributed to the interplanetary dust particles rather than to the electrons.

The observational data on the surface brightness and the polarization of the zodiacal light must be corrected for atmospheric effects and for the background radiation from stars. Once corrected, the data are customarily interpreted in terms of a spatial density and a size distribution for the interplanetary dust particles. Estimates of the size of dust particles considered to dominate in the scattering of sunlight have ranged from dimensions of several hundred microns in the early work down to approximately 0.1 micron as better observational data have been obtained. Allen considered dust particles having radii of 10 microns to produce the F-corona, while van de Hulst used a distribution of sizes by which the major contribution to the scattered light arose from dust particles having radii slightly smaller than 350 microns. In both cases, the dust particles were sufficiently larger than the wavelength of the scattered radiation that classical optics could be employed. Fraunhofer diffraction of sunlight by the dust particles was used to explain the F-corona, while simple reflection was used to explain the zodiacal light.

Subsequent work has usually tended toward interpretation of the zodiacal light in terms of small dust particles having dimensions comparable to or slightly larger than the wavelength of light. Elsasser[19] apparently was the first to use this interpretation, but his later interpretations[20] are more like the early one of van de Hulst. Ingham[21] computed a series of size distributions in order to find the one which best fitted the data obtained by Blackwell and Ingham. Small dust particles having radii as small as 0.3 microns were considered to dominate in producing the zodiacal light. This lower limit on the particle size is somewhat artificial, because Ingham used classical optics in computing a radiation pressure limit on particle size as was done for Table I.

Giese[22] used the Mie theory to compute the intensity and polarization produced by various model distributions of small solid particles. The computed results were applied by Giese and Siedentopf[23] to the interpretation of observations of the zodiacal light, but their selection of a likely model appears to be biased by their assumption that the electron density is approximately 300 electrons/cm^3 at 1 A.U. from the sun. Weinberg has found that his observational data on the surface brightness and polarization of the zodiacal light are fitted closely by the results computed by Giese for small dielectric spheres (water ice) having radii between 0.08 and 2 microns. This size distribution corresponds closely to the one presented by Ingham but includes smaller particles, since Giese used the Mie theory in computing a radiation pressure limit on the particle size. Weinberg also notes, however, that it has not been possible for him to establish a unique size distribution for the dust particles. This negative result suggests that additional computations using Mie theory and best estimates of the distribution of orbits should be made in order to determine whether a unique size distribution can be obtained from the observational data on the zodiacal light.

The distribution of the orbits of small dust particles that produce the zodiacal light and the F-corona presently is unknown. The only available information about the distribution of orbits for these dust particles comes from studies of the shape of the isophotal contours of the zodiacal light and the F-corona. The F-corona appears to be almost spherically symmetric about the sun, but the zodiacal light does not. The isophotal contours show that the zodiacal dust cloud has a lenticular shape, or (more generally) the shape of an oblate spheroid. The axis of rotational symmetry is almost normal to the plane of the ecliptic, and the concentration of dust particles appears to decrease with increasing distance from the sun. At distances less than a few tenths of an astronomical unit, the dust particles would, of course, be vaporized by solar radiation.

The departure of the zodiacal light from spherical symmetry rules against a random distribution of orbits for the interplanetary dust particles. The fact that the zodiacal light does not appear like an edge-on view of the rings of the planet Saturn demonstrates that the dust particles are not predominantly in circular orbits having zero inclination. The sometimes used assumption that most of the dust particles responsible for producing the zodiacal light are in heliocentric orbits having zero or very low inclinations is an idealized model which cannot be justified on the basis of the presently available data. The isophotal contours, as best they can presently be established, indicate a general concentration of the orbital distribution toward the ecliptic but with a sizeable fraction of the dust particles having orbits of moderate and even high inclinations. The action of the Poynting–Robertson effect generally is invoked in arguments that the orbits have fairly low eccentricities.

The determination of the isophotal contours of the zodiacal light is severely hampered not only by the low surface brightness at moderate and high ecliptic latitudes but also by the emission and scattering of radiation by the earth's atmosphere. Weinberg[24] has demonstrated that concurrent measurements of the airglow are essential for use in correcting the observational data on the zodiacal light for the deleterious effects of the earth's atmosphere on the measurements of the surface brightness and polarization of the zodiacal light. Observations of the zodiacal light and the solar corona from an extra-terrestrial station (a spacecraft or the moon) would be free from the effects of the earth's atmosphere. Additional new data about the zodiacal light will appear soon in the published proceedings of the I.A.U. Symposium No. 31 held in Honolulu, Hawaii, 30 January to 2 February, 1967.

DIRECT MEASUREMENTS OF INTERPLANETARY DUST PARTICLES

The advent of spacecraft placed in orbit about the earth or sent into interplanetary space has provided the opportunity for making direct measurements of certain physical and dynamical characteristics of interplanetary dust particles. Dust particle sensors that have been flown encompass the range of particle mass

$$10^{-13}\ \text{gm} \lesssim m \lesssim 10^{-7}\ \text{gm}.$$

This range includes most of the size range of the dust particles that scatter sunlight to produce the zodiacal light and the F-corona.

The most important results obtained so far by the direct measurements technique were rather unexpected. First, the flux distribution measured with Explorer VIII for dust particles in the vicinity of the earth was completely different from what was expected on the basis of either the results from studies of the zodiacal light or the extrapolations of results from observations of meteors. Second, the direct measurements have revealed periods of greatly enhanced flux of small dust particles. Third, the flux of dust particles measured with Mariner IV is in agreement with the flux computed on the basis of results from studies of zodiacal light.

The impacts of more than 10,000 dust particles have now been recorded by means of sensors flown on spacecraft. Most of the impacting dust particles were observed by means of acoustical type sensors (piezoelectric crystals mechanically coupled to exposed surfaces or plates on the spacecraft). The acoustical type sensor measures the impulse imparted by an impacting dust particle and yields a measure of the momentum of the dust particle. A smaller number of impacts of dust particles (of somewhat smaller sizes than those detected by the acoustical type sensors) has been detected by means of the photomultiplier type sensor. This sensor measures the luminous energy released as dust particles impact on a thin metal film which shields the photocathode from background light. The remaining small number of observed impacts has been recorded with a variety of sensors too numerous to describe here. The measurements of flux obtained through the use of both acoustical type and photomultiplier type sensors are shown in Figure 1.

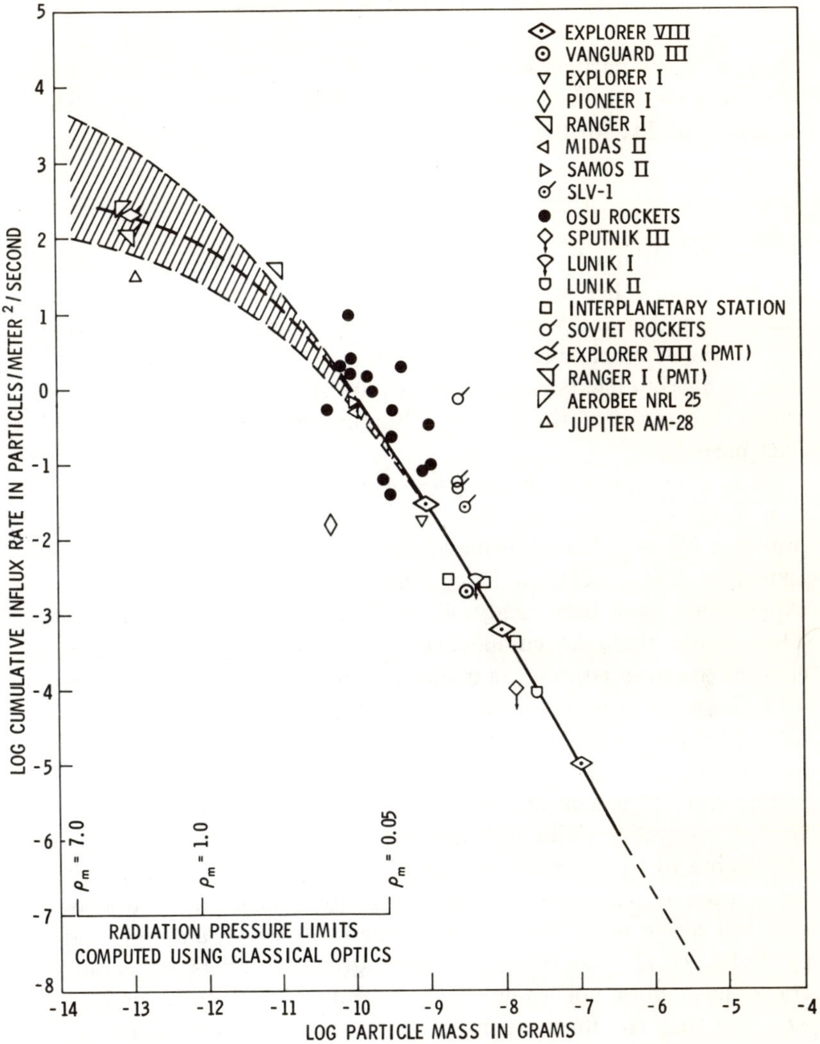

FIG. 1. Cumulative flux distribution measured with acoustical type and photomultiplier type dust particle sensors on spacecraft (after Alexander et al., Ref. 25).

The cumulative flux distribution measured with the acoustical type sensor on Explorer VIII can be expressed in the form of a power law as

$$\log I(m) = -17.30 - 1.70 \log m, \qquad (27)$$

where m is the particle mass in gm, and $I(m)$ the flux in particles/m^2/sec/2π ster for dust particles having masses of m and larger.[26] The range of validity for the foregoing distribution is

$$10^{-10} \text{ gm} \lesssim m \lesssim 10^{-7} \text{ gm}.$$

The results obtained with Explorer VIII provided the necessary basis for analyzing other direct measurements of dust particles. Such an analysis by Alexander et al.[25] showed that good agreement existed among the available direct measurements in 1962. The scatter of the various measurements (some of which are shown in Figure 1) is well within the range of the observed fluctuations in the flux of small dust particles.

As is often the case in analyses of experimental data, an exception arose subsequent to the above analysis. The exception was provided by experiments using penetration type sensors carried on Explorers XVI, XXIII and Pegasus satellites. The results of these experiments have been analyzed by D'Aiutolo[27] and Naumann [28]. They found that the cumulative flux of penetrating dust particles was two to three orders of magnitude lower than the flux measured with the acoustical type sensors. Investigations of this discrepancy have produced several explanations and the following section summarizes two of them.

The simplest explanation is that the acoustical sensors are sensitive to thermal cycling (Nilsson[29]) and thus produce self-generated noise pulses due to the thermal shocks which occur twice in each orbit for a satellite. Therefore, the data from the acoustical sensors consisted of sensor noise rather than impacts. However, another study by Bohn, et al,[30] concerning thermal gradient effects on ceramic transducers does not show similar results. This study finds no evidence that the flux measured by the satellite experiments was the result of thermal noise in the sensors.

A second explanation results from a study of the relations between the observed measureable quantity for the sensors and the physical and dynamic properties of the dust particles. For sensors of the type used, these relations yield an observable quantity R which is associated with the parameters of the impacting particle by a relationship of the form.

$$R \approx C \rho^{i} m^{j} (v \cos \theta)^{k} \tag{28}$$

where C is a constant related to a particular sensor; ρ, m and v are the density, mass and velocity, respectively, of the dust particle; and θ the angle of incidence of the impacting particle. For the penetration experiments, R is the thickness of material required to be penetrated for a theshold measurement. For the acoustical sensor, R is the magnitude of the electrical signal from the piezoelectric ceramic transducer required to produce a threshold signal to the experiment electronics. There is no single unique value of m which relates to a particular value of R for either type of sensor, because values for ρ, v, and θ vary over a large range.

Laboratory hypervelocity experiments using particles with masses between 10^{-2} gm and 1.0 gm and velocities up to about 10 km/sec have provided some information for evaluating the coefficients in the expression for R of the penetration experiment sensors. In presenting the date from the penetration sensors, Naumann used

$$R = C\rho^{0.147} m^{0.352} (v \cos \theta)^{0.667}. \tag{29}$$

In analyzing the acoustical sensor data, only calibrations using particles with low velocities (a few meters/sec) were possible. These studies gave:

$$R = C\, mv \cos \theta \tag{30}$$

In considering the velocity parameter alone, an extrapolation of about four orders of magnitude was necessary to reach meteoritic velocities for data analysis. Recent laboratory experiments by Bohn, et al,[31] using microparticles accelerated to hypervelocities and also theoretical studies by Bjork[32] of hypervelocity impacts indicate that R is not a linear function of v. The exponent k appears to be greater than 1 and possibly to be a function of velocity. These tentative results have a marked effect on the comparisons of the satellite data obtained with the two types of sensors. For example, consider the velocity parameter, where in the case of the penetration sensor, R is varying as a function of $v^{2/3}$; while for the acoustical sensor, R is varying as a function of $v^{f(v)}$. If further laboratory work confirms this general picture, then the discrepancy between the two cumulative flux distributions measured by the types of detectors will decrease. Furthermore, when the dependence of penetration on the structure and shape of the impacting dust particle is known and can be added to the expression for R, the discrepancy may be again reduced. The hypervelocity impact is a very complex physical phenomenon, and the predominant differences between the measurements using the two types of sensors is most probably due to the complexity of equation 28.

In addition to providing measures of the average flux, as shown in Figure 1, the direct measurements have revealed periods of greatly enhanced flux of small dust particles. The most prominent enhancement detected so far was measured with an acoustical type sensor flown on Vanguard III.[33] The average flux during a 10-day interval in mid-November, 1959, was higher by a factor of 10 than the average flux during the remainder of the 80-day lifetime of the experiment. At times, during this 10-day interval, the flux rose by factors of 100 to 1000 above the average background rate. The dust particle sensor with which this period of enhanced flux was observed was omnidirectional, so it was not possible to establish the radiants for the dust particles. The enhanced flux occurred while the earth was passing through the Leonid meteoroid stream. This coincidence in time was the basis of a suggestion that the small dust particles may have come from the Leonid meteoroid stream. There are strong arguments against this suggestion. Small dust particles cannot remain with the Leonid stream for appreciable lengths of time because of the effects of radiation pressure. The hypothesis that the observed dust particles came from larger dust-ball meteoroids which broke up shortly before the observations were made is weakened if the total amount of material which would have been required in the break-up process is considered.

Another period of enhanced flux was observed with Explorer I, and Nazarova and her colleagues have reported that a period of enhanced flux was observed in interplanetary space with Mars I. A highly variable flux of small dust particles seems to be the rule rather than the exception. No evidence exists now for the association of the periods of enhanced flux of small dust particles with the known meteoroid streams.

Measurements of the cumulative flux distribution of dust particles in interplanetary space have been made with experiments on the Mariner II and Mariner IV spacecraft by Alexander, et al.,[34] with detectors using acoustical sensors. These measurements gave the dust particle flux at heliocentric distances ranging from 0.72 to 1.0 A.U. and 1.0 to 1.58 A.U., respectively. The measured flux is in agreement with the flux computed on the basis of results from studies of zodiacal light. Table II gives a summary of the cumulative flux distributions for various heliocentric distances from the Mariner IV measurement. A comparison of these results with the satellite measurements made with similar sensors indicates that the flux in deep space is nearly four orders of magnitude lower than that near the earth. However, on the basis of the recent information concerning the calibration of the acoustical sensors, particularly the sensitivity to velocity found by Bohn et al,[31] the flux of small dust particles in interplanetary space is about two orders of magnitude lower than the flux measured in the satellite experiments using acoustical sensors. This enhancement of the average flux will be treated in greater detail in a subsequent section. Direct measurements of the speeds and radiants of the small dust particles are required before the time-dependent fluctuations in flux can be discussed further.

TABLE II
Cumulative Flux-mass Distribution for Various Heliocentric Distances from the Mariner IV Measurement

N_p (No. of particles)	Δt (days)	psd (probe-sun distance: A.-U.)	flux (N_p/m^2 sec π ster.)	S (slope of cum. flux-mass distribution)
44	100	1.0 –1.25	7.3×10^{-5}	−0.54
43	34	1.25–1.36	2.1×10^{-4}	−0.85
40	20	1.36–1.43	3.3×10^{-4}	−0.57
39	29	1.43–1.49	2.2×10^{-4}	−0.56
49	45	1.49–1.56	1.8×10^{-4}	−0.68
26	57	1.56–1.58	7.6×10^{-5}	−0.81

THE ORIGIN OF INTERPLANETARY DUST PARTICLES

Once the existence of interplanetary dust particles has been established through observations, a question naturally arises concerning the origin of the dust particles. The possible origins of the dust particles are:

(a) primordial debris left over following the formation of the solar system,

(b) interstellar grains streaming through or accreted by the solar system,

(c) dust particles produced by the collisional disintegration of asteroids, and

(d) dust particles released by the evolutionary disintegration of comet nuclei.

A brief discussion of these possible origins is given as a prelude to putting observational data together in the form of a cumulative flux distribution in the following section.

Small dust particles which had not coagulated into larger particulate aggregates possibly existed following the condensation of the primordial solar nebula to form the solar system. The decrease in gas density that must have occurred during the condensation process would have permitted the dust particles to assume heliocentric orbits about the newly-formed sun. Further coagulation of dust particles would have been inhibited by the decrease in density of the gas from which the dust particles were forming; also, the higher relative speeds that could occur for dust particles situated in regions of low gas density would have increased the tendency of colliding dust particles to break apart rather than to coagulate into larger dust particles.

Any dust particle for which the force of radiation pressure exceeded the gravitational force exerted by the sun would have been blown out of the solar system. Additional consideration must be given to those dust particles for which the gravitational force was the larger. Those dust particles which (in the presence of the gravitational and radiation fields of the sun) had total energies less than the parabolic limit would have assumed elliptic orbits about the sun. In the absence of planetary perturbations, such dust particles would then spiral toward the sun under the action of the Poynting–Robertson effect. The dust particles which had energies equal to or greater than the parabolic limit would have begun their exits from the solar system along parabolic or hyperbolic trajectories, respectively. Some of these dust particles may have been recaptured into elliptic heliocentric orbits by the Poynting–Robertson effect, but most would have been blown out of the solar system.

The equation given earlier for the orbital lifetimes permitted by the Poynting–Robertson effect may be used to show that a dust particle having a mass density of 1 gm/cm^3 and a radius of 6 microns would have to have been more than 1000 astronomical units (for a circular orbit), or considerably farther (for an elliptic orbit), from the sun shortly after the formation of the solar system in order to be in the vicinity of the earth's orbit at the present time. Considering the dust particles which produce the zodiacal light and which are being studied through the use of spacecraft to be primordial debris requires that the dust particles are now near the end of a long spiral journey from the outer reaches of the solar system to the sun. The orbits of such dust particles should be nearly circular because of the long time interval required for the Poynting–Robertson effect to bring the dust particles to the inner regions of the solar system. The inclinations of the orbits might be almost randomly distributed like those of the long-period and non-periodic comets. The evaluation of a primordial origin for the small interplanetary dust particles would be very difficult (even if the distribution or orbits of the dust particles were known) because of perturbations by the planets.

The accretion of interstellar dust by the solar system has been considered as a possible source for the interplanetary dust particles. This idea became prevalent following the recognition in 1930 of the existence of interstellar absorption which was attributed to interstellar dust particles. An interstellar origin for meteoroids received further support from visual observations of meteors which indicated that a sizeable fraction of the meteoroids were in hyperbolic heliocentric orbits. This indication was misleading; the application of radar methods to the study of meteors following World War II yielded measurements of the speeds which showed that less than 1% of the observed meteors might have hyperbolic orbits. The observational errors were of such a magnitude that this remaining small fraction could just as well have had elliptic orbits. There is general agreement now that most, if not all, the meteors are produced by meteoroids that travel in closed heliocentric orbits and are members of the solar system. Best and Patterson[35] have shown, however, that the Poynting–Robertson effect can act to capture small interstellar dust particles into closed heliocentric orbits.

Small dust particles produced in asteroidal collisions should be angular fragments and solidified droplets of metals and minerals having mass densities and compositions similar to those of the meteorites. Dust particles having mass densities of 3.5 gm/cm^3, radii less than 500 microns, and initially circular orbits lying in the asteroid zone (roughly 2.8 A.U. from the sun) will spiral into the sun within a time interval of 10 million years. An asteroidal origin for the small interplanetary dust particles therefore requires that asteroidal grinding be a continuing process. Both Piotrowsky and Fesenkov have calculated that the rate of production of small dust particles by the collisional break-up of asteroids appears to be adequate to maintain the cloud of interplanetary dust particles against the loss to the sun.

The hypothesis of an asteroidal origin for the small dust particles seems to depend critically on whether the distribution of the orbits of the dust particles that could be produced by collisional grinding contains sufficiently large numbers of orbits of moderate to high inclination to explain the isophotal contours of the zodiacal light.

According to the icy conglomerate model proposed by Whipple, a comet nucleus consists of grains of metals and minerals imbedded in a matrix of the ices of water, ammonia, carbon dioxide, and methane. Evolutionary break-up of such a comet nucleus under the heating action of solar radiation should produce dust particles that are friable and which have a low mass density. The dust particles could also have irregular shapes and a dendroidal structure. Photographic and radar studies of meteors have indicated that such a model for the meteoroids is indeed a reasonable one. The orbits for the photographic meteors are quite similar to the orbits of the short-period comets. Most of the major meteor streams can be identified with a parent comet through an analysis of the orbits of the meteorids. The observational data support the concept that the meteors observable by visual and photographic methods and the brighter meteors observable by radar have their origin in the short-period comets (periods $<$ 200 years). A question exists concerning the origin of the fainter radar meteors that have orbital inclinations in excess of those for the short-period comets. These meteoroids may have their origin in the long-period and non-periodic comets or (possibly) in the primordial debris of the solar system.

The source of the small dust particles which produce the zodiacal light presently is unknown. Computations have shown that either asteroidal grinding or cometary break-up is adequate to maintain the zodiacal dust cloud, but the relative importance of these two sources cannot be readily established. The relative proportions of cometary and asteroidal dust particles appear to depend markedly on the size of the dust particle. It is difficult to see why the process of asteroidal grinding should again become important for the small dust particles after having been over-shadowed by cometary break-up throughout the range of particle size characteristic of the meteors. A cometary origin for the small dust particles in the zodiacal dust cloud is quite acceptable on the basis of all the presently available data. A knowledge of the orbits of the small dust particles or, alternatively, an accurate knowledge of the isophotal contours of the zodiacal light should help to establish the most probable origin for the small dust particles.

THE MASS DISTRIBUTION OF INTERPLANETARY DUST PARTICLES

The compilation of information about the flux of interplanetary dust particles in order to obtain a mass distribution that extends over a larger range of particle mass than is covered by any one observational technique has been a continuing process. One of the simplest approaches to this problem was taken in 1948 by Grimminger, who used a linear extrapolation of the magnitude distribution for visual meteors as it was given by Watson.[7] Data obtained more recently have been included in the compilations by Kaiser,[36] Alexander et al.,[25] Whipple,[3] and McCracken and Dubin.[37]

The selected data which will be used here in constructing a mass (or flux) distribution curve are given in equation form in Table II. These data apply for average or sporadic conditions; the intervals of enhanced flux which are measured by spacecraft and which are observed during meteor showers must be considered separately. Data

TABLE III

Selected observational data on the mass distribution of interplanetary dust particles

Source (Reference)	Equation of Distribution	Range of Validity
	Meteor observations	
7	$\log N(M_v) = 5.87 + 0.4M_v$	$-3 \leq M_v \leq 10$
4	$\log N(M_v) = 6.3 + 0.4M_v$	$3 \leq M_v \leq 10$
	$\log N(M_v) = 6.0 + 0.50M_v$	$0 \leq M_v \leq 3$
	$\log N(M_v) = 6.0 + 0.57M_v$	$-10 \leq M_v \leq 0$
38	$\log N(M_v) = 4.93 + 0.538M_v$	$0 \leq M_v \leq 4.1$
39	$\log N(M_v) = 4.52 + 0.4M_v$	$-10 \leq M_v \leq -3$
40	$\log N(M_r) = 4.86 + 0.468M_r$	$8 \lesssim M_r \leq 10.8$
	Meteorite collections	
41	$\log N(m) = 3.5 - 0.8 \log m$	$10^4 \lesssim m \lesssim 10^{11}$ gm
	Zodiacal light studies	
19	$\log I(s) = -14.32 - 2.5 \log s$	$2\mu \leq s \leq 0.2$ cm
21	$\log I(s) = -16.38 - 3.0 \log s$	$0.4\mu \leq s$
	Direct measurements (near the earth)	
26	$\log I(m) = -17.30 - 1.70 \log m$	$10^{-10} \lesssim m \lesssim 10^{-7}$ gm

$N(\ )$ = cumulative accretion rate in particles/earth/day.
$I(\ )$ = cumulative flux in particles/m^2/sec/2π ster.
M_v = visual magnitude, M_r = radar magnitude, $M_r \doteq M_v$.
m = particle mass in gm.
s = particle radius in cm.

from studies of the statistics of meteorite falls, from observations of meteors, from studies of the zodiacal light, and from direct measurements made with spacecraft have been included. Other data can be added, and it generally supports the mass distribution given by the data in Table III.

The data given in Table III are plotted as a cumulative flux distribution which is shown as Figure 2. Actually, Figure 2 shows two flux distributions which apply for different regions of space; the probable meaning of the two distributions can be clarified by considering briefly the procedure followed in constructing Figure 2 from the data given in Table III.

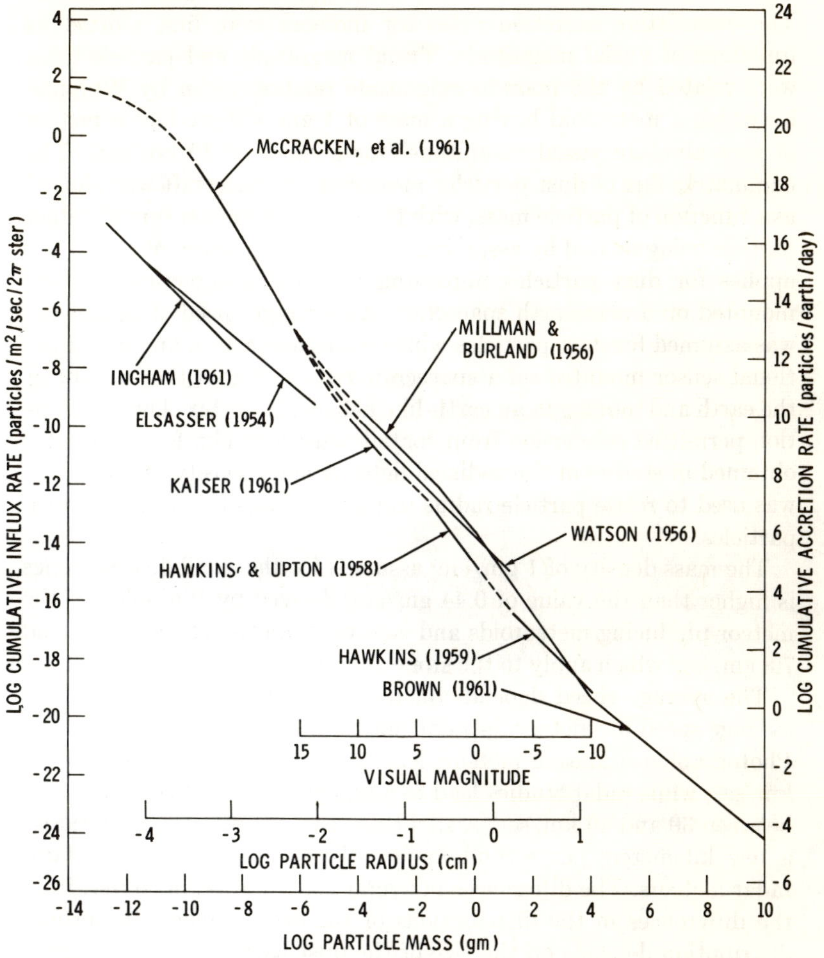

FIG. 2. Cumulative flux distribution based on the results from studies of meteors and meteorite falls, observations of the zodiacal light, and direct measurements. The direct measurement portion of the curve is expected to change when the microparticle hypervelocity calibrations of acoustical sensors are completed. (See Table III for the data expressed in equation form.) (After McCracken and Dubin, Ref. 37.)

The cumulative accretion rate for meteorites was plotted directly as a function of particle mass. (The particle mass is not as fundamental a parameter for the meteorites as might be thought at first, since an estimate of the material lost through ablation must be made.) The cumulative accretion rates for meteors were first plotted as functions of visual magnitude. Visual magnitude and particle mass were related by the mass-to-magnitude relation given by Whipple[3] by which a meteoroid having a mass of 1 gm will produce a meteor of zero absolute visual magnitude (for a speed of 30 km/sec). The cumulative flux of dust particles measured by spacecraft was plotted as a function of particle mass, with the mass and momentum of a dust particle being related by assuming that an average speed of 30 km/sec applies for dust particles impacting on an omnidirectional sensor mounted on a near-earth spacecraft. An average speed of 10 km/sec was assumed for dust particles which would impact on an omnidirectional sensor mounted on a spacecraft removed from the vicinity of the earth and moving in an earth-like heliocentric orbit. This assumption permitted conversion from spatial density to flux for the results obtained in studies of the zodiacal light. A mass density of 1 gm/cm^3 was used to relate particle radius to particle mass for the small dust particles.

The mass density of 1 gm/cm^3 assumed for the small dust particles is higher than the value of 0.44 gm/cm^3 derived by Whipple[3] for the meteor-producing meteoroids and is lower than the values of 3.5 and 7.8 gm/cm^3 which apply to the stone and iron meteorites, respectively.

The average speed depends on the size of the dust particle. The average speed for meteorite-producing meteoroids is 15 to 20 km/sec. Photographic studies of meteors lead to an average speed of 25 to 30 km/sec, while radar studies lead to a higher value that generally falls between 30 and 40 km/sec. A slight decrease in the average speed by a few kilometers per second apparently occurs for the very faint radar meteors. The differences in average speed arise, of course, from the differences in the distributions of the orbits; hence, the orbital distribution depends on the size of the dust particle.

Each of the distributions listed in Table III and used in the construction of Figure 2 applies, at most, for a few orders of magnitude in the particle mass. The use of straight line segments to represent the distributions facilitates expressing the observational data as power law distributions (as in Table III).

Data obtained with spacecraft, from observations of meteors, and from studies of the frequency of meteorite falls establish a cumulative flux distribution that applies for dust particles in the immediate vicinity of the earth. Data obtained from studies of the zodiacal light, together with the direct measurements of flux that were obtained with Mariner II and Mariner IV, can be used to replace the direct measurements made in the vicinity of the earth in order to derive a cumulative flux distribution that applies for regions of space removed from the earth and situated at approximately 1 A.U. from the sun.

The flux of dust particles sufficiently large to produce meteors or meteorites upon colliding with the earth is essentially the same at the earth as elsewhere along the orbit of the earth. Hence the two flux distributions shown in Figure 2 merge for dust particles having masses larger than about 10^{-5} gm. For smaller dust particles, the flux distribution for the vicinity of the earth departs considerably (toward increasing values of flux) from the flux distribution for other regions along the earth's orbit. This measured enhancement of flux constitutes the basis of the numerous assertions that the earth pos-

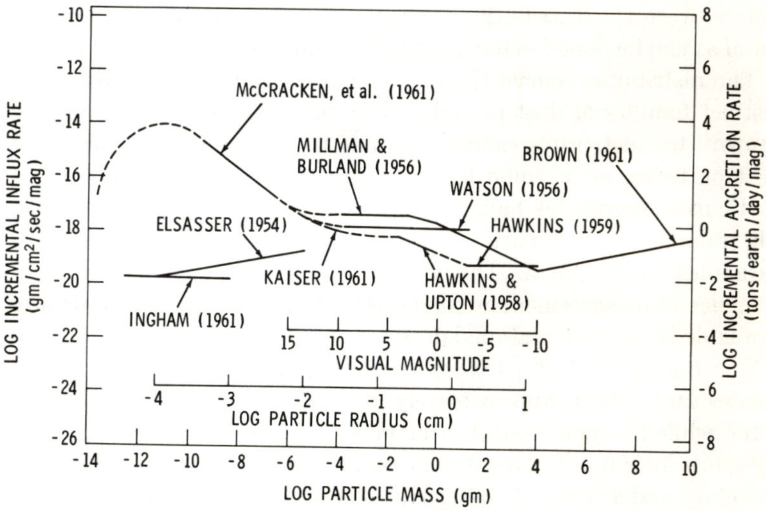

Fig. 3. Incremental mass flux distribution derived from the cumulative flux distribution shown in Figure 2. The direct measurement portion of the curve is expected to change when the microparticle hypervelocity calibrations of acoustical sensors are completed. (After McCracken and Dubin, Ref. 37.)

sess a *dust belt* or a *dust shell*. All that can really be said at present is that an enhancement of flux for dust particles having masses $m < 10^{-7}$ gm has been measured in the vicinity of the earth. It is instructive, before delving further into this subject in the following section, to have a further look at the cumulative mass distribution shown in Figure 2.

The cumulative mass distribution shown in Figure 2 is a segmented distribution that seems to show significant departures from a single straight line extending over the entire range of particle mass encompassed by the observational data. Examination of these departures is made easier by deriving from Figure 2 a second distribution which is just the incremental mass distribution multiplied by the masses of the dust particles. Such a distribution is presented here as Figure 3. The distribution shown in Figure 3 emphasizes the amount of material within a given range of particle size in contrast to the number of dust particles larger than a given size that is emphasized by the distribution shown in Figure 2. Figure 3 shows the flux of mass per unit magnitude (an interval of 0.4 in the logarithm of the particle mass) plotted as a function of the particle mass. Interpolation between the distributions and smoothing (as was done in Figures 2 and 3) can be done freehand entirely within the accuracy of the data.

The distribution curve shown in Figure 3 appears to be revealing distinct families of dust particles. The distribution curve derived by Brown[41] for meteorites cannot continue to rise indefinitely for larger dust particles, or it would lead to an unreasonably large number of particulate aggregates having asteroidal dimensions (see Baldwin[9]). The families of dust particles appear in the form of three inverted parabolas in Figure 3. The appearance of these families of dust particles is in agreement with special information that was already available in the observational data.

The family of dust particles consisting of those particles having masses larger than approximately 10^4 gm is of an asteroidal character, while the meteoroids having masses smaller than approximately 10^4 gm belong to a family of cometary origin. The crossover between cometary and asteroidal dust particles at approximately 10^4 gm agrees well with the results obtained by Jacchia and Whipple in a study of photographic meteors. In addition to these two families of dust particles that exist both in interplanetary space and in the vicinity of the earth, there seems to be a third family that is unique to the vicinity of the earth and which consists of dust particles having masses smaller than approximately 10^{-5} gm.

The foregoing interpretation of the distribution shown in Figure 3 is, of course, quite speculative and depends critically on the degree to which the various distributions given in Table III represent the actual mass distribution of the interplanetary dust particles. Assume, for the purposes of discussion, that the distribution curves in Figures 2 and 3 are reasonably valid, so that we may proceed to the next section for a discussion of the family of dust particles which appears to be unique to the vicinity of the earth.

THE ENHANCED FLUX OF SMALL DUST PARTICLES IN THE VICINITY OF THE EARTH

Two types of enhancement of the flux of small dust particles have been observed in the vicinity of the earth. Assume that the time-dependent fluctuations in the flux can eventually be explained in terms of variations in the supply of small dust particles so that our attention can be directed now primarily toward the enhancement of the average flux above the interplanetary value.

The possible existence of an enhanced flux in the vicinity of the earth was mentioned in 1953 by de Jager, by Öpik,[42,43] by Singer,[44] and by Beard.[45] Whipple[46] suggested, on the basis of a comparison of the directly measured flux with the flux inferred from studies of the zodiacal light, that a *dust belt* existed around the earth. Singer[47] extended his earlier treatment involving the use of Liouville's theorem and suggested that the phenomenon really was a *dust shell*, with the maximum flux occurring at an altitude of about 1000 km. Rather than discussing these and other hypotheses (and they must be considered as such until the appropriate direct measurements can be obtained), let us take a close look at some of the fundamental principles of the problem of the enhanced flux.

There are two aspects to the problem of the enhanced flux in the vicinity of the earth. The first aspect concerns the origin or source of the small dust particles, while the second concerns the mechanism by which the flux is enhanced above the interplanetary value for the flux.

The two possible sources for the dust particles are: (1) interplanetary dust particles entering the gravitational field of the earth, and (2) dust particles ejected from the moon. At this point, one encounters a very basic question. Do the small dust particles have open (parabolic and hyperbolic) or closed (elliptic) geocentric orbits? Data which could be used to answer this question do not presently exist. Dust particles from either source could stream through the vicinity of the earth along open geocentric orbits. If a loss of energy occurs in the vicinity of the earth, dust particles from either source could be captured into closed geocentric orbits. In addition, dust particles ejected from the moon at speeds between 2.4 km/sec (speed required to escape from the moon) and 2.8 km/sec (speed required to escape from both the moon and the earth) could enter directly into closed geocentric orbits.

The energy loss mechanisms which have been suggested include lunar and solar perturbations, solar radiation pressure, atmospheric drag (for grazing encounters), Coulomb drag, and electromagnetic drag. The last two mechanisms depend on the electric charge induced on a dust particle by solar ultraviolet radiation, solar plasma, and the earth's radiation belts and ionosphere. Öpik,[43] Singer,[44] and Shapiro et al[48] (4 papers) attempted to evaluate the importance of these drag forces, but such an evaluation is difficult because of the lack of knowledge concerning the shape, structure, and composition of the small dust particles. Atmospheric drag has been invoked both to trap small dust particles into closed geocentric orbits and to break up dust-ball meteoroids into smaller dust particles which might then be captured by any one of the energy loss mechanisms. The explanation of the enhanced flux in terms of small dust particles trapped in closed geocentric orbits requires long orbital lifetimes.

The authors are in no position to evaluate all the combinations of the possible sources and possible energy loss mechanisms which might be used in attempting to explain the enhancement of the flux of small dust particles in the vicinity of the earth. Only two of the combinations will be discussed here.

First, consider the moon as a source of the small dust particles. Possible ejection mechanisms include lunar volcanic activity, electrostatic expulsion, and ejection through the action of larger dust particles impacting at speeds between 2.4 km/sec and 72 km/sec. The last of these three mechanisms was suggested by Whipple,[46] who has since largely abandoned the hypothesis. Arguments for the lunar impact hypothesis were advanced by Gault et al.,[49] and arguments against this hypothesis were given by McCracken and Dubin.[37]

The lunar impact hypothesis cannot now be evaluated fairly because of the lack of information about the nature of a high-speed impact of a projectile (preferably a dust-ball) on a surface resembling the surface of the moon (whatever that may be, but most probably a porous surface). Probably the ultimate test of the lunar impact hypothesis will be provided by direct measurements of the flux of small dust particles near the surface of the moon. The incident flux of dust particles having masses between 10^4 gm and 10^{-5} gm is known to be low from the studies of meteors and amounts to only about 1 gm/cm^2 per 4.5 billion years.[37] The lack of appreciable color differences on the moon speaks against the removal of large quantities of surface material by any of the possible ejection mechanisms.

Next, consider the possibility that interplanetary dust particles constitute the source and their streaming into the vicinity of the earth constitutes the mechanism for enhancing the flux of small dust particles in the vicinity of the earth.

The geocentric orbital speed v_g of a dust particle at geocentric distance r_g is given (for the case of no loss of energy) by the law of conservation of energy to be

$$v_g^2 = v_\infty^2 + (2\mu_\oplus/r_g) \tag{31}$$

where $\mu_\oplus = GM_\oplus$ and v_∞ is the speed at which the dust particle enters the gravitational field of the earth. The geocentric orbits are parabolic for $v_\infty = 0$ and hyperbolic for $v_\infty > 0$; hence, if no energy loss occurs, the dust particles return to interplanetary space after streaming past the earth.

The angular momentum per unit mass is

$$l = |\mathbf{r}_g \times \mathbf{v}_g| = r_g v_g \sin \zeta. \tag{32}$$

A dust particle that has its perigee passage at an altitude just sufficiently high to avoid being dragged into the earth's atmosphere has (at perigee)

$$l_c = r_p v_p, \tag{33}$$

since $\zeta = 90°$ in this case. Dust particles having angular momenta $l < l_c$ will be accreted by the earth to become micrometeorites.

The scalar flux ϕ at geocentric distance r_g is

$$\phi(r_g) = \rho(r_g) v_g(r_g) \tag{34}$$

where $\rho(r_g)$ is the number density of dust particles at geocentric distance r_g.

Beard[45] assumed $v_\infty = 0$ and assumed zero angular momentum ($l = 0$) in order to derive the dependence of the density enhancement on geocentric distance as

$$\rho(r_g)/\rho_\infty \propto 1/r_g^{3/2}. \tag{35}$$

This derivation involved a dependence of flux on distance given by

$$\phi(r_g)/\phi_\infty \propto 1/r_g^2$$

and a dependence of the geocentric speed on distance given by

$$v_g(r_g) \propto 1/r_g^{1/2}.$$

A flux enhancement of 10^4 would decrease to the interplanetary flux at about $400R_\oplus$, where $R_\oplus = 6378$ km is the radius of the earth. All dust particles that contribute to the enhancement of the flux collide with the atmosphere of the earth for the case $l = 0$. The corresponding accretion rate would be, from Figure 3, about 10^4 tons per day. Such an accretion rate is in fair agreement with the studies of micrometeorites, but is slightly higher than the accretion rate indicated by the studies of sky brightness at balloon altitudes made by Newkirk and Eddy. The assumption of zero angular momentum for the dust particles is unrealistic. If a large fraction of the dust particles have angular momenta $l > l_c$, then they can contribute to the enhancement of flux near the earth without being accreted by the earth to become micrometeorites.

Singer[44] applied Liouville's theorem to the case of interplanetary dust particles streaming through the gravitational field of the earth. The treatment was further extended by Singer[47] and has been developed in considerably more detail by Shelton et al.[50] By Liouville's theorem, the density n of particles remains constant along a dynamic trajectory in phase space. Use of a six-dimensional phase space composed of the geocentric position and velocity vectors $\mathbf{r}_g$ and $\mathbf{v}_g$ permits one to write

$$n(\mathbf{r}_g, \mathbf{v}_g) = n(\mathbf{r}_g', \mathbf{v}_g') \tag{36}$$

for two points $(\mathbf{r}_g, \mathbf{v}_g)$ and $(\mathbf{r}_g', \mathbf{v}_g')$ on the trajectory in phase space.

The density ρ in real space is given by

$$\rho(r_g) = \int_{\mathbf{v}_g} n(\mathbf{r}_g, \mathbf{v}_g)\, d\mathbf{v}_g \tag{37}$$

where the integral is to be evaluated over velocity space. (See Shelton, et al.,[50] for the details of one possible scheme for integration.) If r_g', v_g', ρ_∞, and ϕ_∞ are taken to apply when the dust particles enter the gravitational field of the earth, then the density enhancement at geocentric distance r_g is given by

$$\rho(r_g)/\rho_\infty = f[v_g(r_g)/v_\infty] \tag{38}$$

and the flux enhancement is given by

$$\phi(r_g)/\phi_\infty = f[v_g(r_g)/v_\infty]^2, \tag{39}$$

where f is a correction factor to allow for those dust particles which collide with the earth and are absent from the receding component of the flux.

The density enhancement $\rho(r_g)/\rho_\infty$ and the flux enhancement $\phi(r_g)/\phi_\infty$ both increase without limit, according to the foregoing relations, as the speed v_∞ at which the dust particles enter the gravitational field of the earth decreases toward zero. At the altitudes of near-earth spacecraft, $v_g \lesssim 10$ km/sec, so a flux enhancement of 10^4 requires $v_\infty \lesssim 0.1$ km/sec. The flux enhancement could decrease much more rapidly with geocentric distance than indicated above if the dust particles should prove to be in closed geocentric orbits.

The results obtained through application of Liouville's theorem show that the enhancement of flux depends critically on the geocentric speeds of interplanetary dust particles as they enter the earth's gravitational field. As was shown by Singer,[47] values of 1 to 3 km/sec for v_∞ lead to enhancements of the flux by factors between 10 and 100, as compared to the observed enhancement of 10^4. Lowering the speed as the dust particle enters the gravitational field of the earth helps to obtain larger factors of enhancement. A too common assumption for justifying the required low values of geocentric speed is that the interplanetary dust particles have orbits of predominantly low inclination and low eccentricity (that is, earth-like orbits). This assumption is objectionable from the standpoint of the interpretation of the isophotal contours of the zodiacal light. Additional arguments for large eccentricities were given by Öpik.[43] Let us check whether the assumption of earth-like orbits for the small dust particles may be objectionable for other reasons; if so, the assumption will have to be discarded.

The heliocentric speed of the earth in its orbit, assumed to be circular of radius $r_\oplus = 1$ A.U., is

$$v_\oplus = (\mu/r_\oplus)^{1/2} = \mu^{1/2}, \tag{40}$$

but the heliocentric speed of a small dust particle having an earth-like orbit is

$$v = (\mu_{\text{eff}}/r_\oplus)^{1/2} = [\mu(1 - \beta)/r_\oplus]^{1/2} = [\mu(1 - \beta)]^{1/2}. \tag{41}$$

Notice that $v < v_\oplus$ for $\beta > 0$. The geocentric speed v_∞ of the dust particle as it enters the gravitational field of the earth is

$$v_\infty = | v_\oplus - v | = \mu^{1/2} - \mu_{\text{eff}}{}^{1/2} = \mu^{1/2}[1 - (1 - \beta)^{1/2}]. \tag{42}$$

Thus, $v_\infty = 0$ only for $\beta = 0$. The value $\beta \approx 0$ applies for dust particles much larger than those for which the enhanced flux occurs. Reference to Table I shows that the direct measurements ($m \lesssim 10^{-7}$ gm) apply for dust particles with $\beta \gtrsim 0.01$. The speed v of the dust particle is reduced below the earth's speed $v_\oplus$ by the factor $(1 - \beta)^{1/2}$. The concept that the speed of a small dust particle having an earth-like orbit is lower than the speed $v_\oplus$ of the earth is nothing new—discussions of this effect appear in the writings of Poynting soon after the experimental verification that light falling on a surface exerts a small but finite force on the surface (radiation pressure). The reduction of the speed v of the dust particle by the factor $(1 - \beta)^{1/2}$ yields values of v_∞ which are too large to produce the observed enhancement of flux, as treated through the use of Liouville's theorem. The assumption that the small dust particles measured by spacecraft have earth-like orbits ($a \approx 1$ A.U., $e \approx 0$, $i \approx 0$) may therefore be dropped from all future discussions of the enhanced flux of small dust particles near the earth.

What is needed now is a general expression for the geocentric speed v_∞ of an interplanetary dust particle as it enters the earth's gravitational field, with v_∞ expressed as a function of the heliocentric orbital elements a, e, and i and the parameter β. Such an expression is given simply by the cosine law for adding two vectors $\mathbf{v}$ and $\mathbf{v}_\oplus$ to obtain the magnitude of the vector sum $\mathbf{v} + \mathbf{v}_\oplus$ as

$$v_\infty^2 = v_\oplus^2 + v^2 + 2v_\oplus v \cos \epsilon, \tag{43}$$

where ϵ is the true elongation (the angle between the direction of earth's motion and the direction from which the dust particle entered the gravitational field of the earth). The true elongation ϵ can be expressed in terms of the inclination i of the heliocentric orbit of the dust particle and the angle ψ between the position vector $\mathbf{r}$ and the velocity vector $\mathbf{v}$ of the dust particle to yield

$$v_\infty^2 = v_\oplus^2 + v^2 - 2v_\oplus v \sin \psi \cos i. \tag{44}$$

Now,

$$|\mathbf{r} \times \mathbf{v}| = rv \sin \psi$$

and intercept (entry of the dust particle into the earth's gravitational field) occurs when $r \approx r_\oplus = 1$ A.U. so

$$\sin \psi = |\mathbf{r} \times \mathbf{v}|/r_\oplus v = |\mathbf{r} \times \mathbf{v}|/v, \tag{45}$$

where v is given by

$$v^2 = \mu_{\text{eff}}[(2/r_\oplus) - a^{-1}] = \mu_{\text{eff}}(2 - a^{-1}). \tag{46}$$

But,

$$|\mathbf{r} \times \mathbf{v}| = h$$

is the angular momentum per unit mass of the dust particle at intercept and measured with respect to the sun. Now,

$$h = [\mu_{\text{eff}} a(1 - e^2)]^{1/2} \tag{47}$$

so

$$\sin \psi = [\mu_{\text{eff}} a(1 - e^2)]^{1/2}/v. \tag{48}$$

Substitution of the expressions for $\sin \psi$ and v into the general expression for v_∞ yields

$$v_\infty = v_\oplus\{1 - 2(1 - \beta)^{1/2}[a(1 - e^2)]^{1/2} \cos i + (1 - \beta)(2 - a^{-1})\}^{1/2} \tag{49}$$

in which all the parameters appearing under the radical are now dimensionless and relate to the heliocentric orbit of a dust particle. The use of a high-speed computer permits rapid evaluation of v_∞ as a function of the orbital elements a, e, and i and the parameter β. There are special cases for which the expression for v_∞ reduces to a readily solvable form, and these cases can be treated briefly here.

First, let us examine the expression for v_∞ to see whether there may be cases for which $v_\infty = 0$. The vector v_∞ can be resolved into rectangular components; one component may be chosen to be normal to the plane of the ecliptic, one chosen to be parallel to $\mathbf{r}$, and the third chosen to be parallel to $\mathbf{v}_\oplus$. Any non-zero inclination i will produce a non-zero value for the component of $\mathbf{v}_\infty$ that is normal to the ecliptic plane. The component of $\mathbf{v}_\infty$ that is parallel to $\mathbf{r}$ vanishes only if intercept occurs while the dust particle is making its perihelion passage or (for elliptic orbits) its aphelion passage.

Consider first the case for intercept at aphelion for a dust particle having an orbit of zero inclination. Change variables, replacing a by the aphelion distance q' where $q' = a(1 + e)$ and set $q' = 1$ A.U. to obtain

$$v_\infty = v_\oplus[1 - 2(1 - \beta)^{1/2}(1 - e)^{1/2} + (1 - \beta)(1 - e)]^{1/2} \tag{50}$$

which has the solution

$$v_\infty = v_\oplus[1 - (1 - \beta)^{1/2}(1 - e)^{1/2}]. \tag{51}$$

In order for $v_\infty = 0$, the condition

$$1 = (1 - \beta)^{1/2}(1 - e)^{1/2} \tag{52}$$

must be met. But $e < 1$ requires $\beta < 0$ (which is impossible), so intercept at aphelion cannot yield $v_\infty = 0$.

Consider next the case of intercept at perihelion for a dust particle having an orbit of zero inclination. Change variables, replacing a by the perihelion distance q where $q = a(1 - e)$ and set $q = 1$ A.U. to obtain

$$v_\infty = v_\oplus[1 - 2(1 - \beta)^{1/2}(1 + e)^{1/2} + (1 - \beta)(1 + e)]^{1/2} \quad (53)$$

which has the solution

$$v_\infty = v_\oplus[1 - (1 - \beta)^{1/2}(1 + e)^{1/2}]. \quad (54)$$

In order for $v_\infty = 0$, the condition

$$1 = (1 - \beta)^{1/2}(1 + e)^{1/2} \quad (55)$$

must be met. Solve for β (and denote the value of β for which $v_\infty = 0$ by β_0) to obtain

$$\beta_0 = 1 - (1 + e)^{-1}.$$

The ranges are

$$0 < \beta_0 < 0.5, \quad 0 < e < 1; \quad \text{elliptic orbits}$$
$$\beta_0 = 0.5, \quad e = 1; \quad \text{parabolic orbits}$$
$$0.5 < \beta_0 < 1, \quad e > 1; \quad \text{hyperbolic orbits.}$$

Recall that $\beta \geq 1$ means that the dust particle will be blown out of the solar system, since β is independent of the heliocentric distance. An example showing the dependence of v_∞ on the parameter β for various inclinations of the orbit is given in Figure 4. The left-hand ordinate gives v_∞ in units of $v_\oplus$, while the right-hand ordinate gives v_∞ in units of v_e, the speed of escape from the earth. The orbit chosen as an example has $q = 1$, so a singular solution occurs for $i = 0°$ when $\beta = 1 - \dfrac{1}{1 + e}$.

The foregoing analysis shows that $v_\infty = 0$ whenever $q = 1$, $e > 0$, $i = 0°$, and β is given by

$$\beta = 1 - (1 + e)^{-1}.$$

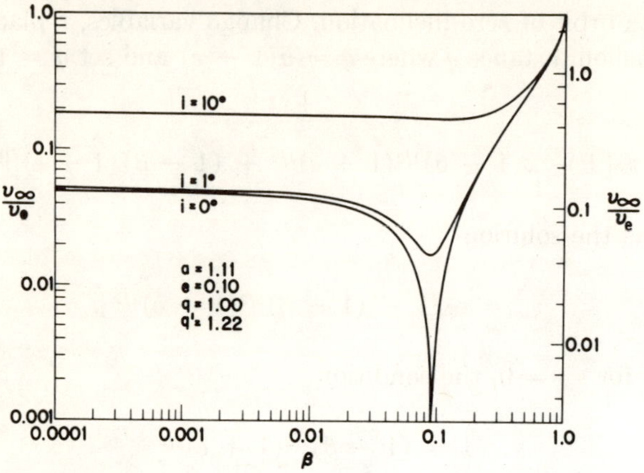

FIG. 4. The dependence on particle size (or β) of the speed at encounter for a planet moving in a circular orbit and dust particles moving in elliptic heliocentric orbits.

This result means that an entire class of orbits of small interplanetary dust particles is subject to intercept with the earth such that the speeds of the dust particles as they enter the gravitational field of the earth are nearly zero. The value $v_\infty = 0$ cannot be used in the expressions for the density enhancement and flux enhancement derived through application of Liouville's theorem. Most of this trouble with boundary conditions results from the use of two-body equations of motion in this simplified analysis. The problem should be treated as a three-body problem with the effects of radiation pressure included. However, the value $v_\infty = 0$ can be used to advantage in locating those values of a, e, i, and β, which will yield very small, non-zero values for v_∞ (see McCracken 51). These orbits can be handled then, through the use of Liouville's theorem, in an approximate manner.

It is not yet known whether the foregoing mechanism for the enhancement of flux will produce the observed enhancement of flux in the vicinity of the earth. The feasibility of the mechanism depends entirely on the distribution of orbits of the small interplanetary dust particles. Evaluation of the importance of the mechanism involves integrating over this distribution of orbits. This integration presents problems, for the greatest contribution to the enhancement of the flux in the vicinity of the earth comes from those orbits which lie in the neighborhood of the singular solutions. Several features of the mechanism make it look much more reasonable than some of the previously suggested mechanisms in which radiation pressure was neglected.

Only one restriction on the eccentricity of the orbits is involved, and this restriction is the very desirable one, $e > 0$. The presence of the $\cos i$ term in the general expression for v_∞ means that the component of v_∞ normal to the ecliptic is non-zero for $i \neq 0$. The presence of the factor $(1 - \beta)^{1/2}$ in the $\cos i$ term reduces the effect of non-zero inclinations on v_∞ so that a larger range of inclinations can yield values of v_∞ smaller than some chosen value when β is non-negligible than for the case $\beta = 0$.

The direct measurements apply for dust particles having masses smaller than 10^{-7} gm, for which $\beta \gtrsim 0.01$, with some of the direct measurements extending to such small dust particles that $\beta \sim 1$ or even $\beta > 1$ (possibly). The case $\beta = 0$ simply does not apply to the small dust particles which produce the zodiacal light and which are being studied through the use of spacecraft. The singular solutions of the equation for v_∞ have occurred only because the effects of radiation pressure were included in the conditions of encounter between the earth and the small particles.

SUMMARY

A knowledge of the distribution of orbits and a better knowledge of the size distribution of the small dust particles which produce the zodiacal light would permit one to check immediately on the degree of enhancement of flux that would be produced by small interplanetary dust particles streaming through the vicinity of the earth. For lack of information about the distribution of orbits (both in interplanetary space and in the vicinity of the earth), one can consider at least three other approaches which might be followed in checking whether interplanetary dust particles streaming past can produce the observed enhancement of flux near the earth.

One approach involves taking the distributions of orbits for large dust particles obtained through photographic and radar studies of meteors, using these distributions as approximations for the distribution of orbits of the small dust particles, and computing the enhancement of flux near the earth. This approach has little chance for success, since the distributions of orbits for the widely different sizes of dust particles most probably are considerably different.

A second, more comprehensive (and admittedly more difficult approach) is as follows: Begin with an assumed distribution of orbits (which might be borrowed from the larger dust particles or even from the comets). Assume a size distribution and a composition for the dust particles (assumed to be spherical) and compute, using the Mie theory, the intensity and polarization of the zodiacal light and solar F-corona which would be produced by the model distributions. Compare the computed model with observational data on the intensity and polarization of the zodiacal light; better observational data are now being obtained by Weinberg in Hawaii. Compare the model also with the direct measurements in interplanetary space, which are presently directed toward the objectives of obtaining measures of the flux of dust particles of a given mass and of obtaining some information about the mass distribution of the small dust particles. Adjust the assumed distributions until reasonable agreement is obtained among the computed results, the observational data on the zodiacal light, and the direct measurements.

The values of β for the dust particles may be computed simultaneously with the light scattering properties of the dust particles, since the Mie theory is required in both calculations. The distributions of orbits, the distribution of particle size, and the values of β can be used then in computing the enhancement of flux near the earth and, in fact, near any other planet. The latitude in the assumptions permitted by the presently available data are such that this approach is not *yet* likely to lead to a satisfactory solution. Repeated iterations among these boundary conditions should lead, with the help of better observational data as it becomes available, to an improved solution.

The third approach involves obtaining direct measurements which will give definitive information about the nature of the enhancement of flux. Dust particle sensors for measuring the speed, direction of motion, momentum, and kinetic energy of an impacting dust particle are under development. Such sensors could give additional information about the mass distribution shown in Figure 1 and about the variation of flux with geocentric distance for dust particles having masses $m > 10^{-13}$ gm. The measurements of the speed and direction of motion of a dust particle would permit one to compute the geocentric orbit of the dust particle if the position, velocity, and orientation of the spacecraft were known. If such direct measurements should show the majority of the small dust particles near the earth to be in closed geocentric orbits, then serious consideration could be given to evaluating the possible energy loss mechanisms and, perhaps, to the possibility that the moon is a significant source for small dust particles in the vicinity of the earth.

REFERENCES

1. H. C. van de Hulst, Light Scattering by Small Particles, John Wiley & Sons, Inc., New York, 1957.
2. S. P. Wyatt, Jr., and F. L. Whipple, The Poynting–Robertson Effect on Meteor Orbits, Astrophys. J. **111,** 134 (1950).
3. F. L. Whipple, On Meteoroids and Penetration, J. Geophys. Res. **68,** 4929 (1963).
4. D. W. R. McKinley, Meteor Science and Engineering, McGraw-Hill Book Company, Inc., New York, 1961.
5. E. J. Öpik, Physics of Meteor Flight in the Atmosphere, Interscience Publishers Ltd., London, 1958.
6. A. C. B. Lovell, Meteor Astronomy, Oxford University Press, London, 1954.
7. F. G. Watson, Between the Planets, Blakiston Company, Philadelphia, 1941; Harvard University Press, Cambridge, 1956.
8. E. J. Opik, The Stray Bodies in the Solar System, Part II. The Cometary Origin of Meteorites, Advances in Astronomy and Astrophysics, Z. Kopal, ed., pp. 301–336, Academic Press, New York, 1966.
9. R. B. Baldwin, Lunar Crater Counts, Astronom. J. **69,** 377 (1964).
10. R. K. Soberman, and C. L. Hemenway, Meteoric Dust in the Upper Atmosphere, J. Geophys. Res., **70** 4943 (1965).
11. N. H. Farlow, M. B. Blanchard, and G. V. Ferry, Sampling With A LUSTER Sounding Rocket During a Leonid Meteor Shower, J. Geophys. Res., **71** 5689 (1966).
12. C. L. Hemenway, R. K. Soberman, and G. Witt, Investigation of Noctilucent Cloud Particles, TELLUS, **16** 84 (1964).
13. O. E. Berg, W. M. Alexander, and L. Secretan, Physical Parameters of Cosmic Dust Obtained from Rocket Collections, to be published in the Proceedings of the Symposium on Meteor Orbits and Dust, Smithsonian Astrophysical Observatory, Cambridge, Mass. 9–13 August 1965.
14. R. A. Schmidt, A Survey of Data on Microscopic Extraterrestrial Particles, University of Wisconsin Research Report Series 63–2, Jan. 1963.
15. W. A. Cassidy, ed., Cosmic Dust, New York Acad. Sci. Ann. **119,** 1 (1964).
16. H. Siedentopf, A. Behr, and H. Elsässer, Photoelectric Observations of the Zodiacal Light, Nature **171,** 1066 (1953).
17. D. W. Beggs, D. E. Blackwell, D. W. Dewhirst, and R. D. Wolstencroft, Further Observations of the Zodiacal Light from a High Altitude Station

and Investigation of the Interplanetary Plasma: I. Introductory Survey and Photoelectric Measurements of Brightness, Month. Not. Roy. Astron. Soc. **127,** 319 (1964); II. Spectrophotometric Observations and the Electron Density in Interplanetary Space, Month. Not. Roy. Astron. Soc. **127,** 329 (1964).
18. J. L. Weinberg, The Zodiacal Light at 5300 Å, Annales d'Astrophysique **27,** 718 (1964).
19. H. Elsässer, Die räumliche Verteilung der Zodiakallichtmaterie, Zeitschrift für Astrophysik **33,** 274 (1954).
20. H. Elsässer, The Zodiacal Light, Planet. Space Sci. **11,** 1015 (1963).
21. M. F. Ingham, Observations of the Zodiacal Light from a Very High Altitude Station. IV. The Nature and Distribution of the Interplanetary Dust, Month. Not. Roy. Astron. Soc. **122,** 157 (1961).
22. R. H. Giese, Light Scattering by Small Particles and Models of Interplanetary Matter Derived from the Zodiacal Light, Space Sci. Rev. **1,** 589 (1962).
23. R. H. Giese, and H. Siedentopf, Optische Eigenschaften Von Modellen der Interplanetaren Materie, Zeit. für Astrophysik **54,** 200 (1962).
24. J. L. Weinberg, White-Light versus Narrow-Band Observations of the Polarization of the Zodiacal Light, Nature **198,** 842 (1963).
25. W. M. Alexander, C. W. McCracken, L. Secretan, and O. E. Berg, Review of Direct Measurements of Interplanetary Dust from Satellites and Probes, *in* Space Research III, pp. 891–917, North-Holland Publishing Company, Amsterdam, 1963.
26. C. W. McCracken, W. M. Alexander, and M. Dubin, Direct Measurement of Interplanetary Dust Particles in the Vicinity of the Earth, Nature **192,** 441 (1961).
27. C. T. D'Aiutolo, Review of Meteoroid Environment Based on Results from Explorer XVI and Explorer XXIII Satellites, to be published in Proceedings of the Fourth International Space Science Symposium of COSPAR, Warsaw, Poland, June 1963.
28. R. J. Nauman, The Near-Earth Meteoroid Environment, NASA TN D–3717, November 1966.
29. C. Nilsson, Some Doubts About the Earth's Dust Cloud, Science, **153**, 1242, September 1966.
30. J. L. Bohn, W. M. Alexander, and A. Wever, Studies of the Thermal Gradient Effects on Ceramic Transducers Used in Cosmic Dust Experiments: Part I, Technical Report, NASA Res. Grant NsG–84–60, March 1967.
31. J. L. Bohn, W. M. Alexander, and A. Wever, Studies of Microparticle Hypervelocity Impacts onto Cosmic Dust Detectors Using Ceramic Transducer Sensors: Part I, Technical Report, NASA Res. Grant, NsG–84–60, April 1967.
32. R. L. Bjork, K. N. Kreyenhagen, and M. H. Wagner, Analytical Study of Impact Effects as Applied to the Meteoroid Hazard, Final Report, NASA–805, SHOCK HYDRODYNAMICS, INC., Sherman Oaks, Cal., Sept. 1966.
33. W. M. Alexander, C. W. McCracken, and H. E. LaGow, Interplanetary Dust Particles of Micron-Size Probably Associated with the Leonid Meteor Stream, J. Geophys. Res. **66,** 3970 (1961).

34. W. M. Alexander, C. W. McCracken, and J. L. Bohn, Zodiacal Dust: Measurements by Mariner IV, Science **149**, 1240 (1965).
35. G. T. Best, and T. N. L. Patterson, The Capture of Small Absorbing Particles by the Solar Radiation Field, Planet. Space Sci. **9,** 801 (1962).
36. T. R. Kaiser, Meteors and the Abundance of Interplanetary Matter, Space Sci. Rev. **1,** 554 (1962).
37. C. W. McCracken and M. Dubin, Dust Bombardment on the Lunar Surface, *in* the Lunar Surface Layer, Materials and Characteristics, J. W. Salisbury and P. E. Glaser, eds., pp. 179–214, Academic Press, New York, 1964.
38. G. S. Hawkins, and E. K. L. Upton, The Influx Rate of Meteors in the Earth's Atmosphere, Astrophys. J. **128,** 727 (1958).
39. G. S. Hawkins, The Relation between Asteroids, Fireballs and Meteorites, Astronom. J. **64,** 450 (1959).
40. T. R. Kaiser, The Determination of the Incident Flux of Radio-Meteors: II. Sporadic Meteors, Month. Not. Roy. Astron. Soc. **123,** 265 (1961).
41. H. Brown, The Density and Mass Distribution of Meteoritic Bodies in the Neighborhood of the Earth's Orbit, J. Geophys. Res. **65,** 1679 (1960); Addendum: J. Geophys. Res. **66,** 1316 (1961).
42. E. J. Opik, Collision Probabilities with the Planets and the Distribution of Interplanetary Matter, Proc. Roy. Irish Acad. **54,** 165 (1951).
43. E. J. Öpik, Interplanetary Dust and Terrestrial Accretion of Meteoric Matter, Irish Astronom. J. **4,** 84 (1956).
44. S. F. Singer, Measurements of Interplanetary Dust, *in* Scientific Uses of Earth Satellites, J. A. van Allen, ed., pp. 301–316, The University of Michigan Press, Ann Arbor, 1956.
45. D. B. Beard, Interplanetary Dust Distribution, Astrophys. J. **129,** 496 (1959).
46. F. L. Whipple, The Dust Cloud about the Earth, Nature **189,** 127 (1961).
47. S. F. Singer, Interplanetary Dust near the Earth, Nature **192,** 321 (1961).
48. I. I. Shapiro, D. A. Lautman, and G. Colombo, The Earth's Dust Belt: Fact or Fiction/, Parts 1–4, J. Geophys. Res., **71**, 5695 (1966).
49. D. E. Gault, E. D. Heitowit, and H. J. Moore, Some Observations of Hypervelocity Impacts with Porous Media, *in* The Lunar Surface Layer, Materials and Characteristics, J. W. Salisbury and P. E. Glaser, eds., pp. 151–178, Academic Press, New York, 1964.
50. R. D. Shelton, H. E. Stern, and D. P. Hale, Some Aspects of the Distribution of Meteoric Flux about an Attractive Center, *in* Space Research IV, P. Muller, ed., pp. 875–907, North-Holland Publishing Company, Amsterdam, 1964.
51. C. W. McCracken, Conditions of Encounter Between Dust and the Planets, to be publ. in the Proceedings of the Symposium on Meteor Orbits and Dust, Smithsonian Astrophysical Obs., Cambridge, Mass., 9–13, August, 1965.

Chapter 12

COSMIC CHEMISTRY

Bertram Donn

INTRODUCTION

One hundred and twenty-five years ago the importance of systematic chemical investigations of geologic material for the study of the earth was first pointed out. The name *geochemistry* was introduced at that time for this study.

A quarter of a centry ago Rupert Wildt proposed a generalization of that idea to include the behavior of matter under all possible conditions in the universe. To this aspect of cosmogony he applied the term *cosmochemistry*. In this review cosmic chemistry is restricted to the chemical analysis of natural objects or the investigation of chemical phenomena that takes place in natural systems. Thus, the origin of the elements, discussed in another chapter, is not included.

At various times regions develop where temperatures, densities and composition are in the right range for chemical forces to compete in importance with nuclear and gravitational forces. These chemical forces are interactions between atoms or clusters of atoms associated with the building of molecules and molecular aggregates. The appearance of such interactions require the study of a wide range of chemical phenomena, from bimolecular reactions to the behavior of solids.

Very few classes of celestial objects can be studied in the laboratory. These are primarily limited to the earth's crust and to meteorites. The recognition of meteorites as celestial objects dates from about 1800 and the first chemical analysis followed almost simultaneously. In recent years, investigations of the structure, composition and age of meteorites have been rapidly expanding.[1]

Still more recently, increasing effort has been devoted to attempts at collecting and analyzing *cosmic dust*. This term refers to fine particulate matter that is collected from the atmosphere, sea sediments or glaciers. A serious obstacle in such research is the great uncertainty of separating true extra-terrestrial particles from those arising from geological or industrial processes or from contamination by the collecting device.

Comets provide considerable opportunities for research in cosmic chemistry. They also provide the primary means of investigating conditions throughout all of interplanetary space and may also be the best guide to the early history of the solar system. Later in this Chapter (pages 511-518), attention will be given to the nucleus of comets and to radicals and ions in the coma.

The largest system in which chemical phenomena make a significant contribution is that composed of the gas and solid particles in interstellar space. Many of the problems encountered here are similar to those of the comets except that the conditions are even more extreme.[2]

Chemistry also has a role in the study of the atmospheres of the cooler stars.[3] For stars with temperatures of 6000°K, comparable to the sun or cooler, diatomic molecules form in the atmosphere. Recently,[4] from observations by a balloon supported telescope, water has been identified in late spectral type red stars with temperatures of about 3000°K.

It also appears that complex molecule and grain formation in such stellar atmospheres may be an important factor for interstellar matter.[5, 6]

In concluding this outline of the various aspects of cosmic chemistry, it is important to call attention to the role of chemistry in the study of the origin of the solar system. This phase of the subject has been investigated most actively by Urey.[7] The composition and structure of primary particles discussed in Chapter 11 are closely related to it.

Generally, in cosmochemical systems, all that is available are the results of complex processes as, for example, the crust of the earth, and the problem is to understand the sequence of events whose consequences we observe. Sometimes, it is the other way and an indication of the initial conditions exists, as with interstellar matter, and we must deduce the reactions that can occur and the nature of the ultimate product, the interstellar grains. Thus, unlike the situation in the chemical laboratory where conditions are nominally well known, in nearly all natural systems the state has to be deduced from other fields, usually geology or astronomy, and cannot be completely specified. This feature is perhaps the main distinction between *ordinary*, i.e., terrestrial, and *cosmic* chemistry.

In addition, cosmic systems are much more complex. They are composed of all elements with their cosmic or geochemical relative abundances and often they involve several phases. The cosmic chemist must take the problem as he finds it and is not free to restrict himself solely to simple definitive experiments. It is often necessary to begin an analysis with a highly simplified system, but the more detailed problem must be considered before conclusions are given any weight.

A characteristic feature of many problems here is the absence of thermodynamic equilibrium. Several factors contribute to this. In some circumstances, as in the interstellar gas, low densities and a disproportionate abundance of ionizing and dissociating radiation, prevent equilibrium from being attained. Low temperatures of comets and interstellar grains can literally freeze a non-equilibrium composition of molecules into a permanent condition.

A third feature of cosmic chemistry in which it differs from normal laboratory chemistry is the ranges of temperature, pressure and composition. These generally are much more extreme than those which have been studied.

Recent advances in several areas of physico-chemical research are beginning to provide the required data. These experiments in turn have stimulated theoretical studies. It is now possible to undertake a study of cosmochemical problems at low temperatures and pressures on a reasonably firm basis. Some of these are described in the next section.

ASTROCHEMICAL PROBLEMS

Studies of chemical phenomena associated with solid particles in the solar system, comets, interstellar matter and molecules in stars differ in fundamental ways from geochemical problems of the earth's crust and meteorites. Such studies are so intimately associated with the astronomical facets of the respective problems that it is appropriate to denote this phase of cosmic chemistry by the name *astrochemistry*. The remainder of this section considers two problems in which solids form the significant phase.

Primary Interplanetary Particles

The observational features of interplanetary grains and their interpretation have been presented in Chapter 11. We are mainly concerned here with the question of the nature and origin of solid particles which formed in the primordial solar nebula. These will be called *primary particles*.

According to present ideas of the formation of the sun and planets, solid particles had a major role. After the formation of the planets, the density of condensable gases was too low for further condensation. All primary grains date from the pre-planetary period of the solar system. Particular interest therefore is associated with primary particles.

The formation of primary grains by condensation requires that some of the constituents of the nebula had been supersaturated to a sufficient degree. It appears rather certain that during some stages this requirement was satisfied. As Urey has emphasized, a supersaturated solar nebula will condense into innumerable grains in a short time.

Particle formation from a supersaturated vapor occurs in two steps. First, there is the formation of the smallest stable molecular aggregate consisting of about 50 molecules. This is the nucleation process.[8] Thereafter, the aggregate, an embryo crystal if it is a solid, increases in size by the mechanism of crystal growth.[8]

If the nucleus develops in the vapor phase in the absence of foreign particles, it is known as *homogeneous nucleation*. For this case a supersaturation of at least 10 is required for nucleation to occur at a significant rate. When nucleation takes place in the presence of foreign particles or ions a lower supersaturation is required.

Theoretical analyses of nucleation are incomplete because of the difficulty of establishing the properties of embryos containing a small number of particles. However, the form of the expression for the nucleation rate, number of nuclei formed per cm^3 per second, has been well established. This is given by:

$$dn/dt = B \exp\left(-A/T^3 \ln^2 \alpha\right), \tag{1}$$

where T is the absolute temperature and α, the supersaturation ratio, is the ratio of the partial pressure of the vapor species to its value at saturation. The major uncertainty in the nucleation theory is the determination of the factor B.

Below a certain critical supersaturation α_c, nuclei form at a negligible rate, and above α_c the rate increases extremely rapidly. As nucleation rapidly occurs throughout the medium, the supersaturation decreases and nucleation ceases.

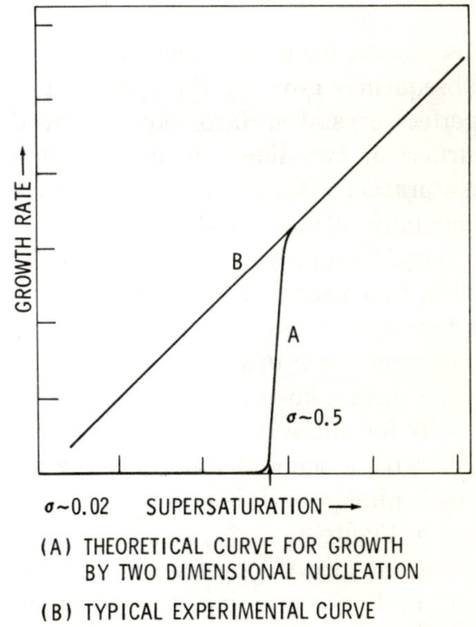

FIG. 1. Crystal growth by two dimensional nucleation: A, theoretical curve; B, typical experimental curve.

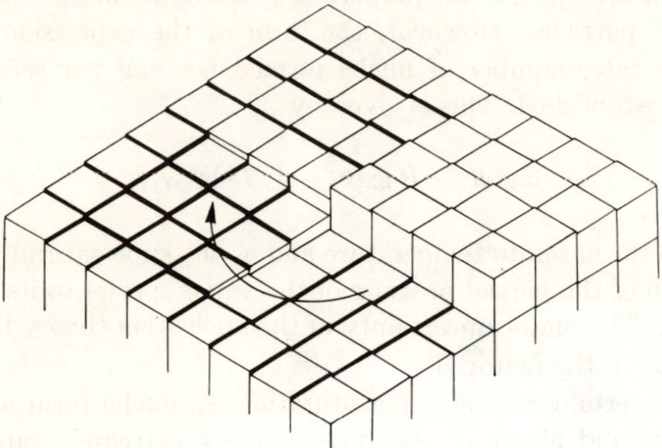

FIG. 2. Intersection of screw dislocation with crystal surface.

The nuclei thus formed have dimensions of the order of 10 Angstroms. These subsequently grow by the appropriate crystal growth mechanism. A perfect crystal surface requires the development of new steps by surface or two-dimensional nucleation. This usually requires a supersaturation ratio of about 1.5 for a sensible growth rate. But experimentally, it was found that crystal growth occurred with α very near unity. Theoretical and experimental rates are shown in Figure 1. In 1949, this discrepancy was explained by F. C. Frank, who pointed out that real crystals do not have perfect surfaces but almost certainly contain *screw dislocations*. These are defects which do not disappear as a surface layer is completed, but rather propagate so as to provide a site for initiation of the next layer without nucleation. Figure 2 illustrates a screw dislocation. The manner in which it leads to continued spiral growth is seen in the electron-micrograph of a crystal surface in Figure 3.

The screw dislocation growth mechanism applies at low supersaturations. In any system containing ions, which are continuously replenished, nucleation would continue until the supersaturation fell to a value near unity. This condition should apply to the primordial solar nebula because of the ionizing effect of both energetic

FIG. 3. Electron micrograph showing spiral growth steps in paraffin, n-$C_{36}H_{74}$ (F. C. Frank, Advances in Physics **1**, 91 (1952), after Dawson and Vand).

charged particles and the much more intense radioactivity then occurring. Consequently, after an initial state of nucleation and growth, further growth of primary particles in the solar system must have been by Frank's dislocation method.

Densities in the solar nebula have been estimated at 10^{-6} to 10^{-12} g/cm^3. Of this, the partial pressure of silicates was about 2×10^{-4}. At a temperature of 100°K, the corresponding silicate pressures would be 10^{-3} to 10^{-9} Torr. Theories of nucleation apply to this pressure range. Experiments on surface nucleation of zinc between 10^{-1} and 10^{-7} Torr showed no significant change in the nucleation process. Another question that must be considered is the effect on particle formation of the complex composition in the solar nebula as compared with the usual case of pure or nearly pure vapors in the laboratory. For binary vapors the form of the nucleation rate equation is unaffected and the numerical values of constants probably not drastically different. If nucleation on ions were the dominant process, the composition should have relatively little effect on nucleation.

A fundamental question for homogeneous nucleation from a vapor of complex composition is the composition of the nucleus. Can embryos of nearly random composition develop or will the vapor condense into aggregates of molecules with similar crystalline properties? The evidence indicates that the latter will tend to be the case particularly when the process occurs at low densities. A rapid nucleation or growth process can trap impurity species but under conditions in space this is not very likely.

In the growth stage, impurities can either inhibit growth or radically change the morphology of the crystal. Generally, only certain crystal faces are poisoned whereas others grow normally leading to a major change in the shape of the crystal. As we propose that the shapes are determined by the dislocation pattern of the nucleus rather than equilibrium crystal properties of the molecules, the effect of poisons would tend to increase the complex, non-equilibrium shapes.

It appears therefore that nucleation and growth processes established under laboratory conditions may be reasonably applied to formation of solids in the solar nebula. The extreme and complex astronomical conditions will modify details of the results but not the major characteristics.

With the understanding that further research will require some modification of our analysis and conclusions, we consider the nature of the solid particles that formed in the solar nebula.[9] The supersaturation required for condensation will modify the condensation sequence from that generally given assuming equilibrium always prevails as Blanders and Katz[10] have emphasized. This effect, which can be very great for homogeneous nucleation will be reduced if condensation on ions or other nuclei occurs.

Nuclei present in the pre-planetary nebula from the initial interstellar cloud are likely to contain screw dislocations. High supersaturations produce nuclei containing many defects which can grow as an array of *whiskers*. Radiation damage offers a possible source of screw dislocations. The presence of such objects allows equilibration of the supersaturated vapor phase with the stable crystal phase and guarantees that the supersaturation rapidly diminishes after nucleation. Only those nuclei containing dislocations can grow and these will tend to develop as whiskers and platelets or various combinations of such forms. The major mass of condensed solids in the nebula must have been the irregular, whisker-type particle.

Several properties of whiskers are important for the further development of the nebula. (1) Whiskers have strength characteristics of perfect crystals. A pure iron whisker has withstood a tensile strain of $1.9 \times 10^6/cm^2$. Ordinary iron has an elastic limit of about $5 \times 10^3/cm^2$. (2) Metal whiskers would generally not break upon collision but plastically deform and tend to intertwine as relative collision velocities were not likely to have been high enough to destroy grains. (3) A whisker has such a high ratio of surface to volume that it would effectively radiate heat generated in a collision and evaporation would be greatly reduced. (4) The large cross section for collision would increase the collision rate compared to equi-axial crystals by a large factor.

These characteristics of primary particles in the form of whiskers are very important for the accumulation of grains into planets.

Particles which grew by the process just described will have the composition of the vapor. The complex silicate minerals such as olivine and enstatite commonly found in meteorites and in particles collected in searches for cosmic dust would have formed by solidification of a melt rather than condensation of a vapor. Such particles are most likely representative of meteorite ablation if spherules or fragmentation if angular and would generally not be primary particles.

FIG. 4. Comet Mrkos, 1957 V, showing straight ion tail and curved dust tail (Mt. Wilson and Palomar Obs.).

A probable source of primary grains are the solid particles in the coma and tail of comets, considered in the next part.

Comets

A fully developed comet consists of three parts as shown in Figures 4 and 5. The characteristic feature of all comets is the head or coma which is an extremely diffuse more or less spherical cloud. It may be mainly gaseous, mainly composed of solid grains or a mixture of both. A tail may or may not be present and if present, may be conspicuous or faint. An *invisible nucleus* is the third component and the one with which we are primarily concerned in this Chapter.[11]

As a comet revolves around the sun, gases are evolved from the nucleus and form the coma and tail. Small solid grains in the nucleus are carried along by the gases. Because of the small mass of a comet this material escapes into interplanetary space and must be continuously replenished for the comet to retain its characteristic appearance. Thus the coma and tail are transient features. Only the nucleus has a permanent existence. Although too small to be seen except possibly for a few distant comets when the coma may be nearly non-existant, the nucleus is the essential part of a comet and the permanent feature which revolves around the sun.

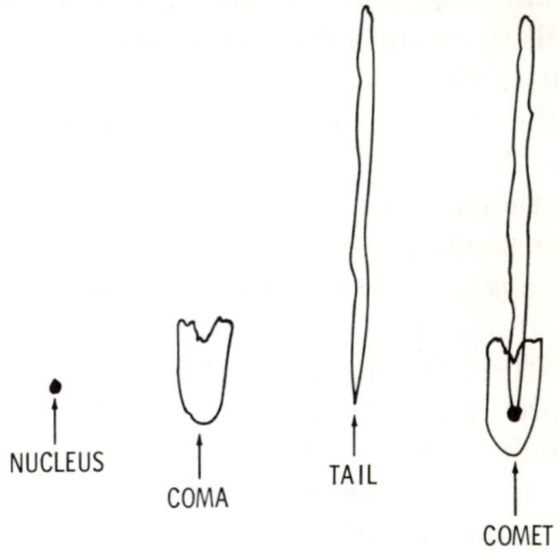

FIG. 5. Structure of a comet (F. W. Watson, Between the Planets, Harvard Univ. Press).

TABLE I

Spectroscopic observations of comets[12]

Coma	Tail
Dust continuum	Dust continuum
CN	CO^+
C_3, NH_2	N_2^+
C_2	CO_2^+
OH, NH, CH	CH^+
Na	OH^+
Fe, Ni, Ca, K	

Since all cometary material emanates from the nucleus, spectroscopic observations of the coma and tail are our source of information about the nucleus. These observations are summarized in Table I. The constituents in the coma are listed in order of appearance as a comet approaches the sun. For the tail, the order is generally that of decreasing intensity in the spectrum.

The continuum shows an absorption spectrum similar to that of the sun and therefore consists of scattered sunlight. Further analysis indicates that the dimensions of the scattering particles must be of the order of the wavelength of light, i.e., about one micron diameter. Usually, as the distance from the sun increases the continuum increases in intensity relative to the molecular emissions.

The intensity distribution within those emission bands which have been examined with sufficient detail shows a close correspondence with the solar continuum and indicates that such emission is produced by resonance fluorescence.

Whipple's icy conglomerate model of the comet nucleus seems to be the most satisfactory for further analysis. As initially proposed, the nucleus was a 1–10 km approximately spherical body, 75 percent of which was frozen gases and the remainder non-volatile meteoric silicate and metallic particles. The gases were chosen as possible parent molecules for the observed radicals and included water, ammonia, methane, carbon dioxide, cyanogen, or hydrogen cyanide. It should be noted that C_2 and C_3 require more complex hydrocarbons or other organic compounds.

The presence of a large proportion of volatile material in the nucleus will maintain a low temperature even at small distances from the sun. Under normal conditions in interplanetary space, comet temperatures result from a balance between thermal solar radiation and the sublimation energy of the cometary *ices*. The temperature is found from the equation:

$$(1 - \alpha)(c/r^2) = H(T)G(T) \tag{2}$$

where α = albedo of nucleus, c = solar constant (energy flux at 1 A.U.), and r = distance of comet from sun in A.U. $H(T)$ is the heat of sublimation available from tables and $G(T)$ is the rate of vaporization given by

$$G(T) = \epsilon(m/2\pi kT)^{1/2}\, P(T) \tag{3}$$

where $P(T)$ is the vapor pressure in $dynes/cm^2$, m the molecular mass, and k is Boltzmann's constant. The sticking coefficient ϵ is the probability that a molecule striking the surface sticks to it. Few precise determinations of ϵ are available but normally it appears to be near unity, although in some cases values as low as 0.001 have been proposed.

For a given distance the product GH follows from Equation (2). This product is calculated for a particular molecule as a function of temperature and the value of T, satisfying Equation (2) determines the temperature of the nucleus. Table II displays the temperatures of a nucleus consisting of a single molecular species at several distances, for $\epsilon = 1$.

TABLE II

Temperatures of icy nucleus

Distance in A.U.	Temperature (°K)			
	Water	Ammonia	Methane	Blackbody
0.1	240	150	60	910
0.5	210	130	55	410
1	200	125	52	290
5	140*	115	48	140
10	110*	110*	45	110

* Temperature from thermal balance.

In Table II, temperatures marked with an asterisk are so low that radiative rather than evaporative cooling determines the equilibrium. The three molecules included in the table cover the range of volatility of molecules that contribute to the equilibrium. Icy cometary nuclei composed of mixtures of molecules therefore have surface temperatures below 200°K, except very close to the sun.

The composition of the nucleus must be such as to lead to the observed radicals of Table I. Several methods of obtaining these radicals may be imagined. Until recently, photodissociation of parent molecules evaporating from the nucleus was accepted. The density distribution of radicals in the coma and occasional rapid changes observed near the nucleus indicate time scales for production of 10^3 sec. This is shorter by one or two as orders of magnitude than photodecomposition of the usual parent molecules appears to yield. Proton impact rather than ultraviolet light has been proposed but according to recent measurements of the solar proton flux, this process also appears inadequate. However, a calculation of photochemically produced radical densities shows quantitial agreement with densities deduced from observations.

Early investigations[2] of trapped radicals at low temperatures led to the suggestion that the cometary radicals may exist in the nucleus. Later measurements of radical concentration, particularly at expected cometary temperatures 50°–150°K, yielded values below 0.1 percent.

TABLE III

Thermodynamically stable compounds in solar nebula, $P = 10^{-3}$ atm.

Element	Temperature	
	298°K	1200°K
H	H_2, CH_4, NH_3, H_2O	H_2, H_2O, H_2S
He	He	He
C	CH_4	CO
N	NH_3, NH_4^+	N_2
O	H_2O	H_2O
Ne	Ne	Ne
Si	SiO_2	SiO_2, SiC
S	FeS	H_2S
Fe	FeS, Fe_3O_4 Fe_2SiO_4	Fe, Fe^3C

This is too low for radical storage to be the primary source but in some cases this process may be important.

The mechanism of formation of radicals consequently is still unsolved. Recent attempts have sought to explain the ionization in cometary atmospheres by means of shock wave interaction with interplanetary plasma or some other type of plasma interaction. These are not well-understood physical processes and outside the scope of this chapter and hence will not be discussed further here. If it turns out that this mechanism can indeed account for the ionization, it can very likely also explain radical formation.

The formation of icy cometary nuclei postulated by Whipple can be fitted into current ideas of the origin of the solar system as a whole. On this assumption, the probable composition can be estimated. In "The Planets" Urey[13] considered the equilibrium of the solar nebula under several conditions and cites earlier work. Since then more extensive investigations have been carried out for arrays of about one hundred molecular species. The most useful for the present purpose of those of Tsuji[5] and of Lord.[14] Table IV summarizes these results and is a modified form of Urey's results.

Several factors can be expected to have seriously modified the composition of the solar nebula from the equilibrium values. Radioactive nuclei were much more abundant at the time of formation of the solar system than they are at present. For nuclei with half life, τ, measured in 10^9 years and present abundance N_0, the abundance five billion years ago was $N_0e^{5/\tau}$. The half life of U^{238} is 7×10^8 years and therefore $N_5 = 1100N_0$. In addition, there is evidence for short lived radioactive elements, $\tau \sim 10^5$ years that were present in the early solar system. Of more significance for producing the non-equilibrium effect could have been the energetic corpuscular radiation which has been proposed as an important feature of the primordial nebula.

A mechanism for the synthesis of the light elements during the early stages of the solar system was formulated by Fowler, Greenstein and Hoyle.[15] Energetic protons from the sun by interaction with heavier nuclei produced the light elements lithium, berryllium and boron and also neutrons which produced heavy radioactive isotopes. In this theory the nuclear reactions occurred in aggregates with dimensions of the order of meters.

Qualitative considerations will indicate that significant chemical effects should have occurred. During this period the energy flux of energetic protons has been estimated at 10^{45} ergs. If all the energy were absorbed by the nebula, it would correspond to 10^{57} eV.

Radiation chemists measure the effects of radiation in terms of a quantity known as the G value for a reaction. The G value is the number of molecules destroyed or formed per 100 eV of radiation absorbed. Although G values show a considerable spread and depend upon the medium and its detailed composition, a reasonable value is $G = 1$. This leads to a conversion of the thermodynamically stable compounds to the extent of 10^{55} molecules. With a mean molecular weight of 20, this is equivalent to 3.5×10^{32} gm or about two solar masses. If only a fraction of the radiation were absorbed, a substantial proportional of the material in the solar nebula would have been transformed.

Some ideas of the ultimate composition of comets may be obtained from experiments on chemical composition of irradiated condensed gases. Closely related experiments are the condensation of dissociated gases. In the latter case, a number of radicals are obtained in addition to more complex and more reactive molecules. Warm up of the condensed films yielded an additional array of molecules. It is not clear in most instances whether these existed as such in the condensed matrix or were recombination products produced as the solid warmed and vaporized.

The most famous of these low temperature compounds is Rice's *blue stuff*.[16] Dissociated HN_3 yields a very unstable blue product which by careful warm up to 148°K becomes ammonium azide, NH_4N_3. Frequently, the experiment suddenly ended with a violent explosion before the blue material was converted, indicating it is a very unstable, energy rich compound. As a measure of the complexity of these problems, it may be noted that the nature of the *blue material* first obtained in 1951 is still unsettled. It has been proposed by Papazian[17] that nitrogen-chained compounds of the form:

$$H{-}N{=}N{-}N{=}N{-}H,$$

and

$$H{-}N{=}N$$

are stabilized at low temperatures. Normally, these occur only as organic derivatives rather than isolated species.

Condensation of discharged methane-nitrogen mixtures has yielded the following molecules[18]: acetylene, C_2H_2; ethylene C_2H_4; ethane C_2H_6; butadiene C_4H_6; propylene C_3H_6; propane C_3H_8; and butene C_4H_8 in addition to the NH radical.

Dissociated water vapor, when condensed in a cold trap, produces a high percentage of hydrogen peroxide, H_2O_2, plus the HO_2 radical. The formation of the super peroxide H_2O_4 has been claimed but recent work does not support such a molecule.

The production of such unstable species as HO_2, HCO, HONO, NH, NH_2 and almost certainly similar yet unidentified compounds by photolysis of simple mixtures in the laboratory again suggests the variety of molecules to be expected in the primordial condensed gases.

Electron and protron irradiation of water, ammonia, methane mixtures have yielded a variety of complex organic compounds as urea and various amino acids.

Much more work devoted to the type of chemical processes just described but with the conditions appropriate for the primordial solar nebula is required. At this stage we can propose as a typical composition for the nucleus of a comet that which is shown in Table IV.

The extent to which hydrogen is trapped would depend upon the temperature of accumulation. If it were as low as 10°K at the distance of Jupiter, as proposed in some investigations,[19] hydrogen would be initially present. However, the central regions of short period comets, especially Comet Encke, should have warmed up sufficiently to have lost all hydrogen. CO is probably the dominant constituent in some comets, as comets Morehouse, 1908 III and Humason, 1961e and should be a major constituent in comets with prominent ion tails. A high CO concentration is consistent with calculations of molecular equilibrium in a nebula of solar composition at temperatures above 500°K. At temperatures about 1000°K, even with a large excess of hydrogen benzene may become important if the CO ratio is near unity. The composition may be roughly estimated as 70 percent volatile, 30 percent non-volatile and less than 1 percent radicals. Molecular species in parentheses possibly occur and may even be a significant fraction in some cases. Although it is reasonably certain that the other types of molecules occur we cannot say what the relative proportion of each specie is. Spectroscopic observations

TABLE IV

Suggested Composition of Nucleus

Type of Material	Examples
Volatile Inorganic	H_2, N_2, Ne, Ar, H_2O, NH_3, CO, H_2O_2, N_2H_4, H_2S
Volatile Organic	CH_4, C_2H_2, C_2H_4, HCN, CH_3HO, (C_6H_6, higher aromatics and complex organic molecules)
Non-volatile, Inorganic	Silicates, metallic oxides (metals, carbon grains)
Radicals	OH, NH, C_2, C_3, CH_2, HO_2

indicate a considerable variation of composition among comets and no single model can represent all comets.

A nucleus of such composition may have a considerable amount of chemical energy stored within it. If the material evaporates without reacting there is no contribution to the formation of the coma. However, rapid heating by any mechanism, such as intense proton bombardment, may cause sufficiently high temperatures for reactions to occur. A reaction such as

$$2\,H_2O_2 + CH_3OH \rightarrow CO + 4\,H_2O \tag{4}$$

produces 111 kcal/mole or 4.8 eV/molecule. Other possible reactions have comparable heats of reaction. This corresponds to 1700 cal/gm of reacting material.

If one fourth of the reactive material does recombine, it corresponds to 10 percent of the nuclear mass. The chemical energy thus adds about 200 cal/gm, which is about equivalent to heats of vaporization of the volatile material. Any heating which leads to reaction before vaporization becomes an important energy source. An estimate of thc effectiveness of solar proton irradiation indicated that normal solar activity has a negligible effect. Enhanced particle ejection associated with solar flares may be important and could be the explanation of outbursts of Comet Schwassman–Wachmann I at 5 A.U. and similar cometary phenomena.

TABLE V

Correlation table for cosmic dust

Source material	Sample material			
	Primary particles	Cometary grains	Zodical cloud	Rocket and Satellite measure-ments
Interstellar grains	×	×	?	?
Primary particles	—	×	×	?
Cometary grains		—	×	×
Asteroidal frag.				
Asteroidal dust			×	×
Meteor residue				
Meteorite ablation				
Lunar ejecta				?

Source material	Sample material			
	High alt. dust layer, noctilucent clouds	Atmosphere and ground collections; sea, snow sediments	Meteors	Meteorites
Interstellar grains	?	?	?	
Primary particles	?	?	?	*
Cometary grains	×	×	×	
Asteroidal frag.				×
Asteroidal dust	×	×	×	
Meteor residue	?	×	—	
Meteorite ablation		×		
Lunar ejecta	?	?		

A cross in columns 2-10 indicates the probable occurrence of the source in that sample, a question mark means its presence is uncertain or unlikely and a blank space indicates its absence.

* The fine grained matrix of carbonaceous chondrites may include only partially modified primary grains. Wood and Suess have proposed that the chondrites are primary material also. The history of carbonaceous chondrites is highly controversial.

(B. Donn, Annals, N.Y. Academy of Sciences **119**, 5, 1964.)

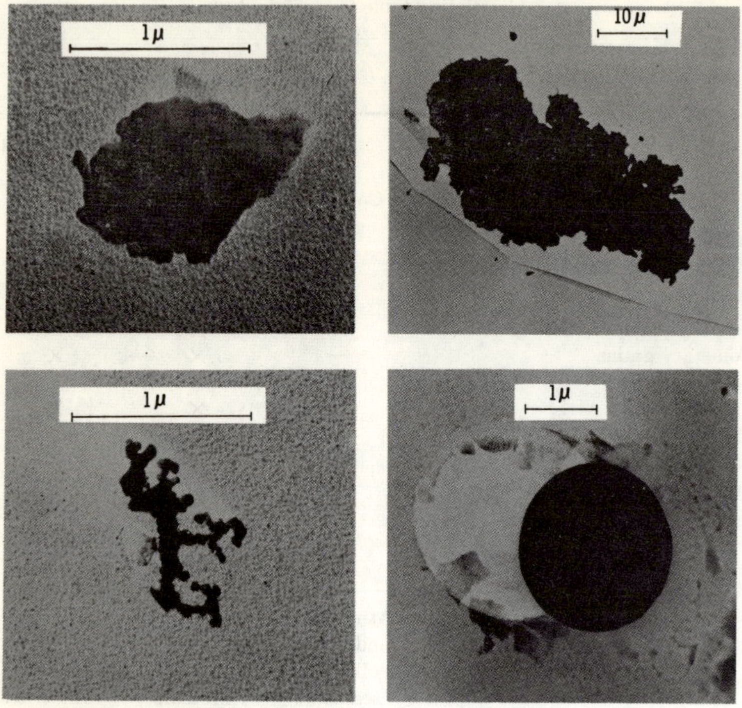

FIG. 6. Types of particles collected by "Venus Fly-Trap" rocket [C. L. Hemenway and R. K. Soberman, Astronom. J. **67**, 256 (1962)].

Space Research

Interplanetary Particles

The collection and analysis of interplanetary particles by the various sampling techniques would be a major step in demonstrating the characteristics of extra-terrestrial solid particles.[20] Table V shows the relationships between collected samples and sources contributing to various samples. It is the mixture of several sources in a given sample that confuses the interpretation. High altitude collections contain fewer sources and above about 125 km, i.e., rocket altitudes, all particles found are forms of interplanetary dust except for contamination.

Analyses of such material should indicate the source of the particles from their morphology and composition according to the analysis presented earlier. The comparison of particle characteristics with the predictions of that analysis and Table V will indicate the validity of the theoretical conclusions about the nature of interplanetary dust.

Collections of sample particles have been made by the Venus-Fly-Trap rocket[21] at altitudes between 88 and 168 km. Three types of particles appeared to be of extra-terrestrial origin and are shown in Figure 6. They have been classified as *spherical, irregular* and *fluffy*. Dimensions of particles ranged from about 0.1 to over 10 μ. Analysis of the particles is being carried out using electron diffraction, electron microprobe analysis and neutron activation.

Only two out of 254 particles gave diffraction patterns. Recondensed material evaporated by the electron beam did produce a diffraction pattern which could have been austenite, taenite or copper. One large irregular particle had a pattern of iron nickel sulfide $(FeNi)_3S_4$.

Electron microprobe analysis indicated the presence of aluminum, silicon, iron, calcium and magnesium in varying proportions among several particles analyzed. Evidence for copper was found by the probe and neutron activation analysis but this may have been contamination.

The results of this rocket collection demonstrate the value of such research for investigating extra-terrestrial particles. Although only one flight has been made, the particles collected appear to be in a general agreement with what has been predicted for the types of particulate matter expected to occur in space. Lack of crystalline structure as indicated by the absence of diffraction pattern requires that the particles were condensed at such low temperatures that an amorphous state was frozen in or they were rapidly quenched from a liquid. In the latter case they would be spheroidal.

Generally, similar results are indicated by collections at lower altitudes or on the ground. However, the large number of particles in such a collection that did not originate outside the atmosphere make the interpretation of results very difficult and uncertain.

Cometary Chemical Release Experiments

Several types of cometary research are suggested by the present

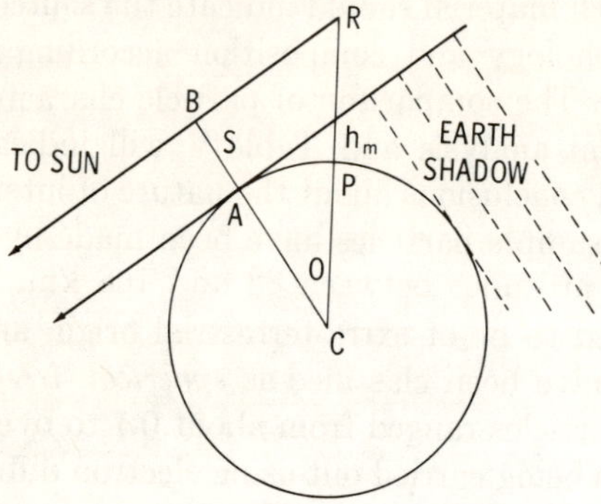

Fig. 7. Geometry of release experiment.

ability to carry chemicals and instruments into near or distant space. Atmospheric investigations have been carried on for several years, through the release of various chemicals at altitudes from 85 to 400 km. Modifications of these experiments to release proposed cometary parent molecules would enable the dissociation and ionization compounds in a natural environment to be studied. Preliminary experiments have been carried out by the European Space Research Organization (ESRO). In the first experiments ammonia was released at 125 km and the cloud observed photometrically and spectroscopically. A tentative identification of NH_2 has been made.

An important factor to consider in chemical releases is the reaction with atomic oxygen because oxygen at 100 km at night is of the order of 10 percent dissociated.

At altitudes in the 100–200 km region, reactions of atomic oxygen with the original molecule or radicals from photodissociation would be quite rapid, the oxygen collision frequency being of the order of 10 per second. At 400 km it has fallen to about 0.01/sec. The effect of collisions is important in the outer zone of the released cloud where photochemical effects also occur.

A second effect is illustrated in Figure 7 which portrays the general features of a release experiment.

Release experiments occur at twilight (morning or evening) when the observer P is in shadow but the point of release R is in sunlight. PR is the release altitude, h and $AB = S$ is the height at which solar radiation illuminating the cloud passes through the atmosphere. The

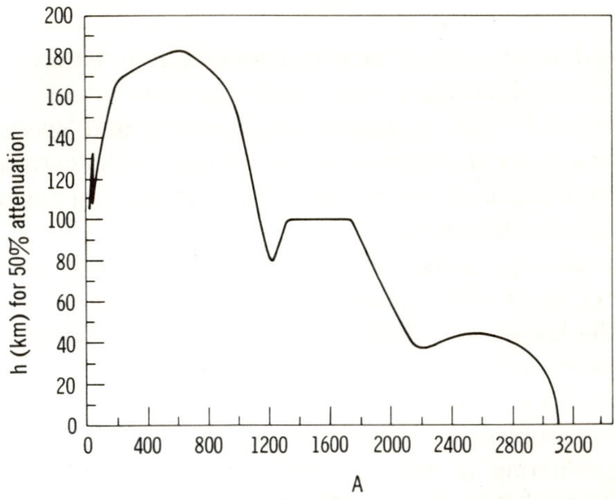

FIG. 8. Attenuation of solar radiation in the atmosphere (L. Goldberg, The Earth as a Planet, G. P. Kuiper, Ed., Univ. of Chicago Press).

minimum release height h_m for a solar depression below the ground horizon, θ is given by

$$\sec \theta = 1 + (h_m/R_0), \tag{5}$$

where R_0 is the radius of the earth. The time of release relative to sunset determines θ. The height of the illuminating radiation through the atmosphere is

$$S = h \cos \theta - R_0(1 - \cos \theta) \tag{6}$$

The shortest wavelength transmitted through the atmosphere to irradiate the released cloud depends upon S in the manner shown in Figure 8. The curve for the transmission of the atmosphere shows that a release at 150 km a half hour after sunset is initially irradiated by wavelengths larger than about 1700 Å whereas at 400 km there is no lower cut-off. With time, θ increases, decreasing S. This leads to a cut-off wavelength increasing with time. The choice of h and θ also determines the darkness of the sky background and the duration of visibility of the cloud, i.e., the length of time the cloud is in sunlight.

Artificial Comet

With the development of powerful rockets it is now possible to put into a terrestrial orbit a small scale version of an icy comet nucleus. In principle the structural, photometric and spectroscopic characteristics of actual comets may be compared with corresponding features for the artificial nucleus as a test of theories of the origin of various cometary phenomena.

The essential requirement for an artificial comet experiment is the production of observable radicals or ions. Preferably, these should be among the known cometary molecules listed previously. A useful study of the icy nucleus model appears possible if sufficient concentration of visible species can be obtained for a few days. One of the major current problems of cometary research is the source and mechanism of production of radicals and ions. The calculated rate of photodissociation for nearly all commonly proposed parent molecules is too low. We do not consider this problem here but will investigate the behavior of an artificial comet as if a means of producing visible species were available.

The rate of disintegration of the artificial nucleus dm/dt is given by:

$$dm/dt = 4\pi R^2 G, \tag{7}$$

where G is the rate of vaporization in gm/cm² and R is the radius. From this the time for the nucleus to decay to a fraction R/R_0 of the original size is

$$t_r = (\rho/G)[1 - (R/R_0)], \tag{8}$$

ρ is the density of the nucleus. If the radius changes sufficiently slowly the luminosity will be proportional to the surface area. Upon expressing the relative decrease in magnitudes we obtain

$$\Delta m = -5 \log (R/R_0). \tag{9}$$

For material having the thermal properties of ammonia, the disintegration rate t is given by Table VI, calculated according to the method described on page 485.

A possible orbit for a synthetic comet is one with a 24 hour period for which the altitude is 36,600 km. If observations are limited to 0°.5 radius, the linear dimension of the coma is 320 km.

TABLE VI

Disintegration of nucleus

Δm	R/R_0	t (days)			
		R_0 50	50	150	150
		ρ 0.3	1	0.3	1
0.5	0.80	0.95	3.2	2.8	9.5
1.0	0.63	1.7	5.8	5.2	17.4
2.5	0.32	3.2	10.7	9.6	32
5.0	0.10	4.2	14.2	12.7	42.5

The luminosity of an artificial comet may be obtained by scaling the luminosity of the equivalent angular region of an actual comet. At 1 A.U. coma would subtend a half angle of 0.44 seconds of arc.

The artificial comet luminosity is L_A and the equivalent comet luminosity is L_C, then

$$L_A/L_C = (R_a/R_c)^2 (\Delta/h)^2, \tag{10}$$

where R is the diameter in meters, h the height above the surface of the earth, and Δ is the distance of the comet from the earth. Estimates of L_C for several comets and the use of equation (10) lead to a predicted value of L_A of about magnitude 13. Only a small fraction of cometary molecules contribute in the visible region whereas the artificial nucleus will be designed for optimum luminosity. It appears that the luminosity of an artificial comet in a 24 hours orbit at 36,600 km would be about tenth magnitude.

With current observing techniques significant measurements could be made. It appears possible to supplement ground observations with space sensors included as part of the comet payload. Finally, it must be realized that there would be considerable technical problems with the construction and launching of such a payload.

Comet Probes

Serious limitations exist in the ability of ground based observations to determine many properties of comets. The development of vehicles for carrying scientific payloads to great distances demonstrated by the planetary probes, suggests their use in the exploration of a comet. In fact much of the instrumentation for studying the interplanetary plasma, magnetic fields and micrometeoroid flux would require little modification for a comet problem.

The orbit of a comet is never known well enough to plan a close intercept. Even for known short period comets the predicted time of perihelion may be a month in error although this could be considerably improved by using all positions for a precise definitive orbit. Because a lead time from launch to intercept of several months is required, only limited time would be available to obtain the orbit the comet is following prior to intercept. Thus, several mid course corrections may be necessary.

The better known periodic comets are considerably fainter than the unexpected new long period comets, the average difference being perhaps 5 magnitudes or a brightness ratio of one hundred. Bright comets with greater space density of particles and larger diameters, making them more favorable for probe analysis have poorer determined orbits and are less favorable to intercept.

It is not appropriate here to present a complete discussion of comet probe experiments. Instead the desirable measurements and the likelihood of successfully performing them are summarized.

The period of penetration of the coma will be at most a few hours, and the relative velocity will be about 30 km/sec.

The simplest scheme for photometry appears to be electronic scanning of an image tube with elements 0.1 mm across. With a 100 cm focal length cassegrain telescope the resolution would be 20 seconds. A 10 km nucleus at 50,000 km subtends an angle of 40 seconds. The resolution and guidance obtainable by the time a comet probe is flown would easily resolve the nucleus and permit its structure to be determined.

Flight pressure gauges can now determine the total density to 10^{-12} Torr or 10^5 molecules/cm^3 at coma temperatures. Neutral particle mass spectrometers have a limit of 10^{-13} Torr or 10^4/cm^3.

Detectors for ions are 10^3 or 10^4 times more sensitive than neutral detectors. Hence, the limiting concentration for ion density is about 1/cm^3. The high velocity of the detector relative to the comet will produce a ram effect which would increase the sensitivity by about a factor of 10.

Observed molecular densities of CN and C_2 in comets are given in Table VII for a comet 1 A.U. from the sun.

TABLE VII

Densities of neutral molecules in comets

Bright comet		Faint comet	
R (km)	N/cm^3	R (km)	N/cm^3
near nucleus	3×10^7	2×10^3	1000
10^4	3×10^3	1×10^4	75
10^5	3×10	5×10^4	4
		1×10^5	1

The intensity of CO^+ relative to the neutral molecules varies. In comet Bester (1948I) the CO^+ densities were estimated at about 30 times the CN densities.

A comparison of particle densities and detector sensitivities indicates that neutral densities could be measured beyond 10^3 km for fainter comets. For bright comets, the corresponding distance approaches 10^5 km. The total particle density is several orders of magnitude greater than for the visible species. The greater sensitivity of ion detectors and mass spectrometers indicates that successful probing could be done to the observable limits of coma and tail.

Calculated dust densities and micrometeoroid detector characteristics lead to values for total impacts of several hundred. As the velocity is known, masses could be found without including velocity measuring sensors.

In addition to these measurements, environmental sensors for solar ultraviolet, interplanetary plasma and magnetic fields would be included.

The determination of plasma and magnetic field characteristics would allow a firm analysis of the interaction of the comet and solar wind.

Measurements at a comet would be critical tests of theories proposed to explain the influence of the solar wind on the heads and tails of comets.

REFERENCES

1. See for example articles on meteorites by E. Anders, J. A. Wood and E. R DuFresne and E. Anders in: Moon, Meteorites and Comets, G. P. Kuiper Ed., Univ. of Chicago Press, Chicago, 1963.

2. B. Donn in Production and Trapping of Free Radicals, N. Bass and H. P. Broida, Eds., p. 347, Academic Press, N. Y., 1960.
3. L. H. Aller, Astrophysics, p. 94, Ronald Press, N. Y., 1963.
4. N. J. Woolf, M. Schwartzschild and W. K. Rose, Ap. J. **140**, 833 (1964).
5. Tsuji, T. Annals Tokyo, Astrophys. Obs., IX, No. 1, 1964.
6. Donn, B., Wickramasinghe, C., Hudson, J. P., Stecher, T. P., to be published Astrophysical Journ. (1967).
7. H. C. Urey, Some Cosmochemical Problems, Pennsylvania State University, University Park, Pa., 1963; Mon. Not. Roy. Ast. Soc. **131**, 191, 1966.
8. J. P. Hirth and G. M. Pound, Condensation and Evaporation, The Macmillan Co., N. Y., 1963.
9. B. Donn and G. W. Sears, Science **140**, 1208 (1963).
10. Blander, M. and Katz, T. L.
To be published Geochemica et Cosmochinich Acta. (1967).
11. See articles on comets by E. Roemer, K. Wurm and F. L. Whipple in Ref. 1.
12. P. Swings, Vistas in Astronomy, Vol. 2, p. 958, A. Beer, Ed., Pergamon Press, London, 1956.
13. H. C. Urey, The Planets, Yale University Press, New Haven, 1952.
14. Lord, H. C., Icarus, 4, 279, 1965.
15. W. A. Fowler, J. C. Greenstein and F. Hoyle, Geophysical J. Roy. Ast. Soc. **6**, 148 (1962); see also W. A. Fowler, Science **135**, 1037 (1962).
16. F. O. Rice in Ref. 2. This book is a good summary of low temperature and free radical chemistry.
17. H. A. Papazian, J. Chem. Phys. **32**, 456 (1960).
18. J. A. Glasel, Proc. Nat. Acad. Sci. **47**, 174 (1961).
19. B. Donn, Icarus 2, 396 (1963), and references cited there.
20. Symposium on Cosmic Dust, W. A. Cassidy, Ed., Annals of N. Y. Academy of Sciences **119**, (1964).
21. C. L. Hemenway and R. K. Soberman, Astronom. J. **67**, 256 (1962).

Chapter 13

ORBITAL MECHANICS

Part I by Peter Musen and Part II by R. K. Squires

PART I—ANALYTICAL TREATMENT

Celestial mechanics is an old science. Substantial contributions to the subject were made by Hansen, Newcomb and Hill in the 19th century and by Brown in the early 20th century. But in spite of the extensive study in this subject, satellite orbits opened up new areas of study. Most of classical celestial mechanics dealt with orbits that were nearly circular of low eccentricity and also of low inclination.

With the launching of artificial satellites a new group of problems in celestial mechanics came into the fore. The theories of motion of artificial satellites had to be developed quite accurately on the basis of all the available observational material in order to separate the unknown effects from the known ones and to find their explanations. One solved problem often created sevaral new problems which had to be solved without any great delay. New methods had to be developed or time honored techniques had to be improved and perfected in a relatively short time.

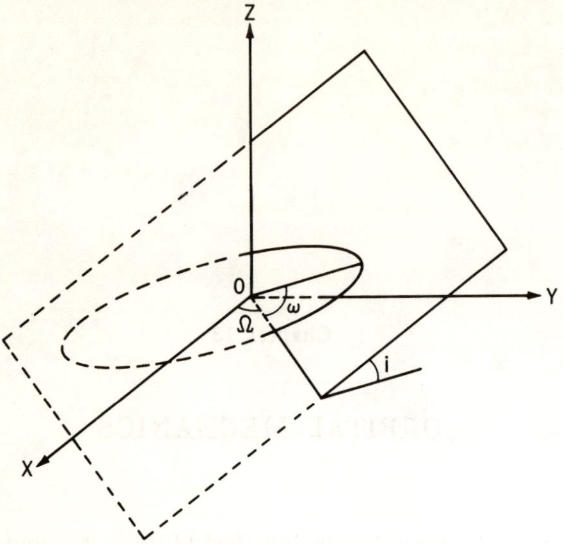

FIG. 1. Elements of a Keplerian orbit.

THE EQUATORIAL BULGE

When the work on artificial satellites was started the effect on the earth's *equatorial bulge* on the motion of the satellite constituted the most important problem.

If the equatorial bulge were non-existent and the earth were a homogeneous sphere, then the orbit of the satellite would be a fixed ellipse. The motion of the satellite in this ellipse would be characterized by six constants—the six *elements.* The most commonly used elements are (Figure 1):

Ω, the longitude of the ascending node,

ω, the argument of the perigee,

i, thc inclination of the orbital plane toward the equator,

a, the semi-major axis of the ellipse,

e, the eccentricity of the ellipse,

τ, the time of the passage through the perigee, or l_0 the mean anomaly at the initial moment of time t_0.

We also shall introduce the period of revolution of the satellite T and the mean motion in its orbits, $n = 2\pi/T$.

Let us designate the gravitational constant of Newton's gravitational law by G, the mass of the Earth by M, the distance of the satellite from the center of the Earth by r, and its coordinates with respect to the inertial system by x, y, and z. We choose the inertial system in such a way that the xy-plane coincides with the equator of the Earth.

There is a simple relationship between n and T (the third Kepler's law) which can be written in the form

$$n = (GM)^{1/2}/a^{3/2}. \tag{1}$$

We shall also make use of the angle

$$l = l_0 + n(t - t_0), \tag{2}$$

which represents the mean position of the satellite in its orbit without influence of the eccentricity and for this reason this angle is called *the mean anomaly.*

If the equatorial bulge is not present then the satellite of the mass m is attracted to the center of the Earth with the force

$$\mathbf{F} = -(GMm/r^3)\mathbf{r}. \tag{3}$$

The acceleration has a *force function* V, where V is defined as

$$V = GM/r. \tag{4}$$

This means that in the inertial system the acceleration itself has the components

$$\begin{aligned} -Gmx/r^3 &= \partial V/\partial x \\ -GMy/r^3 &= \partial V/\partial y \\ -GMz/r^3 &= \partial V/\partial z. \end{aligned} \tag{5}$$

Let us now assume that the bulge is present. Then the force function

V is not given by (1). In general, the force function at P is

$$V = \int [G/(PQ)]\epsilon \, d\tau, \tag{6}$$

where the integral is taken over the total attracting mass, ϵ is the density at the mass element Q, and $d\tau$ is the element of volume.

If r is the distance from the center of gravity of the attracting mass to P, and ρ is the distance from the center of gravity to the mass element Q, then

$$(PQ)^{-1} = [r^2 + \rho^2 - 2r\rho \cos \theta]^{-1/2}, \tag{7}$$

when $\theta = \angle POQ$. This can be expanded by using Legendre polynomials,

$$(PQ)^{-1} = r[1 + (\rho/r)^2 P_2 + (\rho/r)^3 P_3 + \cdots], \tag{8}$$

where $P_2 = \frac{1}{2}(3\,\mu - 1)$, $P_3 = \frac{1}{2}(5\,\mu^3 - 3\,\mu)$, and $\mu = \cos \theta$, and the force function of the disturbing mass M, assuming symmetry about a north–south axis, is given by

$$V = (GM/r)[1 + (1/r^2)\, J_2 P_2 + (1/r^3)\, J_3 P_3 + \cdots], \tag{9}$$

where P_2, P_3, $\cdots$, are the Legendre polynomials with the latitude β of the satellite as the argument and where the J's are constants that are related to the distribution of mass. For the earth the most important term is the dipole-like term involving J_2. This is due to the equatorial bulge of the earth. The term involving J_3 is the one referred to in describing the earth as pear-shaped. The magnitude of this J_3 term was determined first from a study of the orbit of the Vanguard I satellite.[1]

Because of the additional terms in (9) it is clear that the orbit is not a fixed ellipse any more. It will become an ellipse with the elements variable. They are varying slowly, however, because of the smallness of the coefficients J_2, J_3, $\cdots$. The additional contribution to V, containing the factors J_2, J_3, $\cdots$, is called the *disturbing function R* as caused by the presence of the bulge. In higher order theory the contribution done by the moon (the lunar disturbing function), or by the sun (the solar disturbing function), or by the assymetry of the earth must also be considered.

The disturbing function R varies with time, but to find the steady monotonic changes in the orbit elements, the secular perturbations, which are the important changes for our case, we can use the time independent value of R averaged over one cycle of the mean anomaly. If we are concerned with the first order effect involving the equatorial bulge only,

$$R_s = \frac{1}{2\pi}\int_0^{2\pi} R\,dl = -\frac{GMJ_2}{2a^3(1-e^2)^{3/2}}\left(\tfrac{3}{2}\sin^2 i - 1\right). \tag{10}$$

It can be shown that the secular changes in the orbital elements are given directly by derivatives of this average disturbing function. For example,

$$\dot{\Omega}_s = \frac{na}{GM(1-e^2)^{1/2}}\operatorname{cosec} i(\partial R/\partial i) = \frac{-3nJ_2\cos i}{2a^2(1-e^2)^2} \tag{11}$$

and

$$\dot{\omega}_s = \frac{-3nJ_2}{4a^2(1-e^2)^2}(5\sin^2 i - 4), \tag{12}$$

where the equatorial radius of the earth is taken as the unit of length.

Assuming that the satellite, if observed from the North Pole, moves counterclockwise, under the influence of the disturbing effect of the bulge the orbit plane of the satellite rotates clockwise with the constant velocity of $\dot{\Omega}_s$.

This can be understood physically by analogy to a gyroscope. The gyroscope precesses when a torque is put on it. The equatorial bulge of the earth puts a torque on the satellite and tries to rotate the satellits orbit into the earth's equatorial plane. This torque makes the orbit precess as given by equation (11).

In the rotating orbit plane the perigee will rotate counterclockwise with the constant angular velocity of $\dot{\omega}_s$ for $i < 63.4°$. For larger values of i the perigee will rotate clockwise.

These two effects are the main ones among all the effects caused by the bulge. The other effects in the elements are of the periodical nature. Some of them are short periodic with the period T or less and some are long periodic with the period $2\pi/\dot{\omega}_s$ and the long periodicity follows from the smallness of $\dot{\omega}_s$. The values of a, n, e, and i in (2) and (3) are the mean, constant values around which the elements oscillate periodically.

These effects in the elements, which are known as *perturbations*, in the language of celestial mechanics represent a superposition of short and long wave-like terms. They are developable into a multiple Fourier series. Decomposition of the perturbations in the elements into Fourier series constitutes the main, and mathematically the most difficult, problem in the analytical mechanics of satellites.

There are two ways to solve this problem. The purely analytical theories are developed by Brouwer,[2] Garfinkel,[3] and Vinti,[4] and the numerical theory by Musen[5] based on the ideas of Hansen's lunar theory.[6] The arguments of the periodic terms in the development of perturbations are of the form $jl + k\omega$, where j and k are integers. The coefficients of these terms have divisors of the form $jn + k\dot{\omega}_s$. This results from time integrating the original Fourier series containing derivatives of the elements.

RESONANCES

For the inclination $i = 63.4°$ the mean motion $\dot{\omega}_s$ of the argument of the perigee becomes very small according to equation (12) and for the case $j = 0$ the presence of the small divisors in the terms with the arguments $k\omega$ causes these terms to become very large. This effect is known in celestial mechanics as a *resonance* at this *critical inclination*. Generally speaking such resonances cannot be treated by the development into Fourier series. In their nature the resonances can show the same pattern of behavior as the mathematical pendulum does. The change of an element, say of ω, in the case of the critical inclination, can be progressive: that is, ω can change linearly with time and the periodic oscillations are added on the top of this linear change. This is like the pendulum for the case of continuous motion in one direction. The second possibility is that ω simply oscillates between two limits. This oscillation is called *the libration* and it is similar to the oscillatory motion of a pendulum.

Resonances appreciably affect only one element—an angle. The changes in all other elements are relatively small. This statement is valid for the theory of satellites as well as for the motion of other celestial bodies. Extreme accuracy of the observational data is necessary in order to decide about the type of motion in the case of resonance. One more case of the resonance—if the period or revolution of the satellite is nearly 24 hr and the equatorial section of the earth is not a circle but an ellipse, then the near *commensurability* with the period of rotation of the earth will also produce a resonance in the motion of the satellite. The satellite nearly follows the earth in its rotation In its motion it can merely oscillate over a certain point on the surface of the earth (this is the case of libration) or it can drift away further and further from its original position. This case would be similar to the progressive motion of a pendulum. There are two stable locations on the earth's equator for a 24 hour satellite. They are over the Indian Ocean and the Eastern Pacific Ocean. The radial gaps in Saturn's rings at certain radii may be due to such commensurabilities between the planet's period of rotation and the period of that of the ring.

LUNAR AND SOLAR EFFECTS

The long period effects play a dominant role in the problems of the evolution and stability of planetary orbits. We can predict such planetary effects, but we cannot observe them over full periods because these periods are too long.

In the motion of an artificial satellite we are able to observe such effects directly and in a relatively short period of time, as if the time scale had become contracted. From this viewpoint the importance of artificial satellites for testing the theories of celestial mechanics is clearly understood.

The first methods used to solve the problem of determination of the long period lunar and solar effects on earth satellites were based on the development of the lunar part of the potential of the acceleration into a trigonometric series, as it was done by Kozai,[7] Musen *et al.*,[8] Jegorova,[9] and Kaula.[10] If we are interested only in the main characteristics of the long period evolution of the orbit, then the terms of short periods, equal to the period of the revolution of the satellite or less, can be omitted from the development of the disturbing function.

With the remaining terms we can determine the main long period lunar effects in the elements. Such a simplified approach is possible if the perturbations in the longitude of the node Ω and in the argument of the perigee ω contain terms which are linear with respect to time. In this case we say that Ω and ω possess mean motions. This will always be the case for a close satellite, because the source of the mean motion will be the equatorial bulge of the earth. A very small contribution to these motions will also be made by the moon and the sun directly.

In this case the arguments of the trigonometric series are linear combinations of longitude of the nodes and of arguments of the perigee of the satellite and of the moon.

However, such a simplified approach to the problem of lunar and solar perturbations based on the development of the disturbing function into a trigonometric series becomes impossible even for relatively close satellites, if the mean motion of some argument becomes exceedingly small. Then a small divisor appears in the process of integration causing a near resonance condition in the motion of the satellite. The resonances can occur at different orbital inclinations, like 46.4°, 56.1°, 63.4°, 69.0°, and 73.1°.

The first case of these resonances for earth satellites took place for Explorer VI ($a = 4.35R_e$, $e = 0.76$, $i = 47.1°$). For this satellite the mean motion of the longitude of the perigee is very small. Figure 2 shows how the satellite orbit changed with time. The original estimate of its lifetime was about two hundred years, but Kozai has shown that because of the resonance it would be only of the order of two years. By choosing a different time of day for the launch the solar and lunar perturbations on the orbit would be different and Musen has shown that in this case a longer lifetime can be achieved, probably of the duration originally estimated (Figure 3).

By taking the critical argument, which causes the resonance, as a new element we can reduce the resonance problem in first approximation to an equation identical in form to the equation of motion of the mathematical pendulum.

The complete treatment of the resonance, including the higher order effects leads to the extremely complicated formulas even for close satellites. For distant satellites which move in very eccentric

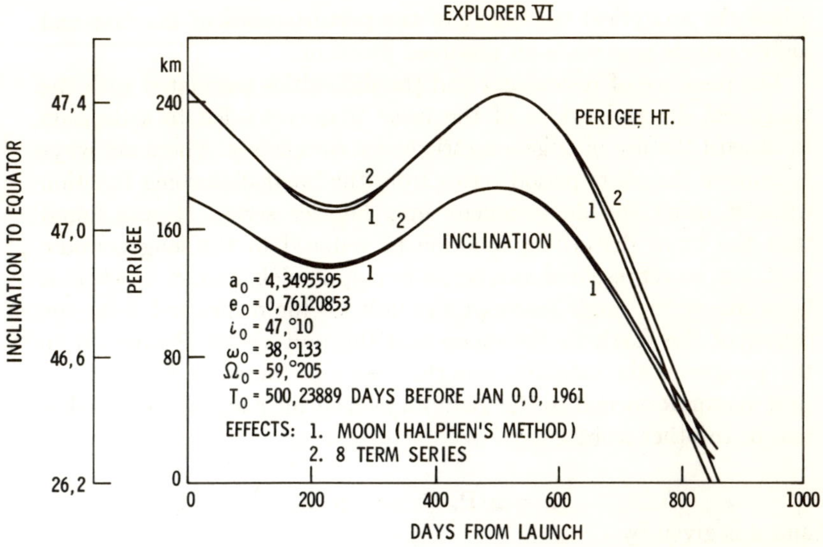

FIG. 2. Calculated inclination and perigee height of the Explorer VI satellite (after Smith, Ref. 17).

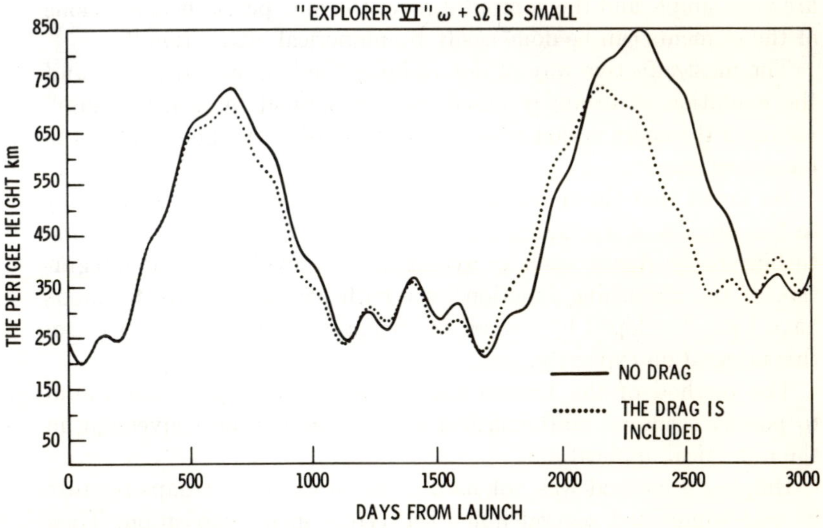

FIG. 3. Calculated perigee height of Explorer VI, assuming the launch time was changed by six hours.

orbits the analytical treatment of the perturbations of the first and higher orders represents an unsolved problem.

The presence of resonances and the difficulties associated with the analytical determinations of the basic frequencies led to a decision to discard the use of trigonometric series completely. There are ways to remove the short period terms from the lunar disturbing function without using the development into Fourier series. It was found that the lunar disturbing function as reduced to the long periodic part, can be represented as a series in polynomials in α, β, γ, where α is cosine of the angle between the unit vector **n** directed from the center of the earth to the moon and the unit vector **P** directed to the perigee of the satellite, γ is the cosine of the angle between the unit vector **R** normal to the instantaneous orbit plane of the satellite and **n**. In other words,

$$\alpha = \mathbf{n}\cdot\mathbf{P}, \qquad \gamma = \mathbf{n}\cdot\mathbf{R},$$

and β is given by

$$\beta = \mathbf{n}\cdot\mathbf{Q},$$

where **Q** is the unit vector normal to **P** and **R**. These polynomials are very simple and the computation of the long period perturbations of the elements can be done easily by numerical integration.[11]

The most effective way of determining the long period effects and the evolution of orbits is based on the method by which Gauss[12] removed the short period effects from the disturbing potential in the planetary case.

He found that the removal of the short period terms is equivalent to the process of averaging over the orbit of the disturbed and disturbing body. Again, such an averaging is equivalent to the integration of the disturbing function over both orbits. A set of formulas have been developed by several mathematicians in order to perform this integration explicitly.

The method of the French mathematician Halphen[13] was found to possess a special mathematical elegance and to be convenient in the numerical applications.

Halphen's method was not used at all originally, perhaps because he had committed several numerical errors in his exposition. They all were discovered by Goriachev[14] from the University of Tomsk in Russia. Published in Russian with a short abstract in German,

Goriachev's brilliant work did not attract much attention outside his own country. The author decided to apply Halphen's and Goriachev's ideas to the determination of the evolution of orbits of artificial satellites under the lunar and solar influence. Certain small modifications were made in the method in order to speed up the convergence of the series. The first applications of Halphen's method showed results which were not anticipated. The eccentricities oscillated wildly in a large interval. The eccentric orbits became almost circular in the course of time and vice versa. One can say that the orbits were *pulsating*. The inclinations were also affected by changes of as much as 20°. Such large changes in a relatively short time seemed to be so unusual that a check was necessary. A program using the Encke method of the perturbation of coordinates confirmed that the effects were real.

An application of Halphen's method to the IMP satellite performed by Shute[15] shows this large variation in the eccentricity (Figure 4). Under the influence of the combined effects of the sun and the moon the duration of the lifetime of this satellite will be about eight years.

It is of interest to note that the long period effects of the type described above do not effect the mean motion and the semi-major axis of the satellite.

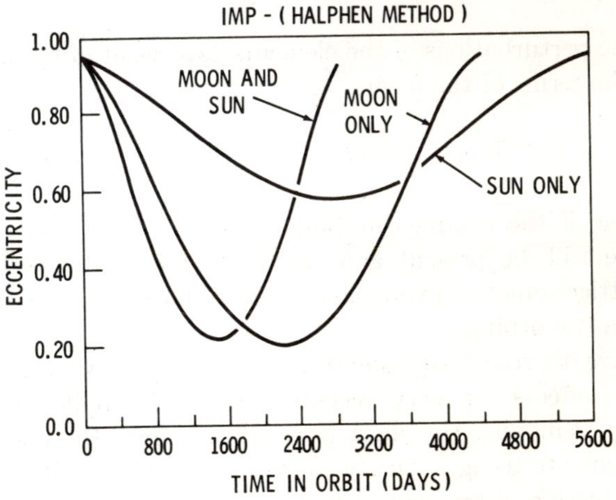

FIG. 4. Calculated variations of the IMP satellite due to solar and lunar perturbations (after Shute, Ref. 15).

One important problem must be mentioned in connection with this statement: if the satellite is moving in the orbit with the large semi-major axis, it may happen that its mean motion will be nearly commensurable with the mean motion of the moon. This will cause a resonance in the motion of the artificial satellite and also all possible *cross effects* between this resonance and other long period effects mentioned before, if the eccentricity and the inclination are large. The removal of the direct influence of the short period terms in the case of the commensurability of mean motions must be performed in a purely numerical way and a future task is to extend and to complete Halphen's method.

Considering the problem of commensurability let l be the mean anomaly and n the mean motion of the satellite l' and n' be the mean anomaly and the mean motion of the moon.

We assume that n and n' satisfy the approximate relationship

$$n'/n \cong i/i',$$

where i and i' are two relative prime integers. In the development of the disturbing functions a long period term appears of the form

$$A \cos (il - i'l') + B \sin (il - i'l)$$

and in the perturbations of the elements as a result of the integration appear the terms of the form

$$[A'/(in - i'n')] \cos (il - i'l') + [B_1/(in - i'n')] \sin (il - i'l').$$

As we see, if the commensurability of n and n' is very sharp, the resonance will be present and the treatment of the perturbations using a trigonometric expression becomes impossible, especially for very eccentric orbits.

We have to resort to numerical integration if we want to treat resonance effects for very eccentric orbits. Liouville[16] invented a method of removing the short period terms from the perturbations of the elements using a purely numerical procedure. It is based on the introduction of two new elements

$$l = i'\sigma$$

and

$$l' = i\sigma - (\theta/i')$$

instead of l and l'. After this transformation all terms containing σ are removed from the derivatives of the elements by applying the numerical integration in the interval $0 \leq \sigma \leq 2\pi$. The reduced expressions for the derivatives of the elements are then integrated step by step with respect to time.

The actual removal of σ, however, requires a subdivision of the orbit of the satellite into a large number of intervals. This type of computation could hardly be done before the advent of electronic machines. Generally speaking, many results which were obtained in the past by the founders of celestial mechanics were never used, precisely for the reason stated above.

One can always find an inspiration in the results obtained by the celestial mechanicians of previous generations. These results can be modified and adopted to suit the modern needs. Liouville's method of the determination of the critical effects is a typical case in which classical theory can serve our needs. One hardly can think of a more general way to treat numerically the effects of the resonances and the long period effects of the secular types combined together in the case of an eccentric orbit. The equations for variation of constants containing such effects do not differ much in form from the corresponding equations in Halphen's method. Liouville's method can be extended to more critical terms of the form

$$\sum_j [A_j \cos j(il - i'l') + B_j \sin j(il - i'l')]$$

or two or more resonances can be treated the same way.

The analytical theory of two coupled resonances is not yet developed in celestial mechanics. It seems that Liouville's numerical method is the only convenient approach if the number of the independent critical terms exceeds one.

There are limitations to the use of the theories of the long period effects as described in this chapter. The prediction of the evolution of earth satellite orbits based on the averaged disturbing function may give satisfactory qualitative results in the interval of approximately 10–15 years. The omitted short period terms will produce higher long period effects, which must be taken into account, if we want to extend the interval beyond these limits. An approximate calculation shows that these omitted effects constitute about 2–3 percent of the total result. An extensive investigation concerning the interval of validity and accuracy of Halphen's method was recently carried out by A. Smith.[17]

An analytical development of perturbations of highly eccentric orbits of satellites has been started. In connection with this, contributions done by Kozai[18] and Kovalevsky[19] especially deserve to be mentioned. They both used Von-Zeipel's method of elimination of short period terms by means of a chain of canonical transformations. Similar results are contained in the older work by Brown,[20] who also used a method similar to Von Zeipel's. These works, if continued and completed, will constitute a significant step toward the analytical solution of the problem of perturbations of a satellite moving in the orbit of the cometary type.

PART II—PROBLEMS

In this section we will consider several numerical problems in celestial mechanics that have been worked on recently that involve artificial satellites and the moon

The easiest problem in celestial mechanics is the two body problem involving particles moving in Keplerian orbits. Even though it may be a fairly crude approximation to what is really a multi-body problem it can frequently give useful and rapid answers. An especially useful approximation to a three-body problem is to use a coupled set of two two-body problems. This method is called patched conics.

PATCHED CONICS

A patched conic approach consists of solving the two-body problem, preferably in closed form coupled with a scheme for switching from one central body to another. The scheme for switching is called the *sphere of influence*. Laplace defined the central body as the one which provides the smaller ratio of disturbing force to the central attraction. Consider a particle moving in the moon–earth system.

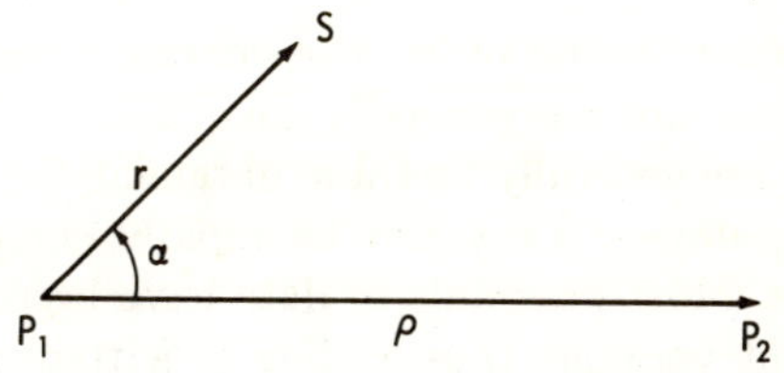

Fig. 5. Sphere of influence variables (after Battin, Ref. 21).

It is on the "sphere of influence" if the relative perturbation of the moon on the particle considered as an earth satellite equals the relative perturbation of the earth on it, considered as a moon satellite. A derivation of the equation as given by Battin[21] is

$$r/\rho = \{(m_1/m_2)^{-2/5}(1 + 3\cos^2\alpha)^{1/10} - \tfrac{2}{5}\cos\alpha[(1 + 6\cos^2\alpha)/(1 + 3\cos^2\alpha)]\}^{-1}, \quad (13)$$

where m_1 and m_2 are the masses of the two bodies, r is the scalar distance to the satellite, ρ the scalar distance between the bodies, and α the angle between the two vectors. The quantities are shown in Figure 5. In the case of the earth–moon system, a two dimensional cross-section of the system is shown in Figure 6.

For other bodies, where $m_2 \gg m_1$ and replacing $(1 + 3\cos^2\alpha)^{1/10} \leq 1.15$ by unity

$$r/\rho = (m_1/m_2)^{2/5} \quad (14)$$

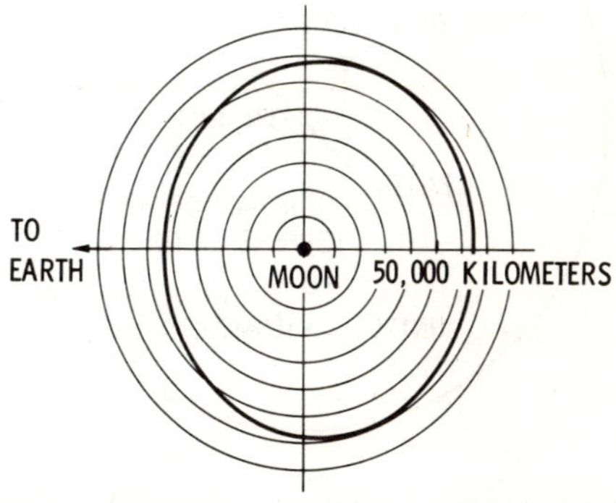

Fig. 6. Surface of influence of the moon (after Battin, Ref. 21).

is thus truly a sphere to first order. Considering other patched conic approximations, first order is generally adequate.

Patched conics are generally useful in obtaining the gross behavior of numerous trajectories. They can be rapidly computed on most digital computers. Most programs to date have been based on time as the independent variable, thus leading to a transcendental equation known as Kepler's equation. Kepler's equation is generally solved by iteration using a Newton–Rhapson technique or some other rapidly converging scheme. However, for many applications it is not at all necessary to use time as the independent variable thereby eliminating the need for iteration and solving the patched conic even more efficiently.

The patched conic was used by Shute[22] to study the dynamical behavior of ejecta from the moon. In this case the first order calculation is considered adequate even though $m_1/m_2 \sim 1/81$, r then is $\sim$ 60,000 km (see Figure 7).

The purpose of the study was to examine the general characteristics of particle trajectories emanating from the surface of the moon. It is of interest to understand what will happen to material ejected from the surface of the moon by the impact of large meteors. Most of the material will fall back on the moon directly but a small fraction of the mass ejected from a lunar impact crater may have velocity

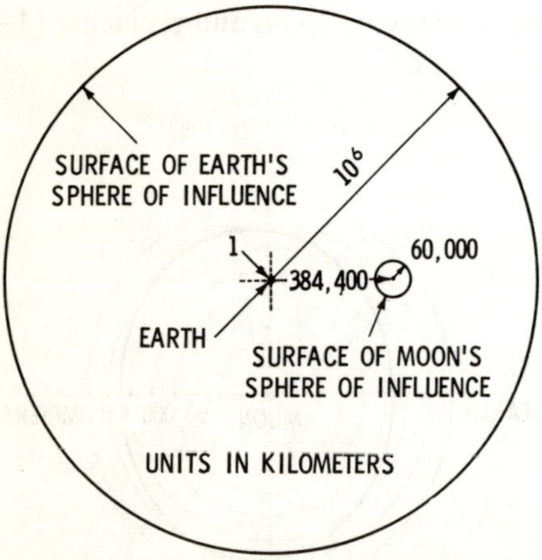

FIG. 7. Approximate spheres of influence of the earth and moon (after Shute, Ref. 22).

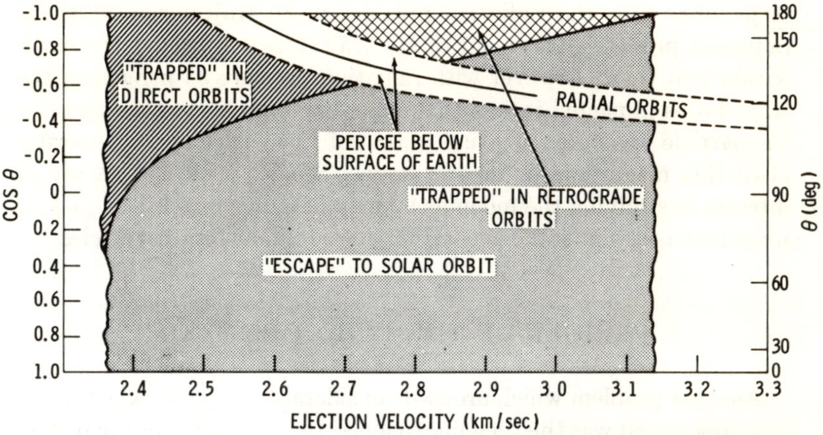

FIG. 8. Analytic approximate geocentric orbits as a function of ejection velocity vs. ejection angle (after Shute, Ref. 22).

$v \geq 2.4$ km/sec and be able to escape the moon's gravitational field. What is the fate of this material? In order to reduce the number of independent variables in the problem, the assumption is made that the position and velocity vectors leaving the lunar *sphere of influence* are parallel. Thus, the major concern is the velocity vector which Shute chose to represent by the magnitude and two angles. The results of the study of lunar ejecta are shown qualitatively in Figure 8. As one may expect, only those particles which leave the *sphere of influence* in the hemisphere opposite to the moon's motion have any chance of being "trapped" in the earth–moon system. *Trapped* in this sense means temporarily in unstable orbits about the earth. (Neglecting energy dissipation processes.) Direct orbits are those in which the particle moves in a counterclockwise rotation when looking down on the North Pole, i.e., the same direction of rotation as the earth. There is a small class of orbits in which a component of the particle velocity cancels the orbital velocity at lunar distance—thus the particle proceeds radially from the earth. Half proceed away from the earth—the others towards the earth with, therefore impact in less than one orbit. Of the trapped orbits, selected ones, dispersed over the angular coordinates and with velocities obtained from the patched conic were numerically integrated, not using patched conics now but including the perturbation of the sun and moon for 50 trajectories plus the radiation pressure in 36 trajectories. Without radiation pressure 10 percent of the trajectories impacted the earth within two years, however with radiation pressure 55 percent of the trajectories impacted in less than one year. The area/mass ratio for the particle was taken as equal to that of Echo since micrometeorites have this magnitude or greater ratios. Studies such as this are of interest in studying the moon as a possible source of tektites, and/or micrometeorites plus the general interest in lunar return vehicles.

PERIGEE OF THE ECHO I BALLOON

Another problem which aroused considerable curiosity when it was first discovered was the then unexplained time variation of the perigee altitude of the Echo I Balloon shown in Figure 9. The perigee oscillated up and down. It is now known that the disturbing force pro-

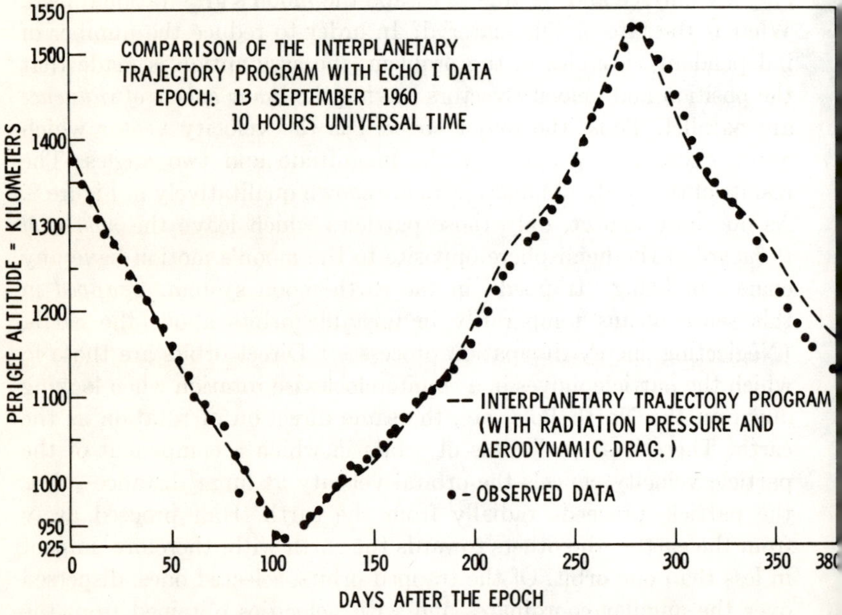

FIG. 9. Comparison of interplanetary trajectory program with Echo I data. Perigee altitude vs. time in orbit.

ducing this oscillation is sunlight. In a balloon type satellite such as the Echo passive communication satellite, the area to mass ratio was sufficiently large that the radiation (light) pressure from the sun played a significant part in the behavior of the satellite's orbit. This simple fact necessitates a fairly elaborate scheme to determine the exact boundaries of the earth's shadow, both umbra and penumbra, refracted by the earth's atmosphere plus a fairly complete knowledge of the atmospheric density to calculate the drag. In addition, the spectral characteristics of the balloon surface were important.

Light quanta emanating from the sun carry a momentum equal to $h\nu/c$, where h is Planck's constant and ν the frequency of the light. When the light impinges on a surface all or part of this momentum is imparted to the surface depending on the spectral characteristics of the surface. For an earth satellite wholly within the sunlight or a perfectly circular orbit, the net acceleration is zero because the contribution is symmetric in the true anomaly, i.e., it adds to the central body acceleration for half the orbit and subtracts for the other half. However for an elliptic orbit which passes through the earth's shadow, the contribution is asymmetric and there is a net change in the total energy. Consider a satellite moving as shown in Figure 10.

Prior to entering the shadow, the solar pressure acceleration a_s has a component which decreases the central body acceleration, a_c, i.e. increases the velocity. Since this velocity increase is lacking in the shadow there is a net decrease in the energy of the orbit. The converse is true when the other half of the orbit is within the shadow and the net energy change is zero when the line connecting apogee and perigee (the line of apsides) exactly coincides with the earth–sun line. Hence, the entire process is cyclic depending on the rate of motion of the line

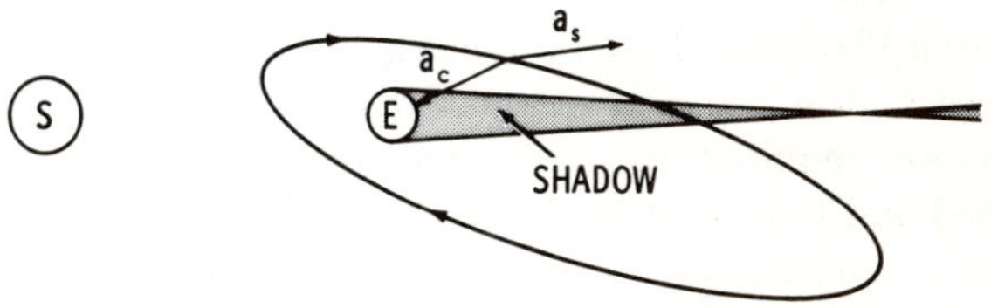

FIG. 10. Acceleration vector notation for radiation pressure and shadowing on an earth satellite.

of apsides with respect to the sun line. During the period of decreasing energy, as shown in Figure 10, the perigee lowers.

The acceleration of a satellite due to absorbed or reflected solar radiation pressure in vector notation is

$$\mathbf{a} = \rho C_r A \mathbf{R}/mr^3 (\text{cm/sec}^2), \quad (15)$$

where ρ describes the spectral characteristics of the satellite, $C_r = 4.6 \times 10^{-5}$ dynes/cm² is the radiation pressure, A is the cross-sectional area (cm²) subjected to the sun, m the mass of the satellite (gm), $\mathbf{R}$ is the vector position of the satellite with respect to the sun, and r its magnitude (cm). It has been shown that a linear interpolation for the refracted shadow boundaries is necessary for earth satellites to avoid an accuracy loss of several degrees in true anomaly after 100 days for Echo type balloons. ρ has been assumed 1.0 for most calculations and $\frac{1}{2}C_r$ was used for the penumbra and obviously $0.0C_r$ for the umbra. The calculations have resulted in reasonable agreement between observation and theory as shown in Figure 9. The observed perigee height vs. time in orbit is compared with the computed perigee height using a numerical integration program (modified Encke Method) with the aforementioned requirements. Considering that numerical integration programs are subject to accumulated effects of round-off error, the agreement is excellent. The uncertainty in ρ and C_r and drag discussed below are too large to warrant further refinements in the present theory.

VERY ECCENTRIC SATELLITE ORBITS

All of the highly eccentric earth satellites (i.e., $e > 0.5$) such as the Eccentric Geophysical Observatory (EGO) and the Interplanetary Monitoring Platform (IMP) are markedly affected by the gravitational forces of the sun and moon. Long term analyses of these perturbations are required and programs such as the Halphen program, modified and improved by Musen[23] are essential.

In addition to the effect on eccentricity and inclination, as discussed by Musen earlier, there is a substantial effect on the perigee height of the satellite. Figure 11 shows the behavior of the perigee height as

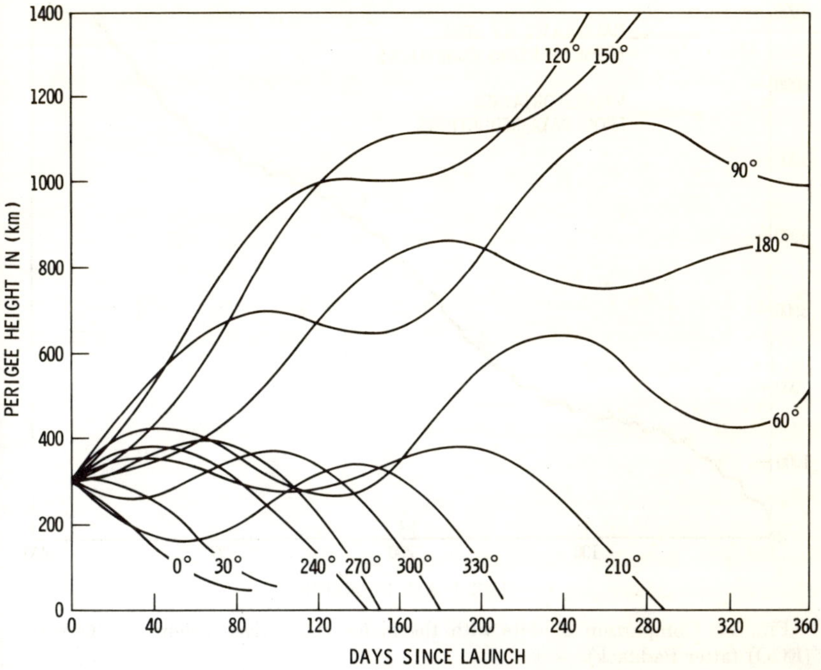

FIG. 11. Solar–lunar perturbation effects on perigee height vs. days since launch as a function of right ascension of the ascending node.

a function of time for an orbit with elements

$$a = 50{,}000 \text{ km}$$

$$e = 0.867$$

$$i = 33.0 \text{ deg}$$

$$\omega = 153.5 \text{ deg.}$$

The third parameter in Figure 11 is Ω, the right ascension of the ascending node, which could also be written as launch-time, all other things being constant. Thus the launch time changes one hour for each 15° of node. For higher apogee satellites the slopes change more radically and vice versa. This fact may be utilized in either direction, to

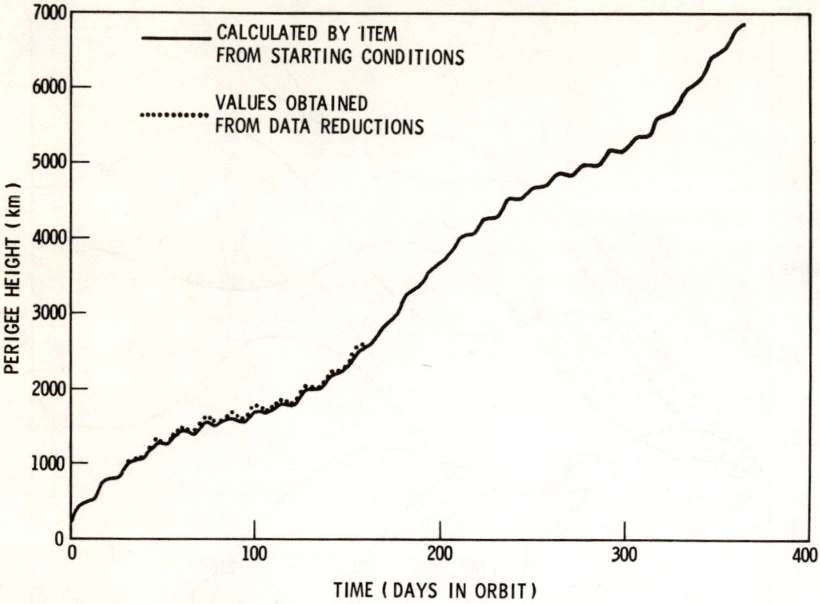

FIG. 12. Comparison of data with theory for the perigee height of OGO–A (EGO) (after Paddack).

raise the perigee when using a low energy booster or to cause the satellite to come back to earth after a predetermined time. How well these may be accomplished depends on how closely the launch time and the injection conditions can be controlled. In addition, the uncertainties in the environment are also important. For Explorer XII an attempt was made to utilize this *perigee driving* as a means of destroying the satellite after a year but it was decided the uncertainty in satellite lifetime due to the uncertainty in injection velocities was too large to try this.

A comparison of theoretical and experimental determination of perigee height using solar and lunar perturbations was performed for the EGO satellite by Paddack. Results for this case are shown in Figure 12. Here, the comparison covers a period of 160 days.

DRAG

The third and final special problem to be discussed in this section is particle drag. Most earth satellites dissipate considerable energy in the atmosphere. The drag force is assumed to be of the form

$$D = -\tfrac{1}{2} C_D \rho A \mid \mathbf{V}_{\text{eff}} \mid \mathbf{V}_{\text{eff}}, \tag{16}$$

where

$$\mathbf{V}_{\text{eff}} = \dot{\mathbf{R}} - \boldsymbol{\omega} \times \mathbf{R} \tag{17}$$

C_D = drag coefficient,
ρ = atmospheric density,
A = effective cross-sectional area,
$\mathbf{R}$ = satellite position vector,
$\dot{\mathbf{R}}$ = satellite velocity vector, and
$\boldsymbol{\omega}$ = sidereal rotation rate vector of the earth.

The minus sign in equation (16) indicates the drag force is opposite to the effective velocity. The effective velocity is the relative wind on the satellite. Implicit in the above equations are the assumptions that the atmosphere is spherically symmetric and rotating with the earth at the same angular velocity. In a computational program, depending on the particular problem, the evaluation of ρ may become quite elaborate or conversely ρ may be deduced through the observed behavior of the satellite. A particularly nice equation for this application was derived by King-Hele[24] using essentially the same assumptions as above plus the following additional assumptions:

1. The air density is independent of time.
2. The air density is proportional to $\exp(-r/H)$, where r is the scalar distance from the earth's center and H is a constant (scale height).
3. The earth's gravitational field is a point-mass.
4. Orbit changes per orbit due to drag are small (second order terms may be neglected).
5. $0.2 \leq e < 1.0$; in practice e must be less than 0.985 to remain in the earth's sphere of influence, and if the satellite is strongly perturbed by the moon e should be less than 0.8 (e is the orbit eccentricity).

$$\rho_p = -(\dot{T}/3\delta)(2e/\pi H a)^{1/2} \times \left[\frac{(1-e)^{1/2}}{(1+e)^{3/2}} \left\{ 1 + \frac{H(8e - 3e^2) - 1}{8 r_p e (1+e)} + \mathrm{O}\,(0.00005) \right\} \right], \tag{18}$$

where

ρ_p = air density at perigee,
$\dot{T}$ = rate of change of the period,
δ = FAC_D/m,
m = mass of the satellite,
F = $[1 - (r_p\omega/v_p) \cos i]^2$,
r_P = perigee distance,
v_p = perigee speed,
ω = magnitude of the angular velocity of the earth,
i = satellite inclination, and
a = semi-major axis.

for an extensive discussion of the limitations of this and other equations see Reference 24.

For very low earth satellites, special programs are generally used incorporating detailed models of the atmosphere to determine the approximate lifetime under the variable effects of drag. This is a field involving dynamic atmospheres and free molecular and slip flow regimes of hypersonic flow. It is often further complicated by the undetermined attitude of the spacecraft. Considerable research is directed toward improving the state of the art in this field but much remains to be done, particularly with regard to other planets where the atmosphere is even more uncertain. Experimentation which would lead to a better understanding of drag phenomena in the free molecular and slip flow regime is somewhat lacking. It is also difficult to simulate space flight conditions in ground test facilities. In free flight, on the other hand, the effects are often masked by other uncertainties in the complete analyses of satellite data. Furthermore, there is no research project specifically designed to research this area in free flight. The researcher in this field must rely on obtaining data from other missions, which may or may not prove fruitful.

Nevertheless much valuable data on atmospheric densities has been obtained in this manner. Chapter 2, Figure 6, shows experimental data on $\dot{T}$ for several satellites that has been used for atmospheric density studies. On this figure one can see the quiet day normal atmospheric drag and also in November 1960 an abrupt increase in the atmospheric drag which occured in connection with a series of solar proton events (see Chapter 10) and can quite likely be linked to increased heating of the upper atmosphere and the resultant expansion upwards.

REFERENCES

1. J., A. O'Keefe, Eckels and R. K. Squires, The gravitational field of the earth, Astronomical Journal **64,** 245 (1959).
2. D. Brouwer, Solution of the problem of artificial satellite theory without drag, Astronomical Journal **64,** 378 (1959).
3. B. Garfinkel, On the motion of a satellite of an oblate planet, Astronomical Journal **63,** 88 (1957).
4. J. Vinti, A new method of solution for unretarded satellite orbits, Journal of Res. of the Nat. Bureau of Standards **63**B, 105 (1959).
5. P. Musen, Application of Hansen's theory to the motion of an artificial satellite in the gravitational field of the earth, J. Geophys. Res. **64,** 2271 (1959).
6. P. A. Hansen, Fundamenta Nova investigationis orbitae verae quam luna perlustrat, 1–331, Gotha, 1838.
7. Y. Kozai, On the effect of the sun and the moon upon the motion of a close earth satellite, Smithsonian Inst. Astrophys. Observ. Rep. No. 22, 7–10, 1959.
8. P. Musen, A. Bailie, and E. Upton, Development of the lunar and solar perturbations in the motion of an artificial satellite, NASA Techn. Note D–494, 1961.
9. A. V. Jegorova, The perturbations in the motion of an artificial earth satellite due to the moon, the sun and earth oblateness, Bull. Inst. Theor. Astr. **93,** 815 (1960).
10. W. M. Kaula, A development of the lunar and solar disturbing function, Astronomical Journal **67,** 300 (1962).
11. P. Musen, On the long period lunar and solar effects in the motion of an artificial satellite, J. Geophys. Res. **66,** 2797 (1961).
12. K. F. Gauss, Determination attractions etc., 1818.
13. G. H. Halphen, Traite des fonctions elliptiques et de leurs applications, Vol. 2, Paris, 1888.
14. N. N. Goriachev, Halphen's method for the computation of secular perturbations (in Russian), Tomsk, 1937.
15. B. Shute, Prelaunch analysis of high eccentricity orbits, NASA Techn. Note D–2530, 1964.
16. J. Liouville, Note sur le calcul des inegalites periodiques du mouvement des planetes, J. Math. Pures et Appliquees **1,** 197 (1836).
17. A. J. Smith, A discussion of Halphen's method for secular perturbations and its application to the determination of long range effects in the motions of celestial bodies, Part II, NASA Techn. Rep. R–194, 1964.
18. Y. Kozai, Motion of a lunar orbiter, Jap. Astr. Journ. **15,** 301, 1963.

19. J. Kovalevsky, Sur le mouvement du satellite d'une planete a excentricite et a inclinaison quelconques, Compt. Rend. **258,** 4435 (1964).
20. E. W. Brown, The stellar problem of three bodies, Monthly Notices **97,** 116 (1937).
21. Richard M. Battin, Astronautical Guidance, McGraw-Hill Publishing Company, 1964.
22. Barbara E. Shute, Dynamical Behavior of Ejecta from the Moon, 1964.
23. Peter Musen, A Discussion of Halphen's Method for Secular Perturbations and its Application to the Determination of Long Range Effects in the Motions of Celestial Bodies, Part I, NASA Technical Report R-176, 1963.
24. D. G. King-Hele, The Contraction of Satellite Orbits Under the Influence of Air Drag, *Proc. Roy. Soc.* A**267,** 541, 557, Part III (1961).

Chapter 14

MAN IN SPACE

Winifred Sawtell Cameron, Lawrence Dunkelman, Jocelyn R. Gill

and Paul D. Lowman, Jr.

INTRODUCTION

The scientific achievements of man himself in space have been modest considering that the manned programs thus far have been necessarily primarily engineering operations toward the goal of reaching the moon, in which the technical achievements have been remarkable. The relatively minor role that science has played has been due to several reasons, among which are the limited space and weight capabilities of the Mercury-Atlas vehicle which was not greatly improved in the Gemini-Titan combination. These two parts of the United States manned space effort have been concerned chiefly with proving the technical feasibility of space flight and science has had to remain a secondary objective. In fact, it was just prior to the first orbital launch that the possibility of scientific experiments was recognized. Those experiments that were scheduled and performed in the Mercury project were therefore simple, and mostly observational. Visual and photographic astronomical observations were made possible by the installation of a multi-panel window; unfortunately, its short wavelength transmission qualities were low and precluded ultraviolet

TABLE I SUMMARY OF MANNED SPACE FLIGHTS

Flight Designations	Astronauts	Launch Date	Revolution	Duration Time	Days	Altitudes (s.m.) (P)	Altitudes (s.m.) (A)	Remarks
Vostok I	Y. Gagarin	4/12/61	1	$1^h\ 48^m$		110	187	1st manned flight
MR–3	A. Shepard	5/5/61	Suborbit	15^m		116		
MR–4	V. Grissom	7/21/61	Suborbit	16^m		118		
Vostok II	G. Titov	8/6/61	17	$25^h\ 18^m$	1	100	159	
MA–6	J. Glenn	2/20/62	3	$4^h\ 56^m$		100	162	
MA–7	M. Carpenter	5/24/62	3	$4^h\ 56^m$		99	167	
Vostok III	A. Nikoyalev	8/11/62	64	$94^h\ 35^m$	4	114	156	tandem orbits
Vostok IV	P. Popovich	8/12/62	48	$70^h\ 57^m$	3	112	158	
MA–8	W. Schirra	10/3/62	6	$9^h\ 13^m$		100	176	
MA–9	G. Cooper	5/15/63	22	$34^h\ 20^m$	1.5	100	166	
Vostok V	V. Bykovsky	6/14/63	81	$119^h\ 6^m$	5	107	146	
Vostok VI	V. Tereshkova	6/16/63	48	$70^h\ 50^m$	3	113	144	1st woman
Voskhod I	V. Komarov, B. Yegorov, K. Feoktistov	10/16/64	16	$24^h\ 17^m$	1	110	255	1st multiple flight, scientists
Voskhod II	P. Belayayev, A. Leonov	3/18/65	17	$27^h\ 2^m$	1	107	307	1st space walk
GT–3	V. Grissom, J. Young	3/23/65	3	$4^h\ 54^m$		100	139	1st man in space 2X, 1st orbit change
GT–4	J. McDivitt, E. White	6/3/65	63	$97^h\ 50^m$	4	100	175	EVA
GT–5	G. Cooper, C. Conrad	8/21/65	120	$190^h\ 56^m$	8	100	189	
GT–6	W. Schirra, T. Stafford	12/15/65	16	$25^h\ 51^m$	1	100	140	1st rendezvous (R)
GT–7	F. Borman, J. Lovell	12/4/65	205	$330^h\ 55^m$	14	100	177	record length (14d.)
GT–8	N. Armstrong, D. Scott	3/16/66	7	$10^h\ 42^m$	0.5	99	147	1st dock (D), R
GT–9	J. Stafford, E. Cernan	6/3/66	46	$72^h\ 21^m$	3	99	144	EVA, R
GT–10	J. Young, M. Collins	7/18/36	44	$70^h\ 47^m$	3	99	145 (475)-	1st dual R, 1st EVA to other craft
GT–11	C. Conrad, R. Gordon	9/12/66	45	$71^h\ 17^m$	3	100	151 (850)	R, D, EVA, altitude record (850 mi.)
GT–12	J. Lovell, E. Aldrin	11/11/66	59	$94^h\ 34^m$	4	100	185	R, D, EVA

(UV) observations. In the Gemini series, UV observations became possible only during the extra-vehicular activities (EVA). The U. S. astronauts did contribute additions and confirmations of theories and made some discoveries. Here we include little of the life sciences and the interested reader is referred to the series on the Life Sciences and Space Research symposia sponsored by COSPAR, the symposium on "The Role of the Vestibular Organs in Space Exploration" sponsored by the National Academy of Sciences, published as NASA SP-77 and the Bioacoustics and Biomechanics symposium sponsored by the National Research Council Committee on Hearing, published as NASA SP-115.

RUSSIAN VOSTOK SERIES

The Russians made substantial advances in the technique of space flight, but made relatively little use of their potential to further the physical sciences in their Vostok and Voskhod manned orbital flight series.

The first two manned orbital flights were made by the Soviets, who called their series Vostok (East), on April 12 and August 7, 1961 respectively. Y. A. Gagarin made one complete orbit around the earth on Vostok I. The only scientific results published on his flight were biomedical data, which may be found in a report by Volynkin, et al.[1] Vostok II made eighteen orbits around the earth with G. S. Titov as the cosmonaut, coming down on land in the Soviet Union as did all their flights.

Vostoks III and IV were the tandem missions flown by A. G. Nikolayev and P. R. Popovich respectively, being launched just a day apart (see Table 1 for flight data). They flew in almost identical orbits and at closest approach, they were reported to be 6.5 km apart and within sight of each other. Vostok III was the first multiple day flight. The cosmonauts got out of their seats and performed tasks and exercises for a total of about three hours in this unrestrained condition.[2] In visual observations of the earth they reported that when the ship was in shadow (at time of full moon—they noted that the earth's surface seemed to be clothed in a gray shroud.

Vostoks V and VI were piloted by V. Bykovsky and V. Tereshkova

TABLE 2. MERCURY EXPERIMENTS

Experiment	Scientific Equipment	Purpose	Results	Mission Numbers
		ASTRONOMICAL		
Dim light phenomena	Unaided eye, camera voasmeter photometer	Observe, increase our knowledge on auroras faint comets near sun, faint magnitude limit of stars, gegenschein kordylewsky libration clouds, meteorite flashes, zodiacal light-corona connection	MA–6 not dark adapted – unsuccessful; MA–9 saw zodiacal light, photographs underexposed, airglow visual and photographs	6, 9
Sunlight and starlight intensity	Voasmeter photometer	Measure atmospheric attenuation	Intensities not measured	6
Normal airglow	Unaided eye, 5577 Å filter	Determine intensity, distribution, structure, variation, color and solar energy conversion	Saw visually on all flights, photographs on MA–9, filter on MA–7	6, 7, 8, 9
Red airglow	Unaided eye	Same as normal airglow	Detected visually on MA–8, confirmed visually MA–9	9
Micrometeorite impact	Visual and amplified inspection	Determine danger and spacecraft protection	One impact found on MA–9 window	6, 7, 8, 9
Refraction of setting objects	Unaided eye, camera	Test and refine theory	Photographs MA–6, MA–7, visual MA–7, MA–9	6, 7, 9
Glenn effect (particles)	Unaided eye, camera	Determine nature and source	Discovered on MA–6, all others saw visually, MA–7 photographs	7, 8, 9
Ultraviolet spectra of Orion stars	Ultraviolet spectograph	Extension of knowledge below atmospheric cutoff	Spectra obtained, window did not transmit to expected wavelength	6
		BIOMEDICAL		
Drinking and eating	Tubes	Possibility and ease	Problems with crumbling and floating crumbs	6, 7, 8, 9
Nuclear radiation	Radiation-sensitive emulsions, lithium-fluoride detectors, ionization chambers	Establish dosage and danger	Dosage well below tolerance level	6, 7, 8, 9
Otolith balance and auto-kinesis phorometer eye measurements	Phorometer	Effect of zero gravity	Not performed due to ASCS malfunction	6
Physiological reactions of man	ECG, body function sensors, 16mm camera monitor	Measure stress and reactions on respiration rate, depth, blood pressure and temperature	Satisfactory reactions on MA–6, MA–8, MA–9, MA–7 had a false reading on body temperature	6, 7, 8, 9
Night vision tests	Eye patch	Test night vision and dark adaptation for navigation	Unsuccessful-patch did not work, did not get fully dark adapted	6
Space suit evaluation	Temperature sensors	Test use and performance	Satisfactory performance on MA–6, MA–9, MA–7 and MA–8 had temperature regulating difficulties	6, 7, 8, 9

ENGINEERING				
Ablation of materials	Several kinds of test materials	Measure ablation properties of test materials	All test materials behaved well	8
Confetti release	1/4 inch plastic disks	Measure drag for reentry conditions	Successful-dispersed rapidly	7
Flashing light beacon	Light	Measure acuity and distance judgment for docking and rendezvous maneuvers	Successful – not seen till next night	9
Ground flare	Several 10^6 Cd lights and a 10^6 Cd Xenon light	Measure visual acquisition and atmospheric attenuation for navigation	MA–9 successful – cloudy on others	6, 7, 8, 9
Liquid behavior	Liquid and glass container	Test theory and to aid in drinking and eating	Successful – liquid behaved as predicted	7
Tethered balloon	Balloon of 5 different colored panels	Measure drag and visual acuity of colors in space for reentry conditions	Partially successful on MA–7 – did not fully inflate, did not deploy on MA–9	7, 9
GEOLOGICAL, GEOPHYSICAL				
General surface observations and photography	Unaided eye, camera	Identify geological and topographical features from high altitude photographs for comparison with planetary surfaces	Photographs obtained on all – quality best on MA–9	6, 7, 8, 9
Specific terrain targets	Unaided eye, camera	Interpretation of photographs from known geological features	Few selected ones obtained, quality fair	8
METEOROLOGICAL				
Albedo intensities day and night sides	Unaided eye voasmeter photometer	Compare to theory and improvement	Not obtained due to ASCS malfunction	6
Cloud structure	Camera, filters of various wavelengths	Compare to Tiros photographs and map forecasts improvement	MA–8 and MA–9 only obtained scheduled ones	6, 7, 8, 9
General weather photographs and observations	Unaided eye, camera	Same as above	All obtained photographs	6, 7, 8, 9
Horizon definition	Camera, red and blue filters	Determine best wavelength for definition of horizon for navigation	Successful – red sharper, blue more stable	7, 9

(the only woman in space to date) respectively. They too, were launched within two days of each other. There have been few reports of data in the physical sciences from these flights. A very interesting report, however, has been published by Rozenberg and Cosmonaut Nikolaeva-Tereshkova[3] on stratospheric aerosols measured from Vostok VI by black and white photography.

Table 1 summarizes all the manned flights of both the Soviet Union and the United States.

UNITED STATES MERCURY SERIES

The Russians made the first manned flight, inaugurating man in space; the United States followed quickly and A. Shepard made the first United States flight less than one month after Vostok I. It was, however, a suborbital one and it was followed by a second suborbital flight by V. Grissom. Both came down in the ocean as did all the United States flights. The orbital missions started with J. H. Glenn's flight MA-6. We shall discuss chiefly the scientific achievements (omitting the unsuccessful experiments). All experiments are listed, however, in Table 2, which summarizes the Mercury flights.

The main scientific results obtained from planned observations for Mercury missions were: (1) confirmation of the predicted flattened sun[4,5,6] effect of refraction on extended objects at sunset or sunrise; (2) confirmation of the expected appearance of the night airglow[6,7] including a heretofore unpublished value of its edge on brightness; (3) finding of small luminious particles, called the *Glenn effect*, and (4) many good photographs of clouds, ocean and land areas.

The theory of refraction indicates that the setting sun, or moon, when seen from above the atmosphere, effectively the situation in orbit, should show a much more pronounced flattening of the disk than that observed from the ground. It should have a sausage shape at sunset, due to twofold refraction for an observer above the atmosphere.[4] Although Glenn looked for it carefully, he did not notice it visually, but he did photograph the effect in a series of sunset and twilight photographs.

Glenn[5] noted and described the normal airglow layer as a tannish or buff band, separate from the horizon band of the lower atmosphere.

He estimated its angular height above the true horizon to be about 7°–8°, which is more than twice its actual height. It was found that all the astronauts similarly visually overestimated angular height at the horizon. Evidently, the moon illusion—exaggeration of the apparent size of horizon objects and distance—operates in space as on the ground. It follows that the moon illusion is not a gravity orientation effect.

The luminous particles referred to as the Glenn effect were first seen by Glenn at his first sunrise, when they appeared to him to resemble a field of stars or fireflies. Sometimes they looked to him like snowflakes, and at other times as if they glowed by their own light. They seemed to drift by very slowly, at about the rate of a man's brisk walk, which indicates that the particles really partake of the capsule's motion and are a part of its environment. They are probably a condensate from the life support system.[5] The Russians did not report such a phenomenon in their Vostok series. An observation reported in the Voskhod series, however, mentioned in that section, suggests they too saw this phenomenon.

Glenn's photographs of extensive cloud areas, showing differences in cloud levels, can be used by weather experts. Those of land areas can be examined for topographic correlations to be used in the study of unknown regions on earth, and the interpretation of planetary photographs taken on space probes.

M. S. Carpenter, on MA-7, tested several theories. The experiment on the behavior of liquid under zero-gravity conditions confirmed predictions that it would fill the capillary tube within a glass sphere.

The Massachusetts Institute of Technology (MIT) horizon definition experiment using a red and blue split-field filter in photographing the horizon confirmed the expectation that the longer wavelength gave somewhat better definition.

Carpenter, in photographing the setting sun, obtained a flattened solar image. (see Fig. 1). He also saw the phenomenon visually.

Carpenter made extensive and definitive observations on the Glenn effect and the airglow layer. After accidentally bumping the side of the capsule and seeing a swarm of particles swirl up, he tried knocking the sides several more times, each time seeing increased numbers of particles. This strongly suggested that they were a part of the capsule environment. He too likened them to snowflakes and to an occasional lathe-shaving piece. The slight difference in velocity from that of the capsule is due to differential drag on the particles.

Carpenter[6] made several estimates of the height of the air glow layer above the true horizon: (1) estimated by eye that it was 8°–10° above it; (2) judged that it was about twice as high as the width of the horizon band, which he estimated to be about 5 times the diameter of the 0.5° solar disk, which would make the air glow layer 5° above the horizon; and (3) noted that the cross bar of the reticle inscribed on the window just spanned the distance from the horizon to the layer. Knowing the distance from his eye to the window and the dimensions of the reticle, the angular distance was computed to be 2.6°. Finally, the most accurate determination was made from Carpenter's timings of the passage of γ Ursae Majoris (Phecda) through the layer. From a knowledge of the orbital parameters the height can be determined accurately.[7] It yielded a height of 91 km, in good agreement with the 90 km obtained from rocket data. This amounts to about 3° angular measure.

FIG. 1. Setting sun photographed by Carpenter on MA–7 showing the pronounced flattening due to refraction of the atmosphere near the horizon. There is some distortion and smearing out of the image from the capsule motion (after Cameron et al., 1963, ref. 4).

W. M. Schirra on MA-8 took about fifteen photographs of cloud areas for the Weather Bureau, with a mosaic of six filters used in connection with a 70 mm Hasselblad camera. Analysis indicated that the yellow and red filters yielded higher contrast than did the others. (see Fig. 2).

Schirra[8] made two new and unscheduled observations. The first was of a patch of light seen on his third orbit when he was just south of Madagascar. He was in drifting flight at the time and did not know his orientation as he did not recognize any of the few faint stars in the field. He noticed that the patch of light was rather wide, brownish in color like smog at dusk, and was high because neither the earth's horizon nor the normal air glow was seen in the window at the time. With this meager information it can only be conjectured as to what the phenomenon was, but it is thought to have been an equatorial arc of the red air glow at about 10°–15° above the horizon.

FIG. 2. Weather experiment photograph taken by Schirra on MA-8. From left to right, the strips were photographed through blue, green, neutral, yellow, red and far red filters respectively. Different cloud levels and cloud types are easily recognized. Note the increased definition of the horizon from blue to red (courtesy of S. D. Soules, of the U. S. Weather Bureau).

FIG. 3. Himalayan mountains photographed by Cooper on MA-9 showing the remarkable detail visible.

The other observation was of the twilight horizon, in which Schirra was the first to see a layered structure in the atmosphere and the first to see a wide, grayish-blue transition area above the normal blue horizon arc. He noted that the blue band was divided by a lighter, narrower blue band. He also observed that the top of the wide gray-blue band, as it faded into the black sky, was ragged "*like underneath a rain cloud*," in contrast to the smooth transition as the other layers blended into each other. This observation suggests turbulence in the band. Schirra also timed the passage of the planet Mercury through some of the layers. This twilight scene is discussed in more detail later.

The last Mercury mission was made by L. G. Cooper in MA-9 on May 15–16, 1963. Cooper confirmed Schirra's observations of the layering of the twilight horizon and its ragged-topped transition region. He also confirmed the horizon definition results of Carpenter's MA-7 mission except that, surprisingly, the blue end gives a better navigational reference as it is more stable and less subject to interference effects of clouds and other atmospheric effects than the red end. He also saw a hazy patch of faint light, brownish in color. From his observations and a review of Schirra's observations it was concluded with reasonable certainty that it was an equatorial arc of the red air glow. Cooper enjoyed superior weather conditions compared to the other flights, and his photographs of the earth can be used effectively for topographical and geological interpretations, (see Fig. 3). He reported seeing buildings in Tibet and a vehicle traveling along a road, raising a dust cloud.

Cooper did not successfully photograph the juncture of the zodiacal light with the solar corona, but observed it visually within 4° of the sun, which is much closer to the sun than any other observations that have been made. He did obtain photographs of the normal air glow.

He was the first to be able to see stars in the daytime.[9] He noted that the threshold of star visibility in the daytime was about two magnitudes less faint than at nighttime, i.e., he could not see stars fainter than 4th magnitude in the daytime.*

RUSSIAN VOSKHOD SERIES

The Russians inaugurated the era of multi-manned missions with Voskhod I (sunrise), when they sent three men up in the one spacecraft. The three-man team was composed of the pilot commander V. N. Komarov, an engineer, K. P. Feoktistov and a medical doctor, B. B. Yegorov. They were in a *shirt sleeve* environment (sea level pressure, room temperature and humidity), and therefore wore only gray sweatsuits instead of cumbersome space suits.[10] They were in orbit 24 hr 17 min, for sixteen revolutions, after which they made a *soft* landing in the capsule. Feoktistov took photographs of the earth's

---

* For more detailed accounts of the Mercury missions the reader is referred to the official publications for each flight and the Mercury Summary. The title of the first is, Results of the First United States Manned Orbital Space Flight, February 20, 1962. The others are similarly titled as the Second and Third, respectively. Cooper's results are included in the Summary, entitled Mercury Projects Summary, Including Results of the Fourth Manned Orbital Flight, May 15-16, 1963. They are published by the Manned Spacecraft Center, National Aeronautics and Space Administration and obtainable for sale by the Superintendent of Documents, U.S. Government Printing Office, Washington, D. C. 20401.

surface and of the Aurora Borealis (Northern Lights), a first view from a space ship. Their description is "First there was the horizon, then the dark sky, then the upper layer of brightness illuminated by the moon and under it rays perpendicular to the horizon 6°–8° high with intervals of the order of 2°. Along the horizon the Aurora Borealis took up the entire field of vision.[11] They also measured the height of the stars along the horizon.

Feoktistov remarked that when the ship was in the earth's shadow he could see a halo of brightness at a height of 60–100 km above the boundary between the earth and the atmosphere, which is visible when the full moon is in line with the ship and the earth.[11] One of Komarov's scientific duties was to take bearings by the stars, probably with the sextant that was part of their equipment.

The cosmonauts also saw glowing particles (Glenn effect?) through their portholes. Experiments were conducted on liquids, plants and Drosophila flies.

Voskhod II was a two-man flight. The cosmonauts were A. A. Leonov and P. I. Belayayev. Information on this flight, as in the previous one, was obtained from press releases. The flight was marked by the first attempt of man to egress from the capsule and to be independently free in space except for a lifeline to the capsule. Leonov exited through an airlock, maneuvered in position to the extent that fifteen feet of rope would allow and took photographs in space for about a ten-minute period. He acknowledged disorientation and difficulty in returning, being surprised by the phenomenon of action and reaction, and admitted being somewhat tired after his exertions. This activity was performed while the spacecraft was over the Soviet Union and Leonov commented about seeing the blue Volga and Yenisey rivers and the gray line of the Ural mountains. He noted that the earth looked flat, that the curvature was only noticeable at the horizon. The sky was completely black and the stars were bright but did not twinkle. The sun was exceptional because it did not have the familiar halo—as if it were welded in black velvet. He searched for the Sochi Sanitarium in the Caucasus mountains, but did not find it. Belayayev obtained movies of Leonov's outside activities which also were televised with on-board TV equipment. They were startled during their flight by observing another satellite estimated to be about one kilometer away. Leonov recorded his observations and impressions, drawing star configurations. Belayayev made a series of astronavigational observations, probably with the sextant that they carried.

UNITED STATES GEMINI SERIES

The Gemini manned flight series sought to extend the capabilities of the United States in space. Like Mercury, it was primarily an engineering and technical series, but, with the prospects of two-men teams and much longer flights, science obtained a planned and increased role as described by Gill and Gerathewohl.[12] Many experiments were scheduled and executed including nineteen scientific ones, of which several were carried on more than one flight. All the experiments are summarized in Table III which is organized similarly to Table II for the Mercury flights, but discussion will be confined chiefly to the scientific ones treating related subjects together. The first manned Gemini flight began with the Gemini III flight with V. I. Grissom and J. W. Young aboard. Grissom was the first person to fly in space twice. The missions are frequently referred to as, GT-3 for example, which is an abbreviation for Gemini-Titan flight 3, where MA-6 meant Mercury-Atlas flight 6.

GT-3 signalled an engineering first when Grissom maneuvered the ship and changed the orbit from an elliptical one with apogee altitude of 122 and perigee of 87 n.m. to a circular one of 90 n.m. Later, prior to re-entry another maneuver reduced the perigee to 43 n.m.

GT-4 was the first flight in which the United States attempted extravehicular activity (EVA). Astronaut White successfully performed a spacewalk (Fig. 4), just a few weeks after that of the Russian Cosmonaut, Leonov. White, too, was connected to the capsule by a tether about 25 feet long but maneuvered himself around with a small jet gun, the first to do so. His exhilarating EVA activities totaled 23 minutes after which he was very reluctant to re-enter the spacecraft. The Americans did not have an airlock system, as the Russians did but instead, depressurized the whole capsule, therefore both astronauts had to be protected from the space environment. White, too, admitted being somewhat tired after his spacewalk, a physical condition experienced by all of the succeeding EVA attempts of the series until GT-12 when body restraints and hand and foot

TABLE 3. GEMINI EXPERIMENTS

Experiment Designations	Experimenters	Experiment	Scientific Equipment	Purpose	Results	Mission Numbers
				DEFENSE		
D–1	Wright-Patterson AFB, Ohio	Visual definition of objects in space	Camera, Questar, Special lenses	Man's ability observe, evaluate, photograph, objects in space	Observed celestial objects, aiming ok, not distant objects due to power difficulties	5
D–2	Wright-Patterson AFB, Ohio	Visual definition of objects in space	Camera, Questar, Special lenses	Get high resolution photos of pad	Power difficulties	5
D–3	AFSC, MSC, Houston	Mass determination	None-data acquisition	Vehicle mass determination by contact	Did determine mass within error limits	8,11
D–4	ACRL, Bedford, Mass.	Radiometric measure	Radiometer, interferometer, spectrometer	0.1–12μ irradiation, spectral information of celestial regions	4000 Å radiometric data on stars Vega, Deneb & moon	5,7
D–5	Wright-Patterson AFB, Ohio	Star occultation navigation	Sextant	Navigation	Equipment failed, some information on Gt–10	7,10
D–6	Wright-Patterson AFB, Ohio	Visual definition terrestrail feature	Same as D–1	Acquisition, track, photograph terrestrial objects	Photographs obtained, not all planned	5
D–7	AFCRL, Bedford, Mass.	Radiometric observation of objects in space	Same as D–4	Extension of D–4 – space objects	4000 Å + 8–12μ radiometric data on Rigel, Sirus and moon	5,7
D–8	Kirtland AFB, N. Mex.	Radiation	Ionization chamber, dosimeter	Accurate measure of rates + total dose in cabin	Low dose levels, no hazard, high rates in South Atlantic Anomaly	4,6
D–9	Wright-Patterson AFB, Ohio	Simple navigation	Filters, sextant	Test sighting procedures for navigation and best features	Procedure worked, daytime stars difficult	4,7
D–10	AFCRL, Bedford, Mass.	Ion-sensing altitude control	Ion-Sensor	Feasibility of orbital pitch, yaw determination by ion-sensing	Were able to determine pitch and yaw	10,11,12
D–12	MSC, Houston, Tex.	EV maneuvering unit	Chest & back pack-flight control	Effectiveness of unit for moving in EVA	Visor fogged, tests incomplete – some tests, equipment worked satisfactorily	9
D–13	Duntley, U. of Cal.	Astronaut visibility	Photometer, visual-tester, black cavity	Measure acquisition of small ground objects	Identified correctly orientation of objects, no significant difference before and after	5,7
D–14	NRL, Wash., D.C.	UHF/VHF Polarization measure	Transmitter, antenna boom	Measure electron structure & variation of ionosphere below craft	Tried 6 times, signals received were 20–30 db lower than expected	8,9
D–15	NADC, Johnsville, Pa.	Night image intensification	TV camera, monitor, camera, recorder	Develop system for nite surveillance of terrestrial features	Abort on (8), on (11) systems functioned well, saw things that men did not	8,11
D–16	Wright-Patterson AFB, Ohio	Power tool evaluation	Power tool, battery, restraints, wrench	Man's ability to work and use tools in zero–G	Abort on (8), premature termination of umbilical EVA (eye irritant)	8,11
				ENGINEERING		
MSC–1	MSC, Houston, Tex.	Electrostatic charge	Potentiometer, sensor electron units	Measure accumulation electrostatic charge on craft	High readings on Gt–4, corrected on Gt–5, both high readings	4,5
MSC–2	MSC, Houston, Tex.	Proton-electron spectrometer	Proton-electron detector	Measure radiation, protons, electrons environment on exterior	Peak rate $= 6 \times 10^4$ el/cm^2–sec and 10^2 protons (both soft) in S.A. anomaly	4,7
MSC–3	MSC, Houston, Tex.	Tri-axis magnetometer	Magnetometer	Direction and amplitude earth's magnetic field to Gemini	Field $= 25{,}000\gamma =$ theoretical electrons penetrated, but safe dosage	4,7,10,12
MSC–4	MSC, Houston, Tex.	Optical communication (laser)	Optical coherent radiator	Evaluate optical communication system atmosphere probe	Cloudy, but laser beam visible	7
MSC–6	MSC, Houston, Tex.	Beta spectrometers	Spectrometer, detector stack	Ability to predict radiation hazards	Peaked electric field and magnetic field as expected. Equipment functioned well	10,12
MSC–7	MSC, Houston, Tex.	Bremsstrahlung spectrometer	10 channel spectrometer	Get γ flux and energy spectrum inside	Spectra obtained, spectrometer functioned as expected	10,12
MSC–8	MSC, Houston, Tex.	Color patch photography	Camera	Fidelity of film on colors photographed in space in EVA	Got 4 photos; film suitable; effect on UV negligible	10
MSC–10	MIT, Cambridge, Mass.	2–color earth limb photos	Camera, filters	Photograph limb in black and white	24 photos, blue profiles more regular than red	4
MSC–12	MSC, Houston, Tex.	Landmark contrast measures	Photometer	Ability to detect and compare contrast of landmarks	Photometer malfunctioned	7,10

M–1	MSC, Houston, Tex.	Cardiovascular reflex	Pneumatic cuffs-inflatable	Determine effect in preventing deterioration	Did protect against deconditioning	5,7
M–2	MSC, Houston, Tex.	Cardiovascular effects	Tilt table, 2 channel	Occurrence and degree of deterioration	No significant difficulties	5
M–3	MSC, Houston, Tex.	Inflight exercisor	Bungee cord, foot crossbar, ECG	Astronauts capacity for work	No significant difficulties, exercise good for men	4,5,7
M–4	MSC, Houston, Tex.	Inflight phono cardiogram	Microphone biosensor, transducer	Amount of heart muscle deterioration	No significant difficulties	4,5,7
M–5	MSC, Houston, Tex.	Biochemical analysis of body fluids	Fluids collect equipment	Hormonal assays for stress reactions	Retention of electrolytes and water immediately postflight consistent with inflight weight losses	7,8,9
M–6	Tex. Women's U., Denton, Tex.	Bone demineralization	X-ray equipment, film	Occurrence and degree of bone density and loss	Exercise cut bone loss	4,5,7
M–7	NIH, Bethesda, Md.	Calcium balance	Waste disposal collector	Rate and amount of calcium loss, effects on muscle	No significant difficulties	7
M–8	Baylor Med. Sch., Waco, Tex.	Inflight encephalogram	EEG electrodes, signal conditioner	Assess alertness, consciousness, sleep depth		7
M–9	Navy	Human otolith function, vestibular effects	Special goggles, visual tester, luminous line	Changes in otolith and orientation capability	Accurate orientation is possible in flight	5,7
				SCIENTIFIC		
S–1	Ney, U. Of Minn.	Zodiacal light photography	Rotating lens, 35 mm camera	Color photos of zodiacal light and airglow	Minimum angle of 16° from sun; photographed gegenschein–diameter 10°	5,8,9,10
S–2	Young, Ames Res.	Sea urchin egg growth	8 chamber cylinder temperature recorder	Effect of Zero-G on growth	Equipment failed	3
S–3	Young, Ames Res.	Frog egg growth	Containers	Effect of Zero–G on growth	Normal growth, no abnormalities attributed to space flight	8,12
S–4	Bender, AEC, Oak Ridge Tenn.	Rad. and Zero-G effects on blood	Container for cells, radiation source	Zero–G, radiation effects on white cells	Mutations, chromosomal aberrations, cell killing	3,11
S–5	Lowman, Tiedemann, GSFC, Md.	Synoptic terrain photographs	70 mm camera, film, filters	High quality small scale photos for resolution	1100 photos including IR, film discovery of unknown features	4,5,6,7,10, 11,12
S–6	Nagler, Soules, USWB, Wash., D.C.	Synoptic weather photographs	Same as S–5	More detail and high quality photos of weather features	Photography and visual cloud information	4,5,6,7,10, 11,12
S–7	Saiedy, USWB, Wash., D.C.	Cloud top altitude spectrum	Spectrometer, camera, IR film	Atmosphere oxygen absorption at .76μ, altitude of Cloud tops	Transmission in A-band at 7600Å higher for high than low clouds	5,8
S–8	Duntley, U. of Cal.	Visual acuity	Photometer, ground targets visual tester, Questar, Camera	Limitations of performance recognition surface objects, Navigation	No significant change from before or after, Resolution ~11" on Gt–5 and 12" (Gt–7)	5,7
S–9	Shapiro, NRL, Fichtel GSFC	Nuclear emulsion	Emulsion pack and container	Quality study of heavy particles; trapped protons, flux of V.A. belts	Exposure time ~ 22^h, expect ~1000 tracks	8,11
S–10	Hemenway, Dudley Obs.	Agena micrometeorite Collection	Collection container	Collect micrometeorites on Agena	Outside plates had some craters; microorganisms survived inside but not out	8,9,10,12
S–11	Koomen, NRL, Wash., D.C.	Airglow horizon photography	70 mm camera, diffraction grating	Emission spectra of airglow	69 photos, variations in altitude and intensity of airglow layer	9,11,12
S–12	Hemenway, Dudley Obs.	Micrometeorite collection	Isolated compartment, Collection container	Collection of micrometeorites	2 kinds of holes found, some microorganisms survived, in 6^h exposure no impacts	9,10,12
S–13	Henize, Northwestern U.	UV astronomical camera	2Å/mm grating, 1000Å/mm prism	UV spectra of 5–10 stars down to 2nd magnitude	Obtained spectra of Sirius, Canopus, γ Velorum field	10,11,12
S–26	Medved, EOS, Pasadena Cal.	Ion wake measures	Electrometer, potential analyser	Measure ion wake and perturbations of craft ambient medium	Electron and ion temperature higher than estimated = 0.3 ev at 200 n.m.	10,11
S–28*†	Dunkelman, GSFC, Md.	Dim and twilight Phenomena	16 mm and 70 mm cameras and visual	Photograph and observe dim objects difficult to get from earth	Twilight bands, clouds, airglow, lightning photos; saw aurorae, meteors	4,6,7
S–29	Morris, USGS, Wash., D.C.	Libration points (L4, L5) photography	70 mm camera f/0.95 lens	Confirm existence and photograph for extent, density	Did not detect them on photos	12
S–30	Ney, U. of Minn. Hemenway, Dudley Obs.	Image orthicon astronomical photographs	Image orthicon tube, camera	Obtain fainter sources and photograph them	Some images almost overexposed	11
S–51	Blamont, CNES, France	Sodium vapor clouds observations	Camera	Observe and photograph vapor cloud set off from ground	Did not see it visually and shutter failed	12
S–64†	Hemenway, Dudley Obs.	UV Dust photography	Camera, spectrograph	Find out chemical nature, quantity of dust	Photos obtained – electrostatic marks nullified scientific value	12
Undesig.†	Gill, et al, NASA – Wash., D.C.	Movies and stills of solar eclipse	16 mm and 70 mm cameras	Atmosphere-free photographs of outer corona	7^s of movie photos. – shutter failure on still camera	12

† Scientific subject of opportunity
* Originally subject of opportunity, later identified as S–28

FIG. 4. Photograph of astronaut White during the first U. S. extra-vehicular activity (EVA) taken by a movie camera mounted outside by White on the GT–4 mission. At left is the nose of this spacecraft from which the umbilical (tether) stretches to the astronaut silhouetted against the earth's limb showing its curvature. At the bottom of the picture is the open hatch through which White egressed.

holds overcame the action-reaction difficulties of the preceding ones. An attempt to maneuver the spacecraft close to the sustainer rocket was not successful. The first actual rendezvous in space came during the GT-6, GT-7 concurrent missions, when Schirra maneuvered his spacecraft to within a foot of the GT-7 spacecraft as described by Lineberry[13] and Stafford, Schirra and Grimm.[14] The first actual docking (joining of two vehicles) was accomplished on GT-8, mating the capsule with its Agena rocket.

Biological Experiments

There were three biological experiments planned for the Gemini series. These were 1) sea urchin growth (S–2); 2) frog egg growth (S–3) and synergism of radiation and zero-G effects on white blood cells (S–4). S–3 and S–4 were carried on later flights also (See table III). The objectives of all three experiments were similar, namely, to ascertain the effects of prolonged radiation and weightlessness on the growth of the eggs and on the chromosomal aberrations of human blood cells. Control samples were maintained on the ground for comparisons. The first scientific experiment in Gemini, the sea urchin egg growth, failed due to leakage and breakage of the package. The frog egg experiment fared better: partial data on GT–8 and complete data on GT–12 were obtained. The package had four chambers containing frog eggs in water with a partitioned section containing a fixative. Handles facilitated activation by the astronauts. On GT–8, early cleavage stages were obtained but the later tadpole stage was not, due to short duration of the flight. All stages of the experiment were completed on GT–12. Results of the latter show that the 10 fixed embryos appeared to be morphologically normal and the five unfixed were live, swimming tadpoles, three being morphologically normal and two abnormal. These were not, however, inconsistent with the controls. It was tentatively concluded that there were no abnormalities that could be ascribed to the flight. The five tadpoles died several hours after the flight for reasons unknown. Histological study may provide the answers.

The S–4 experiment, synergistic (combining or cooperative) effect of zero-G and radiation on white blood cells, was successfully carried out on two flights, and the effect on neurospora, (chosen because of their similar chromosomal behavior to leukocytes) was added to the experiment on the second flight (GT–11). Analyses of the flight leukocyte chromosomes showed no increase in the chromosome deletion frequency over the ground control samples. Bender[15] reports on the results on the neurospora which indicate there was no increase in the frequency of mutations in the flight samples. There seems to be no synergism between radiation and weightlessness on white blood cells. Thus the biological experimental results from the Gemini series are encouraging since harmful effects from radiation or weightlessness, even for four days, seem to be negligible, and man can endure the rigors of the hostile, foreign environment of space.

Synoptic Terrain Photography

By 1964, pictures of the earth taken from sounding rocket, weather satellites, and Mercury spacecraft had indicated the potential value of orbital photographs.[16] Accordingly, Gemini astronauts were scheduled to take color photographs of selected terrain and weather features with hand-held 70mm cameras, as part of two experiments—the Synoptic Terrain Photography (S–5) and the Synoptic Weather Photography (S–6).

The objective of the terrain photography experiment was to obtain high-quality, small-scale (large area) color photographs of selected regions. The astronauts were requested 1) to take vertical or near-vertical strip photographs, small groups, or of individual areas selected for scientific interest; 2) to keep the window in shade, if possible, in order to avoid scattered light, and 3) to keep verbal or written records.

The successful terrain photography experiment during the Gemini missions[17] produced over 1100 usable pictures summarized in Table IV. The areas of best coverage were North Africa and the Middle East because of the geometry of the flight path and the generally good weather. Photography of North America was extensive but of variable quality because of cloud cover. The best American coverage was obtained on Gemini IV[18] when McDivitt and White took a strip of 38 overlapping near-vertical photographs covering Mexico and the southern United States from Baja California to San Angelo, Texas. (Fig. 5). Relatively few good pictures were taken over the southern hemisphere, since the southern part of the flight path was generally in darkness whenever the flight schedule permitted photography; some exceptionally good pictures of southern Africa and Australia were taken on Gemini V and XI, however. Coverage of tropical land areas, such as Central America and equatorial Africa, was generally of poor quality because of the prevailing cloud cover. Even cloud-free pictures were generally of little use for terrain studies because of reduced resolution from strong atmospheric scattering (due to humidity) and the associated degradation of most colors into shades of blue.

The principal problems encountered in the terrain photography were the following: (1) at the times of flight the proportion of global cloud cover was unexpectedly high; (2) ground resolution and contrast

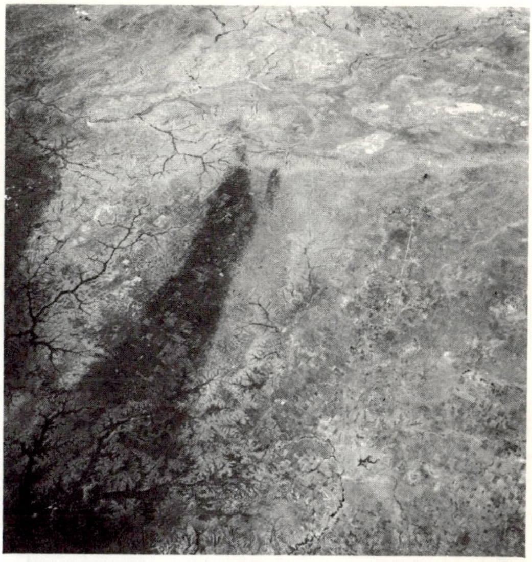

FIG. 5. Central Texas area photographed by McDivitt and White on GT–4. The dark finger-like streak is a large area of ground moistened by rain just before the flight. Aerial mosaics do not show ephemeral features of this sort.

were found to be extremely sensitive to spacecraft orientation (i.e., slant range) and (3) there was some window obscuration (especially on Gemini VII) due to outside residue resulting from second-stage ignition and degassing of interior gaskets.

Regional Geological Mapping

The application of orbital photographs in general, and the Gemini photographs in particular, to the compilation and revision of geological maps is demonstrated in Fig. 6, which shows part of northern Chihuahua, Mexico and southern New Mexico. The large volcanic field at the center of the picture, the Sierra Carizarilla, is clearly a major geologic feature of the area. Aerial reconnaissance and study of geological maps of New Mexico near the Sierra shows that it is a basaltic field of Quaternary age. However, the latest geologic map

TABLE 4. S–5 PHOTOGRAPHY ON GEMINI FLIGHTS

Flight	Camera	Film	No. Usable Pictures	Land Areas Covered
3	80 mm Hasselblad 500C Zeiss Planar	Ektachrome SO 217	7	NW Sonora, Rio Grande valley, Bermuda
4	80 mm Hasselblad 500C Zeiss Planar	Ektachrome SO 217	100	NW Mexico, SW U.S.A., N. Africa, Bahama Islands, Arabian Peninsula
5	80 mm Hasselblad 500C Zeiss Planar	Ektachrome SO 217 Super Anscochrome	175	SW U.S.A., Bahama Islands, South West Africa, Tibet, India, SW Asia, China, Australia
6	80 mm Hasselblad 500C Zeiss Planar	Ektachrome SO 217	60	NW, central and eastern Africa, Australia, Canary Islands
7	80 mm Hasselblad 500C Zeiss Planar	Ektachrome SO 217 Ektachrome IR 8443	250	N. Africa, Arabian Peninsula, India, Caribbean Sea and Adjacent land areas, Brazil, Mexico; infrared film: Gulf Coast, U.S.A. northeast Brazil
9	80 mm Hasselblad 500C Zeiss Planar 38 mm Hasselblad SWC Zeiss Biogon 70 mm Maurer Space Camera 80 mm Zenotar	Ektachrome SO 217	160	N. Africa, northern South America, Caribbean Sea, Mexico
10	70 mm Maurer Space Camera 38 mm Hasselblad SWC Zeiss Biogon	Ektachrome SO 217	75	N. Africa, China, Taiwan, NE South America
11	70 mm Maurer Space Camera 80 mm Zenotar 38 mm Hasselblad SWC Zeiss Biogon	Ektachrome SO 368	102	N. Africa, Arabian Peninsula, S. India, NW South America, Gulf Coast of U.S.A.
12	70 mm Maurer Space Camera 80 mm Zenotar 38 mm Hasselblad SWC Zeiss Biogon	Ektachrome SO 368	160	Southern U.S., N. Mexico, N. Africa, SW Asia, Arabian Peninsula

NOTE: Spacecraft altitudes in Gemini flights ranged from about 100 to 200 statute miles. On the Gemini 11 flight, however, the orbit was changed for 2 revolutions from about 174 statute miles (circular) to 174 (perigee) and 850 (apogee) statute miles. Most of the pictures taken on Gemini 11 were from the two high revolutions, at altitudes of about 400 to 850 miles.

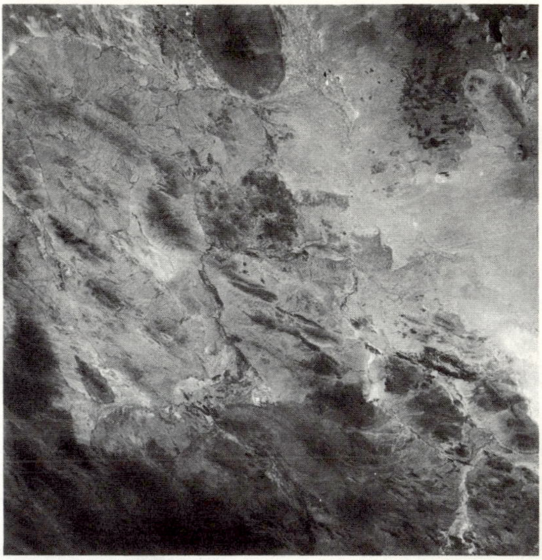

Fig. 6. Northern Chihuahua and southern New Mexico photographed by McDivitt and White on GT–4, U. S.-Mexico border approximately parallel to and one-third of the picture down from the top. Geological features of interest include: Potrillo mountains, volcanoes at upper right; Sierra Carizarilla volcanic field in center; gradation from folded mountains at lower right to block-faulted mountains at upper left. Dimensions of picture approximately 110 miles E-W distance at top.

of North America, published in 1965 shows no Quaternary volcanics in this area. It is obvious, therefore, that this picture—taken on the first flight on which the S–5 experiment was carried—will be of value in revising the map. Other Gemini photographs in this region have been used by the Mexican Government to compile a geologic map covering about 7,000 square miles in Baja California, (e.g., Fig. 7).

The improved knowledge of regional geology, especially in remote areas,[19] will certainly find application to geologic problems such as continental drift, the origin of rift valleys, and the nature of mountain belts. Figures 8 and 9 illustrate these points.

Revision of Topographic Maps

The need for better topographic maps even of settled areas is well-known. During indexing of the Gemini photographs, this need was frequently demonstrated by the numerous discrepancies, omissions, and unmapped areas found on the Army Map Service 1:1,000,000 series. For example, it was sometimes found, in Africa particularly, to

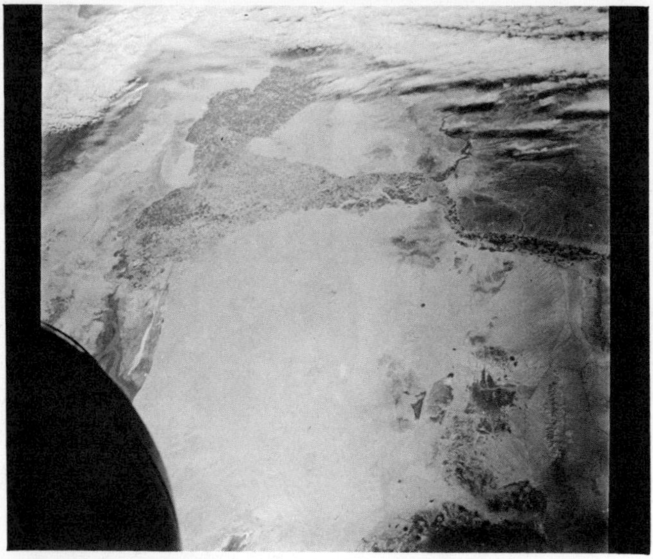

FIG. 7. Southwestern United States photographed by Grissom and Young on GT–3. The Pinacate volcanic complex may be seen in the bottom center. In the center of the picture is the Colorado river delta. Lower left corner is the nose of the spacecraft.

FIG. 8. Southwest Africa photographed by Cooper and Conrad on GT–5. Walvis Bay inside northernmost of three capes. Linear features at bottom are sand dunes in the Namib Desert. Area to north is underlain by complexly folded and faulted Precambrian metamorphic and igneous rocks. Note joint pattern at upper right.

be impossible to match adjoining maps. Also, new roads, urban developments, and canals were noticed in Gemini photographs that were not shown on recent maps issued by the United States.

An incidental advantage of orbital photography came to light in the Gemini map studies. The photographs show *all* the visible aspects of an area, while a map, being an abstraction, shows only selected aspects. Pictures like Fig. 10 (Iran) show topography, geology, something of soil types, and the distribution of cultivated land; consequently, they can supplement maps in the planning of regional projects such as reservoirs, irrigation canals, and new railroads.

Oceanographic Studies

The great speed of satellites and the extremely wide area which can be covered by individual photographs combine to make orbital photography of great potential value for oceanographic research. In several pictures the ocean bottom and river topography can be seen

FIG. 9. Hadramaut Plateau on the southern coast of Arabia photographed by Borman and Lovell on GT–7. Gulf of Aden is at the top. Dendritic pattern of wadis is prominent, with several examples of intersecting wadis resulting from stream piracy. Wadi Hadramaut is the large valley at lower right.

in considerable detail, permitting study of sediment distribution impossible to do by mosaics of air photographs because the dodging and assembling necessary would destroy most of the tonal or color detail.

Photographs of the sun's reflection from the ocean obtained on a number of occasions contain useful information. The glitter pattern frequently contains dark streaks which are slicks (smooth patches) caused, in some cases, by organic matter in the water. Regular patterns which may be caused by internal waves also appear in the glitter pattern.

Man contributed to the success of the experiment (1) by his initiative in taking unplanned photography of interesting features, such as the Richat structures (see Fig. 11), resulting in many valuable pictures beyond expectation; (2) by photographing only adequately lighted and relatively cloud-free terrain in contrast to the unmanned MA–4 mission where the automatic camera continued operating long into darkness, wasting much film; (3) he was able to

FIG. 10. Southeast Iran photographed by Cooper and Conrad on GT–5 with the city of Kerman at left center. Ridges at center are plunging folds in sedimentary rock, truncated by a major fault over 100 miles long at right. Linear features at lower right are ridges in the Dasht-i-Lut (salt desert).

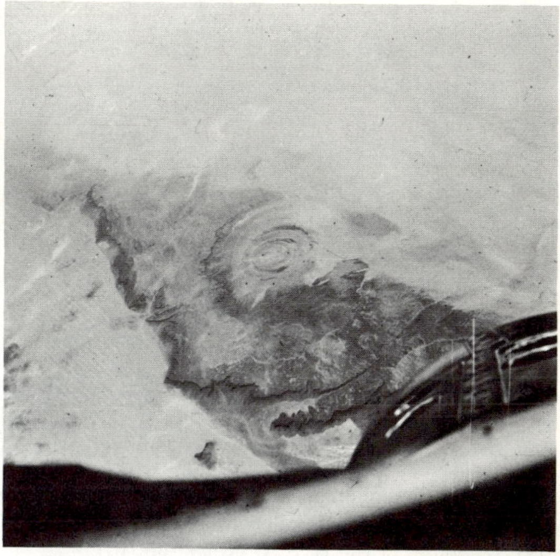

FIG. 11. Dhar Adrar, a rocky plateau in Mauritania photographed by McDivitt and White on GT–4. Circular feature in center is the Richat structure, a dome, probably produced by igneous intrusion but suspected by some geologists of being the root of a former impact crater. Note Semsiyat below and to the right of the Richat. The Richat is about twenty miles in diameter.

supplement the pictures with descriptions of terrain color, atmospheric conditions, location of photographed areas, and reports of equipment operation. It is safe to say that the participation of the astronauts immeasurably increased the success of the photographic experiments. It should be remembered that the photography to date has been done by pilots primarily charged with testing new spacecraft and developing techniques for space operations; improvements in coverage and quality can be expected when missions whose chief objectives are scientific can be flown.

Synoptic Weather Photography

Many photographs were obtained for the Synoptic Weather experiment (S–6). Figure 12 is a view northwest over Camaguey Province

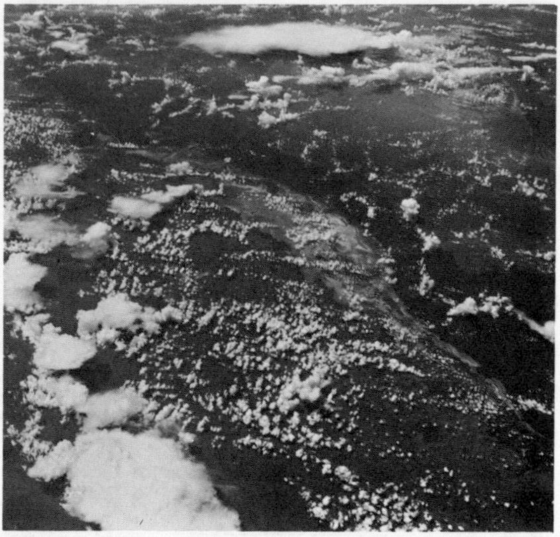

FIG. 12. Camaguey, Cuba photographed by Cooper and Conrad on GT–5, showing polygon-shaped cells and thunderstorms (courtesy of the experimenters K. Nagler and S. D. Soules of the U. S. Weather Bureau).

FIG. 13. Southern India and Ceylon photographed by Conrad and Gordon on GT–11, showing fifty mile wide clear coastal zone and a line of convective clouds several hundred miles off-shore (courtesy of the experimenters K. Nagler and S. D. Soules of the U. S. Weather Bureau).

in Cuba taken on August 23, 1965, on GT–5. A number of thunderstorms are found along the southern coast of Cuba. At the lower left, off the northern coast, some cumulus clouds are arranged in polygon-shaped, open cells. Several are hexagonal with taller cumulus clouds where the cell corners touch. These patterns illustrate a mesoscale cellular convection system that develops when relatively cool air passes over warmer water. Air is tending to sink within the cell and rise near its borders where the cumulus clouds have formed. These open cells would be undetected by a standard satellite TV picture because the cell walls are too thin and their diameter is very small (Hubert,[20] Kuettner and Soules[21]).

Figure 13 is a view of southern India and Ceylon taken from GT–11 on September 14, 1966 with the super wide-angle lens attached to the Hasselblad camera. A clear zone, nearly free of clouds, extends along the west coast of India varying from thirty to fifty miles in width. The zone continues around the southern tip of India and into the Bay of Bengal where a line of convective clouds has formed several hundred miles off shore. Why this clear region should be here is not well understood, but two reasons may be suggested for its extent: (1) the lack of clouds may be the result of drier air subsiding off shore, which would have the tendency to suppress any cloud development. The sea breeze, where the low level winds are moving the air toward land, may cause the air to descend in the clear region; (2) there may be cold water upwelling along the coast. Surface winds are northwesterly along the west Indian coast and southwesterly along the east Indian coast. The northwest winds will transport the surface water southeastward; however, the Coriolis force will tend to deflect the water towards the southwest away from the land. This would permit the upwelling of cooler water along the coast line and the water temperature may be sufficiently low to inhibit the development of cumulus clouds. A surface temperature change of about one degree may be enough to accomplish this. To the east of India where southwest winds prevail, the Coriolis force would act to transport the surface water in an easterly direction. Again this would produce a favorable condition for upwelling of water near the coast. Observations of sea water temperatures from ships are scarce, but the few reports available indicate that the coastal waters were one or two degrees cooler than the sea water temperatures farther west in the Arabian Sea.

Many of the 1900 terrain and weather photographs from the Gemini series are to be published in a series of atlases. The first one was issued as NASA SP–129 by the Government Printing Office in the spring of 1967, entitled "Earth Photographs from Gemini III, IV and V" and were issued in full color which can yield many more bits of information than a black and white photograph alone.

Visual Acuity

The visual ability of the astronauts in the detection and recognition of objects on the earth's surface (S–8) was tested during Gemini V and VII. Data from an inflight tester used on these two flights showed no change in the visual performance of the crews. Results from observations of the ground site near Laredo, Texas (Fig. 14) confirm that the visual performance of the astronauts during space flight as reported by Duntley et al.[22] was within the statistical range of their preflight visual performance and that there was no degradation of the astronaut's visual perception during spaceflight. They successfully acquired and described the orientations of the rectangles laid out. Their resolution capability proved to be about 11–12 seconds of arc, equivalent to about 35 feet from an altitude of 100 n.m. Therefore, it may be concluded that the astronauts did see what they claimed to see, inferred from visual clues.

Micrometeorite Collections

The Agena micrometeorite collection (S–10) from the Agena 8 rocket, which had been there for four months, was recovered during the Gemini X flight. The cover plate to expose the inner collection surfaces had not been opened, but optical scanning by C. Hemenway and his colleagues at the Dudley Observatory of the four stainless steel slides on the outside of the box (protected from launch) have revealed at least four craters larger than 4 microns (μ) which appear to be hyperballistic. Figure 15 shows a crater $\sim$200 μ diameter and this one is named Crater Schweickart, after the astronaut who suggested that there should be an external area on the package which

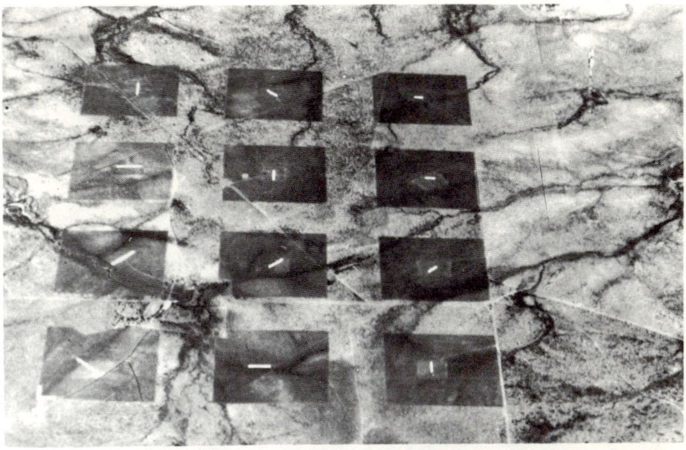

FIG. 14a. Airplane view of the Visual Acuity test patterns at Laredo, Texas.

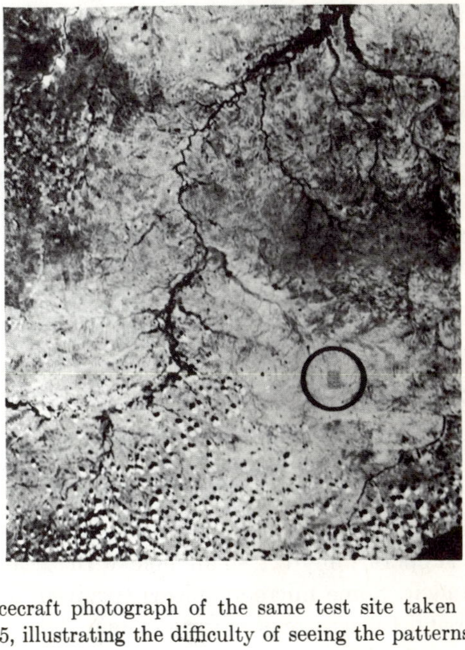

FIG. 14b. Spacecraft photograph of the same test site taken by Cooper and Conrad on GT–5, illustrating the difficulty of seeing the patterns at orbital altitudes.

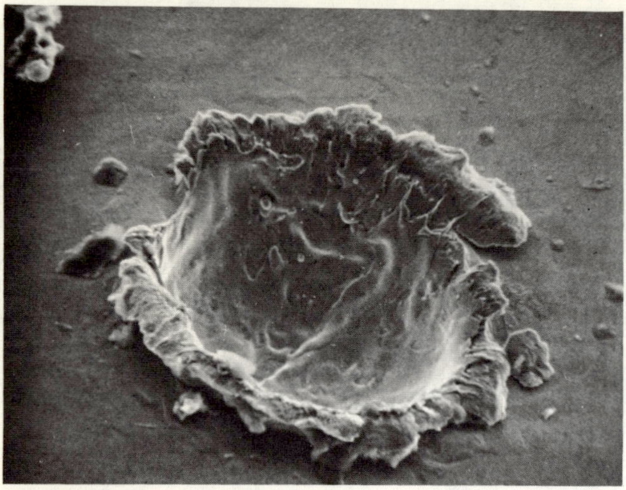

FIG. 15 Crater Schweickart, a pit produced from the impact of a micrometeor in stainless steel in the S–10 experiment on GT–10. Dimensions in microns are 160 diameter, 65 depth and 22 lip height (courtesy of the experimenter, C. L. Hemenway of Dudley Observatory, University of Pennsylvania).

micrometeorites could impact. The astronaut could not approach to open the box during extravehicular activity (EVA) on Gemini X but this was accomplished on Gemini XII. Results of the latter cannot be known since the Agena 12 due to a lack of fuel had to be left in a low orbit which will decay after a few months. The Dudley Observatory has installed a stereoscan electron microscope which will permit scanning the surface in original form, thus minimizing sample contamination. Results of this work will not be known for months.

The Gemini Micrometeorite Collection experiment (S–12) was successfully recovered on Gemini IX during which the experiment had an exposure of more than sixteen hours. Indications are that micrometeorites were collected. Two types of holes were found: (1) a long break in the support film; (2) small hole with a large amount of material around it. Some microorganisms survived (Penicillium, T_1 bacteriophage and tobacco mosaic virus), some did not (Stearothermophilus). For comparison another package was exposed for six hours during the GT–12 flight. For the Gemini XII six-hour exposure, preliminary examination of one square centimeter of the collection surface had revealed no impacts at the time of the astronaut debriefing of Gemini XII. Full analysis has yet to be completed.

Nuclear Emulsion Experiment

For the Nuclear Emulsions experiment (S–9), two groups, Naval Research Laboratory (NRL) under M. Shapiro and Goddard Space Flight Center (GSFC) under C. Fichtel, are working on the scanning of the emulsion stacks. At this time, NRL has finished the initial scan of about $\frac{1}{4}$ of their share and in this part, they have found about 700 tracks of heavy nuclei (charge six and above), which must be sorted out as to their origin either inside the spacecraft or outside during the activation of the experiment. It is estimated that about 200 of this portion will be those of interest to the experiment. If this percentage can be used throughout the analysis of the experiment, then it may be expected that between 1000 and 2000 usable heavy nuclei tracks will have been recorded. At the present time, the experimenters are doing a special kind of scan of the tracks (now about ten percent completed). The detailed scanning to give information on the heavy nuclei will be undertaken next and is expected to take one to two years to complete. The Goddard group is concentrating on the detailed scanning in order to make progress on the analysis of the heavy nuclei which was the main objective of the experiment.

Ion Wake

The objective of the S–26 Ion-Wake Measurement experiment was to measure and confirm the ion- and electron-wake structure and perturbation of the ambient medium produced by the orbiting Gemini spacecraft. The experiment was designed to obtain the following:

(a) A mapping of the spacecraft ion-density wake as a function of position coordinates relative to the reference frame of the spacecraft

(b) A contour mapping of the spacecraft electron-density wake as a function of the same position coordinates

(c) Determination of electron temperature as a function of the position coordinates

(d) Detailed information on ambient ion and electron densities and electron temperature as a function of altitude and diurnal variations from the Gemini Agena Target Vehicle (GATV)

(e) Ionization transients caused by spacecraft thruster firings.

The spacecraft motion is supersonic with respect to the ions and subsonic with respect to electrons which are omnidirectional whereas the ions (and neutral particles) are swept up by the vehicle motion. If the ionospheric constituents were completely at rest, a shadow zone would extend an indefinite distance behind the spacecraft. Random thermal motions of the omnidirectional electrons, by interacting mechanisms will rapidly fill in the shadow or hole region, resulting in a plasma rather than an ion-wake.

Fuel constraints limited the attainment of the experiment objectives on GT–10 but some useful data were obtained on GT–11 despite the nonfunctioning range and position systems at critical periods. Detailed correlations will entail considerable effort by Medved, the experimenter, but he was able to make such early tentative conclusions as (1) thruster firings in the Target Docking Adapter (TDA) south configuration produced a) decrease in ion flux to the outboard ion sensor, b) apparent increase in ion flux to the inboard ion sensor and c) enhanced electron concentration by the outboard electron sensor; (2) visual inspection of the strip-chart data shows that definitive cone angles can be determined, (3) apparent electron distribution, for many cases, follows the ion depletion effects, indicating that this wake is a plasma rather than an ion-wake.

Dim Light Phenomena

We have discussed what were more or less formal experiments. One of the chief characteristics of a manned satellite which distinguishes it from the unmanned type is that observations by the eye (and by photography) of an exploratory or confirmatory nature are possible and indeed desirable. In this section, non-scheduled observations which were performed on a scientific subject of opportunity on both Mercury and Gemini are summarized. Because many of these observations led to and generally provided training and experience for several later formal astronomical and air glow photographic experiments, these visual observations will be reported here, followed by discussion of some typical experiments such as S–11, S–13 and S–29 (see Table 3).

The manned Mercury orbital flights conducted from February, 1962 to May, 1962 established through visual observations by the astronauts the following general features:[9]

1) The night air glow band centered some 90 km above the earth is visible at all times on the night side of the earth. Visual measurements were made of the altitude, width and luminance of the air glow[7] and confirmed by rocket observations.

2) As seen through the Mercury window, the faintest stars observed at night, even under relatively ideal conditions, were described as of the fifth magnitude.

3) With no moon, the earth's horizon is visible to the dark-adapted eye. The earth's surface is somewhat darker than the space just above it, which is filled with the diffuse light of air glow, zodiacal light, integrated starlight and resolved stars. Clouds and coastlines are just barely visible to the dark-adapted eye.

4) With moonlight reflected on the earth, the horizon is still clearly defined, but in this case, the earth is brighter than the background of space. Indeed with moonlight, the clouds can be seen rather clearly and their motion is distinct enough to provide a clue to the direction of motion of the spacecraft.

5) The night sky (other than in the vicinity of the air glow band and horizon) appears quite black with the stars as well-defined points of light which do not twinkle. Lights upon the earth do twinkle when viewed from above the atmosphere.

6) The zodiacal light was successfuly observed by Cooper in the last of the Mercury flights but was not seen during the previous Mercury flights presumably because of cabin lights which could not then be extinguished.

7) A "high air glow" was observed on one occasion on the night side by both astronauts Schirra and Cooper. Schirra described this as a brownish "smog-appearing" patch which he felt was higher and thicker (wider) than the "normal" nightglow layer. Schirra observed such a patch over the Indian Ocean and Cooper while over South

America. It is possible that this phenomenon may have been a tropical 6300Å atomic oxygen emission first reported by Barbier and Glaume.[23]

8) Twilight is characterized by a brilliant, banded, multicolored arc which exists along the horizon in both directions from the position of the sun. On Mercury 8 during twilight an observation for the first time was made of a very remarkable scene. The scene is shown in Fig. 16a which is a black-and-white reproduction of Schirra's description (see ref. 8) of a series of blue bands. Figure 16b is a black-and-white reproduction of one of many 16mm color movie film frames

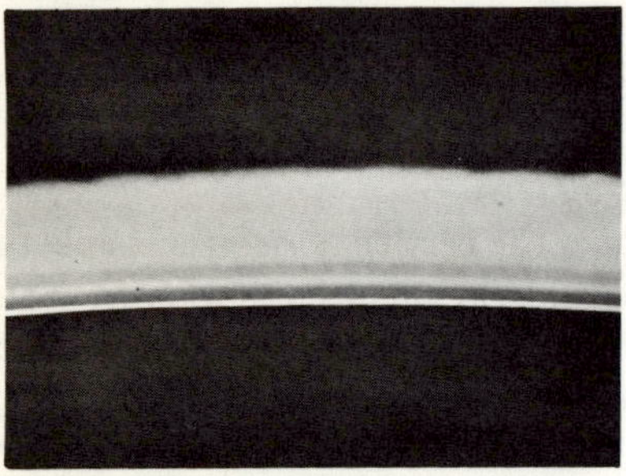

FIG. 16a. Artist's water color conception from Schirra's description of the twilight arc on MA-8, the first observation of structure in this band.

FIG. 16b. Frame from the 16mm movie film taken by McDivitt and White on GT-4 of the horizon arc showing bands thought to be confirmation of Schirra's visual observations (confirmed visually by Cooper on MA-9). The upper light, broad band does not show in the photograph as it is too faint to be recorded in this exposure time.

of this phenomenon photographed by McDivitt and White from GT–4. These color photographs were the first physical proof of Schirra's bands which had also been visually observed by Cooper from MA–9.

9) Finally, during the Mercury flights, the following phenomena were *not* observed:

(a) vertical structure in the nightglow
(b) polar auroras
(c) meteors
(d) comets

Four months before the launching of Gemini IV and subsequently before each mission, there was a significant change in the manner of

FIG. 17. Normal air glow layer photographed by Cooper on MA–9 showing its relation to the true horizon as delineated by the lightning flashes below, at the left of the picture. The altitude of the air glow above the true horizon is about three degrees.

FIG. 18a. Aerobee rocket photograph (Tri-X film, f/1.4 at 3 sec.) of the night air glow layer taken at a height of 184 km on Nov. 30, 1964. Earth's horizon defined by the uppermost bright ground light is about 1500 km distant.

scientific briefings to the astronauts at the Morehead Planetarium, Chapel Hill, North Carolina for star field and flight plan familiarization. The primary crew of McDivitt and White and their back-up crew, Borman and Lovell (who later flew the 14 day GT–7 mission) received many hours of briefings on astronomical and geophysical phenomena. The observations of the Mercury astronauts were reviewed and augmented by ground, rocket and unmanned satellite experimental data, e.g., Ney camera photographs by Cooper in Mercury and more detailed photographs subsequently obtained by Hennes and Dunkelman[24] from an Aerobee rocket experiment as shown in Figs. 17 and 18a,b respectively. Frequently operational

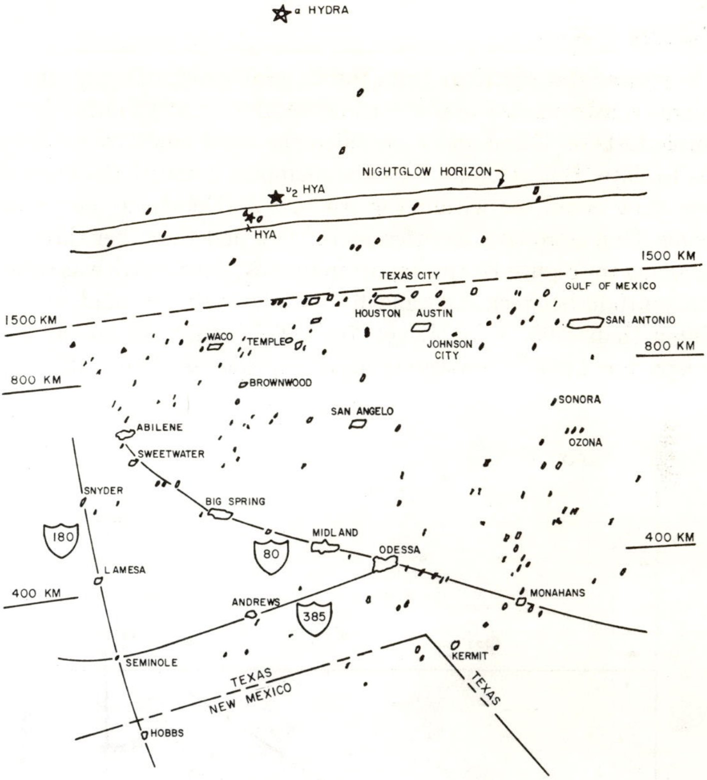

FIG. 18b. A map of the region pictured in a), identifying the brighter stars and cities (after Hennes and Dunkelman, ref. 24).

problems were resolved by the crew and crew support personnel with great appreciation by the experimenters.

From the Gemini flights IV through VII additional information was derived. These included:

1) Specific information on day and night star sightings.
2) Observations of Aurora Australis from Gemini IV and VII.
3) Meteors were first observed by Gemini IV astronauts and again by Gemini V, VI and VII.
4) Vertical structure in the night air glow was first observed and noted in the logbook by the Gemini IV astronauts.

Visibility of Stars

Nighttime star sightings from the Gemini spacecraft indicates that generally, astronauts can observe magnitudes slightly fainter than 6th. Simple tests on GT–6 and 7 provided the most objective evidence of this to date. Both Gemini VI crew members counted the number of stars they could see within the triangle β (Denebola) and δ and θ Leonis. One astronaut reported seeing two and the other three stars. Referring to Figure 19 this report indicates that Schirra was seeing to a magnitude between 6.00 and 6.05 while Stafford could see a star fainter than 6.05. Pilot Lovell on GT–7 could see to only 4.37. Except for Lovell's observation, which was seriously affected by

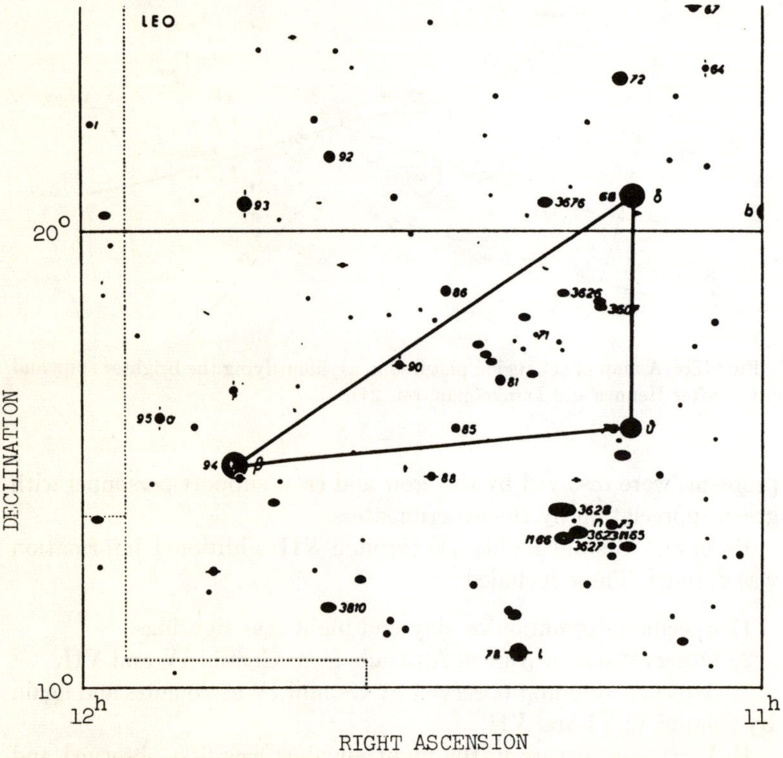

FIG. 19. Test card for nighttime star counting carried on board Gemini VI.

deposits on window surfaces, these compare well with less objective, but nevertheless important, sightings by the Gemini IV crew whose cards show stars in more than five well-known constellations, in which the constellation Corona Australis provided a critical test with stars identified down to 5.95 magnitude. Both McDivitt and White reported that they could easily see all stars on their card, in fact, to stars as faint as about seventh magnitude. All astronauts have made varying subjective comment during scientific debriefings that more nighttime stars are seen from the spacecraft than from ground observations and about the same or slightly more than from a high flying jet aircraft. A vigorous analysis of these results for both night and day star sightings is simply not possible because of the many unknowns that have a great bearing on the result, e.g., subjectivity.

The sighting of stars in the spacecraft *daytime* has been difficult. This is due to many factors, particularly scattered sunlight and earth light on any window. However, there have been several instances where daytime stars were definitely observed, two of these occurring in Gemini V and VIa. The Gemini V sighting occurred when proper precautions were taken during the performance of the S–1 experiment. From papers by Stafford et al.[14] and Roach et al.,[25] daytime sightings of stars prior to and during Gemini VIa rendezvous exercises are analyzed. Schirra was able to see stars as faint as fourth magnitude, as deduced from his observation of several stars in the "Sword" of Orion. Star and planet visibility during twilight has been treated comprehensively by Tousey and Koomen,[26] providing the basis for more complete analyses from future flights wherever simultaneous *photometric observations* of backgrounds and *visual* observations of stars are made.

Aurora Australis

Additional unscheduled observations and data obtained by man in the Gemini series are also of considerable interest. On at least three flights, the astronauts saw the Aurora Australis (southern lights) even though the zone is ordinarily outside the orbital latitude range of the flights. The auroral zone migrates to some extent and can sometimes be seen from equatorial orbital altitudes. McDivitt and White on GT–4 saw auroral sheets combined with the air glow projected against

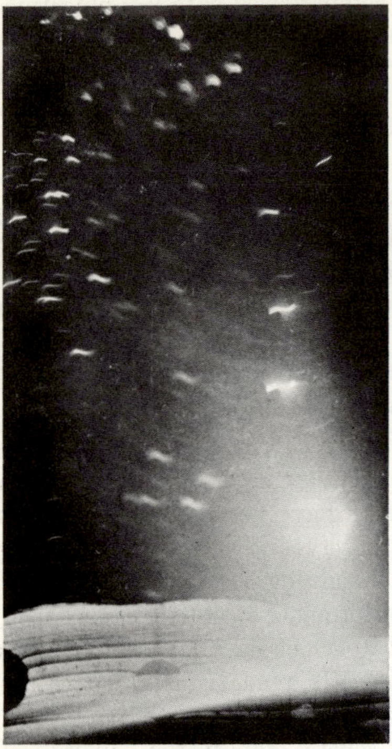

20a.

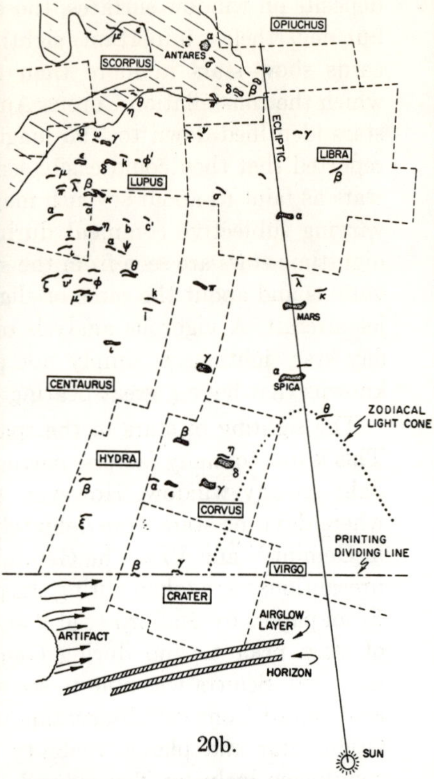

20b.

FIG. 20a. Zodiacal Light photographed by Stafford and Cernan on GT–9 with a 16mm camera, showing also the air glow and star field.

FIG. 20b. Sketch identifying bright stars and parts of constellations (courtesy. of the experimenter, E. P. Ney of the University of Minnesota).

the earth below them (Dunkelman, et al.[27]). Cooper and Conrad on GT–5 gave similar descriptions of auroral phenomena they saw. Borman and Lovell on GT–7 also saw them and drew sketches (see p. 321, Gemini Midprogram Conference[25]).

Meteors

Another phenomenon seen by some of the astronauts was meteors[25] in the earth's atmosphere but seen from above, rather than from below. GT–5 was aloft during one of the annual meteor showers (Perseids in August). GT–6a and 7 were in orbit during the December shower of the Geminids whose flux rate is lower than that of the Perseids. (Meteor shower names are derived from the constellation from which they seem to radiate, called the radiant.) The GT–5 team estimated 20 per hour, compared to about 6 per hour on the other flights.

During scientific debriefing, Aldrin reported that during EVA on GT–12, while in the dark, that when he rubbed his fingers together he would see a vary faint glow between his rubber-gloved fingers. This is probably the familiar phenomenon of static electricity.

Dim Light Photography

The first nighttime photography which led to a variety of Gemini dim light experiments was accomplished by Cooper in the MA–9 flight. The zodiacal light sequence, which is critical timewise, was missed. However, the air glow photography was very successful and F. C. Gillette et al.[28] reported their analyzed results. Figure 17 is a reproduction of one of the exposures showing the nightglow layer. The majority of the light in the visible region of the spectrum arises in a layer some 20 km thick and at an altitude of about 95 km. Densitometry of the MA–9 negatives showed that only 10 percent of the light was reaching the film because of the low transmission of the spacecraft multi-paned window. Throughout the manned program deposits on the windows have shown that effective transmission covered a wide range, sometimes affecting both visual and photographic observations.

The Gemini experiment (S–1) was proposed by E. P. Ney of the University of Minnesota in order to answer the following questions: (a) what is the smallest elongation at which zodiacal light measurements can be made from an earth-orbiting spacecraft above the air glow layer? and (b) can the gegenschein be detected and measured above the air glow layer with an f/1 lens and the fast Kodak Tri-X film, successfully used by Hennes and Dunkelman[24] in their Aerobee rocket air glow photography November 30, 1964? Ney and Huch[29] describe the conduct of the experiment which is typical of many that require acquisition, orientation and guiding. Their results showed that the smallest elongation at which the zodiacal light could be studied on any of the 15 photographs obtained was 16° (compared to about 20° from the ground).

Figure 20a is a print of one frame which shows the zodiacal light; Figure 20b is a sketch of the star field shown in Fig. 20a. The top

FIG. 21. Another photograph of the Zodiacal Light taken by Stafford and Cernan on GT–9. The bright object near the tip of the cone of the Zodiacal Light is the planet Venus whose image shows distortion by the simple lens. The narrow band just above the bright layer (moonlit earth) near the bottom of the picture is the airglow layer (after Ney and Huch, ref. 29).

portion of the frame was printed with 40 times the exposure of the bottom. This ratio brings out the earth limb and the airglow layer. The air glow layer, viewed in profile, has about the same surface brightness as the zodiacal light at 16° elongation. The zodiacal light at this elongation is 4×10^{-12} of the sun's surface brightness. The ratio of profile to zenith brightness of the airglow is about 40, so the inferred zenith air glow brightness is 10^{-13} of the sun's surface brightness. This is about three times brighter than the average nighttime air glow but within the range of variation commonly observed. There were also exposures of thruster firings of use to engineers and scientists. Two of the frames appeared to show the gegenschein having an angular size of about 10°, the center being within a few degrees of the direction opposite to the sun. On GT–9, the S–1 experiment was to have been performed outside the spacecraft, but visor fogging during EVA cancelled it. The astronauts took seventeen good photographs through the window, however, two of which are reproduced as Fig. 21 and Fig. 22. They will be analyzed with an isodensitometer for morning and evening intensity distribution of the zodiacal light and height and intensity of the air glow at various geographic positions.

Another air glow experiment, S–11 was performed on Gemini IXA, XI and XII using a general-purpose 70mm camera with f/0.95 lens. The objective of the experiment was to obtain air glow data on a global scale. In contrast to the S–1 experiment the camera was optically filtered to photograph air glow in the visible bands centered at 5577Å (atomic oxygen emission) and 5893Å (atomic sodium emission). Forty-four air glow photographs were obtained by the Gemini IXA astronauts. The photographs were of excellent quality indicating a stable camera (which was bracket mounted) during the exposure. The photographs show clear variations in altitude and intensity of the night air glow layer. The experiment was flown again using the same camera with additional filtering to isolate the 6300Å atomic oxygen radiation. Many useful air glow photographs were obtained covering northern, southern and eastern air glow horizons. A light leak in the camera system made the 6300Å exposures taken during the high-apogee orbit of unusable quality. Twenty air glow photographs were obtained on GT–10 and analyses by Koomen gave evidence of the

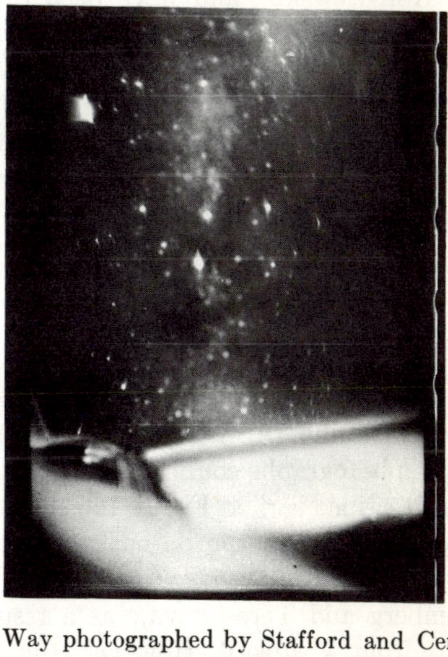

FIG. 22. Milky Way photographed by Stafford and Cernan on GT–9. The group of brightest stars is the familiar Northern Cross (upside down), officially called Cygnus, the Swan (flying up the Milky Way). Courtesy of the experimenter, E. P. Ney, of the University of Minnesota.

existence of a higher air glow layer at 200–300 km as well as the wisps extending upward from the lower air glow layer first photographed on GT–9. Gemini XII produced many good pictures of the sunlight and night air glow in the 5577Å, 5893Å and 6300Å emissions. The twilight photographs with the 6300Å filter, for example, show a low-layer emission band, presumably caused by the OH radical. This is the first photograph of an OH layer from a spacecraft and represents an interesting and useful addition to air glow observations, according to M. J. Koomen of the Naval Research Laboratory (NRL).

Libration Clouds in Earth-Moon System

Among the several astronomical experiments was the Libration Regions Photography (S–29). The purpose was to investigate by photographic techniques the libration points (L–4 and L–5) of the earth-moon system to determine the possible existence of clouds or particulate matter orbiting the earth in these regions. These were first reported by Kordylewski[33] and have been called the Kordylewski Clouds. The Gemini XII mission was the first one on which any libration region was available for photography. These regions (analogous to the relation of the Trojan asteroids to Jupiter) are to be found in the lunar orbit in locations 60° preceding and 60° following the moon. The experiment was performed but the shutter of the camera failed to close and no results were obtained.

Twilight Bands

On Gemini IV McDivitt and White obtained excellent twilight photographs using color film (ASA 64) exposed in a 16mm motion picture camera. The photographs confirmed the existence of the light blue Schirra band[8] described earlier. Figure 16b shows vividly the blue banding originally described in the on-board tape of MA–8.[8] The structure in the foreground (lower left corner) is similar to that reported by Rozenberg and Tereshkova,[3] as a result of black and white photographs from Vostok 6. However, the Schirra band of interest was not recorded on the Vostok black and white photographs. Mateer et al.[30] have analyzed the GT–4 16mm frames and concluded that this band is due to an aerosol layer in the upper atmosphere.

Another interesting result has been reported by Ney and Pepin.[31] They have determined stellar extinction from data obtained on GT–9 in which they found no detectable extinction, due to a high dust or absorbing layer near 100 km suggested by Link.[32]

Ultraviolet Stellar Astronomy

The UV Astronomical Photography (S–13) which used the 70mm Maurer camera obtained usable spectra from the last three Gemini flights. The experimenters report (K. Henize et al.)[34] a fine UV spectrum (region 2200–4000 Å) of the bright star Canopus a solar-type star which was useful for calibration purposes [see Fig. 23a,b taken by Conrad and Gordon on GT–11]. It was possible to compare this spectrum of Canopus with the rocket solar spectra already published by R. Tousey some years earlier.[35] A widened spectrum of γ Cas was obtained on GT–12 which showed no lines. This star is of particular interest for having ejected a shell which subsequently dissipated. Within the past year, there have been indications of hydrogen emission lines. The GT–12 observations provide negative evidence for the return of the shell at this time. A grating spectrum of Sirius (Fig. 24) was also obtained on GT–12 by Aldrin and Lovell. They obtained in addition to these two remarkable grating spectrograms, a number of grating and prism spectrograms of the Orion field. The latter have lower dispersion but give information on a large number of stars. Henize reached stars of fainter magnitude than he anticipated and it is expected that there will be work to be done on the UV energy curves for many months as a result of these photographs. Two of these Orion spectrograms are reproduced in Figs. 25 and 26. These UV experiments made use of the EVA and would not have been possible without it since the spacecraft windows do not admit ultraviolet light. In the chapter entitled *Space Astronomy* an illustrated discussion of this experiment is made.

Comparison of UV and Visible Phenomena

It is of interest here to note with the aid of Table 5, the large variety and intensity range of celestial extended sources. The table is

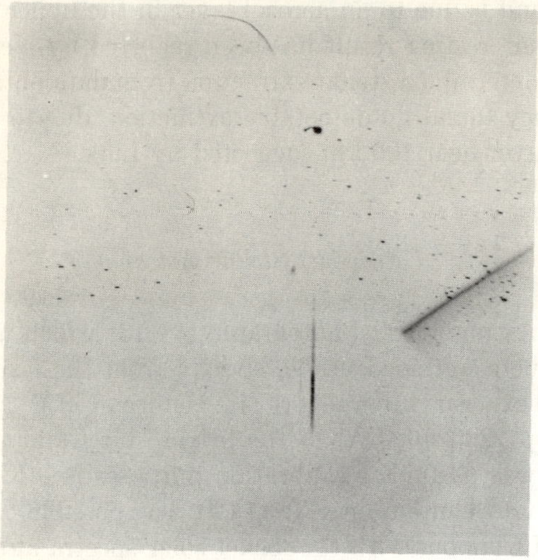

FIG. 23a. Air glow and ultraviolet spectrum of Canopus, photographed by Conrad and Gordon on GT–11 showing the zero order image of Canopus (large star image above center) and its first order spectrum (vertical line).

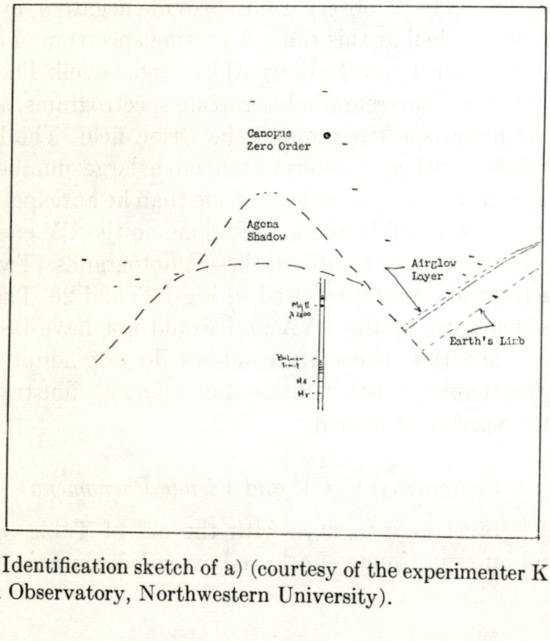

FIG. 23b. Identification sketch of a) (courtesy of the experimenter K. G. Henize of Dearborn Observatory, Northwestern University).

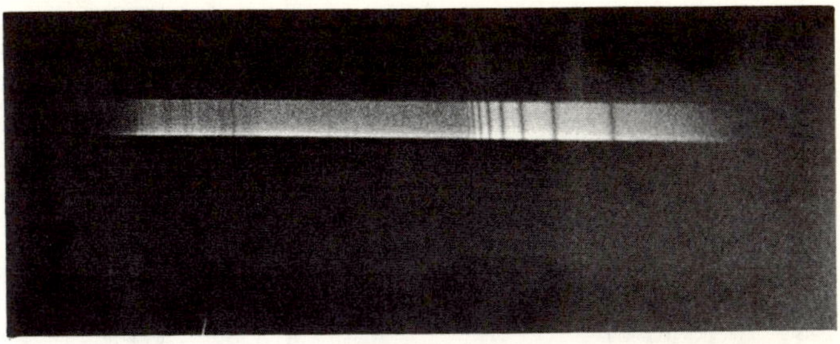

FIG. 24. Ultraviolet grating spectrum of Sirius photographed by Lovell and Aldrin on GT–12 showing the Balmer series of hydrogen lines at the right, and the Mg. II doublet and weak lines of FeII at the left (courtesy of the experimenter, K. G. Henize of Dearborn Observatory, Northwestern University).

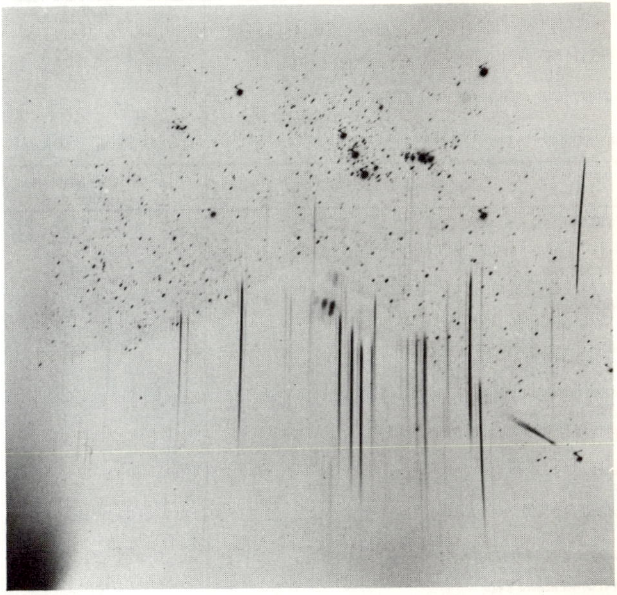

FIG. 25. Ultraviolet *grating* spectra of the Orion field of stars photographed by Gordon on GT–11. Conrad used the coupled Agena vehicle to occult the zero order images of many stars thus minimizing background interference of zero order images with the desired first order spectra. Note how well the spectra are revealed against a clean background (courtesy of the experimenter, K. G. Henize of Dearborn Observatory, Northwestern University).

TABLE 5†

Phenomenon	Luminance (candles cm^{-2})
Milky way, dimmest region, near Perseus	1×10^{-8}
Gegenschein	1.6×10^{-8}
Visible night glow (zenith)	2×10^{-8}
Milky way brightest region, near Carina	4×10^{-8}
Zodiacal light, (30° elongation)	1.2×10^{-8}
Visible night glow (edge-on)	6×10^{-7}
Great Orion nebula M42	5.6×10^{-6}
Full moon	4×10^{-1}
Fluorescent lamp 4500 white	4×10^{-1}
Phenomenon	**Radiance (ergs sec^{-1} cm^{-2} $ster^{-1}$)**
Middle UV night glow (nadir), 2600 A	2×10^{-5}/100 A
Middle UV nightglow (edge-on), 2600 A	5×10^{-4}/100 A
Hydrogen lyman-alpha glow (mean), 1216 A	3×19^{-3}
Sunlit earth albedo (narid), 2600 A	5×10^{-2}/100 A
Phenomenon Visible, Ultraviolet, Combined	**Relative brightness (photon flux)**
Middle UV nightglow (nadir), 2600 A	0.03
Milky way (mean)	0.7
Gegenschein	0.8
Middle UV nightglow (edge-on), 2600 A	0.8
Visible nightglow (zenith)	1
Hydrogen lyman-alpha glow (mean), 1216 A	2
Aurora IBC-I	3
Zodiacal light (30° elongation)	6
Visible nightglow (edge-on)	30
Great Orion nebula M42	300
Sunlit earth UV albedo (nadir), 2600 A	800
Aurora IBC-IV	3000
Full moon	2×10^{7}

†Reproduced from Table I of L. Dunkelman and J. P. Hennes, Jap. Jour. App. Phys. 4, Suppl 1, (1965).

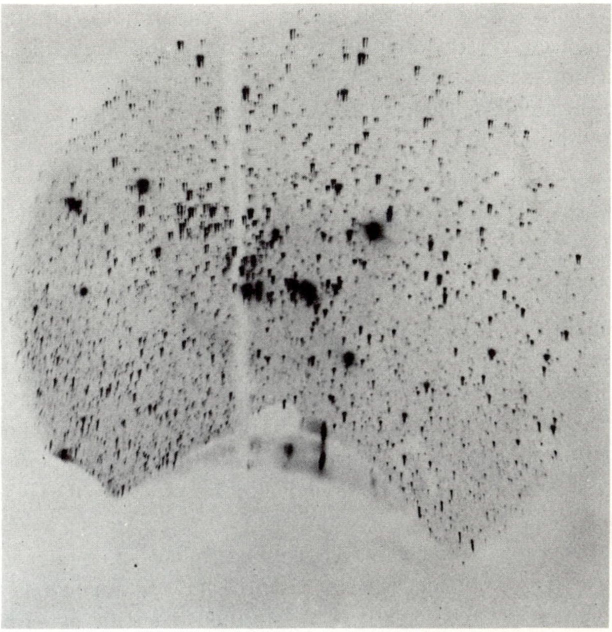

FIG. 26. Ultraviolet *prism* spectra of the same field of Orion as in Figure 25. These latter two figures illustrate the differences, advantages and disadvantages of the two types of spectra. More detailed but fewer spectra are obtained with a grating, while less detail but many spectra can be obtained with a prism spectograph with similar exposure times (courtesy of the experimenter, K. G. Henize, Dearborn Observatory, Northwestern University).

divided into three sections. The first relates to phenomena in the visible region of the spectrum and is intended to provide 1) a convenient reference to objects fairly well known, within our normal experience and 2) a means with which one can compare ultraviolet with visual sources. The astronauts have observed these visible phenomena. Note that the visible nightglow (edge-on) is uniquely visually observed from the vantage point of earth orbital spacecraft. In the first section the luminance (brightness) of each source is given in cd cm^{-2}. In the second section of the table there is listed celestial UV sources, all of which must be measured from above the ozonosphere. Here the radiance (brightness) is given in ergs cm^{-2} sec^{-1}

ster^{-1}. These UV sources will be "seen" by astronauts equipped with UV image converter monoculars now becoming available. Finally, in the third section of the table, some of these sources are listed on a relative brightness scale. The UV fluxes have been converted to the corresponding 5500Å fluxes on a one-for-one photon basis.

SUMMARY

The Gemini program was highly sucessful from the engineering and technological standpoints. Particularly, it demonstrated the laws of physics and motion very well, especially those of inertia, direction (changing orbits) and action and reaction. The latter presented great difficulties in the extravehicular activities (EVA) frequently terminating or aborting them before scheduled tasks were performed or completed and resulting in the cancellation of some of the scientific experiments as well.

The program demonstrated (1) Feasibility of rendezvous and docking of two vehicles; (2) Capacity of men to survive 14 days of weightlessness in good health; (3) Manned scientific experiments in space; (4) Partially controlled re-entry and landing, and (5) Some new technologies, e.g., fuel cells to generate electric power.†

The performances of the astronaut in flight and pilot procedures were outstanding, making the difficult maneuvers of rendezvous and docking seem very easy. The altitude record of 850 miles was also set in this series. Many space firsts were accomplished such as rendezvous, docking, space working, inspecting and recovering articles from other space vehicles. The men's adaptation, calm and resourcefulness in times of stress and crisis were shining examples of the usefulness and advantage of man in space. It helps assure us of attaining our goals to explore the solar system in person. The Apollo program will soon help us to reach the first stop in the solar system—the moon.

† For further information of the Gemini missions, the reader is referred to the Manned Space Flight Experiments Symposium: Gemini Missions III and IV; Oct. 18–19, 1965 and the Gemini Midprogram Conference, Including Experiment Results, NASA SP-121, Feb. 23–25, 1966; and Gemini Summary Conference, February 2–3, 1967, NASA SP-138.

REFERENCES

1. Y. M. Volynkin, et al., translation from Pervyye Kosmicheskiye Polety Chelovska Medico-Biol. Issledovaniya (Moscow), (1962) FTD-II-62-1619/12.
2. A. A. Blaganravov, The Flight of Space Ships Vostok 3 and Vostok 4, The First Group Space Flight in the World, Fourth International Symposium of the Cospar at Warsaw, Poland, June 1963.
3. G. V. Rozenberg and V. V. Nikolaeva-Tereshkova, Stratospheric Aerosole Measured from a Spaceship, J. of Atmospheric and Oceanic Phys. (USSR), 1,386, 1965.
4. W. S. Cameron, J. H. Glenn, M. S. Carpenter and J. A. O'Keefe, The Effect of Refraction on the Setting Sun as Seen from Space in Theory and Observation, *Astronom. Jour.* **68,** 348, (1963); also NASA TN-D-1721, 1963.
5. Results of the First U. S. Manned Orbital Space Flight, February 20, 1962.
6. Results of the Second U. S. Manned Orbital Space Flight, May 24, 1962, NASA S-6.
7. M. S. Carpenter, J. A. O'Keefe and L. Dunkelman, Visual Observations of Nightglow from Manned Spacecraft, Science **138,** 978, (1962).
8. Results of the Third U. S. Manned Orbital Space Flight, October 3, 1962, NASA SP-12.
9. Mercury Project Summary, Including the Results of the Fourth U. S. Manned Orbital Space Flight, May 15-16, 1963, NASA SP-45.
10. Newsweek, October 26, 108, 1964.
11. Press releases from Tass, Pravda, and Izvestia news agencies.
12. J. R. Gill and S. J. Gerathewohl, The Gemini Scientific Program, Astronautics and Aeronautics, November 1964, p. 58-65.
13. E. C. Lineberry, Gemini VI—A Rendezvous Mission Planning, Gemini Midprogram Conference, NASA SP-121, U. S. Gov't. Printing Off., Washington, D. C., 277, 1966.
14. T. P. Stafford, W. M. Schirra and D. F. Grimm, Rendezvous of Gemini VII and Gemini VIA, Ibid., 283, 1966.
15. M. A. Bender, Radiation Research (in press), (1967).
16. P. D. Lowman, Jr., A Review of Photography of the Earth From Sounding Rockets and Satellites, NASA TN-D-1868, Dec. 1964.
17. P. D. Lowman, Jr., The Earth From Orbit, Nat. Geog. Mag. **130,** No. 5, 645-671, Nov. (1966).
18. P. D. Lowman, Jr., J. A. McDivitt and E. H. White, Terrain Photography on the Gemini IV Mission: Preliminary Report, NASA TN (D-3982).
19. A. Morrison and M. C. Chown, Photography of the Western Sahara Desert from the Mercury MA-4 Spacecraft, NASA Contr. Rep't. CR-126, Dec. 1964.
20. L. Hubert, Mesoscale Cellular Convection Meteorological Satellite Rep't. No. 37, May 1966.
21. J. P. Kuettner and S. D. Soules, Organized Convection as Seen from Space, Amer. Meteorol. Soc. **47,** 364-70, (1966).

22. S. Q. Duntley, R. W. Austin, S. H. Taylor and J. L. Harris, Experiment S–8/D–13, Visual Acuity and Astronaut Visibility, Chap. 34, Gemini Midprogram Conference, NASA SP–121, 329, 1966; and S. Q. Duntley, R. W. Austin, S. H. Taylor, J. L. Harris and W. H. Richardson, J. Opt. Soc. Amer. **56,** 538A, (1966).
23. D. Barbier and J. Glaume, Ann. de Geophys. **16,** 319, (1960).
24. J. Hennes and L. Dunkelman, Photographic Observations of Nightglow from Rockets, J. Geophys. Res. **71,** 755 (1966).
25. F. E. Roach, L. Dunkelman, J. R. Gill and R. D. Mercer, Geoastronomical Observations, Chap. 32, Gemini Midprogram Conference, NASA SP–121, 315, 1966.
26. R. Tousey and M. J. Koomen, The Visibility of Stars and Planets During Twilight, J. Opt. Soc. Amer. **43,** 177–183, (1953).
27. L. Dunkelman, J. R. Gill, J. A. McDivitt, F. E. Roach, and E. H. White II, Geoastronomical Observations, Manned Space Flight Experiments Symposium: Gemini Missions III and IV, Oct. 18–19, Wash., D. C., (1965). NASA Publication.
28. F. C. Gillett, W. F. Huch, E. P. Ney and G. Cooper, Photographic Observations of the Airglow Layer, J. Geophys. Res. **69,** 2827–34, (1964).
29. E. P. Ney and W. F. Huch, Optical Environment in Gemini Space Flights, Science, **153,** 297–9, 1966.
30. C. L. Mateer, J. V. Dave, L. Dunkelman and D. C. Evans, Evidence for a High Level Stratospheric Dust Layer from Satellite Twilight Color Photographs, *Roy Meteor. Soc. Quart. J.* (*in press*).
31. E. P. Ney and T. J. Pepin, Observations on Stellar Extinction Obtained in the Gemini IX Space Flight, Nature, **211,** 1382–3, (1966).
32. F. Link, Lunar Eclipses, Chapter 6, Z. Kopal, Ed., Physics and Astronomy of the Moon, Academic Press, N. Y., 1962, p. 161–229.
33. K. Kordylewski, Photographic Investigations of the Libration Point L_5 in the Earth-Moon System, Acta Astron. **11,** 165 (1961).
34. K. G. Kenize, L. R. Wackerling and F. G. O'Callaghan, Ultraviolet Stellar Spectroscopy on Gemini XI, Science **155,** 1407, 1967.
35. R. Tousey, The Extreme Ultra Violet Spectrum of the Sun, Space Sci. Rev., **2,** 3, (1963).

Part III

THE SOLAR SYSTEM AND BEYOND

INTRODUCTION

This part of the book deals with the massive objects in the solar system, and the stars and galaxies. This encompasses much of the classical discipline of astronomy. Space research in this area has started but not progressed very far yet. There are several large satellites planned to gather information in this area in the next few years.

Three Ranger and four Lunar Orbiter spacecraft have taken thousands of detailed photographs of the moon and two Surveyors have soft-landed on the lunar sufrace and tested its properties. Mariner has flown past Venus and Mars to study their atmospheres and surfaces. Unmanned scientific exploration of the planets will be a major effort of the next decade. Very likely, probes will be sent in close to the sun to study its atmosphere and surface. Also, we are likely to see large telescopes put into orbit near the earth to get above the earth's atmosphere.

Chapter 15

ORIGIN OF THE SOLAR SYSTEM

A. G. W. Cameron

INTRODUCTION

The origin of the solar system is an important problem of scientific philosophy that has engaged the attention of leading scientists for a little more than 300 years.[1] Probably the first person to attempt a serious scientific answer to this problem was Descartes, who proposed ideas that bear a remarkable resemblance to some of the schemes that have been seriously advanced in the very recent past. Many first-rate scientific investigators have published serious speculations on the subject during these last three centuries, but the problem still remains essentially unsolved. It is interesting to ask why it has been so resistive to a definite solution.

Until recently, there was very little data upon which to base theories of the origin of the solar system. We know, from studying the relative abundances of the parent and daughter isotopes of radioactive substances contained in meteorites, that the meteorites have remained essentially unchanged for the last 4.5×10^9 years. Similar studies of rocks on the surface of the earth show that the oldest of them have remained unchanged for only some 3.3×10^9 years. Thus the first 10^9 years of terrestrial history have been completely obliterated. During this period, the earth presumably accumulated from smaller bodies in space, became heated in the interior, and became chemically differentiated. The record of these events has been washed away due to erosion by water and air and due to continual shifts in the surface topography of the earth, but it may be that the crust of the earth did not exist at all during the first 10^9 years of the earth's history. This indicates that we cannot investigate the early history of the earth by examination of the rocks that we find upon its surface. We must look outside the earth in order to find bodies which still retain the record of earlier events written upon their surfaces and in their interiors.

Until recently all theories of the origin of the solar system have attempted only to explain certain large-scale regularities of motion and position within the solar system itself. These regularities are of essentially two kinds: the alignment of the angular momentum vectors within the solar system, and the regularity of the increase in the spacing of the orbits of the planets within the solar system, and of the satellites about the major planets. The alignment of the angular momentum vectors within the solar system is demonstrated by the fact that the planets rotate in essentially the same plane about the sun and with the same direction of motion, and that this plane also very nearly contains the equator of the sun, and the sun itself spins in the same direction of rotation. Similarly, the equatorial planes of the planets tend not to be inclined very greatly to the common plane of their motions, and the planets rotate in the same sense. The satellites of the planets also exist in planes that are closely aligned with the equatorial planes of the planets, and the motion of the satellites about their parent bodies also has the same sense. The regularity in the spacing of the planetary orbits has been known for a long time as *Bode's law*, and we can very rapidly characterize it by saying that each orbit has about 1.5 times the radius of the next innermost orbit. The same law is also approximately true for the regular satellites of Jupiter, Saturn, and Uranus.

In recent years many other kinds of evidence about the early history of the solar system have come to light. The foremost source of this evidence lies in the chemical and physical investigations of the properties of meteorites. These bodies occasionally drop on the earth from outer space, and it is generally believed that they represent the fragments of collisions between asteroids, small bodies lying between the orbits of Mars and Jupiter. By studying the details of the chemical and physical texture of the rocky material out of which some of the meteorites are formed, it is possible to deduce much about the actual conditions under which the formation of the material took place. Similarly, some of the meteorites contain gases, trapped in their interiors, which are emitted upon heating, and we can learn very much from the study of the relative amounts of the different elements included within these gases, as well as from the isotopic composition of the elements themselves. The importance of these gases lies in the fact that they represent surviving samples of the solar nebula out of which it is believed that the planetary system has condensed. In the very near future, we shall extend our investigation of the chemical and physical properties of the bodies in the solar system to the moon and the nearer planets. Some of these very interesting investigations in space research should allow us to learn many more and very valuable boundary conditions that must be satisfied by any theory of the origin of the solar system.

THEORIES OF SOLAR SYSTEM ORIGIN

The theories of the origin of the solar system that have been proposed during the last three centuries for the most part fall into two classes: *Dualistic* and *Monistic*. The basic idea of a dualistic theory is, generally, that there is a close passage of two stars in our galaxy, one of them usually being our sun, and it is assumed that tongues of gas are drawn out from each of the stars by the gravitational field of the passing intruder. The planets are then assumed to condense from these tongues of gas.

The first form of a dualistic theory was suggested in 1745 by Buffon. He suggested that a comet hitting the sun might have been responsible for knocking off sufficient material to produce the planets. He suggested that the comet hit the sun somewhat skewly, thus producing the rotation of the planets. In fact, he suggested that the planetary rotation was so fast that rotational instability ensued, and satellite systems were thus spun off from the planets. The collision was also supposed to have produced the rotation of the sun.

This seems nonsense to us today, but it must be remarked that in Buffon's time there were fantastic ideas about comets. Buffon estimated the mass of the comet of 1680 as 28,000 times the mass the earth, which is about the mass of the sun. It is, therefore, clear that Buffon really had in mind encounters like those which are described in dualistic theories in which one star makes a close approach to the sun.

Further work upon dualistic theories was carried on toward the end of the last century. In 1878, Bickerton attempted to develop the theory of an encounter between two stars which also produced a nova-like explosion, with the eruption leading to the formation of the planets. In this century Arrhenius published in 1913 a theory in which there was a head-on collision of two stars leading to one star and a gaseous filament. At about the same time, in 1910, T. J. J. See considered the collision of two *nebulae* leading to one *nebular protosun* which subsequently captured the planets. In the early years of this century, Chamberlin and Moulton developed the *planetesimal theory* which considered an encounter between two stars in which a large tidal protuberance was raised on the surface of the sun facing the intruding star. Because of the expansion of the hot gases in this tidal protuberance, a series of clouds of matter was ejected by great eruptions from the filament. These clouds were supposed to be accelerated into a sideward motion about the sun by the motion of the passing star. The planets were supposed to have been formed by a condensation of liquid drops of matter out of the clouds, and subsequent accumulation of these solid particles into larger bodies. Jeffreys and Jeans later produced modified forms of this theory. They postulated that the passage of the star close to the sun drew out a long cigar-shaped filament of matter which subsequently broke into separate fragments. The larger planets were to arise from the fattest central part of the cigar.

Very few serious investigators still believe in any kind of dualistic theory. When such a theory is subjected to quantitative analysis, it appears to fail on several important grounds. In the first place, gas drawn out of the stars should be so hot that it would tend to disperse in space rather than have condensation of solid material take place in it. In the second place, the gas that is drawn out from one star by the passage of another will have such a trajectory that more than 99 percent of it will fall back onto its parent body, as

Lyttleton has shown recently, and thus it will not be available for planetary formation. Perhaps most important of all we now know that a star like our sun will destroy all the deuterium throughout its interior by means of thermonuclear reactions when it is still in the process of contraction toward its present size. Thus, in a solar system formed by the gases drawn from such a star, the earth and meteorites could not have any deuterium in them; and this is contrary to the observation that both of these objects contain a little more than one part in ten thousand of deuterium relative to hydrogen.

The basis of a monistic theory is that a cloud of gas and dust forms in space about the sun, possibly at the same time as the sun itself is formed. In such a theory, the planets are believed to condense out of this gas and dust cloud, with the bulk of the gas later being dispersed in some fashion. Most investigators believe that some theory of this sort must be the correct one.

The original theory of Descartes was of the monistic type. He published a vortex theory of the formation of the solar system in 1644, before Newton had found the general law of gravitation, and at the time that the scientific method of investigation was being formulated as a method. Descartes postulated that there was a general vortex motion in the primitive ether which set up innumerable more-or-less circular eddies of all sizes. The friction between such eddies would grind down the rough shape of any primordial matter that is present. The small pieces would tend toward the center of the vortex forming the sun, while the larger pieces would stay at larger distances and form the planets. Secondary vortices would form about the planets which would capture additional material to form the satellites.

These ideas remained dormant for more than a century until 1755, when Kant published his cosmogonical ideas about ten years after the publication of Buffon's comet theory. Kant started with the universe filled, more or less uniformly, with gas, and assumed that regions of larger density would act as sinks for matter and that condensations would take place there. Thus the sun would form from a nebular mass which would be flattened due to rotation and, in the disk thus formed, secondary mass concentration would grow which would become the planets.

A different version of such a nebular hypothesis was published in 1796 by Laplace. He assumed a large primitive nebula which gradually cooled and contracted. During the contraction, the rotational velocity increased until centrifugal forces at the equator became larger than the gravitational forces, and part of the matter was left behind in the form of rings. The planets were then supposed to condense out of these rings.

Further developments of the nebular hypothesis did not occur until half a century ago. In 1912, Birkeland introduced a new idea into the problem. He assumed that the sun emitted ions and that the ions spiralled in the sun's magnetic field until they could collect to form the planets. Birkeland was thus the first to introduce magnetic fields into the problem of the formation of the solar system. He was followed by Berlage, who produced a number of theories in which the motions of ions in the magnetic field were supposed to form the planets in some way. In turn, Berlage has given way to Alfven, who more recently has produced a theory of this type, except that the ions are supposed to be formed by the fall of neutral material from space toward the sun, whereupon it becomes ionized, and the ions are then trapped by the solar magnetic field.

The modern revival of monistic theories undoubtedly is due to von Weizacker, who in 1944 introduced ideas concerning the role of turbulence in a primitive solar nebula. He considered a primitive nebula formed about the sun, in which there would exist massive eddies in a hierarchy of sizes. The turbulent motions of these eddies are supposed to bring those particles together which are contained within them, and these particles are thus supposed to accumulate into larger and larger bodies. The largest eddies are supposed to have a regular shape in space, with increasing dimensions outward from the sun corresponding approximately to Bode's law. The planets are supposed to be formed at the boundaries between such larger eddies, resulting in a reasonable spacing.

A few years later, Kuiper was attracted by these ideas and attempted to develop the turbulent eddy picture somewhat further. He rejected von Weizacker's regular system of eddies, and considered that the turbulent motion would be completely chaotic. In Kuiper's theory the main formation of the planets is supposed to arise through large scale condensations in the solar nebula which form protoplanetary bodies. Chemically condensable material settles to form the planets at the center of these protoplanetary bodies, and presumably, the satellites form farther out in the protoplanets, but most of the gases are eventually dissipated by particle streams from the sun.

A further alternative picture to von Weizacker's theory was presented by Ter Haar, who pointed out that there were immense difficulties with the time scale of dissipation of the gas due to turbulence. In subsequent years, some question has arisen as to the role that turbulence can play in the primitive solar nebula, since the particles of gas tend to have restoring forces which prevent their being moved very far from their proper orbital positions. Such forces will presumably prevent the formation of very large eddies within the gas. However, this is a problem which has not yet been properly investigated.

It is worth noting that the difference between the basic character of dualistic and monistic theories makes a tremendous difference in our estimate of the relative probabilities for the formation of planetary systems within our galaxy. A close passage of stars is a very rare event, and perhaps during the entire history of our galaxy we would not have had more than one or two sufficiently close passages of two stars to form a planetary system. On the other hand, a monistic theory leads to the idea that planetary systems must be very common, particularly if the primitive planetary nebula which is formed about the parent star is a natural consequence of the star formation process itself.

Just two decades ago, the curious situation existed that measurements of the distances of various galaxies and the rates of their recession from us, according to the magnitude of their redshift, suggested that all such galaxies were flying away from a common point point at which they had been just 2×10^9 years ago. At the same time, radioactive dating had shown that the surface rocks of the earth were in some instances more than 3×10^9 years old. To the extent that any attempt was made to reconcile these numbers, it was generally believed that the solar system was somehow a product of the very early stages of the development of the universe. This encouraged people to make the assumption that the initial conditions for the formation of the sun were entirely unknown and might be very different from those which are presently observed where star formation takes place.

At the present time this attitude is no longer a tenable one. The age of the earth and of the meteorites has now been increased somewhat over the earlier figure; it is now determined rather precisely from radioactive dating to be 4.5×10^9 years. However, the distance scale of the distant galaxies has undergone a drastic revision, and consequently, when the motions of these galaxies are projected back into the past, they would appear to meet at a common point rather more than 10×10^9 years ago. The present distance scale is so uncertain that this number still cannot be given very precisely. However, much study has also taken place of the age of various objects in our galaxy. In particular, it is possible to assign ages to various star clusters within our galaxy by studying the distribution of the luminosities and surface temperatures of the stars within the clusters and by comparing these descriptions with those which are predicted by modern theories of stellar evolution. The results indicate that the galaxy is at least 10×10^9 years old, and that it may be much older still, perhaps older than 20×10^9 years. Because star formation in our galaxy is believed to have taken place more rapidly in the early stages of its history than in the later stages, it now appears that the conditions in the galaxy at the time our sun was formed cannot have been qualitatively different from those we see today. Thus, it seems very desirable to determine under what conditions one might obtain a primitive solar nebula as part of the star formation process.

INVESTIGATIONS OF INTERSTELLAR GAS

In recent years one of the most interesting subjects that has been investigated, by the techniques of both optical and radio astronomy, is the gas that lies between the stars in our own and other galaxies. About 2 per cent of the mass of our galaxy appears to lie as gas and dust between the stars.

The interstellar gas represents the steady state condition in which there is a balance between the gain of gas due to the destruction of old stars and the loss of gas due to the formation of new stars. Old stars sometimes lose mass into space by means of a steady outward streaming process that can continue for a long time, similar to the loss of gas in the solar wind. Sometimes they lose mass explosively, as in *nova* and *supernova explosions.*

One of the interesting and significant aspects of the supernova explosions is that they should scatter into space some of the products of the nuclear reactions that have been going on in the interiors of the parent stars. This process will be described further in a subsequent chapter. Some of these products are radioactive, and hence we can expect that the interstellar medium will always contain a certain mixture of radioactive elements with a wide variety of half-lives. We shall shortly see how such radioactivity may contribute to our knowledge of the processes of the formation of the solar system.

Studies of the gas in space, both by means of the 21 centimeter radio emission which it emits and by means of the interstellar absorption lines produced in the spectra of very distant hot stars, show that the gas is very irregularly distributed in the space between the stars. It appears that the overall density of the gas near the sun is about 1 hydrogen atom per cubic centimeter. However, there are great irregularities in the density of the gas, and most of the gas seems to be clumped together in clouds of some sort. A typical interstellar cloud may have a gas density of about 10 hydrogen atoms per cubic centimeter, but it must be remarked that there is indeed a wide variation in these densities, with densities up to 10^3 hydrogen atoms per cubic centimeter being seen in some objects.

In order for a star to be formed, some portion of the interstellar gas must become gravitationally unstable and collapse towards high density. In order for this to happen, the region must be dense enough so that the forces of gravitational attraction of the gas on itself exceed the thermal or other forces in the interior which will tend to expand the gas. The gas in the interstellar medium is continually subjected to a mixing and buffering process owing to the fact that supernova explosions and the formation of new hot stars cause the local expansion of gas in their vicinity, and other gas must correspondingly be squeezed together at high densities due to the general turbulent motions that thus ensue. Occasionally the criterion of gravitational instability will be satisfied.

In order to be able to discuss this situation semiquantitatively, let us write down a general expression for the *virial theorem* that applies to the stability of this gas in space. The general form of the virial theorem is:

$$\tfrac{1}{2}(d^2I/dt^2) = 2K_r + 2K_t + 3(\gamma - 1)U + \mathfrak{M} + \Omega - 3pV. \quad (1)$$

Here, I is the moment of inertia of the gas under consideration,

K_r the rotational kinetic energy of the gas, K_t the turbulent kinetic energy of the gas, U the internal thermal energy, γ the ratio of specific heats of the gas (normally 5/3), $\mathfrak{M}$ the magnetic energy contained by the magnetic field within the gas, Ω the gravitational potential energy, intrinsically a negative quantity, and p the pressure exerted over the boundary of the cloud which has a volume V.

THEORIES OF STAR FORMATION

Two theories of star formation have been propounded which represent more or less extreme views as to how the above equation should be interpreted. In each case, in order to get a region of gas, which is initially in a steady state equilibrium, neither contracting nor expanding, to become unstable and undergo a collapse, it is necessary that the left-hand term, $\frac{1}{2}(d^2I/dt^2)$, become negative. To determine the conditions under which this can occur, it is necessary to evaluate the terms of the right-hand side of the expression in order to see whether some or all of the terms will be positive or negative.

One approach has been taken by Hoyle and Fowler.[2,3] They assumed that the starting point for a portion of the interstellar gas to become gravitationally unstable must be represented by gas containing the typical interstellar cloud gas density, or about 10 hydrogen atoms per cubic centimeter. They then asked the question as to how large a region with such a density would have to be in order that $\frac{1}{2}(d^2I/dt^2)$ should become negative. They considered that the region of space should form an isolated system in space, so that there would be no pressure p on the external boundaries of the system. Hence the only negative term on the right-hand side of the above expression would be Ω. Upon assuming that the turbulent and rotational kinetic energies of the gas and the internal magnetic energy should also be negligible quantities, they arrived at the conclusion that the internal energy should be the largest positive quantity in their initial system. Typical interstellar gas temperatures in un-ionized regions are about 100°K. With this value entering into the positive thermal energy of the gas cloud, they arrived at the conclusion that, in order for Ω to have a larger magnitude than the thermal energy, the mass of the cloud must be in the range 10^5 to 10^6 times the mass of the sun. This mass is very much larger than the typical mass of an interstellar cloud, which is about 10^3 times the mass of the sun, and is more characteristic of the most massive congregations of interstellar gas clouds in space, which form a few large massive cloud complexes. Thus Fowler and Hoyle's main hypothesis on star formation is that it should occur in the most massive cloud complexes that are seen in space, in material having an approximate density of 10 hydrogen atoms per cubic centimeter.

At the other extreme is a theory proposed by Cameron,[4,5] whose point of view was based upon the fact that the critical mass for gravitational instability decreases as the initial density increases. He therefore proposed that one should consider a gas cloud having about the largest density which is observed in practice, about 10^3 hydrogen atoms per cubic centimeter, and that one should ask how massive such a cloud must be in order that gravitational instability might occur. He also assumed that there was no external pressure on the cloud, and that the magnetic, rotational, and turbulent energies in the cloud were negligible. These assumptions once again make the thermal energy the dominant positive energy source, and gravitational instability can arise when the mass of gas amounts to about 10^3 times the mass of the sun. This is on the order of magnitude commonly possessed by interstellar gas clouds. Thus, in Cameron's theory a typical interstellar cloud, upon occasionally being compressed to about 10^3 hydrogen atoms per cubic centimeter, will attain gravitational instability and will collapse.

It should be noted that from either point of view the mass of the gas which becomes gravitationally unstable is very much greater than the mass of a typical star such as the sun. Hence we must ask under what conditions fragmentation of the gas into smaller and smaller sub-units may take place during the gravitational collapse of the initial cloud.

The temperature in the interstellar gas represents a steady state condition under which the energy, which is received from various sources and goes into internal heat, is radiated away by various electromagnetic emission processes that can occur within the gas. The most important mechanisms of heating appear to be the absorption of starlight by the interstellar gas and by the interstellar dust particles, and the dissipation of the turbulence of gas motions in space. All of the electromagnetic radiation processes involve inelastic collisions between two particles of some sort. These particles include electrons, hydrogen atoms, other atoms to a smaller degree, molecules of various sorts, particularly hydrogen molecules, various kinds of ions, and dust grains. At the lower temperatures the most effective cooling mechanisms consist of the collisions between electrons and carbon, silicon, or iron atoms which have low-lying excited states. Upon excitation of these states, de-excitation by electromagnetic radiation is the usual result.

In any event it should be noticed that the rates of emission of such radiation depend upon the square of the density of the medium. As the density of an interstellar gas cloud increases, the efficiency of the cooling processes increases even more rapidly. This means that as an interstellar gas cloud undergoes contraction, its internal temperature cannot rise very rapidly owing to the much greater rate at which radiation will stream forth from its interior. Thus we may describe the collapse of an interstellar gas cloud as taking place under roughly isothermal conditions. At the same time the gravitational potential energy of the cloud as a whole is growing extremely rapidly. Soon it will come about that we could consider a small sub-unit of the cloud and find that the gravitational potential energy which is associated with that sub-unit alone will have a greater magnitude than the internal thermal energy of that portion of the cloud. It is evident that such a sub-unit would thus satisfy the virial theorem independently, and hence that it is possible in principle for the cloud to break up into many such small fragments once such conditions are satisfied. While the conditions necessary for the fragmentation process are not known in any detail, it is clear that the basic energy requirement will certainly be satisfied during the collapse of the cloud.

In principle, this fragmentation process can continue until the ultimate size of the fragments is down to the point where the ultimate fragments themselves have become so dense that radiation can no longer readily escape from their interiors. Under these conditions, the energy that is released by the gravitational contraction will no longer be radiated freely into space, but must be stored in the interiors of the clouds. Under such conditions the internal thermal energy term in the virial theorem rises more rapidly in magnitude than the gravitational potential energy of the ultimate fragment, and so eventually the contraction of the fragment must be halted.

In Cameron's theory there is nothing to halt the contraction of the fragments which result from the fragmentation of his interstellar cloud until the fragments have in fact reached approximately solar size, at which point the internal opacity becomes high enough to do the halting.

However this is not the case for the theory propounded by Hoyle and Fowler. Two effects arise which prematurely halt the contraction of their cloud complex: *rotation* and *magnetic energy*.

The rotation of the interstellar clouds takes place too slowly for direct measurement. However, there are some basic considerations which allow us to establish what we should expect for the rate of rotation in an approximate way. There appears to be a general interstellar magnetic field that is associated with all of the interstellar gas. Such a magnetic field is rather closely glued even to the neutral gas in space for the following reason: Starlight, in general, will keep certain atoms in space ionized for by far the greater portion of the time. Near the hottest stars the hydrogen atoms will be fully ionized, and these will succeed in removing from the starlight that penetrates to farther distances all those photons beyond the ionization limit of hydrogen atoms. Thus distant starlight will be unable to ionize any of those atoms which have ionization potentials greater than hydrogen. These atoms include the most abundant atoms in space: helium, nitrogen, oxygen, neon, and so forth. However, there are some common atoms, and practically all of the heavier atoms, which have ionization potentials less than that of hydrogen, and which therefore are susceptible to ionization by starlight. These atoms include carbon in particular. About three quarters of all the ions produced in space will be carbon ions. These carbon ions, or any other ion for that matter, will spiral around the magnetic lines of force of the interstellar medium. From time to time the ions will collide with the neutral atoms, and for this reason there will exist a general friction between the neutral and ionized components of the interstellar medium which will prevent one from being moved very rapidly relative to the other. Consequently, it is very difficult for magnetic lines of force to move through even the regions of space populated by a generally neutral gas.

It is expected that the interstellar magnetic field will play some role in constraining the motions of the interstellar gas. One of the most elementary forms of control is to require that each part of the interstellar gas should very nearly rotate once every time that portion of the gas revolves around the center of the galaxy. In this way the overall configuration of the spiral lines of the galaxy would be preserved. At the distance of the sun from the center of the galaxy, the resulting angular rotation would amount to 10^{-15} radians per second. This is the initial rate of rotation assumed by Fowler and Hoyle.

They found that after their initial cloud complex had contracted to a moderate degree, it was starting to rotate so rapidly that a rotational instability was reached in the equatorial regions. This would halt the general contraction of the cloud. They also noted that after a time the more rapid rotation of the cloud would result in twisting the interstellar magnetic field contained within the cloud complex very greatly with respect to the general surroundings. They noted that this twist would exert a torque on the surroundings which was thus capable of transmitting angular momentum from the cloud complex to the surroundings. They claimed, in fact, that by far the great majority of the angular momentum of the cloud complex could be transmitted to the surroundings in this fashion. Following this, the cloud complex would be free to contract without further hindrance.

However, as the cloud complex contracts, it drags with it the lines of force of the interstellar magnetic field trapped within it. The energy associated with this magnetic field will then vary inversely as the dimensions of the system, just in the same manner that the gravitational potential energy varies. Thus the magnetic energy in the cloud will bear the same proportion to the gravitational potential energy at any time during the contraction that it did at the beginning. Hoyle and Fowler noted that this would place a limit upon the fragmentation process that might take place in the cloud, since with typical values of the initial interstellar magnetic field, only the major subunits of the cloud would be able to undergo fragmentation before the internal magnetic energy placed a lower limit on the size of further fragments. This lower limit is reached when the internal magnetic energy of the sub-unit becomes equal to the gravitational potential energy of the same sub-unit. Hoyle and Fowler claimed that this difficulty would ultimately be overcome when the cloud became dense enough, so that it shielded itself against the incoming starlight in the interior portions. Under these conditions they expected the ionization due to starlight to cease, and the ions in the interior of the cloud to vanish, so that the magnetic field would spring out from the cloud. However, they neglected the process of ionization by internal radioactivities, such as K^{40}, which would impede this process.

It is useful at this point to consider what would happen if a collapsing fragment of solar mass contained no angular momentum, so that it were to form a star directly without any associated solar nebula. There is an upper limit to the size that such a fragment can have while it is in hydrostatic equilibrium. This limit rises from the fact that the temperature in the interior must be high enough to provide sufficient pressure to allow the interior of the star to hold up the overlying layers. Such a high temperature can only be obtained if the gas in the interior, particularly the hydrogen and helium, is ionized. However, it requires a fairly large amount of energy to dissociate the initial hydrogen molecules that the star would contain, and then to ionize the hydrogen, and both singly ionize and doubly ionize the helium atoms. Normally, when a star contracts out of space, half the released gravitational potential energy is stored as internal heat, and the remaining half is radiated away from the surface. In the present situation, however, it is evident that, even if the collapsing star radiates no energy at all into space, half of the released gravitational potential energy must still be stored as internal thermal energy, and the remaining half would then be required to provide the energy necessary to dissociate and ionize the hydrogen and helium. In the case of the sun, insufficient gravitational potential energy will be released to do this until the radius has shrunk to some 60 times the present radius of the sun, which is somewhat less than one-third of an astronomical unit, inside the orbit of Mercury.

It was recently predicted by the Japanese astrophysicist, Hayashi, that the sun would be an extremely luminous object during this early contraction phase. This prediction has been confirmed by recent calculations of Ezer and Cameron,[6] which are reproduced in Figure 1. In this figure are plotted the *luminosity* and *surface temperature* of the sun during its early contraction phase. It may be seen that when the sun first forms a stable object, its surface temperature is nearly as high as it is at the present time, so that its luminosity is some hundreds of times the present luminosity. During this stage the sun is a completely convective object, so that the gravitational potential energy released during further contraction in the interior is rapidly

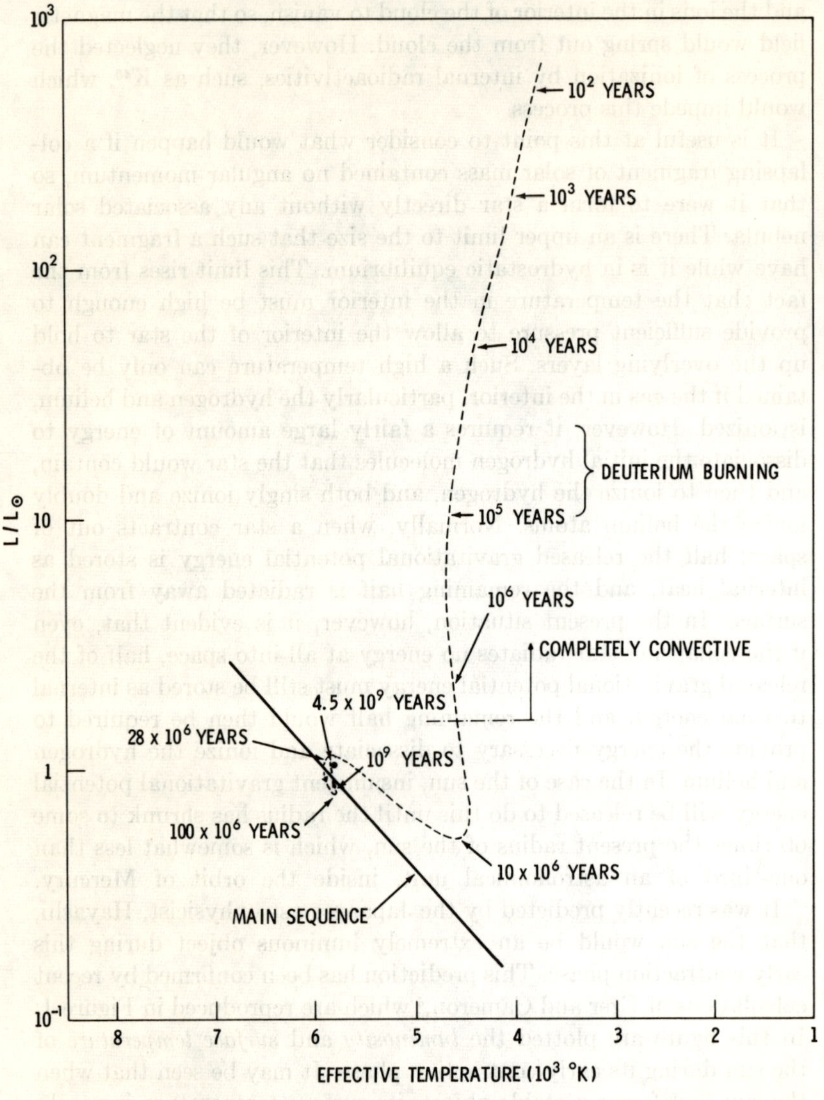

Fig. 1. Luminosity and surface temperature during the early contraction phase of the sun.

convected to the surface, and there radiated away into space. It may be seen from the time scale indicated in Figure 1 that the sun rapidly contracts in these early stages and the luminosity correspondingly decreases, since the surface temperature changes very little during this period, The onset of deuterium burning, which may take place when the sun is some 10 times its present radius, may slow the contraction for about half a million years. During this time the sun remains fully convective, so that the deuterium which is destroyed by central thermonuclear reactions is destroyed throughout all of the star at the same time, owing to the rapid mixing of the interior.

The sun then continues to contract and gradually changes to a structure which is predominantly in radiative equilibrium in the interior, until the central temperature rises sufficiently to halt further contraction while thermonuclear reactions convert hydrogen into helium. The sun is then *a main sequence star*. It is evident that the early high luminosity phase of the solar contraction will have important effects on the temperature environment of any primitive solar nebula which surrounds it.

Now let us consider the effects of angular momentum upon the late stages of the collapse of the protosun. Hoyle, Fowler, and Greenstein assume that the sun will contract and not reach rotational instability in the equatorial regions until the radius is down to the vicinity of that of the orbit of Mercury. They then assume that a small nebular disk will be spun off upon further contraction. Because the protosun continues to trap some of the initial interstellar magnetic field which threaded the interior, the disk which is spun off at this stage will remain magnetically attached to the more rapidly spinning interior protosun. This magnetic linkage will tend to make the disk co-rotate with the inner protosun; and such a condition of co-rotation will impart to the gas in the disk an angular velocity which exceeds that corresponding to Kepler orbital motion. Consequently, the gas in the disk will be accelerated towards larger and larger radii from the central protosun. In this way, Hoyle, Fowler, and Greenstein envisaged that their primitive solar nebula will gradually recede from the central protosun and sweep over the entire region of the present planetary system. They consider that the primitive solar nebula will contain approximately one percent of the mass of the sun, and that the bulk of the angular momentum in the sun will be transferred to this nebula.

Chemical condensations are supposed to form in the nebula, and to grow by accumulation. When the size of the accumulated bodies has reached some tens of meters, they become sufficiently large so that they can no longer be carried along by the gas which is being accelerated outwards in the plane of the ecliptic, and consequently, such metric planetesimals will be dropped behind by the receding gas. It is supposed that in some way they then collect to form the inner planets. According to this view, the outer planets are composed principally of icy materials which attract to them varying amounts of hydrogen and helium.

Hoyle, Fowler, and Greenstein have developed an elaborate theory based upon this picture and designed to explain the abundances in the earth and meteorites of the light elements: deuterium, lithium, beryllium, and boron. In this theory, the process of acceleration of the primitive nebula outwards from the protosun by means of its magnetic linkage to the protosun will result in a great dissipation of magnetic energy. This dissipation is assumed to proceed rather efficiently into the acceleration of cosmic ray particles which will have typical energies of some hundreds of MeV. These energetic particles are then believed to traverse the region of the inner solar system and to bombard, again fairly efficiently, the metric planetesimals which have been deposited in that region. This bombardment thus creates lithium, beryllium and boron, and also a certain amount of deuterium, by *spallation* reactions in the outer skin of the metric planetesimal. Secondary neutrons produced by the bombardment may produce additional deuterium through being captured in any hydrogen that is bonded into the metric planetesimals.

It is perhaps worth mentioning, that in the picture of Hoyle, Fowler, and Greenstein one might not expect a very constant ratio of the isotopes of He^3 and He^4 to be produced within the metric planetesimals. Yet it is observed that primordial gases removed from meteorites have just such a constant ratio. There is a further puzzle concerning the abundances of beryllium in the sun and in the meteorites. Relative to the other light elements, the abundance of beryllium in the meteorites is much less than it is in the sun. The meteoritic abundances of beryllium disagree strongly with those predicted by the Hoyle, Fowler, and Greenstein theory. It might also be expected that there should be some variation between the meteorites and the earth in the isotopic abundances of those nuclides which have very high neutron capture cross sections. No such variations have been found.

Let us now consider Cameron's picture of the collapsing interstellar cloud. Cameron also assumes that his cloud is initially rotating with an angular velocity of 10^{-15} radians per second. However, because of the higher density, this cloud would collapse so rapidly that there would be virtually no chance for any transmission of angular momentum from the cloud to its surroundings through the medium of the interconnecting magnetic field. Consequently, it is necessary to assume that the fragments formed during the fragmentation of the cloud will conserve their angular momentum of rotation and orbital motion at all stages of the collapse. In this picture a fragment of one solar mass will become rotationally unstable at the equator when its dimensions have been reduced to several tens of atronomical units, comparable to the dimensions of the present planetary system. Upon further contraction, the fragment will flatten into a disk. It is noteworthy that there is no central body in hydrostatic equilibrium at the center of this disk. Hence the sun is not a primary product of the collapse in this picture, but must be formed by some secondary process.

Cameron notes that the very massive primitive solar nebula which he finds to be formed in this way will contain chemically condensable material in amounts which are about two orders of magnitude greater than now exist in the planets. He argues that the accumulation of the planetary bodies must take place directly in this massive primitive solar nebula. It is very difficult to understand how small bodies moving independently in a vacuum can accumulate, particularly if their relative velocities amount to just a few kilometers per second. The orbital velocity of the earth about the sun is some 30 kilometers per second, and any body even in a relatively similar orbit is likely to encounter the earth at a velocity of several kilometers per second. If the bodies are only typically some meters in dimension, collisions at such high velocities would be very catastrophic, and would shatter the bodies into many pieces. However, in a massive solar nebula, the solid bodies could not move very rapidly with respect to the gas, and, therefore, they would have to encounter one another rather gently. In this way accumulation may proceed until the body is large enough so that its gravitational field may play some role. Nevertheless, in this picture the formation of the planets still represents a very inefficient accumulation of these solid bodies, amounting to only about one percent of the chemically condensable material that is present.

It is an interesting question as to how this primitive solar nebula is dissipated. The theory of the dissipation has not yet been treated in detail. However, Cameron suggests that two processes may play a significant role. On the one hand, he expects that the primitive solar nebula will be formed hot and partially ionized, so that the interstellar magnetic field will remain trapped in it at all times. Since the primitive solar nebula is likely to contain a relative shear between adjacent layers, such shear will draw out the lines of force of the magnetic field, thus squeezing the lines of force closer together, and increasing the energy of the magnetic field. This increase of energy must come at the expense of the energy of the relative shearing motion, which in turn is maintained by the mass distribution of the nebula. Hence the nebula can make energy available to the magnetic field only at the expense of the redistribution of the mass. This redistribution of mass must take the form of a flow of material inwards towards the center of the nebula, accompanied by some outward flow at the edges in order to maintain an overall conservation of angular momentum of the nebula.

The second mechanism which may play a role is turbulence. To the extent that turbulence is able to promote any mixing between adjacent layers of the nebula, the mixing will tend to make the nebula rotate more like a rigid body, and this also will bring about a redistribution of the mass, with the majority of the mass flowing toward the center of the nebula, and some of the mass near the edges flowing outwards in order to conserve angular momentum. Thus it is apparent that the sun is likely to form as a result of the dissipation of the massive nebula.

In this picture the planets are formed before the sun. As the gas streams inward to form the sun, it will carry with it the great bulk of the chemically condensable material which has not managed to gather into larger bodies. It is possible to estimate roughly how large a body must be in the inner regions of the solar nebula to withstand being swept into the sun by friction with the gas that is streaming inwards. It turns out that the bodies must be larger than the larger asteroids in order to withstand this streaming motion. Hence, Cameron postulates that the present planets were formed in virtually their present size from the solar nebula at this stage.

In Cameron's picture there is also a considerable dissipation of magnetic energy, and some of this energy may go into the acceleration of particles. However, the huge mass of the solar nebula will result in the ready absorption of any accelerated particles, mainly by the hydrogen and helium, so that very little bombardment of the solid materials is likely to take place. Even so, what little bombardment there is may suffice to make sufficient amounts of such radioactivities as Al^{26}, with a half-life of about 1 million years, so that any small bodies that are then accumulated in the solar nebula will contain sufficient short-lived internal energy sources to provide melting. About 1 atom of Al^{26} per million atoms of silicon is sufficient for this purpose. This spallation history is much less violent than that postulated by Hoyle, Fowler, and Greenstein, and consequently, Cameron does not clain that the light elements deuterium, lithium, beryllium, and boron are formed in this fashion. He believes them to have been present in the interstellar medium from which the contraction of the sun commenced, and thus to have been products of *galactic nucleosynthesis*, probably principally produced by non-equilibrium reactions associated with supernova explosions.

PLANETARY DEVELOPMENT

Many other ideas have been proposed concerning the development of the planets out of the primitive solar nebula. Especially noteworthy are those of Urey.[7] During the past several years, Urey has developed a theory of the origin of the planets which depends upon the formation of very large numbers of primitive objects of approximately lunar size. He starts his theory with the assumption of a massive solar nebula. He assumes that gravitational instability can occur in this primitive solar nebula, so that a large number of self-gravitating gaseous condensations can take place. He believes that the total mass of these condensations is likely to be in the range that would be typified by the mass of the moon as augmented by the missing volatile gases. In his picture the chemically condensable solids settle to the center of the gas spheres thus formed, and the solid material is then heated by adiabatic compression of the gas during the subsequent contraction of the spheres. He assumes that this heating will induce the chemical differentiation which has occurred in meteorite-like material. When these gas spheres collide with one another, the gas is assumed to be largely dissipated into space, leaving behind the solids. Most of his primitive lunar bodies are then supposed to accumulate to form the planets, and the earth's moon is assumed to be one of them which was left over from this process, and which was captured by the primitive earth.

One of the interesting questions that arises from these diverse views regarding the origin of the solar system is the nature of the atmospheres that may be formed about various planets. In the Hoyle, Fowler, and Greenstein picture, where the planets are assumed to accumulate from very small bodies in a vacuum, any atmosphere would have to arise entirely by outgassing from the interior. In Cameron's point of view, it is to be expected that the terrestrial planets will capture some of the primordial solar nebula that is streaming by them to form the sun, and this will form a significant primordial atmosphere. It is difficult to see how the gas in Urey's primordial lunar bodies can be completely dissipated by collisions, and, when they accumulate to form the planets, it might also be expected that a primitive atmosphere derived from the solar nebula would then exist. We have significant information concerning these points in connection with the atmosphere of the earth, and apparently now also in connection with other planetary atmospheres.

It has been known for some time that the bulk of the volatile materials above the surface of the earth have outgassed from the interior. These include the oxygen and nitrogen in the atmosphere, the A^{40} in the atmosphere which is derived from the decay of K^{40} in the interior, and the oceans of the earth. Hence if there is to be any evidence for a portion of the earth's atmosphere which was derived from the primitive solar nebula, then this evidence should presumably be sought in the rare gases of the atmosphere, which would not have undergone chemical alteration during the preceding history of the earth. It turns out that we now have some important information regarding the heaviest of these rare gases, the xenon.

Some of the stony meteorites that fall upon the earth's surface contained trapped gases in their interior; these gases appear to be surviving remnants of the original solar nebula. The meteorites with the greatest concentration of these primordial gases are the very rare and fragile carbonaceous chondrites, which have evidently never been heated to very high temperatures, and which also contain water of crystalization and various complicated carbon compounds.

One of the most significant discoveries that has been made in recent years affecting the investigation of the early history of the solar system was that of John Reynolds, of the University of California, Berkeley, who found that xenon extracted from the meteorites has differing isotopic composition. We see the comparison of the isotopic composition of the xenon extracted from three carbonaceous chondrites in Figure 2. The relative abundances of the isotopes of xenon extracted from the Murray meteorite are shown by the solid bars in the bottom of the figure. The abundances of the two lighter isotopes have had to be increased by a factor of ten in order to see them. The differences between the isotopic abundances in the meteorites Mighei and Orgueil and those in Murray are shown at the

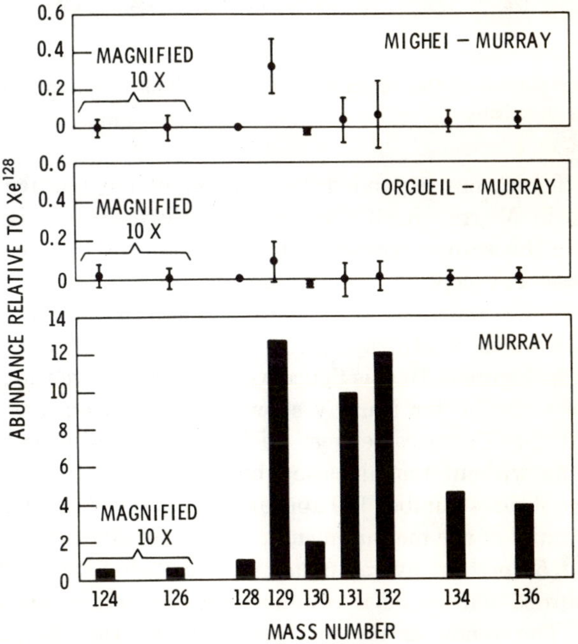

FIG. 2. Comparison of the xenon isotopic composition in three carbonaceous chondrites.

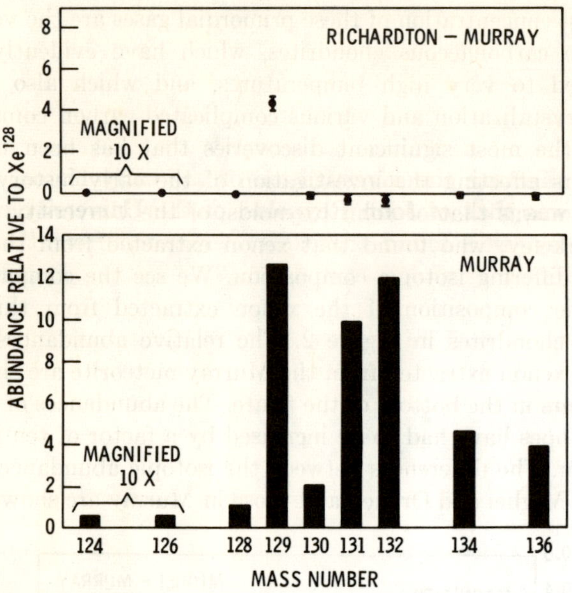

FIG. 3. Comparison of the xenon isotopic composition in an ordinary and a carbonaceous chondrite.

top of the figure, but expanded by a factor of ten relative to the abundances in Murray itself. We see that there are no significant differences in the isotopic composition of xenon in these three sources, with the possible exception of mass number 129.

However, if we compare the xenon extracted from a more ordinary chondrite with that of Murray, then we see a very different picture, as is shown in Figure 3. In this figure we once again have the isotopic composition of xenon for Murray shown at the bottom and at the top we have the differences between Richardton and Murray plotted, but this time without magnification by a factor of 10. All of the isotopes except mass number 129 appear to have the same abundance within the limits of the measurements, but we see that mass number 129 is about 50 percent overabundant in the Richardton meteorite. This has turned out to be a common state of affairs among the meteorites. The xenon extracted from them appears to have essentially the same composition except for these variations in mass number 129. Moreover, the smaller the amount of xenon present per

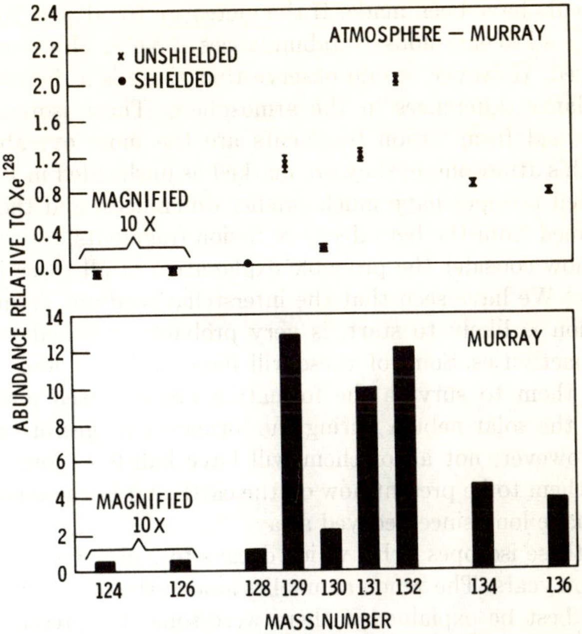

FIG. 4. Comparison of the xenon isotopic composition in the earth's atmosphere and in a carbonaceous chondrite.

gram of meteorite, then the greater is the overabundance of Xe^{129} likely to be.

One now receives a great surprise when one compares the isotopic composition of xenon in the earth's atmosphere with that in the Murray meteorite. This comparison is shown in Figure 4, in which the differences between the atmospheric and Murray compositions are shown at the top magnified by a factor of 5. These two compositions have been normalized at mass number 128, and we see that mass number 128 is the only point that lies on the line of zero differences. All of the other points lie off the line by a minimum of five standard deviations.

We notice at first that the earth does not share the trend that was found in the meteorites. The xenon in the atmosphere is probably the bulk of the xenon associated with the entire earth, and on these grounds the amount of xenon per gram of earth is much less than the amount of xenon per gram of any of the meteorites for which the measurements have been made. If the meteorite trend were followed, this would lead to enormous overabundances of Xe^{129}. This is certainly not observed. However, we do observe that there is a dichotomy of the abundance differences in the atmosphere. Those isotopes that can be formed from fission fragments are the most overabundant in the earth's atmosphere; they are marked as unshielded in Figure 4. The shielded isotopes have much smaller differences, and these cannot be formed from the beta decay of fission fragments.

Let us now consider the probable explanation of all these isotopic differences.[8] We have seen that the interstellar medium, from which condensation is likely to start, is very probably enriched in short-lived radioactivities. Some of these will have half-lives long enough to permit them to survive the formation process itself and to be present in the solar nebula during the formative stages of the solar system. However, not all of them will have half-lives long enough to permit them to be present now on the earth and in the meteorites; they will have long since decayed away.

Among these isotopes is I^{129} which decays to Xe^{129}, with a half-life of 17 million years. The xenon anomalies among the meteorites themselves can best be explained if there were some I^{129} present in the meteorite parent bodies at the time that they had formed and had cooled sufficiently so that the trapped xenon could no longer diffuse throughout the interior. Then the subsequent decay of the I^{129} will result in the enrichment of this mass number in the trapped xenon. The precise amount of enrichment of the Xe^{129} thus provides a tool which can be used to estimate the time of cooling of the meteorite parent bodies relative to one another. It is also possible to estimate how long it was since the material that was due to form the solar system separated itself away from the source of fresh radioactivities in the interstellar medium. Such times come out to be a few tens of millions of years, indicating that the meteorite parent bodies formed and cooled relatively quickly, as would be characteristic of bodies of relatively small size, which had some very short-lived initial heat source in their interiors.

However, evidently we must conclude either that the earth did not contain much I^{129} when it was formed, or else that the xenon decay products of this radioactivity were able to escape from the earth for a considerable period of time after the meteorite parent bodies had formed and cooled. We may summarize either of these possibilities by stating that the earth cannot have retained any Xe^{129} in its atmosphere for 100 million years after the meteorite parent bodies formed and cooled. Apparently the Xe^{129} must either have been lost from the earth's atmosphere, or it had never entered it due to being released too deep in the earth's interior. The latter is a distinct possibility, since iodine may have been carried down and into the iron core to a significant extent.

There is another extinct radioactive isotope, Pu^{244} with a half-life of 76 million years, that has probably been responsible for the formation of the fission products of xenon in the earth's atmosphere. One in every three hundred decays of this radioactive isotope is a spontaneous fission process. The amount of these fission products within the earth's atmosphere is quite consistent with the 100 million year interval, that we have just mentioned.

The abundance differences of the shielded isotopes of Figure 4 are shown in more detail in Figure 5. Here it may be seen that the differences are definitely real and amount to several times the probable errors of the measurements. Cameron has interpreted these abundance differences as resulting from a neutron capture process, in which xenon was exposed to an intense flux of neutrons, probably

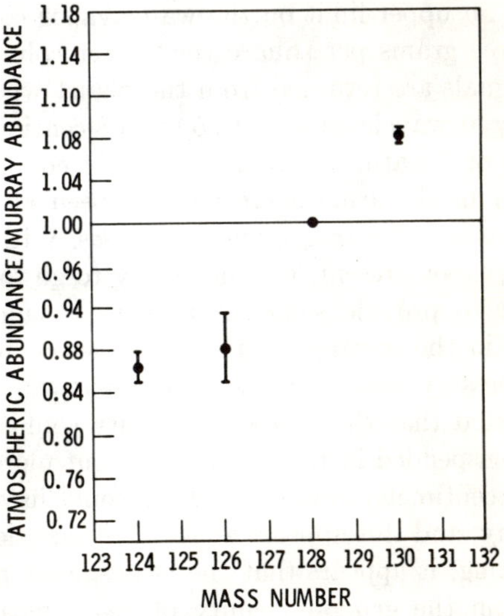

FIG. 5. Expanded portion of Figure 4 showing the shielded isotopes.

during the deuterium burning phase of the sun during its early contraction. The abundance differences observed for the atmosphere are consistent with the conclusion that atmospheric xenon of essentially normal composition almost entirely resided within the sun at one time. The earth has probably captured it from the solar wind throughout terrestrial history. Thus there is no evidence whatever that the earth contains any gases which would have been captured from the primitive solar nebula at the time of the earth's formation. This would seem to be a difficulty with Cameron's theory of the formation of the solar system.

However, it is difficult to understand how the very massive atmosphere of Venus can represent anything other than an accumulation from the primitive solar nebula.[9] As is discussed in the chapter on planetary atmospheres, it now appears certain that the high microwave brightness temperatures of Venus represent temperatures in excess of 600°K at the surface of the planet. These high values of the temperature indicate that the atmosphere of the planet must be extraordinarily opaque in the infrared region, but relatively transparent in the long wavelength microwave region. The latter circumstance rules out the possibility that there can be any substantial amount of water vapor in the atmosphere, and indeed Thaddeus has set an upper limit on the water vapor content of the atmosphere of three grams per square centimeter column from the fact that radar signals are returned from the planet with essentially the same efficiency at wavelengths of 12.5 and 68 centimeters. About five to ten percent of the atmosphere appears to be composed of CO_2, and the remainder of the atmosphere has not been identified. Presumably, it is composed of nitrogen and rare gases. With these small amounts of water vapor present, it is necessary to go to rather high pressures in order to provide sufficient pressure-broadening of the CO_2 bands to fill in the infrared region of the spectrum, and thus to provide the necessary opacity of the atmosphere.

Opik has suggested that the necessary opacity could be provided by dust particles suspended in the atmosphere, but more than three grams per square centimeter column of dust would have to be suspended in this way, and this appears a less likely solution.

Generally speaking, it appears that the atmospheric ground pressure on Venus is in the general vicinity of 100 atmospheres, presumably at least half nitrogen, which is scarce in the surface rocks, and nearly absent in the interior, and consequently it appears that Venus as a whole contains two orders of magnitude more nitrogen than does the earth. The simplest explanation of this situation seems to be that Venus did in fact capture a large primitive atmosphere from the primordial solar nebula. The lack of water in the Venus atmosphere may also imply that there has been, relatively speaking, much less outgassing on that planet than has been the case for the earth.

The atmosphere of Mars represents a very complex situation, and it is unsure to what extent gaseous constituents may have escaped in the past. Consequently, it is not possible to estimate whether the Martian atmosphere was primarily produced by outgassing from the interior or by capture from a primitive solar nebula.

Thus, in Cameron's picture of the origin of the solar system, it seems necessary to provide some mechanism by means of which the earth may have lost its primordial atmosphere. Moreover, this should be a mechanism which would not have operated in the case of the earth's sister planet Venus. These conditions seem to place very stringent requirements on the situation, and the only mechanism that seems possible to produce such a state of affairs would be a period of rotational instability undergone by the earth but not by Venus.

Such a period of rotational instability, during which the atmosphere would be lost to nearby space by centrifugal forces and presumably then swept away by the solar wind, is reminiscent of the theory propounded by George Darwin, in 1898, to account for the origin of the moon. In this theory, it was originally supposed the solar tides could raise large bulges in the earth, at a time when the earth was rotating very rapidly, and this could lead to successive elongation of the earth through a sequence of *MacLaurin spheroids* and *Jacobi ellipsoids*, until the deformation was so great that fission into two bodies would take place. One of these bodies was supposed to be the moon. Darwin's theory failed quantitatively when it was demonstrated that such tidal bulges would be dissipated by internal viscosity long before they could cause such elongations.

However, D. U. Wise recently revived Darwin's theory by proposing that if the earth were initially rotating very rapidly, then the decrease in the moment of inertia associated with the formation of the iron core of the earth might cause the spin to increase sufficiently to trigger rotational instability. This theory has the enormous advantage of providing that the moon should be formed out of material from the outermost parts of the earth which had been sufficiently heated so that the iron would have drained away. This would account for the low density of the moon, which was discussed as a problem in the chapter on planetary structure.

However, there are some difficulties with this picture which merit discussion. Foremost among these is the fact that the earth must be initially rotating rather close to the limit of rotational instability. It must be formed in an initial state which is already a Jacobi ellipsoid, with very great deformation along one of the axes. The angular momentum per unit mass of such a primitive earth is thus much greater than is observed for the other planets, and is much greater than one should expect for an average process of gradual accumulation from much smaller bodies. Cameron has suggested that this situation would require that the earth had accumulated in its final form from two bodies of comparable size, rather than by a process of growth of one large body from any smaller ones. Such a pair of bodies would have to capture one another in the massive primitive solar nebula, so that the nebular gases could act as a viscous medium which could remove energy from the orbital motion of the pair of bodies, thus allowing them gradually to spiral together. When such bodies merged, they would indeed form a rapidly spinning Jacobi ellipsoid. This would presumably occur in the later stages of dissipation of the solar nebula. Then upon melting of the iron in the interior, which from the character of the melting curve is likely to take place first near the surface, there is likely to be a somewhat catastrophic formation of the iron core, and subsequent shattering of the outer parts of the Jacobi ellipsoid, with the ejection of much mass into space.

It is noteworthy that many times the mass of the moon must be ejected by some such process if one wishes to account for the present angular momentum of the earth-moon system, which is very much less than the angular momentum that must be associated with one earth mass in the form of a rapidly-rotating Jacobi ellipsoid. The accumulation of this material to form the moon must presumably take place in vacuum, and for the reasons previously stated it is likely that such accumulation could only take place if it is a very inefficient process, in which a great deal of shattering of material takes place during the more violent mutual collisions. When the moon does begin to grow, its gravitational field will undoubtedly become sufficient to perturb the motions of much of the material present and to eject it from the earth–moon system. This material, which would then be in independent orbit about the sun, may recollide with the earth or the moon, but in so doing the initial angular momentum of spin in the earth–moon system would be converted into angular momentum of orbital motion about the sun. However, much of the material will also be perturbed by the earth into orbits that cross those of the other planets, so that the material will either collide with the other planets or eventually be perturbed out of the solar system entirely by Jupiter.

There is an immediate further consequence of this picture. When the material that is to form the moon or be ejected from the earth–moon system is receding from the rapidly spinning primitive earth due to tidal effects, the earth will be slowed in its rotation and will be relaxed towards the spherical shape. A great deal of gravitational potential energy will be released by this process. Since the interior of the earth will already be considerably heated if the iron has been able to melt and form the core, the additional heat released by the relaxation toward the spherical shape is likely to melt the rocky mantle as well; leading to a completely moltent state for the earth. The possible evolution of such a completely molten earth has been discussed in the chapter on planetary structure. There are a number of aspects of the chemical differentiation in the thermal history of the earth which would seem to be satisfactorily accounted for if the earth did go through such a completely molten stage. It is also likely that the earth would become nearly completely outgassed from the interior during such a completely molten stage. A planet which has not undergone this completely molten state, as perhaps Venus did not, may undergo very little outgassing from the interior, and this may help to explain why the water vapor content of the Venus atmosphere is so low. This hypothesis would indicate that Mars had undergone very little outgassing from the interior, and that its atmosphere is predominantly derived from the solar nebula.

It is evident that the investigation of planetary atmospheres and planetary interiors, as well as the surface and the interior of the moon, will provide us with many checks on these various theories of the formation of the solar system. As knowledge of the characteristics of the solar system has begun to rise dramatically during the last decade, the interest in problems of the origin and history of the solar system has likewise been mounting rapidly. It seems likely that the next few years of space research will result in the final production of answers to this age-old question of scientific philosophy.

BIBLIOGRAPHY

1. R. Jastrow and A. G. W. Cameron, Origin of the Solar System, Academic Press, New York, 1963.
2. W. A. Fowler, J. L. Greenstein and F. Hoyle, Nucleosynthesis During the Early History of the Solar System, Geophys. J. **6**, 148 (1962).
3. W. A. Fowler and F. Hoyle, Star Formation, Roy Obs. Bull. No. 67, 1963.
4. A. G. W. Cameron, The Formation of the Sun and Planets, Icarus **1**, 13 (1962).
5. A. G. W. Cameron, Formation of the Solar Nebula, Icarus **1**, 339 (1963).
6. D. Ezer and A. G. W. Cameron, The Early Evolution of the Sun, Icarus **1**, 422 (1963).
7. H. C. Urey, The Origin and Evolution of the Solar System, *in* Space Science, D. P. LeGalley, Ed., John Wiley and Sons, Inc., New York, 1963.
8. A. G. W. Cameron, The Origin of Atmospheric Xenon, Icarus **1**, 314 (1963).
9. A. G. W. Cameron, The Origin of the Atmospheres of Venus and the Earth, Icarus **2**, 249 (1963).

Chapter 16

THE PHYSICS OF THE SUN

*John C. Brandt and Stephen P. Maran**

INTRODUCTION

The study of the sun and the many aspects of solar activity represents a large field that involves a great many scientists. The existing collection of observational data is staggering, and whole monographs have been written on various aspects of the subject. Therefore, considerable compromise has been necessary in order to make a useful presentation in just a few score pages. Historical material is largely restricted to the introductory parts of this article and subsequent sections outline the present status of solar physics, with emphasis on summary data in the form of representative graphs and tables. Finally, we consider the techniques that may be used to solve the problems that challenge us now and that may well constitute the solar physics of the future.

Man's interest in the sun has a long history. It was first based on a recognition of the sun as our prime source of light and heat. Thus, the sun has played a major role in primitive cultures; among them are the Mayan, Aztec, and Inca Sun Kingdoms of the Americas.

*Kitt Peak National Observatory; Observatory Contribution No. 235.

Astronomers have a strong interest in the sun as a typical star that is conveniently located for observation, and space physicists look to it as the controlling factor for the conditions in the interplanetary medium and the upper atmospheres of the planets. Many stellar quantities are accurately known only for the solar case because the next nearest star is distant some 3×10^5 times the mean sun-earth separation of one Astronomical Unit. Since isotropic emission of photons and particles results in fluxes that decrease as the inverse square of the distance, radiation from the nearest star is already down by $\sim 10^{-11}$ with respect to the corresponding quantities from the sun. Absorption by the interstellar gas is also an important factor. The sun, therefore, is the only normal star that has been observed in the extreme ultraviolet (100Å–1000Å), x-ray ($<$ 100Å) and radio-wave regions of the spectrum. Likewise, the sun is the only star for which we have direct quantitative measurements of corpuscular radiation or out-flowing plasma; this matter is commonly called the solar wind and it has been sampled by apparatus carried on several deep space probes.

The angular size of the moon as viewed from the earth is about the same as that of the sun. Thus, the outer solar atmosphere (chromosphere, prominences, and corona) can be studied under circumstances relatively free of the bright, contaminating radiation of the lower solar atmosphere (photosphere) during eclipse, when the solar disk is occulted by the moon. Such studies are of inestimable value and, in fact, it is only for the sun that we have a large body of quantitative facts on chromospheric and coronal conditions.

Finally, our proximity to the sun and consequent ability to resolve the solar disk have led to the straightforward discovery and subsequent study of (1) the rate of solar rotation as a function of latitude, (2) granulation, (3) limb darkening, (4) sunspots, (5) flares, etc. Most of these phenomena are known unequivocally only for the sun—a fact that results in an understanding of the atmosphere of the sun that is impossible for other stars.

The total solar radiant energy emitted (erg/sec) is called the luminosity. It is determined by measuring the solar constant, which is the radiant energy falling on a unit area (perpendicular to the earth-sun line) per unit time outside the earth's atmosphere at the earth's mean distance; this quantity is difficult to measure because some wavelengths cannot penetrate the earth's atmosphere. Rocket and satellite observations have now helped to determine these additional contributions to the solar constant and the present value is unlikely to be significantly in error. This value is $\approx$ 1.95 cal/cm^2-min; note the unusual units. Taking 1.496 $\times$ 10^{13} cm as the length of the Astronomical Unit, we obtain the solar luminosity, $L_\odot = 3.9 \times 10^{33}$ erg/sec.

A blackbody the size of the sun would have the solar luminosity if it radiated at the solar *effective temperature*, as defined by the Stefan-Boltzmann law,

$$L_\odot = 4\pi R_\odot \sigma T_{eff}^4 \tag{1}$$

Here, the constant $\sigma = 5.7 \times 10^{-5}$ erg/cm^2-sec-deg^4 and the solar radius $R_\odot = 6.96 \times 10^{10}$ cm, so that $T_{eff} = 5750°K$.

The sun appears to be rather typical of the stars in the Milky Way galaxy, as it is intermediately placed in terms of most significant stellar parameters. Known masses of stars range from 0.1 to $50M_\odot$; radii from 0.1 to $200R_\odot$, and effective temperatures from 2000 to 30,000°K. There are stars that are 10^4 to 10^5 times brighter than the sun, and also stars that are fainter than the sun by like factors. The sun's age, at least 4.5×10^9 years, as implied by meteoritic and geological evidence, is believed to be intermediate among those of the stars, as is its fraction of mass in elements heavier than helium ($Z = 0.015$). The apparent normalcy of the sun is of immense value to astronomy; this must be qualified in one important respect—namely, the sun is a single star whereas some two-thirds of the stars in our galaxy are binaries or members of multiple systems. There is no evidence at present that this distinction is serious for stellar physics.

The sun's age, heavy element fraction, and kinematical properties indicate that it is a member of the "disk population" that comprises the stars in the fast-rotating disk of our galaxy. The sun's galactic rotational velocity is $\approx$ 250 km/sec, and its distance from the galactic nucleus is $\approx 3 \times 10^{22}$ cm ($\approx$ 10 *kpc*); its position seems to be at an inner edge of a spiral arm close to the plane of the galaxy.

HISTORY OF SOLAR STUDIES

Modern solar studies can be taken to have begun in 1611 with the essentially simultaneous "discovery" by telescopic observation of sunspots by Fabricius, Galileo, Scheiner, and Harriot. Even their relatively crude observations disclosed that spots occurred in groups and in preferred regions of the disk, and that the solar rotation period was about one month.

The existence of sunspots, however, had actually been known for millennia; an early report (circa 300 B.C.) of a naked eye sunspot is attributed to Theophrastus of Athens; numerous other references can be found in the oriental archives. The lack of references in the occidental literature is sometimes blamed on the Aristotelian dictum stating that the sun is a perfect body and, hence, one without blemish.

Further progress in solar physics in the post-1611 era came slowly and was limited to studies of the spots. In 1769, it was found that the *penumbrae* or outer, less dark regions of sunspots, exhibit a peculiar behavior ("Wilson effect") as spots approach the limb. The part of the penumbra away from the limb decreases in size and disappears while the part toward the limb remains almost unchanged. This phenomenon is generally thought to indicate that sunspots are saucer-like depressions in the solar surface, although some astronomers consider that this is a transparency effect, rather than a geometrical phenomenon.

The variation in the number of sunspots on the sun from a small ("sunspot minimum") to a large ("sunspot maximum") value is now well known ("sunspot cycle"). The periodicity in this quantity was established in 1843 by Schwabe, who found a cycle-length of ten years. In 1848, Wolf collected all available sunspot observations and determined the more accurate value of eleven years. In order to systematize the counts made by observers with widely differing equipment, he also introduced the "sunspot number",

$$R = K\,(10g + f) \tag{2}$$

where f = number of sunspots, g = number of sunspot groups, and K = a constant assigned to an individual observer and his instrument. The sunspot number has proven to be a very useful index of solar activity, and it is still in use today.

The "equatorial acceleration" of the sun (a differential rotation with smaller speeds at higher latitudes) was determined (circa 1860) by Carrington from his sunspot observations; he also discovered that the average latitude of sunspots decreases during the solar cycle. The first known observation of a *solar flare* was made on September 1, 1859 by Carrington and Hodgson, observing independently in white light.

Galileo had noticed areas near sunspots that were brighter than the surrounding photosphere; these (now called *faculae*) regions were the second phenomenon, after sunspots, to be detected on the solar surface. The solar *granulation* was first observed visually by Carrington and, a few years later, photographically, by Janssen. The solar limb darkening was studied in 1851 by Secchi using both daguerrotype images and a thermoelectric pile with the directly projected solar image. He found that the "white light" brightness at the limb was about one-half the brightness at the disk center—in fair agreement with modern determinations.

Considerable progress was made at the beginning of the 20th century following the development of photographic and spectroscopic techniques; most notable was the discovery in 1908 of strong magnetic fields in sunspots by G. E. Hale.

Observation of the sun during eclipse had proven fruitful during the 19th century. The *corona* and *prominences* had been discovered (or rediscovered) and a fundamental problem resolved—namely, do these "appendages" belong to the sun or to the moon? This question arises because at first they were observed only at the time of eclipse. However, the eclipses of 1851 and 1860 showed their solar origin when the moon was seen to move over them. Janssen and Lockyer also confirmed the solar origin in 1868 when they viewed prominences outside of eclipse by using a spectrometer to reduce the scattered light. The corona was observed outside of eclipse in the 1930's using the "coronagraph" developed by Lyot. The chromospheric "flash spectrum" was first observed by Young during the solar eclipse of 1870.

Spectroscopic studies of the sun had begun with the discovery (1802) of the solar absorption lines by Wollaston and independently (1814) by Fraunhofer. The latter made a detailed study of these lines that bear his name today and introduced the familiar alphabetical designation (e.g., D line of Na I) of the principal lines. These spectroscopic studies were continued by Huggins and Secchi. Rowland's Tables of the Solar Spectrum were published in the 1890's, and comparison with laboratory spectra revealed the presence in the sun of many terrestrially known elements. Helium lines had been found in the solar spectrum by Lockyer in 1868, but their "identification" dates from the discovery of helium in mineral samples by Ramsay in 1895. When unfamiliar lines were subsequently discovered in the coronal spectrum, the process of attributing them to a new element (so successful in the case of helium) was repeated, and "coronium" was postulated. We now know that coronium has no place in the periodic table of the elements; it consists only of familiar elements in unfamiliar stages of ionization as established by Grotrian and Edlén, circa 1940.

The first observations of solar radiation outside the visual region of the spectrum appear to be the detection of the solar infrared in the early nineteenth century by John Herschel, who used thermometers and strips of black paper moistened with alcohol as crude energy detectors. Considerable interest in the solar infrared ensued and by 1888, Langley had produced a remarkable map of the solar spectrum from 0.3 μ to 5.3 μ. Solar radio waves were sought as early as 1901 by Deslandres and Lodge, but they were first detected in 1942, when they interfered with wartime radar equipment. The extension of the solar spectrum into the far ultraviolet and x-ray regions became a reality after World War II when rockets made essentially atmosphere-free observations practicable; much of the pioneering work was done by Tousey and his collaborators at the Naval Research Laboratory. The Lyman-α (λ1216Å) line of neutral hydrogen was detected in emission by Rense at Boulder in 1953.

The nineteenth century also saw considerable activity in the theory of the solar interior. The importance of the great pressures expected there led to the conclusion that the interior was not solid as William Herschel had thought. In fact, we now know that the perfect gas law holds throughout the sun. A quantitative model of the solar interior assuming energy transport by convection was derived by Lane in 1869. His model had a central temperature of 10^7 °K, which is not far from the modern value, even though we now know that radiation is the primary mode of energy transfer in the interior. Quantitative techniques for studying radiative energy transport are generally considered to date from the work of Karl Schwarzschild in 1906.

During the nineteenth century, scientists also inquired as to the source of the sun's enormous energy output. A quantitative theory (1853) was developed by Helmholtz and also Kelvin (1887). They showed that the solar luminosity could be maintained from gravitational potential energy if the sun's radius shrank by 75 meters/year. However, this leads to a maximum solar age of "only" 10^7 years, which is about 1% of the value subsequently inferred geologically. Moreover, the fossil evidence strongly indicates a near constant solar luminosity for times in excess of 10^8 years. Eventually, it was realized that only the direct conversion of matter into energy (we recall Einstein's famous equation, $E = mc^2$) could sustain the solar luminosity over times of this length. The essential nuclear reactions that convert hydrogen into helium and yield energy corresponding to the mass defect were discovered by Bethe in 1939.

The specific nuclear reactions involving hydrogen point to another long-standing problem—namely, the composition of the solar interior. Pre-1930 views held the sun to be composed of metallic vapors; the Balmer lines of hydrogen were regarded as mere "skin phenomena." Equations had been produced by Eddington that gave the mass-luminosity-radius relation in terms of the mean molecular weight and opacity; when the opacity was considered well-determined and a mean molecular weight of a little greater than 2 was assumed, the calculated luminosity exceeded the observed luminosity by a factor of more than 10. The solution (which was a possibility considered by Eddington) consisted of a much larger amount of hydrogen which then became the most abundant element; accepting a relatively large abundance for helium is also a fairly recent development.

The importance of the discovery of hydrogen as the most abundant element in stars cannot be overemphasized. It clearly opened the door to the understanding of the correct solar and stellar nuclear reactions. Similarly, after this discovery, Wildt was able to resolve a long-standing problem by suggesting the H^- ion as the source of opacity in the solar photosphere. These discussions form the basis of much of modern astrophysics.

THE STRUCTURE OF THE SUN

Introduction

Studies of the solar interior belong to the discipline of stellar structure and evolution. In this area, theoretical techniques developed for the sun correspond rather closely to those that have been applied to stars in general. However, investigations of the photosphere, or lower atmosphere of the sun, where the mean optical depth is of order unity, are more sophisticated than those attempted by workers in stellar atmospheres, because superior observational data are available for purposes of inspiration and comparison. Besides the obvious fact that the great apparent brightness of the sun facilitates spectroscopic measurements at very high dispersion, the large arc subtended by the solar diameter ($d \approx 32'$) permits the inspection of center-to-limb variations in the solar continuum, the Fraunhofer lines, and indeed the general appearance of solar phenomena. all of which yield detailed information on the structure of the photosphere. For other stars, such data are few and crude, as they arise from discussions of the light curves of eclipsing binaries and from the results of a small number of pioneering interferometric studies. Due to the sparsity of relevant observational material on the outer atmospheres of stars other than the sun, analysis of the solar chromosphere, the region immediately above the photosphere, is virtually a self-contained field. The same is true with respect to the theory of the corona, a tenuous envelope that surrounds the sun and extends to great distances, finally merging into the interplanetary gas or solar wind.

Interior

The observational inputs to model calculations for the solar interior consist of the total mass ($1.99 \times 10^{33}g$, the most precisely known stellar mass) as derived from planetary mechanics, and boundary values of the temperature, pressure, total flux, and chemical composition at one solar radius, determined from photospheric studies. The slow axial rotation of the sun (about 27 days, depending on the latitude) is usually ignored, as is the magnetic field. The chemical composition observed in the photosphere is considered to represent the abundance of the elements in the material from which the sun formed. Using this composition, one constructs an initial model for the march of density, temperature and pressure with radius, using the techniques described by R. F. Cameron in Chapter 21 of the present volume. Subsequent epoch models are then calculated to take account of the changing composition of the interior that results from the processes of nuclear energy generation. Finally, a representation of the contemporary sun is obtained. Results of such work show that the proton-proton chain is the dominant source of energy in the sun, although a contribution of several per cent is made by the CNO cycle. In the model of Weymann[52], shown in figures 1 and 2, a region in radiative equilibrium extends from the center to $r/R_{\odot} = 0.86$, where it is surrounded by a convective zone associated with the ionization of hydrogen. (The ionization of hydrogen, which sets in at $T \approx 10{,}000°K$, provides a source of electrons that can associate with neutral H atoms

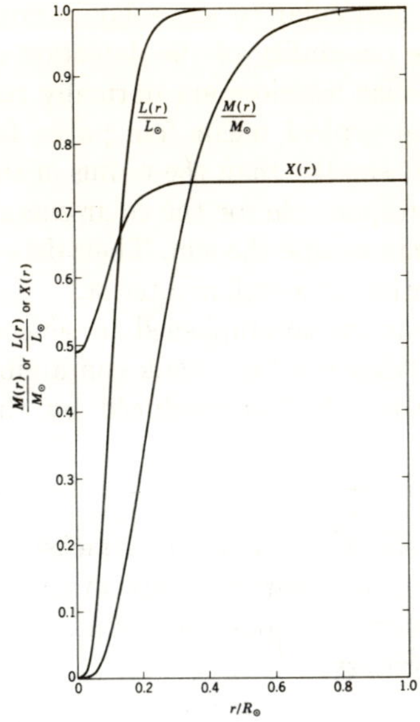

FIG. 1. The run of luminosity L(r)/L⊙ mass M(r)/M⊙ and the mass fraction of hydrogen (X) on the basis of Weymann's solar model. From *Solar System Astrophysics* by J. C. Brandt and P. Hodge. Copyright, 1964, McGraw-Hill Book Company. Used by permission.

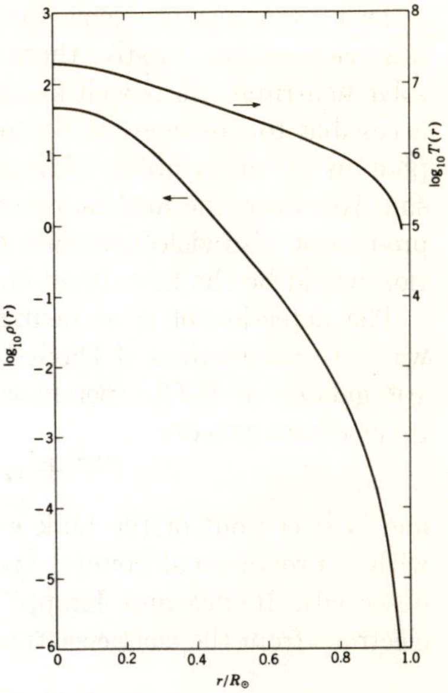

FIG. 2. The run of temperature and density on the basis of Weymann's model of the sun. From *Solar System Astrophysics* by J. C. Brandt and P. Hodge. Copyright, 1964. McGraw-Hill Book Company. Used by permission.

to form H^- ions. These in turn provide enough opacity to significantly impede the flow of radiation, and convection becomes the effective mode of energy transport.) The mass fraction of hydrogen rises from $X \approx 0.5$ at the center, where $T \approx 10^7$ °K, to the photospheric value $X \approx 0.74$, which is already attained at $r/R_\odot \approx 0.3$. In an attempt to derive the solar neutrino flux, Sears[53] considered a number of evolutionary sequences, for various possible initial compositions. His results were not very different from those of Weymann, but included a best-fit value $Y = 0.27$ for the mass fraction in helium. Bound-free transitions constitute the major source of opacity; however, free-free processes and electron scattering may make a numerical contribution, and recently the contribution due to absorption lines has been considered.

Two intriguing problems that are particular to the solar interior deserve mention. Firstly, there is the possibility of the detection of solar neutrinos. As is well known, stellar interiors are normally not accessible to observation because the typical mean free paths for photons are many orders of magnitude smaller than the radius of the sun. However, the nuclear processes responsible for the solar energy production also yield neutrinos that can escape the sun. Their detection would be the first direct observation of a stellar interior.

The detection of solar neutrinos can be accomplished in several ways. In the method of Davis[54], the "detector" is a tank containing 10^5 gallons of C_2Cl_4 (perchlorethylene). Ar^{37} is produced by the inverse beta process

$$Cl^{37} + \nu_e \rightarrow Ar^{37} + e^- \tag{3}$$

and is swept out of the tank with helium, processed and measured with a proportional counter (perhaps 10 events per day might be observed). Reines and Kropp[55], however, propose to count recoil electrons from the elastic scattering process

$$\nu_e + e^- \rightarrow \nu_e + e^-, \tag{4}$$

using a large scintillation detector. In both cases, the "observatory" is located far underground, in order to reduce the effect of cosmic rays.

Such measurements could determine the solar central temperature as follows. A thorough discussion (Bahcall[56]) of the nuclear physics involved indicates that some 90% of the neutrinos detected in the Davis experiment would come from the method of completing the proton-proton chain that involves neutrino production from B^8; this is true despite the fact that neutrinos from B^8 decay constitute only about 0.1% of the solar neutrino flux. The B^8 neutrino flux is sensitive to the central temperature of the sun and a measurement of this flux to 50% would determine the central temperature to within 10%. Furthermore, the observations would provide experimental verification that the proton-proton chain is the dominant source of nuclear energy in the solar interior, and the neutrino flux would provide an additional condition on acceptable models of the solar interior.

The second problem concerns the recent measurement of the solar oblateness by Dicke and Goldenberg[57] who found that

$$\frac{R\,(eq) - R\,(pole)}{R} = (5.0 \pm 0.7) \times 10^{-5}. \tag{5}$$

If this is interpreted in terms of a solar gravitational quadrupole moment, then it must be responsible for an advance in the perihelion of Mercury equal to 10% of the value predicted by general relativity. The latter value appears to be in good agreement with observation, however. One way out is to postulate a scalar-tensor theory of gravitation (*e.g.*, Brans-Dicke cosmology) that would require a smaller advance than that calculated by Einstein. Further, the observed oblateness may be interpreted as an angular momentum effect, implying that the solar core may rotate in a time of the order of a day. This core would supply angular momentum to the solar atmosphere, and a momentum-sink would be required for a steady-state to prevail. Theory and observations pertaining to the orientations of ionized comet tails (Brandt[58,59]) suggest that the correct amount of angular momentum is carried off by the solar wind.

Besides the obvious impact Dicke's hypothesis has on the theory of stellar and solar interiors, it may also bear on the classical problem of angular momentum in the solar system. The planets have 98% of the angular momentum and the sun 2%; it is hard to construct a model of the origin of the solar system with this end result. A first step towards the resolution of this dilemma could be a model in which the sun's angular momentum is much higher than previously supposed, due to a rapidly rotating core beneath the relatively superficial atmosphere where the slow (27-day) rotation is observed.

There are difficulties, however, with the model suggested by Dicke[59a]. Goldreich and Schubert[59b] find that such models are unstable; perhaps there is another cause for the oblateness. Coche[59c] has suggested that turbulent viscosity forces acting on photospheric meridional currents could cause the observed oblateness.

Photosphere

The theory of radiative transfer can be used to derive the variation of effective temperature with optical depth (τ) in the photosphere, given the observed dependence (*limb darkening*) of monochromatic flux on distance from the center of the solar disk. This T-τ relation, together with the condition of hydrostatic equilibrium and an evaluation of the opacity as a function of temperature and pressure for the assumed chemical composition, determines a model photosphere. The fine structure or photospheric *granulation* is usually ignored in this approach, which has been attempted by many astronomers. However, some details of the center-to-limb variation of line profiles appear to require a nonhomogeneous photosphere with temperature fluctuations of a few hundred degrees Kelvin.

Limb darkening can be determined by direct observation of the intensity distribution across the sun. This is accomplished, for example, by "allowing" the sun to drift past the entrance aperture of a spectrophotometer. Such measurements must be corrected for scattered light in the instrument and in the atmosphere, and for the instrumental profile (*apparatus function*). Further, the spatial resolution attained is limited by the finite dimensions of the scanning aperture and the presence of atmospheric seeing. The difficulty of obtaining useful data increases towards the limb, and is particularly severe for $\cos \theta < 0.03$, where θ is the heliocentric angle between the directions to the earth and the point under observation. This method is reviewed by Goldberg and Pierce[20]. Evaluation of scattered light in the atmosphere from measurements of the intensity distribution across the lunar limb at the time of a partial solar eclipse has also received much attention (*cf.* ten Bruggencate *et al.*[60]). An independent estimate of the limb darkening can be obtained by analysis of measurements of the integrated intensity of the exposed crescent during a total eclipse of the sun (*cf.* Kristenson[61]). However, in addition to the error introduced by the irregular character of the edge of the moon, this method suffers from the usual atmospheric limitations. A variation on this technique involves the use of a coronagraph to produce an artificial eclipse (Dunn[62]). Clearly, the acquisition of data unaffected by the terrestrial atmosphere would provide a considerable stimulus to studies of photospheric structure. A promising step in this direction is the analysis of photographs obtained in the stratosphere (Rogerson[63], Gaustad and Rogerson[64]). A representative model photosphere due to Unsöld[65] is contained in Table 1.

The observed limb darkening is analyzed in terms of the standard equation of transfer, *viz.*,

$$\mu \frac{dI_\nu}{d\tau_\nu} = I_\nu - \mathcal{J}_\nu \tag{6}$$

where the opacity is defined by

$$\tau_\nu = \int_z^\infty \rho\, k_\nu\, dz \tag{7}$$

TABLE 1

A MODEL SOLAR ATMOSPHERE CONSTRUCTED BY UNSÖLD*

Height z, km	Solar radii	Temperature T, °K	Gas pressure, dynes/cm², log P	Electron pressure, dynes/cm², log P_e	Electrons/cm³, log N_e	Turbulent velocity V_t km/sec	Layer	Main energy transfer
1,400,000	3.0	$2 \cdot 10^6$	−3.8	−4.1	5.5			
700,000	2.0	$2 \cdot 10^6$	−2.8	−3.1	6.4			
350,000	1.50	$2 \cdot 10^6$	−2.1	−2.4	7.2		Corona	Thermal conduction
42,000	1.06	$2 \cdot 10^6$	−0.9	−1.2	8.4			
20,000	1.03	$2 \cdot 10^6$	−0.8	−1.1	8.5			
		Very inhomogeneous				~15	Transition layer	Mechanical energy
3,000		~4–6000	0.2	−1.7	10.5			
2,000	***	~4–6000	0.5	−1.4	10.8	12	Chromosphere	Radiation
1,000		~4–6000	1.2	−0.9	11.3	7		
	Opt. depth τ_{5000}							
0	0.005	4090	4.1	−0.5	11.7	1–2		
	0.01	4295	4.3	−0.3	12.0			
	0.05	4855	4.6	+0.2	12.4			
	0.1	5030	4.8	+0.4	12.6		Photosphere	Radiation
	0.5	5805	5.1	1.2	13.3	2		
−260	1.0	6400	5.2	1.8	13.8			
	2.0	7180	5.3	2.4	14.4		Hydrogen	
−280	***	10^4	5.3	4.0	15.86	2	Convection Zone	Convection
	Solar radii							
−16,000	−0.02	10^5	9.4	9.1	20.0	0.3		
−140,000	−0.2	10^6	12.3	12.0	21.9	0.0		Radiation

* By permission from A. Unsöld, in "Space Age Astronomy", ed. A. J. Deutsch and W. B. Klemperer, Academic Press, New York, 1962.

Here ρ is the density and k_ν is the scattering coefficient per unit mass; z is measured positive along the outward normal, θ is the angle a particular ray makes with the outward normal, and $\mu = \cos\theta$. The intensity I_ν is in erg/cm^2-sec-steradian-sec^{-1} and $\mathcal{J}_\nu$ is the source function. For the case of the solar atmosphere, the formal solution to eqn. (6) is

$$I_\nu(O, +\mu) = \int_0^\infty \mathcal{J}_\nu(\tau_\nu)\, e^{-\tau_\nu/\mu} \frac{d\tau_\nu}{\mu} \qquad (8)$$

Limb darkening permits a determination of $\mathcal{J}_\nu(\tau_\nu)$ for $\tau \leq 1$. The source function (when known) can be related to a temperature through the assumption of local thermodynamic equilibrium (LTE), which is defined by

$$\mathcal{J}_\nu = B_\nu(T) = \frac{2h\nu^3}{c^2} \frac{1}{[e^{h\nu/kT} - 1]} \qquad (9)$$

Hence, the source function for LTE is the Planck function; such an assumption produces the T-τ relation needed to deduce the structure of the photosphere. While we note that LTE is a good approximation in many discussions of the formation of the solar continuous and line emission, its theoretical (and practical) justification is open to question as has been stressed by Thomas and Athay[24]. A thorough discussion of departures from LTE would be rather complex and far outside the scope of this review. However, the Planck distribution (derived for confined photons) may itself be strictly inapplicable to the solar case which involves photon flow and loss (*i.e.*: we don't have a hohlraum). Waddell[66] has argued that the Kothari-Singh source function (which contains an extra parameter depending on the divergence of the photon flux) may well be more appropriate.

The Fraunhofer absorption lines are formed principally in the photosphere by selective absorption at the characteristic wavelengths of the various elements. The theory of line formation requires the detailed specification of the processes of absorption and reemission. In combination with a model atmosphere, it can be used to determine the abundances of the elements in the photosphere. The profiles (dependence of intensity on wavelength) of individual lines can be used, but the most reliable datum is the "equivalent width". (Definition: if a section of the adjacent continuum is chosen such that its intensity equals that subtracted from the solar spectrum by an absorption line, then the width of this continuum, in Ångstroms, is the equivalent width of the line.) These studies give the abundances of the elements in the photosphere; averages of several determinations are shown in Fig. 3. Note that even though helium was discovered in the solar (*chromospheric*) spectrum, there are no suitable lines for determination of the photospheric helium abundance.

Although radiation is the predominant mode of energy transfer, inspection of white light photographs obtained under conditions of good seeing reveals the presence of variable fine structure, called granulation, in the photosphere. The granules constitute visible evidence of the underlying convection zone, whose presence is suggested by theoretical models of the solar interior. Perhaps the most celebrated work in the recent history of solar physics has been the determination of their properties (Schwarzschild[67], Bahng and Schwarzschild[68,69]) by photographic observations from the Project Stratoscope balloons (see figure 4). Also despite the great difficulties involved in making systematic time-studies of seeing-limited features, a considerable amount of ground-based work has been accomplished on the

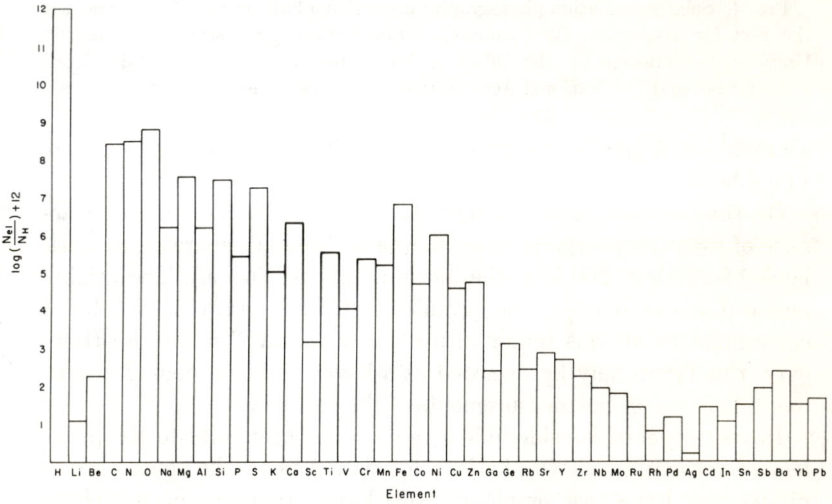

FIG. 3. Solar abundances as determined from studies of the Fraunhofer lines; values given are typical of those in the current literature.

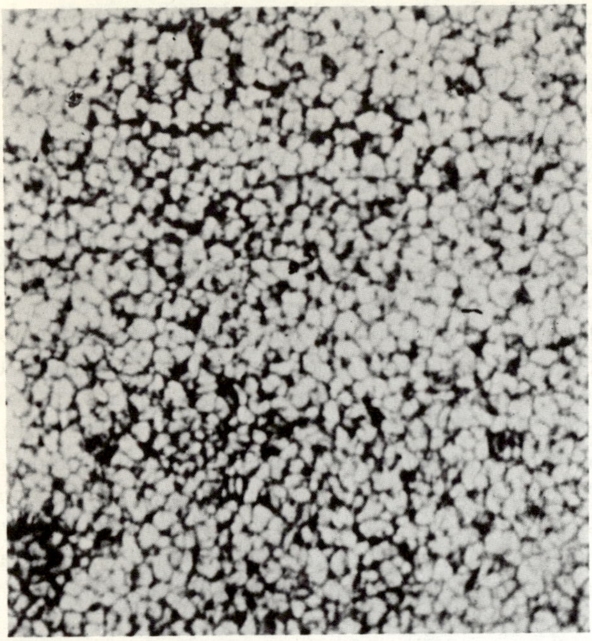

FIG. 4. Solar granulation photograph taken with a balloon-borne telescope; see the text for discussion. By permission from Project Stratoscope of Princeton University, sponsored by the Office of Naval Research, the National Science Foundation, and the National Aeronautics and Space Administration.

granulation. A good summary of these results is provided by Bray and Loughhead[43].

On the best available photographs, the granules have the appearance of irregular polygons, of mean size $\approx$ 700 km, separated by dark lanes of width $\lesssim$ 290 km (the instrumental resolution). The brightness differences between the granules and the intergranular material correspond to an rms temperature fluctuation $\Delta T \approx 100°K$. However, this figure may be regarded as a lower limit, in view of instrumental effects on the measurements of the dark lanes.

In recent years, considerable attention has been paid to the problem of measuring photospheric velocities. Among the several techniques that have been employed, two have proven to be especially fruitful: the *wiggly line* method (see figure 5) which involves the measurement of localized Doppler displacements in an absorption line

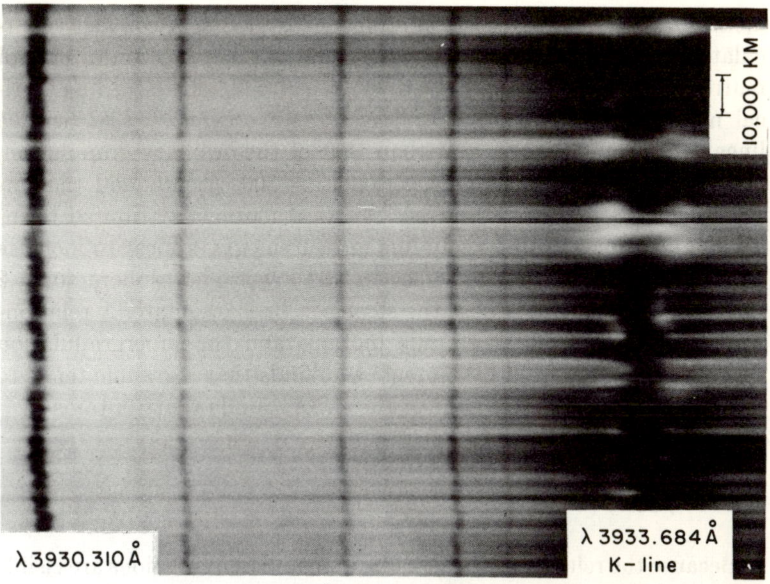

FIG. 5. Spectrogram of the 3933Å region obtained by N. R. Sheeley on June 19, 1966 using the 33-inch solar image of the McMath telescope at Kitt Peak National Observatory. The artificial dark streak along the dispersion indicates a location on the sun where a magnetic field of several hundred gauss was measured approximately 1 minute before this spectrogram was taken. Note that the Ca II K-line has bright K_{232} emission at this position. Also note the general "wiggly" appearance of the lines.

on a high dispersion spectrogram (*cf.* Evans and Michard[70]), and the analysis of *Doppler spectroheliograms* (Leighton, Noyes and Simon[71]; Noyes and Leighton[72]; Simon and Leighton[73]), which provide pictorial representations of the distribution of the radial velocity vector on the solar disk. If observations are made in suitably selected lines, it is possible to build up a model for the velocity field in the photosphere. This follows from two considerations: (1) the strength of absorption is correlated with height, in that weaker lines tend to originate deeper in the photosphere (but absolute values for the heights are very poorly known); (2) radial velocities measured at the disk center correspond to vertical motions, whereas those observed at the limb represent horizontal motions.

Two apparently distinct phenomena are seen on the Doppler spectroheliograms. The *supergranulation* consists of a network of cells of mean size ≈ 32,000 km, with lifetimes ≈ 20 hours. A convective circulation pattern exists in each cell, in the sense of horizontal flow of matter from the center to the rim, at speeds ≈ 0.4 km/sec. The vertical motions are smaller, ≈ 0.1 km/sec. An *oscillatory field,* with a linear scale somewhat greater than that of the ordinary granulation, is also observed. The average period is ≈ 300 seconds and the rms vertical velocity is ≈ 0.4 km/sec. Vertical motions attributed to individual granules, from wiggly line measurements, appear to increase with height, from ≈ 0.3 km/sec in the lower photosphere to ≈ 2 km/sec at the heights where the strongest lines originate. A relationship between horizontal granule motions and the supergranulation flow pattern is reported by Simon[74] who finds that a granule tends to move toward the nearest edge of the supergranule that contains it.

Chromosphere

Because the relevant optical continuum and line emissions are visible only near the limb, the determination of chromospheric structure has been largely based on observational material (*flash spectra,* see figure 8) secured during total solar eclipses. Additional data is derived from spectroheliograms of the disk, taken in or near the cores of strong absorption lines that originate in the chromosphere, from observations of fine structure (*spicules*) at the limb, and from radio measurements.

The eclipse material is analyzed to obtain the variation of intensity with height for the various spectral features. This information then yields the variation of the density scale height in the chromosphere. The technique is summarized by van de Hulst[22] and Pagel.[25] It has long been recognized that these scale heights differ from those predicted by the extrapolation of photospheric models: the observed scale heights are greater than the predicted ones, and they increase with altitude, whereas the reverse is expected. These anomalies suggest that mechanical energy transported from deeper layers is deposited in the chromosphere. The exact interpretation of this phenomenon (*the chromospheric problem*) is clearly basic to our comprehension of the structure of the solar atmosphere, and considerable attention has been paid to it by both observers and theoreticians.

As the radiation field departs from hohlraum conditions, and the mechanical energy flux is significant, useful chromospheric models cannot be calculated with the assumption of local thermodynamic equilibrium. Spectroscopic observations reveal the coexistence at a given height of emission in lines corresponding to transitions that require quite different conditions of excitation and ionization. This suggests that spherical symmetry also has to be abandoned, a result that is supported by photographs of structure in the chromosphere and by millimeter-wave observations (Coates[75]) of the center-to-limb variation of solar brightness temperature. Non-LTE, multi-component models are discussed extensively by Thomas and Athay.[24]

Spectroheliograms taken in the K line of ionized calcium show a pattern (*chromospheric network*) of bright emission features called *flocculi* or *plagettes*. Vertical chromospheric fine structure (see figure 6) is seen on the limb in the form of spicules (described below). The distribution of spicules is apparently not a function of the solar cycle. On the other hand, it is not uniform over the chromosphere; rather, the spicules seem to extend upwards from the flocculi. Simon and Leighton[73] found that the network is associated with the large-scale velocity pattern of the supergranulation; Howard[76] and Leighton[77] had already shown an association between the calcium emission features and the magnetic field. Parker[78] has provided a theoretical basis for an understanding of the enhanced field at the network as a result of the convective process in a supergranule.

Recent studies of spicules, based on H-alpha filtergrams, include both observations at the limb (Dunn[79]) and on the disk (Beckers[80]). A typical spicule has lifetime $\approx$ 15 minutes, diameter $\approx$ 800 km, and length $\approx$ 3000 km (however, there is a disagreement about spicule lengths, with some observers favoring values $\approx$ 10,000 km). The total number of spicules is a quantity of great interest, since it is related to the fractional areas that must be assigned to various components of the chromosphere in model calculations. Sample spicule counts must be corrected for the effects of overlapping features, and several workers have used quite different procedures. Beckers' result is that there are $\sim 4 \times 10^5$ spicules in the chromosphere at a given moment. Zirin and Dietz[81] analyzed the extreme ultraviolet spectral measurements that have been secured by rocket experimenters and concluded that the high-excitation chromospheric lines originate in the spicules. Zirin[5] regards spicules as the sole component of the upper chromosphere; he believes that the coronal gas does not fill the regions be-

(a) SOLAR CORONA (Photograph Courtesy K. Hallam)

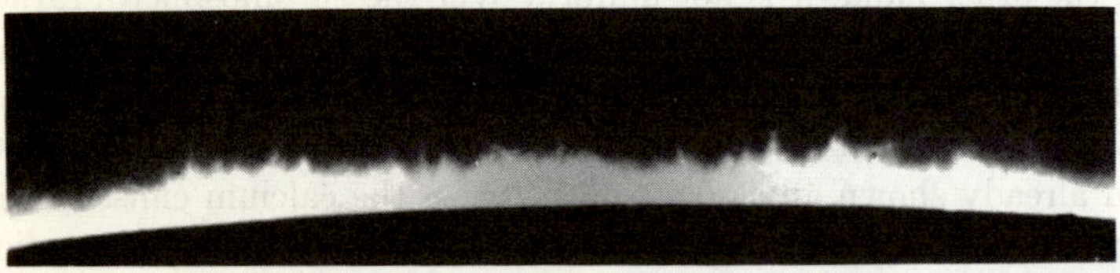

(b) CHROMOSPHERE OF THE SUN SHOWING SPICULES (Sacramento Peak Observatory, Geophysics Research Directorate, AFCRC)

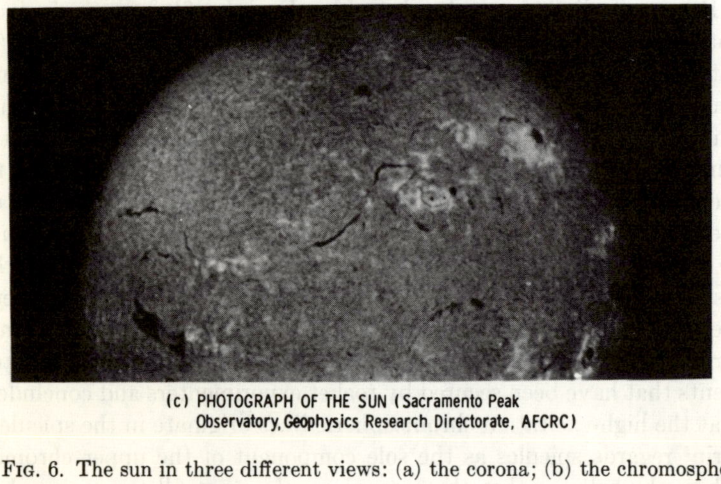

(c) PHOTOGRAPH OF THE SUN (Sacramento Peak Observatory, Geophysics Research Directorate, AFCRC)

FIG. 6. The sun in three different views: (a) the corona; (b) the chromosphere (spicules); and (c) the photosphere.

tween the spicules, but rather extends outward from them in plumes.

The most comprehensive theory for the energy source of the chromosphere is that of Osterbrock[82]. He stressed the importance of hydromagnetic effects and showed that the upward flux of mechanical energy generated in the hydrogen convection zone can account for the radiative energy losses observed in the chromosphere and corona. In particular, the main source of chromospheric heating is identified as the dissipation of shocks that originate as fast mode waves, and the spicules are described as slow-mode waves that travel along the lines of force.

Corona

The halo that is found to surround the sun at the time of total eclipse (figure 6) is called the corona. This tenuous outermost component of the solar atmosphere has a temperature $\approx 10^6$ °K. It is the site of several fascinating phenomena, and is intimately involved with the solar wind that pervades interplanetary space. The physical conditions in the chromosphere go over very rapidly with height to those that characterize the corona, and the separate name of "transition zone" has been advocated to denote this interface region. The corona is commonly taken to begin at 1.03 $R_{\odot}$.

The coronal light is composed of three parts with physically distinct sources: (1) the *K corona* is the "true" coronal light that results from Thomson scattering of photospheric light by free electrons in the corona; (2) the *F corona* results from diffraction of photospheric light by interplanetary dust particles; thus, it must be separated from the K component in order to use measurements of the latter to derive properties of the coronal gas; (3) the *E corona* (sometimes called *L corona*) arises from discrete emission lines due to highly ionized atoms. Many such transitions are observed, but the three famous ones, to which much of the available observational material is limited, are the *red line* (λ6375Å) of Fe X, the *green line* (λ5303Å) of Fe XIV, and the *yellow line* (λ5694Å) of Ca XV. Billings[28] tabulates identified coronal lines.

Coronal densities are given in Table 1; they are derived in a rather straightforward manner from the intensity of the K corona since the emergent photospheric intensities and the cross-section from Thomson scattering are known. The most restrictive assumption usually concerns the geometry because some assumption (*cf.* Brandt and Hodge[4], p. 105) is necessary to interpret the observed two-dimensional intensity distribution in terms of the three-dimensional coronal source. Spherical symmetry is almost always assumed. The rough densities and the variation with heliocentric distance are probably certain; however, the existence of several scales of coronal fine structure greatly complicates *detailed* calculations of coronal densities.

The coronal temperature of $\sim 10^6$ °K is found through several independent arguments. Firstly, density gradients provide an estimate through the temperature needed for support in hydrostatic equilibrium; the systematic expansion implied by the solar wind does not alter these estimates interior to about 3 $R_\odot$ (Brandt, Michie and Cassinelli[83]). Nonetheless, the small-scale, ray-like coronal fine structure must be included in order to obtain an accurate coronal temperature. Physically, this means that the mean density gradients inferred from studies of the K corona that assume spherical symmetry actually reflect the decrease of *two* quantities; these are (a) the actual decrease of density within the "model-rays" and (b) the change of the fraction of space filled by "model-rays." The latter variation can be studied (Brandt, Michie and Cassinelli[84]) through the function $1/X = \langle N_e \rangle^2 / \langle N_e^2 \rangle$ where $1/X$ has a simple interpretation as the fraction of space occupied by matter. Available optical and radio results indicate that $1/X$ decreases outward. This effect changes the coronal temperature inferred from density gradients, which was formerly estimated at 1.5×10^6 °K to 2×10^6 °K or slightly more.

The coronal electron temperature can be inferred from the ionic species present (*cf.* Shklovskii[27]). Now ionizations and recombinations both occur by electron collisions, and, hence, the degree of ionization depends only on the temperature (and not the density). The value accepted for the ionization temperature has been raised recently by the discovery (Burgess[85]) of the importance of dielectronic recombination. In this process, a free electron with energy equal to the excitation energy of *two* bound states excites a bound electron to one of these states; then the first electron has just the right energy to occupy the other bound state. Thus, the recombined atom has two excited electrons that drop to lower states and emit the coronal lines. In hot plasmas like the corona, dielectronic recombination is far more efficient than ordinary radiative recombination and the ionization temperature deduced for the corona is about 2×10^6 °K.

A coronal temperature can be obtained from the widths of emission lines (*cf.* Billings [86]) under the reasonable assumption that the corona is optically thin in the lines in question. The width of a Doppler profile corresponds to the kinetic temperature provided only that turbulent motions are negligible and that the instrumental broadening is known. Thus,

$$T = 1.9 \times 10^{12}\left(\frac{h}{\lambda}\right)^2 \mu \ (^\circ\mathrm{K}) \tag{10}$$

where h is the full width at half-intensity, λ is the wavelength of the line, and μ is the atomic weight. A typical value for the quiet corona is 2.1×10^6 °K from observation of the red line ($h = 0.89$Å). Note, however, that different temperatures are found for the various types of *active* coronal regions.

A temperature can also be crudely estimated from the widths of Fraunhofer lines. Most of these photospheric absorption lines are invisible in the K-coronal spectrum because of the great Doppler broadening due to the high thermal velocity of the coronal electrons. Thus, only the strongest lines, such as H and K of Ca II, can be used; the expected width of these lines can be estimated from eqn. (10) by inserting the atomic weight for an electron, 1/1836. One finds a half-intensity width of about 120Å for a temperature of 10^6 °K. This means that the depression is $\sim 1\%$ of the continuum and is very hard to measure. At present, the widths of Fraunhofer lines in the K corona strictly establish only that $T > 10^5$ °K.

The various arguments described above are consistent with a coronal temperature close to 2×10^6 °K. The observed radio brightness temperature (*cf.* Kundu[38]) of 700,000 − 1,000,000 °K in the meter wavelengths range appears to be an exception to this consensus, but a suggested fundamental change in the applicable equation of transfer (Oster[87], Oster and Sofia[88]) would remove this discrepancy. Ray paths in the corona are computed according to Snell's law and the emergent radio intensity is obtained by integrating along a trajectory. Traditionally,

$$I = \oint_0^{\tau_0} e^{-\tau} B(T)\, d\tau \tag{11}$$

where by τ_o we denote the total optical depth in the corona along the ray path. If the corona is optically thick, the brightness temperature is equal to the kinetic temperature; in the radio wavelength region, the Rayleigh-Jeans approximation is valid, and $B(T) = 2kT/\lambda^2$. The use of the Planck function in eqn. (11) as the source function might be taken to imply that local thermodynamic equilibrium (LTE) is assumed, although it is well known that the coronal gas is not in LTE. However, we are dealing here with absorption and emission of radiation by free-free processes, and the only tenet of LTE required to hold true is that the electrons have a Maxwellian velocity distribution. Collisions in the corona are sufficiently frequent to establish this condition. Hence, the traditional equation of transfer predicts an equality of the kinetic and radio brightness temperatures for the corona; this would place the radio measurements at variance with the other methods of determining the coronal temperature.

Oster's contribution to this problem consisted of attributing the discrepancy to the dispersive nature of the corona as a medium for the traversal of radio waves. His transfer equation, not necessarily accepted by all workers in this field, is

$$I = \oint_{0}^{\tau_o} e^{-\tau}\, n\, B(T)\, d\tau \tag{12}$$

where n is the index of refraction. For the case at hand,

$$n = (1 - x)^{\frac{1}{2}} \tag{13}$$

where

$$x = \left(\frac{f_o}{f}\right)^2 = \left(\frac{e^2 N_e}{\pi m_e}\right) \frac{1}{f^2} \tag{14}$$

Here, f is the radio frequency and f_o is the plasma (or "critical") frequency. When eqn. (12) is applied, the brightness temperature is found to be only $0.25 - 0.6$ the electron kinetic temperature (because $n < 1$). Hence, the observed brightness temperatures in the range $700{,}000 - 1{,}000{,}000$ °K would imply an electron temperature near 2×10^6 °K, and bring the radio value into accord with the other measurements. Thus the new equation of transfer is very attractive. Delano[89] used a Fresnel coefficient formulation to study the intensity of radiation emerging through a "slabified" window (the index of refraction varied in discrete steps) in a hohlraum filled with rarefied gas; his results require the revised equation of transfer. On the other hand, Zheleznyakov[90] argues that the conventional equation is correct; clearly the situation is unsettled. Thus, we conclude at present that most evidence points to a temperature of 2×10^6 °K for the quiet corona.

The energy source required for the maintenance of the corona is presumably the same in broad outline as that of the chromosphere (Osterbrock[82]). Wave motions having their "final cause" in the hydrogen convection zone propagate upward and are dissipated in the form of shocks. The details of this process are, unfortunately, still sketchy. Further, gravity waves have been advocated as an alternative mechanism for coronal heating (Whitaker[91]).

The coronal shape undergoes a systematic variation throughout the solar cycle. At any given time, it is largely determined by the position and number of coronal *fans*, the large streamers having widths $\approx 1R_\odot$. The fans are related to solar activity; in particular, they are associated with the quiescent prominences. At the time of maximum solar activity, the numerous fans cause the circular appearance of the corona. At solar minimum, the small degree of solar activity is concentrated to the equator; this produces a rather asymmetrical corona.

Another major constituent of the coronal fine structure is typified by the narrow streamers called *polar rays*. (Rays are also found within the fans). Their existence is due to the coronal magnetic field, which very effectively inhibits diffusion perpendicular to the lines of force.

The corona is also the site where the solar mass loss occurs in the form of the "corpuscular radiation" long inferred from observations of geomagnetic activity (*cf.* Chapman[92]) and more recently deduced (Biermann[93,94]) from studies of Type I comet tails; finally, it has been directly measured by deep space probes (*cf.* Mackin and Neugebauer[95]). The subject is discussed at length by N. F. Ness in Chapter 8 of the present volume; the hydrodynamic formulation of "solar wind" theory is given in detail by Parker[29], and a valuable review has been written by Dessler[30]. Clauser (see Parker[29], p. 63) discussed the coronal expansion through an analogy with the flow of gas into a vacuum through a de Laval (rocket engine) nozzle. Much attention

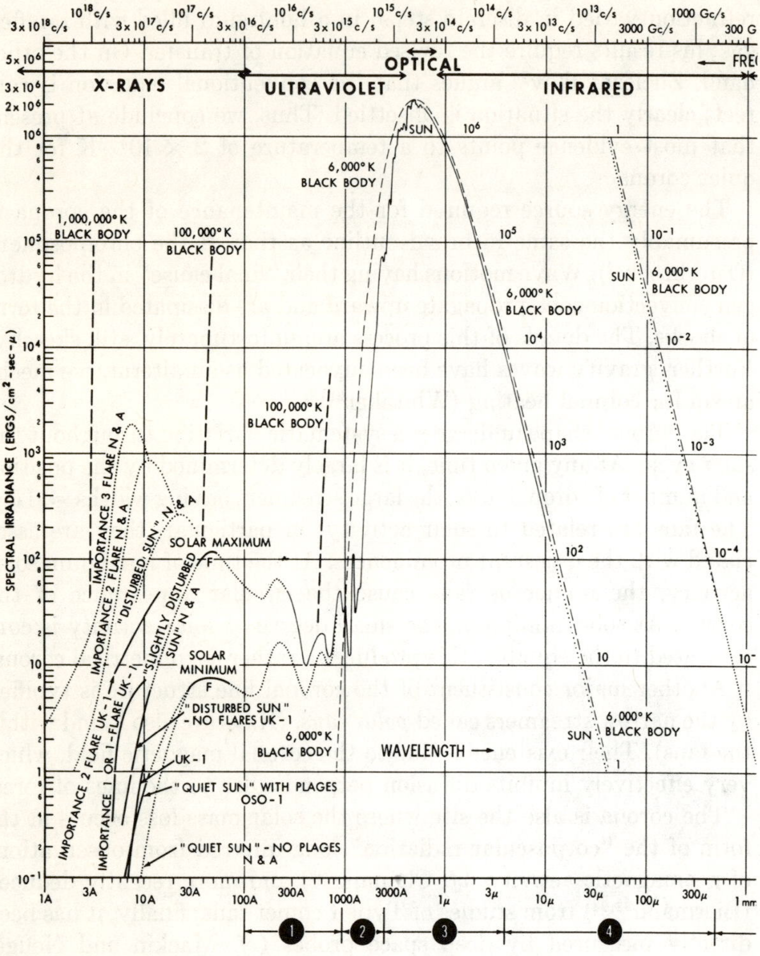

FIG. 7. The solar spectrum from the x-ray to the radio region as observed outside the earth's atmosphere. Compiled by H. H. Malitson.

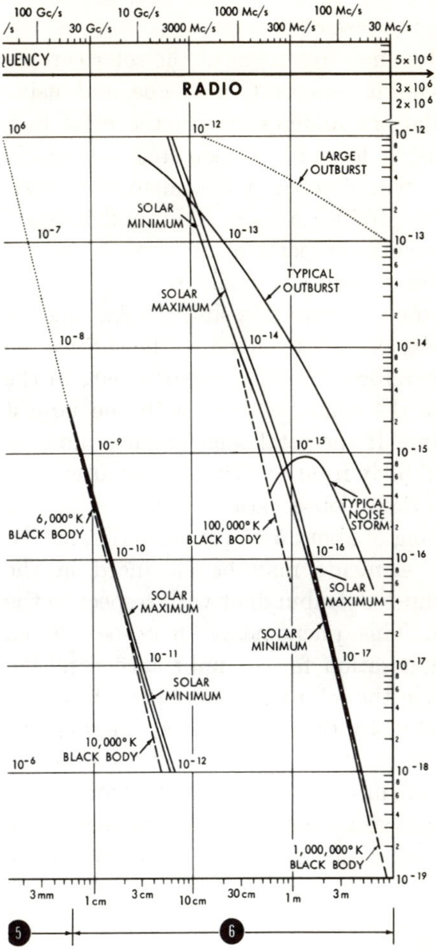

1

MEASUREMENTS BY HINTEREGGER:

(1) IN LILLER, "SPACE ASTROPHYSICS", P. 81, 1961.

(2) "ANALYSIS OF THE SOLAR EMISSION SPECTRUM FROM 1300 TO 250A AS OBSERVED IN AUG. 1961 "(WITH ZIRIN & HALL), 3RD COSPAR MEETING, 1962. ESTIMATES BASED ON ABOVE AND ON OSO-1 DATA IN "TIME CORRELATION OF EXTREME ULTRAVIOLET RADIATION AND THERMOSPHERIC TEMPERATURE" J. GEOPHYS. RES. 69, 4531-4535, 1964. BOURDEAU, R. E., CHANDRA, S., AND NEUPERT, W. M.

2

DETWILER, GARRETT, PURCELL, TOUSEY, "THE INTENSITY DISTRIBUTION IN THE ULTRAVIOLET SOLAR SPECTRUM", ANN DE GEOPHYS. 17 (3), 263-272, 1961.

3

JOHNSON, F. S., "THE SOLAR CONSTANT", J. METEOR. 11 (6), 431-439, 1954. (INCLUDES WORK OF DUNKELMAN AND SCOLNIK, J. OPT. SOC. AM. 49 (4), 356-367, 1959).

4

MURCRAY, F. H., MURCRAY, D. G., AND WILLIAMS, W. J., "THE SPECTRAL RADIANCE OF THE SUN FROM 4μ TO 5μ", APPLIED OPTICS 3, 1373-1377, 1964. SAIEDY, F. AND GOODY, R. M. "THE SOLAR EMISSION INTENSITY AT 11μ", M. N. ROY. ASTR. SOC. 119, 213-222, 1959.

5

6,000° K BLACK BODY

6

ALLEN, C. W., "ASTROPHYSICAL QUANTITIES", LONDON, ATHLONE PRESS, 1963, P. 188.

F = FRIEDMAN, H., "SOLAR RADIATION", ASTRONAUTICS, 14-23, AUGUST 1962 (ROCKET MEASUREMENTS).

N & A = NICOLET, M. AND AIKIN, A.C., "THE FORMATION OF THE D REGION IN THE IONOSPHERE," J. GEOPHYS. RES. 65, 1469-1483, 1960.

OSO-1 = DATA FROM THE FIRST ORBITING SOLAR OBSERVATORY. W.A. WHITE, PRIVATE COMMUNICATION, 1964.

UK-1 = BOWEN, P.J., NORMAN, K., POUNDS, K.A., SANFORD, P.W., AND WILLMORE, A.P., "MEASUREMENTS OF THE SOLAR SPECTRUM IN THE WAVELENGTH BAND 4 TO 14A", PROC. ROY. SOC. A 281, 538-552, 1964. ARIEL 1 SATELLITE DATA.

SOLID LINES REPRESENT MEASUREMENTS; DOTTED LINES, ESTIMATES.

has been paid to the controversy between advocates of the evaporative and supersonic solutions to the expansion equations (summarized by Chamberlain[96] and Parker[97]). It is now clear that the supersonic solution is appropriate to the case of the sun.

For years before the hydrodynamic treatment of the solar corona, mass loss was computed assuming evaporative escape and using exospheric theory. Such calculations underestimated the mass loss. However, the strong velocity dependence of collision cross sections in the corona requires the introduction of velocity dependent exospheric levels. These latter calculations produce a solar wind, and indicate that the solar wind is a fundamental property of the corona and not of the method of treatment (Brandt and Cassinelli[98]).

The solar wind has a reaction on the sun as a sink of solar angular momentum because the outgoing plasma is linked to the sun by the frozen-in magnetic field and therefore attempts to participate in the solar rotation; it is clear that the solar particles with the largest angular momentum are escaping. If the total solar angular momentum is computed assuming solid body rotation, theory, measurements of comet tail orientations, and satellite observations agree in assigning an e-folding time for solar braking of about 5×10^9 years (Brandt[99]).

The abundances of certain elements may be enhanced in the corona; iron appears to be 20 times overabundant with respect to the photosphere (*cf.* Pottasch[100]). This phenomenon may be viewed (Brandt[101]) as a coronal compensation for an imbalance resulting from a mass-dependent process in the solar wind; note that the entire corona above 1.1 $R_{\odot}$ is replenished by the solar wind in times on the order of weeks. Thus, the corona reacts quickly to an imbalance. An equivalent view pictures the enhancement as a result of diffusion; computed drift velocities are approximately equal to the solar wind velocities in the low corona, and hence an enhancement would be expected.

THE SOLAR SPECTRUM

Infrared, Visible, and Near Ultraviolet

This is the region of the solar spectrum that has been the subject of traditional studies and the amount of data available is enormous (*cf.* Moore-Sitterly[102]). From roughly 3000Å to about 1 cm in wavelength, the broad variation in the solar spectrum is given by a black body spectrum (Planck curve) at a temperature of 6000 °K ("color temperature"). Local variations in the intensity are due to line absorptions and also the absorption continua near various series limits (*e.g.*, the Balmer continuum at $\lambda < 3646$Å). The broad variation is shown in figure 7 (due to Malitson[103]). The measurements of absorption lines constitute the raw material for the determination of photospheric abundances. A section of photospheric and chromospheric spectrum that includes the strong H and K lines of Ca II is shown in figure 8. Note the reversal in the chromospheric *flash spec-*

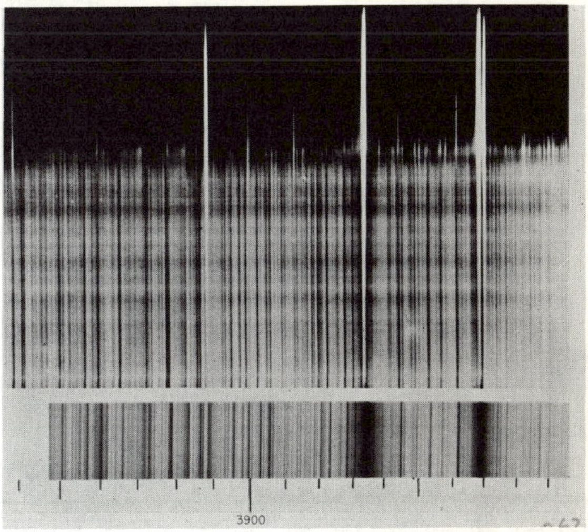

FIG. 8. Chromospheric or flash spectrum (top) near 3900Å with the photospheric spectrum included for comparison (bottom). Notice the reversal (positive print). Courtesy of Lick Observatory.

trum: strong absorption lines in the photosphere become strong emission lines in the chromosphere.

The relative contributions to the solar constant from various spectral regions are: ultraviolet ~ 9%, visible ~ 40%, infrared ~ 51%, so that the trace amounts in the EUV and x-ray regions (10^{-3}%) and the radio region (10^{-10}%) do not contribute to the determinations of the solar luminosity and the effective temperature (5750 °K). Variations in the total solar luminosity are presently impossible to deter-

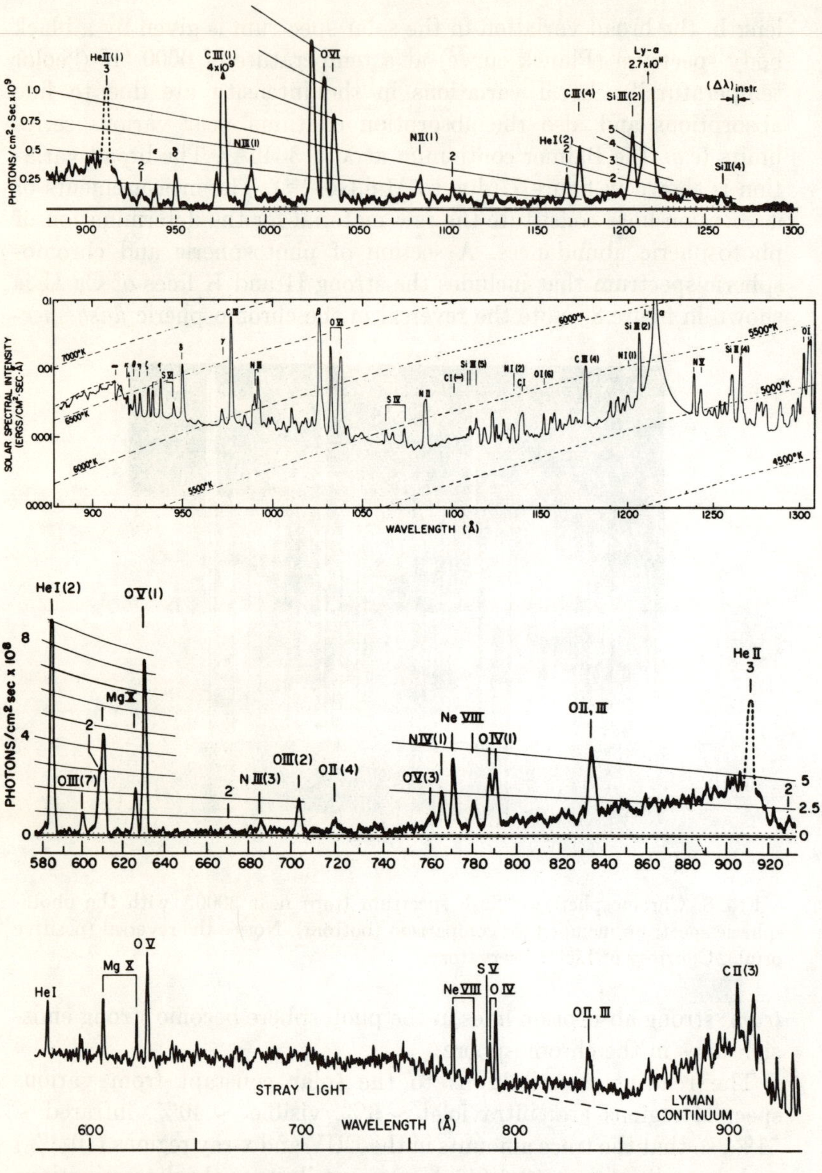

FIG. 9. Comparison of the solar spectrum from about 600Å to 1300Å as obtained by Hinteregger and co-workers (top and third segments) and Tousey and co-workers (second and bottom segments).

mine from direct solar observations, and normally one seeks them through careful photometry of objects that reflect sunlight, such as the outer planets (Hardie and Giclas[104]). In any case, such variations are certainly very small.

Extreme Ultraviolet (EUV)

Here we discuss the solar spectrum in the "rocket ultraviolet" that extends down to about 100Å from the atmospheric cutoff near 3000Å; refer again to figure 7. The region 500–2000Å is relatively straightforward and will be described first. Recent work in this wavelength range has been done by NRL and AFCRL groups led by Tousey and Hinteregger, respectively. The NRL group (Tousey[33], Tousey *et al.*[105]) has flown double-dispersion grating spectrographs, while the AFCRL group has used photoelectric monochromators at grazing incidence (Hinteregger[34,106]). Their results are in excellent agreement as shown by the comparison of data in figure 9.

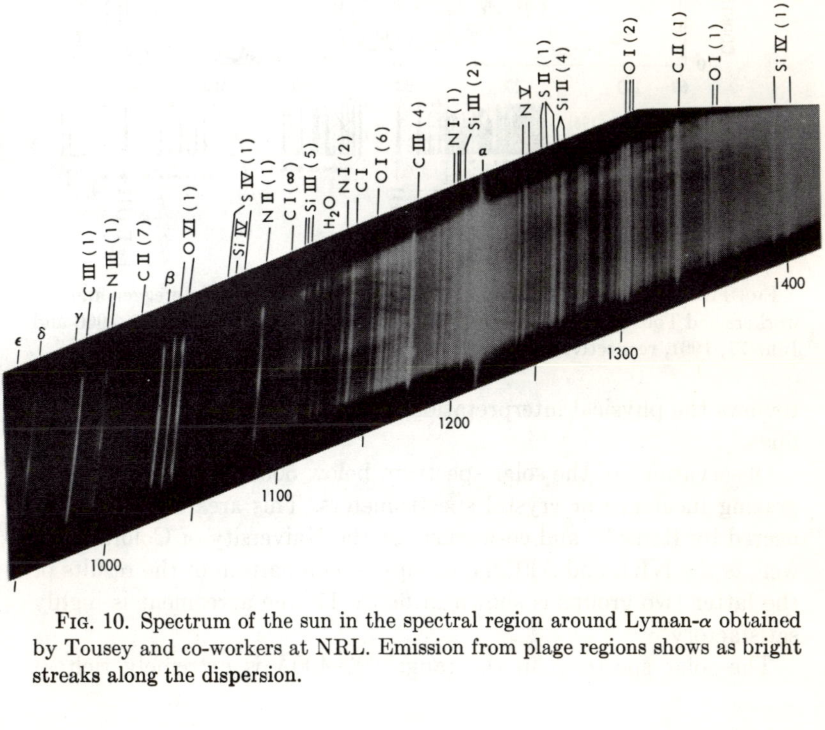

FIG. 10. Spectrum of the sun in the spectral region around Lyman-α obtained by Tousey and co-workers at NRL. Emission from plage regions shows as bright streaks along the dispersion.

The "true" solar continuum ends near 1800Å and the solar spectrum shortward of this wavelength is composed almost exclusively of discrete emission lines (*e.g.*, hydrogen Lyman-α at 1216Å) and emission continua, such as that beyond the Lyman limit at 912Å. A reproduction of a spectrum showing the region around Lyman-α is given in figure 10. This line is one of the most important ones in the solar ultraviolet spectrum, with a flux at the earth of 5–6 erg/cm^2-sec. It is believed to originate in the "transition zone" between the chromosphere and corona at a temperature of about 10^5 °K. Pottasch[107]

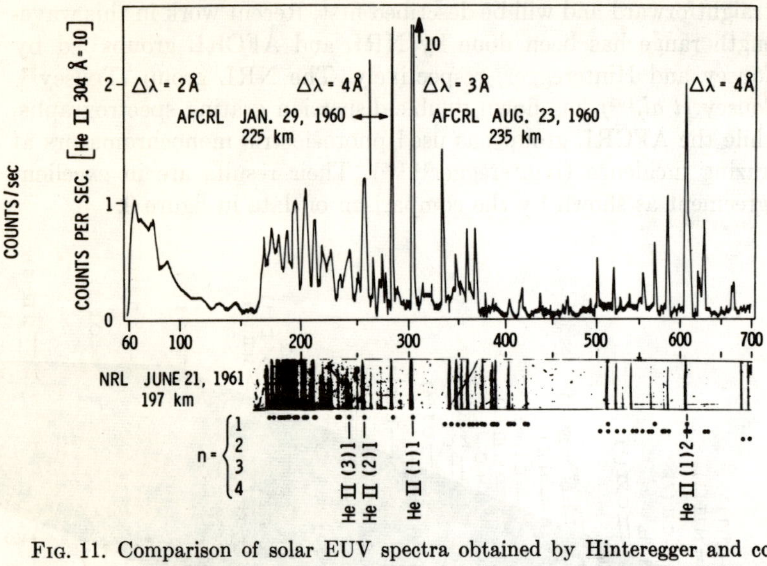

Fig. 11. Comparison of solar EUV spectra obtained by Hinteregger and co-workers and Tousey and co-workers, from rocket flights on January 29, 1960 and June 21, 1961, respectively.

reviews the physical interpretation of observations of the solar EUV lines.

Observations of the solar spectrum below 500Å involve the use of grazing incidence or crystal spectrometers. This area has been pioneered by Rense[108] and co-workers at the University of Colorado, as well as the NRL and AFCRL groups. A comparison of the results of the latter two groups is shown in figure 11; the agreement is highly satisfactory.

The solar spectrum in the range 200–400Å is extremely rich in

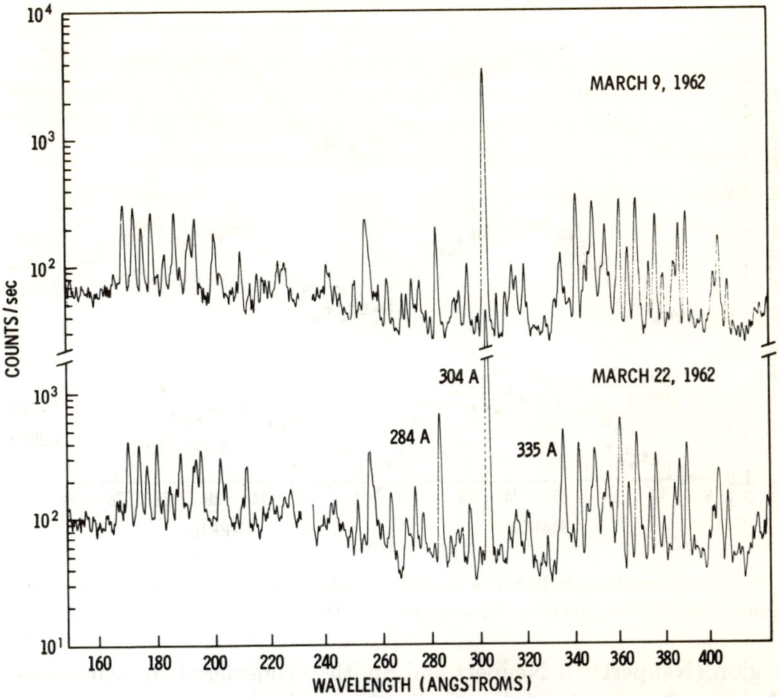

FIG. 12. Typical EUV spectrum obtained from the spectrometer flown on OSO-1 for a quiet sun on March 9, 1962 and for an active sun on March 22, 1962.

emission lines as shown in spectra obtained (Neupert, Behring and Lindsay[109]) with a grazing-incidence scanning monochromator flown on the OSO-1 (Orbiting Solar Observatory) satellite (figure 12). The optical arrangement permitted resolution of lines separated by only 0.85Å and the detector was a windowless photomultiplier with a tungsten photocathode of minimal response to wavelengths longer than 1500Å. These spectra are in good agreement with the results from the AFCRL and NRL rocket experiments. Further, the OSO–1 results reveal enhanced emission, in certain lines, that is associated with solar activity. For example, the Fe XVI (335Å) and Fe XV (284Å) lines show an increase for the active sun with respect to the neighboring lines.

Besides the astrophysical interest implicit in these variations, they are of value in making line identifications in this crowded spectral

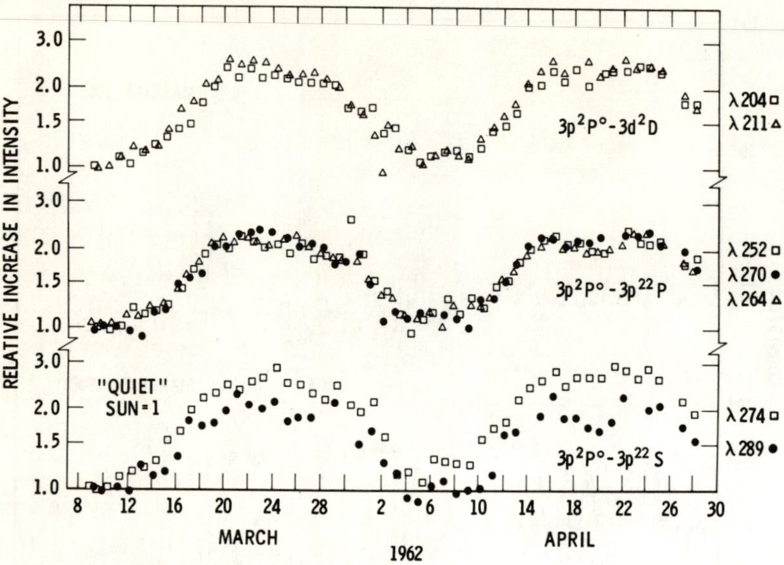

FIG. 13. Variations in flux as a function of solar activity of spectral lines attributed to Fe XIV as given by Neupert and Smith.

region (Neupert[110]). Neupert and Smith[111] considered the variations due to solar activity for optically-thin multiplets. If the excitation cross-section is only weakly temperature-dependent, then lines arising from the same or several closely-spaced upper levels will vary in unison. Observed variations of this type, for permitted lines of Fe XIV in the range λ200–300Å, are shown in figure 13. The identification process is thus aided by classifying the observed solar lines according to their amplitudes of variation. Further, the spectrum of the discharge from the "Zeta" toroidal plasma pinch experiment has shown a remarkable correspondence with Tousey's solar spectrum in the region λ170–212Å (Fawcett, Gabriel, Griffin, Jones and Wilson[112]) as shown in figure 14.

Finally, the combination of data for lines due to five ionic species of iron, three of silicon and one of magnesium shows that the relative amplitudes of their variations are correlated with the electron temperatures corresponding to their maximum relative abundances (Neupert[110]). This phenomenon exists in the sense that lines of the "cooler" ions have much smaller amplitudes of variation.

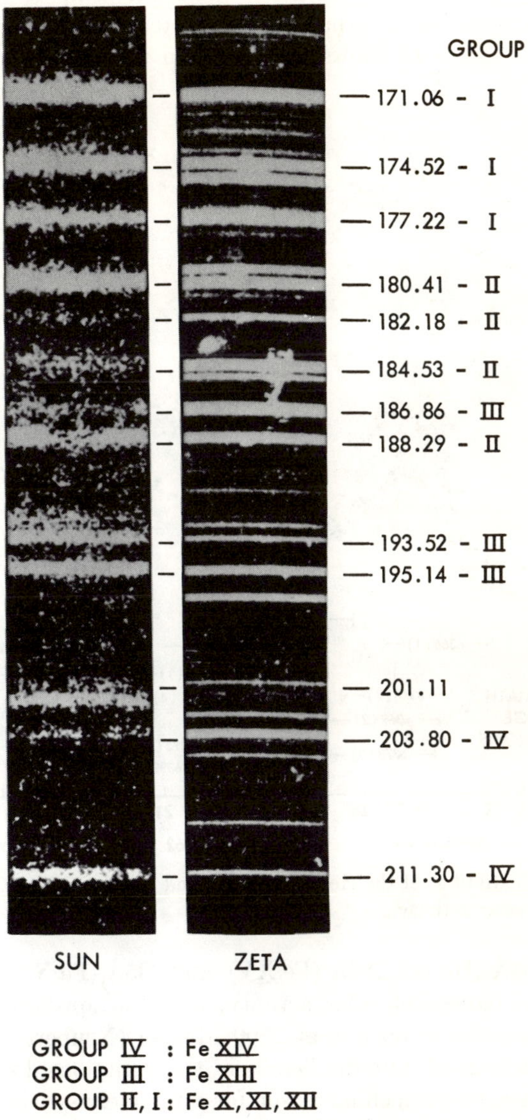

FIG. 14. Corresponding spectra obtained from the sun and from the Zeta pinch showing tentative grouping and identification of some of the more prominent spectral lines (ref. 112).

Now we turn to the question of the relationship to solar activity of the aforementioned variations in the intensities of solar EUV emission lines. Figure 15 shows the variation in counting rates for the

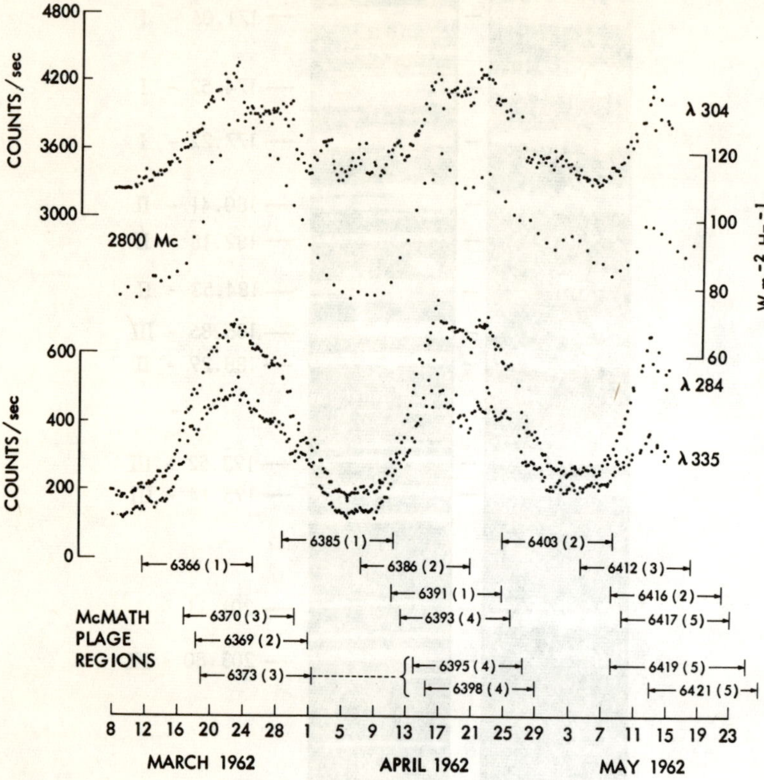

FIG. 15. Variations of the He II 304Å line and the Fe XV and XVI lines as a function of solar activity.

lines λλ304Å (He II), 284Å (Fe XV), and 335Å (Fe XVI) versus two traditional indices of solar activity, *viz.*, the appearance of plage regions, and the radio flux at 2800 Mc/sec (Neupert[110]); the gross correlation is good, but the details in the variations do differ. However, there is no compelling reason for completely synchronous variation of the various indices of activity, and detailed comparisons may require the introduction of a coronal model for a center of activity. Neupert[110] also found that the fluxes in the coronal lines of Fe XIV, Fe XV, and Fe XVI are associated with plages. This is illustrated for

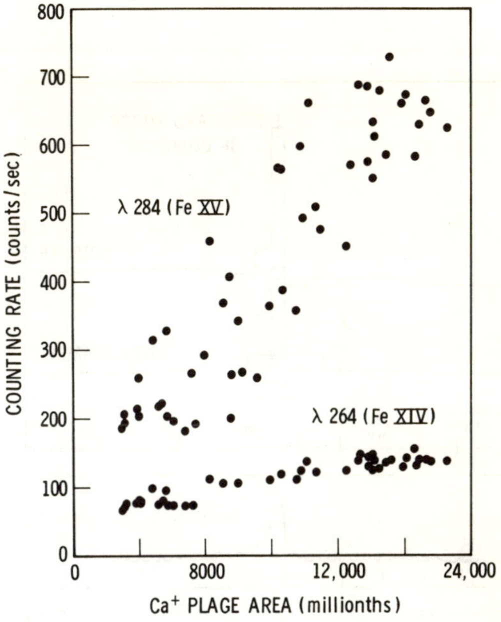

FIG. 16. Comparison of the Ca II plage areas with the Fe XIV and Fe XV fluxes.

λ284Å (Fe XV) and 264Å (Fe VIX) in figure 16. Note that linear extrapolation of the λ284Å intensities yields a non-zero y-intercept, that presumably can be identified as the quiet sun emission in this line. Now it is well known that the Ca II plages do not simply disappear, but rather tend to break up, become patchy, and ultimately merge into the normal chromospheric network. Hence, the residual, quiet sun radiation in λ284 (Fe XV) may well result from the extension into the corona of the "activity" associated with the chromospheric network.

The experiments described above have delineated a certain correlation between EUV radiation and solar activity, primarily through comparison of the radiation from the entire disk with indicators of solar activity such as plage area; by inference, low temperature ions such as Fe X, Fe XIII, Si VIII-X, Mg VIII, and Mg IX show little association with active regions while high temperature ions such as Fe XV and Fe XVI show a strong association with plages and/or active regions. It is not clear how the various physical factors (such

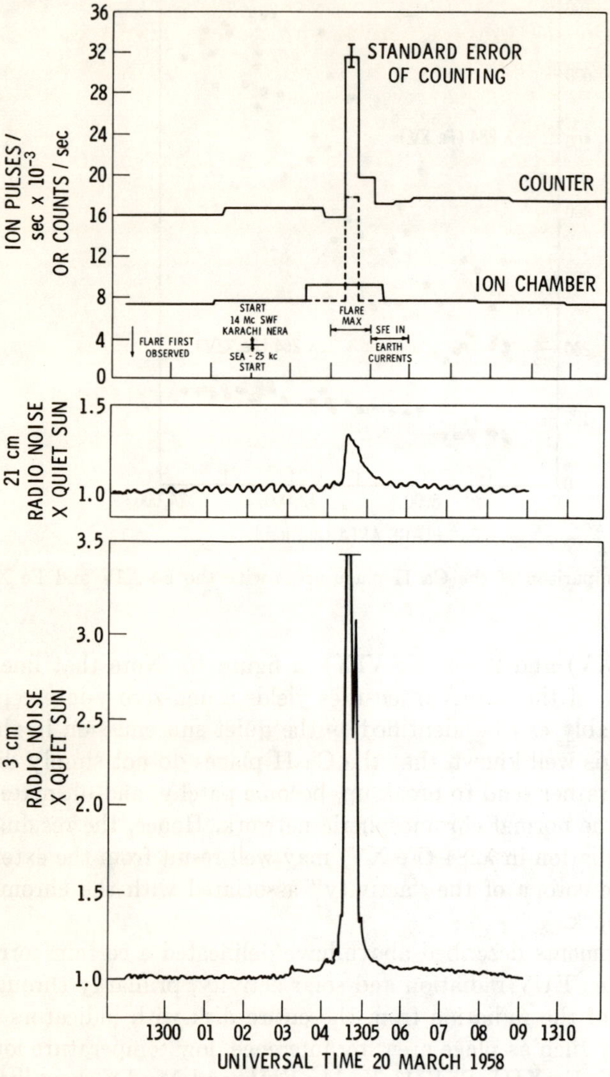

FIG. 17. X-ray data obtained from a balloon-borne ionization chamber and Geiger counter by Peterson and Winkler for the solar flare event of March 20, 1958.

as density, temperature, and area) combine to produce the enhancements, nor can the distribution of enhanced elements over the disk be determined. Needed are observations with spatial *and* spectral resolution. Such observations in the form of EUV spectroheliograms with one arc-minute resolution were obtained by Purcell, Garrett and Tousey[178] (see also Austin, Purcell, Tousey and Widing[113]) for eight spectral features in the range λ300–630Å. They found that the λ304Å emission occurs over the entire disk, and is enhanced in the plages. On the other hand, Fe XV, XVI emission was recorded only in the plage areas. Filtergrams taken in x-ray wavelengths can also be used to study the spatial distribution of high energy (photon) phenomena (see below).

X-Rays

1. Hard X-Rays

A somewhat arbitrary convention is used in discussing x-rays; soft x-rays are those with wavelengths above 1Å and hard x-rays can originate from electronic transitions involving inner shell electrons (line emission) or from Bremsstrahlung or free-free (continuous) emission. Currently, one avoids speaking of very hard x-rays as "gamma rays" because the latter term implies an unproven nuclear origin.

Hard solar x-rays have been detected to date only in bursts with time scales on the order of minutes. The record of the first such detected burst (Peterson and Winckler[114]) is shown in figure 17 and was associated with an Importance 2 flare that occurred on March 20, 1958. High energy x-ray bursts show a strong correlation with centimeter wavelength radio bursts (Kundu[115]). Figure 18 shows two sets of spectral measurements for a hard x-ray burst that accompanied an Importance 2^+ flare on August 31, 1959 (Chubb, Friedman and Kreplin[116]). The data, obtained with a rocket-borne scintillation spectrometer, refer to times several minutes later than the maximum H-alpha intensity of the flare.

Hard x-rays have not yet been detected from the non-flare sun. For example, Frost[35], from data obtained on OSO-1, determined an upper limit of 3.4 ± 1.0 photons/cm^2-sec for the flux in the energy range 20-100 Kev. Frost, Rothe and Peterson[117] have determined an upper limit to the 160-800 Kev solar flux at Earth of 0.05 photons/cm^2-sec. The relationship of certain bursts to flare activity is treated in the section on solar activity.

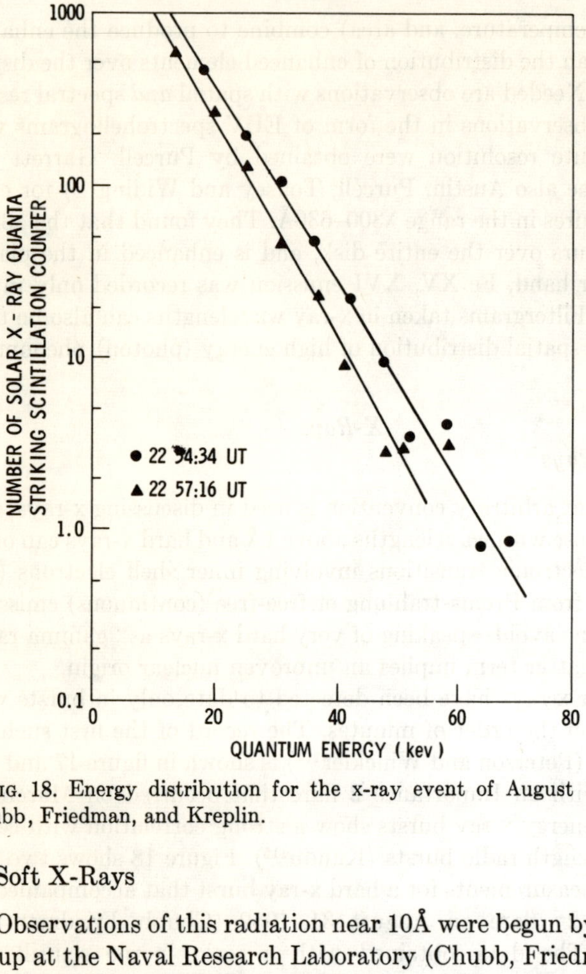

FIG. 18. Energy distribution for the x-ray event of August 1959 found by Chubb, Friedman, and Kreplin.

2. Soft X-Rays

Observations of this radiation near 10Å were begun by Friedman's group at the Naval Research Laboratory (Chubb, Friedman, Kreplin and Kupperian[118]); subsequent studies include those made at the University of Leicester, the Goddard Space Flight Center, Los Alamos, Kitt Peak National Observatory, and in the Soviet Union. The early work is reviewed by Friedman[32]. Near continuous monitoring of solar x-rays has been made possible by experiments on OSO-1 (White[119]). Such data are shown in figure 19, together with an indication of predictions based on plage data and 2800 Mc/sec flux data.

The ordinate of figure 19 requires a word of caution. Its calibration involved an assumption concerning the energy spectrum of the

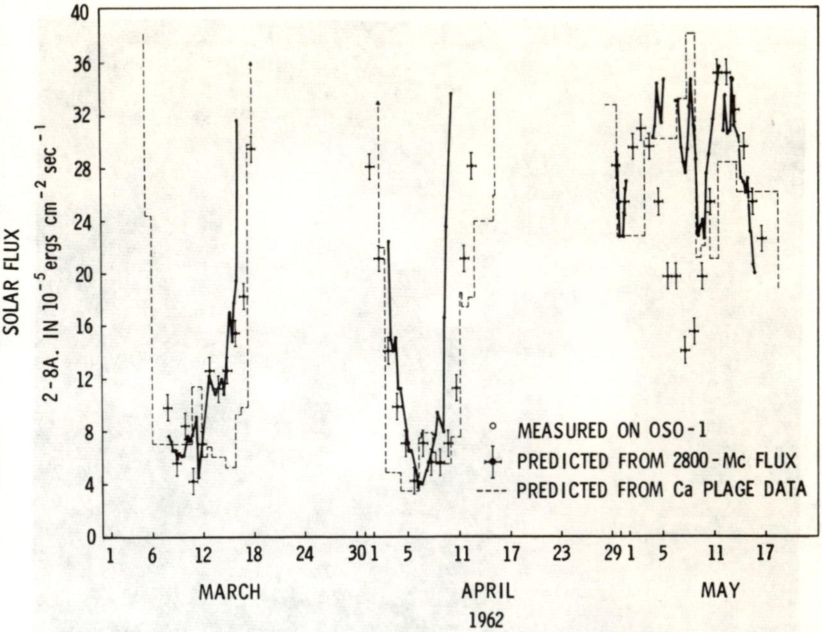

FIG. 19. X-ray (1–11Å) data obtained by White from an ionization chamber flown on the first Orbiting Solar Observatory compared with 2800 Mc/sec radio data and plage data.

x-rays. At the time of the analysis, most energy spectrum data depended on measurements by the Ariel satellite, and the spectral shape as measured by Ariel (Culhane, Willmore, Pounds and Sanford[120]) on a specific date was taken as appropriate; this particular spectral distribution is consistent with a 2.8×10^6 °K plasma for wavelengths in the range 3–11Å. Theoretical calculations show that it is difficult, if not impossible, to reproduce the measured fluxes with thermal emission; even very unreasonable models produce fluxes that are too low. The explanation appears to be the importance of line emission in this region, a conclusion previously reached by Elwert[121]; a possibility is emission in the Lyman series of Fe XXVI near λ2Å. There are probably no important lines shortward of about 1.5Å, because iron is the heaviest relatively abundant element.

The association of soft x-ray emission with active regions is clearly shown in figures 20 and 21 given by Underwood and Munney[122]; figure 20 is an x-ray photograph of the sun taken from a rocket, and

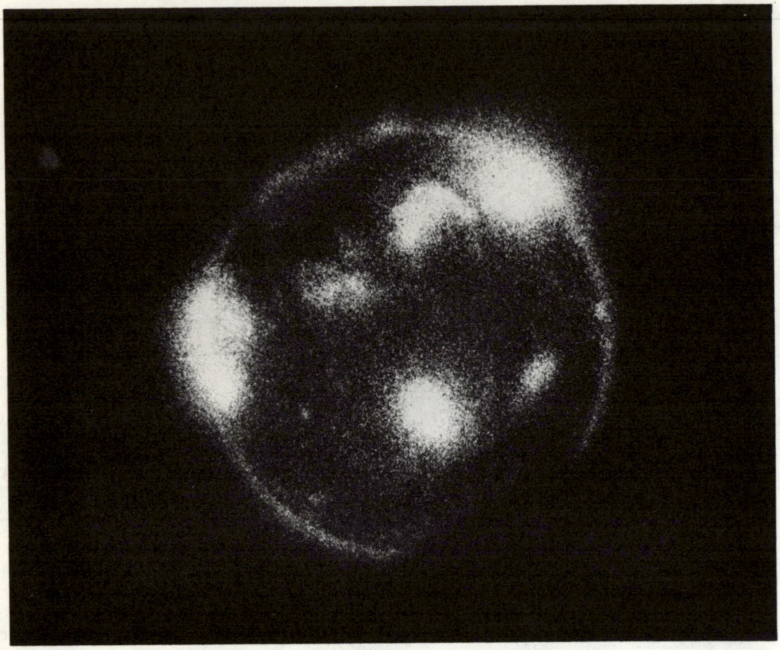

FIG. 20. X-ray photograph of the sun taken from an Aerobee rocket with filters passing radiation in the range 3–22Å and 44–60Å. Compare with figure 21. Courtesy of J. H. Underwood and W. S. Muney, Solar Physics Branch, NASA Goddard Space Flight Center.

figure 21 is an H-alpha filtergram taken on the same day at the Goddard Solar Physics Branch Observatory.

Radio Waves

1. Quiet Sun

To reach the terrestrial surface, radio waves must pass through a "window" bounded at short wavelengths by molecular absorptions and on the long end by reflection in the ionosphere. The total energy in radio wavelengths is a very small part of the total solar luminosity, and the variation with wavelength is shown in figure 7. At λ 1 cm, the apparent radio brightness temperature of the quiet sun is close to 6000°K and this value increases with wavelength to about 1,000,000°K at three meters. These correspond *roughly* to the kinetic temperatures

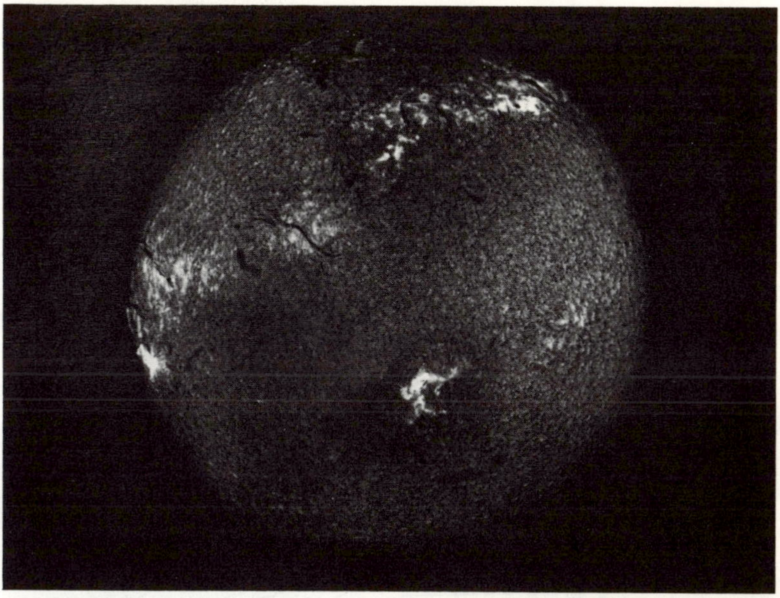

FIG. 21. H–α filtergram of the sun taken at approximately the same time as the x-ray photograph in figure 20. Courtesy P. F. Knittel, Solar Physics Branch Observatory, NASA/Goddard Space Flight Center.

in the appropriate layers of origin, which range from the photosphere to the corona. (The effective emitting region for radiation at a given frequency lies near and above the critical layer for that frequency.) However, some interpretation is necessary in relating the observed radio brightness temperature to the kinetic temperature of the emitting gas; see the discussion above concerning the allowances for a dispersive medium in studying the corona.

High resolution solar radio observations are highly dependent on modern engineering technology because the resolution depends on the ratio of the wavelength used to a characteristic linear dimension of the antenna; an example is the 32-element crossed-grating interferometer (Bracewell and Swarup[123]) at Stanford University. This instrument is used to make "spectroheliograms" with a three arc-minute pencil beam at λ 9 cm. A problem of interest for the theory of the quiet sun is the wavelength dependence of the distribution of brightness temperature with distance from the disk center, shown (after

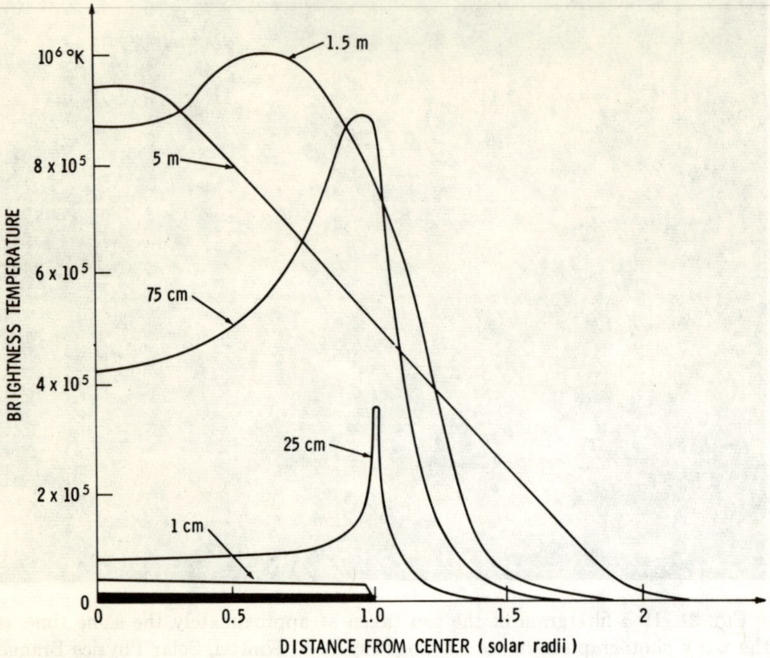

FIG. 22. The variation of brightness temperature with distance from the center of the solar disk at various wavelengths.

Steinberg and Lequeux[124]) in figure 22. The limb brightening shown in this figure is expected for an atmosphere with a temperature that increases with height.

2. Active Sun

Maps of the radio brightness of the sun at centimeter wavelengths show bright regions that are closely associated with the Ca II faculae; these areas (called "radio condensations") may last for several solar rotations and are described in some detail in the next section in connection with the development of a "center of activity." "Noise storms" (observed on meter wavelengths) may be associated with the extension of the radio condensation into the corona (typical heights $\approx \frac{1}{2} R_\odot$ above the photosphere). The storm radiation is strongly circularly polarized and may last for hours or days. The individual features or fine structure that are superimposed on the noise storm continuum are referred to as type I bursts. Noise storms are always associated with large sunspot groups, but many large sunspots are not accompanied by storm radiation. No generally accepted mechanism is available for the production of noise storms. There are also "isolated bursts" that have no clear-cut association with specific solar phenomena; however, as data collection and analysis improves in quality and completeness this class of radio events is decreasing. The subject is reviewed in detail by Kundu[38].

Finally, there is the "outburst" consisting of the multitude of radio phenomena associated with a solar flare. These are the initial type III and type V bursts that follow the onset of the flare, and then the somewhat longer-lived type II and type IV bursts; an idealized

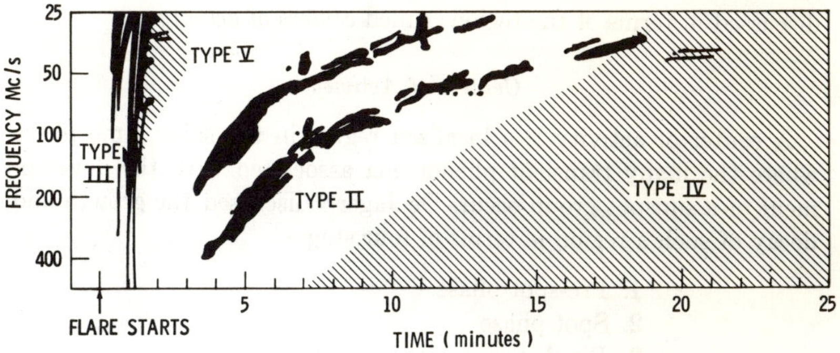

FIG. 23. An idealized representation of dynamic spectra showing the general characteristics of radio bursts at meter and decameter wavelengths.

sequence (after Wild[125]) is shown in figure 23. We return to a discussion of outbursts in the description of a "center of activity."

SOLAR ACTIVITY

Introduction

The existence of transient solar phenomena has been recognized for centuries. Activity is observed in one form or another at all latitudes on the disk and at all levels in the atmosphere, and it is manifest in all accessible regions of the electromagnetic spectrum. As we have seen in Parts I and II of the present volume, it affects the distribution of magnetic fields and charged particles in the interplanetary medium and has important consequences for the physical state of the earth's upper atmosphere and for geomagnetic activity.

Casual examination of the photospheric image in white light is sufficient to reveal localized dark (*sunspots*) and bright (*faculae*) areas that change in appearance from day to day or even hour to hour. In the chromosphere, the most dramatic events are the *flares*, normally observed in the light of H-alpha. *Prominences*, *streamers* and *condensations* are found in the corona. In general, all features of the active sun for which sufficient data are available appear to be spatially related to variable magnetic field structures. The general solar field is cyclic in time, and the frequencies of occurrence of the various species of solar activity are correlated with this cycle. Although isolated events do occur, the general pattern of solar phenomena is best discussed in terms of the regions called *centers of activity*.

Centers of Activity

A center of activity is a localized region in the solar atmosphere where the many transient phenomena associated with the development of a sunspot group occur. De Jager[42] discussed the growth and decay of such a region in terms of four stages:

1. Pre-spot phase
2. Spot phase
3. Bipolar magnetic, post-spot phase
4. Unipolar magnetic phase.

Evolutionary summaries are also given by Kiepenheuer[41], Righini[126], and Bray and Loughhead[43]. Observations of the development of individual centers are described by Dodson and Hedeman[127] and Christiansen *et al.*[128]

The formation of a center of activity begins with the appearance of adjacent areas of opposite magnetic polarity, referred to as a *bipolar magnetic region.* This event is suggestive (Babcock[129]) of the emergence of loops of a toroidal magnetic field from beneath the visible photosphere. The bipolar regions follow the polarity rules that were found for sunspots by Hale and Nicholson[130]: the preceding* components of all regions in the Northern Hemisphere have the same sign, which is identical to that of the following components of regions

* Preceding in the sense of the solar rotational motion.

in the Southern Hemisphere, and the sense of these polarities reverses in each eleven-year cycle. When the field strength in the center exceeds about two gauss, the *plages*, also called *faculae*, appear.* The development of these bright emission regions, as observed on Ca II spectroheliograms, parallels the increase in strength of the bipolar magnetic field. Howard[76] found that the plages correspond very closely to the 10-gauss contours on magnetograms.

Above the plages, a coronal condensation (or "green patch"), recognized in coronagraph observations by the enhancement of λ 5303Å (Fe XIV) has begun to form. The condensations can also be observed in other emission lines, and in the continuum. Models largely based on optical data, in some cases obtained during solar eclipses, are presented by Athay and Thomas [131], Newkirk[132], Hiei[133], Aly, Evans and Orrall[134], and Nishi and Nakagomi[135]. Newkirk inferred from continuum measurements with a K-coronameter that the condensation is purely a density effect, *i.e.*, that the temperature of the condensation equals that of the undisturbed corona at the same height. On the other hand, differences in the temperatures deduced from the widths of various line emissions imply the presence of both relatively hot and relatively cool structures within a condensation (Billings and Lehman[136]). Billings[28] has given typical temperatures for certain features of the active corona; they are summarized in Table 2.

TABLE 2

Temperatures of Coronal Regions
Based on Line Widths

(After Billings, ref. 28)

Region	Temperature (10^6 °K)	Remarks
Red line (λ 6374Å)	1.2	Early days of new cycle, little green line enhancement.
Moderate enhancement	1.4–2.0	Both green and red line enhanced.
Active enhancement	3.0	Over active sunspots, enhanced green line, little red line, and some yellow line.
Yellow-line (λ 5694Å) condensation	4.5	Intense yellow line, diminished green line, and red line absent.

* Terminology in this subject is particularly perplexing; see Kuiper[1], p. 744.

TABLE 3

Model for Electron Density in a Coronal Condensation
(Inner Corona)

(Aly, Evans and Orrall, ref. 134)

r (distance from axis)	N_e
0 km	35.0×10^8 cm^{-3}
10,000	29.5
20,000	24.5
30,000	19.2
40,000	13.9
50,000	8.8
60,000	6.0
70,000	4.1
80,000	2.9
90,000	2.1
100,000	1.8
110,000	1.3
120,000	0.9

Aly *et al.* assumed axial symmetry about the vertical direction in a model based on eclipse measurements of the optical continuum intensity in a condensation at the limb. Their values of the electron density as a function of distance from the axis are given in Table 3.

Within a few days from the birth of the bipolar magnetic region, the first sunspot appears, and the development of the center of activity enters the second stage. Radio emission associated with the center consists of brief events, the *radio bursts*, and the long-term (lifetime ≈ center of activity lifetime ≈ several months) radiation known as the *slowly-varying component*. The source of the slowly-varying component is identified with the coronal condensation; pertinent discussions include those of Christiansen *et al.*[128], Kawabata[137], Kakinuma and Swarup[138], Swarup *et al.*[139], and Kundu[38]. The radio spectrum of a condensation peaks near 10 cm, where the brightness temperature may attain several million degrees. If the possibility of a non-thermal contribution to this radiation can be neglected, then this implies the presence of "hot spots" in the corona. Strong sources subtend two or three minutes of arc in the wavelength range 3 to 10 cm, and are slightly larger at 21 cm; their heights are typically 10,000 to 20,000 km at 3 to 10 cm and 40,000 to 90,000 km at 21 cm. The disk-projected position of a source is close to the center of the associated sunspot group, and the 10 cm brightness temperature is well correlated with sunspot area and magnetic field strength. Circular polarization of the radiation is observed (see Table 4); the sense of polarization is opposite over the two areas of a bipolar spot group and corresponds to that expected for the extraordinary wave. Mechanisms suggested for this radio emission include simple thermal models (Waldmeier and Müller[140], Newkirk[132]), a magnetoionic theory (Piddington and Minnett[141]), and the theory of absorption at the gyrofrequency and its harmonics (Kakinuma and Swarup[138], Zheleznyakov[142]). At present, this last model seems to be in the best accord with the observational material.

TABLE 4

Circular Polarization in a Strong Source of the Slowly Varying Component

(Swarup *et al.*, ref. 139)

Wavelength	Polarization
3 cm	>30%
10	~10
21	< 2

Noise storms observed on meter wavelengths (and discussed above in the section on the solar radio emissions) may represent the radio emission from the very extended portion of the coronal condensation (at distances $\sim 0.5R_\odot$ above the surface).

The spot phase represents the most active stage in the life of an activity center. An immense literature exists on such observational properties of the sunspots as their numbers, areas, relative motions and orientations, disk locations, spectra, and associated magnetic and velocity fields. The reader should consult Bray and Loughhead[43] for the most comprehensive recent survey of these data. In Table 5 (condensed from their book, but largely based on Michard's results for a spot of diameter 24 arc sec), we summarize the physical conditions in a sunspot, as contrasted with the state of the undisturbed photosphere. Note that the gas pressure is much less in the sunspot than in the surrounding photosphere; a magnetic field of ≈ 3000 gauss is needed to make up the difference. Other than this summary, we content ourselves here with a description of the relation of the spots to the evolutionary history of the center of activity.

The first indication of sunspot activity is the appearance of one or more small dark regions (*pores*). Frequently, they subtend only a few arc sec and disappear in a day or two; at any event, the pores always represent both the first and the last phases in the history of a spot group. In the case of a major group, the pores assume a bipolar configuration, then extended, less dark areas (*penumbrae*) develop around the larger spots, and the structure of these spots becomes complex, while additional smaller spots appear around and between the two

TABLE 5

Physical Conditions in a Sunspot
(after Bray and Loughhead, ref. 43)

	Optical Depth (λ 5000Å)	Umbral Value	Photospheric Value
Geometrical Depth	0.1	733 km	132 km
	1.0	2120	300
Gas Pressure	0.1	1.3×10^4 dynes/cm²	3.5×10^4 dynes/cm²
	1.0	3.6×10^4	9.8×10^4
Electron Pressure	0.1	0.45 dynes/cm²	2.9 dynes/cm²
	1.0	2.7	58.0
Radiation Temperature	0.1	4060 °K	5010 °K
	1.0	4870	6450
Effective Temperature	0.67	4160 °K	5785 °K
Excitation Temperature	—	4000 °K	5000 °K
Degree of H Ionization	0.1	1.2×10^{-8}	5.0×10^{-6}
	1.0	2.0×10^{-6}	5.1×10^{-4}
Degree of Metal Ionization	0.1	0.25	0.83
	1.0	0.59	0.94
Electrical Conductivity	0.1	5.2×10^{10} e.s.u.	1.1×10^{11} e.s.u.
	1.0	1.1×10^{11}	7.3×10^{11}

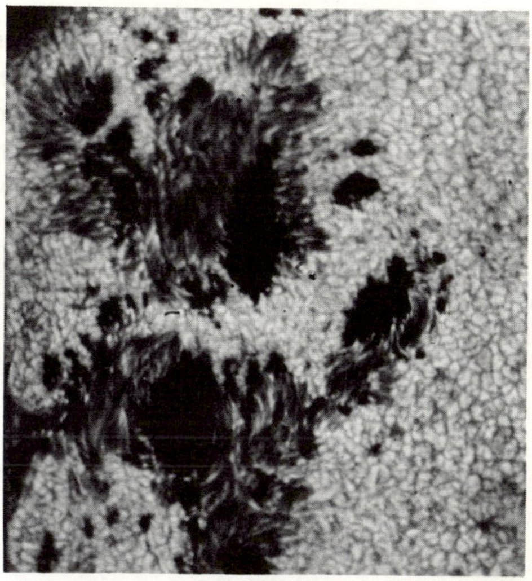

FIG. 24. Balloon photograph of sunspots showing the umbra and penumbra as described in the text. By permission from Project Stratoscope of Princeton University, sponsored by the Office of Naval Research, the National Science Foundation, and the National Aeronautics and Space Administration.

main ("preceding" and "following") *umbrae* (see figure 24). At the peak of its extent, the group may stretch over more than 15° in solar longitude and the distribution of polarities among the spots can be quite complex. As the group wanes, the smaller spots disappear first; eventually only a unipolar remnant of the preceding spot is left; it declines to the final (pore) stage. On rare occasions, great spot groups can attain areas $\sim 10^{-3}$ of the photospheric surface; however, 95% of all groups never exceed 2.5×10^{-4} of the photosphere. The maximum intensity of the magnetic field in a group rises steeply at first, remains constant during most of the spot lifetime, and finally decreases at a rate somewhat slower than that of the initial rise; in the greatest spots, it can surpass 4000 gauss.

Until recently, a sunspot has been taken to be in radiative equilibrium (Minnaert[143]) but with the total net flux (carried beneath the surface by convection) reduced due to the constrictive action of the spot's magnetic field. The prime datum for this interpretation is the limb darkening for sunspots, which is, however, difficult to determine. The construction of sunspot models in radiative equilibrium has not been successful, and now it appears that some other form of energy transport is required. This conclusion is reinforced by observations of the umbral granulation (Bray and Loughhead[144]). The so-called "flux deficit" is also of interest in this connection. It represents the additional flux presumably available to the photosphere outside a spot due to the blocking action in the sunspot. It has been argued that the *bright rings* (Waldmeier[145]) surrounding sunspots are visible evidence of this phenomenon. Such bright rings are not observed around all sunspots and whether or not they are related to the flux deficit is debatable; certainly they do not account for all of it (Sweet[146]). In any event, there is a growing indication of the importance of non-radiative energy in the spots. The interaction of convection with the magnetic field might produce a transport of energy in the form of Alfvén waves (Musman[147]). The propagation of this substantial amount of energy in "mechanical" form upward into the chromosphere and corona could be responsible for many of the phenomena associated with a center of activity.

In a sunspot penumbra, the action of the field in arresting convection is apparently less efficient or it differs because the magnetic field (traditionally thought to be nearly vertical in the *umbra*) approaches a horizontal configuration. Danielson[148,149] has shown that with a horizontal field the convective element is elongated with the long axis along the field ("convection roll"); however, Bray and Loughhead[43] have summarized conflicting evidence that indicates a horizontal field orientation in the umbra and a vertical one in the penumbra.

The flare events in a center of activity are observed during the spot phase. Flares are defined as transient brightenings of discrete chromospheric regions, observed in the light of H-alpha. They are visible in other lines as well, but are rarely seen in white light. Frequently, they are accompanied by the emission of x-ray and radio bursts, and by the ejection of charged particles from the sun. Some good summaries of the extensive work that has been done on flares and related phenomena appear in Smith and Smith[45], Wild, Smerd and Weiss[37], Hess[47], Kundu[38] and Lüst[48].

A flare is described in terms of its *Importance*, a scheme of classification that is based on the flare area (corrected for projection effects) at the time of maximum H-alpha intensity.

Flare durations and H-alpha intensities are correlated with flare areas, but the dispersion in these distributions is rather large. The majority of all listed flares are subflares; further, there are many

TABLE 6

Classification of Solar Flares by Area

Importance	Area*	Remarks
3^+	>1200	
3	600–1200	
2^+	250–600	#
2	250–600	
1^+	100–250	#
1	100–250	
1^-	<100	*Subflare*

* Unit = 1 millionth of solar hemisphere.
# denotes flare of unusual intensity or Doppler broadening for an event of its area.

much smaller events (*microflares*) that are not regularly reported. In fact, determinations of the flare occurrence rate, and of the relative frequencies of flares of the different Importance classes, are strongly affected by the completeness of the observational material for small flares. In any case, for the larger flares, the probability of occurrence is enhanced in long-lived active regions, and especially in those regions that have already experienced a large flare. The maximum flare rate in an activity center is associated with the stage in which the sunspot magnetic fields are most complex. Further, 60% of the flares in a given center occur within 40,000 km of the associated spot group, and the flare-spot separation exceeds 10^5 km in less than 2% of the cases. Often, flares recur at almost exactly the same position in the active region (*homologous flares*). Magnetograms show that flares are often located near apparent neutral points in the magnetic field. The field gradients near these points are usually smaller after the flares.

Typically, a flare is first seen in H-alpha as one or more rapid, pointlike brightenings in a pattern of dark, fibrillar chromospheric structures. The flare expands, and assumes a flattened, horizontal appearance that may involve intense "ropes" of emission. After maximum light, the flare intensity decreases less rapidly than it grew. Subflares may expire in a minute or two, but great flares may persist for several hours. Filtergrams taken at 10-second intervals following the initial rapid brightening (*explosive phase*) have shown wave-like phenomena propagating horizontally away from flares at speeds $\sim 10^3$ km/sec. *Filaments* (dark, threadlike structures that represent prominences projected against the disk) situated $\sim 10^5$ km from a flare have been found to move up and down ("wink") by amounts $\sim 10^4$ km in response to such disturbances.

Matter may be visibly ejected from flares in the form of *surges* and *spray prominences*, with velocities $\sim 10^2$ km/sec and as much as 10^3 km/sec, respectively. Further, solar cosmic rays are sometimes emitted (see Chapter 10). Both surges and sprays appear to rise with a twisting motion, suggesting magnetic confinement. Surges are observed to fall back along their paths of ascent, whereas much of the spray material escapes from the sun. Often, surge trajectories follow the loops or arches of the *active prominences* that form above sunspot groups.

At meter waves (see figure 23) type III ("fast drift") burst groups, with durations of several minutes, can be observed simultaneously with the onset of the flare. They are interpreted as emission from plasma oscillations excited in the corona by electron streams moving upwards from the flare at speeds near one-half the velocity of light. At the same time, equally brief ("impulsive") bursts of microwave synchrotron radiation and "hard" ($\lambda < 1$Å), probably nonthermal x-rays may be produced in the chromosphere. Sometimes, "soft" x-ray events are observed at 1Å $< \lambda <$ 10Å; they are attributed to temporary heating of the chromosphere to $T \sim 10^6$ °K, and are accompanied by microwave bremsstrahlung that may last several dozen minutes (*gradual burst*).

Immediately after the type III event, a brief ($\sim$ minutes) meter wave continuum, called a type V burst, is sometimes observed, and within a few minutes the type II meter-wave event (also ascribed to coronal plasma oscillations) may occur. In this case, however, the measured velocity (from the frequency drift rate, or from interferometer observations of the motion) is only ~ 1000 km/sec, and the oscillations are considered to be excited by the passage of a shock caused by the flare. Type IV meter-wave continua sometimes follow the type II events, and may persist for many hours; they are caused at least in part by synchrotron radiation from temporarily-trapped electrons, and are strongly correlated with solar cosmic rays. A schematic representation, due to Wild[150], for a model of the totality of these events, is given in figure 25.

Satellite spectrometer observations show that the solar extreme UV radiation, *e.g.*: $\lambda\lambda$ 304Å He II, 284Å Fe XV, and 335Å Fe XVI,

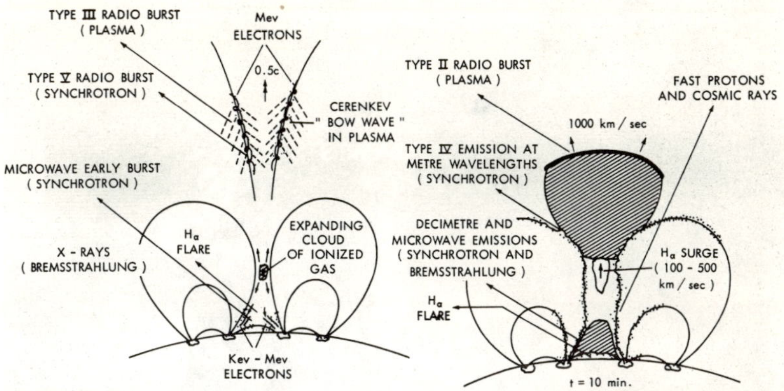

FIG. 25. A pictorial representation (due to J. P. Wild) of the physical processes leading to the origin of the two phases of a solar outburst. It is postulated that the initial flare event is due to some self-annihilation of opposing magnetic fields above a sunspot group (from ref. 150).

is enhanced during flares, and also, the ratio I(λ335Å)/I(λ284Å) increases temporarily.

The spot phase has lasted perhaps a few weeks, and as the sunspots die out and flare activity ceases, the center of activity enters the bipolar magnetic post-spot phase, which lasts for a month or two. The chief activity is in the corona now, where the *quiescent prominences* are still observed. They are denser and cooler ($T \sim 10^4$ °K) than the surrounding corona, and have lifetimes of a few months. Examined at high resolution, they are seen to possess filamentary structure, and it is apparent that downward streaming of matter is occurring, with velocities of several km/sec. These prominences are located between fields of opposite polarity, and apparently represent matter that has condensed along horizontal lines of force. Some recent discussions of the processes involved appear in Evans[26], Shklovskii[27], Zirker[151], and Zirin[5]. (In contrast, the *active prominences* occur during the spot phase of the center of activity and have lifetimes of the order of days.) The magnetic field decreases in strength and becomes unipolar; no optical phenomena are present, except possibly for streamers in the corona. This last stage in the life of the center of activity may persist for several months; the multifarious optical aspects of a "center of activity" are summarized in Table 7. Small differences among such

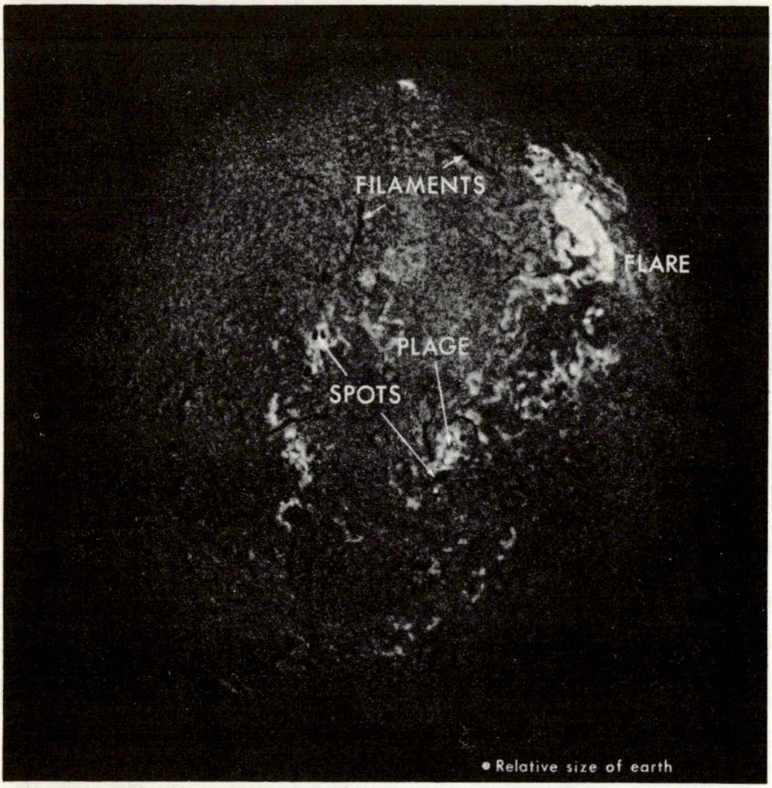

FIG. 26. The solar surface in H–α, showing various types of solar activity (Courtesy of the Sacramento Peak Observatory, Geophysics Research Directorate, Air Force Cambridge Research Laboratories).

summaries serve to emphasize their schematic nature. Figure 26 is an H-alpha photograph showing several aspects of solar activity.

The Solar Cycle and the General Solar Field

The most straightforward manifestation of the solar cycle is the quasi-periodic change in the numbers of sunspots. Approximately 11.0 years elapse between successive sunspot maxima or minima; however, the cycle is asymmetrical since the time from maximum to minimum is 6.5 years while it is only 4.5 years from minimum to maximum (Bray and Loughhead[43]). Sunspots are part of the bipolar

TABLE 7

Life History of Typical Center of Activity*

DAY	SPOTS	MAGNETIC REGION	FACULAE	FLARES	PROMINENCES (Filaments)	CORONAL REGIONS
1	Spots may originate a few hours after faculae appear, but usually only pores.	Assumes bipolar character.	Systematic alignment of chromospheric fine structure around area which becomes bright facular speck. Speck develops east-west** orientation.			Green patch forms after spots appear. Parallels facular development. Same extent as facular region.
2	West end of facular regions shows first p spot.	Characteristic dimension magnetic region ~50,000 km.	Area increases in size and brightness.	First signs of activity.	Small unstable filaments appear around faculae.	Green line increases.
5	f spot appears at east end of faculae; other small spots in between; main spots are continually separating.			Flares between f and p spots.	Sunspot prominences appear.	
11	Maximum development.	Field irregular and variable.	Size and brightness increases; dimensions $\sim 10^5$ km.	Peak of flare activity.		Yellow-line brightness with flare activity.

TABLE 7 (continued)

DAY	SPOTS	MAGNETIC REGION	FACULAE	FLARES	PROMINENCES (Filaments)	CORONAL REGIONS
27	Usually only p spot remains; the f spot disappears by being "bridged" by bright, normal photosphere.	Magnetic flux reaches maximum.	Facular area still increases in size.	Flares rare.	Stable filament forms on poleward side of facular area, makes angle of some 40° with meridian and points toward p spot.	Maximum green-line brightness.
54	No spots.		Brightness decreases and area cut in half by filament.		Filament now $\sim 10^5$ km in length and starting to swing toward east-west orientation.	Green line decreases in brightness.
81		Field strength declines.	Area constant, dissolves into bright facular network (still brighter than normal chromosphere).		Length increases and orientation continues toward east-west.	
108			Faculae dissolved.		Maximum length achieved and nearly east-west (parallel to equator).	

DAY	SPOTS	MAGNETIC REGION	FACULAE	FLARES	PROMINENCES (Filaments)	CORONAL REGIONS
135		Area occupies one-twentieth of disk.	CA undetectable in photosphere or chromosphere except for some fine structure oriented around filament.		Length decreases; prominence starts poleward migration.	
162–270		BMR fills one-fifth of disk and changes into UMR.			Filament changes structure and joins polar crown of filaments.	Coronal streamers, associated with UMR.

** Note: East and west are as the sun is seen from the earth.

* From "The Sun and Stars" by J. C. Brandt, copyright 1966, McGraw-Hill Book Company, used by permission.

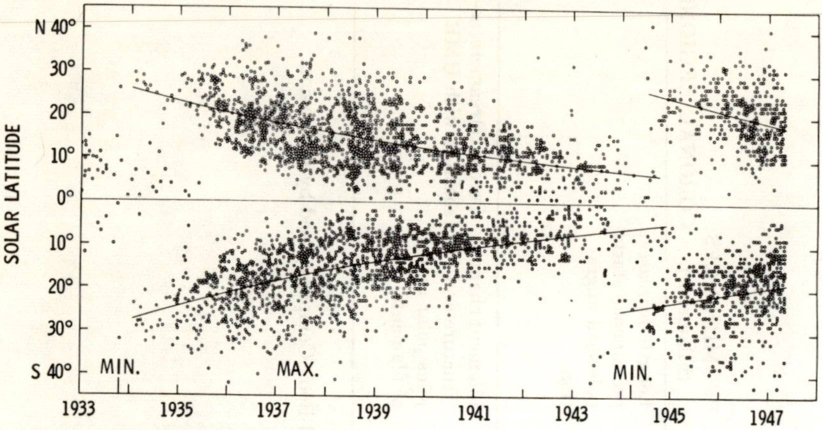

FIG. 27. Butterfly diagram showing the latitude variation of sunspot locations over the solar cycles (*Sky and Telescope*).

magnetic regions discussed in the section on centers of activity, and, hence, obey the polarity laws formulated for BMR's (actually, the polarity laws were *first* discovered for spots and later generalized to BMR's). Thus, with the change of the fundamental polarity in each solar hemisphere after eleven years (Babcock[152]) one can think of a "true" solar cycle of length 22 years.

Because sunspots are intimately connected with centers of activity, most indices of solar activity vary, on the average, synchronously with the sunspot number. This includes the facular area, number of prominences, brightness of active coronal regions, occurrence frequency of flares, etc. However, the distribution of sunspots (and related phenomena) shows a pronounced change in heliographic latitude through the solar cycle; this behavior is shown in the Maunder *butterfly diagram* (figure 27). This migration of the zones occupied by sunspots is called *Spörer's Law.*

Belief in a general solar field relies primarily on its direct measurement. It is commonly believed that most of the early reports of general field measurements are spurious, and that technological advances incorporated into the solar magnetograph enabled the first reliable measurements of a general field of ~ 1 gauss (Babcock and Babcock[129]). However, this result appears to be undergoing a reinterpretation. Sheeley[153] has found (see the caption to figure 5) that the "general field" is approximately 100 gauss in the plagettes that make up the chromospheric network (as seen in Ca II K_{232}) and essentially zero elsewhere. The bunching of the field lines in the chromospheric network can be qualitatively understood because the motions observed in the supergranulation would tend to convect any magnetic field to the boundary (Parker[78]), which is roughly coincident with the chromospheric network. The Babcocks' result of $\sim$ 1 gauss is then to be interpreted as a gross spatial average.

A satisfactory theoretical reason for the existence of the general solar magnetic field is very hard to give. The older literature is summarized by Cowling[50]. Some suggestions are:

(1) The sun's field is a natural result of its process of formation—a fossil (*cf.* Mestel[154]). Estimates of the decay time for the field in a static sun are some 10^{10} years for a field that pervades the entire sun. If, however, the field is confined near the surface of the sun, a time of 10^7 years is more appropriate. The field observed at the *surface* is clearly not a fossil, in any case.

(2) Currents flowing in meridional planes (Biermann[155]) can produce an azimuthal magnetic field; the currents can be produced by the solar rotation and gradients of the electron pressure. However, the result is an azimuthal field while the more difficult poloidal field is the goal.

(3) Dynamo theories, in which motion of material across existing field lines is used to sustain the field, have been the object of considerable study (*cf.* Cowling[156]). The dynamo theory has neither been proved nor disproved.

(4) A process roughly analogous to equipartition could amplify a small field through interactions with turbulent motions (Batchelor[157]). This mechanism produces an irregular, small-scale field that does not correspond to a general field.

Thus, we clearly do not have even a working hypothesis concerning the general solar magnetic field. Next we turn to the question of the solar cycle where the general field is presumed. There are many attempts at a solar cycle and sunspot theory (*e.g.*, Allen[158], Hoyle and Wickramasinghe[159]) but the most successful is that of H. W. Babcock[160]. In this model, the solar differential rotation winds up a pre-existing, sub-photospheric poloidal field. Then, magnetic buoyancy or instabilities cause a flux loop to protrude through the surface, producing a BMR and resultant effects; this mechanism naturally satisfies the various polarity laws. It is assumed that the flux loop continues to expand outward into the chromosphere and corona. As this process continues, the field can neutralize part of the general field by severing and reconnection. This is a major feature of the model and on this picture, the reversal (albeit in a piecemeal fashion) of the general field can take place. The observations taken in 1957–58 provide a good example of this piecemeal process, because reversals took place in the two solar hemispheres about a year apart (H. D. Babcock[152]).

Leighton[161] has greatly extended our understanding of the solar cycle and Babcock's theory through the conceptual introduction of a diffusion process in which the supergranulation convection currents move the magnetic field lines about in a random-walk process. Sunspots disappear because the BMR continually expands; since the net magnetic flux remains constant, the field strength required to suppress and/or greatly modify convection (and hence produce the spot) does not exist for long. Leighton also shows that buffeting of solar field lines by the supergranulation motion can increase the areas of BMR's at the known rate of 10^4 km^2/sec. Many features of the solar cycle can then be reproduced by following the various field lines and assuming that the field pattern is due to an *algebraic* sum at that time of the fields associated with previously existing spot groups. Such algebraic bookkeeping accounts for the field reversal in a natural way. In particular, the "rush to the poles" of the *polar filaments* (Waldmeier[162]) that occurs at sunspot maximum may be identified with the model-deduced latitude variation of the position of zero magnetic field (Hyder[163]), and Sheeley[164] has pointed out that the number of *polar faculae* is inversely correlated with the sunspot number.

THE FUTURE OF SOLAR ASTRONOMY

In every region of the spectrum, the possibility exists of making instrumental advances that can contribute significantly to our understanding of the sun. For example, much of what we know about the x-ray appearance of the sun is based on a small number of photographs recovered from sounding rocket experiments. The desirability of taking such pictures with a long-lived orbiting instrument is clear. Data on x-ray *bursts*, on the other hand, have largely been collected with satellite-borne counters that were sensitive to rather wide frequency bands and that accepted radiation from the entire facing hemisphere of thc sun. Here, the prime need in second generation experiments is for greater spatial and spectral resolution. In this connection, the image-forming slitless spectrometer proposed by Gursky and Zehnpfennig[165] appears to have considerable potential for solar x-ray astronomy.

In the normal spectral range covered by most ground-based solar telescopes, atmospheric *seeing* constitutes a serious limitation. Three lines of approach are being explored in this area: to detect and take maximum advantage of brief moments of unusually good seeing (Bray, Loughhead and Norton[166]), to develop electro-optical methods of compensation for seeing effects, and to locate the telescope above the altitudes where the seeing originates, by use of stratospheric balloons (Schwarzschild[67], Kiepenheuer and Mehltretter[167]). The last approach, in particular, is very expensive in terms of dollars per observing second, although it offers the best possibility of obtaining data at high spatial resolution. Also, it does not lend itself very well to long-term solar studies. A large diffraction-limited orbiting telescope, usable over the greatest possible wavelength range, is to be preferred in the long run, although the problems involved are very great. Certainly, occasional astronaut rendezvous would be necessary, for purposes of maintenance and instrument change. Also, the information content of the records obtained would be very great and might place considerable strain on the telemetry system. Thus, recovery of photographic material is also a possible astronaut function. Two recent approaches to the problem of reducing *instrumental* seeing at ground-based solar observatories are described by Kiepenheuer[168]—a domeless coudé refractor—and Dunn[169]—an evacuated tower telescope.

Basic patrols of solar phenomena have tended to emphasize white light photographs and records (*e.g.*: H-alpha and Ca II K-line spectroheliograms and filtergrams) that are taken in just a few monochromatic bandpasses. A further step, of considerable value to astrophysicists, would be the development of a solar patrol in five to ten bands. Their wavelengths would correspond to radiation that originates at different, selected levels in the solar atmosphere. Lyot-type monochromators might be used, and for some of the bands, image intensification equipment is indicated, in order to keep the exposures sufficiently brief.

The manned space program has a vital requirement for rapid

detection, or, if possible, reliable *predictions* of those solar flares that give rise to energetic particle events at the earth. Further, the question of flare forecasting is of great importance to global communications services. The considerable efforts undertaken in this area by a bewildering array of U.S. agencies are discussed in the news report by Pay[170]. From the astrophysicist's point of view, the prime needs in the study of solar flares are for (1) improved, quantitative studies of the manner in which flares evolve in space and time and (2) more spectra of flares (and magnetograms of flare regions) taken with better time resolution, from the *onsets* of the events.

An instrumental system capable of making a major contribution to the study of flares, with respect to the requirements for improved astrophysical data and also the practical needs of the space program, is under study at Kitt Peak National Observatory (Maran and Sullivan[171]). A block diagram of this ASO ("Automatic Solar Observatory") system is shown in figure 28. Basically, ASO consists of two rather different telescopes and their associated instrumentation, all under the control of a large computer. One telescope ("Flare Patrol Telescope") would be a relatively small refractor, similar to those presently in use at many of the existing flare patrol observa-

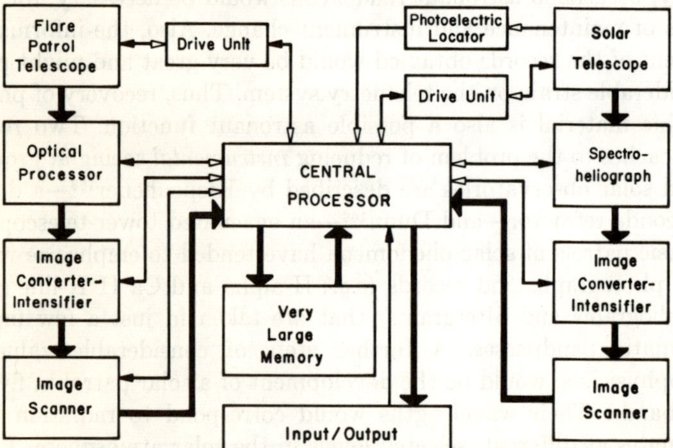

FIG. 28. Block diagram for a proposed Automatic Solar Observatory (Maran and Sullivan). The heavy lines indicate the flow of astronomical information; the thin lines indicate the transmission of control signals (filled arrowheads) and status monitor data (empty arrowheads).

tories. Its output, filtered by the "Optical Processor", would yield a monochromatic image in any (for example, H-alpha) of several wavelengths. The intensity of this image would be stepped up (to allow rapid time resolution) by an electronic image converter whose output is imaged on a vidicon-type tube that is scanned by the computer. The digital television read-out system under development at Socorro by Colgate[172], for use in an automatic supernova detection system, might be appropriate for this function. The appropriate software would be provided to enable the computer to immediately determine whether a flare was in progress and, if so, to determine its total intensity, area, brightness distribution and disk location. If certain criteria were satisfied with respect to some of these parameters and possibly also their time dependencies, a "proton event" warning would be issued. The refractor and associated instruments would continue to furnish filtergram-type data on the evolution of the flare region in H-alpha and other wavelengths of interest. In the meantime, the "Solar Telescope", a tower instrument like the McMath Solar Telescope (Pierce[173]; this equipment is already partially automated) shown in figures 29 and 30, may have been taking relatively routine spectroscopic observations of the photosphere prior to the flare. However, discovery of the flare would cause the computer to command the tower telescope heliostat to slew over so that the flare's coordinates (determined from the Flare Patrol Telescope observation) would be positioned on the slit of a spectroheliograph. The spectroheliograph grating would simultaneously be slewed into position for observations of the wavelength region of interest for the flare spectra program. (The spectroheliograph would convert to an "ordinary" spectrograph upon command.) Spectroscopic observations of the flare could then commence. This system would be easily capable of obtaining spectra during the first one or two minutes when the so-called "explosive phase" of a flare usually occurs. A number of independent lines of investigation suggest that this is a stage of the greatest astrophysical importance in the development of the flare and associated solar phenomena, such as the ejection of solar cosmic rays. The "Flare Patrol Telescope" component of the ASO would be similar in concept to the vidicon flare patrol system developed by Dunn (Evans[174]), which, however, has a man in the loop.

In the radio frequency range, one of the most promising developments (James[175]) has been the construction of radar facilities capable

FIG. 29. The McMath Solar Telescope of the Kitt Peak National Observatory.

of providing velocity data on density irregularities such as shocks propagating through the corona. The effort in this field deserves to be expanded; in particular, simultaneous observations at several frequencies are desired, as is greatly improved resolution in angle and time. Exciting prospects are also offered by several planned and proposed space experiments. Radiometers carried on board the Alouette II (Nelms *et al.*[176]) and OGO-B (Haddock[177]) satellites have already detected solar radio bursts at frequencies below the ionospheric cutoff. The data obtained are important for the study of the coronal gas at great distances from the sun. Further, these experiments constitute only the first generation of effort in this direction; the next two stages are represented by the Radio Astronomy Explorer satellites and such very large instruments as the 10 km orbiting radio telescope.

The possibility that the sun possesses a measurable gravitational quadrupole moment, as suggested by Dicke's oblateness experiment (discussed in the section on the Interior) needs to be investigated. Because of the ambiguities involved in relating a photospheric brightness distribution to the physical state of the interior, and because of the possible relevance of this question to fundamental gravitational theory, through the interpretation of the perihelion advance of Mercury, it is vital that an unambiguous study of the solar gravitational field be made. The best way to do this would appear to be by observation of the motion of an artificial satellite of the sun. Further, such a close-in satellite could be equipped with plasma detection equipment suitable for delineating the gross shape of the solar particle beams associated with the "M regions" postulated in the 1930's by J. Bartels to account for recurrent geomagnetic storms. Theoretical studies are also needed to understand the path of the enhanced particle emission under the influence of the solar rotation and the interaction with the magnetic field and ambient material; this is a problem in fluid dynamics (see below).

A great deal of theoretical support will be required in future solar investigations and much of it will follow traditional lines. Analyses of absorption lines based on atomic parameters and the fundamental physics of absorption and emission as reflected in the equation of transfer will undoubtedly occupy a great deal of the theoretical effort and will require a good deal of concomitant laboratory work. Theoretical studies concerned with nuclear reactions, such as the prediction and/or analysis of the solar neutrino flux, should be actively pursued. However, it seems that much of the future of solar theoretical studies lies in fluid dynamics and, in particular, fluid dynamics in the presence of a magnetic field; the fundamental physics in this area is just now becoming clear.

The general area of magneto-fluid dynamics is clearly of importance *throughout* the sun. If the sun has a rapidly rotating core as Dicke has suggested, then this phenomenon is of considerable interest in studies of general stellar structure and evolution, and the mechanism for effectively isolating the core and envelope must be investigated. In the same area, the problem of the general solar magnetic field remains to be solved, as does the possibility of a solar meridional circulation. Close to the surface, we find that the physical application of fluid dynamics to the hydrogen convection zone is urgently required even though some progress has been made through the method of

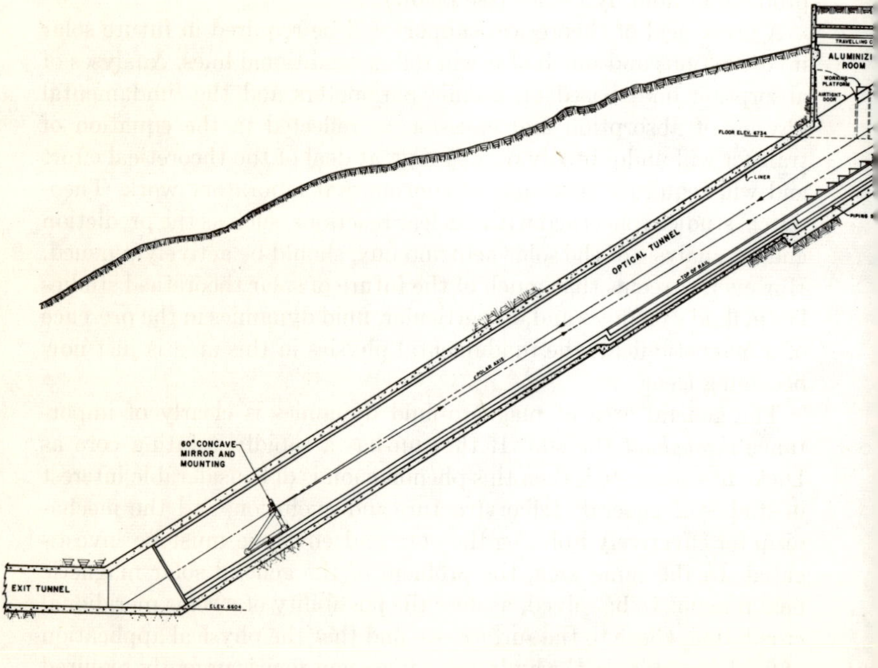
TRAVELLING C
ALUMINIZI ROOM
WORKING PLATFORM
AIRTIGHT DOOR
FLOOR ELEV. 6734
LINER
PIPING S
OPTICAL TUNNEL
TOP OF RAIL
POLAR AXIS
60" CONCAVE MIRROR AND MOUNTING
EXIT TUNNEL
ELEV. 6604

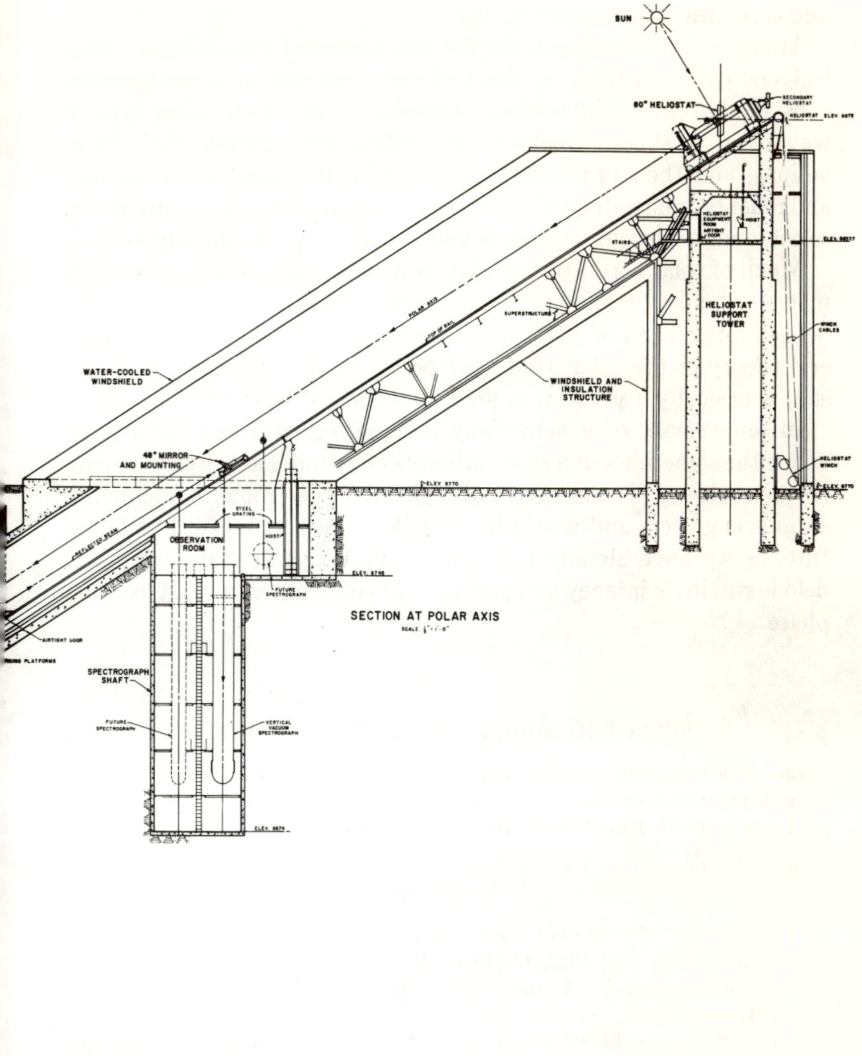

FIG. 30. The optical and geometrical layout for the McMath Solar Telescope of the Kitt Peak National Observatory.

laminar modes. The inhibition of convection motions by a strong magnetic field also requires study.

Analysis of the heating of the chromosphere and corona by mechanical energy originating in the hydrogen convection zone deserves considerable effort; the method of noise generation, the exact type of waves and/or shocks involved, the mode of propagation, their interactions, and their processes of attenuation and dissipation are not, and must be, understood. Of particular interest is the physics by which select areas in the chromosphere and corona are preferentially heated.

Much of solar activity may also require an explanation in terms of magneto-fluid dynamics; examples are the theories of flares, of prominence motions and of the Evershed effect in sunspots. Also, the general theory of the solar wind, and especially its interaction with the sun, is basically a problem in fluid dynamics, at least near the sun.

In some sense, solar astronomy may be regarded as a focal point where the sometimes divergent interests of astronomers, hydrodynamicists, nuclear-, plasma- and radio-physicists, astronauts, communications engineers and would-be neutrino observers meet. The literature, as we have already indicated, is vast, but in many ways this field is still in its infancy and perhaps has yet to enter its most exciting phase.

SELECTED BOOKS AND REVIEW ARTICLES

(Note: A new journal, *Solar Physics* began publication in January, 1967, as the present paper was completed; the editors are C. de Jager and Z. Svestka and the publisher is the D. Reidel Company, Dordrecht-Holland.)

Introduction

The following references cover essentially the entire subject area of this article:

1. Kuiper, G. P. (ed.) 1953, *The Sun*, Chicago; University of Chicago Press.
2. Ambartsumyan, V. A. (ed.) 1958, *Theoretical Astrophysics*, New York; Pergamon Press.
3. Flügge, S. (ed.) 1958–1959, *Handbuch der Physik*, **LI, LII** Berlin; Springer-Verlag.
4. Brandt, J. C., and Hodge, P. W. 1964, *Solar System Astrophysics*, New York; McGraw-Hill Book Co.
5. Zirin, H. 1966, *The Solar Atmosphere*, Waltham, Mass.; Blaisdell Publ. Co.

For elementary surveys see:

6. Menzel, D. H. 1959, *Our Sun*, Cambridge, Mass.; Harvard University Press.

7. Kiepenheuer, K. O. 1959, *The Sun*, Ann Arbor; University of Michigan Press.
8. Brandt, J. C. 1966, *The Sun and Stars*, New York; McGraw-Hill Book Co.

Historical Introduction

Consult:

9. Goldberg, L. 1953, in *The Sun*, p. 1, (ed. G. P. Kuiper), Chicago; University of Chicago Press.
10. Strömgren, B. *ibid.*, p. 36.
11. Strömgren, B. 1951, in *Astrophysics*, p. 1, (ed. J. A. Hynek), New York; McGraw-Hill Book Co.
12. Strömgren, B. 1951, *ibid.*, p. 172.
13. Clerke, A. M. 1908, *A Popular History of Astronomy During the Nineteenth Century*, London; A. and C. Black.
14. Struve, O. and Zebergs, V. 1962, *Astronomy of the 20th Century*, New York; The Macmillan Co.
15. Pannekoek, A. 1961, *A History of Astronomy*, New York; Interscience.

and references (7) and (8).

The Structure of the Sun

Introduction: *see* references (1) through (8).

Interior: *See*

15a. Schwarzchild, M. 1958, *Structure and Evolution of the Stars*, Princeton, N. J.; Princeton University Press.
16. Wrubel, M. H. 1958, *Handbuch der Physik*, **LI**, 1.
17. Aller, L. H. and McLaughlin, D. B. (eds.), 1965, *Stellar Structure*, Chicago; University of Chicago Press.

Photosphere: *see*

18. Minnaert, M. 1953, in *The Sun*, p. 88, (ed. G. P. Kuiper), Chicago; University of Chicago Press.
19. Unsöld, A. 1955, *Physik der Sternatmosphären*, Berlin; Springer-Verlag.
20. Goldberg, L. and Pierce, A. K. 1959, *Handbuch der Physik*, **LII**, 1.
21. Leighton, R. B. 1963, *Ann. Rev. Astron. Astrophys.*, **1**, 19.

Chromosphere: *see*

22. van de Hulst, H. C. 1953, in *The Sun*, p. 207 (ed. G. P. Kuiper), Chicago; University of Chicago Press.
23. de Jager, C. 1959, *Handbuch der Physik*, **LII**, 80.
24. Thomas, R. N., and Athay, R. G. 1961, *Physics of the Solar Chromosphere*, New York; Interscience.
25. Pagel, B. E. J. 1964, *Ann. Rev. Astron. Astrophys.*, **2**, 267.

Corona: *see* references (21), (22), and

26. Evans, J. W. (ed.) 1963, *The Solar Corona*, New York; Academic Press.
27. Shklovskii, I. S. 1965, *Physics of the Solar Corona*, Reading, Mass.; Addison-Wesley Publ. Co.
28. Billings, D. E. 1966, *A Guide to the Solar Corona*, New York; Academic Press.

29. Parker, E. N. 1963, *Interplanetary Dynamical Processes*, New York; Interscience.
30. Dessler, A. J. 1967, *Rev. Geophys.*, **5**, 1.

The Steady Solar Spectrum

This entire section is covered in

31. de Jager, C. (ed.) 1965, *The Solar Spectrum*, Dordrecht-Holland; D. Reidel Publ. Co.

Other reviews may be consulted as follows:

Analysis of the visible spectrum—the Fraunhofer lines:

See the general references and the photospheric, chromospheric, and coronal references.

The Extreme Ultraviolet (EUV)

32. Friedman, H. 1963, *Ann. Rev. Astron. Astrophys.*, **1**, 59.
33. Tousey, R. 1963, *Space Sci. Rev.*, **2**, 3.
34. Hinteregger, H. E. 1965, *Space Sci. Rev.*, **3**, 461.
35. Lindsay, J. C., Neupert, W. M., and Stone, R. G. 1965, in *Introduction to Space Science* (1st Ed.), p. 585 (ed. W. M. Hess), New York; Gordon and Breach.

X-rays: *see* the EUV references and

36. Mandel'štam, S. L. 1965, *Space Sci. Rev.*, **4**, 587.

Radio Waves: *see*

37. Wild, J. P., Smerd, S. F., and Weiss, A. A. 1963, *Ann. Rev. Astron. Astrophys.*, **1**, 291.
38. Kundu, M. R. 1965, *Solar Radio Astronomy*, New York; Interscience.
39. Aarons, J. (ed.) 1965, *Solar System Radio Astronomy*, New York; Plenum Press.

For a popular exposition, see

40. Smith, A. G. 1967, *Radio Exploration of the Sun*, Princeton, N. J.; D. Van Nostrand.

Solar Activity

Introduction: *see* the general references, reference (23) and

41. Kiepenheuer, K. O. 1953, in *The Sun*, p. 322 (ed. G. P. Kuiper), Chicago; University of Chicago Press.

Centers of Activity: *see*

42. de Jager, C. 1961, *Vistas in Astronomy*, **4**, 143.
43. Bray, R. J., and Loughhead, R. E. 1965, *Sunspots*, New York; John Wiley and Sons, Inc.
44. Babcock, H. W. 1963, *Ann. Rev. Astron. Astrophys.*, **1**, 41.
45. Smith, H. J., and Smith, E. v. P. 1963, *Solar Flares*, New York; The Macmillan Co.
46. Severny, A. B. 1963, *Ann. Rev. Astron. Astrophys.*, **2**, 363.
47. Hess, W. N. 1964, *The Physics of Solar Flares*, Washington, D.C.; National Aeronautics and Space Administration.

48. Lüst, R. (ed.) 1965, *Stellar and Solar Magnetic Fields*, Amsterdam; North-Holland Publ. Co.

See also the various references for the chromosphere, corona, and the solar spectrum.

Solar activity in the context of solar-terrestrial relations is beyond the scope of this article; however, see, for example

49. Athay, R. G. and Warwick, C. S. 1961, *Advances Geophys.*, **8,** 1.

The Solar Cycle and the General Solar Field: see the general references, references (44), (48) and

50. Cowling, T. G. 1953, in *The Sun*, p. 532 (ed. G. P. Kuiper), Chicago; University of Chicago Press.

The Future of Solar Astronomy

51. Space Science Board, National Academy of Sciences, National Research Council, 1966, *Space Research, Directions for the Future* (Part 2), Washington, D.C.; NAS-NRC.

REFERENCES

52. Weymann, R. 1957, *Ap. J.*, **126,** 208.
53. Sears, R. L. 1964, *Ap. J.*, **140,** 477.
54. Davis, R. 1964, *Phys. Rev. Lett.*, **12,** 303.
55. Reines, F. and W. R. Kropp 1964, *Phys. Rev. Lett.*, **12,** 457.
56. Bahcall, J. N. 1964, *Phys. Rev. Lett.*, **12,** 300.
57. Dicke, R. H. and Goldenberg, H. M. 1967, *Phys. Rev. Letters*, **18,** 313.
58. Brandt, J. C. 1966, *Ap. J.*, **144,** 1221.
59. Brandt, J. C. 1967, *Ap. J.*, **147,** 201.
59a. Dicke, R. H. 1964, *Nature*, **202,** 432.
59b. Goldreich, P. and Schubert, G. 1967, *Science*, **156,** 1101.
59c. Coche, W. J. 1967, private communication.
60. ten Bruggencate, P., Gollnow, H. and Jager, F. W. 1950, *Zeitschrift für Astrophysik*, **27,** 223.
61. Kristenson, H. 1960, *Arkiv. für Astronomi*, **2,** 315.
62. Dunn, R. B. 1959, *Ap. J.*, **130,** 972.
63. Rogerson, J. B. 1959, *Ap. J.*, **130,** 985.
64. Gaustad, J. E. and Rogerson, J. B. 1961, *Ap. J.*, **134,** 323.
65. Unsöld, A. 1962, in *Space Age Astronomy*, p. 161, (eds. A. J. Deutsch and W. B. Klemperer), New York; Academic Press.
66. Waddell, J. 1965, *Ap. J.*, **142,** 326.
67. Schwarzchild, M. 1959, *Ap. J.*, **130,** 345.
68. Bahng, J. and Schwarzchild, M. 1961, *Ap. J.*, **134,** 312.
69. Bahng, J. and Schwarzchild, M. 1961, *Ap. J.*, **134,** 337.
70. Evans, J. W. and Michard, R. 1962, *Ap. J.*, **135,** 812.
71. Leighton, R. B., Noyes, R. W. and Simon, G. W. 1962, *Ap. J.* **135. 474.**
72. Noyes, R. W. and Leighton, R. B. 1963. *Ap. J.*. **138. 631**.

73. Simon, G. W. and Leighton, R. B. 1964, *Ap. J.*, **140**, 1120.
74. Simon, G. W. 1966, in *The Fine Structure of the Solar Atmosphere*, p. 47, (ed. K. O. Kiepenheuer), Wiesbaden; Franz Steiner Verlag.
75. Coates, R. J. 1958, *Ap. J.*, **128**, 83.
76. Howard, R. 1959, *Ap. J.*, **130**, 193.
77. Leighton, R. B. 1959, *Ap. J.*, **130**, 366.
78. Parker, E. N. 1963, *Ap. J.*, **138**, 552.
79. Dunn, R. B. 1965, *Contrib. Sacramento Peak Obs.*, No. 87.
80. Beckers, J. M. 1963, *Ap. J.*, **138**, 648.
81. Zirin, H. and Dietz, R. D. 1963, *Ap. J.*, **138**, 664.
82. Osterbrock, D. E. 1961, *Ap. J.*, **134**, 347.
83. Brandt, J. C., Michie, R. W. and Cassinelli, J. P. 1965, *Ap. J.*, **141**, 809.
84. Brandt, J. C., Michie, R. W. and Cassinelli, J. P. 1965, *Icarus*, **4**, 19.
85. Burgess, A. 1964, *Ap. J.*, **139**, 776.
86. Billings, D. E. 1959, *Ap. J.*, **130**, 961.
87. Oster, L. 1963, *Ap. J.*, **138**, 761.
88. Oster, L. and Sofia, S. 1965, *Ap. J.*, **141**, 1139.
89. Delano, M. 1967, private communication.
90. Zheleznyakov, V. V. 1967, *Ap. J.*, **148**, 849.
91. Whitaker, W. A. 1963, *Ap. J.*, **137**, 914.
92. Chapman, S. 1964, *Solar Plasma, Geomagnetism and Aurora*, New York; Gordon and Breach.
93. Biermann, L. 1951, *Zeitschrift für Astrophysik*, **29**, 274.
94. Biermann, L. 1952, *Zeitschrift Naturforsch.*, **7a**, 127.
95. Mackin, R. J. and Neugebauer, M. (eds.) 1966, *The Solar Wind*, Pasadena; Jet Propulsion Laboratory.
96. Chamberlain, J. W. 1965, *Ap. J.*, **141**, 320.
97. Parker, E. N. 1965, *Ap. J.*, **141**, 322.
98. Brandt, J. C. and Cassinelli, J. P. 1966, *Icarus*, **5**, 47.
99. Brandt, J. C. 1967, *Ap. J.*, **148**, 905.
100. Pottasch, S. R. 1963, *Ap. J.*, **137**, 945.
101. Brandt, J. C. 1966, *Ap. J.*, **143**, 265.
102. Moore-Sitterly, C. E. 1965, in *The Solar Spectrum*, p. 89, (ed. C. de Jager), Dordrecht-Holland; D. Reidel Co.
103. Malitson, H. H. 1965, *Sky and Telescope*, **29**, 162.
104. Hardie, R. H. and Giclas, H. L. 1955, *Ap. J.*, **122**, 460.
105. Tousey, R., Purcell, J. D., Austin, W. E., Garrett, D. L. and Widing, K. G. 1964, *Space Research*, **4**, 703.
106. Hinteregger, H. E. 1965, in *The Solar Spectrum*, p. 179, (ed. C. de Jager), Dordrecht-Holland; D. Reidel Co.
107. Pottasch, S. R. 1964, *Space Science Rev.*, **3**, 816.
108. Rense, W. A. 1961, in *Space Astrophysics*, p. 17, (ed. W. Liller), New York; McGraw-Hill.
109. Neupert, W. M., Behring, W. E. and Lindsay, J. C. 1964, *Space Research*, **4**, 719.
110. Neupert, W. M. 1965, *Ann. d'Astrophysique*, **28**, 446.

111. Neupert, W. M. and Smith, E. v. P. 1964, *Astron. J.*, **69**, 554.
112. Fawcett, B. C., Gabriel, A. H., Griffin, W. G., Jones, B. B. and Wilson, R. 1963, *Nature*, **200**, 1303.
113. Austin, W. E., Purcell, J. D., Tousey, R. and Widing, K. G. 1966, *Ap. J.*, **145**, 373.
114. Peterson, L. E. and Winckler, J. R. 1959, *J. Geophys. Res.*, **64**, 697.
115. Kundu, M. R. 1961, *J. Geophys. Res.*, **66**, 4308.
116. Chubb, T. A., Friedman, H. and Kreplin, R. W. 1960, in *Space Research I*, p. 695, Amsterdam; North-Holland Publ. Co.
117. Frost, K. J., Rothe, E. D. and Peterson, L. E. 1966, *J. Geophys. Res.*, **71**, 4079.
118. Chubb, T. A., Friedman, H., Kreplin, R. W. and Kupperian, J. E. 1957, *J. Geophys. Res.*, **62**, 389.
119. White, W. A. 1964, in *The Physics of Solar Flares*, p. 131, (ed. W. N. Hess), Washington; NASA.
120. Culhane, J. L., Willmore, A. P., Pounds, K. A. and Sanford, P. W. 1964, in *Space Research IV*, p. 741, (ed. P. Muller), Amsterdam; North-Holland Publ. Co.
121. Elwert, G. 1961, *J. Geophys. Res.*, **66**, 391.
122. Underwood, J. H. and Muney, W. S. 1967, *Solar Phys.*, **1**, 129.
123. Bracewell, R. N., and Swarup, G. 1961, *IRE Trans. Antennas Propagation*, **AP-9**, 22.
124. Steinberg, J. L. and Lequeux, J. 1963, *Radio Astronomy*, New York; McGraw-Hill.
125. Wild, J. P. 1963, in *The Solar Corona*, p. 115, (ed. J. W. Evans), New York; Academic Press.
126. Righini, G. 1964, in *Space Exploration and the Solar System*, (ed. B. Rossi), New York; Academic Press.
127. Dodson, H. W. and Hedeman, E. R. 1956, *M.N.R.A.S.*, **116**, 428.
128. Christiansen, W. N., Mathewson, D. S., Pawsey, J. L., Smerd, S. F., Boischot, A., Denisse, J. F., Simon, P., Kakinuma, T., Dodson-Prince, H. and Firor, J. 1960, *Ann. d'Astrophysique*, **23**, 75.
129. Babcock, H. W. and Babcock, H. D. 1955, *Ap. J.*, **121**, 349.
130. Hale, G. E. and Nicholson, S. B. 1938, *Publ. Carnegie Inst.*, No. 498.
131. Athay, R. G. and Thomas, R. N. 1957, *Ap. J.*, **125**, 788.
132. Newkirk, G. 1961, *Ap. J.*, **133**, 983.
133. Hiei, E. 1962, *J. Phys. Soc. Japan*, **17A-II**, 227.
134. Aly, M. K., Evans, J. W. and Orrall, F. Q. 1962, *Ap. J.*, **136**, 956.
135. Nishi, K. and Nakagomi, Y. 1963, *Pub. Astron. Soc. Japan*, **15**, 56.
136. Billings, D. E. and Lehman, R. C. 1962, *Ap. J.*, **136**, 258.
137. Kawabata, K. 1960, *Pub. Astron. Soc. Japan*, **12**, 513.
138. Kakinuma, T. and Swarup, G. 1962, *Ap. J.*, **136**, 975.
139. Swarup, G., Kakinuma, T., Covington, A. E., Harvey, G. A., Mullaly, R. F., and Rome, J. 1963, *Ap. J.*, **137**, 1251.
140. Waldmeier, M. and Müller, H. 1950, *Zeitschrift für Astrophysik*, **27**, 58.
141. Piddington, J. H. and Minnett, H. C. 1951, *Australian J. Sci. Res. A*, **4**, 131.

142. Zheleznyakov, V. V. 1962, *Soviet Astron.* **6**, 3.
143. Minnaert, M. 1953, in *The Sun*, p. 722 (ed. G. P. Kuiper), Chicago; University of Chicago Press.
144. Bray, R. J. and Loughhead, R. E. 1959, *Australian J. Phys.*, **12**, 320.
145. Waldmeier, M. 1939, *Astr. Mitt. Zürich*, No. 138.
146. Sweet, P. A. 1955, *Vistas in Astronomy*, **1**, 675.
147. Musman, S. 1967, *Ap. J.*, **149**, 201.
148. Danielson, R. E. 1961, *Ap. J.*, **134**, 275.
149. Danielson, R. E. 1961, *Ap. J.*, **134**, 289.
150. Wild, J. P. 1963, in *Radio Astronomy Today*, (eds. H. P. Palmer, R. D. Davies, and M. I. Large), Cambridge, Mass.; Harvard University Press.
151. Zirker, J. B. 1965, in *Stellar and Solar Magnetic Fields*, p. 208, (ed. R. Lüst), Amsterdam; North-Holland Publ. Co.
152. Babcock, H. D. 1959, *Ap. J.*, **130**, 364.
153. Sheeley, N. R. 1967, *Solar Physics*, **1**, 171.
154. Mestel, L. 1965, in *Stellar and Solar Magnetic Fields*, p. 420, (ed. R. Lüst), Amsterdam; North-Holland Publ. Co.
155. Biermann, L. 1950, *Z. Naturforsch.*, **5a**, 65.
156. Cowling, T. G. 1955, *Vistas in Astronomy*, **1**, 313.
157. Batchelor, G. K. 1950, *Proc. Roy. Soc. London A*, **201**, 405.
158. Allen, C. W. 1960, *Observatory*, **80**, 94.
159. Hoyle, F. and Wickramasinghe, N. C. 1961, *M.N.R.A.S.*, **123**, 51.
160. Babcock, H. W. 1961, *Ap. J.*, **133**, 572.
161. Leighton, R. B. 1964, *Ap. J.*, **140**, 1547.
162. Waldmeier, M. 1960, *Zeitschrift für Astrophysik* **49**, 176.
163. Hyder, C. L. 1965, *Ap. J.*, **141**, 272.
164. Sheeley, N. R. 1964, *Ap. J.*, **140**, 731.
165. Gursky, H. and Zehnphennig, T. 1966, *Applied Optics*, **5**, 875.
166. Bray, R. J., Loughhead, R. E. and Norton, D. G. 1959, *Observatory*, **79**, 63.
167. Kiepenheuer, K. O. and Mehltretter, J. P. 1964, *Applied Optics* **3**, 1359
168. Kiepenheuer, K. O. 1966, *Sky and Telescope*, **31**, 256.
169. Dunn, R. B. 1964, *Applied Optics*, **3**, 1353.
170. Pay, R. 1967, *Technology Week*, **20**, 20.
171. Maran, S. P., and Sullivan, N. J. 1967, unpublished.
172. Colgate, S. 1967, private communication.
173. Pierce, A. K. 1964, *Applied Optics*, **3**, 1337.
174. Evans, J. W. 1966, *Astron. J.*, **71**, 818.
175. James, J. C. 1966, *Ap. J.*, **146**, 356.
176. Nelms, G. L. *et al.* 1966, *Canadian J. Phys.*, **44**, 1419.
177. Haddock, F. T. 1967, private communication.
178. Purcell, J. D., Garrett, D. L. and Tousey, R. 1964, *Astron. J.*, **69**, 147.

Chapter 17

THE MOON

John A. O'Keefe

INTRODUCTION

In the present chapter we shall attempt to penetrate to an understanding of the structure and origin of the moon. We shall attempt to work from the outside inward and from the present toward the past. This will permit us to begin with the study of the geometrical facts about the moon, proceed to its morphology, its surface properties and geophysical problems and conclude with the chemical problems, and the problem of its origin.

GEOMETRICAL FACTS

Distance

The orbit of the moon is, to a very poor approximation, an ellipse, with the earth at one focus. Because the shape of the ellipse is strongly perturbed by the sun, it is necessary to define carefully what is meant by the *mean distance* of the moon, and astronomers have chosen for this purpose that value of a which satisfies the equation

$$n^2a^3 = G(M + m). \tag{1}$$

In this equation n is the moon's average angular velocity in its orbit; G the universal constant of gravitation; M, mass of the earth; and m, the mass of the moon. It is important to note that the real average value[1] of the distance is $a(1 + \beta')$, where β' is a constant which takes account of the effect of the constant portion of the perturbations of the sun.

This equation has a good deal of importance in that it relates several of the fundamental constants of the earth-moon system. β' can be determined from theory with all the accuracy required. n is likewise known with all necessary accuracy; it represents the mean length of the month, and this is a constant which is exceedingly well-determined as the result of observations going back for centuries. If we write the equation in the form

$$n^2a^3 = GM(1 + \mu), \tag{2}$$

where

$$\mu = m/M,$$

then we see that as soon as an accurate value of a has been determined (for example, by radar measurements of the moon's distance) we have an important relation between two quantities. The second of these quantities, $1 + \mu$, involves the ratio of the mass of the moon to that of the earth and will be discussed further below. If this quantity can be fixed accurately, it is clear that we will have an accurate knowledge of GM.

The value of GM is important for measuring the size of the earth. There is a theorem of Gauss relating to inverse square forces such as gravity which can be formulated by saying that every gram of matter puts out $4\pi G$ lines of force. The number of lines of force traversing a given area can be measured. It is not precisely what geophysicists call gravity, since that includes centrifugal force, but it can be found from measures of gravity because the allowance for centrifugal force can be made with great accuracy. As a result, it is possible to determine with precision the average number of lines of force traversing 1 cm^2 of the earth's surface, as a function of latitude, from the equator to the pole. We may call this vector $\mathbf{g}^*$ to distinguish it from the ordinary value of gravity. Gauss's theorem asserts that

$$\iint_s \mathbf{g}^* \cdot \mathbf{n}\, ds = 4\pi GM, \tag{3}$$

where s is any surface enclosing the mass, and $\mathbf{n}$ is the unit vector normal to s. Thus GM is determinable, provided that we can find the quantity ds. On a sphere, ds would be

$$r^2 \cos\phi \, d\lambda \, d\phi,$$

where ϕ, λ, are the latitude and longitude, respectively. There is an analogous expression for ellipsoid which it is not necessary to state here. The essential point is to see that $\mathbf{g}^*$ is related to GM through the radius of the earth. Hence this equation can be employed to measure the size of the earth.

It is interesting to note that when Newton first employed this general approach to test his theory of gravitation, he inserted an incorrect value of the radius of the earth. When the answer did not check, he concluded that the theory was not correct, and he put the *Principia* on the shelf. When news reached him of Picard's measurement of the size of the earth by means of the arc from Dunkirk to Paris, he saw that the correction would be in the right direction, and it is said that his hands shook so that he had to have a friend perform the computation.

A recent determination of the mean distance of the moon by Fischer[2] gives

$$a = 384{,}400 \pm 2 \text{ km}.$$

When combined with the recent measurements of the mass of the moon, this distance implies a value for the radius of the earth which is in agreement with other recent measurements, near 6,378,166 meters.

Mass

The determination of the mass of the moon is important for its own sake as well as in order to fit into the above equation. The principle which is used here is that the earth and the moon revolve around their common center of gravity at distances which are inversely proportional to their masses. It follows that the mass ratio, μ, is the reciprocal of the ratio of the mean distances from the center of gravity. So long as we are inside the earth–moon system, almost everything goes as if the earth had a mass of

$$M + m$$

and were unmoving, while the moon went around it as we have hitherto assumed. It is only possible to observe the motion of the earth around the *barycenter*, as we call the center of mass, through its effect on the apparent motion of other systems. The observations of the angular displacement of the planet Eros have been used in this way. An accurate determination was made by the Jet Propulsion Laboratory by radar observations of Mariner II. They find $\mu = 1/81.30$, a value confirmed by direct measurements of the acceleration produced by the moon on Ranger VI through IX.

Radius and Density

Two values of the radius of the moon are commonly used. One value is employed for the calculation of eclipses. For a total eclipse of the sun, it is important that no portion of the solar disc should be exposed, since it will ruin the effectiveness of the eclipse from the point of view of seeing the corona and any other faint phenomenon. The *eclipse radius* is therefore the radius to the lowest point of the valleys which notch the moon's edge. It is ordinarily given as the fraction

$$0.272274$$

of the earth's radius.[3]

For most purposes the value which is taken for the radius of the moon's disc is the average value around the limb. The accepted figure[3] is

$$0.272446.$$

Taken in conjunction with the above mass of the moon, the implied density is

$$3.36 \text{ gm/cm}^3.$$

Orbit

The problem of the orbit of the moon is the most important single problem in celestial mechanics. Its study goes back to Greek times when a few of the larger perturbations were noted; Newton made some important contributions; and during the 18th century much effort was devoted to it. The effort was, in part, stimulated by a prize of about $100,000 offered by the British Government for a method of determining longitude. It was hoped that very accurate tables of the moon would permit a sea captain to determine his longitude (by measuring the angular distance between the moon and the sun, and hence determining his Greenwich time, from which the longitude is easily obtained). (In the end, the invention of the marine chronometer by J. Harrison won the prize.)

During the 19th century, three fundamentally different approaches to the calculation of the orbit of the moon were developed.[1] The Danish mathematician, P. Hansen, started from the assumption of a fictitious moon moving in a Keplerian orbit with a steady motion of the node and perigee. He then determined the actual motion of the moon in terms of oscillations in longitude, latitude, and time around this fictitious body. The French mathematician C. Delaunay developed a theory based on the Hamilton–Jacobi method. The original variables, for example, the coordinates and velocities, are first replaced by elements which are modifications of the Keplerian elements. Next, by a special type of transformation the Keplerian elements are replaced by new elements which are defined in such a way that they retain the properties of canonical variables, but they now include some of the perturbations. By successive transformations of this kind, it is possible to include more and more of the perturbations. The process is somewhat lengthy; a device to increase the rapidity of convergence was invented by von Zeipel some years ago and has been further developed by Brouwer. The third theory is that of G. W. Hill, in which the initial orbit is a periodic orbit but not a Keplerian ellipse. Instead it takes into account the most serious of the solar perturbations, called the *variation*. Perturbations of the so-called *variational orbit* were studied from the theoretical viewpoint by Hill and were developed into a complete theory and a set of tables by E. W. Brown. Brown's tables form the basis for the existing ephemerides.

MORPHOLOGY

Craters

When the external form of the moon is examined through the telescope, the most striking feature is a large number of circular objects, generally called *craters*, (*Fig. 1*) of all sizes from 200 km across down to the smallest size that can be distinguished. The smaller craters are cup-shaped, with such smooth walls that they look as though they were turned out by a machine. Around these craters there is ordinarily a low rim, generally so small that, contrary to the usual opinion, it would not approach a filling of the crater if its material were dumped in. In the craters above 12 km in size, a more complex type of bottom is observed, which is nearly always flattened in the center. Often there is a small central peak or small cluster of peaks. The walls, instead of being simple and smooth, are complicated piles of irregular shape. Many of these craters show a system of concentric walls, and in some, such as Posidonius and Taruntius (see Figures 2 and 3), there is a system of narrow internal valleys of the type described below as rilles, some of which may parallel the external walls of the crater.

Opinion is divided about the origin of these craters between those who favor an origin by the impact of an asteroid or a comet and those who favor an origin by vulcanism. It is not possible to doubt that some of the large craters on the moon are the result of impact,

FIG. 1. The region of Maurolycus (center right) and Stöfler (center left), in the lunar highlands. North at the top. Rays from Tycho traverse Stöfler. Maurolycus is 120 km in diameter. Courtesy of Lick Observatory. Compass directions in all figures are given in the I.A.U. convention.

FIG. 2. Posidonius and a portion of Mare Serenitatis with a prominent wrinkle ridge. North at the top. Note the rill inside Posidonius on the west side which turns into a ridge in the southwest corner. Posidonius is 100 km in diameter. Courtesy of Lick Observatory.

FIG. 3. The crater Taruntius, 60 kilometers in diameter, with concentric rilles and ridges in the interior. Orbiter I photograph. North at the left. (NASA).

since the impact origin of several large craters on the earth has been demonstrated by the discovery of the high-pressure minerals *coesite* and especially *stishovite* in the walls.[4] Among the largest of the known terrestrial craters is the Ries Kessel in Germany, some 25 km in diameter, in which Chao discovered both coesite and stishovite in 1960 and 1961. The pressure required to form coesite is 16,000 atm; that required to form stishovite is over 100,000 atm. There is no evidence that such pressures as these can be generated in a volcanic outburst. Volcanic outbursts are generally in the nature of a steam boiler explosion; that is to say, a contained gas finally bursts its container and erupts. It is not believed possible for the earth to contain pressures more than a few hundred atmospheres at moderate depths; thus these indications of pressures hundreds of times higher are logically blamed on meteorite impact following the example of Meteor Crater. If such impacts occur on the earth, then they must also occur on the moon.

A great deal of work has been done on the mechanics of cratering, both by Baldwin,[5] who has compared the shapes of the craters with those of terrestrial impact formations, and by the group at the U.S. Geological Survey and the NASA Ames Research Center who have worked out the relationship of lunar craters to experimental impact and explosion craters. An important result of this work is that the diameter of a crater varies with the energy, w, required to produce it approximately as $w^{1/3.4}$.

On the other hand, Fielder[6] and others have pointed out that there are alignments of craters such as those between Abulfeda and Almanon which are very probably volcanic, not only because we can hardly imagine a linear arrangement of bodies flying in from space, but especially since the alignment conforms to the general pattern of joints and faults in the lunar surface which is indicated in other ways.

The ring of hills about 100 kilometers in diameter, which includes the crater Flamsteed in its interior was photographed by the probe Orbiter I (See Fig. 4). The hills resemble flows of viscous magma, though some geologists think they are shaped by rock creep.

The similarity of the great majority of lunar craters to one another, however, suggests that one or the other of these processes predominates, and it is on this point that the greatest amount of dispute

FIG. 4. A portion of the Flamsteed Ring, showing light-colored rounded hills, possibly acid lavas. North at the left. The heavy horizontal lines are 1600 meters apart. Note the relative deficiency of craters on the hills. (NASA).

has taken place. The advocates of impact point to the remarkable similarity of the majority of craters, to their immense size as compared to terrestrial craters, and to the lack of obvious evidence of lava flows emerging from them. On the other hand, the adherents of a volcanic origin point to the fact that in many cases the central peaks themselves have small craters, to some evidence of shaping of the craters into polygonal forms as through faulting, and to the systems of concentric walls as evidence of a volcanic origin. Jeffreys and some others have pointed out that there is a well-known group of geologic formations which are called ring dikes, huge circular systems of faults through which magmas have been extruded to the surface. He compares these with the lunar craters.

It does not seem to be impossible that both parties are partly right. The presence of large asteroids such as Eros, in orbits which approach that of the earth, makes it reasonably certain that impacts of very large size occur both on the earth and on the moon with fair frequency. From experiment and theory it appears that the immediate result is to excavate a cup-shaped crater. On the other hand, the strength of the materials of the earth or of the moon is probably insufficient to support this cup shape for any length of time in a large crater, and a hydrostatic rise of the interior is nearly certain to take place. It seems possible, at least to me, that some of the concentric walls of the craters are due to the extrusion of magma along the circular faults produced in this way. At the same time, there may be a down-faulting of blocks belonging to the rim. (See figure 5) A crater may thus evolve from a pure impact feature to one in which subsequent volcanism is responsible for nearly everything which we see. It may thus be in a certain sense, both volcanic and impact in its origin.

Some further light has been cast on this problem by the photographs taken from the lunar probes Ranger VII and Ranger VIII. It is found that the sequence of crater size continues beyond the lower limit of visual resolution around 200 meters down at least as far as craters about 20 cm in diameter. The general shape is similar except for craters between about 20 and 200 meters, for which there is a strong tendency toward a very shallow, saucer-like form. The tendency disappears for the smaller craters. The reason

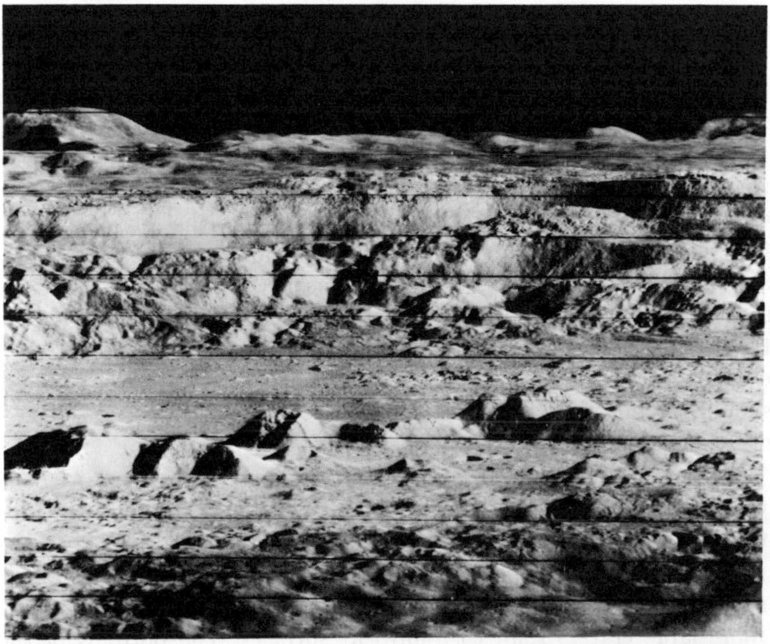

FIG. 5. View looking north across the crater Copernicus, 90 km in diameter by Orbiter II. Note the downfaulted blocks on the far side of the crater; also the central peaks. (NASA).

for this shape is controversial. Some authorities consider that the craters have been filled in by the debris from the erosion of the moon. The erosion is attributed to the impact of meteors and of smaller bodies. Another group of students consider it possible that the moon's surface is overlaid with beds of volcanic ash from 20 to 50 meters thick, which mantle and soften the craters in the 100 meter range. In this case the craters of smaller size evidently result from impacts subsequent to the deposition of the ash.

Maria

The dark patches which are visible to the naked eye on the surface of the moon turn out, in the telescope, to be extensive level plains. The early telescopic astronomers considered these plains as seas, and gave them names, such as Mare Imbrium (the Sea of Showers), Mare Serenitatis (the Sea of Serenity), and the like; in the plural, they are called the *maria*.

The surfaces of the maria are not only darker and more level than the non-mare regions (the uplands); they also have an entirely different distribution of craters. As seen through the telescope (i.e., for craters ½ km in size or larger) the uplands have about 10 times as many craters per unit area as the maria. On the other hand, as seen by moon probes, such as Ranger IX or the Orbiters (See figure 6) the maria are like the previously-mentioned floors of the craters, and contain between 10 and 100 times as many craters. The possibility should be kept in mind that the majority of these smaller craters are of volcanic origin (fumaroles).

Around the shores of most of the maria there are craters in about the same abundance as elsewhere; and many of these continue into the mare itself. The shore craters are usually tilted, so that they are partly submerged on the mare side. The abundance of shore craters is such that it is quite certain that the bottoms of the maria had many more large craters than their present surfaces. The bottoms of the maria may be of approximately the same age as the uplands. Mare Imbrium appears to be an exception to these statements; it has few evidences of submerged craters, and its shores have been greatly reshaped.

The average level of the mare surfaces is lower than that of the surrounding upland in every case; but, unlike the terrestrial seas, the levels of the maria are not the same either within a single mare, or from one mare to another. Numerical data are hard to give, and not particularly significant, since of the three coordinates of the moon's center of gravity, only one, the north-south coordinate, can be determined, and this is almost a kilometer different from the N–S coordinate of the center of the visible face; but the general situation is unmistakable.

Most writers have regarded the maria as huge flows of basaltic lava, like the flows of the Deccan, in southern India, or the Columbia Basin flows. In favor of this idea are their relatively dark color and their smoothness, the latter suggesting a fluid lava rather than a viscous melt such as an obsidian. Kuiper[8] regards the lavas as ex-

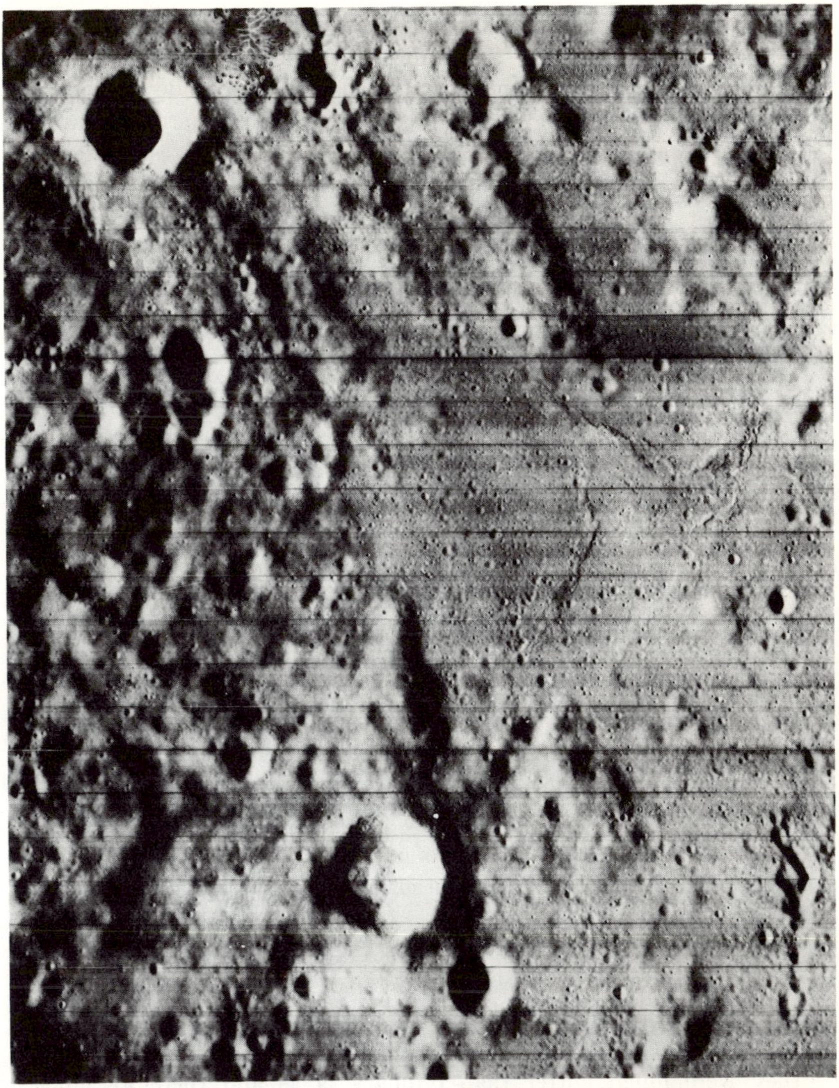

FIG. 6. A portion of the lunar surface west of the crater Gambart, at 17° 15′ W longitude. 0° 45′ N latitude. North at the top. Note the difference in crater distribution between highland and mare ground: large craters in the highlands and small craters in the lowlands. The heavy lines are at intervals of 1.6 kilometers.

truded through cracks in the crust, while Urey feels that they were formed at the surface by the heating action of an impacting body.

Gold[9] has emphasized the important observation made by Shaler[10] that the maria, unlike terrestrial lava flows, do not normally end in *scarps*, (steep slopes) 5 to 10 meters in height. The terrestrial scarps result from the fact that even a molten lava is normally not completely fluid; it carries on its surface a load of solidified stone. At the lava front, the stone makes a kind of retaining wall; and the lava flow makes progress by repeatedly bursting through the wall with streams of liquid rock which then congeal. When the lava flow finally stops, the last of these walls remains in place as a scarp. On the moon, such scarps are not seen, although at low sun angles, of the order of 1 degree, a 10-meter scarp would cast a shadow some 600 meters in breadth, and hundreds of kilometers in length, even if the wall is no higher than on the earth. The shadow would be conspicuous; it has been repeatedly sought in vain, by Shaler, Gold, and by the writer. It is fair to assume that the wall is really missing; and that its absence indicates a degree of fluidity in the material filling the maria which exceeds that of terrestrial lavas. It is particularly surprising because the hydrostatic pressures which cause the bursting of the scarps on earth should be only one-sixth as great on the moon for the same wall height.

Shaler also noted that the remarkably gentle slopes of the maria and the absence of any swellings which would indicate the location of the volcanic vents were likewise indications of an extremely fluid material. The maria are so nearly level that occasional dome-like structures with maximum slopes of only one or two degrees are clearly distinguished from the rest of the surface. The average slope of Mare Crisium is 1 or 2 km in a distance of several hundred kilometers. Even the basaltic lavas are not as fluid as this.

To explain this extraordinary fluidity, Shaler and some others have postulated that the lavas were produced by gigantic impacts which also scooped out the mare basins. According to this theory, they were much hotter than ordinary lavas and therefore much more fluid. Against this idea is the fact that as noted above the mare bottoms come from a considerably earlier time in the moon's history than the surfaces of the maria. The coastal craters, which clearly precede the mare material itself, should have been scraped away by the impact which produced the mare flow if the two events were nearly simultaneous.

Gold contends that in order to meet this problem we must suppose that the maria are filled in, not with basaltic lava, but with fluidized dust, produced by erosion, especially in the upland areas, and conveyed to the mare regions by the successive hops which result from the electrostatic charging of the dust by cosmic rays of the solar flux. This theory explains the absence of scarps around the maria. On the other hand, it suffers from the very serious difficulty that it seems to require too much erosion. It is possible to form an idea of the erosion of bodies in space from the isotopes which are produced by cosmic rays in meteorites. The cosmic rays do not penetrate more than a meter or so, and if we find that the level of the cosmic-ray-produced isotopes in a stony meteorite corresponds to an exposure of one hundred million years, as sometimes happens, then it is clear the erosion in space cannot be much greater than ten meters per billion years or thereabouts. At this rate only 45 meters would have been taken off the uplands and, allowing for the fact that the maria occupy one-third of the total area of the visible face of the moon, we could not account for an average depth of the maria greater than about 150 meters. The total depth of the maria, however, must exceed several km since they have drowned the great majority of the craters which once underlay them. It is only possible to rescue the theory by supposing a very vigorous erosion by micrometeorites during an early stage of the moon.

Russian photographs of the back side of the moon cast significant light on the problem. A number of large basins are found to which they have given the name thalassoids. The thalassoids appear to be mare basins without the fill of dark mare material, which in fact is much scarcer on the far side of the moon. The thalassoids have about as many of the larger craters as the uplands. It is difficult to reconcile these facts with the origin of the mare material by erosion.

Another explanation that has been offered for the maria is that they are *ash flow tuffs*, i.e., masses of volcanic ash which have been emplaced by flow under the process of *fluidization*.[10a] This process results from the action of gases generated within the hot ash; the gases cushion the ash particles from each other, and thus destroy the internal friction so that the whole mass becomes a sort of mobile liquid. In terrestrial flows of this kind the ash has moved with speeds of one or two miles per minute. Terrestrial ash flows are known to cover hundreds of thousands of square km. Since they are in effect fluidized dust, they conform to the requirements which Gold has pointed out.

It is a well-known property of ash flow tuffs that as the porosity is gradually squeezed out, the tuff surface comes to imitate, on a reduced scale, the surface of the basin in which it lies because the porosity was greatest in the deepest parts and hence the sinking can be greatest here. This "exhumation of the topography" may be responsible for the curious softened appearance of many of the craters seen on the Ranger photographs as well as on photographs from Orbiter I and II. A further morphological argument in favor of the ash flow hypothesis has been made by W. Cameron,[11] who pointed out that the *sinuous rilles* of the moon's surface, which resemble river beds, may be channels through which a portion of the ash flowed from the initial craters to the maria.

Bright rays

A third remarkable feature of the moon's surface is the bright streaks which radiate from some of the craters in all directions across the moon's surface. These *bright rays* (See Fig. 1) come from the craters which are the most recent. That is to say, if a ray crater and a non-ray crater overlap, it is always found that the ray crater bites into the side of the non-ray crater rather than the reverse. It has long been noted that the rays appear most conspicuous, not when the sun is vertically above the ray, but at full moon, i.e., when the sun and the earth are in line as seen from the surface of the moon. Conceivably this might be the result of dispersing small glass spheres over the surface of the moon in somewhat the same manner as on a beaded glass screen, but the majority of students disagree. It is difficult to disentangle the problem photometrically, and the most accurate observations do not show a marked difference in photometric properties between the rays and the other portions of the moon's surface.

Other Features

In many portions of the surface of the moon there are long narrow depressions of the surface called *rilles* (See Fig. 7). The arrangement of the rilles over the moon's surface has been described by Fielder,[12] under the name of the *lunar tectonic grid*. The rilles are clearly related

FIG. 7. Mare Humorum (lower half) and Gassendi. North at top. Note fault at lower left, rilles at center left. Courtesy of Lick Observatory.

in some way to the maria, since for example there is a system of rilles radial to Mare Imbrium and another which follows the boundaries of Mare Humorum. The fault which bounds Mare Humorum on its N.W. side (Figure 7) appears to belong to the same system as the rilles, and to suggest that the rilles are related to faults. In some cases the bottoms of the rilles are visible. They are typically either level or, occasionally, slightly convex. Some of the wide rilles such as Ariadaeus present the appearance of *grabens*, i.e., narrow trenches bounded by faults. In Posidonius or Taruntius (Figs. 2 & 3) the small internal rilles concentric with the walls appear to pass over into ridges. In at least one case a rille, namely the Hyginus rille, has a group of craters along it.

The *lunar tectonic grid* is more than the above-mentioned system of rilles; it is in fact a system of directions on the moon which is followed by most of the predominant features other than craters in each part of the moon. The grid consists principally of two systems, the A system (Fielder's notation) running NE-SW, and the B system, running NW-SE; there also are lines radial to Mare Imbrium, and other minor systems. A conspicuous manifestation of the lunar tectonic grid is seen near the Haemus Mts. and the crater Julius Caesar, where the whole upland surface is made up of ridges which parallel the lunar tectonic grid. It seems at least reasonable to suppose that the lunar tectonic grid represents a system of faults on the moon's surface. The faults have, in some cases, produced narrow grabens; in other cases they have localized small systems of volcanoes; and in yet other cases they have localized ridge-like extrusions of viscous lava (as was suggested by N. S. Shaler).

A similar kind of extrusion may perhaps be responsible for the fields of small knob-like bodies which are so common around the shores of Mare Imbrium. These are often, but, I think, naively, considered to be the debris of the impact which produced the mare itself. They are more probably extrusions of viscous and probably acid lava.

In some places the surfaces of the maria are swollen by broad structures 5 to 10 km in diameter which are called *domes*. Although they are circular in plan, they are very far from being hemispherical in elevation; in fact, the domes have angles of only a few degrees. The domes have been compared by some authorities, at least, with *laccoliths*, i.e., swellings of the surface of the earth produced by the intrusion of a viscous lava into a layered structure, and the mushrooming of the body there.

On the surface of the maria, there appear the so-called *wrinkle ridges* (Fig. 2) only a few hundred meters in height, and often in the shape of a festoon. Sometimes the crests of the wrinkle ridges are marked by narrower and steeper features which look like extrusions through a crack on the crest. The wrinkle ridges are usually considered to be the moon's nearest equivalent to the folded mountain ranges of the earth.

During the last 200 years observers have repeatedly noted orange spots on the moon, particularly around the crater Aristarchus. Although the observers included some of the most famous names in astronomy, it was customary to dismiss these observations as illusory until 1958, when Kozyrev obtained a spectrogram of a bright spot in Alphonsus. The spectrogram was interpreted as that of C_2. Since 1963, a large number of visual observations have been made which have placed the reality of the observations beyond doubt. A later spectrogram by Kozyrev indicates H_2 coming out of Aristarchus.

The explanations offered for these phenomena include:

a. Luminescence of rock under solar ultraviolet or particle emission. Since the emissions are often seen against the sunlit face, the energy source should be comparable with sunlight, which is hard to reconcile with the solar UV or particles.

b. Decomposition of acetylene.

c. Volcanic phenomena, such as ash flows, accompanied by lightning.

It is an interesting fact that the morphological explanations which seem natural for many of the moon's features are consistent with the idea that much of it is an acid lava. For example, ash flows are normally acid; the glass beads of which the rays might be composed would indicate acid rock; the steep-sided ridges, if extrusions, are best explained as acid lavas because these are more viscous than the basaltic lavas; for the same reason the laccoliths, like those on earth, are probably granitic, i.e., acid, in character and the H_2 observed by Kozyrev might conceivably come from water, if a source of high excitation exists. Students of the subject are by no means agreed on this point; for example, Kuiper regards the maria as basaltic flows, despite the absence of scarps; but it is interesting to see that explanations in terms of acid lavas are at least possible.

SURFACE PROPERTIES

The surface of the moon is one of the most curious substances known to man. Although the moon is usually considered to be white, it is actually black. Its surface acts toward light as if it were almost perfectly rough, while it acts toward low-frequency radar as though it were almost perfectly smooth. The surface temperature as judged by infrared radiation ranges through hundreds of degrees, whereas the temperature judged by radio waves is almost constant. This bundle of paradoxes can be explained in a reasonably consistent way only by the assumption that the moon's surface is porous as we shall show at the end of this chapter.

Optical Properties

The reflective power of the moon's surface is extremely low, as may be directly verified by holding up a piece of paper or a stone so that we see it alongside the moon at a time when the sun is also above the horizon. Then we see the stone by the light which has come through the atmosphere and been reflected from the stone; we see the moon by the light which has been reflected from the moon and then comes through the atmosphere, so that the two are reasonably comparable. It turns out that we get a fair equality with the brightness of the moon only when we use something as black as a domino. A piece of paper is much too white.

The color of the moon is approximately that of ordinary sunlight; but its surface absorbs somewhat more light toward the blue end of the spectrum, so that it is slightly brown. There is a tendency for this brownness to be correlated with the general darkness of the surface so that a plot of absorption vs. wave length is very nearly linear. Determined attempts have been made to find deviations from these linear relationships, but the deviations are small and uncertain.

If we follow the reflected light of the sun into the infrared we find that it can be detected up to 3 or 4 microns. Up to this point the moon's spectrum appears in detail to be just the reflected solar spectrum. Some observers have claimed that the dark lines, the so-called *Fraunhofer lines* of the spectrum, are not as black in the lunar spectrum as in the solar spectrum. If true, this might imply some source of light on the moon, such as fluorescence. At present the observations need further confirmation.

It is an extraordinary fact that the full moon appears to be of almost uniform brightness across its disk. Any ordinary spherical object held up in such a way that the eye sees only the illuminated hemisphere will show a pronounced dark edge where the sunlight falls obliquely and the surface is obliquely illuminated. No trace of this effect is visible on the full moon; the edge is, if anything, slightly brighter than the center. On the other hand, when we observe the moon at first quarter, the illuminated half disk again appears to be nearly uniformly bright except close to the dark half, but the brightness is one-ninth and not one-half of the brightness of full moon. The difference is conspicuous; the full moon casts conspicuous shadows of objects in the earth, while at first quarter the shadows are nearly invisible. Yet at first quarter we can see portions of the moon which from the sun's direction must look just as bright as the full moon looks to us. It is clear from these examples and from some very detailed photometric studies that have been made, that the moon's surface as a whole has somewhat the same properties as those which in previously described for the bright rays, though not quite so strongly. The moon's surface is strongly *retrodirective*, like some types of highway markers which return the light preferentially in the direction from which it comes. It is fairly certain that this property is not the result of any distribution of glass spheres, however. It is probably related to a phenomenon seen from an airplane which passes over a forest. Around the shadow of the airplane, as seen from the air, the forest appears to be much brighter. This is because the observer who looks at the shadow is looking away from the sun. Consequently he does not see the shadows cast by the leaves. However, as soon as he turns his eye in any direction away from the shadow of the airplane, he is looking into the shadows cast by the leaves on the ground. For this reason the forest appears to be dark everywhere except just around the shadow of the plane. This property of strongly retrodirecting the light is one which is possessed by rough materials in proportion to their roughness. Ordinary roughness, like a pile of sand or even like a piece of pumice, is not sufficient. Hapke[13] has further shown that we must assume that the material is dark; otherwise the internal reflections in the material will make the shadows insufficiently black.

When we pass toward the infrared beyond about 5 microns the energy radiated by the moon exceeds the energy reflected. In this portion of the spectrum the moon behaves, so far as present studies go, like a black body. No structure has been found in the spectrum except that produced by atmospheric absorption and radiation, principally by water vapor and ozone. This region extends from about 5 microns to about 25 microns. The intensity of the radiation from the moon varies through the month to a remarkable degree. As gauged by the heat emitted by a point at the center of the disk, the moon goes from +100°C at full moon to −150°C at new moon. This large range of temperature is undoubtedly connected with the absence of an atmosphere. What is even more extraordinary is the sudden drop[14] during a lunar eclipse from +100°C to −80°C within 1 hour. The cause of this very rapid response of the moon's surface to changing solar illumination is undoubtedly the very low amount of heat stored in the regions to which the heat penetrates. The stored heat is clearly related to: (a) the heat capacity c of the surface material; (b) the density ρ of the material since the heat capacity is measured in calories per gram; and (c) the thermal conductivity k, since this regulates the depth to which the heat penetrates. From observation of eclipses, the quantity $(k\rho c)^{-1/2}$ can be determined. It is found to be about 500 to 1,000 c.g.s. units, compared to 20–50 for ordinary rocks. Recent detailed studies of the moon's thermal behavior show that around the ray craters, the thermal properties are notably different; the surface heats and cools more slowly, as if some bare rock were exposed. Orbiter II photographs confirm the existence of fields of boulders around some craters. See also Fig. 10.

Radio

In radio frequencies the moon's brightness is about one-twentieth that of the sun. As a consequence of the Rayleigh–Jeans law, which says that for very low frequencies the intensity of radiation is proportional to the temperature, it is possible to make reasonably accurate measurements of the temperature of the moon by using radio frequencies. The result which emerges is that the temperature is constant within 5° at a wavelength of 10 centimeters. These radio waves are known to come from less than a meter below the surface, and they thus reconfirm the conclusion which was obtained from the thermal properties that the moon's surface is an extraordinarily efficient insulator.

As we go from these longer radio waves to shorter waves, the fluctuations of the moon's temperature rise somewhat, and at the very shortest usable radio waves, around 1.5 mm, a range of temperature over 200° is found, thus approaching the situation which is seen for the heat waves.

Radar

The fourth tool for the study of the surface is radar. When the first radar pulses were sent to the surface, the receivers were tuned for an echo which was expected to last for several thousandths of a second, as the returns came in, first from the portion of the moon nearest the earth and then from those portions which were further away. The very weak total returns which were then obtained discouraged experiments, as indicating that the moon was a poor reflector. Some years later, however, it was discovered that the moon gives a sharp echo because the reflection from the part of the moon nearest the earth is stronger than of that from the rest of the moon. At the longer wavelengths this reflection predominates over all others as if the moon were a polished surface which reflected only a highlight. The contrast between this behavior and behavior at optical frequencies is extraordinary. As radar techniques have been improved, it has been possible to mitigate the strangeness of this contrast. For the very shortest radar waves the moon's surface begins to show a more nearly uniform reflectivity.[15] More recently, Pettengill and his associates have shown that there are peaks in the reflecting power around some of the ray craters such as Tycho and Copernicus apparently due to the above-mentioned fields of boulders. The radar reflectivity is found to be nearly independent of frequency which indicates that the conductivity of the surface is low. Some observers are inclined to think that this excludes materials such as chondritic meteorites on the moon's surface.[16] It is also found that the general reflectivity is low, which indicates a low value of the dielectric constant, around 1.8 according to Hagfors.[17] It suggests a siliceous material in a finely divided state.

Polarization

The question of polarization of light by the moon's surface has been studied. It is found to be less than that for any normal rock surface. This constitutes a clear indication that there are no bare surfaces of rock exposed on the moon's surface. The polarization is so extraordinarily small that it appears that we cannot even account for it on the assumption of coarse powders. Using the so-called *ashen light*—that is, the reflection of earth light on the moon—Lyot and his successor, Dollfus[18] have studied the ability of the moon's surface to depolarize light at visible frequencies. Dollfus concluded from this study that the moon's surface should be identified as a powdery opaque substance. In radio frequencies it is found that the radar highlight at the center of the disk moon's surface does not depolarize a transmitted radar beam, which indicates that the surface does not possess roughnesses on the order of the wave lengths of the radar, or about 10 centimeters. It is further found that the (much weaker) reflections from points near the edge of the moon's disk are depolarized, which indicates that they are due to individual facets of the surface which happen to face the earth.

Interpretation

We must now attempt to synthesize from these data a picture of the moon's surface. From the properties of the radar reflections it is clear that the majority of the moon's surface must be smooth when we are thinking of things a few cm in height. From the transition to poorer reflectivity, it is clear that roughnesses begin to set in at about 1 cm. These roughnesses increase so that by the time we have reached about 20 microns the moon's surface has become wildly irregular. From the Surveyor images (See Figures 8 and 9) we see that the mare surfaces, apart from rocks which cover about 1/10 of the surface, consist of some finely-divided material, with a grain size less than 500 microns, and a consistency like loam. The surface

FIG. 8. Footpad No. 2 of the Surveyor 1 spacecraft. Note the impression made in the lunar soil. The pad is 30 cm in diameter, (NASA).

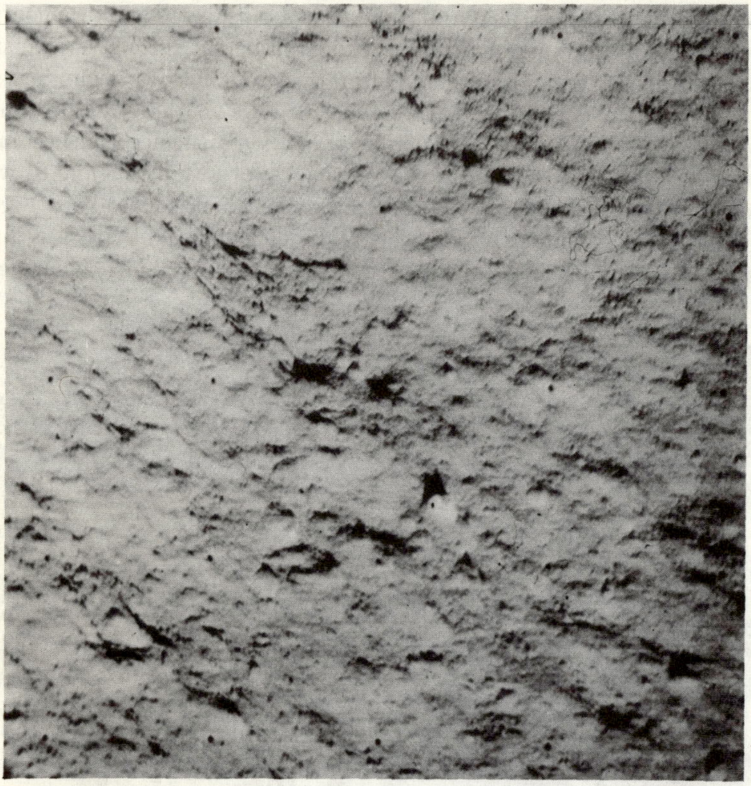

FIG. 9. Undisturbed lunar soil near the Surveyor I spacecraft. Note the system of lineation. The width of the picture is approximately 15 centimeters (NASA).

is pitted by roughly-circular craters down to about 2 cm across, below that size the irregularities, though very abundant, are no longer circular. The low thermal conductivity of the moon's surface is adequately explained by such a material. The low dielectric constant implies either a siliceous material and a porosity on the order of 60%, or a more basic rock, and a higher porosity (up to 80% for chondritic material).

On the Surveyor photography, Shoemaker has found that craters less than 1 meter in depth are not surrounded by rocks while those over 1 meter in depth are. (Fig. 10)

GEOPHYSICAL PROBLEMS

The question of the response of the surface of the moon to imposed loads is one of the most important ones for a study of its physics. From the dynamics of the Surveyor spacecraft, it has been found that over areas of the order of a square meter, the surface will support about 300,000 dynes per cm^2. This bearing capacity is like that of soft soil.

Our next datum relates to much larger areas and loads, namely those which are produced when a crater is formed. No matter what we think about the relative numbers of impact craters vs. volcanoes among the large lunar craters, we must recognize that the existence of moderately large asteroids such as Eros in the vicinity of the

FIG. 10. The near rim of a crater several hundred meters in diameter. The far rim is visible against the horizon. Surveyor I image. (NASA).

earth implies the production of very large impact craters from time to time. The mechanism of formation by impact indicates that the crater will originally have a hemispherical shape since this is what we expect from the hydrodynamical analogy and since the analogy becomes increasingly valid as we go toward larger craters. Their present shape is therefore the result of some kind of readjustment. One of the forces which tends to readjust the shapes of craters is the slumping of the crater walls. The other is the hydrostatic rise of the bottom of the crater under the forces produced by the difference of load on the moon's interior. No careful studies have yet been made of this point, and we cannot say more at this time than that it appears possible that the hydrostatic rise of the bottom may have played an important part not only in producing the flat bottoms of the craters, but also perhaps in lowering the height of the rim and in producing ring faults around the inside of the crater just below the rim. Note that several of the large Canadian impact craters show concentric rings of faults.

On a larger scale many students, especially Alter and Baldwin, have detected evidence of isostatic adjustment in the relation of the maria to the uplands. As a whole the maria are lower than the uplands. If they had been scooped out mechanically as the result of a great impact on a homogeneous moon, then hydrostatic forces would tend to bring the bottoms of the maria up to a level with the surrounding region. This tendency would persist even at the present time, since the maria are now lower than the rest of the moon. If, however, the maria are like the terrestrial oceans, which are underlain by denser rocks than the continents, then we might expect that the mare basin was older than the infilling material and might expect that the effect of the inflow would be to cause the level to fall. We find, in fact, that the maria have been sinking with respect to the surrounding countryside; this is particularly evident in the case of Mare Humorum (See Fig. 7) which is surrounded by a concentric pattern of rilles and which has on its N.W. side a scarp which is clearly the result of faulting, the downthrow side being toward the mare. In addition to the fault scarps, there is evidence of the sinking of Mare Humorum from the inclination of the craters around the border, inward toward the mare. Similar indications are found around other maria, especially Mare Crisium and Mare Nectaris.

We note that if we are to interpret these adjustments as isostatic in nature, then we are probably compelled to suppose that the moon's surface is differentiated; that is to say that the material which underlies the uplands is a lighter rock than that which underlies the maria. If that were not true, then the maria would tend to continue to rise until the difference in level between the two was abolished. To explain how isostatic adjustment can take place we are further led to suppose that, as on the earth, there exists a plastic layer at some depth in the moon in which the flow occurs. It may be going too far to suppose that the rings of raised terrain which are seen around a few of the maria, especially Mare Crisium, are caused by the outflow of material in this plastic layer as a result of the loading of the mare surface.

At a deeper level it appears that the material of the moon possesses approximately the same ability to sustain loads as the lower mantle of the earth. It is known that the second harmonics of the moon's gravitational field do not have their equilibrium value. They are mathematically related to the moments of inertia; and the values to be expected for a fluid body can be calculated in a remarkably definite way, provided the moon's internal stratification is reasonable.

The data on the actual moments of inertia comes from perturbations of the moon's rotation. The difference between the polar moment of inertia and the mean of the two equatorial moments can be estimated, as with the earth, from the precessional motion of the moon's axis. The difference between the two equatorial moments can be measured because the axis of least moment of inertia points, on the average, toward the earth. As it swings back and forth across the earth–moon line of centers for purely kinematical reasons (the *optical librations*) restoring forces are called into play which produce the barely detectable *physical librations*.

We find that the actual values are not what we would expect for a fluid mass subjected to the centrifugal force of the moon's rotation and the tide-raising action of the earth. It is true that the polar axis of the ellipsoid which best fits the moon's gravitational equipotential surface is the shortest of the three axes, and that the one which points to the earth is the longest. This relation is the one which would be expected for a fluid under the action of gravitation, but it is also the relation to be expected for an irregular solid having enough internal friction so that the tides can damp out oscillations and bring it to an equilibrium position. The magnitudes of the distortions are several times those which could be expected for a fluid, and they clearly indicate the presence of stress differences in the interior of the moon. If the moon is solid to its center, then, as Jeffreys[7] pointed out, the implied stress differences are on the order of 20 bars. Much the same stresses are indicated in the interior of the earth by the non-equilibrium values of the spherical harmonics of the earth's gravitational field discovered from satellite motions.

In the case of the earth, we can follow Jeffreys in supposing that these stresses are supported by ordinary mechanical strength. The calculated melting temperature in the interior of the earth increases faster with depth than the actual temperature, because of the increase in pressure. In the case of the moon, we need one-fourth as fast a rise in temperature to carry off the heat generated if the moon has the same abundance of radioactive elements; but the pressure near the center of the moon is an order of magnitude less than that at the same depth (1700 km) in the earth. It is therefore more difficult to explain how the moon's interior can have enough strength to resist the calculated stresses.

Two significant features of the earth's surface appear to be missing on the surface of the moon. One of these is *wrench faulting*, in which there is horizontal movement between the two sides of the fault. Such faults are conspicuous and of large extent on the earth, and their absence on the moon indicates some fundamental difference in the movements. The second is the absence of folded mountain ranges. The nearest representatives of these are, as has been mentioned before, the small wrinkle ridges of the mare surface.

CHEMISTRY OF THE MOON

The most fundamental fact about the moon's chemistry is the fact that the problem is within our present grasp. It has been shown by Elsmore[19] and later workers that the moon not only lacks an atmosphere in the ordinary sense, but it even lacks the very tenuous gases which make up an ionosphere. The very small number of atoms which circulate around the moon at great heights are totally unable to arrest an incoming particle, however small, or an escaping secondary fragment. It follows that material thrown from impact craters, large and small, must be constantly escaping. The escape velocity of the moon is only 2.3 km/sec, while a typical asteroid moves at 10–15 km/sec with respect to the earth. A typical impact at these speeds would be expected to expel a large amount of ejecta from the moon with velocities greater than the escape velocity. This expectation is strongly supported by impact studies which have been made by Gault, Shoemaker, and others. It appears that in a typical meteorite impact the quantity of matter escaping the moon is several times the quantity which is brought in. Calculations by Öpik lead us to believe that a major amount of this material will later be swept up. Hence the earth's surface contains a detectable portion of matter from the surface of the moon. The problem is really how to recognize and put our hands on it.

Several theories have been put forward on this subject. Urey[20] draws attention to the fact that *chondritic meteorites* have conspicuously less age-in-space as judged by the cosmogenic isotopes which they have accumulated than iron meterorites. Some of them have ages less than a million years. He suggests that chondritic meteorites originate by impact on the surface of the moon, especially by the impact of iron meteorites. This idea suffers from the difficulty that there exists at least one accurately observed orbit for a chondrite, which entered the atmosphere with the kind of velocity to be expected in a body which had originated in the asteroid belt. This orbit is not easily reconciled with a lunar origin. Moreover, from the fact that meteorites are more often observed to fall in the afternoon than in the morning, there is a suggestion that they travel around the sun in the same direction as the earth's orbit, but with somewhat greater speed, so that they overtake the earth. Such orbits would correspond to bodies coming from the asteroid belt. In addition, the *petrographic marks of strong shock* are rarely observed in chondritic meteorites, as would be expected if they were impactites. The possibility that some of the *achondrites* come from the moon is a promising one which is just beginning to be explored at this

FIG. 11. A tektite of the Muong Nong type (type A) from Thailand. Length 50 centimeters.

time. It is especially hopeful for the basaltic achondrites, which do, in fact, show the marks of shock.

The other suggestion relates to *tektites,* which are centimeter-size glassy black or green pebbles, found in certain well-defined areas of the earth's surface called strewn fields. There are three general types of tektites: Type A, sometimes called the Muong Nong type, which are chunky in outline (See Fig. 11) and usually have an internal structure consisting of parallel layers, often marked by bubbles; type B, sometimes called splash-form tektites, which are usually roundish in outline (spheres, ellipsoids, teardrops, dumbbells, etc.); and type C, the australite forms, which apparently started from shapes like those of type B, especially the spheres, but were modified by ablation on one side.

It appears from the data available up to 1963[21] that tektites originated by meteorite impact either on the earth or the moon. Since that date, L. S. Walter[22] has discovered coesite in some tektites of type A. This is especially significant because at temperatures of the order of 1700°, coesite changes to cristobalite in a few seconds at ordinary pressures. It is hard to imagine a pressure-temperature regime which would permit the formation of glass from typical terrestrial materials, yet retain the coesite. Moreover, calculations of the rates of diffusion of water suggest that the production of a glass as homogeneous and free of vesicles as some tektite glass, even of type A, is difficult to explain in the brief time of a terrestrial impact. Hence tektites, especially those of type A, should also be considered as possible samples of the lunar surface. Fig. 12 shows a lunar rock as seen from Surveyor I; the evidence of layering in it (the nearly-vertical, slaty cleavage) is to be compared with Fig. 11.

Type A tektites (which appear to be the primitive form) have apparently been produced by the welding together of small bits of material. When a terrestrial ash flow tuff collapses, the separate shards become welded together to form what is known as a *welded tuff*. If a similar lunar process is the origin of the type A tektite material, then it appears that the individual bits of glass were round in shape—spherical or ellipsoidal—instead of spiny, like the *shards* of terrestrial volcanic ash.

The chemical comparison between tektites, chondritic meteorites and the earth's crustal material is shown in Fig. 13. The distance between each point plotted in Fig. 7 and the central line gives the logarithm of the ratio of the abundance of some element in tektites or meteorites to its abundance in the crust. We see at once that the black triangles which represent the tektites cluster much more closely around the central line which represents the earth's crust than do

FIG. 12 Rock, 0.5 meters long, observed southwest of Surveyor I spacecraft. Note vertical slate-like layering.

the crosses which represent the meteorites. The few cases in which the triangles are at a considerable distance from the central line can usually be explained as the effect of differential volatilization, as in the case of phosphorus, germanium, zinc, copper, and lead.

Urey and others reject the whole idea that tektites come from the moon on two grounds. First, tektites are found only in certain regions of the earth. If they result from an earth impact, the regionality is to be expected; while if they are of lunar origin, the distribution over a region can only be explained by the breakup of a temporary satellite

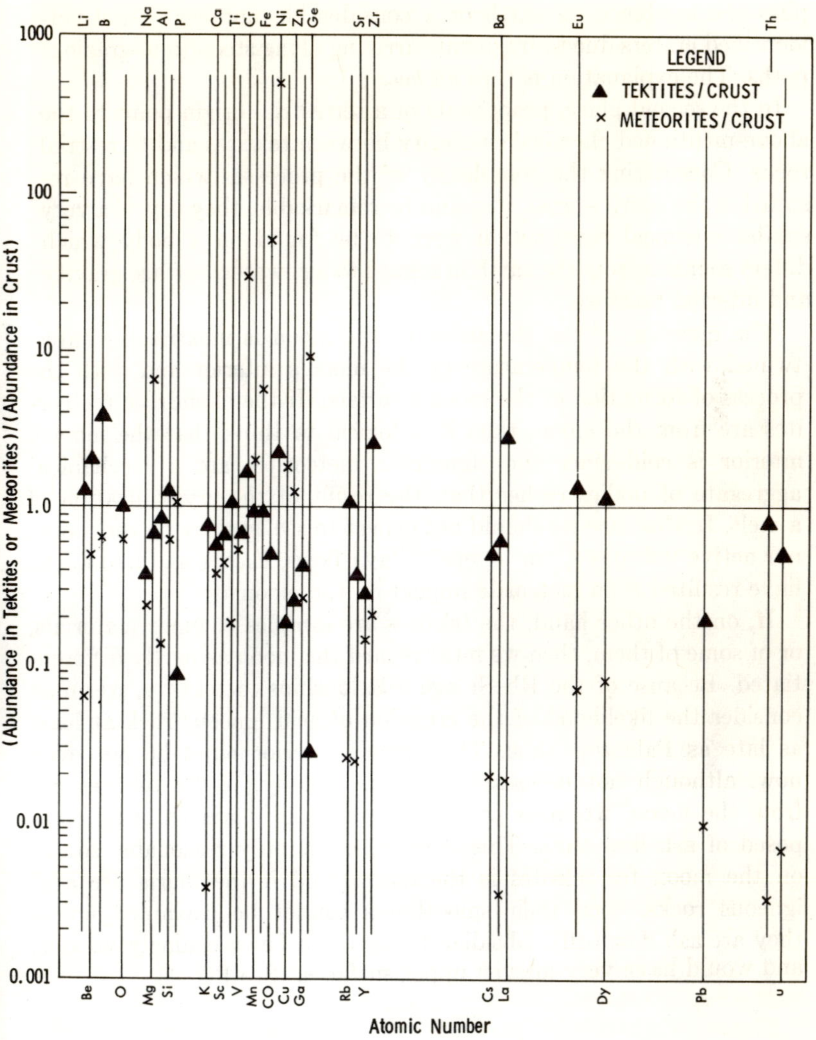

FIG. 13. Abundances of the elements in tektites compared with their abundances

in an elliptical orbit which grazes the atmosphere. This clearly implies the existence on earth of a considerable number of not-yet-identified craters due to fragments arriving along steep (non-grazing) paths. The explanation is thus *ad hoc*.

In the second place, proponents of a terrestrial origin point to the above-mentioned chemical similarity between tektites and terrestrial rocks. Considering the complexity of the processes which have resulted in the earth's crust, it would be remarkable, they say, if a very similar chemical constitution were to be found on a body which differs so much from the earth in regard to air, water, surface gravity and internal pressure.

The question of the chemistry of the moon is inextricably intertwined with the temperature of the moon's interior and with the process of formation of the moon's surface. If the chondritic meteorites are from the moon, then it is logical to think that the moon's interior is cold since the chondritic meteorites are a mechanical aggregate of bodies rather than the result of the crystallization of a melt. In this case we should not expect to see isostasy on the moon, nor active volcanoes, nor extensive lava flows, except as these might have resulted from meteorite impact on the moon.

If, on the other hand, the tektites are samples of lunar materials, or of some of them, then we must regard the moon as partly differentiated. Because of the Rb–Sr age relationships in tektites, we must consider the likelihood of the eruption of acid igneous rock at least as late as Paleozoic time. The moon's interior must be hot even now, although not necessarily molten. Finally, if the tektites are from the moon, we must probably regard the maria as being composed of ash flow tuffs. This is because the only reasonable source on the moon for tektites is the maria, and if the maria are acid igneous rocks, then their smoothness cannot be explained unless they are ash flow tuffs. Obsidian flows would be exceedingly viscous, and would have very uneven upper surfaces. (See Fig. 4)

ORIGIN OF THE MOON

The effects of *tidal friction* in the earth–moon system have changed it radically during the past few billion years. Tides raised in the seas and the solid earth can be thought of as a wave, which goes around the earth with its crest vertically under the moon. Owing to the earth's rotation in the same direction, the wave is dragged forward a few degrees with respect to the moon. This results in a couple which at the same time retards the earth's rotation and feeds energy into the moon's orbit. From a comparison of annual and diurnal growth rings in corals, J. W. Wells has deduced that there were 390 days per year in the Devonian, some 400 million years ago. Scrutton has found evidence of a monthly cycle in corals, from which he has deduced corresponding changes in the orbit of the moon.

Allowing for tidal evolution, three general theories have been put forward for the genesis of the moon. They are:

(a) Capture
(b) Formation alongside the earth
(c) Fission of the earth.

The first of these, the capture theory, is difficult from the point of view of celestial mechanics. Any ordinary body of the moon's size which passes close by the earth ought to escape again. It is only under very special circumstances that a close encounter would lead to capture. If the two bodies come so very close that a substantial portion of the orbital energy of the moon is dissipated as the result of huge tides raised in the earth and the moon during the encounter, then the moon's velocity may be so reduced that capture can take place. In this case the moon will begin with a highly elliptical orbit. If the orbit is direct, that is, if the moon moves around the earth in the same direction as the one usual in the solar system, which is counterclockwise, as seen from the north, then the subsequent evolution of the moon will be extremely difficult to understand. The action of the earth will be to accelerate the moon at each passage near the earth. This will enlarge the orbit, but it will also make the orbit more eccentric so that the present nearly circular orbit is very difficult to understand.

In 1956, Gerstenkorn pointed out that capture in a retrograde orbit was more likely than capture in a direct orbit. In this case the effect of the subsequent interaction with the earth is to cause the inclination of the orbit to become nearly polar, then finally to turn over so that the orbit becomes direct and thereafter to enlarge. This mechanism is now being actively studied.

If the moon has been captured in any way, then we should expect that its chemistry would be quite different from that of the earth, especially since there is good reason to think that the earth's internal chemistry differs from that of its surface. The idea of capture is logically associated with Urey's suggestion that the surface of the moon is composed of essentially the same material as the chondritic meteorites. The density of the moon is approximately equal to the density of the lightest chondritic meteorites, about 3.4 against 3.8. The difference is significant, however, and a few percent of some lighter material is required in order to bring about agreement. Urey has suggested graphite or water.

The second hypothesis, that of formation alongside the earth, is very difficult to reconcile with the known difference in average density between the moon and the earth. If two bodies were really formed in the same way, they should have been formed from the same materials. A suggestion was made by Ramsey in 1948 that the core of the earth is really composed of the same material as the mantle. If this suggestion were correct, and if the core were really a product of the great pressures prevailing in the interior, then we could understand how the moon might have been formed alongside the earth. Unfortunately, F. Birch has shown[23] that this theory is difficult to reconcile both with the extrapolation of the velocity-density curve for earthquake waves and with calculations of the equation of state. The velocity-density curve is, according to Birch, nearly the same for a given chemical composition regardless of its physical state. The data for the core agree with what we expect for iron, but not at all with what would be expected for mantle material. A confirming argument is derived from calculations of Thomas-Fermi degeneracy. These calculations represent the extreme limit of the compressibility of mantle material; and it is found that the densities do not reach those of the core. Birch is of the opinion that if any theory can be disproved in geophysics, it is the theory that the core of the earth is a phase change of the mantle.

From the point of view which has been explained above, it will be clear that if we consider the third alternative, namely the formation of the moon by the fission of the earth, we must suppose the fission to have taken place after the formation of the core. If it takes place prior to the formation of the core, the two materials will be chemically approximately identical, and the difference of mean density will be inexplicable. The mantle of the earth, on the other hand, has approximately the density which we require for the moon. Some recent calculations have shown that it is barely possible that the earth may have divided as a result of the formation of the core. For example, if it had been rotating as a homogeneous body with a period of $2^h\ 39^m$, it would be on the verge of a certain kind of instability. If then the core had formed, by the sinking of iron to the center, the period of rotation would have shortened to $2^h\ 11^m$. But with a core, the limiting period of rotation, for the same kind of instability is $2^h\ 18^m$; hence the earth would have become unstable.

If the earth and the moon were combined in a single body with their *present* angular momentum the body would rotate in a period of 4 hr and would fail by a considerable factor to possess the angular momentum necessary for instability. Hence, if the moon originated by fission then we must suppose that the subsequent evolution of the earth-moon system was accompanied by some loss of angular momentum to the system as a whole. This is quite different from the tidal problem by which the earth and the moon interchange angular momentum. It is conceivable but unlikely that there exists some kind of magnetic interaction between the earth and the interplanetary field by which the earth-moon system can give up angular momentum.

A difficult feature of all theories about the origin of the moon is the fact that the calculation of the rate of change of the moon's distance under tidal interaction indicates that a crisis of some sort, either fission or capture, apparently occurred approximately 2 billion years ago. The geological record is explicit in informing us that nothing of this violence occurred at that time. G. J. F. MacDonald was thus led to postulate the formation of the moon by the mutual capture of a considerable number of satellites in orbit around the earth.

A decisive point in this situation was made by Öpik[24] who pointed out that the lunar craters are nearly circular. Had the moon been bombarded by impacts at a time when it was so close to the earth that it possessed a large tidal bulge, then the craters produced would have been nearly circular. After collapse of the tidal bulge, however, they would have become nearly elliptical, and the ellipticity would have varied systematically from place to place on the moon. Since the ellipticity is the same on nearly all parts of the moon, Öpik considers that the moon's face, as we see it, was formed when the moon was 30,000 km away or more. This holds equally for theories of capture and for theories of break-up by tidal fission. Whatever the earliest history of the moon may have been, it is not revealed by the morphology which we see on the surface.

Summing up, we may trace two main threads in the interpretations of lunar surface data. For one group of students the moon's surface is dead. The craters are the result of mechanical impact only, the maria are impact lavas, or perhaps erosional dust, and the moon has been captured. For the other school of thought, the moon's surface is much like that of the earth, being igneous rock. The craters have been extensively modified by igneous instrusions. The maria are ash or lava flows; and the moon is a body which somehow split off from the earth when the earth was young. We may hope that the difference between these two views may be cleared up by experiments directly on the moon during the next few years.

REFERENCES

1. E. W. Brown, An Introductory Treatise on the Lunar Theory, Cambridge University Press, 1896.
2. I. Fischer, Parallax of the Moon in Terms of a World Geodetic System, Astrophys. J. **67,** 373 (1962).
3. U.S. Nautical Almanac Office, American Ephemeris and Nautical Almanac, pp. 494, 499 (1965).
4. E. M. Shoemaker, Interpretation of Lunar Craters, Chap. 8, Physics and Astronomy of the Moon, Z. Kopal, Ed., Academic Press, New York, 1962.
5. R. Baldwin, The Measure of the Moon, University of Chicago Press, Chicago, 1963.
6. G. Fielder, Lunar Geology, Lutterworth Press, London, 1965.
7. H. Jeffreys, The Earth, Cambridge University Press, 1962.
8. G. P. Kuiper, The Surface of the Moon, Chap. 15, in Space Science, Donald P. Le Galley, Ed., p. 630, John Wiley and Sons, New York, 1963.

9. T. Gold, Processes on the Lunar Surface. Chap. 39, The Moon, Z. Kopal and Z. Mikhailov, Eds., Academic Press, London, 1962.
10. N. S. Shaler, A Comparison of the Features of the Earth and the Moon, Smithsonian Contributions to Knowledge, Vol. 34, pt. 1, 1903.
10a. J. A. O'Keefe and W. S. Cameron, Evidence from the Moon's Surface for the Production of Lunar Granites, Icarus, **1,** (3) 271 (1962).
11. W. S. Cameron, An Interpretation of Schröter's Valley and other Lunar Sinuous Rills, J. Geophys. Res. **69,** 2423 (1964).
12. G. Fielder, Structure of the Moon's Surface, Chap. 11, p. 161, Pergamon Press, London, 1961.
13. B. W. Hapke and H. Van Horn, Photometric Studies of Complex Surfaces with Applications to the Moon, J. Geophys. Res. **68,** 4545 (1963).
14. E. Pettit, Planetary Temperature Measurement, Chap. 10, in Planets and Satellites, Vol. III of The Solar System, G. P. Kuiper and B. M. Middlehurst, Eds., p. 400, Chicago University Press, Chicago, 1961.
15. G. H. Pettengill and J. R. Henry, Radar Measurements of the Lunar Surface, Chap. 50, The Moon, Z. Kopal and Z. Mikhailov, Eds., Academic Press, London, 1962.
16. V. S. Troitsky, Radio Emission of the Moon, Its Physical State and the Nature of Its Surface, Chap. 45, The Moon, ed. Z. Kopal and Z. Mikhailov, Eds., Academic Press, London, 1962.
17. T. Hagfors, Radar Evidence that the Moon is Covered by a Tenuous Surface Layer, Science, **153,** 1153-1155 (1965).
18. A. Dollfus, Polarization Studies of the Planets, Chap. 9, of Planets and Satellites, Vol. III of The Solar System, G. P. Kuiper and B. M. Middlehurst, Eds., p. 343, University of Chicago Press, Chicago, 1961.
19. B. Elsmore, Radio Observations of the Lunar Atmosphere, Phil. Mag. 8th Series **2,** 1040 (1957).
20. H. C. Urey, Origin and History of the Moon, Chap. 13, of Physics and Astronomy of the Moon, Z. Kopal, Ed., p. 481, Academic Press, New York, 1962.
21. J. A. O'Keefe, ed., Tektites, Univ. of Chicago Press, Chicago, Ill., 1963.
22. L. S. Walter, Coesite discovered in Tektites, Science **147,** 1029 (1965).
23. F. Birch, Composition of the Earth's Mantle, Geophysical J. **4,** 295 (1961).
24. E. J. Öpik, Tidal Deformations and the Origin of the Moon, Astronomical J. **66,** 60 (1961).

Chapter 18

PLANETARY ATMOSPHERES

R. Jastrow and S. I. Rasool

INTRODUCTION

The properties of the atmospheres of the planets depend on the abundances of the elements in the solar nebula out of which planetary bodies accumulated, and on the physical conditions which prevailed during the subsequent history of the planets. The study of these conditions permits us to place our detailed investigation of the terrestrial atmosphere, and our comparison of the differences among the planets, within the broader context of the evolving solar system.

The cosmic abundances of the elements are fairly well known. Representative data are shown in Figure 3 of Chapter 23 (Nucleosynthesis) in this volume. Four groups of elements have the highest abundances:

H, He | C, N, O | Mg, Al, Si | Fe.

Hydrogen constitutes roughly 90 percent of all matter of the universe, and helium approximately 10 percent. The remaining elements make up a few percent. The original compositions of the atmospheres of the planets must therefore have been dominated by hydrogen and helium plus smaller amounts of such compounds as

$$CH_4,\ NH_3,\ H_2O,\ N_2,\ O_2,\ CO_2,\ CO.$$

GRAVITATIONAL ESCAPE

During the history of a planet, the composition of its atmosphere is modified by the differences in the rates of escape of various gases from its gravitational field. The rate of escape from the gravitational field of a planet is readily estimated.[1–3] At any level in the atmosphere a number of atoms will be moving upward with speeds in excess of that required for escape from the gravitational attraction of the planet. These atoms are in the high-velocity tail of the Maxwell distribution. The minimum velocity for escape, v_e, is that for which the total energy of the particle is zero:

$$E = T + V = \tfrac{1}{2}mv_e^2 - (mMG/R) = 0,$$

or

$$v_e = (2MG/R)^{1/2}$$

For the earth and other planets

$$v_e = 11.3 \text{ km/sec} \times (M/R)^{1/2}$$

in units of the earth's mass and radius. For the moon, Mercury, Mars, and Jupiter the values of v_e are 2.3, 4.3, 5.0, and 61.0 km/sec, respectively.

The escape of gases can take place effectively only at high altitudes where the density is low and the probability of collisions is small. These requirements define the region of the atmosphere known as the *exosphere*, a region in which the probability for collision of a particle moving in a radially outward direction is substantially less than unity. Atmospheric particles moving in the exosphere execute ballistic trajectories in the gravitational field of the planet.

Density Distribution in the Exosphere

In the lower atmosphere the density distribution is determined by the law of hydrostatic balance. Let ρ, P, and T be the density, pressure, and temperature at a given level, and m the average mass per particle. Then the gas law is

$$P = (\rho/m)kT, \tag{1}$$

and on a small layer of cross section 1 cm² and thickness dh, the balance of forces gives

$$dP = -\rho g\, dh, \tag{2}$$

hence

$$dP/P = (d\rho/\rho) + (dT/T) = (mg/kT)\, dh, \tag{2a}$$

and

$$P = P_0 \exp\left[-\int_0^z (mg\, dh/kT)\right], \tag{2b}$$

where P_0 is the surface pressure. If T is linear in z, then

$$\rho = \rho_0 \exp\left\{-\int [1 - (k\beta/mg)](dh/H)\right\}, \tag{2c}$$

where

$$\beta = dT/dh \qquad \text{and} \qquad H = kT/mg.$$

To the extent that variations in T can be neglected,

$$P = P_0 \exp[-(mg/kT)h] = P_0 \exp(-h/H), \tag{2d}$$

and

$$\rho = \rho_0 \exp(-h/H). \tag{2e}$$

Since $\rho = nm$ we have

$$n = n_0 \exp(-h/H). \tag{3}$$

The quantity $H = kT/mg$ is called the *scale height* of the atmospheric constituent with molecular weight m; it determines the rate at which the pressure of that constituent decreases with increasing altitude, and also the rate of decrease of the density when changes in T can be neglected.

For air near the ground, for example, with a mean molecular weight of approximately 29 and an average temperature of 240°K,

$$H = (1.4 \times 10^{-16} \times 240)/(29 \times 1.6 \times 10^{-24} \times 980) = 7.2 \text{ km.}$$

In the upper atmosphere, near the base of the exosphere, with $\mu \sim 16$ (atomic oxygen) and $T \sim 1500°\text{K}$, we have $H_e \sim 100$ km.

Equation (3) is known as the *barometric formula.* It is usually derived, as we have done here, from the hydrostatic law, which equates the pressure gradient with the gravitational force. In the exosphere the molecules do not collide with each other, and the pressure is no longer isotropic. However, it may be shown that the barometric formula is still valid in the exosphere to a high degree of approximation, within an earth's radius of the surface. Moreover, the temperature is observed experimentally to be nearly constant at exospheric altitudes. In the exosphere, therefore, the density is

$$n(r) = n_e \exp [-(h - h_e)/H_e], \tag{4}$$

where h_e is the altitude of the base of the exosphere and n_e is the density at that level.

To calculate the altitude at the base of the exosphere, we note that a particle moving radially outward from the base of the exosphere will reach the radius r with probability

$$P(r) = \exp \left[- \int_{R_e}^{r} \sigma n(r) \, dr \right] \tag{5}$$

where σ is the cross section for collisions, R_e is the radius of the planet at the base of the exosphere and $n(r)$ is given by equation (4). We define the base of the exosphere as the altitude for which $P(r) = 0.5$. Performing the integration, we obtain

$$P(r) = \exp (-n_e \sigma H_e) \quad \text{for} \quad (r - R_e) \gg H_e.$$

Therefore,

$$n_e = 0.7/\sigma H_e \tag{5a}$$

where H_e is the scale height at the base of the exosphere. Taking $\sigma = 10^{-15}$ cm² and $H_e = 100$ km, we have

$$n_e \approx 7 \times 10^7 \text{ particles/cm}^3.$$

According to satellite observations on the properties of the upper atmosphere this density occurs at an altitude of ~500 km.

Flux of Escaping Particles

At the base of the exosphere the outward flux of escaping particles is

$$F = (4\pi R_e^2) n_e \int_0^{2\pi} d\phi \int_0^{\pi/2} \sin\theta\, d\theta \int_{V_c}^{\infty} (V\cos\theta) f(V) V^2\, dV, \tag{6}$$

where R_e is the radius of the base of the exosphere, n_e is the atmospheric particle density at that level, and $f(V)$ is the Maxwell velocity distribution function,

$$f(V) = (m/2\pi kT)^{3/2} \exp(-\tfrac{1}{2}mV^2/kT) = A \exp(-\beta V^2)$$

$$(\beta = m/2kT). \tag{6a}$$

The integral is

$$\int_0^{2\pi} d\phi \int_0^{\pi/2} \sin\theta \cos\theta\, d\theta \int_{V_c}^{\infty} V^3 f(V)\, dV$$

$$= \pi A \int_{V_c}^{\infty} V^3 \exp(-\beta V^2)\, dV$$

$$= (\pi A/2\beta^2) \exp(-\beta V_c^2)(1 + \beta V_c^2) \tag{6b}$$

and the escaping flux is

$$F = 2\sqrt{\pi} R_e^2 n_e (2kT/m)^{1/2} \exp(-mV_c^2/2kT)[1 + (mV_c^2/2kT)] \quad \text{particles/sec.} \qquad (7)$$

In order to assess the importance of this escaping flux we must compare it with the total number of particles in the exosphere which is, from (4),

$$N = 4\pi R_e^2 \int_{n_e}^{\infty} n(h)\, dh = 4\pi R_e^2 n_e H_e. \qquad (8)$$

The decrease in N in time dt by escape from the gravitational field of the planet is

$$dN = -F\, dt. \qquad (9)$$

From (7), (8), and (9) we have

$$\frac{dN}{N} = \left[(1/2H)(2kT/\pi m)^{1/2} \exp(-mV_c^2/2kT)\left(1 + \frac{mV_c^2}{2kT}\right)\right] dt, \qquad (10)$$

and integrating (10),

$$N = N_0 \exp(-t/t_e),$$

where N_0 is the number of particles per cm² originally contained in the exosphere, and

$$t_e^{-1} = (2H)^{-1}(2kT/\pi m)^{1/2} \exp(-mV_c^2/2kT)[1 + (mV_c^2/2kT)].$$

The quantity t_e measures the lifetime of the gas, i.e., the time in which a fraction $1 - e^{-1}$ of the original content is lost by gravitational escape.

Using the relations

$$H = kT/mg \quad \text{and} \quad g = MG/R^2,$$

and the mean thermal velocity, $\bar{v} = 2(2kT/\pi m)^{1/2}$ we have

$$t_e \approx \frac{4H}{\bar{v}} \frac{\exp (R_e/H)}{R_e/H} \tag{11}$$

assuming $H/R_e \ll 1$ as is the case in all examples of interest. The quantity t_e gives the time for depletion of the amount of gas contained in the exosphere.

The time required for escape of a constituent from the entire atmosphere is much larger than t_e, because the amount of gas in the exosphere is only a part of the total atmospheric content. If B is the ratio of the total amount of a constituent to the amount in the exosphere, then the lifetime of the constituent is

$$t_e \approx B \frac{4H}{\bar{v}} \frac{\exp (R_e/H)}{R_e/H} . \tag{12}$$

No general prescription exists for calculating B. The total amount of a constituent in an atmosphere will be influenced by a number of factors including not only its primary abundance but also the rate of exhalation from the interior of the planet, chemical reaction with the crust and photochemical reactions in the upper atmosphere. Table I presents values of B, estimated by Richard Stewart in a private communication, for a few representative gases for the Earth, Mars, and Venus. Table II and Fig. 1 show atmospheric lifetimes for constituents of interest calculated from (12) and Table I. B is

TABLE I

Correction factor B for terrestrial planets (after R. Stewart)

	Mars	Earth	Venus
H	10^2	5×10^5	3×10^3
He	10^3	2×10^6	3×10^4
O	10^6	10^8	3×10^7
A	10^{11}	10^{11}	10^{13}

TABLE II

Lifetimes of representative constituents of planetary atmospheres (in years)

	Moon	Mars	Earth	Venus	Jupiter
H	10^{-2}	10^{3}	$10^{3\pm0.5}$	$10^{3\pm1}$	$10^{400\pm100}$
He	10^{-1}	10^{6}	$10^{8\pm1}$	$10^{8\pm3}$	$>10^{400}$
O	10^{2}	10^{20}	$10^{32\pm3}$	$10^{32\pm6}$	$>10^{400}$
A	10^{10}	10^{49}	$10^{70\pm7}$	$10^{70\pm14}$	$>10^{400}$

taken as unity in the case of the Moon, which has only an exosphere and no lower atmosphere. For Jupiter the times of escape turn out to be so large in comparison with the age of the solar system that it is possible to conclude that all gases are bound on that planet in spite of the uncertainty in the value of B.

The exponent in equation (12) is very much greater than unity in most cases of interest. For this reason the calculation of the life-

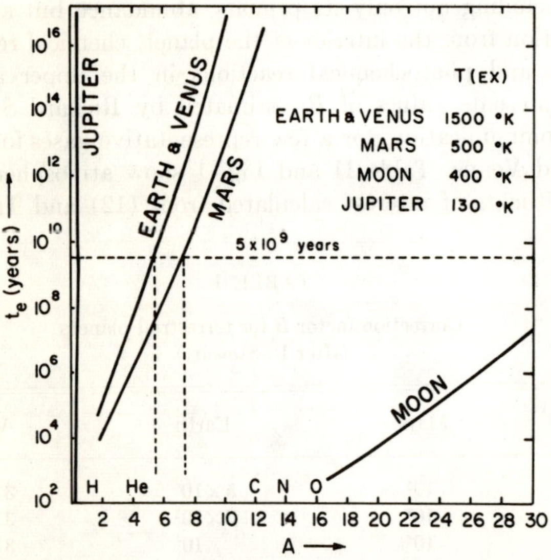

FIG. 1. Effective time of escape of gases as a function of atomic weight A for Earth, Venus, Mars, Jupiter, and Moon.

time is extremely sensitive to the exospheric temperature which enters directly in the exponent through the exospheric scale height.

In the case of the earth we adopt the observed average value of temperature at the base of the exosphere, $T = 1500°K$ and because of the lack of reliable information on Venus, the same temperature has been assumed. For Mars, the recent measurements by Mariner IV and various theoretical calculations suggest that the exospheric temperature may lie between 100°K and 600°K. Escape times for Mars are presented for $T = 500°K$. We estimate a 10 percent uncertainty in the exospheric temperature of the earth and a 30 percent uncertainty in the values given for Venus. The effect of these variations in the exospheric temperature on the lifetimes is indicated in the table.

The exospheric temperature of Jupiter has recently been estimated by Gross and Rasool[5] to be as low as 130°K. This considerably smaller increase in temperature on Jupiter, as compared with the exospheric temperature of the Earth, which is of the order of 1500°K, results because (a) the intensity of solar radiation at Jupiter is 27 times less than at the distance of the Earth; (b) the conductivity for the gaseous mixture on Jupiter is higher than for the O_2–N_2 mixture for the Earth; and (c) the density scale heights in the upper atmosphere are ~ 5 times less than in the upper atmosphere of the Earth.

Atmospheric lifetimes shown in Table II and Fig. 1 have been calculated for the lightest gases, hydrogen and helium; also for atomic oxygen, representing the C, N, O group; and finally argon, as a representative of the heavier gases, but occurring in special abundance by virtue of the steady source provided throughout the life of the planet by the decay of radioactive potassium in the crust.

Inspection of Table II and Fig. 1 permits us to draw interesting conclusions regarding the constituents to be expected in the atmospheres of the planets. We note that the age of the solar system is approximately 10^{10} years, hence lifetimes appreciably less than 10^{10} years imply escape of the gas, while lifetimes $\gg 10^{10}$ years indicate effectively permanent retention. On the basis of this criterion we draw the following conclusions:

(1) The Moon must have lost all but a trace of its atmosphere early in its history; what atmosphere it has must consist of argon, and perhaps the still heavier gas, xenon.

(2) Hydrogen escapes rapidly from all the terrestrial planets. Also methane and ammonia are readily dissociated at the top of the atmosphere by solar ultraviolet radiation, which leads to depletion of these gases by loss of their hydrogen component. This circumstance explains the scarcity of these gases in the atmospheres of the terrestrial planets, although they were probably present in abundance in the primitive atmospheres of these planets. The presence of these gases in primitive terrestrial atmospheres is essential to current ideas on the evolution of living organisms out of inorganic matter in the early history of the earth.

(3) Helium escapes from Mars, but may be partly retained on the Earth and Venus. The uncertainty in the exospheric temperature of the Earth, while small, is still sufficiently large to cover the range of possibilities from complete loss to fractional retention; this circumstance was exploited by Nicolet a few years ago in calculations designed to deduce the temperature of the exosphere by fitting the escape formulas to the observed amounts of He^3 and He^4 in the atmosphere; we now know the upper atmosphere of the Earth has a substantial layer of helium, this having been detected very recently in the exosphere at the 1600 km level, by instruments carried in the Explorer VIII satellite.[6]

(4) Jupiter probably retains all the gases of its primitive atmosphere, including hydrogen: hence its atmosphere may be expected to have, as is observed, large amounts of such compounds as CH_4 (methane) and NH_3 (ammonia) and probably H_2O as well.

The depletion of water is different from that of other hydrogen compounds, at least on the earth, because it exists primarily in liquid form at the temperatures prevailing near the ground. The temperature falls to a minimum of 200°K at an altitude of 10 km, which constitutes a *cold trap* for the water contained on the surface and in the lower atmosphere. Most of the water vapor condenses out of the atmosphere in liquid form below this altitude, and the amount left above 10 km is only the minute concentration corresponding to the saturated vapor pressure of water at the low temperature of 200°K in the cold trap. The ultraviolet light, which dissociates the H_2O and leads to escape of its hydrogen, is absorbed at high altitudes by photodissociation reactions in O_2 and other constituents, and essentially none of it penetrates to the layer below an altitude of 10 km, to which the water vapor is confined. This special combination of circumstances involving the high boiling point temperature of water, the low temperature of Earth's tropopause and the presence of free oxygen in the upper atmosphere preserves the water on our planet.

SURVEY OF ATMOSPHERIC STRUCTURE

Temperature, *density*, and *composition* are the basic atmospheric data. Of these, temperature is the most significant quantity because it directly reflects the processes of absorption of solar energy which determine atmospheric structure. The temperature profile of the

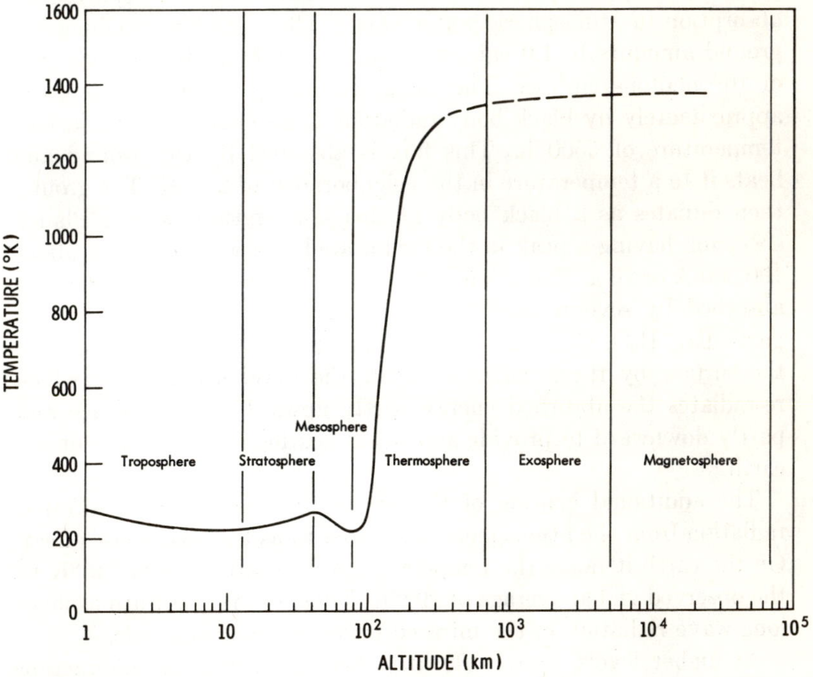

FIG. 2. Temperature profile of Earth's atmosphere (after Jastrow and Kyle, 1961).

earth's atmosphere being fairly well determined (Fig. 2), we first undertake to develop a qualitative understanding of the factors which control the temperature of the terrestrial atmosphere. In our discussion we will consider the regions of the atmosphere, in sequence, from the surface of the planet out to the interplanetary medium, introducing in each case the sources of energy and mechanisms of energy transport which determine the structure of the region.

In the lower atmosphere the ground, heated by the absorption of solar visible radiation, is the source of energy, and the important transport processes are radiation and convection.

About 40 percent of the solar energy incident on the earth is reflected by clouds or back scattered in the atmosphere. Of the remaining 60 percent, most is in the visible region and is transmitted to the ground without attenuation. More precisely, 45 percent reaches the ground, and 15 percent lying in the near infrared is removed by absorption in atmospheric water vapor. The solar flux reaching the ground amounts to 1.0 cal/cm^2/minute, or 0.7×10^6 ergs/cm^2/sec, or 10^{18} photons/cm^2/sec. The energy frequency distribution is given approximately by black body radiation at an effective solar surface temperature of 5500°K. This flux is absorbed by the ground and heats it to a temperature in the neighborhood of 245°K. The ground then radiates as a black body at this temperature, with a Planck spectrum having a peak in the far infrared at a wavelength of about 100,000 Å or 10 μ. Radiation in this region of wavelengths is strongly absorbed by several constituents of the terrestrial atmosphere—in particular, H_2O, CO_2, and O_3. The absorption of the radiation from the surface by these molecules heats the lower atmosphere, which re-radiates the absorbed energy partly upward to outer space and partly downward to provide additional heating of the surface of the earth.

The additional heating of the surface by the return of infrared radiation from the atmosphere is referred to as the *greenhouse* effect. On the earth it raises the temperature of the surface from 245°K to the observed global average of 290°K. Thus, in the lower atmosphere long-wave radiation in the infrared plays the dominant role.

At higher levels, above 100 km, there is little or no interaction with infrared radiation, because the constituents of the atmosphere at great heights are atomic or diatomic particles without absorption spectra in the infrared region. The principal role is played by the short-wave components of the solar spectrum—ultraviolet radiation and x-rays. These components are absorbed in processes of *photodissociation* and *photoionization* and provide the heat source for the upper atmosphere. They carry only a small fraction of the total solar energy flux, there being less than 1 percent in the near ultraviolet and less than 0.01 percent in the far ultraviolet and beyond. However, the cross sections for photodissociation and photoionization are so large in the far ultraviolet that this band of wavelengths is removed from the incident solar flux by the time it has penetrated to a height of 100 km. The absorbed energy is transported downward to lower altitudes by conduction.

At the highest altitudes an important role is played by electrically charged particles and their interactions with the geomagnetic field.

The distribution of temperature in the terrestrial atmosphere can be explained very readily in terms of these energy sources and processes of energy transfer. We refer to Fig. 2 showing in schematic form the observed terrestrial temperature distribution. It was noted above that the surface of the Earth is maintained at a temperature of about 290°K by the absorption of sunlight in the visible region plus the return of infrared from the atmosphere to the ground. As we move away from the heated surface the temperature of the air falls at a rate of 7°K/km determined by combined radiative and convective transport. At 12 km and a temperature of ~200°K the decrease halts and the temperature rises again to a maximum of ~270°K at 50 km. The increase in temperature between 12 km and 50 km is the result of the heating produced by the absorption of near ultraviolet radiation in the ozone layer, the band of absorbed wavelengths extending from 2000 to 3000 Å.

Between 50 km and 80 km the temperature falls again, reaching a minimum of ~200°K at 80 km. This minimum is associated with an energy sink provided by CO_2 and O emission in the far infrared.

Above 80 km the temperature rises rapidly, reaching a roughly constant level of ~1500°K above 300 km. A part of this increase is the product of heating produced by photodissociation and photoionization of oxygen and nitrogen by solar radiation in the far ultraviolet, with wavelengths extending from 2000 to 100 Å and the remainder results from the interaction of solar particle streams with the atmosphere. The division of the heat source between particles and radiation is not yet known, but radiation is believed to be the more important of the two.

Convective transport, occurring in the lower atmosphere, leads to a thorough mixing of all atmospheric constituents. Although the atmosphere should be stable against convection between 12 km and 50 km, and again above 100 km, these being regions of positive temperature gradient, it is found experimentally that the region of mixing extends from the ground to an altitude of about 110 km. Below 110 km the atmosphere is homogeneous and the proportions of atmospheric constituents are independent of altitude, the density of each major constituent falling off exponentially with altitude, as in (3), but with a scale height determined by the average weight per particle, i.e., $\mu = 29m_H$. Above 110 km the various constituents begin to settle out of the atmosphere in proportion to their weights, again in accordance with the *law of hydrostatic balance* (3), but now with the scale heights determined by the separate weights; i.e.,

$$n_i = n_{i_0} \exp\left[-\mu_i gh/kT\right], \tag{13}$$

in which n_i is the density of a constituent with molecular weight μ_i.

The region of the atmosphere in which the relative concentrations are given by (13) is known as the *domain of diffusive separation*. At the boundary of this domain, at a height of 110 km, nitrogen and oxygen are the major constituents, with N_2 dominating in the ratio 4:1, as on the surface. At higher altitudes the relative concentration of oxygen increases, and by 300 km the surface proportions are nearly reversed, with oxygen now dominating in the ratio of 3:1. The reason for this circumstance is that O_2 can be dissociated by absorption of solar ultraviolet, with relatively high probability, yielding atomic oxygen with half the weight per particle, and therefore with twice the scale height in (13) and a correspondingly smaller rate of decrease of concentration with increasing altitude. Molecular nitrogen also undergoes photodissociation, but with a much smaller probability than for oxygen; hence it remains in primarily molecular form up to greater altitudes. At 300 km less than 1 percent of the N_2 is in atomic form, while 99.5 percent of the oxygen is atomic.

For this reason atomic oxygen dominates the composition of the upper atmosphere from 200 km to 1000 km. Above that level the lightest gases, hydrogen and helium, emerge as the principal constituents because of their very large scale heights, although they are present only in trace amounts at lower altitudes. At the greatest altitudes all gases except hydrogen have settled out, and this gas dominates until the atmosphere merges into the interplanetary medium.

Temperature

The *effective temperature* (T_e) of a planet is defined by equating the energy received from the sun to the energy which would be radiated from the entire surface of the planet if the solar energy were distributed uniformly over it: $4\pi R^2 \sigma T_e^4 = \pi R^2 S(1 - A)$. Hence

$$T_e = [(S/4\sigma)(1 - A)]^{1/4}, \quad (14)$$

where R is the *radius of the planet*, S the *solar constant or radiation* flux received at the top of the atmosphere, A the *planetary albedo* averaged over the whole solar spectrum, and σ the *Stephan–Boltzmann constant.* On the Earth $S = 1.4 \times 10^6$ ergs/cm²/sec, $A = 0.4$, and $T_e = 245°K$. This result should be valid for the global average of a rapidly rotating planet. For a synchronously rotating planet and at latitude λ, the effective temperature will be

$$T = [(S/\sigma) \cos \lambda(1 - A)^{1/4}. \quad (15)$$

Table III contains values for T_e and other constants pertaining to the nine planets.

The *greenhouse effect* increases the temperature by an amount which can be estimated from the theory of *radiative transfer*. For an atmosphere whose optical thickness in the infrared is τ, the Eddington approximation to the equation of radiative transfer gives the temperature T_G at the surface of the planet as

$$T_G = T_e(1 + \tfrac{3}{4}\tau)^{1/4}. \quad (16)$$

This result is based on the assumption that the atmosphere is in radiative equilibrium. It neglects energy transfer by convection and circulation, and the effects of latent heats of condensation.

In the case of Earth, for $T_e = 245°K$ and the observed value of

TABLE III

Planet	Distance from Sun, $R_\oplus = 1$	Mass, $M_\oplus = 1$	Radius, $a_\oplus = 1$	Density, gm/cm³	Albedo	g, cm/sec²	T_e, °K
Mercury	0.3871	0.0543	0.39	5.05	0.07	350	616
Venus	0.7233	0.8136	0.966	5.21	0.73	877	235
Earth	1.	1.	1.	5.52	0.39	980	245
Mars	1.5237	0.1080	0.520	3.90	0.26	391	209
Jupiter	5.2028	318.35	10.97	1.33	0.47	2592	105
Saturn	9.5388	95.3	9.03	0.714	0.46	1145	78
Uranus	19.19	14.54	3.72	1.56	0.45	1070	55
Neptune	30.07	17.26	3.50	2.22	0.49	1381	43
Pluto	39.52	0.033(?)	0.45(?)	2.0(?)	0.16(?)	160(?)	42

Distance of Earth from the Sun $= 1.4968 \times 10^{13}$ cm
Mass of the Earth $= 5.975 \times 10^{27}$ g
Mean radius of the Earth $= 6370.97$ km
Solar constant $= 1.395 \times 10^{6}$ ergs cm^{-2} sec^{-1}.

TABLE IV

Planet	T_e (°K)	T_G (°K)	ΔT (°K) Greenhouse	Optical thickness of atmosphere in the infrared
Mercury	616	613	—	∼0
Venus	235	600	365	55.4
Earth	245	290	45	1.4
Mars	209	230	21	0.6
Jupiter	105	150	45	4.3

surface temperature of 290°K, the total optical thickness of the atmosphere from equation (16) is 1.4. It is observed that about 20 percent of the surface radiation actually passes through the earth's atmosphere, corresponding to an optical thickness of $\ln 5 = 1.6$. For this optical thickness (16) yields a surface temperature of 298°K. The discrepancy is removed when allowance is made for the additional cooling effect of convective transport of energy from the ground to the atmosphere.

Table IV gives the values of the effective *blackbody* temperatures for some of the other planets and also the observed surface temperatures.

In the case of Mercury T_e and T_G refer to the subsolar point on the planet. For Jupiter the conditions below the clouds are unknown, and Table IV lists T_G at the cloud top. The differences between T_G and T_e give the magnitude of the greenhouse effect for each planet which in turn provides an estimate of the optical thickness of the atmosphere in the far infrared.

For Mercury we notice that the observed temperature of the ground is approximately the same as the effective temperature, indicating that the atmosphere of Mercury is devoid of molecules which would provide opacity in the infrared. The magnitude of the greenhouse effect on Venus suggests that this atmosphere is almost completely opaque to the far infrared radiation emitted by the surface of the planet. *Polyatomic gases,* which have strong absorption bands in the infrared, should be abundant on Venus. The optical thicknesses of the terrestrial and Martian atmospheres are far smaller than that of Venus. The concentration of polyatomic molecules in the atmosphere of Mars should be comparable to that on the earth.

According to Table IV the atmosphere of Jupiter above the clouds should have an optical thickness of 4.3 in order to explain the cloud top temperature of 153°K. The optical thickness listed for Jupiter indicates a very high opacity (99 percent absorption) in the Jovian atmosphere above the clouds. Gross and Rasool[5] find that the observed amounts of polyatomic gases, methane and ammonia, are not sufficient to provide the required amount of absorption in the far infrared and suggest, following an earlier idea of Öpik, that the relatively high temperature observed at the cloud top of Jupiter may be accounted for by primitive heat released from the interior of the planet.

ATMOSPHERES OF THE PLANETS

We now turn to a detailed discussion of the observed properties of the atmospheres of the planets.

Mercury

One of the most astounding discoveries about the planet Mercury was made in 1965 when Pettengill and Dyce[7] made radar measurements of Mercury and found that the planet is *not* in synchronous rotation, but rotates much faster with a period of 59 ± 5 days (Figure 3). This new result was in complete disagreement with the widely accepted value of 88 days and for a time, the reason for this discrepancy was difficult to understand. The original estimates of an 88-day rotation period were based on the visual observations of Mercury made during the last century. A reanalysis[8] of these data has indicated that because most of the observations were made at a particular phase angle and, by chance, only at particular time intervals, the rotation period was erroneously derived as 88 days. When the time interval between observations is large (several months or years), a number of other rotation periods could be possible. It has now been shown[9] that most of the existing observations could be reconciled with a rotation period not only of 88 days, but also of 58.65 days, which is within the range of the values given by radar. At the same time, Colombo[10] has pointed out that because 58.65 is

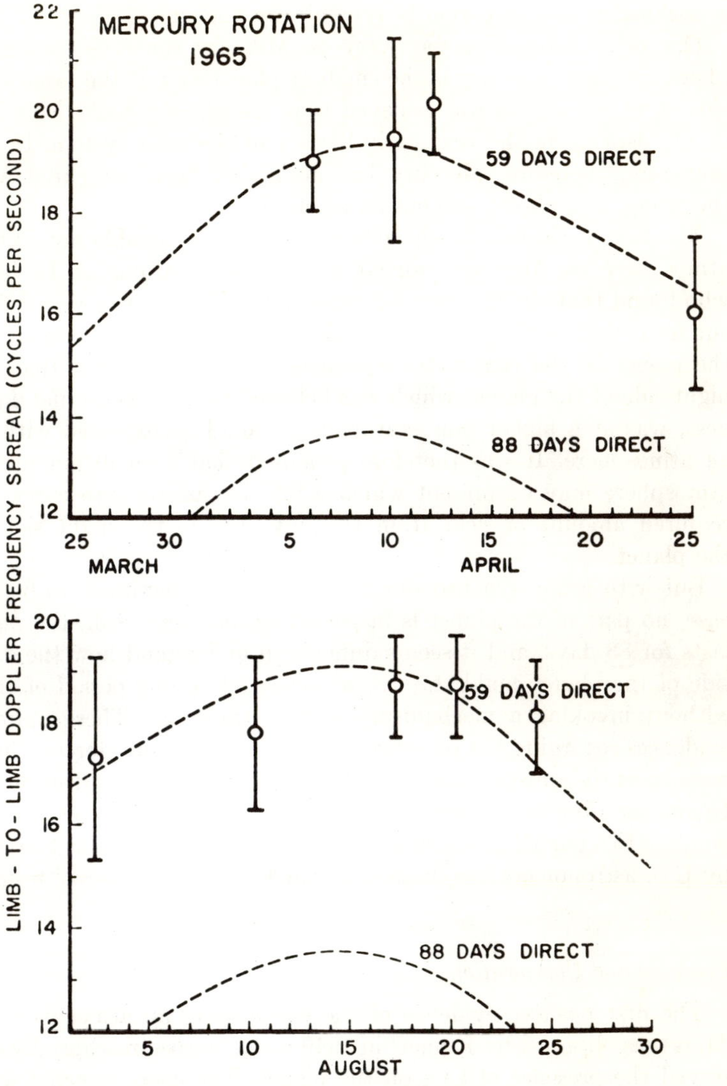

FIG. 3. Plot of the apparent rotational angular velocity of the planet Mercury versus date for several values of rotation during the inferior conjunctions of April and August, 1965. The values inferred from the measurements are shown with their estimated errors (after Pettengill and Dyce, 1966).

exactly two-thirds of 88, which is the orbital period of the planet, a 'locking in' at this period is dynamically plausible.

One other interesting discovery on Mercury concerns its atmosphere. Because Mercury is the smallest planet and at the same time closest to the sun, it was believed that the planet should have lost its atmosphere in the very early history of the solar system. In the literature, therefore, Mercury has invariably been compared with the moon, as a planet without an atmosphere.

The first indication of the presence of a measureable amount of atmosphere on Mercury came from the observations of Dollfus[11] who found that the polarization properties of the planet were different from that of the moon. Also, the temperature measurements of the planet in the centimeter wavelength region indicated that the night side of the planet, which was believed to be in perpetual darkness, was at a higher temperature than could be expected without an atmosphere. It was therefore proposed that a small amount of atmosphere may be present which would be sufficient to carry the required amount of heat from the day side to the night side of the planet.

But with a non-synchronous rotation, as we now know to be the case, no part of the planet is in perpetual darkness. Still, the night lasts for 88 days, and it seems difficult to understand how the dark side of the planet could stay warm over such a long period of time without invoking a mechanism for heat transport. These indirect evidences regarding the presence of an atmosphere on Mercury have been recently substantiated by the fact that Moroz in the Soviet Union has observed a small amount of CO_2 on the planet, corresponding to a partial pressure of $2 - 3$ mb. Search for CO_2 on Mercury by U.S. astronomers has, however, failed to confirm Moroz' results.

Mars

Pressure and Composition

The first positive evidence of the existence of an atmosphere on Mars was obtained by Kuiper[1] in 1949 when he spectroscopically observed the presence of CO_2 on the planet. The measurements were made in the 1.6μ region of the near infrared spectrum. Because these CO_2 bands are highly pressure-dependent, the only way to deduce the amount of CO_2 from these measurements is to first determine the atmospheric pressure by another technique.

An independent method for estimating the atmospheric pressure on another planet was first developed by Lyot in 1929. By studying the polarization properties of the sunlight reflected from different parts of the planetary disk, one can deduce the total thickness of the atmosphere. This method of determining the surface pressure is, however, only valid if the atmosphere of the planet is completely free of dust and other solid particles. If not, the measurements give only an upper limit on the pressure.

Such measurements for Mars have been repeated a number of times since 1929, and the most widely-accepted value for the total atmospheric pressure of Mars has been 85 mb. This pressure value, when combined with the measured intensity of the CO_2 absorption bands, had indicated that the amount of CO_2 makes up only 2% of the total atmosphere.

In 1964 Kaplan, Münch, and Spinrad, measured the absorption of CO_2 on Mars in the 8700 Å region. Here the strength of absorption is directly proportional to the amount of the gas and independent of pressure. From this study, in conjunction with other spectroscopic measurements in the 1.6μ region, it was found that the atmospheric pressure on Mars may be considerably lower than the previously quoted value of 85 mb, viz, in the neighborhood of $20 - 25$ mb. This result also implied that CO_2 may be a major constituent of the atmosphere.

Recently, Kliore, et al,[12] performed an experiment on Mariner IV to determine the surface density and the atmospheric structure of Mars. Just before the occulation of the spacecraft by the planet, the radio signals at 12 cm wavelength, which were used in tracking the spacecraft, passed through the Martian atmosphere. The atmosphere refracted the ray and the longer path length was observed as a change in phase in the signal. From the analysis of this phase change, the authors deduced that the atmospheric refractivity is 3.6 N units, that is, the refractive index of the atmosphere near the surface is 1.0000036, which for atmospheric temperatures of the order of 200°K corresponds to surface pressure of as low as 6 ± 1 mb. It was also found that the scale height of the lower atmosphere may be only

8 – 10 km, contrary to the previous belief that it was of the order of 15 km. This new result suggests that the mean molecular weight of the atmosphere may be ~ 40. From these measured parameters, the authors have deduced the various possible models for the atmosphere of Mars which are summarized in Table V.

Table V Summary of occulation experiment.

Atmosphere	
Surface refractivity	3.6 ± 0.2 N units
Scale height	8 to 10 km
Surface number density	
100% CO_2	$1.9 \pm 0.1 \times 10^{17}$ mol/cm^3
Up to 20% A or N_2, or a mixture	$2.1 \pm 0.2 \times 10^{17}$ mol/cm^3
50% A	$2.5 \pm 0.15 \times 10^{17}$ mol/cm^3
Surface mass density	
100% CO_2	$1.43 \pm 0.1 \times 10^{-5}$ g/cm^3
Up to 20% A or N_2, or a mixture	$1.5 \pm 0.15 \times 10^{-5}$ g/cm^3
50% A	$1.75 \pm 0.10 \times 10^{-5}$ g/cm^3
Temperature	
100% CO_2	180 ± 20°K
Up to 20% A or N_2, or a mixture	175 ± 25°K
50% A	170 ± 20°K
Surface pressure	
100% CO_2	4.1 to 5.7 mb
Up to 20% A or N_2, or a mixture	4.1 to 6.2 mb
50% A	5.0 to 7.0 mb
Ionosphere	
Maximum electron density ($\chi = 70°$)	$9 \pm 1.0 \times 10^4$ el/cm^3
Altitude of maximum	120 to 125 km
Electron scale height above maximum	20 to 25 km
Temperature	< 200°K at 120 to 200 km

More recent spectroscopic measurements made by Belton and Hunten in the near infrared tend to confirm the result that the pressure at the surface of Mars may actually be between 5 and 10 mb, and that CO_2 is the major constituent of the atmosphere of Mars.

Search for the presence of other gases in the atmosphere of Mars has failed to give positive results. The upper limits on their possible abundances are listed in Table VI.

TABLE VI

GAS	m −atm
O_2	< 0.7
N_2O, NO_2, or N_2O	< 2.0
NH_3	< 0.2
CH_4	< 0.1
O_3	< 5 × 10 −4
SO_2	< 3 × 10 −5

Upper Atmosphere and Ionosphere

The occulation experiment also provided information on the ionosphere of Mars. Figure 4 shows the electron density profile measured in the upper atmosphere.

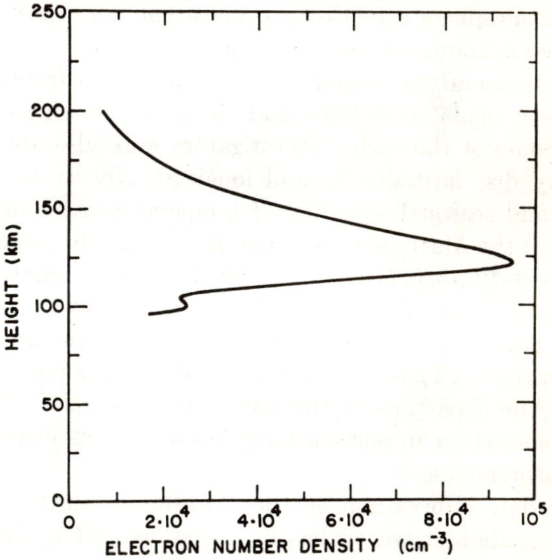

FIG. 4. Electron number density versus altitude above Electris at the time and location corresponding to immersion into occulation (50°S latitude, 177°E longitude, 1300 local time, later winter, solar zenith angle 67°( )after G. Fjeldbo, W. C. Fjeldbo, and Von R. Eshelman, J. Geophys. Res. **82,** 2307, 1966).

Based on these measurements, and on the assumption that the main ion in the upper atmosphere is O^+, it has been suggested that the upper atmosphere of Mars may be isothermal at a temperature of only 85°K. On the other hand, theoretical calculations of the equilibrium temperature in the upper atmosphere of Mars made independently by two groups of workers, suggest that the temperature should be in the neighborhood of 500°K. No satisfactory explanation for this discrepancy has yet been offered.

Surface Temperature

Planetary temperatures are usually estimated by measuring the infrared radiation emitted by the planet. A large part of the infrared spectrum is, however, absorbed by the water vapor and CO_2 present in the Earth's atmosphere. Ground-based observations of the planets in the infrared are therefore confined to the 8–12 μ region where the Earth's atmosphere is relatively transparent. This spectral region in the infrared is known as the atmospheric *window*.

Extensive temperature measurements of Mars by infrared radiometry have been made since 1926, and the results to date are quite consistent. Some of the earlier investigators were also able to scan the planetary disc latitudinally and longitudinally so as to obtain the diurnal and seasonal variation of temperature as a function of latitude. Since the Martian atmosphere should also be largely transparent in the 8–12 μ *window*, it is assumed that the radiation intensity measurements in this wavelength region refer to the surface of the planet. This will be true if the surface of Mars radiates as a blackbody in the 8–12 μ spectral inverval, which may not necessarily be the case. Any departures in the emissivity from unity will require an upward correction in surface temperature values deduced from infrared measurements.

The up-to-date information on the variation of surface temperature with latitude and season has been summarized by Mintz and is reproduced in Fig. 5.

Vertical Distribution of Temperature

The atmosphere of Mars is optically thin in the infrared, and therefore the vertical temperature structure can not be determined from radiometric measurements of the planet because a major part of the observed radiation originates from the surface. But if the composition of the atmosphere is known, then from the knowledge

Polar Caps

The polar caps are rather reliably known to be made of a layer of H_2O frost a few centimeters thick deposited on the surface. These caps are observed to expand to lower latitudes during fall and winter and to recede to high latitudes during spring and summer. The observed evaporation of the polar caps in summer and the almost immediate transport of the water vapor towards the winter pole across the equator is a special phenomenon peculiar to Mars.[13] In

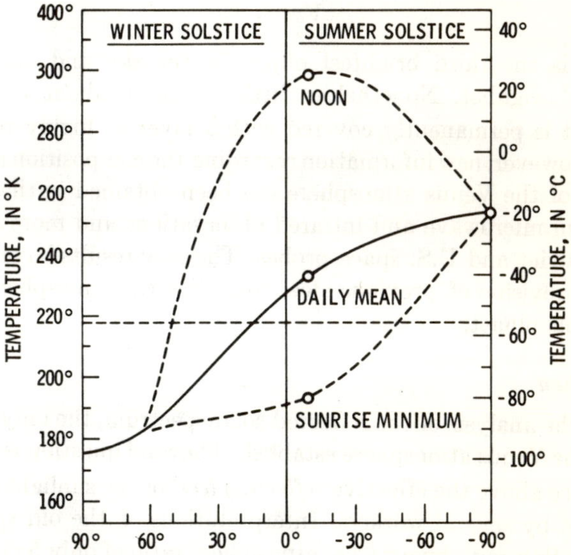

FIG. 5. The surface temperature on Mars at the northern hemisphere winter solstice. These temperatures were derived from the observed temperatures near the equator and the observed temperature variation along the noon meridian (after Mintz, 1961).

the Earth's atmosphere the mixing between hemispheres is extremely small.

The average speed of this *humidity wave* across the planet has been estimated to be about 45 km/day.

Blue Haze Layer

Photographs of Mars taken through blue filters indicate no surface features, and it is therefore believed to be covered with a haze layer known as *blue haze*, absorbing at these wavelengths. Sometimes it clears over certain regions and surface features become observable in the blue. These *blue clearings* usually last a few days. Conflicting arguments have been extended to explain the nature of the blue haze, but an explanation which satisfies all the observed features is yet to be found.

Venus

Venus is the third brightest object in the sky and our nearest planetary neighbor. Nonetheless, little is known about this planet because it is permanently covered with a layer of highly reflecting clouds. However, new information regarding the composition and temperature of the Venus atmosphere has been obtained in the last few years from microwave and infrared observations and more recently by the Soviet and U.S. space probes. The new results have led to a complete revision of previous ideas regarding the atmospheric structure of this plnaet.

Composition

From the analysis of the reflected solar spectrum, the only constituent of the Venus atmosphere established beyond question is CO_2. Its abundance above the effective *reflecting level* of the sunlight has been estimated by several workers. Interpretations of the old spectra of Venus by Spinrad[14] give a CO_2/atmosphere ratio of only 5 percent by mass.

New high resolution spectrograms of Venus in the region obtained and analyzed by Goody, Belton and Hunten indicate that CO_2 may be a major constituent of Venus atmosphere. These findings have been spectacularly confirmed by the Venus–4 and Mariner–5 observations of the planet on October 18 and 19, 1967, respectively. Preliminary results announced from both these missions indicate that CO_2 may constitute as much as 90% of the Venus atmosphere. The amount of nitrogen and rare gases on Venus is probably 10%.

Estimates of water vapor in the atmosphere of Venus vary by several orders of magnitude. Goody *et al.* have presented convincing spectrographic evidence that the mixing ratio of water vapor above the cloud level is less than one part in 10^5. However, Strong *et al.* argue that the clouds are made of ice crystals and that there is a large quantity of water vapor below. The recent Russian Venus–4 measurements give an upper limit of 1.4% for water vapor in the lower atmosphere.

Temperature and Pressure

Cloud Top.—Sinton and Strong have repeatedly measured the planetary emission from the cloud top in the 8–12 μ region and find a temperature value of 235° ± 10°K. This temperature probably refers to the cloud top level of the Venus atmosphere. Also, Murray and Wildey have made extensive observations of infrared emission from Venus across the disc of the planet and obtain temperature estimates varying between 205° and 220°K at different places on the planet. Mariner II also observed Venus in the 8.4 μ and 10.4 μ regions and the results indicate a temperature of 234°K.

However, Goody has pointed out that the interpretation of infrared measurements to give cloud top temperatures may be incorrect because the emissivity of the clouds in the infrared may not be unity. He thus concludes that the cloud top temperatures on Venus may be as high as 275°K. New high resolution spectra in the near infra-red also indicates a temperature of 270°K at the reflecting level of 1 μ photons. The width of individual lines of the CO_2 bands indicate a pressure of 0.1 to 0.2 atmosphere at this level.

Surface.—Determinations of the surface temperature of Venus by optical astronomy have been prevented by the cloud cover. However, the small amount of thermal radiation emitted by the planet in the microwave region would penetrate through the clouds without significant attenuation, and therefore can be used to determine the temperature of the surface of the planet.

First attempts to measure radiation from Venus at centimeter wavelengths were made in 1956 by the radio-telescope of the Naval Research Laboratory. The temperature inferred from the measured radiation intensity was of the order of 600°K. Repeated measurements in the following years have confirmed this result.

The observed temperatures are plotted as a function of wavelength in Fig. 6. At longer wavelengths the brightness temperature is observed to increase when the Venus-sun-earth angle becomes appreciable, indicating that the illuminated hemisphere is significantly hotter than the dark side. (The 10 cm brightness temperature may exceed 700°K on the bright side.) Venus–4 which landed on the night side of planet on October 18, 1967, measured a temperature of 550°K at the surface.

Several models of Venus have been discussed in connection with these measurements but the one most consistent with all the observations seems to be the greenhouse model for the atmosphere. The *greenhouse model*, originally discussed by Sagan, assumes that the 600°K temperature is maintained by a very effective greenhouse effect. The solar radiation in the visible portion of the spectrum penetrates up to the surface; the planet, thereby being heated up to a temperature T_e, emits in the infrared; but due to the presence of such triatomic molecules as CO_2 and H_2O which have strong absorption bands in the infrared region, most of the radiation remains trapped in the atmosphere and heats up the surface.

Jastrow and Rasool have pointed out that in such a model an extremely high infrared opacity of the atmosphere (with an optical thickness of 55, and a transmission of 10^{-24}) will be required in order to obtain a Venus ground temperature of 600°K.

The required opacity can result from pressure broadening of absorption bands if the atmospheric pressure at ground level is many tens or hundreds of atmospheres. Spinrad and Thaddeus have offered

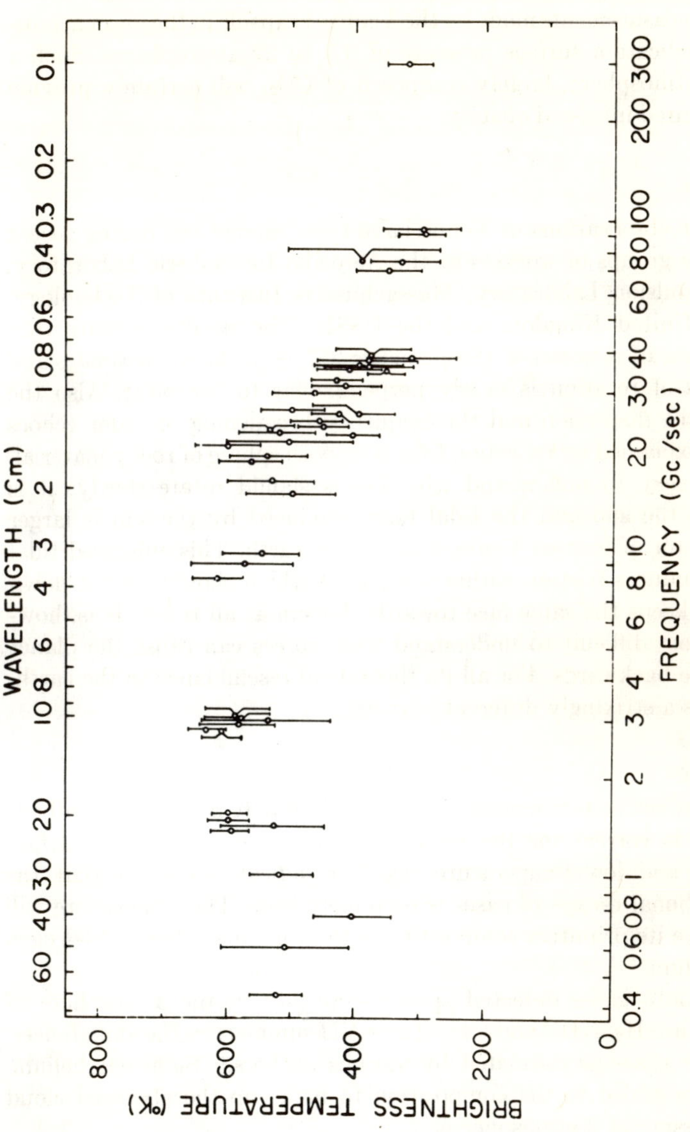

FIG. 6. Radio brightness temperatures of Venus at wave-lengths from 0.1 to 100cm.)after W. Ho, I. A. Kaufman, and P. Thaddeus, J. Geophys. Res. **82,** 5091, 1966).

independent arguments for a ground pressure in this range. Barometric measurements made by the Venus–4 capsule in the lower atmosphere indicate a surface pressure of 15 to 22 atmospheres. Such a heavy atmosphere, largely composed of CO_2, will certainly provide the required infra-red opacity.

Rotation

Radar observations of Venus have been carried out during recent years by groups of workers at the Aerocibo Ionospheric Laboratory, Jet Propulsion Laboratory, Massachusetts Institute of Technology, in the United Kingdom and the USSR. The results indicate that the period of rotation of the planet is 257 $\pm$ 5 days *retrograde,* and the axis of rotation is nearly perpendicular to the orbit. Also the pulse-time dispersion and the frequency broadening of radar echoes can be explained by reflection from a smooth sphere of rocky material.

It is easy to understand why Venus should rotate slowly; it is close to the sun and the tidal force produced by the sun is larger by a factor of four on Venus than on the earth. This enhanced tidal force produces friction within the planet which should slow it down until it keeps the same face towards the sun at all times. It is, however, most difficult to understand what forces can cause the planet to rotate backwards. For all its theoretical resemblance to the earth, Venus is a strikingly different planet.

Jupiter

The effective temperature of Jupiter is 105°K if the visible albedo of 0.47 is correct for the entire solar spectrum. The high surface gravity and low temperature together indicate (Fig. 1) that the gravitational escape of gases is extremely slow. The atmosphere will still have its primitive composition with large quantities of hydrogen and helium.

The only gases detected spectroscopically in the atmosphere of Jupiter are H_2, NH_3, and CH_4. Table VII summarizes the abundances of three gases as estimated by various authors.[17] Sufficient helium has been added to the composition to make up the observed cloud top pressure of 3 atmospheres.

Table VII indicates that the Jovian atmosphere is not as rich in hydrogen as expected from the discussion of gravitational escape.

TABLE VII

Gas	Percentage by volume
H_2	60
He	36
Ne	3
CH_4 and NH_3	<1

Measurements of the scale height of the atmosphere of Jupiter have been made by observing the occultation of σ-Arietis by Jupiter. These observations yield a scale height of $\sim$8 km. For a mean atmospheric temperature of 120°K, the corresponding molecular weight would be $\sim$4. This requires more helium than Table VII indicates. However, according to Gallet, the relative abundance of H and He given in Table VII could also account for a mean molecular weight of 4 if the Jovian atmosphere had a substantial amount of neon.

Temperature

Spectroscopic measurements of infrared emission from Jupiter indicate a temperature of 130°K which probably refers to an intermediate level above the clouds. The cloud top temperatures have been estimated at 153°K on the assumption that the clouds are made of NH_3 crystals and that they are in phase equilibrium with the NH_3 vapor above the clouds.

Gallet has made extensive theoretical analyses of the atmospheric structure of Jupiter below the clouds. Again it is assumed that the visible clouds are made of NH_3 crystals, which condense at the temperature and density of the cloud top. With allowance for the heat released in this condensation, Gallet calculated a relatively slow increase of temperature going down from the clouds to the surface, and thus arrives at high densities and pressures at low temperatures. He has also suggested that the observed cloud layer may be NH_3 of thickness 50 km, below which there is another deck of clouds composed of H_2O. Between these two cloud layers is a region where NH_3 rain storms occur. In such a model the surface of Jupiter would be a few hundred km below the clouds, at a temperature of ~1000°K and a pressure of several thousand atmospheres.[18,19]

Red Spot

The most prominent feature on Jupiter is the great red spot first observed in 1831, which is 40,000 km in length and 13,000 km in width. The intensity and color of this spot are variable. It reached its highest intensity in 1880, when its color was pink. The red spot drifts in longitude as though it were not attached to the surface. Recently, Hide has advanced an explanation in which the surface of the planet in that region would be a plateau only a few kilometers high. Hydrodynamical theories suggest that because Jupiter rotates so rapidly, the effect of a shallow topographical feature on the general circulation of the atmosphere will be attenuated very slowly with height. Thus, the feature will make its presence manifest at the level of the cloud.

REFERENCES

1. G. P. Kuiper, Planetary Atmospheres and Their Origin, *in Atmospheres of the Earth and Planets*, G. P. Kuiper, Ed., Chap. 12, p. 306, University of Chicago Press, Chicago, 1952.
2. H. C. Urey, The Atmospheres of the Planets, in *Handbuch der Physik*, Vol. 52, p. 363, Springer-Verlag, Berlin, 1959.
3. S. I. Rasool and R. Jastrow, The Atmospheres of Mars, Venus, and Jupiter, *in Life Sciences and Space Research II*, M. Florkin and A. Dollfus, Eds., North-Holland Publishing Company, Amsterdam, 1964.
4. J. W. Chamberlain, Upper Atmosphere of the Planets, Astrophys J. **136,** 582 (1962).
5. S. H. Gross and S. I. Rasool, The Upper Atmosphere of Jupiter, Icarus **3,** 311 (1964).
6. R. Jastrow and A. G. W. Cameron, Space: Highlights of Recent Research, Science **145,** 1129 (1964).
7. G. H. Pettengill and R. B. Dyce, Nature **206,** 1240, (1965).
8. W. E. McGovern, S. H. Gross, and S. I. Rasool, Nature **208,** 375, (1965).
9. S. I. Rasool, S. H. Gross, and W. E. McGovern, Space Science Review **5,** 565, (1966).
10. G. Colombo, Nature **208,** 575, (1965).
11. A. Dollfus, Polarization Studies of Planets, in *Planets and Satellites*: Solar System III, G. P. Kuiper and B. N. Middlehurst, Eds., Chap. 9, p. 343, University of Chicago Press, Chicago, 1961.

12. A. J. Kliore, D. L. Cain, G. S. Levy, V. R. Eshleman, F. D. Drake, and G. Fjeldbo, Science **149,** 1243, (1965).
13. G. de Vaucouleurs, *Physics of the Planet Mars*, Faber and Faber, London, 1954.
14. H. Spinrad, Spectroscopic Temperature and Pressure Measurements in the Venus Atmosphere, Publ. Astron. Soc. Pacific **74,** 187 (1962).
15. C. Sagan and W. W. Kellogg, The Terrestrial Planets, Annual Rev. Astron. Astrophy. **1,** 235 (1963).
16. S. I. Rasool, Structure of Planetary Atmospheres, A.I.A.A. Journ. **1,** 6 (1963).
17. P. J. Brancazio and A. G. W. Cameron, Eds., *The Origin and Evolution of Atmospheres and Oceans*, John Wiley and Sons, Inc., New York, 1964.
18. A. G. W. Cameron, Physics of the Planets, in *Space Physics*, D. P. LeGalley and A. Rosen, Eds., Chap. 5, p. 127, John Wiley and Sons, Inc., New York, 1964.
19. R. Wildt, H. J. Smith, E. E. Salpeter, and A. G. W. Cameron, The Planet Jupiter, Phys. Today **16,** 19 (1963).

Chapter 19

PLANETARY STRUCTURE

A. G. W. Cameron

There are several planets in the solar system and also one satellite: the moon, whose interiors we shall discuss in this chapter. We have detailed information about the interior of only one of these bodies: the earth. We may expect similar information to be forthcoming about most of the other bodies during the eventual course of the lunar and planetary exploration program of the NASA. Hence what we shall have to say about the other bodies of the solar system must necessarily be mostly speculation at the present time, but for the most part it will be realistic speculation based upon our knowledge of the earth's interior. When we come to the giant planets: Jupiter, Saturn, Uranus, and Neptune, we shall encounter structures that are expected to be very much unlike that of the earth, and our speculation will be on much less firm ground.

We can learn some extremely important facts about planets by studying a few simple characteristics of their exteriors. Knowledge of the size of the planet and of its mass gives us its *mean density*, and this is an extremely important clue to its chemical composition. In the case of the earth, the geologists tell us that special varieties of rock have been brought up with little alteration from the region below the earth's crust, and these give us some indication of the composition at those depths.

The characteristics of a planetary magnetic field can be expected in the long run to give us information about the deep interior where there exist currents that give rise to such a field. At the present time we do not have a very precise knowledge of the mechanisms that give rise to such currents in the core of the earth.

THE INTERIOR OF THE EARTH

Most of our information about the interior of the earth has, however, come from study of the seismic waves which are transmitted through the interior. These waves are generated by earthquakes and consist of several types. Two types of waves can propagate through the deep interior of our planet. These waves are *compressional waves*, or P waves, in which the displacements are longitudinal or parallel to the direction of propagation of the waves; and also *shear*, or S waves, in which the displacements are perpendicular to the direction of motion of the waves. Two other kinds of waves can propagate around the surface or the earth. These are called *Rayleigh* and *Love waves*. The Rayleigh waves consist of a combination of compressional and transverse vertical motions. Love waves consist of transverse horizontal motions. These waves give information about the crust of the earth. In addition to these waves, very intense earthquakes can give rise to oscillations of the earth as a whole, similar to the ringing of a bell. These oscillations have both a radial and a torsional character, in which the motions respectively lie parallel to and perpendicular to the earth's radial direction. These motions give us information about the deep interior of the earth.

In the following discussion we shall present mostly the information that has been obtained from the analysis of the P and S waves which propagate to the deep interior. These waves have been the basis upon which specific models of the earth's interior have been constructed. When these waves, while propagating through the interior, encounter a discontinuity of some kind, then a complex set of reflections and refractions can take place. When any one type of wave, S or P, encounters such a discontinuity, in general both S and P

waves can be produced by the *reflection* and *refraction* phenomena. Thus many different types of waves may be received at one location from any given earthquake, consisting of S and P waves which travel directly from the earthquake center to the observer as well as secondary waves which have resulted from reflection and refraction from discontinuities in the deep interior. In addition there may be surface waves.

The basic properties of *seismic waves* may be derived from the classical theory of elasticity.[1–3] In particular, their velocities may be written in the following form

$$V_p = \left[\frac{K + 4\mu/3}{\rho}\right]^{1/2} \tag{1}$$

$$V_s = [\mu/\rho]^{1/2}. \tag{2}$$

Here K is the *bulk modulus*, or the reciprocal of the compressibility, μ the *rigidity*, and ρ the *density*. We note that P waves travel faster than S waves, and therefore in general they will be the first to arrive from a given earthquake. The theory of elasticity also gives us the important information that in a liquid the rigidity μ is zero and hence the S waves will not be propagated in a liquid. Since S waves have not been observed to traverse the core of the earth, it is generally concluded that the core of the earth has liquid properties.

The basic piece of information available to the seismologist is the time required for various of the seismic waves to reach points on the circumference of the earth at various angular distances away from the earthquake center. From such a set of travel time data the seismologist attempts to reconstruct the paths taken by the waves through the interior, and hence to obtain information about the velocity of the seismic waves as a function of depth in the earth's interior. These curves of velocity versus depth then become the raw material with which the earth model builder works. Figure 1 shows the velocity curves as a function of depth which were obtained by Gutenberg.

We note that in Figure 1 both the P and S waves undergo a slight drop in the very outermost regions of the earth near the surface, and then after traversing a region called the low velocity layer, the velocities of both types of wave rise continuously, although more

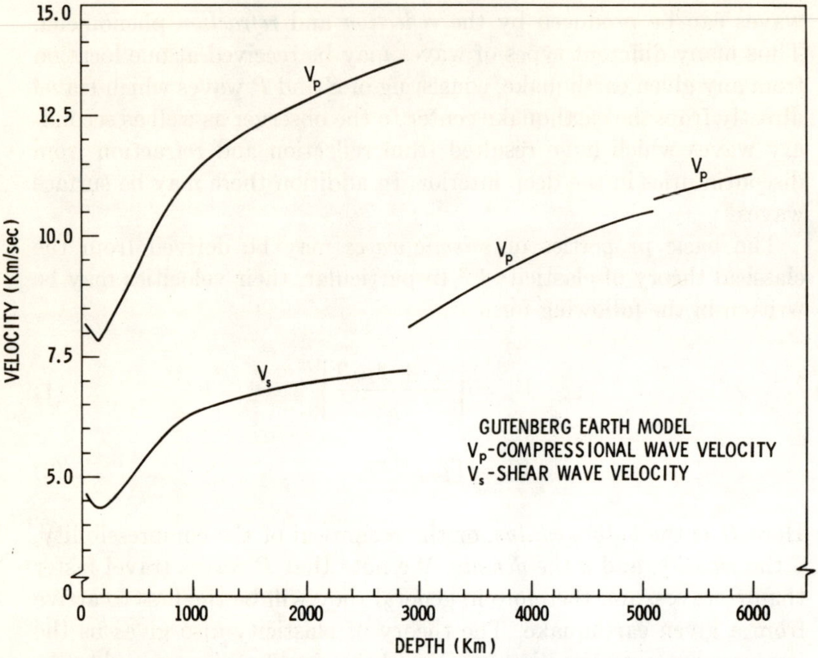

FIG. 1. Seismic velocities as a function of depth in the earth (after Gutenberg).

and more slowly, until a depth of 2900 kilometers is reached. At this point there is a sharp discontinuity and *S* waves are no longer propagated. The outer region of the earth in which the *S* waves propagate is called the *mantle*, and the inner region is called the *core*. At the core boundary the velocity of the *P* waves undergoes a sharp decrease, as can be expected in general from the equation which gives the velocity of the *P* waves, and then the *P* wave velocity once again increases gradually with depth. Further there appears to be a slight discontinuity in the *P* wave velocity at a depth of about 5000 kilometers. Here the *P* wave velocity suddenly jumps to a higher value, and then once more it increases slightly with depth. This innermost region is called the *inner core*, and because the *P* wave velocity undergoes a sharp jump at the boundary of the inner core, it has been speculated that the inner core has essentially solid properties.

We shall now consider how the density distribution in the interior of the earth can be obtained from a knowledge of the seismic wave velocities in the interior. We start with the basic equation of hydrostatic equilibrium of a fluid in a gravitational field, which states that the change in pressure with a small change in radius is equal to the force of gravity acting upon the material which is contained within that interval of radius:

$$dp/dr = -g\rho. \tag{3}$$

Now from our equations for the seismic wave velocities we can obtain the following expression for the bulk modulus

$$K = \rho(V_p^2 - 4V_s^2/3). \tag{4}$$

The bulk modulus is defined as

$$K^{-1} = -v^{-1}(\partial v/\partial p)_T, \tag{5}$$

where v is the *specific volume* equal to the reciprocal of the density ($v = 1/\rho$), or

$$K = \rho(\partial p/\partial \rho)_T. \tag{6}$$

Thus if we neglect temperature effects upon the compressibility we may write

$$dp/d\rho = V_p^2 - \tfrac{4}{3}V_s^2 = F(r), \tag{7}$$

where $F(r)$ is a function that may be determined directly from the velocity curves as a function of depth. Now

$$d\rho/dr = (dp/dr)/(dp/d\rho). \tag{8}$$

We write the local acceleration due to gravity in the form

$$g = GM(r)/r^2, \tag{9}$$

where G is the *universal constant of gravitation* and $M(r)$ the mass interior to the radial distance r. This assumes that we are dealing with a spherical distribution of matter, which is a good assumption for the earth. In order to know the total mass interior to any radial distance r, it is necessary to know the density distribution interior to that point. In practice this simply indicates that we must integrate two simultaneous differential equations, one expressing the change of the density with distance, and the other expressing the change of $M(r)$ with distance. The differential equation which expresses the latter relationship is simply

$$dM(r)/dr = 4\pi r^2 \rho. \tag{10}$$

The final form taken by the equation expressing the change of density with distance now becomes

$$d\rho/dr = -G\rho M(r)/F(r)r^2. \tag{11}$$

It should be noted that these two equations may be integrated in regions where the seismic velocities change smoothly as a function of distance. Clearly it would not be correct to perform these integrations across boundaries in which there is a sudden discontinuity in the velocities. Because there are such discontinuities in the velocity of the seismic waves in the interior of the earth, it is clear that a unique earth model cannot be obtained by a simple integration of these equations away from the center of the earth. What is needed are some additional constraints which allow the discontinuities in density in the interior of the earth to be inferred. Such constraints are provided by our knowledge of the total mass of the earth and by the earth's moment of inertia. The latter quantity is

$$I = 0.3307MR^2, \tag{12}$$

where M is the total mass of the earth and R is its average radius.

A number of attempts have been made to construct a proper density distribution for the interior of the earth. One such solution, after Gutenberg, is shown in Figure 2. The greatest uncertainties that exist in this and other models have to do with the precise density in the inner core of the earth, and this must be regarded as extremely uncertain. However, it may be noted that the typical density of the earth varies from a little less than 4 gm/cm^3 at the surface to nearly 6 gm/cm^3 at the mantle-core boundary. In the core itself the density jumps to about 11 or 12 gm/cm^3, and still higher values seem to

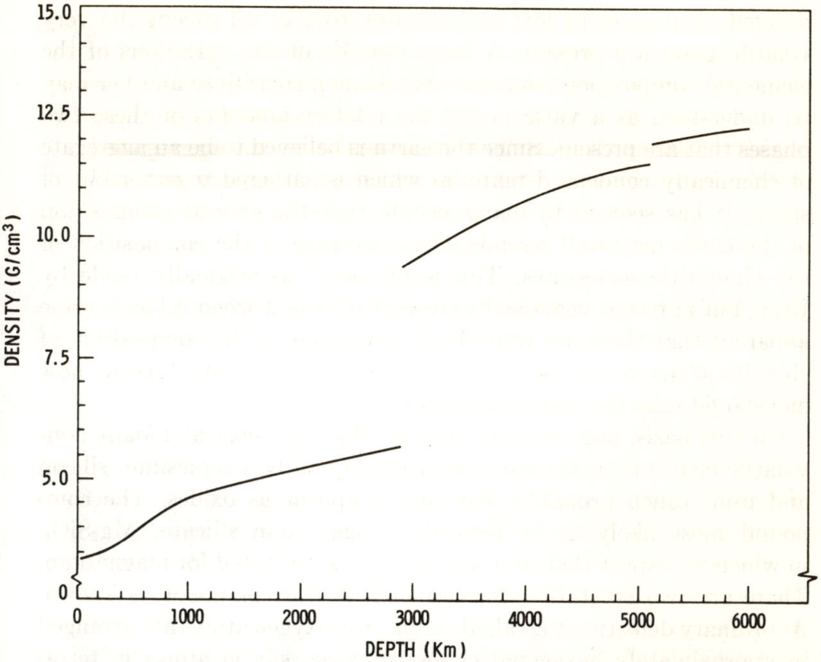

FIG. 2. Density as a function of depth in the earth (after Gutenberg).

exist in the inner core. This density distribution challenges us to see if there can be a reasonable explanation in terms of the chemical composition of the earth's interior.

Composition

One of the most common assumptions in recent years is that the composition of the earth as a whole is probably very similar to the composition of the *chondritic meteorites*.[4–6] These meteorites are stony meteorites which contain little round inclusions, or *chondrules*, a millimeter or so in dimension, which may represent chemical condensations from some primordial solar nebula. Recent work by Anders suggests that the material of these stony meteorites falls into two general classes. In one class it appears that material has been derived from very hot gases, so that only the most refractory materials are present. In the other case it appears that the material has been derived from gases at low temperature, so that all except the very volatile gases are present. A large number of the variations of the elemental composition from one chondritic meteorite to another may be understood as a variation of the relative amounts of these two phases that are present. Since the earth is believed to be an aggregate of chemically condensed material which is gathered together out of space, it has seemed to many people that the overall composition of the earth may well resemble some average of the composition of the chondritic meteorites. This suggestion was originally made by Urey, but in recent years he has retreated from it when it has become apparent that there are quite large variations in the composition of chondritic neteorites, so that great uncertainties arise as to how one should take the proper average.

On this basis, one can then assume that the most abundant non-volatile elements in the earth as a whole will be magnesium, silicon and iron, which probably will largely appear as oxides. The compound most likely to be formed is magnesium silicate, Mg_2SiO_4, in which we expect that iron will often be substituted for magnesium. There are two crystalline forms in which this compound is known. At ordinary densities it is called *olivine*: the oxygen atoms are arranged in approximately hexagonal closest packing, silicon atoms in tetrahedral coordination, and magnesium or iron in octrahedral coordination. At zero pressure the density of olivine is 3.2 gm/cm^3 in the form Mg_2SiO_4, and 4.3 gm/cm^3 in the form Fe_2SiO_4. Another form, or *polymorph*, of this compound can exist which is called *spinel*. This form has a more complicated arrangement of the atoms with oxygen atoms in approximately cubic closest packing. The *olivine–spinel* transition is accompanied by an increase in density of about 10 percent, which will occur at the pressures associated with a depth in the earth of about 200 to 300 kilometers.

As the pressure increases still further, a decomposition into oxides becomes probable, such as

$$Mg_2SiO_4 \text{ (spinel)} \rightarrow 2\,MgO \text{ (periclase)} + SiO_2.$$

At zero pressure SiO_2 is what we call *quartz*, but at higher pressures it forms *coesite* and at still higher pressures ($\sim 1.7 \times 10^{11}$ dynes/cm^3 = 1.7×10^5 bars) coesite forms *stishovite*. In quartz and coesite each silicon atom is surround by a tetrahedron of oxygen atoms, but in stishovite each silicon atom is surrounded by six oxygen atoms at the corners of an irregular octrahedron. Until recently there was no evidence of these higher pressure forms existing in nature, but in recent examinations of the meteor crater in Arizona, small crystals of coesite and also of stishovite have been found which were evidently the result of high pressures produced by the impact. Stishovite was only a theoretical probability until two Russian workers, Stishov and Popova, produced it experimentally in 1961. The density of MgO is 3.59. The densities of the three forms of SiO_2 are 2.65 (quartz), 3.02 (coesite), and 4.35 (stishovite). The combination 2 MgO plus stishovite has the average density of 3.89.

The conversion of olivine to spinel gives an 11 percent volume decrease, while the conversion of olivine to MgO plus stishovite gives a 21 percent volume decrease. These phase transformations will account for much of the increase in density that occurs as one goes deep into the earth's mantle. An additional increase in density will arise from the compression of the materials due to the high pressures present.

There has been much speculation in the past as to whether the large increase in density which occurs at the mantle-core boundary may also represent another *phase change* of some sort in basically olivine material. There are two general arguments which can be put forward to show that this cannot be the case. The first argument consists of theoretical considerations of the equation of state of material at high density. When the densities become very high, then the electronic structure of matter is sufficiently altered, but fairly good representations of the matter can be obtained by using the Thomas–Fermi–Dirac model of the atom. In such a model the electrons are assumed to form a compressed distribution around the central field of the atom, and as the external pressure increased the electrons are forced into a smaller and smaller volume about the atom, and the internal pressure associated with their higher internal velocities correspondingly increases in order to match the high values of the boundary pressure. The higher the pressure, the better is the Thomas–Fermi–Dirac model of the atom. Such a model allows us to determine an average value of atomic charge which must exist for the atoms in the core of the earth in order to correspond to the high densities found there. The average value so found is approximately 24. This suggests strongly that iron, with an atomic number of 26, must form the main constituent of the core, with probably some admixtures of lighter atoms such as silicon and sulfur. It has long been considered that the existence of iron meteorites and of iron phases in stone meteorites constitutes an argument in favor of such a composition.

The second principal argument against a phase change to form the core consists of a comparison of the sound wave velocities that have been determined in various materials at the high pressures generated by shock wave techniques with the sound velocities inferred for the interior of the earth. The velocity of sound in the interior of the earth is a function of the quantity $dp/d\rho$, and we have already seen how this quantity can be obtained in the interior of the earth as a function of the measured values of the seismic wave velocities there. It has been found that the sound wave velocity in the mantle is characteristic of such atoms as Mg or Al, while that in the core is characteristic of such atoms as Cr or Fe. This demonstration gives strong support to the conclusion that the interior of the earth has undergone a chemical differentiation, with the denser Fe flowing toward the center to form the core of the earth, and carrying with it such substances as Ni, Co, and sulfides of various types. The low densities of the rocks in the crust of the earth indicate that apparently a second type of chemical differentiation has occurred in which the material with low melting temperature and small density has risen to the surface to form most of the continental structure of the earth.

We may obtain further insight into this differentiation process by considering the twin questions of the thermal history of the earth and the distribution of radioactivity in the interior. The flow of heat from the interior of the earth clearly depends upon this distribution of the heat sources, or radioactivity, and it also depends upon the initial heat distribution in the earth when it was formed.

Since the general structure of the core and mantle of the earth seems to correspond to the more abundant elements that are found in the chondritic meteorites, let us see what happens when the heat production calculations for the earth are also made upon the assumption that the overall composition is that of the chrondritic meteorites. The principal heat-producing radioactivities are those of U, Th, and K. If we should know the present rate of radioactive heating in the earth, by assuming that all of the heat flow from the interior to the surface is due to the radioactivity, then as we go back in time we find that greater amounts of radioactive heat are produced. Since K has the shortest half-life of the dominant heat producing radioactivities, its heat production goes up most rapidly as we go further into the past.

It is found that we need to know the age of the earth fairly well. This can be found in principle by measuring the growth of a daughter substance relative to the abundance of an initially pure parent. Knowing the radioactive decay constant of the parent, we can readily determine the age from the time when the parent was chemically isolated. The relative abundances of the two lead isotopes which are the end products of the alpha decay chains of U^{238} and U^{235} can give a date for the chemical separation of lead from uranium without knowledge of the abundance of uranium present. Thorium decays to another lead isotope.

By measuring the amount of potassium in a rock, and the amount of argon, particularly of A^{40}, assuming that there is no outgassing and that all of the A^{40} is from the decay of K^{40}, then the date of the last chemical differentiation can be determined.

The oldest rocks associated with the crust seem to have ages of about 3.3 billion years, as determined by these procedures. This means that 3.3×10^9 years ago these rocks solidified and became chemically static systems. Similar techniques show that meteorites have ages of approximately 4.55×10^9 years. Russell has recently used a somewhat complex procedure to determine that the age of the mantle of the earth is also approximately 4.5 billion years. His procedure is based upon an examination of the variations in lead abundance ratios in a large number of rock samples which are believed to be derived more-or-less directly from the mantle. This age presumably corresponds to the time since the primordial lead was removed from the mantle, presumably to be carried to the core as lead sulfide.

Thermal History

If the earth has a chondritic composition, then the abundances of potassium, uranium, and thorium are

$$K = 8.0 \times 10^{-4} \text{ per gram,}$$

$$U = 1.1 \times 10^{-8} \text{ gm/gm,}$$

and

$$\text{Th} = 4.4 \times 10^{-8} \text{ gm/gm.}$$

Using the previous estimates of the age of the earth, the heat production at the present time due to radioactivity in the earth is 3.01×10^{20} ergs/sec, or 59.0 erg/cm^2 sec at the surface, assuming that the heat produced in the interior appears instantaneously at the surface. For the original earth, 4.5 billion years ago, the heat production would be 2.43×10^{21} ergs/sec, or 476 erg/cm^2 sec at the surface.

The measured rates of heat flow at the surface are close to 65 ergs/cm^2 sec. There are local variations but in general the values are remarkably uniform over continental and oceanic areas. The general agreement between theoretical estimates and the measured heat flow has been presented as evidence for a chondritic composition of the earth. It is observed, however, that the uranium and thorium abundances in surface rocks are particularly high, and it appears that the bulk of the uranium and thorium of the earth is probably within these rocks.

For potassium this does not seem to be the case. The equality of heat flow rates over oceans and continents, taken with the fact that the bulk of the heat production is due to potassium, indicates that most of the potassium is in the mantle, but with possible additional strong concentrations in the crust. From geochemical considerations of the chemical properties, none of these radioactive substances is expected to concentrate in the iron phase, and hence the bulk of the earth's radioactivity is assumed to be in the mantle or above.

The problem of the *primordial heat* of the earth must still be considered. This may certainly not be negligible, since if the gravitational potential energy obtained from assembling the earth in its present form were suddenly released in the interior, it would melt the earth completely.

In order to determine the *thermal history* of the earth, it is necessary not only to know the distribution of the heat sources, but also to know the thermal conductivity of the medium. The thermal conductivities of rocks show remarkably little sensitivity to composition, a typical value being 0.03 joules/cm sec °C at room temperature. It appears that the ordinary thermal conductivity decreases with increasing temperature, according to the measurements that are available. However, at high temperatures the thermal conductivity coefficient will increase greatly owing to the onset of the mechanism of energy transport by radiation. A precise determination of such radiative conductivity is very difficult, since it requires a knowledge of the solid state structure of the infrared lattice vibrations at high pressures. There will be certain frequencies at which the lattice vibrations will readily absorb transmitted radiation, while at other frequencies the absorption will be much less. However, it seems safe to say that the radiative transfer in the interior of the earth will be a sufficiently efficient mechanism of energy transport that large temperature gradients are unlikely exist in the deep interior.

G. J. F. MacDonald has made extensive calculations of the thermal history of the earth. If the earth started as a relatively cold body with uniform distribution of radioactivity and no differentiation of material, then it appears that over the course of 4.5×10^9 years the temperature will be raised to the point where everything will melt, from the deep interior up to the surface. This conclusion led MacDonald to try various other possible histories. For instance, the earth might be assumed to have been initially cold and uniform and after a few billion years to heat to the melting point, at which time chemical differentiation occurred. Subsequent cooling then resulted in the present situation. On the other hand it might be assumed that the earth was initially differentiated and cold, or initially differentiated and hot. The differentiated cold earth turns out to be barely possible; if the radioactivities are sufficiently near the surface the present rate of heat flow may be reproduced. However, if differentiation occurs later in the earth's history, then there is too much heat flow at that time, so that the earth must start with a remarkably low temperature. This is an uncomfortable hypothesis because heating is expected from the release of gravitational potential energy upon contraction, and from the compression of the interior regions by gravitation. The formation of the core also presents problems; if the iron is originally distributed uniformly over the volume of the earth, then a subsequent concentration in the core releases gravitational potential energy because iron is much denser than the bulk of terrestrial material. This is sufficient to the raise of the temperature of the interior by about 2000°K. If there is a significant amount of initial heat, and if we add the radioactive heating previously calculated, then the surface heat flow would be considerably in excess of that observed.

MacDonald drew the conclusion that the difficulty with these models lies in the combined assumption that the earth had a significant amount of primordial heat and an overall chondritic composition with respect to radioactivities. He suggested that lowering the abundance of radioactivities by a factor of 2 might produce acceptable solutions.

It is helpful at this point to consider the characteristics of a completely molten model of the earth. The melting temperature in the interior is a function of pressure. It is customary to represent the melting relations in terms of the *Simon semiempirical equation.* The pressure and temperature along the fusion curve are related by

$$P = (a/B)[(T/T_0)^B - 1], \tag{13}$$

where a and B are empirical constants, and T_0 is the melting temperature at zero pressure. Figure 3 shows the schematic representa-

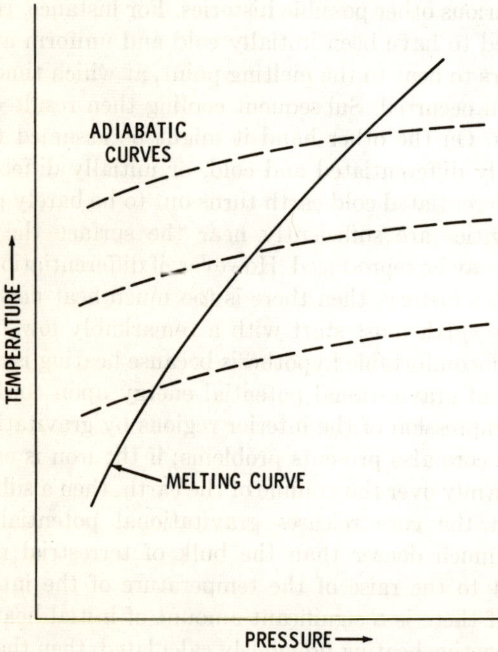

FIG. 3. Schematic adiabatic and fusion curves for solid materials.

tion of this relation. A characteristic feature is that this fusion curve is steeper than the adiabatic gradient. The adiabatic curve represents the temperature rise that will occur in a substance as it is compressed adiabatically without possiblity of exchanging heat with the surroundings.

In a completely molten earth the temperature gradient in the interior could not significantly exceed the adiabatic temperature gradient. Such superadiabatic gradients in the material will drive rapid convective motions in a fluid, and it is evident that the temperature distribution in the interior would rapidly adjust itself to be very close to the adiabatic curve. Figure 4 shows a speculation by Jacobs concerning a possible temperature history of a completely molten earth which could provide the present structure. It may be assumed that initially both mantle and core are completely molten, and the resulting flat temperature distribution would result in a very high temperature at the surface of the earth. This would lead to cooling which would take place rapidly, and the entire temperature distribution in the interior would be lowered from one adiabatic curve to another. Thus from the completely molten adiabatic curve marked 1 we would progress to the adiabatic curve marked 2. Also shown in Figure 4 is a schematic fusion curve for the mantle and the core, and we see that the curve labelled 2 now crosses the fusion curve for the iron core, and that it would produce a solid inner core at this phase. Upon further cooling, to the adiabatic curve marked 3, we see that the mantle might start to solidify before the iron core has become completely solid. It should be emphasized that this fusion curve of Figure 4 is entirely hypothetical, and its only merit lies in the schematic representation of a solid mantle combined with a liquid outer core and a solid inner core. Because heat transport by conduction in the interior of the earth is very slow, while that by fluid convection would be very rapid, the further history of the molten earth would consist of a rapid solidification of the entire mantle, leaving behind a partially molten iron core. This would leave the fusion curve for the mantle as a possible initial temperature distribution that one could assume for a thermal history calculation of the earth. MacDonald, using this assumption, found that a significant fraction of the presently observed heat flow could be due to this primordial heat.

Regardless of whether or not the earth went through an initially

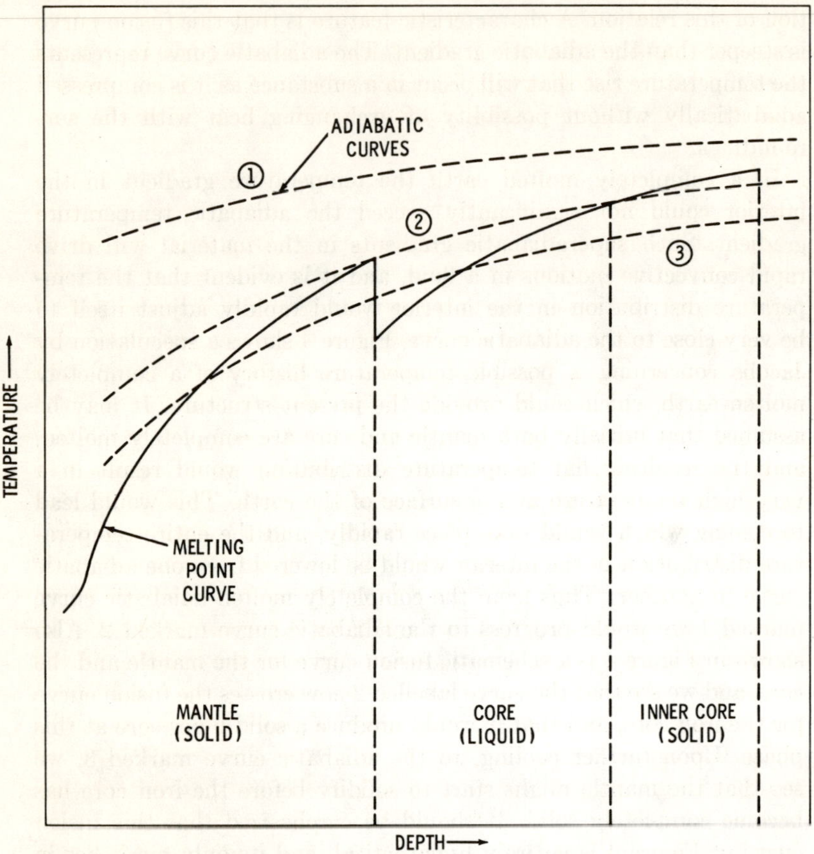

FIG. 4. Possible cooling history of the earth (after Jacobs).

molten stage, the indications are that a fairly large initial temperature is very likely. Thus it seemed to MacDonald that probably one should reduce the amount of radioactivity in the earth's interior, and in practice that means in particular that the amount of potassium present must be reduced relative to chondritic composition. There are several pieces of evidence which support this contention. A number of years ago Gast observed that the abundance of Sr^{87} relative to other isotopes was characteristically different for rocks derived from the mantle than for material in the meteorites. He interpreted this difference to indicate that the bulk content of rubidium in the earth was lower by a factor of 4 than in the meteorites. This conclusion is based upon the fact that the isotope Rb^{87} decays to Sr^{87}. Because rubidium is similar chemically to potassium, one might expect the earth to be underabundant in potassium also. This is also supported quantitatively from extensive analysis of potassium in rocks derived from both crustal and mantle materials which indicate that the potassium content of the earth as a whole is unlikely to be nearly as great as that in chondritic meteorites.

The question of the existence of the geomagnetic field is also related to the thermal and compositional history of the earth. The theoretical problem of the origin of the field is extremely complex. Many years ago it was believed that because the earth had an iron core, and since iron is ferromagnetic, then the magnetic field would have a natural explanation. However, the iron is well beyond the Curie point, and hence we are faced with a more complex problem of generating a magnetic field as the result of the motion of a conducting fluid. A proper solution of this problem would require the simultaneous solution of the equations of fluid dynamics and of Maxwell's equations of the electromagnetic field. While such a solution has not been properly obtained, special solutions corresponding to special assumptions about fluid motions have been made which suggest that the geomagnetic field might be built up as a result of a kind of a self-exciting dynamo action. In this the motion of the conducting fluid would interact with an initial or seed magnetic field in such a way as to convert the kinetic energy of fluid motion into field energy, and thus to reenforce the seed field. Cowling showed some time ago that no axially symmetric flow is capable of building a magnetic field in this way. More recently, a non-axially symmetric flow has been shown to enhance the magnetic field under certain special conditions.

Two types of field configurations are associated with the self-exciting dynamo. The externally observed field is poloidal, and in the case of the earth predominantly dipolar. In the core there is supposedly a toroidal field which is wrapped like a doughnut. The fluid motions are assumed to convert the toroidal field into a poloidal field and back again. There is a pattern of irregularities in the earth's dipole field which is observed to be drifting westward at a constant rate, which has been going on steadily for the last few centuries, and this is presumed to indicate a rotation of the core relative to the mentle.

To derive convective motions in the core, an energy source is needed. Such an energy source must presumably be based upon some form of heating at the center of the earth in order that convective motions can be driven in the fluid core. We have already mentioned the geochemical expectation that the long-lived natural radioactivities should not exist in the core of the earth. One possible driving mechanism might exist if the earth has cooled from an originally completely molten state as is pictured in Figure 4. In this picture the gradual cooling of the iron core of the earth will lead to a gradual increase in the size of the solid inner core, and the heat released upon solidification would then exist as a driving mechanism for convection in the outer molten part of the core. However, this suggestion is very speculative and an open mind must be maintained on this subject.

Malkus[8] has recently suggested a special interaction that may be responsible for building the earth's magnetic field to a much greater extent than one would expect on the other planets. Tidal perturbations of the earth by the moon are remarkably large as compared to those which may exist for any of the other planets. The earth is rotating relatively fast, and consequently it has an equatorial deformation. Because the core is much denser than the mantle, its spheroidal deformation is much less, resulting in a different shape, and, under the influence of the moon's gravitational field, the mantle and core have exerted on them different *precessional torques*. This tends to make the core move relative to the mantle. The westward drift of the main geomagnetic field is taken as evidence of this move ment. Although ordinary viscosity is much too small, magnetic viscosity due to a strong magnetic field in this general region might produce an interaction between the core in the mantle which is strong enough to keep the relative core-mantle motion down to the observed slow westward drift. The fluid motion at the core-mantle interface in the presence of the magnetic viscosity forces is taken to be the principal non-axial motion which is required to build the overall external geomagnetic field. Malkus still needs the basic central energy source to drive convective motion, and he chooses the continued crystalization of the central solid core for this purpose. A consequence of these ideas is the expectation that Venus and Mars should have smaller magnetic fields than the earth.

THE MOON AND PLANETS

We now turn our attention outwards from the earth and try to summarize what can be said about the moon and planets based upon such direct measurements as we can make of their bulk properties from the earth and upon theoretical considerations derived from our knowledge of some aspects of the earth's interior.

We begin with the moon. The mean density of the moon is 3.34 gm/cm^3. In order to compare this density with those of other substances commonly encountered, corrections for compressional effects and thermal expansion must be made. Urey suggests a value of about 3.4 gm/cm^3 for an uncompressed moon at room temperature. The mean density of the earth is 5.5 and that of the chondritic meteorites is in the range 3.57 to 3.76, so that the mean density of the moon appears to be definitely lower than any of these. The maximum pressure reached at the center of the moon would be about 46,000 bars, which is not sufficient to produce any phase transitions in the mantle of the olivine-to-spinel type. Thus the mean density of the moon is directly comparable with typical densities of rocks measured at low pressures. The low mean density of the moon has been a principal problem in the cosmogony of the solar system, and in particular in accounting for the origin of the moon in the past history of the earth–moon system.

Two principal ways have been suggested in which the non-chondritic composition of the moon might be obtained. The first involves the addition of large amounts of light components to chondritic material. Extremely large quantities would be needed; for water the fraction would be 2 or 3 percent and for carbon compounds about 10 percent of the lunar mass. The second suggestion is that there has been a depletion of iron in the moon relative to chondritic composition.

The present figure of the moon does not correspond to hydrostatic equilibrium; the large inequalities in it present clues concerning the history and present condition of the moon. Information about the moments of inertia of the moon is obtained by observing perturbations of the moon's motion. We denote the greatest moment of inertia, about the axis of rotation, by C; the least moment of inertia, A, is about a radius vector pointing toward the earth; and the intermediate moment, B, is about an axis at right angles to the radius vector and the axis of rotation. These quantities are not measured as such, but instead their differences and ratios are found. In particular, the best determined ratio, β, is defined as

$$\beta = (C - A)/B,$$

which depends on the inclination of the axis of rotation to the perpendicular to the plane of the orbit. β is found to be 0.0006279 ± 0.0000015. γ is defined as

$$\gamma = (B - A)/C,$$

which can be derived from observations of the *forced physical libration* of longitude which arises from the attraction of the earth for the moon's equatorial bulge. The numerical value is $\gamma = 0.0002049 \pm 0.0000009$. This involves a slight assumption that will not be discussed here. From observations of the secular motion of the node and perigee of the moon's orbit, one may deduce the quantities

$$(C - A)/mR^2 \quad \text{and} \quad (B - A)/mR^2,$$

where m and R are the mass and radius of the moon, respectively. These determinations are extremely uncertain because the major contribution to the secular motion of the perigee is solar, and the effect sought is $<10^{-5}$ of the solar effect. Consequently the resulting numbers must be considered unreliable. A straightforward reduction gives

$$B/mR^2 = 0.58$$

and

$$C/mR^2 = 0.35.$$

For a homogeneous sphere both of these ratios should be 0.4. At face value, these numbers imply that the outer part of the moon is denser than the inner, which seems very unlikely. When it is possible to establish a satellite in orbit about the moon it will be possible to determine these numbers with much greater precision, and hence to draw more reliable conclusions about the density distribution in the interior of the moon.

Attempts have been made to determine the various radii of the moon accurately; the mean radius of the visible face is 1737.85 ± 0.07

kilometers. Alexandrov has assumed the physical surface to be at equal potential and has then used the mean radius with the mechanical ellipticities to deduce that the three radii are

$$a = 1738.57 \text{ kilometers}$$

$$b = 1738.21 \text{ kilometers}$$

$$c = 1737.49 \text{ kilometers},$$

where the letters correspond to those used previously with the moments of inertia.

The observed values of β and γ differ from what one would expect for hydrostatic equilibrium. The theoretical values would be $\beta_{th} = 0.0000375$, and $\gamma_{th} = 0.0000281$, if the moon were in hydrostatic equilibrium, and thus the observed β is 17 times larger, and the observed γ is 42 times larger, than these values. This led both Urey and MacDonald to note that the interior of the moon is capable of supporting stress differences on the order of 20 bars, in order to preserve these deviations from hydrostatic equilibrium for long periods of time. The stress differences which the moon is capable of supporting are of the same order of magnitude as the stress differences which the earth appears to support in its outer mantle. Jeffreys proposed that the departures from equilibrium would correspond to a solidification of the moon in an equilibrium shape at a time when it was much closer to the earth, but he later abandoned this suggestion.

The moon is known to be receding from the earth gradually due to tidal action which slows the rotation of the earth and transfers angular momentum into the orbital motion of the moon. This may be deduced by comparison of the observed position of the moon in ancient eclipses with that position which would be expected assuming no tidal slowing. It appears however that the observed departures from equilibrium of the figure of the moon do not correspond to an equilibrium figure at any rate of rotation. The large stress differences that are supported in the interior of the moon thus indicate that the temperature in the interior of the moon should not be near or exceeding the melting point because this would render the material incapable of supporting large stress differences over geological time periods. This raises questions concerning the thermal history of the moon.

MacDonald has made an extensive study of the thermal history of the moon as a corollary to his study of the thermal history of the earth. In discussing the thermal history of the earth we noted that because of the rapid increase in pressure with depth, there was a corresponding rapid rise in the melting points of silicate and iron materials. It is because of this increase in melting point in the great depths of the earth to many thousands of degrees that the mantle of the earth can be solid. In the case of the moon the maximum pressure is only 46,000 bars, and consequently there is virtually no elevation of melting point in the deep interior of the moon. If the moon had chondritic composition with respect to radioactivities, about the same amount of heat would be generated per gram as in the case of the earth, so that the interior of the moon would be easier to melt, other things being equal. MacDonald drew the conclusion that if the initial temperature of the moon were 1200°K, then the amount of potassium present must be less than 40 percent of the chondritic value in order that the moon should not be melted. For an initial temperature of 600°K the upper limit for the potassium content is 70 percent of chondritic material, and only for 0°K does it become 100 percent. These conclusions were based upon use of the melting point of *diopside*, a typical material that might be characteristic of the moon's interior. A mixture of materials is likely to give a lower melting point and hence these conclusions are likely to be conservative. Thus it appears that the potassium content of the moon must also be less than that which would be characteristic of chondritic material, if the large stress differences supported by the interior of the moon indicate that the moon has never come anywhere near the melting point over most of the interior.

The Inner Planets

Let us now turn our attention to Mars. The mass of Mars is reasonably well determined from the orbital constants of its satellites, Phobos and Deimos. The radius of Mars is uncertain because it is determined by measuring the apparent size of the disk. This is difficult to do to an accuracy of 1 or 2 percent, particularly because there are refraction effects at the edge of the disk owing to the Martian atmosphere. The orbital motion of the satellites provides an accurate measure of the quantity

$$(C - A)/mR^2,$$

which is defined in an analogous way to the same quantity for the moon. Such information is not available for planets without satellites, such as Venus and Mercury. Estimates of the radius of Mars vary about 3 percent, corresponding to an uncertainty in the mean density of 9 percent. It is hoped that the Martian fly-by experiments will provide a better value for the radius of Mars. A recent determination of the radius of Mars by Dollfus indicates that the mean density is approximately 4 gm/cm^3. This mean density is thus apparently greater than that of the chondritic meteorites, but on the other hand Mars is large enough to have phase changes in its interior.

From the motion of Phobos we have

$$(C - A)/mR^2 = 0.00197,$$

where R is the equatorial radius. If Mars were in hydrostatic equilibrium this number would be ~ 0.00051. In the case of this planet we therefore find a significant departure from hydrostatic equilibrium once again. The question arises: to what extent does the deviation of the figure imply large stress differences? MacDonald has given a dimensional argument to show how these stress differences will scale from one planet to another. The result for Mars indicates that inequalities of figure 6.8 times as large as that of the earth could be supported if the basic strength of materials is the same. The earth supports a deviation of the above ratio,

$$(C - A)/mR^2,$$

which is 1.11 percent of the equilibrium value. For Mars this implies that the structural strength could support a flattening as large as 0.0053, which is comparable to the flattening deduced from the motion of Phobos. This suggests that indeed the mantle of Mars supports stress differences of the same order as the mantle of the earth.

The flattening of the figure of Mars which has been observed by optical techniques amounts to

$$(C - A)/mR^2 = 0.0117.$$

Even if the material within Mars were 10 times as strong as that of the earth, and capable of supporting 20,000 bars stress difference, the maximum observed flattening would only rise to 0.007. At the present time it seems more likely that the visual observations of the flattening are in error than that the material in the mantle of Mars can have a structural strength several orders of magnitude greater than that in the mantle of the earth.

The construction of a density model for Mars requires the assumption of a surface density and a law of density variation with pressure. In order to meet the requirements provided by the total mass and the moment of inertia calculated on the basis of hydrostatic equilibrium, an additional degree of freedom is required. This free parameter can be the radius of the chemically distinct core for a given surface density.

MacDonald has made detailed calculations on possible density models for Mars, assuming that the equation of state for the mantle is that followed by materials in the earth's mantle. This means that phase transitions and major changes in density occur in silicates at pressures around $[1.0 \text{ to } 1.5] \times 10^5$ bars. The principal parameter he used was the size of the iron core and the extent to which iron was still uniformly distributed in the mantle. He concluded that Mars probably has an internal structure intermediate between that of the moon and the earth. He concluded that the differentiation has probably proceeded sufficiently to form a small metallic core, but that the proportion of mass in the core is less than that for the earth. These conclusions are to be taken as extremely tentative.

MacDonald has also calculated models of the thermal constitution and history of Mars, and has arrived at conclusions which are to be expected on the basis of our preceding discussion. The effect of pressure on the melting curve is intermediate between that of the moon and the earth. A model assuming chondritic composition of radioactivities indicates that radioactive heating would cause iron to melt and gravitational differentiation would form an iron core composed of about 10 percent of the total mass of the planet. The resulting surface density would be too low to account for the observed mass and moment of inertia. MacDonald therefore concludes that it is likely that Mars also is depleted in potassium relative to chondritic composition.

Öpik has attempted to calculate deceleration of Phobos due to tidal interaction with Mars. Tidal accelerations vary as a sixth power of the distance of the object from the planet and as the sine of the angle of the phase lag that the tides raise. For a perfectly elastic body there is zero phase lag. As the distance between Phobos and Mars is well known, the uncertainty lies mostly in choosing the phase lag. If the elasticity of material in Mars is similar to that in the earth, the predicted deceleration of Phobos would cause it to collide with Mars in approximately 30 million years from now. This could be interpreted to mean that by chance we are observing Phobos during the last 1 percent of its life, assuming that it has been orbiting Mars for 4.5×10^9 years. On the other hand, this result could be interpreted as meaning that the phase lag for Mars is not the same as that of the earth, and that the outer layers of Mars are probably more elastic than those of the earth. Decreasing the phase lag would greatly increase the prospective lifetime of Phobos. If indeed the material is more elastic, this would seem to indicate that the outer layers of the planet must be farther from the melting point than is the case for the earth, a situation which agrees with MacDonald's suggestion that the composition of Mars is quite different from the chondrites insofar as radioactivity is concerned.

We consider next the interior planets Venus and Mercury. Neither of these planets possesses satellites and hence nothing is as yet known about their figures. The mean density of Venus is very nearly equal to that of the earth, although slightly lower, and the mean density of Mercury is also similar to that of the earth. Because of the close similarity in size of Venus and the earth, similar mean densities are expected if the compositions are roughly the same. This is not the case for Mercury because Mercury is much too small to have any extensive phase changes in the interior; in this respect it is intermediate between the moon and Mars. It seems that Mercury must have a striking composition difference relative to the earth, probably containing more material with higher densities (presumably the metallic phases). The available data seems consistent with a gradation of composition from Mercury to the asteroid belt, the moon being an anomaly. Mercury rotates at a rate close to its orbital angular rate at time of perihelion passage. Venus is not rotating rapidly according to radar measurements, and is rotating slowly in the retrograde sense. All of the other bodies in the solar system are rotating more-or-less rapidly and, with the exception of Uranus, which is tilted slightly more than 90°, they rotate in the same sense. It seems likely that Venus and Mercury were once rotating rapidly and were slowed by solar tides. If we assume that the angle of the tidal bulge on Venus and Mercury is the same as that on the earth, then we can calculate how rapidly slowing will occur due to solar tides. It is then easily found that the couple exerted by the tides on Mercury and Venus would be inadequate to slow them down in 4.5×10^9 years. This suggests that tidal dissipation has been greater in each of these planets that it is in the earth, and therefore it is likely that much higher temperatures must have been present in the upper layers of those planets than is the case for the earth, so that the viscous effects are greater. The ground temperature of Venus is in excess of 600°K, implying that the temperature throughout the upper layers of the planet may be much more elevated than would be the case for the earth. The ground temperature on the sunlight side of Mercury is similarly high.

The Giant Planets

Finally, we turn our attention to the giant planets of the solar system. The mean densities of Jupiter and Saturn are not very different from that of water, and those of Uranus and Neptune are not very much greater. These low densities, taken together with the fact that the masses of the giant planets are much greater than those of the interior terrestrial planets, indicate that the composition of the giant planets is drastically different from that of the terrestrial planets. Indeed, there is now evidence that the composition of Jupiter closely resembles the composition of the sun. The sun has a composition consisting essentially of three-fourths by mass hydrogen, one-fourth by mass helium, and only 2 percent of the mass consisting of all of the other elements. Most of the remaining 2 percent of the mass actually exists in the form of the elements carbon, nitrogen and oxygen, and in a cold planet such as Jupiter these are likely to be present as the compounds CH_4, NH_3, and H_2O. The rocky materials such as the earth is predominantly composed of are likely to constitute only one-fifth as much as the mass of the carbon, nitrogen and oxygen. It may well be that Jupiter has some enrichment of the CH_4, NH_3, and H_2O (which will collectively be called *icy materials*) and of the rocky materials, and it may well be that the increasing densities of the outermost planets indicate still higher contaminations in these bodies of icy and rocky materials, but these points are still matters of speculation.

Most models of Jupiter and Saturn have been based on the assumption that they are completely cold planets, and that the densities throughout most of their bulk do not deviate significantly from the values they would have at a temperature of 0°K. Because the principal constituents are hydrogen and helium, the principal theoretical question involved in the construction of models of these planets must be concerned with the equation of state of cold hydrogen and helium.

To determine the equation of state for cold hydrogen and helium it is necessary to extrapolate data measured at low densities to regions of high densities, guided by a theoretical model such as the Lennard-Jones potential or eventually a Thomas–Fermi–Dirac model of the atom. These models are not very good for hydrogen and helium, so that the exercise of much intuition is necessary, with all of its resulting pitfalls.

This approach indicates that hydrogen is likely to form a cold solid molecular phase which would form the mantle of Jupiter and Saturn, but in the deep interior the high pressures involved will cause hydrogen to form a metallic phase. Our knowledge of the metallic phase of hydrogen is subject to some uncertainties; it was first analyzed by Wigner and Huntington in 1935.

A completely cold model of Jupiter and Saturn is as unlikely to exist as a completely cold model of the earth is to exist. Since these planets are much more massive than the earth, a much larger amount of heating would have occurred when the material was collected together to form them. This gives rise to some question as to whether the mantle of these planets actually consists of a solid state of molecular hydrogen with admixture of helium, or whether in fact it will actually form a fluid state. With the temperature reasonably large in the mantle, more than a couple of thousand degrees at the base of the mantle, then melting is likely to occur. It may well be that the basic models of these planets should consist of metallic hydrogen cores, in which the temperature distribution is nearly isothermal, because of the high thermal conductivity, surrounded by fluid hydrogen and helium mantles along which there is an adiabatic temperature gradient connecting them to the surface of the planet. This is a type of temperature distribution which would be formed by a hot gaseous body which forms a stable structure and cools when the pressure due to internal molecular forces becomes high enough. Such cooling may still be occurring, and there is some evidence indicating that some of the infrared radiation being emitted by the surface of Jupiter must represent a heat flow from the interior rather than mere reradiation of energy received from the sun. It is clear that a great deal of work remains to be done to determine more about the interiors of the giant planets.

The knowledge of the composition and structure of the interiors of the planets and satellites of the solar system is greatly to be desired in order to help unravel the complex history which gave rise to these bodies. Such knowledge will provide important boundary conditions which will apply to any theory of the origin of the solar system. Most of this information can only be obtained by bringing scientific measuring instruments of one kind or another close to the surfaces of the planets and satellites, so that such phenomena as seismic wave propagation can be measured. Many detailed analyses of the compositions of materials at the surfaces of these planets and materials derived from deep drilling will also be required in order properly to elucidate the situations that are present. These investigations should provide an important facet of the future investigations of the planets within the space program.

BIBLIOGRAPHY

1. K. E. Bullen, An Introduction to the Theory of Seismology, 3rd ed., Cambridge University Press, Cambridge, England, 1963.
2. L. D. Landau and E. M. Lifshitz, Theory of Elasticity, Addison-Wesley, Cambridge, Massachusetts, 1959.
3. B. Gutenberg, Physics of the Earth's Interior, Academic Press, New York, 1959.
4. G. J. F. MacDonald, On the Internal Constitution of the Inner Planets, J. Geophys. Res. **67,** 2945 (1962).
5. G. J. F. MacDonald, Calculations on the Thermal History of the Earth, J. Geophys. Res. **64,** 1967 (1959).
6. G. J. F. MacDonald, The Internal Constitutions of the Inner Planets and the Moon, Space Sci. Rev. **2,** 437 (1963).
7. J. A. Jacobs, Some Aspects of the Thermal History of the Earth, Geophys. J. **4,** 267 (1961).
8. W. V. R. Malkus, Precessional Torques as the Cause of Geomagnetism, J. Geophys. Res. **68,** 2871 (1963).

Chapter 20

SPACE ASTRONOMY

Albert Boggess III and C. E. Fichtel

INTRODUCTION

Astronomy is one of the most interesting and yet most challenging fields of science. It is concerned with the physics of the whole universe, and the problems to be investigated involve diverse processes occurring on scales which cannot be imagined in the laboratory. Moreover, the information at our disposal is severely restricted. One cannot touch a star; neither can one taste it or smell it. Experiments cannot be performed in the classical sense, where all variables are controlled except the one under study. All of our information about stars and the space they inhabit must be inferred from observations of their electromagnetic radiations. Studies of the kinetics of stellar systems require only measurements of the directions and brightnesses of the stars, but investigations of the fundamental physical processes occurring in stars and in the interstellar medium require a detailed examination of the electromagnetic spectrum over a wide range of frequencies. Quantitative measures of individual lines and of the overall shape of the spectrum provide excellent clues to the nature of the source of the radiation. If the distance to the source can be deduced, the total amount of energy released may be calculated. In general, the lower frequencies or longer wavelengths in the radio and optical regions can reveal the nature of atomic processes in stars and interstellar material, while the high frequency radiations in the x-ray and γ-ray regions provide information on nuclear processes and particle acceleration and deceleration.

Since the only means of learning about a star is to detect radiation from it, the way the intervening medium affects and modifies the stellar radiation is of particular importance. In the case of observations from the surface of the earth, the dominant portion of that medium is the earth's own atmosphere, which acts like an unstable filter, variable in transmission, color, and optical resolution. These properties place severe restrictions on the kinds of observations that can be made through the atmosphere. Of course, daytime observations of all but the brightest planets are impossible because of the blue sky produced by Rayleigh scattering of sunlight. This scattering is so strong that even the moon at night can illuminate the sky enough to prevent accurate photometry. Not only does the atmosphere scatter radiation incident upon it, but it emits light itself as a result of photochemical reactions, primarily involving oxygen, nitrogen, sodium, and the hydroxyl radical. This airglow is produced in a ten to twenty kilometer thick zone at an altitude of about 100 km, and is intense enough to be easily detected by the naked eye on a dark night away from city lights. Under these conditions, the brightness of the sky increases markedly from the zenith to the horizon because of the increasing line of sight through the emitting zone as one looks toward the horizon. The presence of the airglow places a limit on how faint an object can be seen from the earth's surface, since it will ultimately fog a photographic plate or result in an unusable photoelectric signal-to-noise ratio.

In addition to producing background radiation by scattering and emission, the atmosphere is an absorber. In Fig. 1 the height at which incident radiation is attenuated by 50% is shown. Below 0.3μ the atmosphere is virtually opaque. Most of the ultraviolet and all of the x-ray and γ-ray interval is excluded from ground observations, except for very crude measurements of the highest energy γ rays made by studying secondary effects resulting from interactions in the atmosphere. Beyond 0.7μ there are many broad absorption features pre-

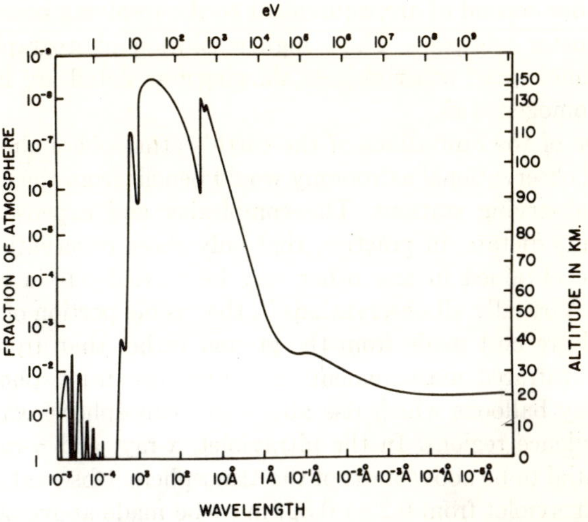

FIG. 1. Curve showing the altitude at which the electromagnetic radiation incident on top of the atmosphere is reduced to half its intensity as a function of both energy and wave length.

dominantly due to water vapor and carbon dioxide. Thus, the atmosphere can be considered transparent only in the near ultraviolet and visible, from 0.3 to 0.7μ, and in discrete windows in the infra-red. It becomes transparent once more at a few millimeters wavelength extending on through the radio-frequency spectrum. Even in the region where the atmosphere is optically transparent, the actual transmission is highly variable due to haze conditions and clouds, and it is also a function of the total air mass in the line of sight. Thus, at 0.5μ, 85 percent of the light from a star at the zenith reaches the earth's surface, but only 72 percent is received from a star thirty degrees above the horizon because the light traverses twice as much air. This fact greatly complicates even a simple measurement of the relative brightness of two stars in different parts of the sky, and an absolute determination of the brightness of a particular star requires many laborious steps in order to extrapolate the observations to outside the earth's atmosphere.

Finally, observations made from the surface of the earth are affected by thermal convection and turbulence in the troposphere. These motions limit the photographic resolution normally attainable to about one second of arc, equivalent to the resolving power of a six inch diameter telescope. Thus, high resolution photographs of the moon, planets, and other objects showing fine detail are limited by these inhomogeneities.

Because of the limitations of the earth's atmosphere almost every branch of observational astronomy would benefit from the use of high altitude observing stations. The complexity and expense of space experiments dictate, in practice, that only those observations which cannot be obtained in any other way be carried out in rockets or satellites. Virtually all observations in the visible portion of the spectrum still are best made from the ground rather than from a space platform. Infrared measurements and high resolution photography can employ balloons which rise above the atmospheric water vapor and turbulence regions. In the ultraviolet, x-ray, and γ-ray regions it is essential to be above most of the atmosphere. Observations in the middle ultraviolet from 0.2 to 0.3μ, must be made above 50 km, the top of the ozonosphere. At that altitude one must still look through the bright ultraviolet airglow due to the Herzberg bands of molecular oxygen. This radiation interferes up to 100 km. To make photometric measurements in the far ultraviolet between 0.1 and 0.2μ, a height of 200 km must be attained before severe absorption is eliminated. In x-ray and γ-ray astronomy, some valuable information can be obtained from large high altitude balloons which reach 40 km, but the ultimate experiments again have to be performed in rockets and satellites.

In the next sections, some of the problems which have been examined in the different spectral ranges will be discussed. In each case some background for the problem is given, the observational techniques are described, and the observations together with their possible interpretation is discussed. In the final section there is a brief summary of the present status of space astronomy, indicating a few of the outstanding research problems yet to be examined.

ULTRAVIOLET AND INFRARED RADIATION

The range of problems that can be studied in the ultraviolet is very broad. The strongest spectroscopic lines of many elements lie in the ultraviolet; in particular, the two most abundant elements in the universe, hydrogen and helium. A number of these transitions are sufficiently strong that they can play a significant role in the energy balance of certain stars. Measurements of intensities and profiles of these lines would give important information concerning temperature distributions within stellar atmospheres. Important subjects for study are the colors and spectral energy distributions of the hotter stars. Stars with effective temperatures greater than 10,000°K radiate most of their energy in the ultraviolet. In effect, our knowledge of these stars is extrapolated from ground observations of the long wavelength tails of their black body distributions, where the information content of the radiation is relatively low. Measurements near the peak and on the short wavelength side of their energy distributions would specify the temperatures and other properties of their atmospheres much more reliably than has been possible in the past. These hotter stars are of particular significance since they evolve rapidly and it is possible to study a wide range of evolutionary types. Moreover, the intense ultraviolet radiation from these stars couples strongly with the gas in interstellar space and has great influence on the physical state of the interstellar medium.

The interstellar medium is itself an important subject for study. Star formation is visualized as a process of condensation out of the surrounding medium, and there is enough material in interstellar space to exert a strong influence on the kinematic properties of the Galaxy. This material is composed of gas plus an admixture of atoms and molecules frozen together to form particles or grains. It can be detected optically in several ways. The gaseous component produces absorption lines in stellar spectra, and, under the influence of stellar ultraviolet radiation, it flouresces to produce emission nebulae. Spectroscopic observations in the ultraviolet would help in determining atomic abundance ratios in the interstellar gas and in studying excitation mechanisms in emission nebulae. The interstellar grains scatter stellar radiation, and measurements of this scattering are important not only for the interpretation of stellar flux observations but also because knowledge of the ultraviolet scattering can give valuable clues as to the composition of the grains themselves.

Just as ultraviolet observations are required of hot stars, the coolest stars are best studied where they radiate most efficiently, in the infrared. In addition, many molecular constituents in cool stars and in the atmospheres of planets can be observed by their prominent infrared spectra. In the far infrared it may be possible to measure thermal radiation from the interstellar grains, or even to detect energy from cold planet-like bodies outside of the solar system.

Infrared Balloon Observations

The only high altitude infrared data to be obtained so far has been in the near infrared where the primary requirement is to rise above the atmospheric water vapor. For this purpose balloons reaching altitudes of 20 to 30 kilometers are satisfactory, and provide fairly stable platforms for periods of a few hours.

Balloon observations of the infrared spectrum of Venus have been made by Strong and his associates[1,2]. An f/1.5 Schmidt telescope of 30 cm aperture is pointed toward the planet by an automatic star tracker. Two types of observations were made. In one case, observations were made at twenty-one wavelengths which corresponded to particular absorption lines in the water vapor band at 1.13 μ. These absorption lines were compared to the adjacent continuum in order to determine the amount of water vapor in the Venusian atmosphere. The second type of instrumentation scanned the spectrum from 1.7 to 3.4μ with 0.08μ resolution. The resulting data, when compared with similar scans of the sun, yield the reflectivity of the cloud layer in this spectral region. These observations detected small but measurable amounts of water in the Venus atmosphere. However, this result is at variance with a recent observation by G. P. Kuiper, who used an infra-red spectrometer in high-flying airplane to measure the Venusian spectrum and was unable to find any trace of water.

The most ambitious astronomical balloon program that has been carried out to date is the Stratoscope II project[3,4]. Stratoscope II consists of a thirty six inch telescope and an infrared prism spectrometer that scans from 0.08 to 3.2 μ. Guidance is actively controlled from the ground by observing a television display and transmitting commands back to the telescope. Observations of Mars with this instrument give an indication of the presence of water vapor, with an upper limit of 40 μ of precipitable water in the atmosphere. The exact amount of water requires an assumption concerning the total pressure, one of the most elusive parameters of the Martian atmosphere.

In addition to planetary observations, Stratoscope II has been used to measure the infrared spectra of several cool giant stars. These stars produce most of their energy in the infrared and the nature of their spectra at these wavelengths is of significance in the overall structure of their atmospheres. The continuous spectrum of these cool stars is determined largely by continuous absorption due to the negative hydrogen ion in the stellar atmosphere. Ionizations of H^- produce a maximum absorption near 0.8 μ which decreases rapidly to larger wavelengths, becoming zero at 1.64 μ. On the other hand, energy losses during near collisions of hydrogen atoms and passing electrons result in a rapidly increasing absorption to wavelengths longer than 1.6 μ. As a result, the atmosphere has a maximum transparency near this wavelength, permitting energy from the hotter lower atmosphere to escape. This prediction is confirmed for stars with surface temperatures from 2200° to 3500°K, which show emission peaks near 1.6 μ when compared with black bodies at the surface temperatures. In addition, the Stratoscope II spectra show an absorption band at 2.35 μ due to CO, and a number of absorption bands of H_2O. These water vapor bands become progressively more intense for cooler stars. The coolest star observed is *o* Ceti, with a surface temperature of 2200°K. In this object the water vapor bands absorb about one-fourth the total stellar energy. Absorptions of this magnitude will significantly affect the structure of a star's atmosphere and will have to be considered in deriving theoretical atmospheric models.

Ultraviolet Instrumentation

The first problem which is encountered in constructing ultraviolet optical systems is the scarcity of useful optical materials at these wavelengths. There are perhaps half a dozen substances suitable for lens manufacture, ranging from fused silica which becomes opaque at 0.16 μ down to lithium fluoride which becomes opaque at 0.11 μ. No lens material is known which can transmit to shorter wavelengths. As a result, while simple transmission systems can be made to work above about 0.18 μ, below this wavelength reflecting optics must be employed. Even mirrors have limited efficiency in the ultraviolet. By coating aluminum mirrors with thin evaporated films of magnesium fluoride, reflectivities as high as 80 percent have been achieved near 0.12 μ. But a conventional reflecting telescope and spectrometer might employ five reflecting surfaces, permitting only $(0.8)^5 = 0.33$ of the light to reach the detector. Since observations of stars and nebulae are invariably limited by the amount of energy available, even under the most favorable conditions, the design of the optical system becomes an important factor.

The method of detection is the next point to be considered. Special photographic films, sensitive to ultraviolet light, are available. Photography has the supreme advantage of recording large quantities of data easily and simply. However, film is not capable of high photometric precision, and, in any case, it is generally difficult to recover film except from manned satellites. Photomultipliers are very satisfactory for spectroscopic observations where the spectrometer provides the required spectral resolution, but if a simple photometer is being built to measure stellar colors, a filter must be used to define the spectral band to be studied. Filter materials in the ultraviolet are as difficult to find as lens materials[5,6]. Interference filters have been produced covering wavelength bands down to about 0.18 μ, but at shorter wavelengths they are less successful. One of the most useful detectors on the 0.10 to 0.15 μ range has been the ion chamber, filled with a gas which ionizes under the influence of radiation at wavelengths shorter than the ionization limit of the gas. The chamber window is transparent only to those wavelengths longer than the absorption limit of the window material. By choosing suitable fill gases and window materials, the spectral sensitivity of the ion chamber can be changed to suit the exact requirements of the experiment.

An important aspect of any photometric device is its calibration. This step is made more difficult in the ultraviolet because oxygen becomes opaque below 0.2 μ, so that calibration procedures must be carried out in vacuum. Moreover, radiation standards are generally not available. It is possible to calibrate gas discharge tubes as monochromatic radiation standards above about 0.2 μ, using thermocouples. The mercury line at 0.2537 μ is convenient for this purpose. At shorter wavelengths thermocouples become very insensitive, and it is difficult to make sure that the sensitive surface of a thermocouple is sufficiently gray to provide a meaningful calibration. Instead, it has become customary to use the known ionization efficiency of nitric oxide at 0.1216 μ, the Lyman-α line of atomic hydrogen. Eighty one percent of the Lyman α quanta absorbed by nitric oxide produce ion pairs, and this fact permits a nitric oxide-filled ion chamber to be used to calibrate a lamp at that wavelength. Once radiation standards are established at hydrogen Lyman α and at the mercury line, it becomes possible to interpolate the calibration between these wavelengths by means of a phosphur such as sodium salicylate, which has a flourescent quantum efficiency virtually independent of wavelength throughout this spectral region.

Of course, the greatest problem in obtaining ultraviolet observations is that of providing a suitable observing platform above the atmosphere. Very little data is available yet from satellites containing astronomical instrumentation. A series of Orbiting Astronomical Observatories, employing telescopes with optics as large as thirty-six inches in diameter, is under construction.[7] The first of these, launched in the Spring of 1966, failed due to a faulty power supply. The second OAO is scheduled for launch in 1968. Most of the data that has been obtained has come from sounding rocket flights. The most commonly used vehicle is the Aerobee, a 15 inch diameter rocket capable of lifting a 250 pound payload to an altitude of 200 km and providing nearly five minutes observing time above the airglow layer at 100 km. Until recently it was not possible to orient the rocket or point the instrumentation in any particular direction. Under these circumstances, telescopes are mounted looking out the side of the rocket so that they can scan the sky as the rocket spins and precesses. A typical flight will survey 50 or 60 percent of the sky above the horizon in this fashion. There are several disadvantages to this procedure: on the average, a telescope will spend half the available time looking back at the earth instead of above the horizon; rather than choosing the targets to be studied, one must accept whatever data the rocket motion provides; the post-flight identification of which stars were recorded is very laborious. In spite of these difficulties, several kinds of instruments have been flown very profitably.

The simplest type of observation to obtain from an uncontrolled rocket is filter photometry. The object here is to measure stellar energies in a specific wavelength interval defined by the bandpass of a filter or by the spectral characteristics of the detector. Above 0.20 μ, where filters and lens materials are available, instruments like the one in Fig. 2 are used. Here a 60 mm diameter quartz lens is the objective. At the focal plane, inside the photometer housing, a field stop limits the field of view to a square piece of sky four degrees on a side. All the light coming from that four degree square passes through a filter and is focused on a photomultiplier. On a successful flight, photometers like these will record perhaps a hundred stellar signals, which can

FIG. 2. A flight photometer. Light passes through a filter inside the case and is focused on a photomultiplier. The electrometer amplifier is mounted on top.

then be identified and analyzed to determine the brightnesses of the stars in the wavelength intervals employed. The most successful satellite instrumentation to date has also consisted of filter photometers. A. M. Smith used an uncontrolled satellite to scan the sky with a small reflecting telescope and detector sensitive to radiation from 0.13 to 0.16μ. During the 84 day life of the experiment, he obtained high quality data on 96 stars.

Uncontrolled rockets have also been used to obtain low resolution spectra. Fig. 4 shows a diagram of a scanning spectrometer in which radiation from a star strikes one of two gratings which disperse the light. A parabolic mirror then focusses the spectrum on a slit. The instrument is pointed out the side of the rocket so that the rocket rotation moves the spectrum across the slit. The photomultiplier signal is again amplified and transmitted to ground. The entrance baffle is designed to limit the field of view so that unwanted background radiation does not enter the spectrometer. Stellar spectra with 50 Angstrom resolution have been recorded in this way[8].

Recently, control systems have been developed which can point the rocket toward a particular star, making it possible to study individual stars in more detail. Fig. 5 shows a telescope and slit spectrometer used for ultraviolet spectrophotometry. This instrument is designed to fill an Aerobee payload. The primary mirror is over twelve inches in diameter. Another type of instrument currently in use consists of a small telescopic camera with a grating mounted in front to disperse stellar light so that a spectrum can be photographed. Cameras of this type have been flown in rockets and obtained stellar spectra with one Angstrom resolution. Another version, shown in Fig. 3, has been used by astronauts to photograph stellar spectra during Gemini Flights (see Chapter 14).

FIG. 3. Camera used for stellar spectroscopy on Gemini flights. The 70 mm camera has a 73mm focal length, f/3.3 lens achromatized for the ultraviolet. The reflection grating is mounted in front of the lens and provides a dispersion of 180 Angstroms/mm (Henize and Wackerling, Ref. 10).

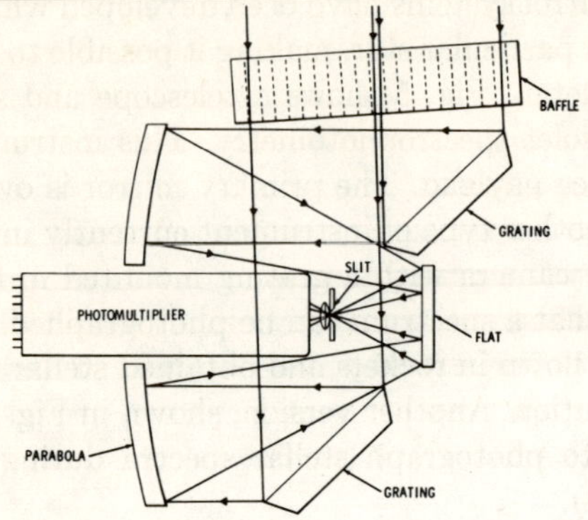

FIG. 4. An unguided scanning spectrometer. Stellar radiation traverses the straylight baffle, is dispersed by the gratings and focused at the slit. Rotation of the instrument produces a scan of the spectrum (Stecher and Milligan, Ref. 8).

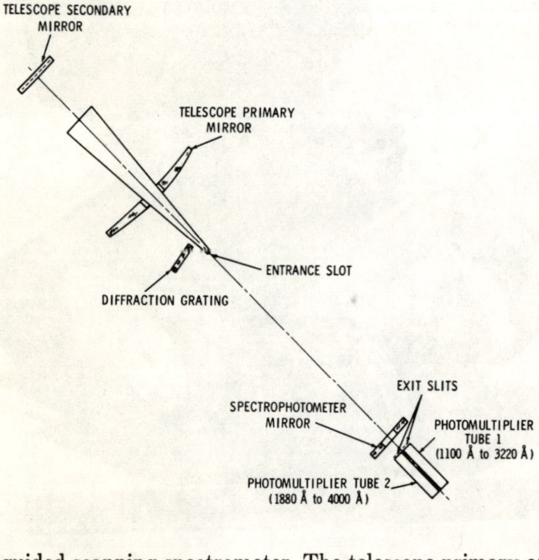

FIG. 5. A guided scanning spectrometer. The telescope primary and secondary mirrors focus the stellar image on the entrance slot of the spectrometer. The spectrophotometer mirror collects the star light and reflects it back to the grating, which disperses the light onto the exit slits. Rotation of the grating produces a spectral scan. (Wright, Ref. 9).

These are some of the instruments now being flown to obtain astronomical data. Many others are under development which will permit increased accuracy, better resolution, and extension of the observations to shorter wavelengths. However, those now in use have already yielded much valuable information.

Ultraviolet Spectra of Planets

The dominant features in the spectrum of Jupiter are methane and ammonia, which produce a number of heavy absorption bands throughout the red portion of the spectrum[11]. It is generally assumed that the atmosphere is composed primarily of hydrogen and helium, the exact amounts depending on the manner in which the Jovian proto-planet formed and evolved. Quadrupole bands of molecular hydrogen have been detected in the spectrum, but analyses of their intensities to derive the amount of hydrogen above the cloud layer yield widely discordant results, ranging from 5 to 270 kilometers of H_2 reduced to S.T.P.

Additional light can be thrown on this problem by considering the scattering properties of the atmosphere, especially in the ultraviolet. When radiation of intensity I_0 falls upon a scattering medium, the amount scattered out of the beam is given by

$$I = I_0 [1 - \exp(-\sigma\chi)], \tag{1}$$

where χ is the thickness of the medium and σ the scattering coefficient. In the case of the Jovian atmosphere, as with the earth, the scattering coefficient, σ, is inversely proportional to λ^4, the proportionality constant being determined by the scattering material. As a result, shorter wavelengths are scattered much more strongly than long wavelengths, producing the earth's blue sky. In the ultraviolet, the measured reflectivity of Jupiter should increase towards shorter wavelengths as a larger and larger percentage of the incident light is scattered back out of the atmosphere. This reflectivity obviously can be related to the quantity $(\sigma\chi)$, so that a value for the amount of scattering material, χ, can be derived.

Several ultraviolet spectra of Jupiter have been obtained both photoelectrically and photographically[12]. The ultraviolet reflec-

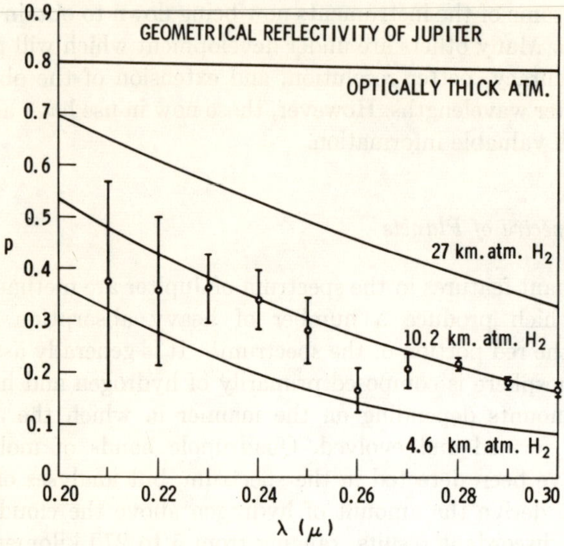

FIG. 6. Geometrical reflectivity of Jupiter. Observations are compared to theoretical curves for different amounts of H_2. The absorption at 0.26 μ may be due to photodissociation of H_2 (Stecher, Ref. 12).

tivities derived from these spectra are shown in Fig. 6, together with computed curves assuming different amounts of hydrogen in the Jovian atmosphere. It is seen that the data follow the curve for 10.2 km of H_2 at S.T.P. These curves were computed assuming a pure hydrogen atmosphere. However, molecular hydrogen is so much more efficient at scattering than helium, the only other significant constituent, that even very large amounts of helium would not drastically change the result. It is concluded that Jupiter's atmosphere contains between 8 and 10 km of molecular hydrogen at S.T.P., the exact amount depending on the amount of helium present.

A similar spectrum of Venus has been obtained. In this case, the planet is too small to retain any appreciable atmosphere of hydrogen or helium, and the dominant constituent is expected to be molecular nitrogen. However, instead of the reflectivity curve that would be expected from a purely scattering atmosphere, the spectrum shows significant absorption below 0.30 μ. This may be interpreted as due to minute amounts of ozone existing just above the cloud layer. Analysis of the data indicates that 0.03 cm of ozone, with a maximum concentration occurring near the top of the clouds, would account for the observed reflectivity. The ultraviolet bands of ozone are a very sensitive indicator of the presence of oxygen, and the amounts of oxygen suggested would not have been detected with previous techniques.

Ultraviolet spectra of Mars[13] show no large absorptions; the spectrum appears to be due to Rayleigh scattering, but through a very thin atmosphere rather than a thick, dense atmosphere like that of Jupiter. The scattering curve corresponds to an atmosphere with only 6 to 8 mm total pressure. It is possible that a cloud or haze layer opaque to ultraviolet radiation prevents one from seeing deeply into the atmosphere. However, the value of the surface pressure determined from the occultation experiment on Mariner IV[14] agrees exactly with that determined from the ultraviolet spectra, indicating that the atmosphere is transparent in the 2000 to 3000 Angstrom region. The resulting high level of germicidal radiation at the surface of the planet is another example of the unfavorable environment existing on Mars.

Ultraviolet Stellar Observations

Stars are classified according to the nature of their absorption line spectra. The state of excitation and ionization of the atomic constituents in the stellar atmosphere depends on the temperature and electron density in the atmosphere, so that, as these parameters vary from one star to the next, the appearance of the absorption spectrum changes in an orderly fashion. This makes it possible to set up an empirical classification scheme as a qualitative description of the stars.[15] It has long been recognized that these classifications are correlated with intrinsic stellar luminosity, and when spectral type is plotted against absolute magnitude, as in Fig. 7, the vast majority of the stars fall along a well-defined main sequence. The fact that some stars fall outside this primary correlation is attributed to evolutionary processes which are discussed in a later chapter.

One of the most straightforward, yet most fundamental, observations that can be made of a star is its color. A color is defined in terms of stellar magnitudes as

$$C(\lambda_1, \lambda_2) = 2.5 \log \int E_\lambda s_1(\lambda)\, d\lambda - 2.5 \log \int E_\lambda s_2(\lambda)\, d\lambda + K, \quad (2)$$

where E is the irradiance due to the star at wavelength λ, s_1 and s_2 are response curves of filters having effective wavelengths λ_1 and λ_2,

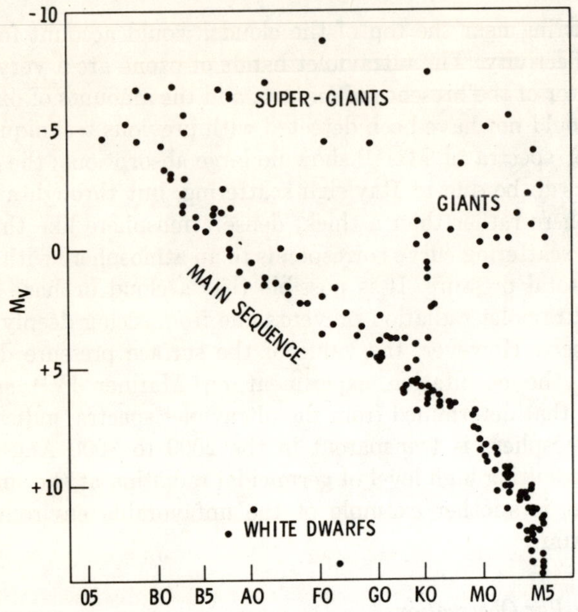

FIG. 7. Schematic plot of stellar luminosity versus spectral type. The vast majority of stars fall in the main sequence, other regions of the diagram being populated by evolutionary processes (modified from Keenan, Ref. 15).

and K is chosen to make $C = 0$ for stars of spectral type A0. Thus a color requires a brightness measurement at two wavelengths and, in general, the numerical value depends upon which wavelengths are chosen. In ground observations three bandpasses, defined as the U, B, V system[16], are most widely used. Their effective wavelengths are: U, 0.36 μ; B, 0.44 μ; and V, 0.55 μ; and the colors measured on this system are referred to as U–B and B–V. Positive colors indicate stars bright at long wavelengths; negative colors indicate stars bright at short wavelengths. Clearly, these colors must be correlated in some way with stellar temperatures, blue stars with negative color being hotter than the positive color red stars, and they must therefore be related to spectral class. An understanding of the exact nature of this correlation requires a thorough knowledge of the physical conditions and radiative properties of the stellar atmosphere.

The atmosphere of a star is defined as those regions from which measurable amounts of radiation reach the observer directly; in other words, the outer, relatively transparent layers of the star. It can be specified in terms of the star's effective temperature, T_e, its surface gravity, g, and its chemical composition, A. In deriving a model of the atmosphere, the temperature, T, and total pressure, p, are computed as functions of the optical depth, τ. τ is equivalent to the exponent, $(\sigma\chi)$, in equation (1), except that it is defined in terms of the total absorption coefficient per gram of material, k:

$$d\tau = k\rho d\chi, \tag{3}$$

$$k = k(p, T, A), \tag{4}$$

and the density, ρ, is given by the perfect gas law:

$$p/\rho T = \text{constant}. \tag{5}$$

The requirement of hydrostatic equilibrium, i.e., that the pressure at each point in the atmosphere be sufficient to support the overlying layers, requires that

$$dp = g\rho \, d\chi. \tag{6}$$

Finally, the theory of radiative transfer determines the variation of temperature with optical depth:

$$T = T_e f(\tau), \tag{7}$$

where the effective temperature, T_e, is related to the total stellar radiance through the Stefan-Boltzmann law:

$$\mathscr{F}_i = \sigma_R T_e^4, \tag{8}$$

σ_R being the radiation constant. Equations (3) through (7) represent a set of simultaneous relations which determine p, ρ, and T as a function of depth in the atmosphere when T_e, g, and A are specified.[17,18]. Once the model is obtained, the monochromatic absorption coefficients, $k_\nu(p, T, A)$, can be computed for each point in the atmosphere, and it is then possible to calculate the monochromatic fluxes emerging at the top of the atmosphere. These fluxes may be used to compute theoretical colors from Equation (2) for comparison with observations.

The relative abundances of the atmospheric constituents are essentially the same for large numbers of stars, so that, in principal, measurements of only two colors such as $(U\text{–}B)$ and $(B\text{–}V)$ suffice to determine the two basic parameters of the atmosphere: effective temperature, and surface gravity. The difficulty with this concept is illustrated in Fig. 8, where flux curves predicted from model atmospheres are shown for a G0 star, only slightly hotter than the sun, and for a B0 star with effective temperature roughly 25,000°K. It may be seen that, while the U, B, V wavelengths span a significant portion of the cooler star's radiation, their base line is not long enough to adequately describe the hot star's energy distribution. It is for the hot O and B stars that ultraviolet observations are particularly needed.

Useful data for this purpose have been obtained with filter photometers having effective wavelengths at 0.22 and 0.26 μ[20]. When colors derived from these photometer observations are compared with conventional $(B\text{–}V)$ colors, a correlation like that in Fig. 9 is obtained. Additional valuable information comes from observations made with the scanning spectrometer described in Fig. 4.[21] These observations may be compared with fluxes predicted from various model atmospheres in order to find values of T_e and g which give the best fit. Fig. 10 shows spectral scan data of the $B3$ star η Ursae Majoris, together with the flux curve from a model atmosphere with $T_e = 18{,}000$°K and $\log g = 4.9$.

It may be seen that the data agree very satisfactorily with the energy distribution of the model except at the shortest wavelength where the sensitivity of the spectrometer is quite low. Using observations of this type and the filter photometry observations, a provisional temperature scale has been derived based on the available ultraviolet data. This scale is shown in Fig. 11 together with the temperature scale derived by Harris from ground observations[22]. The ultraviolet observations yield temperatures systematically lower than those obtained at longer wavelengths. This difference is particularly significant in terms of Equation (8), which states that the integrated emergent flux of a star varies as the fourth power of the temperature. Thus, the lower temperatures correspond to a reduction in total luminosity by as much as a factor of two. This reduction comes almost entirely in the ultraviolet and is scarcely noticeable to ground observers, but the overall energy balance of the star is considerably altered.

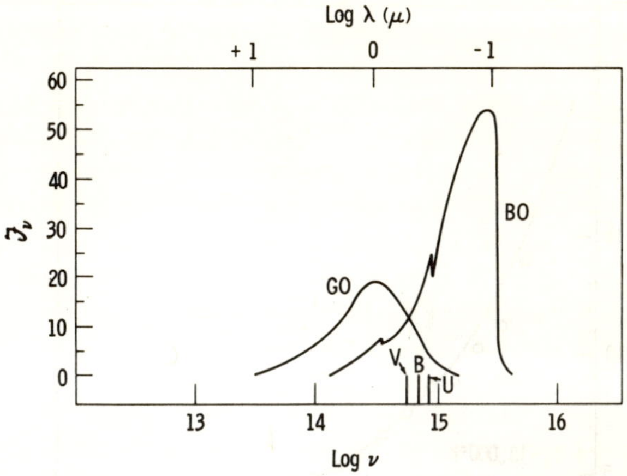

FIG. 8. Emergent fluxes from G0 and B0 model atmospheres. The effective wavelengths of the ground based U,B,V photometry are indicated (modified from Code, Ref. 19).

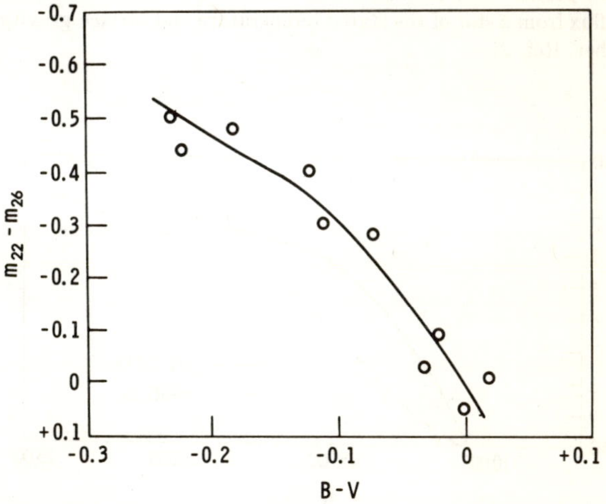

FIG. 9. Correlation of ultraviolet colors with (B-V). The spectral types range from B2 to A0 (Boggess, Ref. 20).

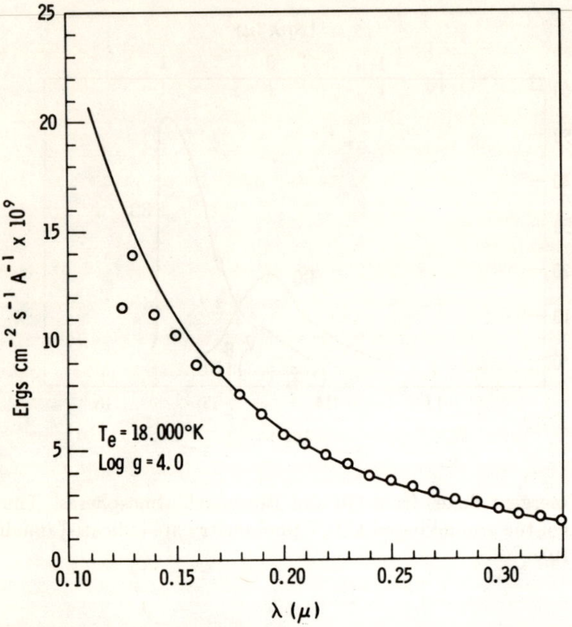

FIG. 10. Spectral observations of the B3 star η UMa. The solid curve is the predicted flux from a star of the stated temperature and surface gravity (modified from Stecher, Ref. 21).

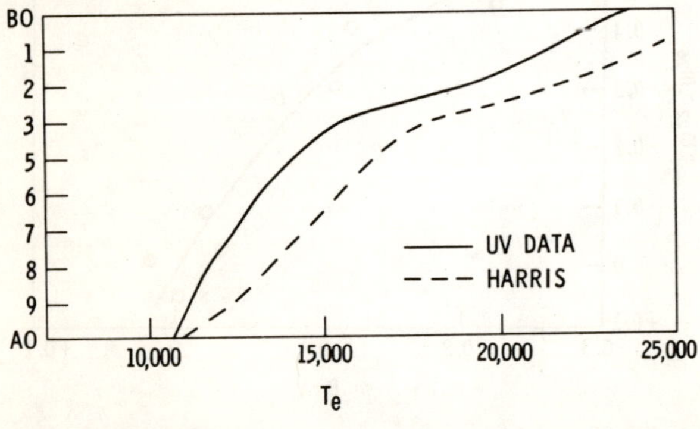

FIG. 11. A provisional effective temperature scale based on rocket observations. The scale derived by Harris (Ref. 22) is shown for comparison.

Figure 9 of Chapter 14 shows spectra obtained during Gemini flights XI and XII. The low dispersion spectra in Fig. 9d are useful for classification, while the remaining spectra have sufficient resolution to study individual absorption lines. The enlarged spectrum of the A0 star Sirius[23] extends from about 0.50 to 0.23μ. The hydrogen Balmer series is visible in the blue and near ultraviolet. A MgII doublet is prominent at 0.28μ, and the lines at shorter wavelengths are predominantly Fe.

In the hotter stars, the abundant atmospheric constituents become more highly ionized. Their spectra should be relatively featureless in the 0.2 to 0.3μ region. However, at shorter wavelengths, from 0.10 to 0.15μ, they should be dominated by very strong lines due to CII, III and SiIII, IV[24]. These expectations are generally confirmed by the spectra obtained so far, but spectra of very bright O and B supergiants have shown an unanticipated anomaly. In these stars emission lines of carbon and silicon have been observed, flanked by absorption edges to shorter wavelengths[25]. These features are explained by assuming that gas is being ejected from the star in all directions. The gas travelling perpendicular to the line of sight emits bright spectral lines at the laboratory wavelengths. The gas expelled in the direction of the observer absorbs the starlight passing through it to produce an absorption line Doppler shifted by an amount corresponding to the velocity of approach. These velocities are found to be on the order of 2000 km/sec, and the rate of mass ejection is estimated to be 10^{-6} to 10^{-4} solar masses per year. The mechanism responsible for this sizeable mass loss is not fully understood, but undoubtedly results from the supergiant's attempts to maintain a stable configuration during the rapid evolution of its interior.

Interstellar Scattering

A serious difficulty in the interpretation of stellar flux measurements is the effect of scattering by interstellar grains. This scattering reduces the observed irradiance and, since the scattering is not gray, may seriously distort the spectral distribution of the radiation. In order to evaluate this effect the scattering coefficient, σ, must be determined as a function of wavelength. The amount of energy received from a star at a particular wavelength, expressed in magnitudes, is given by

$$m_\lambda - M_\lambda = 5 \log (r/10) + \sigma_\lambda r, \tag{9}$$

where m_λ and M_λ are the observed and intrinsic magnitudes of the star at that wavelength and r is the distance of the star in parsecs. If measurements are made at two wavelengths, then the observed color. $C(\lambda_1, \lambda_2)$ is related to the intrinsic color, $\Gamma(\lambda_1, \lambda_2)$, by

$$C(\lambda_1, \lambda_2) - \Gamma(\lambda_1, \lambda_2) = r[\sigma(\lambda_1) - \sigma(\lambda_2)]. \tag{10}$$

The left-hand side of Equation (10) is called the color excess, $E(\lambda_1, \lambda_2)$. Finally, if the colors of two stars, a and b, of identical spectral type are compared, the intrinsic color is the same for each star, so that

$$C_a(\lambda_1, \lambda_2) - C_b(\lambda_1 \ \lambda_2) = K[\sigma(\lambda_1) - \sigma(\lambda_2)], \tag{11}$$

$$K = r_a - r_b. \tag{12}$$

K may be considered an arbitrary constant to normalize the scattering coefficient to any convenient distance. Evidently if the colors of two suitable stars are measured over a wide range of wave lengths, Equation (11) will yield the shape of the interstellar scattering curve. It can be shown that at very large wavelengths $\sigma \to 0$, so that the zero point of the curve is established by extrapolating to $\lambda = \infty$.

Ground observations have shown that in the visible the scattering curve is nearly linear with frequency, or reciprocal wavelength, but flattens out in the infrared as it tends toward zero[26]. The shape of this curve provides most of the information available as to the size and composition of the scattering particles. Extensive theoretical calculations have been made of the scattering properties of small particles of different types [27,28]. Calculations of scattering efficiency are made in terms of the complex index of refraction of the particle and a size parameter.

$$x = 2\pi a/\lambda, \tag{13}$$

a being the particle diameter. The objective is to reproduce the observed scattering curve by suitable choices of these variables. The best fit has been found by assuming that the grains are dielectric, i.e., the index of refraction is real and produces pure scattering with no absorption, and that they have sizes of the order of a few microns. However, the effects of different refractive indices are most pronounced at large values of x, that is, at short wavelengths. Thus the determination of interstellar scattering coefficients in the ultraviolet is of interest in its own right, in addition to being necessary for the interpretation of stellar flux measurements.

In order to derive ultraviolet scattering coefficients normalized to the same scale as those obtained from the ground, Equation (10) is modified to include three wavelengths. Then B and V observations may be combined with rocket data at an ultraviolet wavelength, λ, to give.

$$[C(\lambda, V) - \Gamma(\lambda, V)]/[C(B, V) - \Gamma(B, V)] = [\sigma(\lambda) - \sigma(V)]/[\sigma(B) - \sigma(V)]. \quad (14)$$

Since both the observed and intrinsic $(B-V)$ colors are well known from ground measurements, the only unknowns in Equation (14) are the intrinsic color, $\Gamma(\lambda-V)$, and the ultraviolet scattering coefficient, $\sigma(\lambda)$. Rocket data is available for enough stars to perform a least squares solution for these unknowns[29,30], and the resulting $\sigma(\lambda)$ are plotted with error flags in Fig. 12. In this figure, ground observations from 0.33 μ to 2.0 μ are also shown and are compared with the scattering curve of dielectric grains as computed by Van de Hulst. It is evident that this theoretical curve which appeared to fit the ground data does not satisfy the ultraviolet observations at all. This discrepancy could be removed by introducing a small complex term in

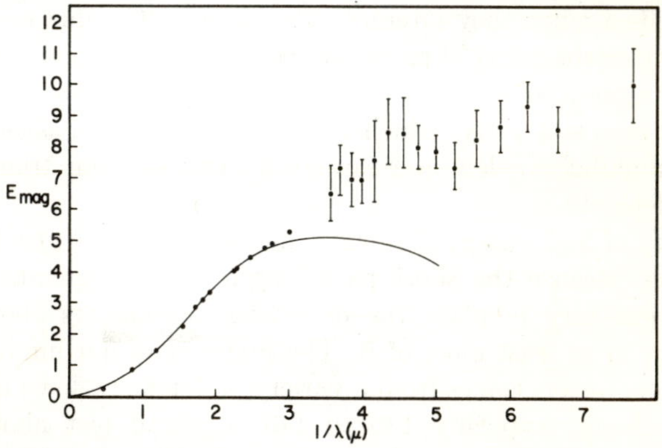

FIG. 12. Interstellar scattering curve from 2 μ to 0.13 μ. Van de Hulst's theoretical curve for dielectric grains is shown for comparison (modified from Stecher, Ref. 30).

the index of refraction to permit some absorption of radiation within the grain. The difficulty with this proposal is that the long wavelength end of the curve would then become depressed and disagree with the infrared observations. A possible solution may be to postulate a grain with an absorbing core, covered by a thin dielectric mantle which would act as a pure scatterer at long wavelengths but permit some energy to be absorbed in the middle ultraviolet. This model is attractive because of the possibility that the atmospheres of some cool giant stars may produce graphite particles which would then be available in interstellar space as nuclei for grain formation. Donn and Stecher[31] have suggested that the graphite flakes themselves might produce the scattering, different orientations of the strongly anisotropic flakes being responsible for different portions of the scattering curve.

X AND γ RADIATION

High energy astronomy is of great significance because it provides a means of detecting many of the major energy transfers occurring in the universe. High energy photons ($E > 1$ KeV) are intimately related to high energy non-thermal electrons and nuclear processes. The energetic electrons radiate x-rays and γ-rays in the form of synchrotron radiation as they are bent by magnetic fields, in the form of bremsstrahlung as they pass through matter, and by the inverse Compton effect when they interact with photons. Gamma rays may also result from the decay of pions formed in the interaction of cosmic rays with interstellar matter. Finally, there are a number of characteristic x-rays and γ-rays which arise in processes such as electron-positron annihilation, deuteron formation, and electron transitions in heavy elements.

The study of high energy photons is necessarily a very new field of astrophysics because the short path length of these quanta in air makes it necessary to place the observing instruments above the atmosphere, or at least most of it. The majority of our information on x-rays and γ-rays comes from a variety of detectors flown on high altitude balloons (typically 130 to 140 thousand feet above the surface of the earth) and sounding rockets (Small rockets capable of carrying payloads to from 80 to 200 miles). A few satellite experiments have been completed, and it is hoped that within the next five years satellite borne instruments will greatly enhance our knowledge of high energy photons.

At this time, a number of specific x-ray sources have been found as well as a general x-ray background. As yet no γ-ray (E > 1 MeV) point source has been seen for certain; however, several significant upper limits to possible point sources have already been set. Of the x-ray sources found, only a few have been identified with known stellar objects, and very little is understood about the nature of the origin of most of the x-rays. Information on energy spectra, size, and time variations are now becoming available however, and this data will be summarized later in this section along with a discussion of its possible interpretation.

Production Mechanisms:

Attention will be given here to the physical processes which can lead to the emission of high energy photons. Possible regions in space where these mechanisms might be significant will be indicated, and in some instances estimates of the intensity which might be expected will be made along with a discussion of the shape of the energy spectra.

1. Synchrotron Radiation

As a charged particle spirals around a magnetic field it emits electromagnetic radiation which is strongly polarized and collimated in the instantaneous direction of motion of the particle, especially for higher energy particles. The rate of energy loss in the synchrotron process is given by the relation:

$$\frac{dE}{dt} = \frac{2}{3}\frac{e^4H^2}{m_e^2\,c^4}\left(\frac{E}{m_ic^2}\right)^2 \tag{15}$$

where H is the magnetic field strength, m_e the mass of the electron, E and m_i are the energy the mass of the particle of interest. The emitted quanta have frequencies which are centered around a characteristic frequency which is given approximately by:

$$w_m = \frac{e}{c}H_\perp\left(\frac{E}{m_ic^2}\right)^2 \tag{16}$$

From these last two equations, it is clear at once that electrons are expected to be a much more important source of high energy radiation than protons because the factor $1/m_i^2$ appears both in the equation for the rate of radition and in the one for the characteristic frequency. If the electrons have a differential energy spectrum which is proportional to $E^{-\gamma}$, the resulting electromagnetic frequency spectra is proportioned to $\nu^{-\alpha}$, where $\alpha = \frac{\gamma-1}{2}$.[32]

The synchrotron radiation is of considerable importance in explaining radio frequency measurements. Electron and radio measurements seem now to agree with the hypothesis that electrons exist in the galaxy in sufficient number to explain the general radio flux and that α has a value of about 2 to 2½. However, present estimates indicate that unless the number of high energy electrons is considerably greater than is expected or unless the magnetic fields are much greater than present estimates, synchrotron radiation is possibly not as significant in the x-ray or γ-ray frequency interval. Estimates of the fluxes made by Gould and Burbidge[33] from the galaxy and intergalactic space are shown in Fig. 13. Possible exceptions to this statement may be the early stages of a supernovae explosions and the Crab nebulae, which is unique in many respects. The possibility of discrete sources being the result of synchrotron radiation is discussed in more detail by Ginzburg and Syrovatskii.[34]

2. Bremsstrahlung

In traversing matter, electrons pass close to other charged particles, and, as a result of the Coulomb interaction, energy is radiated. The probability of emitting a photon in a given frequency interval, $d\nu$, by this process is given by the expression:

$$\phi d\nu dX = \frac{d\nu}{\nu}\frac{dx}{X_0} \qquad (17)$$

where X_0 is the radiation length which is given by the equation:

$$1/X_0 = \frac{4e^2\,n}{\hbar\,c\,A}\,Z\,(Z+1)\,r_e^2\,\ln\,(183Z^{-1/3}), \qquad (18)$$

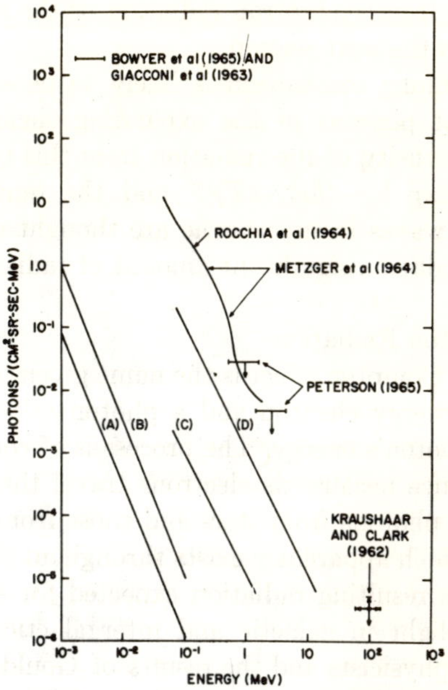

FIG. 13. Expected fluxes of x-rays and γ-rays from galactic and intergalactic space as calculated by Gould and Burbridge (Ref. 33) compared to measured fluxes and upper limits set by Kraushaar and Clark (Ref. 35), Peterson (Ref. 36), Metzger et al. (Ref. 37), Rocchia et al. (Ref. 38), Giacconi et al. (Ref. 39), and Bowyer et al. (Ref. 40). Curves A and B are the calculated fluxes for synchrotron radiation from the galaxy and intergalactic space respectively, and curves C and D are the calculated fluxes for the inverse Compton radiation from galactic and intergalactic space respectively.

and has a value of about 62 g/cm^2 for hydrogen. For this process to be important the medium through which the electrons pass, must be thick enough to produce a detectable number of high energy photons, but not so thick that they, and the electrons which produce them, are absorbed.

Since there is atomic hydrogen in the galaxy and in intergalactic space and also presumably molecular hydrogen, possibly in about the same abundance, there is the possibility of radiation from electrons passing through this material. Garmire and Kraushaar[41] and Gould and Burbidge[33], as well as others, have estimated that the expected intensity from this source in the galaxy and intergalactic space is less than from sychrotron radiation or the inverse Compton effect discussed in the next section.

The bremsstrahlung mechanism is likely to be significant in the optically thin hot plasmas of the expanding shells of supernovae remnants. The intensity of the radiation from this type of plasma is proportional to $\exp[-(h\nu)/(kT)]$[42] and the temperature of the expanding shock waves in supernovae are thought to be sufficiently ($>10^7$ °K) to produce a significant amount of radiation in the x-ray region.

3. Inverse Compton Radiation

The "inverse" Compton effect is the name given to the interaction between a high energy electron and a photon in which there is an increase in the photon's energy. The process is of considerable astrophysical significance because as electrons travel through space they interact with the photons from stars and those from a general black body radiation which apparently exists throughout the universe. The calculation of the resulting radiation expected for electrons passing through the starlight in galactic and intergalactic space has been made by several physicists and the results of Gould and Burbidge[33] are shown in Fig. 13. Again the uncertainty of the parameters which must be used in the calculation is fairly large, but perhaps less uncertain than for the bremsstrahlung calculation. A calculation for the x-ray flux expected for electrons interacting with the 3 °K black radiation has been made by Fazio et al.[43] The results are necessarily even more uncertain than the previously mentioned calculation, but the estimated flux probably lies somewhat below curve D of Fig. 13.

4. π° Decay

Mesons are produced in cosmic ray nuclear collisions with interstellar matter. By far the most common mesons are the π-mesons, which can be either charged or neutral. The neutral π-mesons, π°'s, decay into two gamma rays each of which has 70 MeV in the rest frame of the π°, and an energy between (70 MeV) $\sqrt{\frac{1-\beta}{1+\beta}}$ and (70 MeV) $\sqrt{\frac{1+\beta}{1-\beta}}$ in the laboratory system where β is the π°'s velocity in terms of the speed of light. The present experimental data on nuclear interactions with hydrogen is concentrated below 20 BeV, and the theory of multiple meson production at high energies has been checked only very approximately in the study of interactions of cosmic rays. However, since the number of pions produced seems to increase only slowly with energy, approximately as the total energy of the primary to the $\frac{1}{4}$ power and since the primary cosmic ray energy spectra decrease rapidly with decreasing energy approximately as the total energy to $-5/2$ power, a fairly accurate calculation of the expected π° production can be made for cosmic ray interactions with hydrogen. The number of gamma rays resulting from interactions in the galactic and intergalactic space can then be made with the most uncertain parameter being the density of hydrogen in these regions. These calculations indicate an expected flux of the order of the upper limit presently existing for γ-rays of 70 MeV or more with a peak in the energy spectrum around 10^2 MeV.

π°-mesons can also be produced in interactions of protons with photons in space if the proton energy exceeds 10^{17} ev. [44]; however, in view of the small fraction of cosmic rays with energies greater than 10^{17} ev, this is not a significant process for astrophysical γ-ray production.

Another process for the production of neutral π-mesons is the annihilation of a nucleon and antinucleon. In a typical proton-antiproton annihilation there are about five pions produced, of which some are neutral. If the particles have relatively small rest energies, the resulting γ-rays will have energies concentrated around 10^2 MeV. This process is of considerable interest in the steady state theory of the universe, wherein the spontaneous creation of particles and antiparticles at the rate of about $10^{-22}/\text{cm}^3$ sec is postulated in order to keep the mean density of the universe constant as it expands. One consequence of this theory is that there should be a considerable amount of anti-matter still present with the expected annihilation rate. Using the best available cross section data, Gould and Burbidge[33] have calculated the expected flux of 10^2 MeV photons in terms of the mean ratio of anti-matter, to matter $\bar{\alpha}$, and find it to be about $\bar{\alpha}$ photons/cm^2-sec assuming a particle velocity corresponding to a temperature of 10^5 °K

5. Characteristic X-rays and γ-rays

There are several physical processes which lead to x and γ rays of a specific energy. One of these is the emission of a photon of discrete energy when an electron makes a transition to an empty atomic inner shell of a high-charge atom. Another example is the ½ MeV gamma rays produced in the annihilation of electrons and positrons. The majority of the positrons found in the galaxy should essentially come to rest before combining with an electron to form two γ-rays of equal energy. If the steady state picture is assumed again, Gould and Burbidge[33] have calculated that there should be $\bar{\alpha} \times 10^4$ photons/cm² sec with energies of about ½ MeV. There are, of course, many other sources of positrons, of which the most important is the π^+ meson formed in nuclear interactions which decays first into a μ^+ and then an e^+.

Another characteristic γ-ray line could exist at a somewhat higher energy as the result of deuterium formation from a neutron and proton. This γ-ray would have an energy of 2.23 MeV. This interaction and other nuclear synthesis processes may be important at the time of a supernova outburst. If even a small fraction of the gamma radiations resulting from the hydrogen burning after an implosion can penetrate through the surrounding material a very large flux should be observed during the first hour or so. However, the probability of occurrence of a supernova is very small, about one in a hundred years in our galaxy. (γ-rays of other energies should also be present in large numbers early in a supernova's lifetime resulting from sychrotron radiation, π° decay, and $e^+ - e^-$ annihilation).

Detection Techniques

The detecting systems in high energy astrophysics are generally more similar to particle detectors than optical devices. This situation results from the high frequency of the radiation which precludes the use of reflection and diffraction techniques, except for the lower energy x-rays, and the high energy of each individual photon which makes it possible to use scintillators, Geiger counters, proportional counters, and even spark chambers, for the γ-rays.

In the lowest energy range, approximately 1 to 10 or 20 KeV the detecting system has normally been constructed around thin window Geiger counters or proportional counters, although there has been some effort directed towards using grazing incidence optics. The first x-rays were detected with Geiger counters with several individual mica windows. Each Geiger counter was placed in a well formed by an active anticoincidence shield composed of scintillating material to reduce the cosmic ray background.[45]

Subsequent measurements in this energy range have usually been made with arrays of Geiger counters or proportional counters normally again placed inside a well of active scintillator and covered with a collimator of some type, such as a honeycomb structure, to improve the directionality. The low energy limit is set by the material, and thickness of the window and the high energy limit by the type of gas. The first measurements, and all those which followed in this energy range, were made using rockets because even the high altitudes obtainable by present scientific balloons are not sufficient to overcome the combined problems of primary x-ray absorption in the atmosphere and secondary production of x-rays by cosmic rays in the residual air.

Above approximately 20 KeV, it is possible to perform x-ray experiments on balloons, at least for the stronger sources whose fluxes are greater than the cosmic ray induced background. The principal advantage of a balloon exposure over a sounding rocket exposure is the longer observing time possible, several hours compared to several minutes. The basic approach to the detector system in the higher energy x-ray interval is to use a scintillator counter or more complex counter system and a collimation device which is either a passive material or an active scintillator used in anticoincidence with the primary detector. The first balloon experiment to detect x-rays[46] consisted of a scintillator counter employing a 97 cm^2 Na I (tl) crystal and a collimator of brass slats. A typical example of this type of detector system is shown in Fig. 14a. Subsequently much larger arrays of such detectors have been used by several research groups.

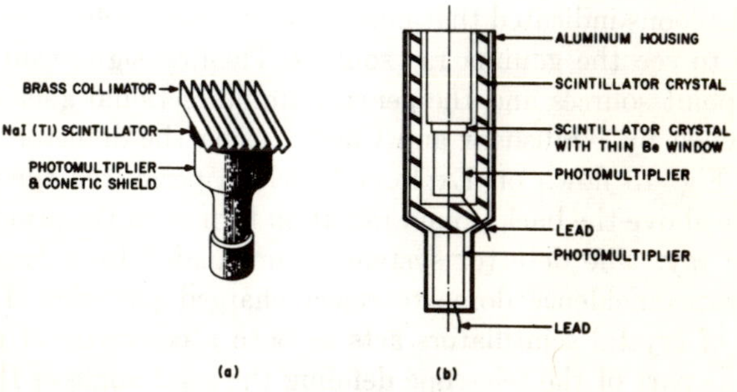

FIG. 14. (a) A schematic diagram of the passive collimeter type x-ray detector (Ref. 46, 69).

(b) A schematic diagram of an active collimeter type x-ray detector.

The other approach mentioned above, using a collimator made out of scintillation material which is an active part of the detector system, has the advantage of eliminating much of the background since x-rays produced in the collimator are discriminated against. This approach is especially valuable in the highest energy x-ray region where the primary x-rays are most penetrating and least numerous. Fig. 14b shows a schematic diagram of this type of detector.

Balloon observations have generally been made either by mounting the detector system in a gondola which orients the detector as desired or by mounting the detector in a vertical position and letting the source pass overhead. Orientation information has normally been obtained with the use of magnetometers and balloon position information. In a satellite experiment, Metzger[37] et al. used a scintillator spectrometer placed at the end of a six foot boom on Ranger III. The boom reduced the number of secondary x-rays reaching the detector from the spacecraft.

In the γ-ray interval above about 20 MeV the only efficient interaction process for gamma rays is the creation of an electron positron pair. The angle between the pair tends on the average to decrease with energy, since it is primarily determined by electron scattering. What information that can be obtained about direction and energy must be derived from this pair because the penetration power of γ-rays and their low flux relative to other possible masquerading events makes shielding difficult and generally undesirable, since it can be the source of these unwanted background events. Early experiments using single counters systems and nuclear emulsions flown on balloons indicated that more sophisticated techniques would be needed to see the gamma ray sources. The first significant upper limits to point sources and the general omnidirectional galactic flux were obtained by Kraushaar and Clark[35] using the detector system shown in Fig. 15 flown on Explorer II which permitted the experiment to be above the background radiation formed in the atmosphere by cosmic ray. The detector system is surrounded by a large very efficient anticoincidence dome to reject charged particles. The top sandwich of crystal scintillators acts as both a converter of gamma rays and as part of the telescope defining the solid angle of the incident gamma ray. The other half of the telescope is the lucite Cerenkov detector, which responds to one or both electrons and is in coincidence with the scintillator sandwich.

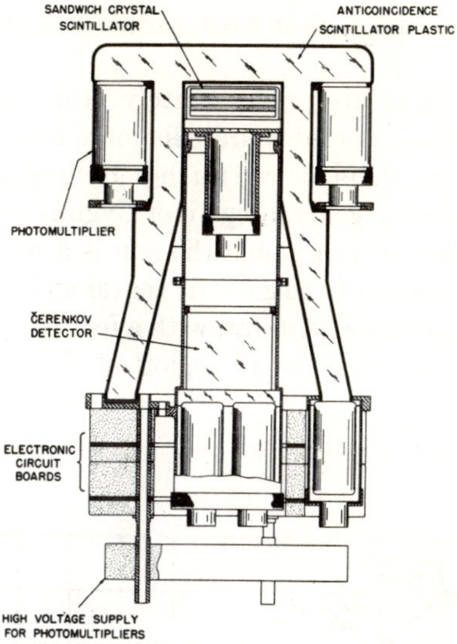

FIG. 15. Detector system flown by Kraushaar and Clark (Ref. 35) on Explorer XI to look for celectral γ-rays.

In view of the very low flux rates of gamma rays as evidenced by the failure of these first experiments to detect any fluxes with certainty, several investigators at present have turned to the spark chamber as the heart of the detector system. The spark chamber has the advantage of being a picture type device which allows the experimentalist to separate the electron positron pairs from the other neutral events—a feature which is of utmost importance because the fluxes are so low, and yet it can be quite large. The spark chamber is generally triggered by a scintillator-Cerenkov counter telescope of some type and surrounded by an anticoincidence system. Detector systems of this type are currently being flown on balloons using similar orientation systems to those of the X-ray balloon experiments.

Observations and Interpretations

Over the last five years high energy astronomy has blossomed in a truly spectacular fashion. At the beginning of 1962, there were no observed x-ray sources, and now there are over two dozen. Further, there are already energy spectral measurements on a few sources, some size estimates, and a clear time variation determination on at least one source.

The observed sources are predominately near the galactic plane; only a very few of the reported sources lie more than 15° from it. The sources within the glactic plane are further grouped into two regions. The first of these is in the general region of Cygnus and Casseopia and could be in the spiral arm of which the sun is a member. The other is in the general direction of the galactic center and includes Scorpius. Hence, the sources observed thus far, with a few exceptions seem to be concentrated in a relatively small region of the sky, as shown in Fig. 16.

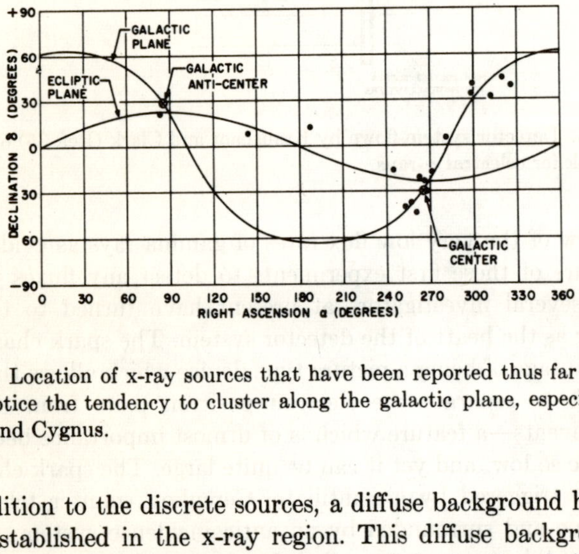

FIG. 16. Location of x-ray sources that have been reported thus far (Ref. 40, 47-49). Notice the tendency to cluster along the galactic plane, especially near Scorpius and Cygnus.

In addition to the discrete sources, a diffuse background has been clearly established in the x-ray region. This diffuse background is, however, not at all even. The number of small fluctuations in intensity from the average is significantly above the number expected purely from statistical arguments[47]. This result suggests that as experiments become more refined, many weaker x-ray sources may be found.

Of the sources which have been found only three are reasonably positively identified with optical objects, namely Taurus A, Scorpius XR-1 and Cygnus XR-2. A few others, Cassiopia A and Tycho Brahe supernovae, and the possible extragalactic sources Cygrus A and M87, have been tentatively identified, but there is either a fairly large uncertainty in the position ($\sim 1\frac{1}{2}°$) or there is some doubt about the existence of the source. In the case of these identified sources the x-ray flux is always one or two orders of magnitude greater than the flux from the radio portion of the spectrum.

The Taurus A source has a finite size, which has been determined to be approximately the same as the optical size of the Taurus A supernovae remnant (Crab Nebulae). The size of the Crab nebulae x-ray source was determined by Bowyer et al.[50] during its occultation by the moon. The only other source for which an accurate size determination has been made is Scorpius X-1. Gursky et al.[51] found an upper limit to the angular dimension of 20″ by using a complex wire grid collimator system flown on a rocket. This small size seems to remove the possibility that Scorpius X-1 is an old remnant of a relatively close supernovae, such as the crab.

Following this accurate determination of the position of Scorpius X-1, a thorough search was made for a possible source in this region of the sky that would have a relatively flat frequency spectrum in the visible and hence appear as a "blue" star. The flat spectrum was expected on the basis that the emission process was probably bremsstrahlung from a hot gas or synchrotron radiation since the spectra did not fit that of a black body[51]. Such a star was found with a reasonable magnitude, 13th, and optical spectrum, and the results were reported by Sandage et al.[52] A "blue" stellar object of 16th magnitude has now also been found for the Cygnus XR-2 source[53]. The weakness of these stars suggests a possible explanation of why not very many x-ray sources have yet been identified. Others may soon be identified with optical objects once their positions are better known provided the "seeing conditions" are reasonably good in the part of the sky where they are located.

Another interesting aspect of the Scorpius X-1 source is that the optical object associated with it is highly variable, changing by more than half an order of magnitude in less than three hours in one case. There have been several indications that many of the x-ray sources may be variable, but most of these result from measurements made by different groups with instruments of different design; so they are uncertain. There is, however, one result which is quite unambiguous. In two flights June 1964 and April 1965, Byram et al.[47], using the same instrument, saw a decrease in the intensity of Cygnus XR-1 by a factor of 4 between the two flights while the intensity of Cygnus XR-2 remained unchanged. In fact, apparently as the result of the decrease in Cygnus XR-1, a new weak source was seen—Cygnus XR-3, which had presumably been hidden in the tail of Cygnus XR-1 distribution on the first flight.

Before discussing the characteristics of a few of the better observed x-ray sources in a little more detail, the results of the γ-ray studies will be mentioned. At present there has been no certain measurement of a flux of high energy γ-rays from any point source. Upper limits now exist for essentially all of the most likely sources including Taurus A, the Cygnus region, Cassiopia A, the Scorpius region, the galactic center, and Virgo. The upper limits for fluxes above 30 MeV now range from 2×10^{-4} to 1×10^{-5} (cm^2. sec.)$^{-1}$, [54–58] and above 5×10^{12} eV limits of (0.5 to 1.0) $\times 10^{-10}$ γ/(cm^2-sec.) exist for most of the same sources.[59,60] Kraushaar et al.[54] in their Explorer XI experiment saw no positive evidence for a general celestial γ-ray flux which was anisotropic and set a general upper limit of 3×10^{-4}/(cm^2-sr-sec) for gamma rays above 50 MeV. Further, studies in the γ-ray region would be of great value since even these upper limits are of considerable significance in that they provide upper limits to the energy spectra from the sources and to the radiation from the galaxy.

Using the result of Burbridge and Gould[33] mentioned earlier, the upper limit to the gamma ray flux above 50 MeV. just mentioned places an upper limit of 4×10^{-3} on $\bar{\alpha}$, the mean ratio of antimatter to matter.

Going now to a discussion of individual sources, Taurus A (the Crab Nebulae) will be discussed first. This unique supernovae with its strongly polarized light, large angular dimensions, and strong radio signal was naturally one of the first targets for x-ray studies. An x-ray flux from this source has been measured by several different observers, and the results seem to give different intensities and somewhat different spectra particularly in the appearance of a high energy cut-off in some measurements.[40,46,61–63] Whether this apparent time variation is real or the result of experimental uncertainties is not yet known. Some of the results are shown in Fig. 17. If the x-ray source is the result of the same synchrotron process which is generally believed to be responsible for the optical spectrum, a time varying high energy spectrum and cut-off are certainly possible and a power

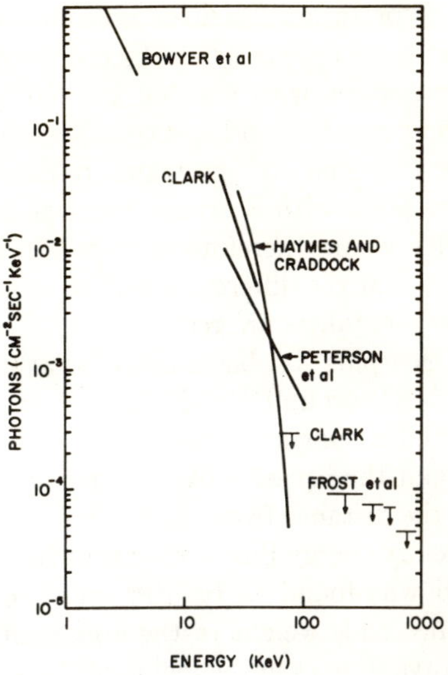

FIG. 17. Energy spectral measurements of Crab Nebulae in the x-ray region by Peterson et al. (Ref. 61), Clark (Ref. 46), Bowyer et al. (Ref. 40), Haymes and Craddock (Ref. 62), and Frost et al. (Ref. 63).

law spectrum in energy would be expected. Peterson et al.[61] did, in fact, see a power law spectrum in total energy in the range from 16 to 120 KeV during their flight when the high energy cutoff observed by others apparently did not exist. The angular size measurements of the Taurus A x-ray source have already been mentioned, and it is consistent with the interpretation mentioned above, but more measurements are needed before a definite conclusion is possible.

In the case of Scorpius X-1, the results of the various groups again disagree on absolute flux, but there seems to be generally good agreement on a frequency spectra of the form

$$I(\nu)\, d\nu = C \exp\,(-\, h\nu/kT)\, d\nu \tag{19}$$

with kT being about 4 KeV ($T \approx 5 \times 10^7\ °K$) at least in the energy range from 1 to 50 Kev.[64–66] This type of energy spectrum is consistent with that which would be expected from a bremsstrahlung emission process occurring in an optically thin cloud of fully ionized gas as mentioned earlier. At present, the limited data permit the possibility of other interpretations; so no definite conclusion can be drawn at this time. However, as noted earlier, the visible continuum measurements are also consistent with the hot gas interpretation—in fact remember that the x-ray flux and spectrum were used to predict the magnitude of the star and its spectrum. Remember also that the optical object associated with Scorpius X-1 was highly variable and a similar variability would therefore be expected in the x-ray region making it plausible that the different absolute intensities observed by the different experimentalists are real.

On the basis of Scorpius X-1 being detected at wavelengths greater than 44A by the NRL group[67] Rossi[68] has set an upper limit on the amount of interstellar matter between this source and the earth. From this result and the conservative assumption of 0.5 atoms/cm^3, an upper limit to the distance from the earth to Scorpius X-1 is set at 100 parsecs. The x-ray energy flux in the range from 1 to 10A can then be calculated and was found to be 10^{35} ergs/sec. The small upper limit to the distance adds weight to the argument, suggested by the lumping of the x-rays sources in the region of the sky of our own spiral arm, that most of the x-ray sources observed so far may be within 1 or 2 kiloparsecs. It is also worth noting that the flux measured in the 44 to 55A region is too large by a factor of 3 to agree with the smooth curve extrapolation between the 1 to 5A region and the optical region. Whether this result represents a real constant effect, a time variation, or an overall instrumentation calibration problem is not yet clear.

Present data seem to suggest that Cygnus XR-1 may have a spectrum similar to Taurus A in the region from 2 to 100 Kev (Bowyer *et al.*[40] and McCracken[69]). As was mentioned earlier Cygnus XR-1 is clearly established as a variable source, and hence, it would be consistent for its radiation to be the result of synchrotron radiation, but at present this hypothesis must be regarded as highly speculative.

Both Serpens X4 and Cygnus X2 seem to have energy spectra in the region from 1 to 180 KeV which are consistent with the form of equation (19) and hence characteristic of a bremsstrahlung mechanism with a Maxwellian distribution of electron energies[40,70] although here again other interpretations are possible.

In addition to the x-ray stellar objects, there also seems to be a diffuse radiation whose existence was first indicated by the Ranger III observations[37]. The combination of balloon, rocket, and satellite experiments are shown in Fig. 14. As mentioned earlier, there is some indication from the rocket results that this radiation may be partly due to a large number of weaker individual sources. From their work, Metzger *et al*[37] have also set upper limits of 0.014/cm^2 sec. for the 0.51 MeV gamma ray line and 0.005/cm^2 sec. for the 2.23 MeV line. Using the result discussed earlier, the limit on the 0.51 MeV gamma ray flux places an upper limit of 10^{-6} on the mean ratio of antimatter to matter. This limit is considerably more severe than the one mentioned earlier from the absence of an 100 MeV peak.

It is not yet clear what the origin of the diffuse radiation is. It could be a summation of thermal sources or a general flux resulting from inverse Compton or synchrotron radiation as the electrons interact with the photons and fields of the galaxy as a whole or even extragalactic space. The apparent isotropic nature of the radiation would argue strongly in favor of the radiation coming from the universe as a whole. The recent evidence which seems to indicate that the universe contains black body radiation at 3°K[71], thus providing a high density of photons, gives added weight to the inverse Compton process as a possible mechanism.

FUTURE RESEARCH PROBLEMS

In the previous sections the progress which has been achieved thus far in space astronomy has been summarized and its significance has been discussed. Clearly, the many significant discoveries made to date are only a beginning. The initial research has been unusually rewarding in terms of the additional knowledge gained and the number of new phenomena which have been discovered in a few short years. The future holds even more promise as these phenomena are studied in depth and still new horizons are opened. In the final few paragraphs, some future research problems are presented together with what it is hoped will be learned from them.

The more sophisticated instruments and improved guidance systems that are now coming into use enlarge the research possibilities tremendously in ultraviolet astronomy. One important experiment that can be carried out now is a program of flux measurements in galactic star clusters. These objects are groups of 20 or 30 to several hundreds of stars which are moving together through space, have identical chemical compositions, and presumably identical ages and common histories. Photometric comparisons among stars in such a homogeneous group avoid the uncertainties arising from different distances, different amounts of interstellar scattering, and different abundance and age effects which may be encountered among stars in general. Only the youngest clusters contain hot stars of the types which have been considered here. These stars evolve rapidly because of the high rate at which they produce and radiate energy. As the hydrogen is exhausted in their interiors they migrate from the Main Sequence, illustrated in Fig. 7, to the Giant region of the diagram. The adjustments occurring in the atmosphere at this time are only generally understood, and observations in the ultraviolet, where the radiative changes are most profound, should prove most helpful. The migration proceeds so rapidly that the probability of finding a star in transit is low. However, if color-color arrays similar to the plot in Fig. 9 are prepared for various clusters, differences may be expected as some clusters begin to show precursors to the formation of a Giant Sequence.

Systematic measurements of absorption line intensities in ultraviolet spectra are needed to provide data on the abundance of atoms in certain ionization states which are undetectable from the ground. The equivalent width of an absorption line, that is, the spectral width of a completely black line removing the same amount of energy from the spectrum as does the actual line, can be expressed in terms of the number of atoms capable of exciting the line, the excitation temperature of the gas, and the kinetic temperature of the gas. If a group of lines, all arising from the same ion, are analyzed together, each of these parameters can be derived. In addition to the abundance information this gives, the temperatures appropriate to that part of the atmosphere producing the lines is valuable for calculation of atmospheric models.

In cooler stars, coronal and chromospheric emission lines should be observable in the ultraviolet. In the sun, these lines are masked at longer wavelengths by the continuous radiation from the photosphere, but in the ultraviolet, where photospheric radiation is very small, the emission lines become the dominant features of the spectrum. Because chromospheric and coronal radiations are unobservable from the ground except in the case of the sun, where light from the solar disk can be masked, there is no knowledge of the extreme outer layers of the other stars. Extrapolation of solar phenomena to stars of other spectral types is very uncertain.

It will be of interest to attempt to record stellar spectra in the extreme ultraviolet below 0.10 μ. Although the hottest stars radiate considerable amounts of energy at these wavelengths, this radiation must pass through the interstellar hydrogen, which is ionized by wavelengths less than 0.09 μ. The large abundance of interstellar hydrogen ensures that the optical depth at this wavelength is extremely high. However, the absorption coefficient varies as λ^3, so that the optical depth should decrease rapidly below the absorption limit. It may be that at 0.05 μ, for instance, the opacity is low enough for nearby stars to be detected. If so, the most interesting observation will be simply the appearance of the spectrum. Many hot stars are surrounded by regions of ionized gas maintained by the extreme ultraviolet radiation from the central stars. In determining the temperature and other physical parameters of these ionized regions, an assumption must be made concerning the spectral distribution of the ionizing stellar radiation. Normally it is assumed that the stellar energy below 0.09 μ is continuous and is as computed from the model atmosphere. However, it is quite possible that a significant fraction of the extreme ultraviolet energy is lumped into emission lines. A simple determination of whether an O star spectrum was continuous or discrete at these wavelengths would be of great value to the theory of emission nebulae.

Ultraviolet spectra of these emission nebulae will be valuable for improving chemical abundance calculations for the interstellar medium. Fig. 18 shows a predicted spectrum for a high excitation nebula computed by Daub.[72] Measurements of these line intensities would yield more accurate temperature and density data than is possible from the ground. Relative intensities of some of the low level forbidden lines should indicate the significance of collisional exictation in the nebula. The hydrogen Lyman series will probably not be observable from a nebula because of strong resonant scattering by interstellar hydrogen, but it will be enormously strong in absorption against stellar spectra. Other interstellar absorption lines, such as those of carbon, oxygen, and nitrogen, can be observed in the ultraviolet. From the ground, lines of only a few interstellar gases can be detected. This is because most elements are so rare in space that only the strongest resonance lines produce measurable absorption. The resonance lines of most elements lie in the ultraviolet. Of particular interest is the detection of the Lyman bands of molecular hydrogen below

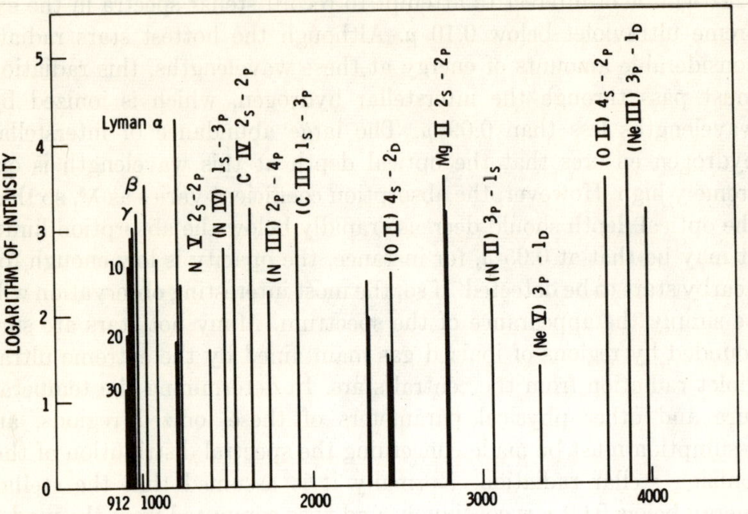

FIG. 18. Predicted ultraviolet spectrum of an emission nebula. Identifications of some of the more interesting lines are shown (Code, Ref. 72).

0.16 μ. Hydrogen molecules are expected to form on the surfaces of interstellar grains, but current estimates as to the reaction rate are very unreliable and there is no way of observing interstellar molecular hydrogen from the ground. The photodissociation limit of molecular hydrogen is roughly the same as the ionization limit of atomic hydrogen. As a result, they must compete for the same ultraviolet quanta which are mostly used up in the immediate vicinities of hot stars. Once a hydrogen molecule forms, therefore, it may have quite a long lifetime, and an appreciable amount of interstellar gas may be tied up in this way. Since the mass of interstellar material is an important factor in the gravitational potential of the Galaxy, a determination of the amount of molecular hydrogen present would be of great interest.

In x-ray and γ-ray astronomy, we see a field which is just emerging, but in a spectacular fashion. Now that the existence of x-ray sources is established, there is a strong need to perform a systematic sky survey over as much of the energy range as possible using large x-ray detector systems flown on satellites. This survey should reveal the number of sources, their location, and the distribution in intensity.

The x-ray source positions can then be compared with optical and other data to try to determine if the detectable x-ray sources are predominantly clustered in a small region of the galaxy close to the earth or if there are a significant number of sources in other parts of the galaxy and, even more interesting, beyond our galaxy.

At the same time it will be very important to measure the energy spectrum of the detected sources and determine the nature of any time variations in the shape of the energy spectrum. These data together with polarization line spectra and size measurements should then provide the basic information with which a reasonable deduction concerning the nature of the origin of the radiation can be made.

The understanding of the origin of the radiation is particularly important both because of the large amount of energy radiated in the x-ray region and because of the insight which can be obtained into the properties of high energy particles and nuclear processes in stellar objects. To obtain the experimental information, the present x-ray detectors described previously will be supplemented by experiments using x-ray optics and refined polarization techniques.

In the γ-ray region, the future path is obvious, but difficult. Since no source of γ-rays ($E > 10$ MeV) has yet been detected, bigger and more sensitive detectors with excellent background discrimination will have to be built and flown on satellites above the earth's atmosphere. The significance of γ-ray astronomy in that it can both reveal the high energy nuclear interactions of charged particles through π° decay and at the same time provide the highest energy information on bremsstrahlung, synchrotron, and inverse Compton effects remains; thus, efforts in this area of research must continue, since even the upper limits of fluxes now being obtained are beginning to be quite significant.

Finally, in the x-ray and γ-ray energy interval, the nature of the general celestial flux must be determined. Is it really primarily the result of many unresolved point sources, or is it principally due to the charged particles interacting with the material, magnetic fields, and photons in space? If the latter is true, the next step will be to determine whether the galaxy or extragalactic space contributes most to the radiation. If it is the former, the radiation should be markedly anisotropic; whereas, if it is the latter, it should be isotropic. Finally, energy, flux and polarization measurements should be made to determine the origin of this general high energy radiation.

REFERENCES

1. M. Bottema, W. Plummer, and J. Strong, Water Vapor in the Atmosphere of Venus, Astrophys. J. **139,** 1021 (1964).
2. M. Bottema, W. Plummer, J. Strong, and R. Zander, Composition of the Clouds of Venus, Astrophys. J. **140,** 1640 (1964).
3. R. Danielson, J. Gaustad, M. Schwarzschild, H. Weaver, and N. Woolf, Mars Observations from Stratoscope II, Astronom, J. **69,** 344 (1964).
4. N. Woolf, M. Schwarzschild, and W. Rose, Infrared Spectra of Red Giant Stars, Astrophys. J. **140,** 833 (1964).
5. L. Dunkelman, W. Fowler, and J. Hennes, Middle Ultraviolet Photoelectric Detection Techniques, *Space Research III*, W. Priester, Ed., p. 1174, North-Holland, Amsterdam, 1963.
6. J. Hennes and L. Dunkelman, Ultraviolet Technology, The Middle Ultraviolet: Its Science and Technology, AES. Green, Ed. p. 304, John Wiley & Sons, 1966.
7. J. E. Kupperian, Jr., and R. R. Ziemer, Satellite Astronomy, International Science and Technology, March, 1962.
8. T. P. Stecher and J. E. Milligan, Stellar Spectrophotometry from above the Atmosphere, Astrophys. J. 136, **1** (1962).
9. D. U. Wright, Jr., An Ultraviolet Stellar Spectrometer, Japanese Journal of Applied Physics, 1965.
10. K. G. Henize and L. R. Wackerling, Stellar Ultraviolet Spectra from Gemini 10, Sky and Telescope **32,** 204 (1966).
11. T. Dunham, Jr., Spectroscopic Observations of the Planets at Mt. Wilson, Atmospheres of the Earth and Planets, G. P. Kuiper, Ed., p. 288, Univ. of Chicago Press, 1952.
12. T. P. Stecher, The Reflectivity of Jupiter in the Ultraviolet, Astrophys. J. **142,** 1186 (1965).
13. D. C. Evans, Ultraviolet Reflectivity of Mars, Science, **149,** 969 (1965).
14. A. Kliore, D. L. Cain, G. S. Levy, V. R. Eshleman, G. Fjeldbo, F. D. Drake, Occultation Experiment:Results of the First Direct Measurement of Mars' Atmosphere and Ionosphere, Science **149,** 1243 (1965).
15. P. C. Keenan, Classification of Stellar Spectra, Chap. 8, Basic Astronomical Data, K. Aa. Strand, Ed., p. 78, Univ. of Chicago Press, 1963.
16. H. L. Johnson, Photometric Systems, Chap. 11, ibid, p. 204.
17. L. H. Aller, The Atmospheres of the Sun and Stars, Chap. 5, Ronald Press, New York, 1963.
18. G. Munch, The Theory of Model Stellar Atmospheres, Chap. 1, Stellar Atmospheres, J. L. Greenstein, Ed., p. 1, Univ. of Chicago Press, 1960.
19. A. D. Code, Stellar Energy Distribution, Chap. 2, ibid, p. 50.
20. A. Boggess III, B Star Colors Between 2000 and 3000 Angstroms, Ann. d'Ap. **27,** 805 (1964).
21. T. P. Stecher, Ultraviolet Stellar Spectrophotometry, Ann d'Ap. In Press.
22. D. L. Harris III, The Stellar Temperature Scale and Bolometric Corrections, Chap. 14, Stellar Atmosphere, J. L. Greenstein, Ed., p. 263, Univ. of Chicago Press, 1960.

23. K. G. Henize, L. R. Wackerling, F. G. O'Callaghan, Ultraviolet Stellar Spectroscopy on Gemini II, Science, in press.
24. D. M. Mihalas and D. C. Morton, A Model for a BIV Star with Line Blanketing, Astrophys. J. **142,** 253 (1965)
25. D. C. Morton, Far Ultraviolet Spectra of Six Stars in Orion, Astronom. J. **71,** 172 (1966).
26. S. Sharpless, Interstellar Reddening, Chap. 12, ibid, p. 225.
27. H. C. van de Hulst, Light Scattering from Small Particles, Rech. Obs. Astr. Utrecht **11,** part 2 (1949).
28. J. M. Greenberg, Interstellar Grains, Annual Review of Astronomy and Astrophysics, Vol. 1, L. Goldberg, Ed., p. 267, Annual Reviews, Palo Alto, 1963.
29. A. Boggess III and J. Borgman, Interstellar Extinction in the Middle Ultraviolet, Astrophys. J. **140,** 1636 (1964).
30. T. P. Stecher, Interstellar Extinction in the Ultraviolet, Astrophys. J. **142,** 1683 (1965).
31. B. Donn and T. P. Stecher, On Graphite and Interstellar Extinction, Astrophys. J. **142,** 1681 (1965).
32. V. L. Ginzburg and S. I. Syrovatskii, Origin of Cosmic Rays, AN SSSR, 1963.
33. R. J. Gould and G. R. Burbidge, High Energy Cosmic Photons and Neutrinos, Annales D'Astrophysique **28,** 171 (1965).
34. V. L. Ginzburg and S. I. Syrovatskii, Nature of Discrete Sources of x rays, Doklady A.N. USSR, Astronomiya, Tom **158,** 808 (1964).
35. W. L. Kraushaar and G. W. Clark, Search for Primary Cosmic Gamma Rays with the Satellite Explorer XI, Phys. Rev. Letters **8,** 106 (1962).
36. Peterson, L. E., Upper Limits to the Cosmic Gamma Ray Flux from OSO 1, Sixth COSPAR International Space Science Symposium, Buenos Aires (1965).
37. A. E. Metzger, E. C. Anderson, M. A. Van Dilla, and J. R. Arnold, Detection of an Interstellar Flux of Gamma-Rays, Nature **204,** 766 (1964).
38. Rocchia, R., R. Rothenflug, P. Boclet, G. Ducros, and J. Labeyrie, 7th International Space Sci. Symp. Vienna (1966).
39. R. Giaconni, H. Gursky, F. R. Paolini, and B. B. Rossi, Further Evidence for the Existence of Galactic X-Rays, Phys. Rev. Letters **11,** 530 (1963).
40. S. Bowyer, E. T. Byram, T. A. Chubb and H. Friedman, Cosmic X-Ray Sources, Science **147,** 394 (1965).
41. G. Garmire, and W. L. Kraushaar, High Energy Cosmic Gamma Rays, Science, Rev. **4,** 123 (1965).
42. C. W. Allen, Astrophysical Quantities, 2nd Edition, The Athlone Press (1963).
43. G. G. Fazio, F. W. Stecher, and J. P. Wright, Cosmic Black-Body Radiation, High-Energy Electrons, and the Origin of Isotropic X-Ray and Gamma Radiation, Astrophys. J. **144,** 611 (1966).
44. S. Hayakawa, and Y. Yamamoto, Ultra-High Energy γ-Rays from Intregalactic Space, Prog. of Theor. Physics **30,** 71 (1963).
45. R. Giacconi, H. Gursky, F. R. Paolini, and B. B. Rossi, Evidence for X-Rays from Sources Outside the Solar System, Phys. Rev. Letters **9,** 439 (1962).
46. G. W. Clark, Balloon Observation of the X-Ray Spectrum of the Crab Nebula Above 15 Kevs, Phys. Rev. Letters **14,** 91 (1965).

47. E. T. Byram, T. A. Chubb, and H. Friedman, Cosmic X-Ray Sources, Galactic and Extragalactic Science, **152,** 66 (1966).
48. Fisher, P. C., H. M. Johnson, W. C. Jordan, A. J. Meyerott, L. W. Acton, Astrophys. J. **143,** 203 (1966).
49. G. Clark, G. Garmire, M. Oda, M. Wada, R. Giacconi, H. Gursky and J. R. Waters, Positions of Three Cosmic X-Ray Sources in Scorpio and Sagittarius, Nature **207,** 584 (1965).
50. S. Bowyer, E. T. Byram, T. A. Chubb, H. Friedman, Lunar Occultation of X-Ray Emission from the Crab Nebula, Science **146,** 912 (1964).
51. H. Gursky, R. Giacconi, P. Gorenstein, J. R. Waters, M. Oda, H. Bradt, G. Garmire and B. V. Sreekantan, A Measurement of the Angular Size of the X-Ray Source SCOX-1. Astrophys. J. **144,** 1249 (1966).
52. Sandage, A. R., P. Osmer, R. Giacconi, P. Gorenstein, H. Gursky, J. Waters, H. Bradt, G. Garmire, B. V. Sreekantan, M. Oda, K. Osawa, and J. Jugaku, Astrophys. J. **146,** 316 (1966).
53. R. Giacconi, X-Ray Observations from Rockets, Proceedings 1967 Texas Symposium on Relativistic Astrophysics, Jan. 23-27, 1967.
54. W. Kraushaar, G. W. Clark, G. Garmire, H. Helmken, P. Higbie, and M. Agogino, Explorer XI Experiment of Cosmic Gamma Rays, Astrophys. J. **141,** 845 (1965).
55. J. G. Duthie, R. Cobb and J. Stewart, Evidence for a Source of Primary Gamma Rays, Phys. Rev. Letters **17,** 263 (1966).
56. G. M. Frye, Jr., and L. H. Smith, Upper Limits to Cosmic γ-Rays in the Range 30 to 500 MeV, Phys. Rev. Letters **17,** 733 (1966).
57. Kniffen, D. A., T. L. Cline, C. H. Ehrmann, C. E. Fichtel and R. W. Ross, Gamma Ray Astronomy with A. Digitized Spark Chamber, AAS Symposium on Ultraviolet, X-ray, and Gamma-Ray Astronomical Observations (1966).
58. G. M. Frye, Jr., and C. P. Wang, Reported Cosmic Gamma-Ray Source in Cygnus, Phys. Rev. Letters **18,** 132 (1967).
50. A. E. Chudakov, V. I. Zatsepin, N. M. Nesterova, and V. L. Dadykin, A Search of Photons with the Energy of 10^{13} eV from Discrete Sources of Cosmic Radiation., J. Phys. Soc. Japan, **17,** Suppl. A-III, 106 (1962).
60. J. H. Fruin, J. V. Gelley, C. D. Long, N. A. Porter, and T. C. Weekes, Flux Limits for High-Energy γ-Rays from Quasi-Stellar and Other Radio Sources, Physics Letters **10,** 176 (1964).
61. Peterson, L. E., A. S. Jacobson, and P. M. Pelling, Spectrum of Crab Nebula X-Rays to 120 keV, Phys. Rev. Letters **16,** 142 (1966).
62. R. C. Haymes and W. L. Craddock, Jr., High Energy X-Rays From the Crab Nebula, Department of Space Science, Rice University (1966).
63. K. J. Frost, E. D. Rothe, and L. E. Peterson, A Search for the Quiet-Time Solar Gamma Rays from Balloon Altitudes, J. Geophys. Res. **71,** 4079 (1966).
64. R. Giacconi, H. Gursky, and J. R. Waters, Spectral Data from the Cosmic X-Ray Sources in Scorpius and Near the Galactic Center, Nature **207,** 572 (1965).
65. G. Chodil, R. C. Jopson, H. Mark, F. D. Seward, and C. D. Swift, X-Ray

Spectra from Scorpius (SCO-XR-1) and the Sun Observed Above the Atmosphere, Phys. Rev. Letters **15,** 605 (1965).

66. R. J. Grader, R. W. Hill, F. D. Seward, A. Toor, X-Ray Spectra from Three Cosmic Sources, Sciences **152,** 1499 (1966).
67. H. Friedman, Naval Research Laboratory X-Ray Observations from Rockets, Proceedings 1967 Texas Symposium on Relativistic Astrophysics, Jan. 23-27, 1967.
68. B. B. Rossi, Recent Advances in X-Ray Astronomy, MIT preprint (1966).
69. K. G. McCracken, A Celestial Source of X-Rays in the Energy Range 20 to 58 KeV, the 9th International Conference on Cosmic Rays, London (1965).
70. D. Brini, U. Ciriegi, F. Fuligni, E. Moretti, and G. Vespignani, Cosmic X-Ray Sources in the 20 to 180 KeV Energy Range, University of Bologna preprint (1966).
71. A. A. Pensias and R. W. Wilson, A Measurement of Excess Antenna Temperature at 4080 Mc/s, Astrophys. J. **142,** 419 (1965).
72. A. D. Code, Astronomy from Space Vehicles, Astronom. J. **65,** 278 (1960).

Chapter 21

STELLAR EVOLUTION

Robert C. Cameron

THE NEW ASTRONOMIES

We may speculate with some hope of success on the formation of the Solar System of which we form a part, for it presents to us numerous perfectly well-known phenomena, susceptible perhaps of giving proof of its true immediate origin. But what, on the other hand, could possibly form a rational basis for our conjectures on the formation of other suns? How confirm or disprove by the evidence of phenomena any cosmogonical hypothesis when no phenomena of such a kind are known, nor, doubtless, are even knowable?

These words, written by A. Comte in his *Cours de Philosophie Positive* (1830–42), were quoted at the dedication of the Victoria Telescope of the Royal Observatory at the Cape of Good Hope by His Majesty's Astronomer, Sir David Gill, in 1901. He continued:

In other words, the philosophic dictum of sixty years ago was that the chemical composition of other systems than our own is a subject which, from the nature of things, must be regarded as unknowable. But the dis-

covery of the lines in the Solar Spectrum by Fraunhofer [1814], *the interpretation of the meaning of these lines by Kirchhoff and Bunsen* [1859], *and the application of the spectroscope to other celestial objects has upset that philosophic conception of the unknowable and given to us the new astronomy.*

The *old astronomy* was essentially that of position and magnitude, and hence of motion, mass, distance, and spatial distribution. The two astronomies have since proceeded side by side, often seemingly as if on separate tracks, but Gill had the insight (not universally shared) that the two cannot be divorced. At propitious times, after periods of amassing of data and subsequent incubation, far-reaching syntheses have taken place. For example, there is the demonstration that stars with similar spectroscopic characteristics are apt to share similar dynamic characteristics of velocity, eccentricity, and inclination to the galactic plane in their course through the galaxy—characteristics which are related to the date and local conditions of birth of the stars in question. The present day is one of such syntheses, not only within astronomy itself, but syntheses with adjacent disciplines as diverse as geology and the physics of elementary particles.

The dark and bright lines of the stellar spectra were of course due to the absorption or emission of energy by atmospheric atoms or molecules at discrete frequencies, the wavelength and intensity of which depend on the element (or compound) involved and the physical conditions of its environment. It will be recalled that helium was discovered in the solar spectrum 26 years before it was produced in the laboratory.

Now a powerful means was at hand for inferring not only the contents and conditions inside the stars, but also the details of their history and future. As Gill so aptly put it 60 years ago:

The different types of star spectra form such a complete and gradual sequence (from simple spectra resembling those of nebulae, onwards through types of gradually increasing complexity) as to suggest that we have before us, written in the cryptograms of these spectra, the complete story of the evolution of suns from the inchoate nebulae onwards to the most active sun (like our own) and then downwards to the almost heatless and invisible ball. The period during which human life—nay, even life of any kind—has existed on our globe is probably too short to afford ob-

servational proof of such a cycle of change in any particular star, but the fact of such evolution, with the evidence before us can hardly be doubted.

With the comparatively recent advent of radio and space astronomy, and the potentially important neutrino astronomy, there is today a multiplicity of new astronomies. New cryptograms provided by radio waves, by radiation in hitherto inaccessible infrared, gamma-ray, and X-ray regions of the spectrum, and by cosmic rays, have only begun to be deciphered. The study of stellar evolution itself has entered an exciting evolutionary phase whose outcome remains to be seen. Nevertheless, there exists a solid and hard-earned basis for the present tackling of the information explosion. One of the principal concerns in this introduction to stellar evolution is to convey some feeling for the subject by examining briefly the fundamentals, by bringing the story up to date, and by indicating the nature of unsolved problems.

Stars of representative masses and chemical compositions will be examined in the following evolutionary phases: (1) birth (initial condensation by gravitational contraction), (2) lifetime (energy production by nuclear fusion) and (3) death (exhaustion of nuclear fuel). Occasionally it will be the case that when general statements are made exceptions will exist under unusual conditions. Certainly the main-sequence and white dwarf phases have been the most thoroughly studied. Other phases have received detailed investigation only within the last decade: here we are at the forefront of the study of the evolution of the stars. [For the reader interested in a more comprehensive introduction, the following books are recommended: for general textbooks including the details of stellar model computation, Refs. 1 and 2; for general review articles, Refs. 3 to 5; for the basic physics of the stellar interior, Refs. 6 to 10; and for symposium reports, Refs. 11 to 15.]

GALAXIES AND STAR POPULATIONS

First, a quick look at the galaxies in which these stars are embedded is in order.

In photographs made with the largest telescopes, in directions relatively free of local obscuring matter (interstellar gas and dust in our own galaxy), the large-scale features of the universe are evident: hosts of galaxies, sometimes more numerous than the foreground stars of our galaxy. The prominent features of Figure 1 are the spiral and elliptical galaxies; and careful examination shows that even some of the apparently star-like images are actually of galaxies—detectable by their irregular or fuzzy appearance.

Two nearby spiral galaxies are shown in Figures 2 and 3 which illustrate in a general way the appearance of our own galaxy if it could be seen from outside. In Fig. 2, NGC 4594 is seen edge-on, showing the typical heavy concentration of stars and interstellar obscuring matter in a plane or disc, surrounded by a less dense spherical halo of stars. In M81 (Fig. 3), which appears in an oblique aspect, the halo is not evident but the spiral-arm structure is well-delineated and a dense central bulge can be seen. Contrary perhaps to the photographic appearance, there is ample space between stars. It has been estimated that in the case where one galaxy passes through another, there is an expectation of about one direct collision between stars.

There are about 100 billion stars (10^{11}) in our galaxy. The sun is about two-thirds of the way out to the edge, at a distance of 30,000 light-years from the center, in or near a spiral arm, and essentially in the central galactic plane. (One light-year = 5.88×10^{12} miles = 9.46×10^{17} cm.) The component stars are in orbit about the center of mass of the galaxy. At the sun's distance, the period of revolution is 200 million years, with stars whose mean distances are less than that of the sun revolving in a shorter period, and vice versa. Hence the sun, with an age of 4.5 billion years, has made the circuit more than 20 times and has gradually departed from its inner and outer neighbors.

The concentration of stars and obscuring matter (dark clouds) in the galactic plane (Milky Way) is seen in Fig. 4, photographed in the direction of the center of the galaxy; the central galactic bulge and other galaxies are not seen, however, because of the effectiveness of interstellar matter in obscuring visible radiation.

Twenty years ago, W. Baade introduced the concept of the two stellar populations, the ramifications and refinements of which are still under discussion.[15] To oversimplify somewhat, the stars of Population I are those in spiral arms and are to be found in association

FIG. 1. Galaxies of various types in a relatively unobscured region in Hercules. Some of the faintest images are of galaxies (Palomar 200-inch photograph).

FIG. 2. The Sombrero Galaxy (NGC 4594) in Virgo, a spiral galaxy seen nearly edge-on, showing the galactic halo and the concentration of obscuring matter in the galactic plane. Its distance is 43 million light-years (Palomar 200-inch photograph).

FIG. 3. The spiral galaxy M81 in Ursa Major, at a distance of 8.6 million light years (Palomar 200-inch photograph).

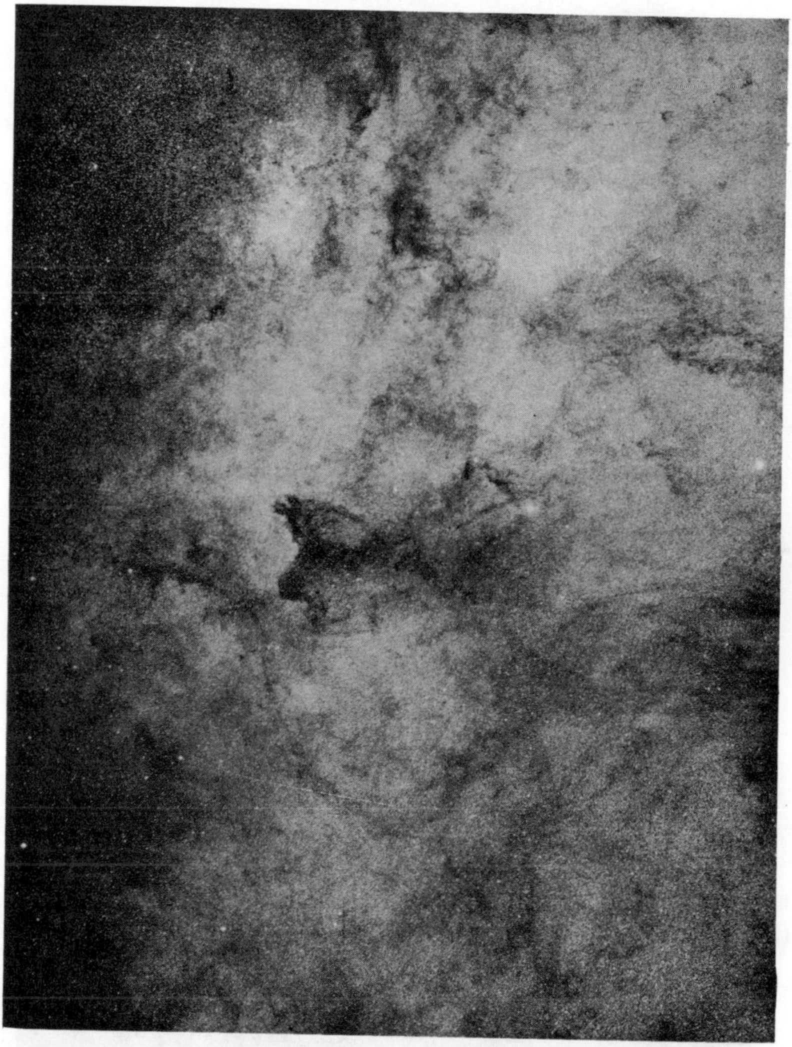

FIG. 4. Concentration of stars and obscuring matter in the galactic plane (Milky Way) in the region of Sagittarius, toward the center of our galaxy (Palomar 48-inch Schmidt photograph).

with the interstellar gas and dust which are plentiful there. Supergiants and other intrinsically bright stars are typical members as well as the so-called galactic clusters, lying in the galactic plane, with populations of up to a few thousand stars. The sun and those other stars which are visible to the naked eye are members of Population I, as are the stars being born today, formed out of the gas and dust.

Essentially, everything else constitutes Population II. Typical members are the halo population of the spiral galaxies with its globular clusters, and the high-velocity stars (with respect to the local standard of rest) in their appreciably elliptic orbits.

Figure 5 shows the cluster M13, with a population of about 100,000 stars, representative of the 125 or so globular clusters in our galaxy.

The most luminous (and most massive) stars of Population II are about 100 times fainter than those of Population I, a circumstance which was a prime factor in the delay in the recognition of Population II. The Population II stars are found in regions relatively free of

FIG. 5. The globular cluster M13 in Hercules, at a distance of 25,000 light-years (Palomar 200-inch photograph).

interstellar gas and dust and are old stars, formed at a time when turbulent velocities in the galaxy were high, and with maximum masses of perhaps only about 5 times that of the sun. The ages of the globular clusters are of the order of 10^{10} years, as compared with 10^7 years for an aged massive supergiant. The fractional abundances (by mass) of elements other than hydrogen and helium are extremely low in Population II stars. Typical initial abundances adopted in stellar model computations, (where X, Y, and Z refer to hydrogen, helium, and all other elements, respectively) are

$$X = 0.71, \quad Y = 0.27, \quad Z = 0.02 \quad \text{(Population I)}; \tag{1a}$$

$$X = 0.90, \quad Y = 0.099, \quad Z = 0.001 \quad \text{(Population II)}. \tag{1b}$$

It now appears that the complex elements have been built up from primordial hydrogen through nuclear processes taking place in the stars, and that, for example, a Population I star like the sun, with a good supply of heavy elements, was formed out of matter which had been processed in the interior of one or more stars and expelled either gently or through a cataclysmic incident into the interstellar medium. Thus the everyday objects with which we are familiar may be regarded as souvenirs of a stellar interior. Chapter 23 gives a description of the element-building process.

THE HERTZSPRUNG–RUSSELL DIAGRAM

It is convenient in stellar evolution computations to give the mass, luminosity, and radius of a star in units of those of the sun (the same symbols without the solar subscript will be used for a given star):

$$M_\odot = 1.987 \times 10^{33} \text{ gm}, \tag{2a}$$

$$L_\odot = 3.90 \times 10^{33} \text{ ergs/sec}, \tag{2b}$$

$$R_\odot = 6.957 \times 10^{10} \text{ cm}. \tag{2c}$$

Before discussing the Hertzsprung–Russell diagram, which is an important meeting-place for observational and theoretical results, a brief discussion of luminosity and surface temperature is in order.

Stellar brightness or luminosity is discussed in terms of magnitude. Apparent magnitudes are those measured at the earth (corrected for the influence of the earth's atmosphere); absolute magnitudes are intrinsic magnitudes and are those that would be observed for a given star if it were located at a standard distance of 10 parsecs (1 parsec = 3.26 light-years). The absolute magnitude, M, and the apparent magnitude, m, are related by

$$M = m + 5 - 5 \log d - A, \tag{3}$$

where d is the distance of the star in parsecs, A the amount of interstellar obscuration expressed in magnitudes, and log d means $\log_{10} d$. The (logarithmic) magnitude scale is such that a difference of 5 magnitudes between two stars corresponds to a factor of 100 in luminosity, *with the brighter star being assigned the lower magnitude.* The bolometric absolute magnitude refers to energy output at all wavelengths:

$$M_{\text{bol}} = -2.5 \log (L/L_{\odot}) + 4.60. \tag{4}$$

The absolute photovisual magnitude, M_v, is derived from observations in the yellow region of the spectrum. The two absolute magnitudes differ by the bolometric correction, which is largely a function of the effective (surface) temperature:

$$M_{\text{bol}} = M_v + \text{B.C.} \tag{5}$$

Estimates of surface temperature are obtained in several ways. The spectral classifications, O, B, A, F, G, K, M are indicators, originally empirical, as suggested by the order of the letters pertaining to the spectral classes, from hot to cool above. The sun is a G2 star.

A more accurate modern indicator is the difference between the blue and photovisual apparent magnitudes, B–V, determined photoelectrically with standard filters.

In theoretical computations, the effective temperature, T_e, defined by Stefan's law for blackbody radiation,

$$\log T_e = 3.76 + 0.25 \log (L/L_{\odot}) - 0.50 \log (R/R_{\odot}), \tag{6}$$

is determined for a stellar model from its luminosity and radius. The effective temperature of the sun is about 5800 degrees. (All temperatures in this chapter are given on the absolute or Kelvin scale, i.e., centigrade temperature plus 273 degrees.)

Fifty years ago, E. Hertzsprung and H. N. Russell, on the basis of the limited data available at that time, put forward the fundamentally important two-dimensional classification scheme, now commonly called the H–R diagram.[5,14] The abscissa corresponds to surface temperature (increasing to the left) and the ordinate to intrinsic brightness (increasing upwards). In the common modern variant, spectral class is replaced by the $B-V$ color as the abscissa; the ordinate is usually the absolute photovisual magnitude, M_v.

In theoretical stellar models, the corresponding quantities arrived at are $\log T_e$ for the abscissa, and either M_{bol} or $\log L/L_{\odot}$ for the ordinate. These quantities can then be converted to $B-V$ and M_v by means of Fig. 6 and Eqs. (4) and (5) for comparison with observation.

All three types of diagram are of similar appearance and, for convenience, all will be referred to here as H–R diagrams.

In Fig. 7, we see at a glance the regions of the H–R diagram in which a few of the important types of stars are to be found, along with an indication of their membership in Population I or II. The gross features of the diagram are closely identified with the operative energy source.

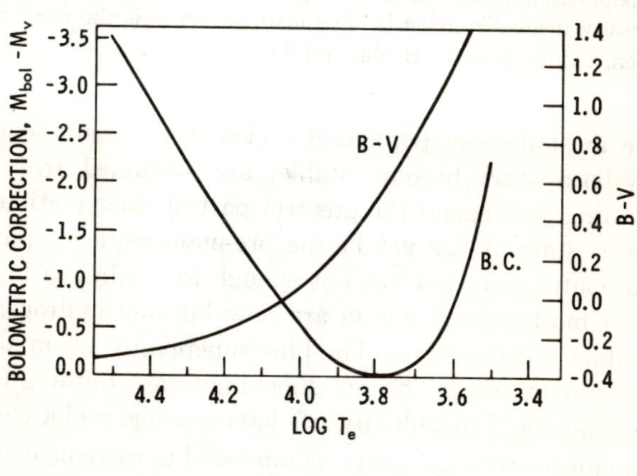

FIG. 6. The bolometric correction and $B-V$ color as a function of effective (surface) temperature (data by H. C. Arp[5]).

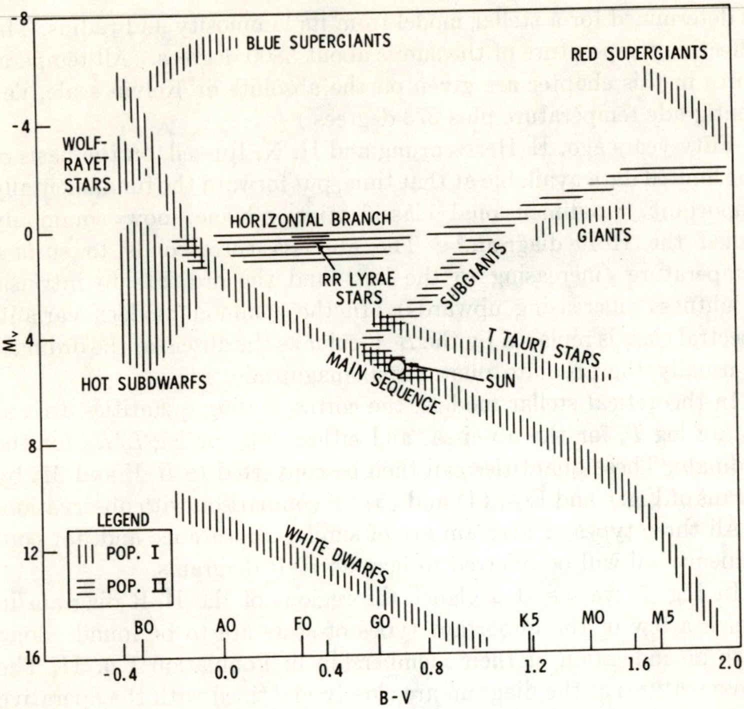

FIG. 7. The Hertzsprung–Russell diagram, showing various classes of stars according to population type. Surface temperature increases to the left and intrinsic luminosity increases upwards. The main sequence is the most densely populated region. (After Hayashi, Hoshi, and Sugimoto.)

By far the most densely populated region is the main sequence, where newly-born stars become stable, are sustained by central hydrogen-burning, and spend the greatest part of their lifetime.

The T Tauri stars are as yet in the pre-main-sequence phase of gravitational contraction, not yet hot enough for hydrogen burning. Subgiants and most normal giants are now burning hydrogen in a shell surrounding a helium core. The blue supergiants are in a phase of central helium burning. The red supergiants are burning carbon and heavier elements. The white dwarfs have reached a phase of final decay with no nuclear energy source, compressed to extreme densities, and cooling off to invisibility.

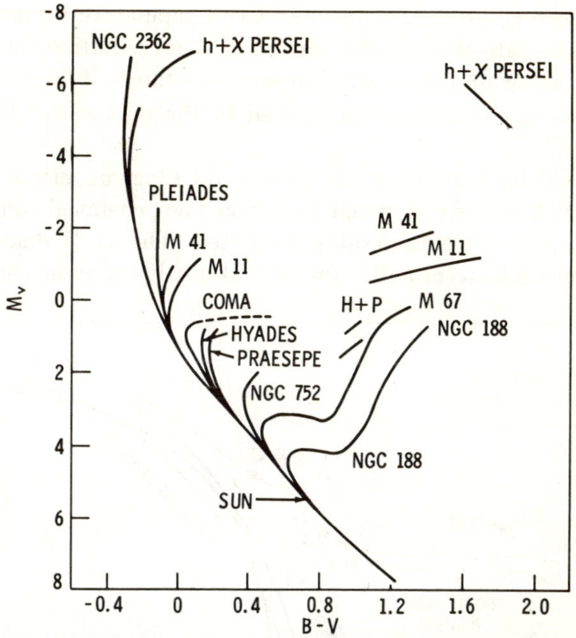

FIG. 8. A composite schematic H–R diagram of several well-known galactic (Population I) clusters. The youngest is NGC 2362; the oldest, NGC 188. The gap between the main sequence and the giant branches is called the Hertzsprung gap and corresponds to a phase of rapid evolution. (After A. Sandage.)

The scant preview presented above is the fruit of a century of study of stellar evolution.* In principle, however, the entire future life of a single star is deducible on physical principles when the star's initial mass and composition are known.

Computations of stellar models should attain sufficient accuracy so that the emitted spectrum for a star of any given mass, age, and initial chemical composition can be predicted. Such a calibration would then allow the determination of these basic parameters, and

* The study of stellar evolution began with the publication in 1869 by I. J. Homer Lane, a sometime U.S. Patent Office employee, of a paper with the rather self-sufficient title, "On the Theoretical Temperature of the Sun under the Hypothesis of a Gaseous Mass Maintaining Its Volume by Its Internal Heat and Depending on the Laws of Gases Known to Terrestrial Experiment."

also the radius, luminosity, and effective temperature for any star, through accurate photoelectric sampling of its radiation by means of a small number of narrow-band interference filters. Just such pioneer work is going forward as exemplified in the research of Strömgren and Kelsall.[16,17]

A star cluster is a microcosm wherein the range of certain variables relevant to the study of stellar evolution (age, chemical composition, and distance) are considerably restricted. The H–R diagram of a cluster gives a locus of points for stars of nearly the same composition

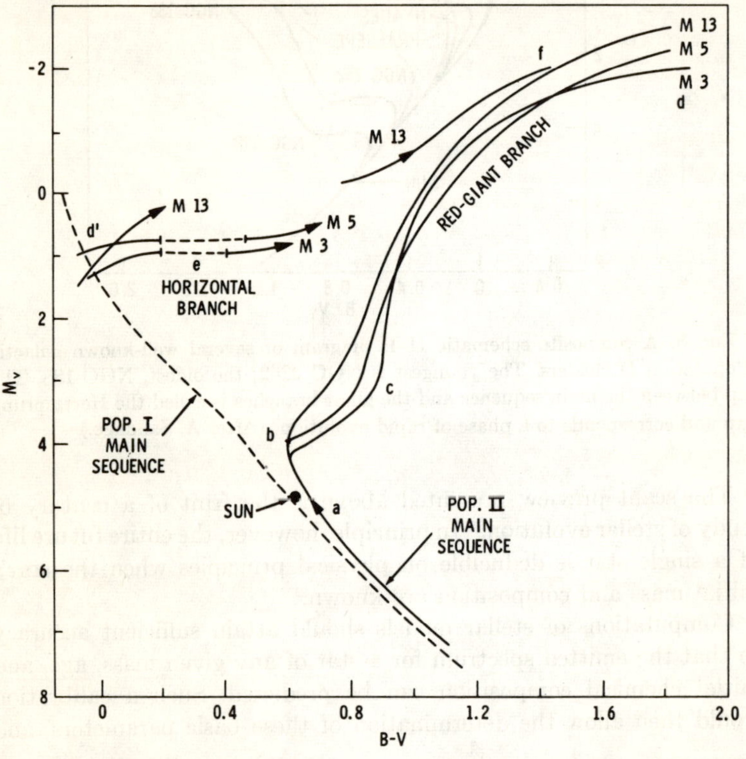

FIG. 9. A composite schematic H–R diagram for three well-observed globular (Population II) clusters. The sun and the Population I main sequence are shown for comparison. After following the track *a–d*, a star is thought to proceed rapidly to point *d′* where it again evolves to the right to point *f*. The dashed regions *e* (RR Lyrae gap) are occupied exclusively by pulsating RR Lyrae stars. (After A. Sandage.)

for a range of masses. Such a diagram resembles, in its rapidly evolving giant branches of narrow mass-range, the computed evolutionary track for a star of given mass and composition throughout its lifetime. Here we have a powerful observational check on theoretical computations.

Sandage has compiled schematic composite H–R diagrams for Population I (galactic) clusters (Fig. 8) and Population II (globular) clusters (Fig. 9).

A prominent feature of these diagrams is the *turn-off point* where the star departs from the main sequence to the right; this just corresponds to the exhaustion of hydrogen in the central region. The stars involved in the turn-off will be the most massive main-sequence stars—and also the most luminous and the most profligate consumers of their energy store. Thus the galactic cluster h and χ Persei, which still retains stars of masses up to about 15 solar masses on its main sequence, is only about 18 million years old, whereas NGC 188, whose stars at the turn-off point are of the order of one solar mass, is about 1000 times older. In short, with proper calibration, the age of a cluster is given by the absolute magnitude of its turn-off point.

THE STATE OF THE STELLAR INTERIOR

Under the conditions generally prevailing within the stellar interior, the matter acts as a perfect gas, i.e., it satisfies Boyle's Law,

$$P_g = k\rho T/\mu H, \tag{7}$$

where P_g is the local gas pressure in dynes/cm^2, $k = 1.380 \times 10^{-16}$ erg/deg is the Boltzmann constant, ρ the local density in gm/cm^3, T the local absolute temperature, and μ the mean molecular weight (average mass per particle) in units of the mass of the hydrogen atom, $H = 1.673 \times 10^{-24}$ gm.

Initially, at least 98 percent of the stellar mass is composed of the two simplest elements, hydrogen and helium, whose electrons are easily stripped off at the high temperatures involved. Thus the effective volume of a gas particle is not that of an atom, but of the tremendously reduced volume corresponding to a nucleus or a free electron. For example, the central density of a star of 15.6 solar masses at the onset of helium-burning is 1600 gm/cm^3, or approximately 200 times the density of iron; yet the central temperature of 160 million degrees insures that the perfect gas relation [Equation (7)] will hold with sufficient accuracy.

In the case of complete ionization, the reciprocal of the mean molecular weight is expressed by

$$\mu^{-1} = \sum \{[(z + 1)/A]X_z\} \tag{8}$$

summed over all the elements present, where z is the atomic number and $(z + 1)$ is merely the number of particles involved per atom (i.e., there are z electrons and one nucleus); A is the atomic mass in units of the hydrogen atom; and X_z is the fractional abundance by mass of each element. Although the heavy elements are not completely ionized in the stars, neither are they abundant. For abundant elements, $(z + 1)/A \simeq z/A \simeq \frac{1}{2}$ is a good approximation, with the exceptions of hydrogen and helium. Taking the fractional abundances of hydrogen, helium, and heavy elements as X, Y, and Z, respectively (so that $X + Y + Z = 1$),

$$\mu^{-1} = 2X + \tfrac{3}{4}Y + \tfrac{1}{2}Z, \tag{9}$$

of which

$$\mu_i^{-1} = X + \tfrac{1}{4}Y \tag{10}$$

is due to ions (nuclei),

$$\mu_e^{-1} = X + \tfrac{1}{2}Y + \tfrac{1}{2}Z = \tfrac{1}{2}(1 + X) \tag{11}$$

is due to free electrons, and

$$\mu^{-1} = \mu_i^{-1} + \mu_e^{-1}. \tag{12}$$

The limits of the mean molecular weight are thus $\frac{1}{2}$ (pure hydrogen) and 2 (heavy elements alone).

The total pressure is just the sum of the gas and radiation pressures:

$$P = P_g + P_{\text{rad}}. \tag{13}$$

The radiation pressure is effectively that of an isotropic radiation field, i.e., it is one-third of the energy density:

$$P_{\rm rad} = \tfrac{1}{3}aT^4 = 2.52 \times 10^{13}(T/10^7)^4 \text{ dynes/cm}^2, \tag{14}$$

where $a = 7.57 \times 10^{-15}$ erg cm^{-3} deg^{-4} is the radiation density constant.

The representation of the gas pressure as the sum of the ion and electron pressures,

$$P_g = P_i + P_e, \tag{15}$$

is of importance when the perfect gas equation (7) fails, e.g., at ultra-high densities when the electron gas becomes degenerate, while the ions (nuclei) continue to constitute a perfect gas.

Degeneracy does not arise from the packing of particles so tightly that they make contact, but is a consequence of the *Pauli exclusion principle*, which states that no two electrons can occupy the same quantum state; i.e., a cell of volume h^3 (where h is Planck's constant) in the accessible 6-dimensional position-momentum (phase) space can be occupied by at most two electrons of opposite spin. In the limiting case of complete degeneracy, all of the lowest quantum states are occupied and all of the electrons have momenta less than a certain maximum value, p_0, the Fermi threshold momentum, which is related to the electron pressure (which is now independent of the temperature), in the case of complete non-relativistic degeneracy, by

$$P_e = (8\pi/15mh^3)p_0^5 = 9.91 \times 10^{12}(\rho/\mu_e)^{5/3}. \tag{16}$$

In the case of complete relativistic degeneracy, i.e., in the case where the Pauli exclusion principle compels the electron velocities to approach the speed of light,

$$P_e = (2\pi c/3h^3)p_0^4 = 1.231 \times 10^{15}(\rho/\mu_e)^{4/3}. \tag{17}$$

In these expressions, $m = 9.107 \times 10^{-28}$ gm is the rest mass of the electron, $c = 2.998 \times 10^{10}$ cm/sec is the velocity of light, and the value of Planck's constant is $h = 6.624 \times 10^{-27}$ erg–sec.

The most notable case wherein the theory of degeneracy is applicable is that of the white dwarf stars, of which the companion of

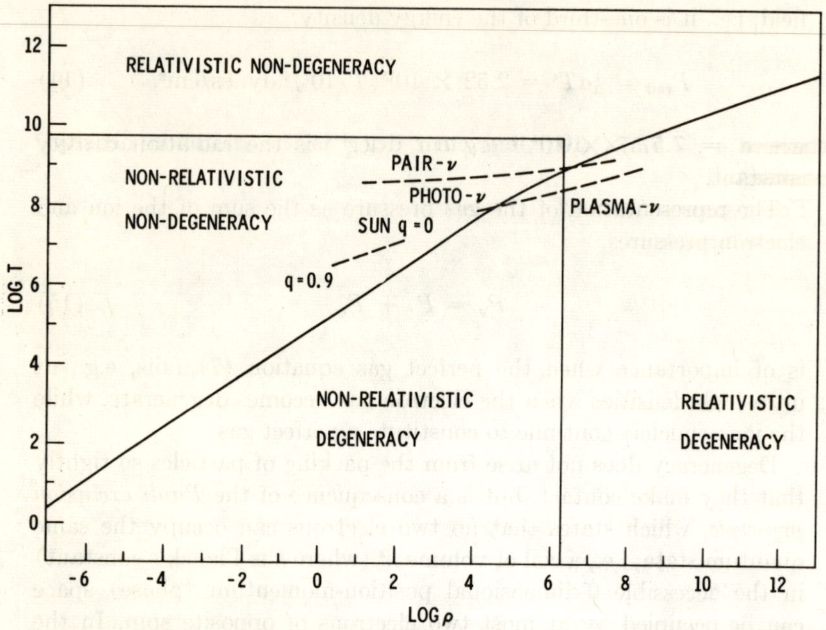

FIG. 10. Regions of relativistic and non-relativistic degeneracy and non-degeneracy in the ρ–T diagram (after G. Wares). The transition from one region to the next is gradual. Non-degenerate regions follow the perfect gas law. The locus of the inner 90 percent (by mass) of the sun is indicated. For temperatures above about 100 million degrees, the dominant neutrino-emission process, arising from the universal Fermi interaction is indicated, according to data by H. Reeves.[11]

Sirius, as massive as the sun, but with only $\frac{1}{300}$ of the solar luminosity, is an example. In these stars, mean densities of 10^6 to 10^8 gm/cm^3 (1 to 100 tons/cm^3) are encountered.

Varying degrees of partial degeneracy for which the limiting equations (16) and (17) are approximately true can be met with in the cores of the less massive stars. The regions of the ρ–T diagram pertaining to relativistic and non-relativistic degeneracy and non-degeneracy are shown in Fig. 10. It is to be noted that in the relativistic non-degenerate case, the perfect gas relation is still applicable. For regions near the boundaries indicated in Fig. 10, certain transition formulae are required.[6]

The basic observational data at our disposal are the mass, luminosity, radius, and chemical composition of the surface layers. The problem of inferring the physical variables throughout the interior of a star would be quite intractable were it not for the saving condition of the relative slowness of the evolutionary changes taking place; i.e., at each point in its evolutionary track, a star is effectively in a condition both of dynamic and thermodynamic equilibrium throughout its structure, hence certain simplifications in the mathematical treatment are justifiable. For example, in the stellar interior there is a balance between the weight of overlying layers and the elasticity of matter below, and a star is said to be in hydrostatic equilibrium under its own gravitation. This condition is expressed for a nonrotating gaseous sphere by two basic differential equations,

$$dP/dr = -(GM_r/r^2)\rho, \tag{18}$$

and

$$dM_r/dr = 4\pi r^2 \rho, \tag{19}$$

where r is the distance in centimeters of an interior point from the center of a star, M_r the mass in grams interior to a shell of radius r, P the local total pressure in dynes/cm^2, ρ the local density in gm/cm^3, and $G = 6.67 \times 10^{-8}$ gm^{-1} cm^3 sec^{-2} is the universal gravitational constant. Equation (19) expresses the conservation of mass.

NUCLEAR ENERGY

By rather general arguments it can be shown that temperatures of the order of 10^7 degrees are to be encountered in the interior of the sun (as contrasted with its surface temperature of 5800 degrees) and that its observed luminosity would be produced for a period of the order of only 10^7 years if its energy derived entirely from gravitational contraction.

From these two considerations, we deduce that at such temperatures essentially complete ionization of the constituent elements prevails, and that an energy source more potent than gravitational contraction will be required to mesh with the geological evidence for essentially constant solar radiation during the last billion years.

Furthermore, with the aid of the Wien displacement law,

$$\lambda_{max} T = 0.29, \tag{20}$$

it is seen that, at the temperatures indicated, the radiation to be dealt with is primarily that of X-rays in the angstrom region (1 Å = 10^{-8} cm). An early doubt as to the applicability of laboratory results for X-rays to the interior of a star was ultimately dissolved in 1932 with the quantum-mechanical modification by Gaunt and by Strömgren of the classical Kramers absorption theory necessary for the prevailing highly ionized stellar conditions, and with the realization, by Strömgren and by Eddington, that the stellar interior consists very predominantly of hydrogen.

Nuclear transformations can be produced, for example, by proton bombardment, and for transmutations to occur, the proton need not have the energy to surmount the potential barrier surrounding the target nucleus.

This is an impossible result in classical mechanics, but is amenable to quantitative treatment in quantum mechanics, which was not developed until 1925. Here is the crux of the disagreement of the theorists of the early 1920's with Eddington's essentially correct guess as to the source of stellar energy—on the classical theory, stellar temperatures (and hence proton energies) were just too low to permit nuclear fusion and the accompanying enormous release of energy. In spite of the energy difficulty, Eddington persisted in believing that helium was being produced in some hot place, the foremost candidate being the interior of a star. And to his antagonists he suggested that they go and find a *hotter place.*

Detailed knowledge of the relevant nuclear processes remained hidden until the researches of Bethe, Critchfield, and von Weizsäcker, beginning in 1938. The two processes found to be important were the carbon-cycle and the proton-proton chain.

Two versions of these hydrogen-burning reactions, of which there are variants, depending on the physical conditions in the stellar interior, proceed as follows:

pp-Chain

$$H^1 + H^1 \rightarrow D^2 + e^+ + \nu \qquad \text{(twice)} \tag{21a}$$

$$D^2 + H^1 \rightarrow He^3 + \gamma \qquad \text{(twice)} \tag{21b}$$

$$He^3 + He^3 \rightarrow He^4 + 2\,H^1; \tag{21c}$$

CN-cycle

$$C^{12} + H^1 \rightarrow N^{13} + \gamma \tag{22a}$$

$$N^{13} \rightarrow C^{13} + e^+ + \nu \tag{22b}$$

$$C^{13} + H^1 \rightarrow N^{14} + \gamma \tag{22c}$$

$$N^{14} + H^1 \rightarrow O^{15} + \gamma \tag{22d}$$

$$O^{15} \rightarrow N^{15} + e^+ + \nu \tag{22e}$$

$$N^{15} + H^1 \rightarrow C^{12} + He^4. \tag{22f}$$

The net result in each case is the combination of 4 protons (hydrogen nuclei) to produce an alpha particle (helium nucleus), with the restoration in the final step of a constituent of the first step (H^1 and C^{12}, respectively). The net mass loss, 4.8×10^{-26} gm, is converted to energy amounting to 4.3×10^{-5} ergs, according to $E = mc^2$, which may be regarded as Einstein's reply to a query by Newton (*Opticks*, 1704): *And amongst such various and strange mutations, why may not Nature change bodies into light and light into bodies?*

Although the sun is converting a prodigious 4.8 million tons of its mass into energy each second, extrapolation of this rate for the 4.5 billion years of its life shows that it has suffered a loss of only about 1/3000 of its mass.

It strikes one's attention that it is somewhat odd that a series of reactions as apparently complex as the CN-cycle should play such a basic role in stellar evolution. The explanation approximates the following. For nuclear reactions to take place frequently enough, light nuclei must be involved. But the light elements deuterium, lithium, beryllium, and boron tend to be rapidly destroyed (converted to He^3 or He^4) early in the life of a star and are not important as energy sources. There remains the possibility of constructing a chain of reactions involving light nuclei heavier than boron, but not destroying them. The CN-cycle is the process satisfying such a criterion that goes forward at the lowest temperature. Equilibrium abundances of the four stable isotopes, C^{12}, C^{13}, N^{14}, and N^{15} are eventually established and subsequently each isotope is created as fast as it is destroyed.[2]

Modern results show that for main-sequence stars more massive than about $1.5M_{\odot}$ (Population I) or $4M_{\odot}$ (Population II), the CNO-cycle predominates over the *pp*-chain; in terms of central temperature, the dividing line is at about 20 million degrees.

As we shall see, a star in equilibrium has a luminosity requirement that depends mainly on the stellar mass ($L \propto M^3$) and more particularly on the structure of the innermost regions. As the star evolves, the luminosity requirement increases rather gradually (as long as nuclear energy is available), except in certain cases of rapid core evolution, as exemplified by the red giants.

When a nuclear fuel fails to maintain the required luminosity, gravitational contraction in the central region commences and increases the nuclear output by raising the temperature of the nuclear fuel. Usually a slight temperature increase will suffice because of the sensitivity of the nuclear energy generation rate to the temperature. In a like manner, over-production of energy is accompanied by expansion and cooling of the central region. Thus gravitational and thermal energy are available as buffers against deviations from the luminosity requirement and act effectively as thermostats in controlling the production of nuclear energy.

When exhaustion of the central nuclear fuel is imminent, gravitational core contraction begins in earnest and its role becomes temporarily that of the principal energy source (rather than a goader of the disappearing nuclear source), until it heats the core to the ignition temperature of the next available nuclear fuel, which is, following hydrogen-exhaustion, the helium which has been produced in the hydrogen-burning phase. [According to one of the great generalizations of physics, *the virial theorem*, in the case of negligible radiation pressure, precisely half of the gravitational energy released becomes available for heating and half for surface radiation.] Helium-burning takes place according to Salpeter's *3α-process*, wherein three helium nuclei are converted into a carbon nucleus:

$$He^4 + He^4 \rightleftarrows Be^8 \tag{23a}$$

$$Be^8 + He^4 \rightleftarrows C^{12*} \tag{23b}$$

$$C^{12*} \rightarrow C^{12} + \gamma. \tag{23c}$$

C^{12*} is the second excited state of the C^{12} nucleus. The ignition temperature is about 150 million degrees.

Central helium-burning is accompanied by a second nuclear source, hydrogen-burning, which has previously shifted to a shell surrounding the dehydrogenized core. Hydrogen-burning is reactivated by the substantially increased temperature of the central regions; it is effective only in a thin shell because of the temperature decrease outward (temperature gradient) and the temperature sensitivity of nuclear sources, previously mentioned, just as a central source is effective only very near the center of a star.

As the sequence of central nuclear fuel exhaustion, gravitational contraction, new nuclear burning, and added shell source proceeds (as will be discussed later), the stellar structure becomes more and

TABLE I

Nuclear burnings in stars.*

Burning	Main reactions	Main products**	Temperature (°K)	Energy release (10^{17} ergs/gm)
H^1	$4H^1 \rightarrow He^4$	He^4	$(1\text{–}4) \times 10^7$	60
He^4	$3He^4 \rightarrow C^{12}$, $C^{12}(\alpha, \gamma)O^{16}$	C^{12}, O^{16}	$(1\text{–}3) \times 10^8$	$5.8 \sim 8.6$***
C^{12}	$2C^{12} \rightarrow Ne^{20} + He^4$, $Na^{23} + H^1$	O^{16}, Ne^{20}, Mg^{24}	$(6\text{–}7) \times 10^8$	~ 4
Ne^{20}	$Ne^{20}(\gamma, \alpha)O^{16}$, $Ne^{20}(\alpha, \gamma)Mg^{24}$	O^{16}, Mg^{24}	1.1×10^9	~ 2
O^{16}	$2O^{16} \rightarrow Si^{28} + He^4$, $P^{31} + H^1$	Mg^{24}, Si^{28}, S^{32}	1.3×10^9	~ 4
S^{32}	$S^{32}(\gamma, \alpha)Si^{28}$, $Mg^{24}(\alpha, \gamma)Si^{28}$	Mg^{24}, Si^{28}	1.6×10^9	
Mg^{24}	$Mg^{24}(\gamma, \alpha)Ne^{20}$, followed by Ne- and O- burnings	Si^{28}	1.8×10^9	~ 3
Si^{28}	$Si^{28}(\gamma, \alpha)Mg^{24}$, followed by Mg-, Ne- burnings, etc., to form Fe	Fe^{56}	2.0×10^9	

* As given by Hayashi et al. (Ref. 1). The notation $C^{12}(\alpha, \gamma)O^{16}$ means $C^{12} + \alpha \rightarrow O^{16} + \gamma$, etc.

** Unprocessed elements are included.

*** The two values correspond to the cases where the final product is all carbon and all oxygen, respectively.

more complex. A summary by Hayashi, Hōshi, and Sugimoto,[1] concerning the details of the complete series of nuclear burnings, ending in iron core formation and supernova explosion for a massive star, is given in Table I, where it is to be noted that hydrogen-burning provides by far the greatest energy release. The sequence is shorter as a star is less massive. A star failing to ignite the next nuclear fuel, then cools down to obscurity.

ENERGY FLUX

The third basic differential equation [along with equations (18) and (19)] for the study of stellar structures expresses the conservation of energy in a star in thermal equilibrium:

$$dL_r/dr = 4\pi r^2 \rho(\epsilon + \epsilon_g). \tag{24}$$

Here ϵ is the energy release due to nuclear sources in ergs/gm-sec, and ϵ_g that due to gravitational contraction, which is prominent in the short-lived evolutionary phases when nuclear sources are ineffective. L_r is the net outward rate of energy flow across a spherical surface of radius r, in ergs/sec, i.e., it represents the leakage of energy which will eventually be radiated into space from the stellar surface, where L_r becomes just the stellar luminosity, L. It is to be noted that radiant energy is darting about in all directions in the stellar interior and in all cases a reference to energy flux means net outward flux.

Typical expressions for ϵ will be given in a later section (pages 790–791). At very high temperatures ($T \sim 10^9$ degrees), neutrino emission is important. The neutrinos escape into space with essentially no further interaction with the stellar interior and it is necessary to include appropriate negative terms in ϵ, representing the neutrino loss.

The expression for energy generation by gravitational contraction (also suitable for the case of energy generation by thermal cooling, as in the case of the white dwarfs) is given, in the simple case when radiation pressure is negligible, by

$$\epsilon_g = -\tfrac{3}{2}\rho^{2/3}(\partial/\partial t)(P/\rho^{5/3}). \tag{25}$$

The general case is treated by Hayashi, Hōshi, and Sugimoto.[1]

The outward progress of the energy flux depends essentially on the temperature gradient, whether the energy transport is provided by radiation, convection or conduction.

The fourth and final basic differential equation is the general expression for the temperature gradient, expressed in the useful form,

$$dT/dr = (n + 1)^{-1}(T/P)(dP/dr), \tag{26}$$

where n is the local polytropic index corresponding to the proportionality, $P \propto T^{n+1}$, when the density is eliminated. The properties of spherically symmetric configurations with constant n throughout (polytropes) are well-known.[10] The case $n = 0$ corresponds to the absence of central condensation (constant density) and $n = 5$ to infinite central condensation, with the special case $n = \infty$ corresponding to an isothermal situation. The polytropic index can be helpful in understanding the mathematical behavior within limited regions of a complex stellar model.

In the stellar interior, local thermodynamic equilibrium obtains. The material is so opaque to radiation that points separated only by a few centimeters are effectively screened from each other. Over this distance, the temperature change is only of the order of 10^{-3} degrees and the radiation field is effectively isotropic. Thus the treatment of the problem is simplified, but it may be noted that the infinitesimal degree of anisotropy present is ultimately responsible for the outpouring of radiation from the stellar surface.

If the temperature gradient is not so great as to cause convective mass motion of the stellar material in a given region, radiative equilibrium obtains, and a mass element is stable against perturbation, i.e., an upward (or downward) adiabatic displacement will be counteracted by a density increase (or decrease) and a return to the equilibrium position.

In this case, the dependence of L_r on the temperature gradient is given by

$$dT/dr = (-3\kappa\rho/16\pi r^2 acT^3)\, L_r, \tag{27}$$

where κ is the local radiative mean mass absorption coefficient in cm^2/gm. Combination with Eqs. (18) and (26) yields

$$(n + 1)_{\mathrm{rad}} = 16\pi c(aT^4/3P)(GM_r/\kappa L_r). \tag{28}$$

A perturbed mass element which persists in its motion is said to be unstable against convection and the adiabatic temperature gradient is appropriate. In this case (convective equilibrium),

$$(n + 1)_{\text{conv}} = (32 - 24\beta - 3\beta^2)/(8 - 6\beta), \tag{29}$$

where $\beta = P_g/P$.

The appropriate value of $(n + 1)$ to be used in Eq. (26) in a given layer is just the larger of $(n + 1)_{\text{rad}}$ and $(n + 1)_{\text{conv}}$, e.g., the outer boundary of a convective core is identified with the layer at which $(n + 1)_{\text{rad}}$ first exceeds $(n + 1)_{\text{conv}}$.

Generally speaking, stars operating on a central nuclear process other than the proton-proton chain will have a convective core whose existence is to be associated with the very strong temperature dependences involved, e.g., $\epsilon \propto T^{16}$ for the carbon cycle, as compared with $E \propto T^4$ for the *pp*-chain. The other principal region where convection prevails is a surface convection zone, which in red giant and supergiant stars may become quite deep, and is to be associated with the incomplete ionization of hydrogen atoms and the high opacity of the cool outer layers. Unfortunately the preceding precepts for determining n can break down here because of the lower density and reduced *mixing length*, the mean distance a convective mass element moves before dissolving into its surroundings, and recourse must be had to an approximate theory.

Convective regions in stars are turbulent and well-mixed and do not upset the hydrodynamical equilibrium. The realization that mixing of the stellar material is limited essentially to regions in convective equilibrium, i.e., that both meridional circulation induced by rotation and diffusion are generally unimportant, represented an important step forward in the study of stellar evolution.

OPACITY

An important controlling factor in the course of the evolution of a star is the radiative opacity, κ, introduced in Eq. (27), which arises from the interaction of radiation and matter. In stars, the radiation is an X-ray photon and the matter is an electron.

The local radiative mean mass absorption coefficient, κ, also called the opacity, is defined by

$$-dI/I = \kappa\rho\, ds, \tag{30}$$

where I is the specific intensity of a pencil of radiation which is reduced by an amount $-dI$ over the distance ds in the direction of the incident radiation.

The principal opacity-producing processes are absorption and electron scattering. In the first of these, the photon is absorbed and the electron, whether in an elliptic (bound) or hyperbolic (free) orbit, is sent into a more energetic state. The three possible absorption processes are described according to the initial and final states of the electron: *bound–bound, bound–free, and free–free.*

The bound–bound process, which produces the spectral lines in the stellar atmosphere and which requires a photon of the correct frequency for a discrete transition, is of the least consequence and has ordinarily been neglected in stellar interiors, partly because of the computational labor involved. It will not be considered further. In the bound-free process, the energy requirement on the photon is only that it exceed the ionization energy; in the free–free process, there is no specific energy requirement, and the electron transition is from one hyperbolic orbit to a more energetic one.

In the stellar interior, matter and radiation are in equilibrium with each other and each absorption event is balanced by its inverse, an emission event, when the entire budget of such processes is considered.

In the case of absorption followed by induced emission, the directions of the absorbed and emitted photons are the same and there is no impedance to the radiation and hence no contribution to the opacity. Spontaneous emission, however, can send a photon in any direction and the net result of absorption followed by spontaneous emission is equivalent to isotropic scattering. Here the outward flow of radiation, with which we are principally concerned, is in effect weakened by its redistribution in direction.

The lengthy details of the computation, combination, and proper averaging by frequency of the basic absorption coefficients to obtain a single mean coefficient, κ, are discussed by Menzel et al.[6] for the non-degenerate case and Hayashi et. al.[1] for the degenerate case.

For illustrative purposes, the following approximate expressions for the non-degenerate bound-free and free–free opacities are useful:

$$\kappa_{bf} = 4.34 \times 10^{25}(\bar{g}/t)Z(1 + X)(\rho/T^{3.5}), \tag{31}$$

$$\kappa_{ff} = 3.68 \times 10^{22}\bar{g}_{ff}(1 - Z)(1 + X)(\rho/T^{3.5}). \tag{32}$$

Since hydrogen and helium are completely ionized, they will be the major contributors to the free–free opacity, whereas the heavy elements will produce the bound–free opacity (but with an indirect assist from hydrogen and helium, which are abundant suppliers of free electrons whose presence affects the number of bound electrons in the heavy atoms).

If we take the Gaunt g-factors of Eqs. (31) and (32) as equal and a typical guillotine factor, $t = 10$, in Eq. (31), it can be seen that for Population I stars ($Z \approx 0.02$), it is the bound-free opacity which dominates, while for Population II stars ($Z \approx 0.001$) it is the free–free opacity. In massive stars the combination of low density and high temperature results in low bound–free and free–free opacities, and the opacity becomes due mainly to electron scattering.

An isolated free electron, uninfluenced by the coulomb field of a nucleus or other electron, is incapable, by considerations of the relativity principle,[18] of absorbing or emitting radiation by processes already discussed. However, in classical terms, a portion of the energy of the incident radiation goes into setting isolated electrons into vibration. These then act as transmitters and send out, in all directions, electromagnetic radiation of their own. The principal effect is again a weakening of the outward flux by a partial redistribution in direction. The consideration that the stellar photon energies are very large as compared with the ionization energies of abundant elements means that even bound electrons may be regarded as isolated in computations of their scattering effects, since their potential energies are essentially negligible.

The absorption coefficient associated with electron scattering is:

$$\kappa_s = (8\pi/3\mu_e H)(e^2/mc^2)^2 = 0.20(1 + X), \tag{33}$$

where $e = 4.802 \times 10^{-10}$ E.S.U. is the electron charge, e^2/mc^2 is the classical electron radius, and κ_s is related to the scattering cross-section, σ_s, by $\kappa_s = \sigma_s/\mu_e H$. Here we have an especially simple expression for the opacity, independent of frequency (in the nonrelativistic case) and depending only on the hydrogen concentration X.

In dense degenerate matter, energy transport is due mainly, but not entirely,[1] to thermal conduction by electrons. The conductivity can be expressed in terms of opacity. As the density increases, the opacity decreases.

In Fig. 11, which pertains to a Population I star, are identified the dominant opacity-producing processes in various regions of the ρ–T diagram.

The most thorough treatment of the opacity problem is that of A. N. Cox,[3,11] who has for many years generously supplied computations—made on the Los Alamos computers—to astronomers on request.

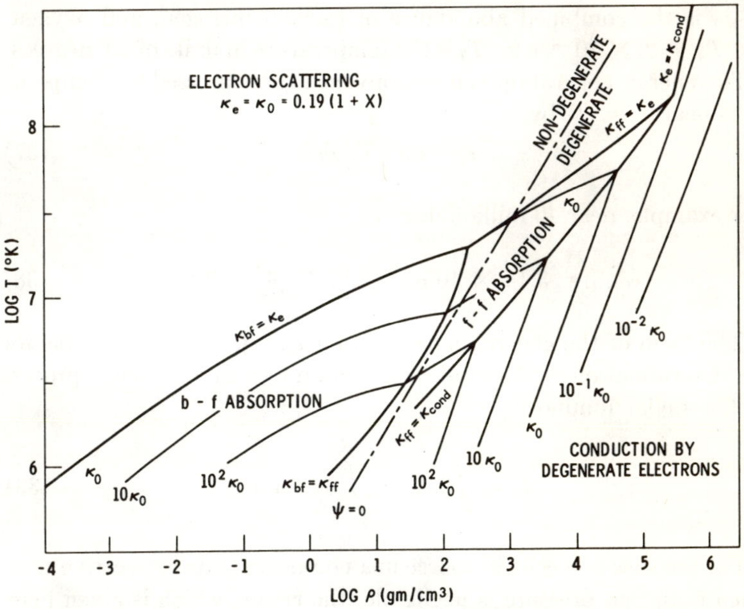

FIG. 11. The dominant opacity-producing processes—electron scattering, bound-free absorption, free-free absorption, and electron conduction—in the interior of a Population I star, as a function of density and temperature. The unit of the opacity is $\kappa_0 = 0.19(1 + X)$. (After Hayashi, Hōshi, and Sugimoto.)

STELLAR LUMINOSITY

The forms of the expressions for the luminosity, L, as derived from the energy generation rate, ϵ, will be illustrated in this section for representative cases.

The explicit energy generation expressions (in erg/gm-sec) for the three most common processes, the *pp*-chain, the carbon-cycle ($4H^1 \rightarrow He^4$), and the 3α-process ($3\, He^4 \rightarrow C^{12}$), are

$$\epsilon_{pp} = 4.4 \times 10^5 \rho X^2 T_7^{-2/3} \exp(-15.69 T_7^{-1/3}), \tag{34}$$

$$\epsilon_{\mathrm{CNO}} = 1.73 \times 10^{27} \rho X_{\mathrm{CNO}} X T_7^{-2/3} \exp(-70.5 T_7^{-1/3}), \tag{35}$$

$$\epsilon_{3\alpha} = 3.5 \times 10^{11} \rho^2 Y^3 T_8^{-3} \exp(-43.2/T_8), \tag{36}$$

where X and Y are the hydrogen and helium abundances, respectively, X_{CNO} is the combined abundance of carbon, nitrogen, and oxygen, and $T_n = T \times 10^{-n}$, e.g., T_7 is the temperature in units of 10^7 degrees.

The energy generation can be conveniently expressed for temperatures near $T = T_0$ by

$$\epsilon = \epsilon^0 \rho^k (T/T_0)^s. \tag{37}$$

For example, near 40 million degrees,

$$\epsilon_{\mathrm{CNO}} = 3.14 \times 10^7 \rho X_{\mathrm{CNO}} X [T/(4 \times 10^7)]^{14.2}. \tag{38}$$

The form of the general expression for ϵ [Eq. (37)] is suitable for the determination of the total contribution by a given nuclear process to the stellar luminosity:

$$L = \int dL_r = \int 4\pi r^2 \rho \epsilon \, dr. \tag{39}$$

For the case of a central source in a non-degenerate convective core, when radiation pressure is negligible, the result, which is given here just to provide some feeling for the way in which the basic quantities enter the luminosity expression, is

$$L_{\mathrm{core}} = \epsilon^0 \rho_c^{k+1} (T_c/T_0)^s \left[\frac{3(n+1)_c}{(k+1)n'_c + s} \frac{P_c}{2G\rho_c^2} \right]^{3/2}, \tag{40}$$

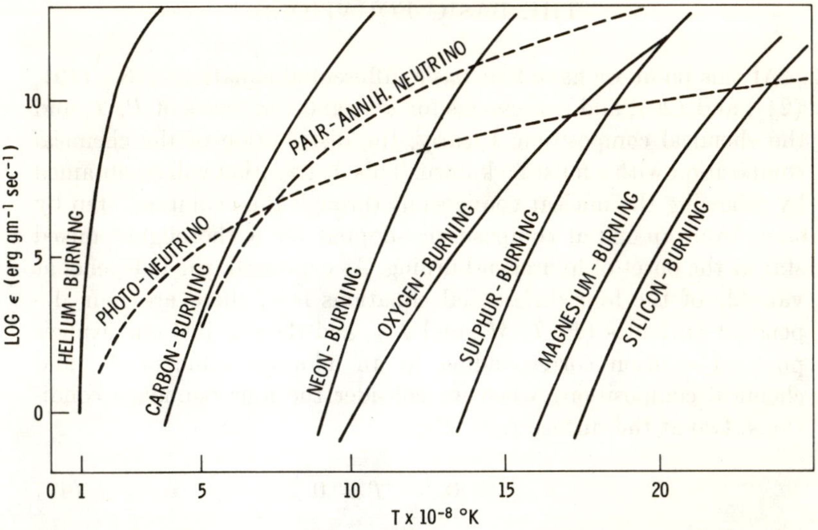

FIG. 12. Energy generation in ergs/gm-sec as a function of temperature for nuclear burnings after hydrogen-burning (Table I) and energy loss for the photo-neutrino and pair-annihilation neutrino processes in the non-degenerate case. The concentration of each fuel is taken as 100 percent, and its density as $\rho = 10^5$ gm/cm³, except in the case of helium-burning ($\rho = 10^4$ gm/cm³). For the neutrino processes, $\rho = 10^5$ gm/cm³ and $\mu_e = 2$. (After Hayashi, Hōshi, and Sugimoto.)

where $n' = (24 - 21\beta)/(8 - 6\beta)$, $\beta = P_g/P$, and n is given by Eq. (29).

Energy generation as a function of temperature for the later nuclear burnings of Table I is plotted in Fig. 12. These curves are based on Eq. (36) and the following generalized expressions,[1] which may be used in connection with Eqs. (39) or (40) for the luminosity computation.

$$\epsilon_{\mathrm{C}} = 2.3\rho X_{\mathrm{C}}^2(T_9/0.6)^{32.6}, \tag{41a}$$

$$\epsilon_{\mathrm{Ne}} = 10^3 X_{\mathrm{Ne}}(T_9)^{65.2}, \tag{41b}$$

$$\epsilon_{\mathrm{O}} = 6 \times 10^{-2}\rho X_{\mathrm{O}}^2(T_9/1.25)^{41.4}, \tag{41c}$$

$$\epsilon_{\mathrm{S}} = 2 \times 10^7 X_{\mathrm{S}}(T_9/1.7)^{64.9}, \tag{41d}$$

$$\epsilon_{\mathrm{Mg}} = 10^7 X_{\mathrm{Mg}}(T_9/2)^{67.3}, \tag{41e}$$

$$\epsilon_{\mathrm{Si}} = 2 \times 10^5 X_{\mathrm{Si}}(T_9/2)^{72.9}. \tag{41f}$$

THE BASIC EQUATIONS

At this point we have four basic differential equations (18), (19), (24), and (26), and expressions for ρ, κ, and ϵ in terms of P, T, and the chemical composition. Clearly, the distribution of the chemical composition with r must be known. This distribution will be obtained by following the nuclear conversions through the evolution, step by step, from an initial composition adopted for a thoroughly mixed star at the onset of hydrogen-burning. As expressed, the independent variable of the four differential equations is r; there are four dependent variables (P, T, M_r, and L_r); and there is just one explicit physical solution corresponding to an adopted value of R (and chemical composition) when we consider the four boundary conditions: two at the surface ($r = R$),

$$P = 0, \qquad T = 0 \tag{42}$$

and two at the center ($r = 0$),

$$M_r = 0, \qquad L_r = 0. \tag{43}$$

As will be seen later, a more rigorous surface condition is necessary for stars of low surface temperature. Otherwise Eq. (42) will be adequate for the integration, since, for example, $T_e/T_c < 10^{-3}$ and the actual non-zero value of the surface temperature will ultimately be given by Eq. (6).

The solution obtained will determine and be characterized by the central values of the temperature and pressure, T_c and ρ_c, and the surface values of M_r and L_r, namely, M and L.

The preceding argument is generally given in the following equivalent form: If the mass and distribution of chemical composition are given, the stellar structure (including the luminosity and radius) is uniquely determined (Vogt–Russell Theorem). Bearing in mind Stefan's law [Eq. (6)] relating L, R, and T_e, it is seen that the evolutionary track of a star in the H–R diagram is completely defined by its mass and initial composition.

The Runge–Kutta method is well-suited for numerical integrations performed on high-speed computers. Details concerning the solutions of the equations are given by Hayashi et al.,[1] Schwarzschild,[2] and Wrubel.[4]

Certain helpful insights can be gained by a general examination of the basic equations. [See also the more detailed treatments by Frank-Kamenetskii[7] and Stein.[11]]

Our earlier general stellar temperature estimate was obtained by setting the thermal energy equal to half of the gravitational potential energy, as demanded by the virial theorem. In order of magnitude,

$$(k\bar{T}/\mu H)M = GM^2/R, \tag{44}$$

which gives a mean temperature,

$$\bar{T} = 2.3 \times 10^7 \mu[(M/M_\odot)/(R/R_\odot)], \tag{45}$$

of the correct order of 10^7 degrees for the sun.

Our estimate of about 10^7 years for the length of time that the sun's present luminosity could be maintained by gravitational contraction alone was obtained by dividing half of the gravitational potential energy, GM^2/R, by the luminosity, L.

Main-sequence stars, slowly burning hydrogen, have roughly similar structures; e.g., the proportionality factors between mean and central pressure are similar. Taking the mean pressure gradient, dP/dr as of magnitude P_c/R, since P ranges from 0 to P_c, and r from 0 to R, Eq. (18) gives $P_c \propto \bar{\rho}M/R$. But

$$\bar{\rho} = M/\tfrac{4}{3}\pi R^3, \tag{46}$$

hence

$$P_c \propto M^2/R^4. \tag{47}$$

By Eqs. (7), (46), and (47),

$$T_c \propto M/R, \tag{48}$$

which corresponds to the result obtained in Eq. (45).

With the mean temperature gradient taken as T_c/R, Eq. (27), together with Eqs. (46) and (48), yields

$$L \propto M^3. \tag{49}$$

This mass-luminosity relation, in which the radius has dropped out, and which was derived without any knowledge of nuclear processes, shows that a star in equilibrium demands a luminosity which depends mainly on its mass; it further demonstrates why the very massive stars are extraordinarily luminous and why their lifetimes,

$$t \propto M_{\text{core}}/L \propto 1/M^2, \tag{50}$$

are correspondingly short.

TABLE II

Evolution of the Sun (initial composition $X = 0.708$, $Y = 0.272$, $Z = 0.020$)*

Age, yr.	L, erg sec^{-1}	R, cm	T_c, °K	ρ_c, gm cm^{-3}	X_c	T_e, °K
(10^9)	(10^{33})	(10^{11})	(10^6)			
0	2.78	0.66	13.7	90	0.708	5470
0.5	2.87	0.66	13.8	94	0.673	5520
1.5	3.08	0.67	14.2	105	0.601	5610
2.5	3.31	0.67	14.6	118	0.525	5680
3.5	3.58	0.68	15.1	135	0.445	5750
4.5	3.90	0.69	15.7	158	0.359	5820

Structure of the present sun (age 4.5×10^9 years)

M_r/M	ρ, gm cm^{-3}	P, dynes cm^{-2}	r, cm	T, °K	L_r, erg sec^{-1}	X
		(10^{15})	(10^{11})	(10^6)	(10^{33})	
0.0	158	252	0	15.7	0	0.359
0.1	83	133	0.08	12.8	2.13	0.584
0.2	59	87	0.10	11.3	3.09	0.648
0.3	43	58	0.13	10.1	3.55	0.679
0.4	31	38	0.15	9.0	3.77	0.694
0.5	22	25	0.17	8.1	3.86	0.702
0.6	15	15	0.20	7.1	3.90	0.705
0.7	9.4	7.9	0.23	6.2	3.90	0.707
0.8	5.0	3.5	0.26	5.1	3.90	0.708
0.9	1.8	0.97	0.32	3.9	3.90	0.708
1.0	0	0	0.694	0	3.90	0.708

* Data by R. L. Sears.

In spite of the crudeness of the derivation of the mass-luminosity relation, observation confirms the result. On the main sequence,

$$L \propto M^{4.0} \quad \text{for} \quad M \gtrsim 0.5\, M_{\odot};$$
$$L \propto M^{2.8} \quad \text{for} \quad M \lesssim 0.5 M_{\odot}. \tag{51}$$

Finally, it may be helpful to examine in Table II, by R. L. Sears, both the actual run of variables in the interior of the present sun (age: 4.5 billion years) and the evolution of selected physical variables during the sun's hydrogen-burning history.

BIRTH OF A STAR

Stars are born in regions where material between the stars is plentiful, as exemplified by the region of the Orion Nebula, beautifully portrayed by N. U. Mayall and H. W. Babcock in Fig. 13, where the interstellar medium is illuminated by nearby hot stars.

Initially, a vast proto-star is formed and begins to contract under its own gravitation.[19] When sufficient gravitational energy has been released, it emerges as a luminous object somewhere to the right of the main sequence.

In the first models computed for such stars, radiative equilibrium was taken to prevail throughout the interior. In this case, the decrease in radius is accompanied by increasing surface temperature and a rather slow increase in luminosity; i.e., the evolutionary track slopes gradually upward to the left until the star reaches the main sequence, where the first significant energy-producing nuclear process, hydrogen-burning, takes place.

Recent studies by Hayashi[1,11,20] have shown that, prior to following the radiative track, the stars are wholly convective* and follow a downward vertical track of nearly constant effective temperature (Fig. 14); i.e., an early luminous phase is followed by a *diminution* in luminosity. For blackbody radiation, Stefan's law [cf. Eq. (6)] states that

$$L = 4\pi R^2 (ac/4)\, T_e^4, \tag{52}$$

* This result supports the correctness of the usual assumption of initial uniformity of the chemical composition in computations for main-sequence stars.

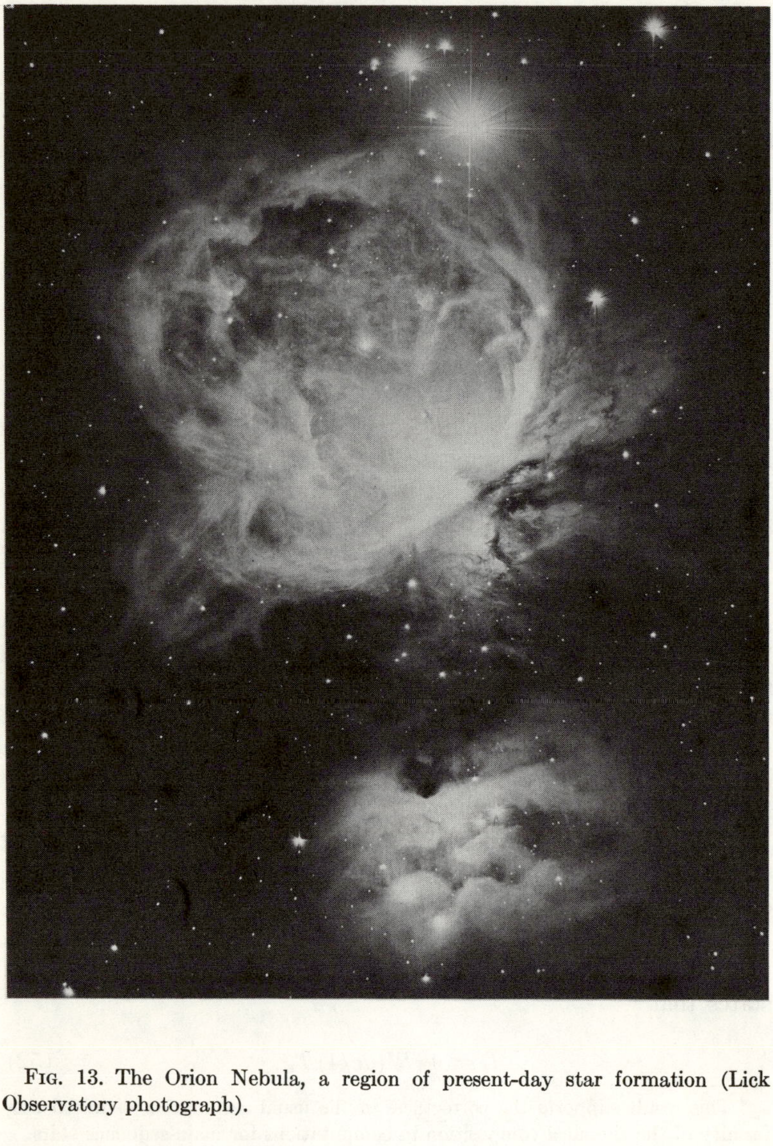

FIG. 13. The Orion Nebula, a region of present-day star formation (Lick Observatory photograph).

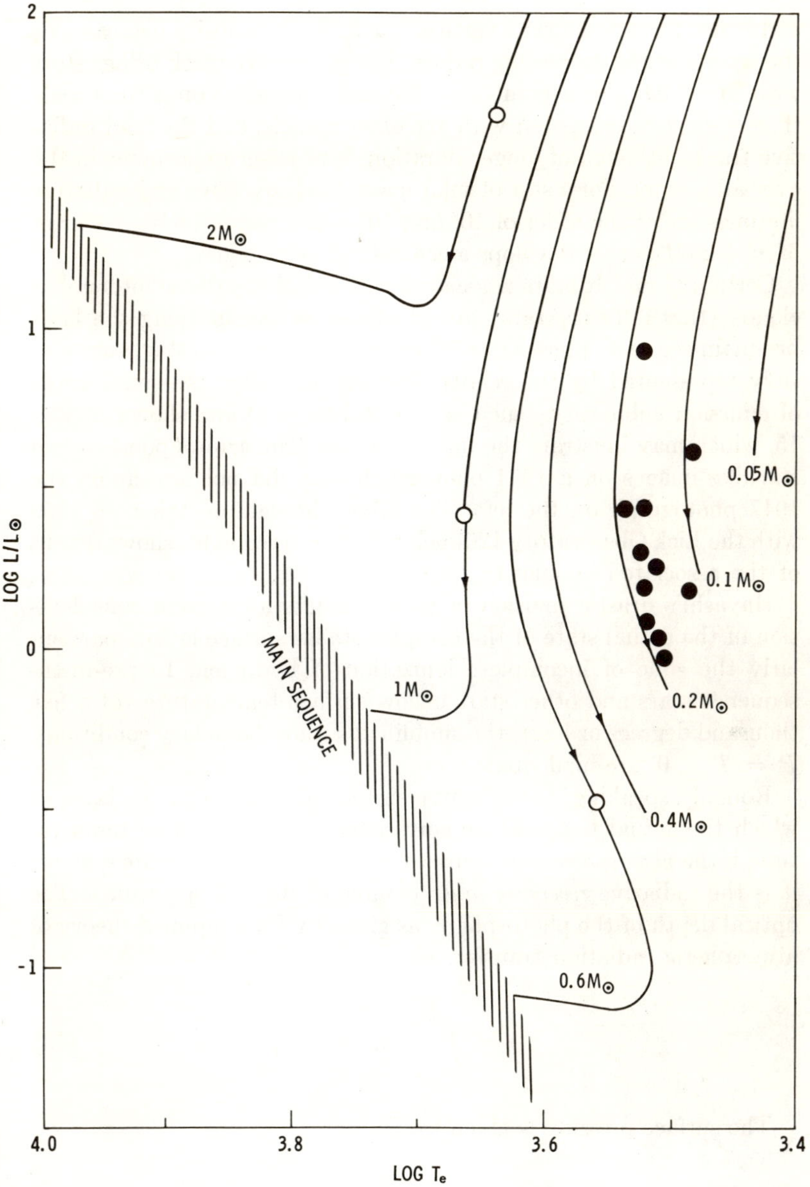

FIG. 14. Pre-main-sequence evolutionary tracks derived by Hayashi for stars of low mass. The open circles indicate the end of the wholly convective phase. The filled circles represent M-type stars observed near the Orion Nebula by Blanco.

and hence for a downward vertical track, the luminosity decreases as the square of the decreasing radius. The convective track brings stars with $M \lesssim 2M_\odot$ to a point near the main sequence on a time scale that is short as compared with the older models, and the final radiative track, although of longer duration, is of lesser importance in the over-all picture. For a star of solar mass, the convective and radiative lifetimes are of the order of 10^6 and 10^7 years, respectively. A star of $M < 0.26M_\odot$ never develops a central radiative region.

Certain G and K stars associated with relatively dense interstellar clouds (the T Tauri stars) are identified as exemplifying the later pre-main-sequence phases.[12,15,21] Very early contraction stages are possibly represented by the related Herbig–Haro objects—small knots of emission nebulosity—such as No. 2/12a, in Orion, shown in Fig. 15, which may illustrate the birth of a star. The arrows point to two star-like images on a 1954 photograph that did not appear on the 1947 photograph on the left. The large photograph, taken in 1959 with the Lick Observatory 120-inch reflector in red light, shows details of the associated nebulosity.

Hayashi's drastic revision of earlier results arose from consideration of the actual state of the complex stellar surface layers, particularly the zone of incomplete ionization of hydrogen. In pre-main-sequence stars and other stars of low surface temperature (of a few thousand degrees or less), the simplified surface boundary conditions, $P = T = 0$, are inadequate.

Roughly speaking, the effective surface of a star, i.e., the layer at which the actual temperature corresponds to the effective temperature, is the *photosphere* (the source of the continuous spectrum); above it is the radiative *reversing layer* (source of the line spectrum). The optical depth of the photosphere, as given by the simplified theory of atmospheric radiation transfer, is

$$\tau = \int_R^\infty \kappa\rho\, dr = \tfrac{2}{3}. \qquad (53)$$

The surface pressure is given by

$$P = (GM/R^2) \int_R^\infty \rho\, dr, \qquad (54)$$

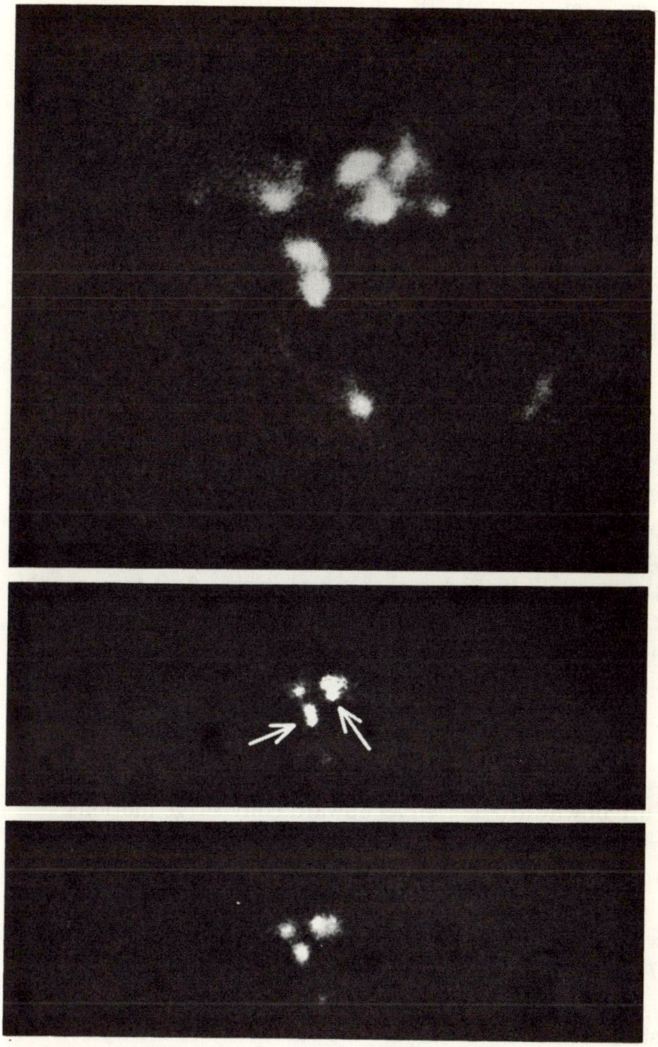

Fig. 15. Herbig–Haro object No. 2/12a in Orion, illustrating perhaps an example of stellar birth. The two star-like images (indicated by arrows) in the middle photograph, taken in 1954, do not appear in the 1947 photograph on the left. The large-scale photograph on the right, made with the Lick 120-inch reflector in red light in 1959, shows minute details of the stellar aggregate and the associated nebulosity (Lick Observatory photographs).

where GM/R^2 is the surface gravity in cm/sec². Taking κ as constant and eliminating the integral expression between Eqs. (53) and (54), we obtain an approximate, but more realistic surface condition, as $r \rightarrow R$ and $T \rightarrow T_e$,

$$P \rightarrow \tfrac{2}{3}(GM/\kappa R^2). \tag{55}$$

Hayashi found that, at least in the early phases of pre-main-sequence contraction, convective high-luminosity models are required to satisfy the surface condition when the diversion of energy to the ionization of hydrogen in a sub-photospheric region is considered, and the opacity in the surface regions is due to absorption of photons by the H^- ion (the hydrogen atom with a captured electron) and neutral hydrogen atoms. He also considered that the relation,

$$(n+1)_{\text{conv}} > (n+1)_{\text{rad}}, \tag{56}$$

no longer guarantees the existence of convective equilibrium at the low densities and temperatures of the surface layers, since convective motion may not be fast enough to transport the surface energy flux, $L/4\pi R^2$ ergs/cm²-sec. Although convective velocities in the outer regions of the star will be higher than usual in an effort to carry the energy flux, there exists an upper limit, viz., the velocity of sound, which is, in order of magnitude,

$$V_s \sim (5kT/3H)^{1/2} = 0.12T^{1/2} \text{ km/sec}. \tag{57}$$

The *effective* upper limit is βV_s, where β is usually taken as 0.2 or 0.3, since there are downward as well as upward convective motions.

The hydrogen ionization zone, though thin, is effective in *refracting* the pressure-temperature adiabat (locus of constant entropy). The convective envelope below it is a polytrope of index $n = 1.5$, i.e., $P \propto T^{5/2}$, with $P = T = 0$ serving as a sufficiently accurate boundary condition. *Polytropic convective envelopes* are characterized by a certain parameter, $E \propto P/T^{5/2}$, which has a maximum value for the existence of a quasi-static solution; this is just the value for the wholly convective models. At a certain point, however, as the star approaches the main-sequence E will decrease and a radiative zone will grow from the center until finally the evolutionary track changes direction

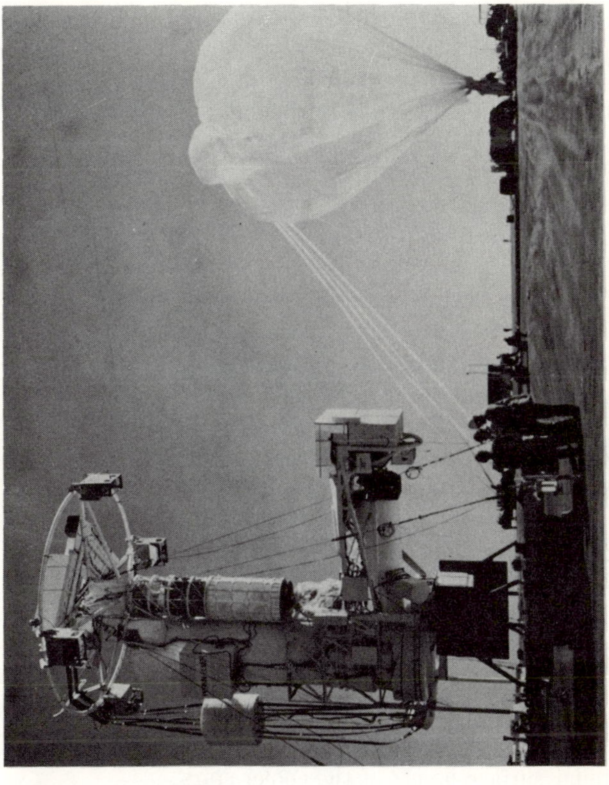

FIG. 16. The remote-controlled and balloon-borne Stratoscope II 36-inch telescope in preparation for a flight. Unique stellar spectrographic observations in the infrared, bearing on questions concerning the surface condition of the red giant stars, were made with this telescope in 1963. The Balloon Project, under the leadership of M. Schwarzschild and R. E. Danielson of the Princeton University Observatory, is jointly sponsored by the National Science Foundation, Office of Naval Research, and National Aeronautics and Space Administration (NSF photograph).

(Fig. 14) and the star is almost entirely in radiative equilibrium ($E \approx 0$).

There is increasing evidence along different lines for the general correctness of the Hayashi results. For example, the light elements lithium and beryllium, produced by spallation in the solar surface layers, would have been mixed into deeper and hotter layers of the early sun. According to Hayashi, the temperatures at the bottom of the surface convection zone were such that lithium would burn and beryllium would not. Solar spectroscopic evidence shows just this effect: normal beryllium and depleted lithium.

The Hayashi models have forced a re-examination of the theories concerning the origin of the solar system because of the difficulty of explaining the presence of liquid H_2O in some meteorites if their parent bodies were formed during the sun's early high-luminosity phase.

Although the following consideration does not alter the over-all evolutionary picture of the pre-main-sequence contracting stars, it represents one of the weakest links in our understanding of the physics of stellar structure. For example, current theories of turbulence and convection do not specify the value of the convection efficiency parameter β.

An invaluable basis for further theoretical studies has been provided in recent years by the remarkable balloon observations of Schwarzschild and his associates, whose classic photographs of the sun's surface granulation[22] and infrared spectra of the cool red giant stars were made at an altitude of about 80,000 feet from large and ingenious unmanned telescopes (Fig. 16). The Princeton team provided the first definitive picture of the sizes and shapes of the solar convective cells and defined the extent of the gigantic infrared H_2O absorption bands in the atmospheres of the red giants. These two important results contributed significantly to the understanding of the complex stellar surface layers of the cooler stars.

LIFETIME OF A POPULATION I STAR

A massive Population I star illustrates the case where a star can proceed through the entire sequence of central nuclear burnings (Table I) up to iron core formation and supernova explosion. The

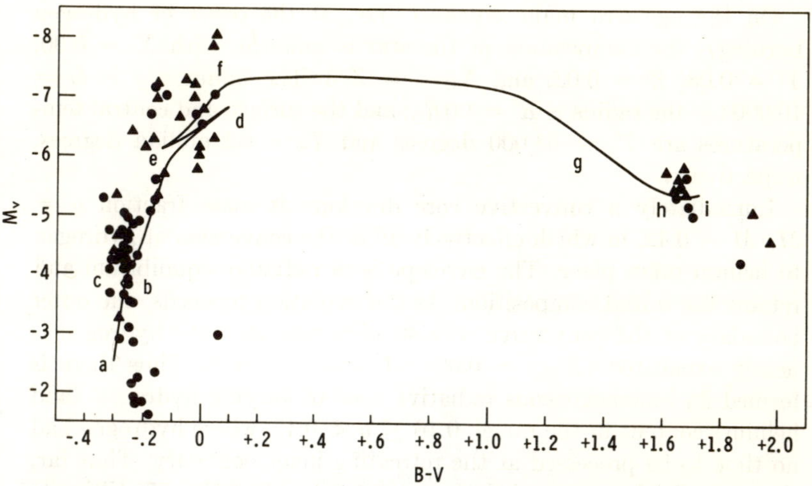

FIG. 17. The evolutionary track of a Population I star of 15.6 solar masses, superimposed on the H-R diagram of the double cluster, *h* and χ Persei. Track *a–b* represents hydrogen-burning; *b–c*, hydrogen exhaustion ($X_c \leq 0.02$); *c–d*, core contraction; *d–g*, helium-burning; *g–h*, helium exhaustion ($Y_c \leq 0.016$) and core contraction; *h–i*, carbon-burning. The segments *b–d* and *f–h* are traversed rapidly. A hydrogen-burning shell-source becomes prominent in segment *c–d*. Track *a–b* is by Sakashita, Ôno, and Hayashi; track *b–i*, by Hayashi and R. Cameron, Astrophys. J. **136**, 166 (1962).

evolution considered here will be that of a star of 15.6 solar masses (Fig. 17); its evolution was the first to be carried out beyond hydrogen-burning for a Population I star. The advanced evolutionary phases, computed by C. Hayashi and R. C. Cameron,[23] continued the evolution of the hydrogen-burning models of S. Sakashita, Y. Ôno, and Hayashi. This research was the first to explain theoretically the observed blue supergiant branch (helium-burning),* the Hertzsprung gap (helium exhaustion), and the red supergiant branch (carbon-burning and later phases up to supernova explosion) of massive stars in open clusters such as h and χ Persei. In this star the opacity is due to electron scattering. Radiation pressure is important, and hydrogen-burning occurs via the CNO version of the carbon-cycle.

* The point in the H-R diagram corresponding to the onset of helium-burning, and the subsequent trend toward the left were found by Hayashi, J. Jugaku, and M. Nishida in a preliminary exploration of the helium-burning phase.

On the age-zero main-sequence (i.e., at the onset of hydrogen burning), the composition of the star is uniform, with $X = 0.90$, $Y = 0.08$, $Z = 0.02$, and $X_{CNO} = Z/3$. The luminosity is $L = 16{,}000 L_{\odot}$; the radius is $R = 4.0 R_{\odot}$; and the surface and central temperatures are $T_e = 33{,}000$ degrees and $T_c = 32$ million degrees, respectively.

Immediately a convective core develops at mass fraction $q = M_r/M = 0.42$, in which effectively all of the conversion of hydrogen to helium takes place. The envelope is in radiative equilibrium and retains the initial composition. As the evolution proceeds, the outer boundary of the convective core steadily retreats until hydrogen is nearly exhausted ($X_{\text{core}} = 0.02$) when $q_{\text{core}} = 0.16$. Thus there is formed an inhomogeneous radiative zone of varying hydrogen (and helium) content in the range, $0.16 \leq q < 0.42$, where hydrogen had no time to be processed at the retreating inner boundary. Thus far, the evolution has proceeded along the track *a–b* of Fig. 17. Ultimate hydrogen exhaustion to $X_{\text{core}} = 0.00001$ results in a slight kink in the track, *b–c*. The radius and luminosity both are now slightly more than double their age-zero values.

To maintain the stellar luminosity, gravitational contraction commences and the subsequent heating of the central regions is sufficient to cause effective hydrogen-burning in a relatively thin shell at the boundary of the dehydrogenized core, accompanied by expansion of the envelope. This relatively rapid excursion (track *c–d*) takes the star in 70,000 years through a luminosity increase of 50 percent and a drastic increase of the radius to $R = 94 R_{\odot}$ to the *right end* of the blue-supergiant branch, where the central temperature is high enough (166 million degrees) to ignite helium.

Helium-burning takes place in a new convective core ($q = 0.08$). There is now a radiative helium zone between the core and the hydrogen-burning shell source, and evolution proceeds in another 800,000 years to the left end of the blue-supergiant branch (track *d–e*) where helium is depleted to $Y = 0.35$ in the growing core ($q = 0.15$).

The central and shell sources are now contributing equally to the star's luminosity, having been respectively increasing and decreasing in importance. Since $L_{3\alpha} \propto Y_c^3$, maintenance of the stellar luminosity requires an accelerated increase in the central temperature during helium exhaustion; this increase is now provided by gravitational contraction, accompanied by envelope expansion. The direction of the track reverses taking the star back to the right end of the blue-supergiant branch in 300,000 years (track *e–f*); it continues across the Hertzsprung gap to the vicinity of the red-supergiant branch in another 20,000 years, with helium nearly exhausted ($Y_c = 0.016$) in the core (track *f–g*). Further significant increase in the stellar radius, which has now ballooned to $R \sim 1000R_\odot$ is suppressed by a growing surface convection zone which eventually extends downward nearly to the shell source.

At point *g*, there is a high degree of central condensation: the carbon–oxygen core, with mass fraction $q = 0.16$, is contained within the radius fraction, $r_{\text{core}}/R = 0.0002$. The central density is $\rho_c \sim 10^4$ gm/cm^3 and the central temperature $T_c = 317$ million degrees. The luminosity has increased slightly (about 25 percent) during helium-burning, to $L \sim 65{,}000L_\odot$. Radiation pressure now accounts for 25, 16, and 32 percent of the total pressure at the center of the star, the edge of the core and the hydrogen-burning shell, respectively. Total helium exhaustion and another gravitational contraction phase (track *g–h*) take the star the short distance to the red-supergiant branch where carbon burns (track *h–i*) and where the star remains nearly stationary until supernova explosion. Furthermore, the remaining evolution takes place rapidly enough that a knowledge of the internal structure through carbon-burning gives a substantial basis for projection of further developments (Fig. 18). As one central fuel expires, a short-lived gravitational contraction phase heats the central regions to the ignition temperature of the next fuel. Throughout the red-supergiant track, there is, however, the following troubling consideration.

The theory of the universal Fermi interaction is based on the coupling constants of the three weak interactions between fermions (i.e., particles of spin $\frac{1}{2}$), neutron β-decay, muon β-decay, and muon capture by a proton being equal within experimental error. In the formalism, for example, of Feynman and Gell–Mann (1958), it becomes possible quantitatively to account for observed processes and to predict new ones. Transition probabilities are based on terms involving combinations of certain fermion pairs and the corresponding antiparticle pairs. Thus far (1967) only cross-term processes have been observed in the laboratory; the cross-sections of the square-term processes are so small as to make their confirmation exceedingly

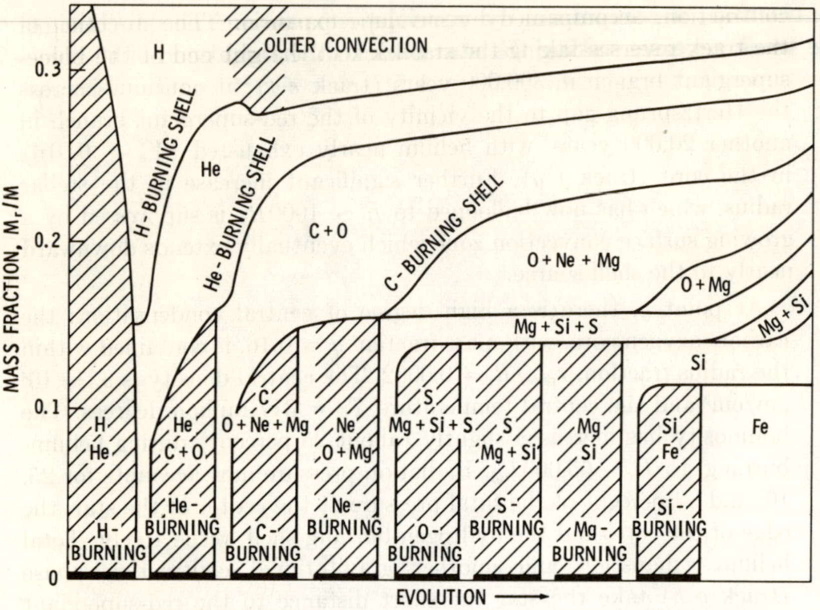

FIG. 18. Schematic depiction of the evolution of the central region of a Population I star of 15.6 solar masses. The shaded regions are convective. Nuclear fuels and their main products are indicated (cf. Table I). Burning phases are separated by gravitational contraction phases. The structure through carbon-burning is based on detailed computations by Hayashi and his associates; the later structure is an extrapolation estimated by Hayashi. Supernova explosion follows iron core formation.

difficult. The square term $(e^+, \nu)(\bar{\nu}, e^-)$ predicts the occurrence of a direct interaction between electrons and neutrinos, even in the absence of nucleons. A number of processes arising from this term have been studied by Chiu[11,24] and his colleagues, by Levine,[25] and by others (see also the discussions by Reeves,[3,11] and by Hayashi et al.[1]). In our context, the most important processes seem to be the photoneutrino process,

$$\gamma + e^- \rightarrow e^- + \nu + \bar{\nu}, \tag{58}$$

and the pair-annihilation neutrino process,

$$e^- + e^+ \rightarrow \nu + \bar{\nu}. \tag{59}$$

They begin to be important at very high temperatures as shown in Fig. 10, where the domains of the dominant processes (including the plasma neutrino process*) are indicated.

The substantial dissipation rate by neutrino loss—as compared with the energy production rate of nuclear burnings at $T \gtrsim 10^9$ degrees—is indicated in Fig. 12. The neutrinos carry energy through the body of the star into space almost as if it were transparent, and do not contribute to the luminosity, L; they represent an energy sink rather than source. Gravitational contraction proceeds to raise the temperature of the central regions and to stimulate the nuclear burning to make good the neutrino loss, which itself depends on a high power of the temperature. The resulting evolution is so extremely accelerated that it seems unlikely, on quite general grounds, that the red-supergiant lifetime can endure for any significant length of time in the presence of neutrino emission.

Table III gives the lifetimes, with and without neutrino loss, as derived by Hayashi and R. Cameron[23] for the star of 15.6 solar masses.

A comparison of the relative lifetimes and the relative numbers of stars in the corresponding H–R diagram locations for the clusters h and χ Persei and NGC 330, whose most massive stars are of the order of 15 solar masses, gives satisfactory agreement only when neutrino emission is omitted.

Briefly stated, if the neutrino processes under discussion actually occur as predicted, there are about 10 times too many red supergiants observed in the clusters mentioned, i.e., observation suggests that the electron–neutrino interaction proceeds with a coupling constant that is at least an order of magnitude smaller than that given by theory, and may be forbidden. On the other hand, detailed studies by Fowler have shown that the terrestrial iron-group isotopic abundance ratios imply the existence of a loss mechanism in massive stars with a rate quite comparable with that predicted for the pair-annihilation neutrino process. Other evidence, both for and against

---

* The relatively new plasma neutrino process of Adams, Ruderman, and Woo (1962),

$$\text{plasmon} \rightarrow \nu + \bar{\nu},$$

appears to be effective at densities higher than those encountered in the star of 15.6 solar masses. The coupled system arising from the interaction of electromagnetic waves with an ionized gas has a normal mode (plasmon) which acts like a particle with non-zero mass.

TABLE III

Lifetime of a Population I star of 15.6 solar masses, with and without neutrino-loss, by evolutionary phase*

H–R diagram location	Energy source	Lifetime without ν-loss (10^5 years)	Lifetime with ν-loss (10^5 years)
Pre-main-sequence	Gravitational contraction	0.5	0.5
Main sequence	Central hydrogen-burning	157	157
Transition	Grav. contr. and H-burning shell source	0.7	0.7
Blue-supergiant branch and Hertzsprung gap	Central helium-burning and H-burning shell source	11.6	11.6
Red-supergiant branch	Grav. contr. and H- and He-burning shell sources	0.6 } ~9	0.6 } 0.8
	Central carbon-burning and H- and He-burning shell sources	2.3	0.2
	Later burnings up to supernova explosion	~6	0
Total lifetime		179	171

* Data by C. Hayashi and R. Cameron.

the theory, has been adduced. Whichever way this impasse is ultimately resolved, the resulting information will be of importance both to stellar evolution and to the theory of weak interactions.

Not all neutrino emission is expected to arise from the electron-neutrino interaction, as we have already seen on the main sequence [Eqs. (21) and (22)]. Impressive experiments are now underway to detect the neutrino flux from space, representing the birth of a new science—neutrino astronomy[26]—which promises to provide a direct look into the deep interiors of the stars.

LIFETIME OF A POPULATION II STAR

When the mass of a main-sequence star is less than about $1.5M_\odot$ (Population I) or $4M_\odot$ (Population II), nuclear energy generation is due mainly to the proton–proton chain and there is no convective core. These conditions apply to the stars comprising the upper main sequence and red-giant and horizontal branches of the globular cluster H–R diagram (see Fig. 9). Their present masses are estimated to lie between 1.0 and 1.5 solar masses. The puzzling red-giant branch was first accounted for in detail by the stellar models derived by Hoyle and Schwarzschild in 1955.

Because of the relatively low temperature dependence of the proton–proton chain ($\epsilon \propto T^4$), energy production becomes negligible only in the outer 60 percent of the stellar mass. The region of total conversion of hydrogen to helium (which tends to be isothermal,

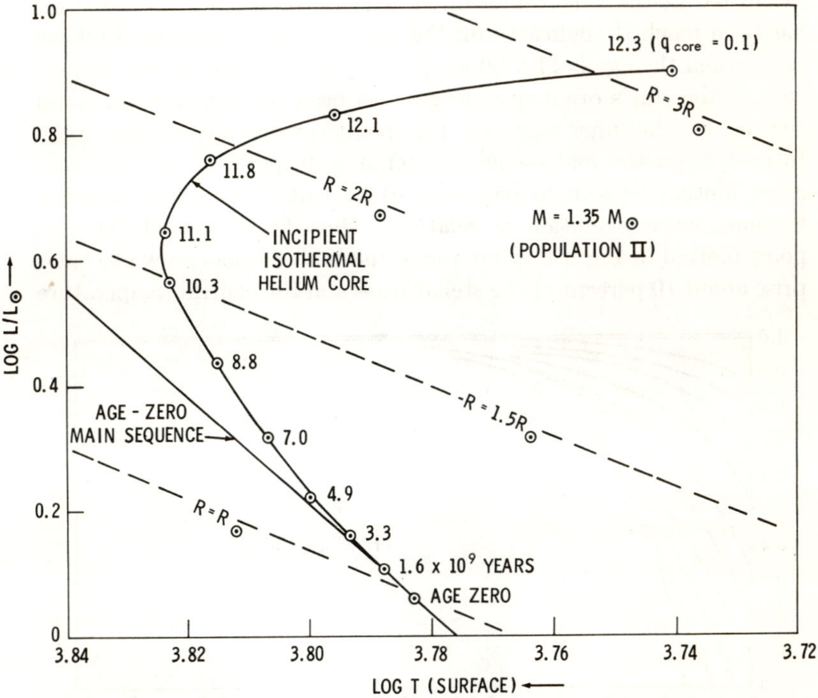

FIG. 19. The evolutionary track of a Population II star of 1.35 solar masses and initial composition, $X = 0.990$, $Y = 0.009$, $Z = 0.001$, as computed by Hoyle. The points plotted (with ages given in billions of years) are to be identified with the eleven stages of hydrogen depletion indicated in Fig. 20. This track corresponds approximately to the track *a–c* of Fig. 9.

since it no longer furnishes nuclear energy) begins as a central point and grows outward to include the cooler central regions, where the conversion has been less rapid.

The early evolutionary track of a Population II star of 1.35 solar masses and initial composition, $X = 0.990$, $Y = 0.009$, and $Z = 0.001$, as derived by Hoyle (1959), is shown in Fig. 19, where the age in billions of years is given for each point, and the corresponding state of the depletion of hydrogen can be found in Fig. 20. The track plotted in Fig. 19 corresponds to the segment *a–c* of Fig. 9.

It is to be noted that the early track is nearly parallel to the main-sequence and the initial slow, steady pace of evolution becomes gradually accelerated. It can be seen in Fig. 20 that hydrogen is not exhausted at the very center of the star until after the eighth stage has been reached (indicated on Fig. 19 as *incipient isothermal helium core*) when the age is 11.5 billion years. (By this time, about 15 percent of the star's original hydrogen has been converted into helium throughout the inner regions.) This point corresponds closely to the turn-off from the main-sequence, (cf. Fig. 9, point *b*) and the red-giant lifetime is seen to be about 10 percent of the main-sequence lifetime, since evolution is relatively short-lived beyond the last point plotted in Fig. 19, where the isothermal core has grown to comprise about 10 percent of the stellar mass and the central temperature

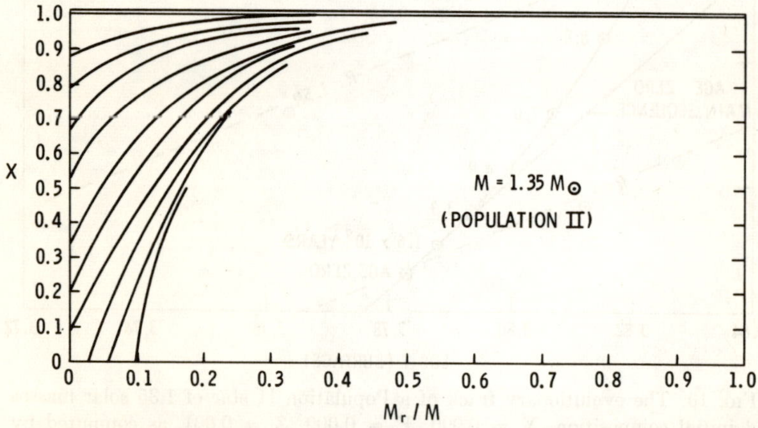

FIG. 20. Hydrogen-depletion as a function of mass fraction for a Population II star of 1.35 solar masses and initial composition, $X = 0.990$, $Y = 0.009$, and $Z = 0.001$, computed by Hoyle. The eleven evolutionary stages indicated are to be identified with the points of Fig. 19.

has reached 34 million degrees. Here the carbon-cycle, with its high temperature dependence ($\epsilon \propto T^{16}$) prevails, and further hydrogen-burning may be regarded as taking place in a shell source surrounding the helium core. Furthermore, the surface condition (discussed in an earlier section) becomes important and the track bends steeply upward to much greater luminosities (and radii) and slightly cooler surface temperatures to the tip of the red-giant branch (track *c–d*, Fig. 9).

In this portion of the track the surface condition controls both the stellar radius (almost independently of the inner structure) and the depth of the growing surface convection zone.

We now reconsider the main-sequence stages and examine the reasons for the accelerated evolution of the star under consideration. As it comes off the main-sequence, there is a gradually increasing luminosity requirement for equilibrium models, which depends largely on the distribution of the chemical composition in the inner regions, and more particularly, on the mass of the core. At the turn-off point, depletion of hydrogen is sufficiently advanced that energy production tends to lag, and gravitational contraction of the central regions (accompanied by expansion of the stellar radius) becomes effective in raising the temperature of the nuclear energy source so that the luminosity requirement will be met. The direct contribution of gravitational contraction to the luminosity is negligible because ϵ_g (for a given rate of contraction) is necessarily smaller as the degree of degeneracy is greater[1] and because of the difficulty of establishing a large temperature gradient in a core that is gradually becoming more and more degenerate. As the degree of degeneracy progresses, the opacity due to electron conduction becomes very low (see Fig. 11), the radiative opacity becomes negligible, since

$$\kappa^{-1} = (\kappa_{\text{cond}})^{-1} + (\kappa_{\text{rad}})^{-1}, \tag{60}$$

and the isothermal approximation for the dehydrogenized core becomes more valid.

Now a shell source is less effective than a central source in sustaining a star, i.e., it requires a higher temperature to maintain a comparable flux of energy. At these high temperatures, the change in the inner structure (growth of the core) becomes rapid, the luminosity requirement becomes greater, more gravitational heating is required, and the cycle repeats at an accelerating rate.

The numerical values given from here on refer to a slightly different Population II star of $M = 1M_\odot (X = 0.900, Y = 0.099, Z = 0.001)$.

At some point in the upper reaches of the giant branch, when the central temperature reaches about 80 million degrees, helium-burning at the very center of the star begins to be noticeable. At this point the inner region out to mass fraction $q \approx 0.4$ is degenerate and hence effectively isothermal. The thin, outer, non-degenerate portion of the core has a temperature gradient, produced by gravitational contraction, such that the temperature at the hydrogen-burning shell source is $\sim$ 50 million degrees, and energy production is sufficient to sustain the envelope.

As we have seen, the onset of central helium-burning in a massive Population I star causes heating, accompanied by an opposing effect, expansion and cooling of the non-degenerate core. Degenerate matter, however, behaves differently when heated; the above-mentioned opposing tendency is absent in the degenerate core of the star of 1 $M_\odot$ since the pressure is independent of the temperature. Here we have the second cause of accelerated evolution, a thermal runaway, known as the *helium flash*, which lasts until helium-burning raises the central temperature sufficiently to dissolve the degeneracy (at the tip of the red-giant branch). Expansion of the core then opposes any further heating and, at the same time, contraction of the envelope takes place. The star slides rapidly back down the giant branch toward a relatively stable equilibrium configuration.

Schwarzschild and Härm [*Astrophys. J.* **145**, 496 (1966)], using the powerful Henyey computational method, have continued their heroic efforts to follow the evolution of their star in detail through the extraordinary conditions of the helium flash. The rapidity of change of the physical variables is so great at the peak of the flash that the age difference between consecutive models, of which hundreds were computed automatically, in this ultimate case was 10 minutes. Furthermore, at this point, which corresponds to disappearance of degeneracy in the core and takes place at the tip of the giant branch, the interior helium-burning luminosity is at a (very temporary) maximum of $10^{11} L_\odot$,* i.e., the rate of energy production is roughly that of an entire galaxy. The thermal blanket of non-degenerate matter, however, allows practically none of this energy to be radiated from the surface. At the peak of the helium flash, the stellar luminosity

* This represents an increase by a factor of 50 billion in 26,000 years.

is 2700 $L_\odot$; the radius, 106 $R_\odot$; and the central condensation such that the (now convective) core, containing 0.43 of the stellar mass, occupies the inner 0.00016 of the stellar radius.

It is expected that further evolution will take the star to a less frantic helium-burning phase at the *left* end of the horizontal branch (point d' of Fig. 9), which will be explained by an evolutionary track proceeding from left to right, along the lines indicated by Hayashi et al.[1] for a star of $0.7M_\odot$ with a helium core of $0.5M_\odot$. According to Sugimoto, the peak of the helium flash occurs whenever the helium core mass reaches a critical value, $0.5M_\odot$, which is nearly independent of the total stellar mass. Since the main-sequence turn-off depends on the conversion of a certain *fraction* (roughly, 10 to 15 percent) of the stellar mass into helium, the relative lifetime in the giant branch is greater as the star is less massive. Thus, the older the cluster, the greater the giant population as compared with the upper main-sequence population. This situation is in agreement with observation.

Although the general picture for the giant branch seems quite satisfactory, evolutionary questions regarding the transition from the helium flash to the slow phase of helium-burning are still under investigation.

Although the evolution outlined here has been given for Population II stars, the picture is much the same for Population I stars of comparable mass (e.g., the sun). At point *c* of Fig. 9, where surface convection becomes important, the Population I star diverges to the right is a less steep upward direction, so as to be about 3 magnitudes fainter at surface temperatures comparable to that of the tip of the Population II red-giant branch. It is the effect of the heavy-element abundance on the surface condition that determines the divergence.

DEATH OF A STAR

Occasionally a star is observed to appear suddenly where none was evident. The brightest of these novae, *the supernovae*, reach maximum visual luminosities of 10^7 to 10^{10} times that of the sun and then fade away to obscurity. Thus some extragalactic supernovae appear as bright as the galaxy in which they are situated. [Zwicky[5] estimates that there are at any one time about 200 million supernovae near maximum light within a distance of 10^9 parsecs, which is roughly the ordinary limit of the 200-inch telescope.]

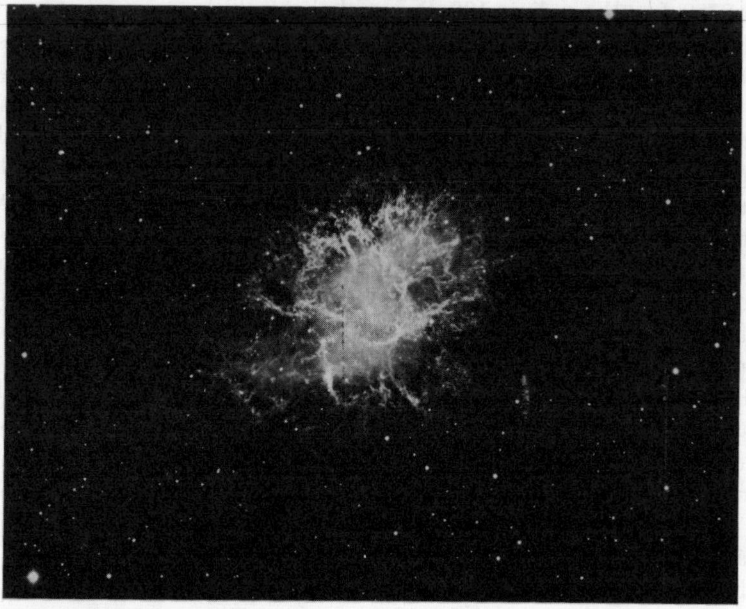

FIG. 21. The Crab Nebula (M1) in Taurus, site of the supernova of July 4, 1054 A.D., photographed in red light (Palomar 200-inch photograph).

The supernovae are understood as explosive events. The Crab Nebula (Fig. 21) was the site of the Type I supernova which appeared on July 4, 1054 A.D. It was recorded by Chinese and Japanese astronomers and was observable in daylight. The naive interpretation of the nebulosity as the expelled surface layers of the stars involved is challenged by Hoyle, Fowler, Burbidge, and Burbidge.[27] They propose that the filaments are of interstellar origin and that the filament expansion (~1000 km/sec) is due to the continuing influence, through emission of high-energy particles, of the central supernova remnant on the interstellar medium. This remnant has not yet been identified photographically.

The enigmatic Crab Nebula is also one of the strongest radio sources in the galaxy; and it is the site of one of 40 known discrete galactic X-ray sources, recently discovered through rocket observations. A spectacular and important result was achieved on July 7, 1964, in the precisely-timed flight of a rocket flown by scientists of the Naval Research Laboratory, who observed the X-ray intensity during occultation of the Crab Nebula by the moon—an event not scheduled to recur until 1972. Interpretation of the gradual diminution of the X-ray intensity showed that a broad source— about 1 light year in diameter—was involved, and not a hot neutron star (point-source) as had been proposed. In 1966, another strong X-ray source, Sco X-1, was identified with an intense ultraviolet stellar object of $V \approx 13$ mag., whose flickering activity and peculiar colors are characteristic of old novae. From the results of another epoch-making flight of May 17, 1967, H. Friedman and E. T. Byram, of the Naval Research Laboratory, established with near-certainty that the well-known quasar 3C273 (to be discussed later in this chapter) is an X-ray source. Furthermore, the same flight confirmed the existence of X-ray flux from the peculiar radio galaxy M87, which is the brightest elliptical galaxy in the Virgo Cluster and one of the intrinsically brightest galaxies known. Clearly, X-ray astronomy has come of age within the space of just a few years.

In the following description of the progress of Type I and Type II supernova explosions, we follow the sequence developed by Hoyle and Fowler,[28] based in part on earlier researches by Burbidge, Burbidge, Fowler, and Hoyle.[29]

A Type I supernova occurs when a Population II star of 1.2 to 1.5 solar masses, at an advanced evolutionary stage develops in its intermediate regions, where C^{12}, O^{16}, and Ne^{20} have been synthesized, (1) relativistic electron degeneracy, and (2) explosive temperatures of 1.5 to 2.5×10^9 degrees appropriate to the above-mentioned light nuclear fuels, which can then enter into sudden fusion processes (e.g., $C^{12} + C^{12}$ and $O^{16} + O^{16}$ reactions) on a time scale of 1 to 100 seconds, liberating the 10^{50} ergs (or more) necessary for supernova explosions. The central temperature just before the explosion is $T_c \sim 3.5 \times 10^9$ degrees and the central density, $\rho_c \sim 10^8$ gm/cm^3. The violence of the resulting explosion is sufficient to shatter the whole stellar mass, almost all of which is degenerate at the time of the outburst, during which certain heavy nuclei are synthesized by neutron capture.

A Type II supernova occurs when the succession of nuclear fuels has finally synthesized iron in the nondegenerate core of a massive Population I star. Conditions are such that strong neutrino emission accelerates the increase in central temperature and density. At the appropriate central densities and a temperature of $T \sim 6 \times 10^9$ degrees, iron tends to dissociate in the endoergic phase change,

$$Fe^{56} \rightarrow 13\ He^4 + 4n. \tag{61}$$

Further significant temperature increase is arrested and density increase accentuated by the lack of available energy required to dissociate iron. (It exceeds the thermal energy of the material.) The core becomes unable to support the star without a temperature increase, and gravitational collapse ensues. The vast amount of energy released in the collapse goes almost entirely into the dissociation of Eq. (61) and the core temperature rise is too small for the re-establishment of hydrostatic equilibrium. The resulting condition is free-fall of the core, which has a time scale,

$$t \sim (G\rho)^{-1/2}. \tag{62}$$

With $\rho \sim 10^7$ gm/cm^3, the time of collapse is of the order of 1 second. In another second, the light nuclei, C^{12}, O^{16}, and Ne^{20} of the intermediate stellar regions, having fallen as a result of the collapse, are heated to their explosive temperatures, $T \sim 2 \times 10^9$ degrees. Outer layers are blown off at 5000 km/sec and in a region between the implosion and the explosion, iron and its neighbors are formed. The details concerning the further evolution of the dense remnant are unknown.

S. Colgate and his associates, through detailed consideration of the shock waves generated, explain cosmic rays as material spalled from the surfaces of Type II supernovae.

It is a necessary condition for the occurrence of a Type II supernova explosion that gravitational collapse occurs so quickly that there is no time available for readjustment of the intermediate layers toward normal and non-explosive conditions, for example, the refrigerating effect of the iron decomposition is essential for promoting the free-fall condition, and it may not be available to a star of $M \lesssim 10M_\odot$. Hence not all stars become supernovae.

But what is the fate of the supernova remnants and of stars whose nuclear fuel is effectively exhausted before a supernova explosion can take place? As these stars cool down, they can be supported by the pressure of degenerate electrons indefinitely if the mass is less than the Chandrasekhar limit;[10,30] i.e., if

$$M < 1.44M_\odot(2/\mu_e)^2. \tag{63}$$

Usually, if not always, $\mu_e \sim 2$.

The observed white dwarfs (see Fig. 7), with densities of the order of 10^6 to 10^8 gm/cm^3, are in this category. It can be demonstrated that the energy source for these low-luminosity *A* stars is just thermal cooling, which is effective over long periods of the order of billions of years. Spectral lines originate in a thin atmosphere (mainly of hydrogen) floating on a degenerate core of heavy elements. The evolutionary track of the transition to the white-dwarf phase has been studied by Hōshi and Hayashi.[1]

W. J. Luyten has recently discovered a small number of *pygmy stars* : which appear to populate a sequence roughly 4 magnitudes below that of the white dwarfs in the H–R diagram.

For densities $\rho \gtrsim 10^{12}$ gm/cm^3, matter is composed mainly of neutrons. The mass and radius of a star in equilibrium at zero temperature, consisting of a neutron core and an envelope of ions and electrons, depends critically on the central density, as shown by Table IV, due to T. Hamada and E. E. Salpeter.

There remains the question of the ultimate state of a star more massive than the indicated limiting masses of white dwarf and neutron stars. Hoyle, Fowler, Burbidge, and Burbidge,[27] in a general reexamination of the gravitational collapse problem, remind us that, so far as anyone knows, massive stars may just continue to contract through an event horizon. They vanish from sight. According to the theory of general relativity a sufficiently dense body can warp spacetime in such a way as to isolate itself from the outside universe, although it will continue to communicate through its gravitational field. The critical or Schwarzschild radius for disappearance is

$$R = 2GM/c^2 = 2.96(M/M_\odot) \text{ km.} \tag{64}$$

TABLE IV

Selected parameters for zero-temperature stars with neutron core and envelope of ions and electrons*

ρ_c (gm/cm^3)	$M_{core}/M_\odot$	R_{core} (km)	$M/M_\odot$	R (km)
0.3×10^{14}	0.043	17	0.66	880
3.44×10^{14}	0.047	8	0.049**	460
6.0×10^{14}	0.271	8	0.271	10

* Derived by Hamada and Salpeter.

** For this star, $M/M_\odot$ reaches a minimum value.

At this radius, radiation is no longer visible to an outside observer, having been redshifted to infinite wavelength (zero energy).

Can the critical radius be reached when the star is rotating? As a rotating body contracts, conservation of angular momentum demands an increase in angular velocity (and centrifugal force) which can arrest the collapse. The implosion of a discrete body, including rotational effects, has not been solved in general relativity. Hoyle et al., however, give reasons why rotation may not necessarily forestall ultimate collapse: e.g., a sufficiently small angular velocity can be *crushed* by a strong gravitational field. This is a non-Newtonian result.

We now consider briefly the collapsing supermassive stars of $M \sim 10^8 M_\odot$, which have been postulated to exist in order to explain the quality and quantity of radiation emanating from certain strong radio sources, including the recently identified *quasars* (pron. kwā′ zarz; an acronym for *quasi-stellar radio sources*). These point-image objects were at first regarded as stars in the ordinary mass-range. Their identification as very extraordinary objects followed M. Schmidt's[27] realization in 1963 that the unidentified spectral lines were familiar lines with large redshifts, implying distances in the billions of light-years. The later discovery of substantial light variations in quasars in intervals as short as one day has presented difficulties in the interpretation of the quasar redshifts as cosmological. The argument is that such variations cannot be produced by an object with dimensions greater than one light-day, whereas the spectral observations seem to require a larger object (at the great distances under consideration). The possibility remains, however, that only a part of the object is varying.

Alternate interpretations for the quasars are currently being closely re-examined:[31,32] e.g., that the quasars are at moderate distances or relatively nearby, with the redshifts due to ejection from a galaxy (perhaps our own) at relativistic velocities, or due to the Einstein gravitational redshift (associated with superdense objects). Furthermore, alternate exotic energy sources, such as would result from the formation of familiar elementary particles from primordial *quarks*, are under consideration.[31] [Quarks are hypothetical massive, fractionally-charged particles; three quarks are required to create a baryon, and a quark-antiquark pair to create a meson.]

The already intense interest among astronomers and physicists concerning the quasars has been enhanced by the announcement by

A. R. Sandage in 1965 of the discovery of the radio-quiet quasi-stellar objects, or *quiesars** (pron. kwi′zärz; fr. L. *quiescere*, to keep quiet; *quies*, rest, quiet). Quiesars are identified through their unusually strong ultraviolet radiation and their large redshifts. They are essentially comparable with the quasars in all respects but one—the absence of detectable radio emission (at least down to a low threshold). It is estimated that the quiesars outnumber the quasars by about 100 to 1.

Attempts are currently being made to establish a genetic relationship not only between the quasars and quiesars, but also between these objects and the radio galaxies (extended sources known to be cosmological), whose maximum observed redshift ($z \approx 0.5$) is substantially less than that of the quasi-stellar objects ($z \approx 2.2$). [The redshift z is related to the wavelength λ by the following expression: $\lambda(\text{observed}) = (1 + z) \cdot \lambda(\text{laboratory})$.]

Until such time as a single blueshifted quasar or quiesar is discovered, or some other contrary crucial observation is made, it seems that the cosmological interpretation of these objects—which will be considered in what follows—suffers the least objection. The quasi-stellar objects bear little or no resemblance, however, to previously known cosmic objects, and the ultimate explanation of their origin and evolution may involve surprising elements.

The quasars (of which more than 150 are now known) not only match the radio energy output of the strongest extended radio sources ($\sim 10^{44}$ ergs/sec), but are also the optically brightest objects in the universe, with luminosities $L \sim 10^{46}$ ergs/sec, about 100 times the luminosity of a large galaxy.

The total initial energy store in the form of magnetic field and high-energy electrons has been estimated for quasars as up to 10^{60} ergs, which is so great as to defy description. This is the energy that would be released by the annihilation of more than 1 million suns. The comparable figure for extended radio sources may be 100 times as large. How are such energies to be achieved?

It appears that we shall require the highly-efficient energy generation process, gravitational collapse, which can, in principle, liberate

* The name *quiesar* is here proposed to fill a frequently mentioned need for a suitable short, pronounceable name for these objects. The most common name in current use is *QSG* (for *quasi-stellar galaxy*); other designations are *interloper*, *BSO* (for *blue stellar object*), and *QSO* (*radio-quiet*).

nearly the entire rest-mass energy of $E/M = c^2 = 9 \times 10^{20}$ ergs/gm. Half of this amount is liberated in a collapse to the event horizon. However, there still remain questions as to the mechanisms involved in converting the dynamic energy of a supermassive star into the relativistic particle energies responsible for the radio emission.

Actually, Hoyle and Fowler have accounted for the luminosity (thermal radiation) of strong radio sources, including quasars, just by hydrogen-burning in a wholly convective supermassive star, which releases only 0.7 percent of the rest-mass energy, and which occurs early in the collapse phase and temporarily slows it down. For example, hydrogen-burning in a star of $M = 10^8 M_\odot$ can provide 2×10^{46} ergs/sec (the output of the quasar 3C48) for about 10^6 years. The radio flux, however, requires the previous and essentially total collapse of a similar supermassive star to provide the relatively undegraded energy of the relativistic electrons.

A major consideration in the study of the evolution of supermassive stars is that very early in the collapse, when central densities are quite low ($\rho_c \sim 4 \times 10^{-10}$ gm/cm^3 for $10^8 M_\odot$) general relativity becomes important, as clearly demonstrated by Fowler,[27] and the evolutionary progress is substantially different from that of ordinary stars, which, incidentally, are unstable above $M \sim 100 M_\odot$.

The supermassive stars have instigated a variety of exciting researches as to their origin, evolution, stability, and ultimate collapse. Explicit theories outlined here are to be regarded not as definitive, but as illustrative of the new modes of thinking engendered by the novel concept of the supermassive star.

In regard to the final evolutionary stages, one possible picture that emerges is the following. After nuclear burning, implosion proceeds and in at least some cases a titanic explosion ensues. There is observational evidence of such explosions in the jet extending 150,000 light-years from the quasar 3C273 (see Fig. 22) and jets associated with some ordinary radio galaxies. A close look at a perhaps similar but less energetic event is provided by a remarkable photograph by Sandage of the nearby galaxy M82 (see Fig. 23), where hydrogen filaments emanating from the galactic nucleus extend above and below the galactic plane by about one galactic radius. Such explosions (including one which may be now going on in the nucleus of our galaxy) have been proposed as sources of cosmic rays.

The central remnant of the explosion of a supermassive star is now

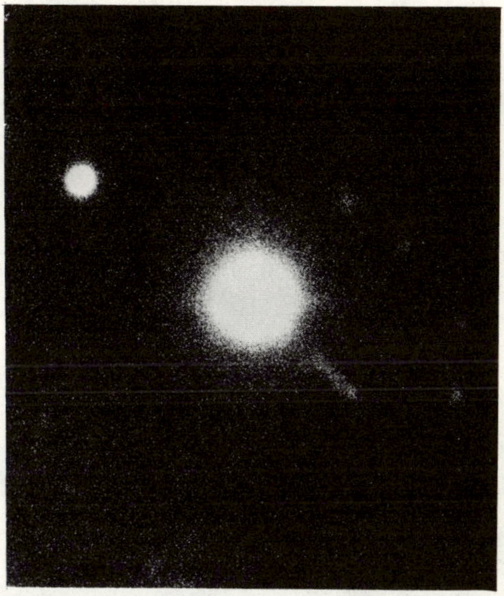

FIG. 22. The brightest image in this photograph is that of the quasi-stellar radio source 3C273, which is the most intrinsically luminous object known ($M_V \approx -25.6$). Its nominal cosmological distance is about 1.5 billion light-years. A radial filament in position angle 223° begins 11″ and ends 20″ from the main image (which resembles that of a star of $m_V \approx 12.6$). This jet may be indicative of an explosive event somewhat similar to that seen in the nearby galaxy M82 (see Fig. 23). (Palomar 200-inch photograph.)

presumed to pass beyond the event horizon. The possibility, in these circumstances, of gravitational radiation, for example, via emission of gravitons of spin 2 is under investigation. Supposing that it continues to radiate, how does the radiation get *out*?

In the steady-state cosmology, the Hoyle–Narlikar theory predicts that such a star, rather than reaching a singularity, will eventually oscillate between the event horizon and a location far inside it where densities $\rho > 10^{30}$ gm/cm^3, corresponding to interparticle separations of 10^{-18} cm or less, are attained. A particle emitted from the surface, for example, at maximum compression, will stay ahead of the surface during the ensuing expansion and will eventually be shoved through the event horizon into the outside universe. Furthermore, Faulkner,

Fig. 23. The nearby exploding galaxy M82, photographed in the red light of hydrogen (Hα, 6563 Å) to show faint filaments extending more than 10,000 light-years above and below the center of the galactic disk. The diameter of M82 is only about one-half of that of its neighbor in space, M81 (see Fig. 3). (Palomar 200-inch photograph.)

Hoyle, and Narlikar show that a particle emitted during the expansion phase will be blueshifted on receipt.

Nuclear physics itself is not, however, at present equipped to describe in detail the happenings under these extreme conditions. However that may be, we have here, on reasonable hypotheses, a qualitative explanation for the continuing injection of high-energy particles, which seems to be required, for example, even in the case of the Crab Nebula, involving a low-mass collapse.

It is frankly uncertain whether supermassive stars or neutron (or hyperon) stars exist and whether collapse beyond the event horizon can in fact occur. Such a reconnaissance, however, focuses attention upon the possibility of the existence in the cosmos of substantial amounts of invisible matter (a situation subject to indirect verification) and upon new vistas in physics, astronomy, and cosmology, open for exploration of the nature of the universe.

REFERENCES*

1. C. Hayashi, R. Hoshi, and D. Sugimoto, Evolution of the Stars, Progress of Theoretical Physics Supplement No. 22, pp. 1-183, Kyoto University, 1962.
2. M. Schwarzschild, Structure and Evolution of the Stars, Princeton University Press, 1958.
3. L. H. Aller and D. B. McLaughlin, Eds., Stellar Structure (Stars and Stellar Systems, Vol. **8**; G. P. Kuiper, Gen. Ed.), University of Chicago Press, 1965.
4. L. Gratton, Ed., Star Evolution, Academic Press, New York, 1964.
5. S. Flügge, Ed., Astrophysics II: Stellar Structure (Encyclopedia of Physics, Vol. **51**), Springer-Verlag, Berlin, 1958.
6. D. H. Menzel, P. L. Bhatnagar, and H. K. Sen, Stellar Interiors, John Wiley and Sons, Inc., New York, 1963.
7. D. Frank-Kamenetskii, Physical Processes in Stellar Interiors, Moscow, 1959; translated and published for the National Science Foundation and the National Aeronautics and Space Administration by the Israel Program for Scientific Translations, Jerusalem, 1962; available from the Clearing House for Federal Scientific and Technical Information, Springfield, Virginia 22151.
8. L. H. Aller, Nuclear Transformations, Stellar Interiors, and Nebulae, Ronald Press Co., New York, 1954.
9. B. Strömgren, The Sun as a Star, *in* The Sun, G. P. Kuiper, Ed., pp. 36-87, University of Chicago Press, 1953.
10. S. Chandrasekhar, An Introduction to the Study of Stellar Structure, University of Chicago Press, 1939; reprint, Dover Publications, Inc., New York, 1957.
11. R. F. Stein and A. G. W. Cameron, Eds., Stellar Evolution, Plenum Press, New York, 1966.
12. J. Sahade, Ed., Symposium on Stellar Evolution, National University of La Plata, Argentina, 1962.
13. University of Liège, Stellar Models and Stellar Evolution, *in* Mémoires de la Société Royale des Sciences de Liège, 5th Series, Volume **3**, 1960. See also the 1953 Liège Colloquium, Nuclear Processes in Stars.
14. J. L. Greenstein, Ed., The Hertzsprung-Russell Diagram, International Astronomical Union Symposium No. 10, reprinted from Annales d'Astrophysique, 1959.
15. D. J. K. O'Connell, S. J., Ed., Stellar Populations, North-Holland Publishing Co., Amsterdam, and Interscience Publishers, Inc., New York, 1958.
16. B. Strömgren, Comparison of Observed and Theoretically Calculated Intensities in the Continuous Spectra of Main-Sequence B Stars, Reviews of Modern Physics **36,** 532 (1964).

* A valuable bibliography of research on stellar interiors published between 1958 and mid-1966 (JILA Report No. 88, University of Colorado, Boulder, 1966) has been compiled by E. Langer, M. Herz, and J. P. Cox.

17. T. Kelsall and B. Strömgren, Calibration of the Hertzsprung-Russell Diagram in Terms of Age and Mass for Main-Sequence B and A Stars, *in* Vistas in Astronomy, Vol. **8**, pp. 159-178, A. Beer and K. Aa. Strand, Eds., Pergamon Press, Oxford, 1966.
18. A. S. Eddington, The Internal Constitution of the Stars, Cambridge University Press, 1926; reprint, Dover Publications, Inc., New York, 1959.
19. G. R. Burbidge, R. Ebert, S. von Hoerner, S. Temesváry, and F. D. Kahn, Die Entstehung von Sternen durch Kondensation diffuser Materie, Springer-Verlag, Berlin, 1960; in English or German.
20. C. Hayashi, Evolution of Protostars, *in* Annual Review of Astronomy and Astrophysics, Vol. **4**, pp. 171-192, L. Goldberg, D. Layzer, and J. G. Phillips, Eds., Annual Reviews, Inc., Palo Alto, Calif., 1966.
21. G. H. Herbig, The Properties and Problems of T Tauri Stars and Related Objects, *in* Advances in Astronomy and Astrophysics, Volume **1**, pp. 47-103, Z. Kopal, Ed., Academic Press, New York, 1962.
22. M. and B. Schwarzschild, Balloon Astronomy, Scientific American **200**, No. 5, 52 (1959).
23. C. Hayashi and R. C. Cameron, The Evolution of Massive Stars. III. Hydrogen Exhaustion through the Onset of Carbon-Burning, Astrophys. J. **136**, 166-192 (1962); Astron. J. **67**, 577 (1962) and **69**, 140 (1964); and unpublished research (reported by Hayashi in Refs. 1 and 11).
24. H.-Y. Chiu, Neutrinos in Astrophysics and Cosmology, *in* Annual Review of Nuclear Science, Vol. **16**, pp. 591-618, E. Segrè, G. Friedlander, and H. P. Noyes, Eds., Annual Reviews, Inc., Palo Alto, Calif., 1966; see also H.-Y. Chiu, Neutrino Astrophysics, Gordon and Breach, New York, 1965.
25. M. J. Levine, Neutrino Processes of Significance in Stars, Thesis, California Institute of Technology, 1963.
26. P. Morrison, Neutrino Astronomy, Scientific American **207**, No. 2, 90 (1962); see also F. Reines and J. P. F. Sellschop, Neutrinos from the Atmosphere and Beyond, Scientific American **214**, No. 2, 40 (1966).
27. I. Robinson, A. Schild, and E. L. Schücking, Eds., Quasi-Stellar Sources and Gravitational Collapse, University of Chicago Press, 1965.
28. F. Hoyle and W. A. Fowler, Nucleosynthesis in Supernovae, Astrophys. J. **132**, 565 (1960).
29. E. M. Burbidge, G. R. Burbidge, W. A. Fowler, and F. Hoyle, Synthesis of the Elements in Stars, Reviews of Modern Physics **29**, 547 (1957).
30. E. Schatzman, White Dwarfs, North-Holland Publishing Co., Amsterdam, 1958.
31. G. Burbidge and F. Hoyle, The Problem of the Quasi-Stellar Objects, Scientific American **215**, No. 6, 40 (1966).
32. A. G. W. Cameron, Ed., Proceedings of the 1967 "Texas" Symposium on Relativistic Astrophysics [see also the Proceedings of the 1964 Texas Symposium, E. L. Schücking, Ed., 1967], Gordon and Breach, New York, in preparation.

Chapter 22

EXTRAGALACTIC RADIO SOURCES

Wolfgang Priester and Johan Rosenberg*

INTRODUCTION

The most challenging research problem in the field of radio astronomy at the present time is the attempt to understand the physical processes which cause certain extragalactic objects to emit *radio radiation*, the total power of which is up to six orders of magnitude greater than the radio emission of our own galactic system or of other "normal" galaxies. There are two kinds of those extremely strong extragalactic radio emitters known: the *radio galaxies* and the *quasi-stellar radio sources*. For convenience, the latter name is now abbreviated "quasar", following a proposal by Chiu.[1]

The radio sources called radio galaxies often show a double or multiple structure where the radio emission is observed to originate

* National Academy of Sciences—National Research Council Senior Research Associate with the Goddard Institute for Space Studies; on leave from Bonn University.

in large areas far outside of the optically observed galaxy. However, there are additional objects wherein a single area of radio emission is nicely centered around the optically-observed galaxy. There are two basic problems in connection with the radio galaxies: first, to explain the observed brightness distribution, in particular in those cases where there are several, or at least two, centers of radio emission separated by distances which are much greater than the diameter of the optically-observed parent-galaxy; and, secondly, to determine the physical process which is capable of providing the extremely large amount of total energy required for the production of the observed radio power.

The second problem is also the main problem for the quasi-stellar radio sources. Here the difficulties are increased due to the comparatively small size of the areas of radio and light emission, and to the evidence that the total power emitted from these objects in the range of optical wavelengths is approximately two orders of magnitude larger than the total light emission from the brightest giant galaxies. Further, the radio emission of the quasars is equal to that of the brightest radio galaxies. Moreover, some quasars show a double-source distribution of radio emission similar to the radio galaxies.

With regard to the terminology of radio sources, when Hey discovered a point source in the constellation Cygnus in 1946, he called it a *radio star*. In the early 1950's about 100 radio sources were observed, but none of them was identified with either a peculiar star or another optically-known object. Then, in 1952, Baade and Minkowski[2] identified the source in Cygnus and a few others with somewhat peculiar galaxies. Since the sources were distributed isotropically on the sky (if we disregard the few supernovae remnants in our own galaxy which are observable radio sources) it became apparent that the great majority of the radio sources were indeed galaxies with abnormally high radio emission and henceforth the term *radio star* was replaced by *radio galaxy*. To date, about 300 sources have been identified with galaxies, the radio emission of which is several orders of magnitude larger than the one from a normal galaxy, such as our own or M31 (Andromeda nebula). The term *radio galaxy* is now generally used for these strong radio emitters.

The term *radio star*, however, reappeared again at the 107th meeting of the American Astronomical Society in New York in 1960 when Allan Sandage announced that the radio source 3C 48 (No. 48 of the

3rd Cambridge catalog of radio sources) was identified with a 16th magnitude *star* in Triangulum by the perfect coincidence between optical and radio position. The identification was backed by the extremely small diameter of the radio source; less than 1″. On a 90-minute exposure with the 200 inch telescope, the *star* is accompanied by a faintly luminous nebulosity measuring about 5″ × 12″, which, however, by no means resembles a galaxy. The spectrum was considered to be very peculiar since no unequivocal identification with lines of known atoms or ions could be made immediately. By early 1963, five radio sources (3C 48, 147, 196, 273, and 286) were identified with similar star-like objects. All these radio sources have extremely small angular diameters of the order of one second of arc. However, the term *radio star* was abandoned once again and replaced by *quasi-stellar radio source* when Maarten Schmidt[3] discovered that the observed optical emission line spectrum of the source 3C273 is explainable if one accounts for a redshift of $\Delta\lambda/\lambda = 0.158$ corresponding to a nominal recession speed of 47,400 km/sec. This pointed strongly towards an extragalactic object at a great distance and, therefore, there was no longer a valid reason to believe that the actual size of these objects was comparable to the size of a normal star. By early 1967 about 150 rather certain identifications of quasars have been made. For 93 of these objects redshifts are known.

In this chapter we report essentially on the observational results on radio galaxies and quasars. Since there are no generally-accepted views for the physical interpretation on the origin, evolution and energy supply of both kinds of objects, and since the proposed theories are likely to be subject to changes or improvement, we decided to leave out a detailed account of the proposed theoretical interpretations for the energy source of the objects. We include a section explaining the basic physical nomenclature. A detailed account of the physical processes which are responsible for the generation of continuous radio emission is given in our report on "Extragalactic Radio Sources".* Both the radio galaxies and the quasars show typical non-thermal radio spectra which are generally believed to originate from magneto-bremsstrahlung of relativistic electrons spiraling in a magnetic field ("synchrotron emission").

* W. Priester and J. Rosenberg, Extragalactic Radio Sources, NASA Technical Note D-2888, July 1965.

BASIC DEFINITIONS

Intensity and Flux

The *radiation intensity* $I_\nu(\theta, \phi)$ in a radiation field is defined as the amount of radiant energy, passing into the direction specified by (θ, ϕ) through a unit area A, perpendicular to this direction, during one time unit into a solid angle of one steradian within a frequency band of 1 c/s. Therefore, in c.g.s. units the dimensions of intensity are:

$$[I_\nu(\theta, \phi)] = \text{erg cm}^{-2}\text{ ster}^{-1}\text{ sec}^{-1}\text{ (c/s)}^{-1}.$$

Although not explicitly mentioned, $I_\nu(\theta, \phi)$ also will depend in general upon time and the position of A. In Fig. 1, then, the energy dE_ν, passing through the area $d\sigma$ into the solid angle $d\Omega$ during the time

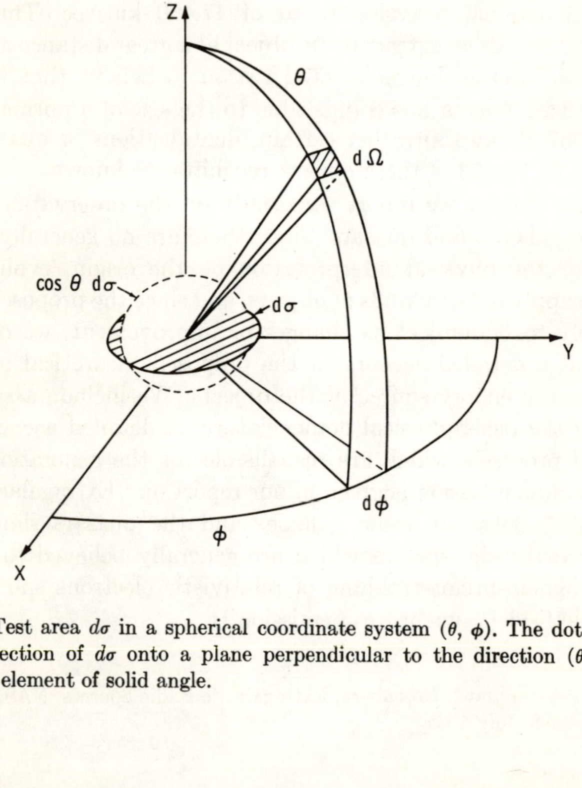

FIG. 1. Test area $d\sigma$ in a spherical coordinate system (θ, ϕ). The dotted area is the projection of $d\sigma$ onto a plane perpendicular to the direction (θ, ϕ). $d\Omega$ denotes an element of solid angle.

dt with frequencies between ν and $\nu + d\nu$, is:

$$dE_\nu = I_\nu(\theta, \phi) \cos\theta \, d\sigma \, dt \, d\Omega \, d\nu. \tag{1}$$

The *radiation flux* S_ν is defined as the amount of energy passing through an area of 1 cm^2 in all directions during one time unit with a frequency bandwidth of 1 c/s. Hence the relation between flux and intensity is given by

$$S_\nu = \int_O I_\nu(\theta, \phi) \cos\theta \, d\Omega \tag{2}$$

with the dimensions

$$[S_\nu] = \text{erg sec}^{-1} \text{cm}^{-2} (\text{c/s})^{-1}.$$

The symbol O denotes the whole sphere over which the integration has to be extended. If one expresses the element $d\Omega$ in spherical coordinates (θ, ϕ) (see Figure 1), we have $d\Omega = \sin\theta \, d\theta \, d\phi$ and henceforth the flux is

$$S_\nu = \int_0^{2\pi} \int_0^{\pi} I_\nu(\theta, \phi) \sin\theta \cos\theta \, d\theta \, d\phi. \tag{3}$$

The average intensity F_ν of the radiation coming out of the unit area into one hemisphere, denoted H, is given by

$$F_\nu = \int_H I_\nu(\theta, \phi) \cos\theta \, d\Omega \bigg/ \int_H \cos\theta \, d\Omega. \tag{4}$$

If the test area is situated in a field of *isotropic radiation*, i.e. if $I_\nu(\theta, \phi)$ is independent of (θ, ϕ), then $S_\nu = 0$. If there is no radiation coming in and the radiation going out is isotropic, we have

$$F_\nu = I_\nu(\theta, \phi) = S_\nu/\pi. \tag{5}$$

The *total flux* over all frequencies is given by

$$S = \int_0^\infty S_\nu \, d\nu. \tag{6}$$

Its c.g.s. units are $[S]$ = erg cm^{-2} sec^{-1}.

Brightness Temperature

A relation between the emitted intensity at a certain frequency and the physical quantities determining the nature of the source will be sought. For this purpose it is convenient to compare the observed intensity with the intensity emitted by a black body at the same frequency. A *black body* is defined by the property that it absorbs all incident radiation at all frequencies. According to *Planck's Law* a black body having a temperature T radiates with an intensity given by:

$$I_\nu \equiv B_\nu(\nu, T)$$
$$= (2h\nu^3/c^2)[\exp(h\nu/kT) - 1]^{-1} \text{ erg sec}^{-1} \text{ cm}^{-2} \text{ ster}^{-1} \text{ (c/s)}^{-1}, \tag{7}$$

where

Planck's constant: $h = 6.6256 \pm 0.0005 \times 10^{-27}$ erg sec

Boltzmann's constant: $k = 1.38053 \pm 0.00006 \times 10^{-16}$ erg deg^{-1}

velocity of light: $c = 2.997925 \pm 0.000002 \times 10^{10}$ cm sec^{-1}.

The radiation of a black body does not depend on its material but only on the frequency and the temperature. The total radiant flux of a black body is given by *Stefan–Boltzmann's Law*

$$S = \int_0^\infty S_\nu \, d\nu = \sigma T^4, \tag{8}$$

where the Stefan-Boltzmann's constant: $\sigma = 5.6697 \times 10^{-5}$ erg cm^{-2} sec^{-1} deg^{-4}.

In radio astronomy, the temperatures of the sources will normally be over 100°K. The frequencies in use will in general be less than 3×10^4 Mc/s (corresponding to $\lambda \geq 1$ cm). Hence the quantity $h\nu/kT$ is generally <0.015. So we can safely approximate exp $(h\nu/kT)$ by the first two terms of the *Taylor expansion*:

$$\exp(h\nu/kT) - 1 \approx h\nu/kT. \tag{9}$$

With this we obtain the *Rayleigh–Jeans approximation* of Planck's Law for $h\nu/kT \ll 1$:

$$B_\nu^{\mathrm{RJ}} = (2h\nu^3/c^2) \cdot (kT/h\nu) = 2\nu^2 kT/c^2$$

$$= 2kT/\lambda^2 \text{ erg cm}^{-2} \text{ sec}^{-1} \text{ ster}^{-1} \text{ (c/s)}^{-1}. \tag{10}$$

This approximation is very important in radio astronomy and will be used throughout this chapter. Note that the quantum theoretical factor h has disappeared. This formula was actually derived by Lord Rayleigh, using classical methods. Inserting appropriate values for k and c we find:

$$B_\nu^{\mathrm{RJ}} = 3.07 \times 10^{-25} \nu^2 T \text{ erg sec}^{-1} \text{ ster}^{-1} \text{ (c/s)}^{-1} \text{ cm}^{-2} \tag{10a}$$

or

$$B_\nu^{\mathrm{RJ}} = 3.07 \times 10^{-28} \nu^2 T \text{ watt m}^{-2} \text{ ster}^{-1} \text{ (c/s)}^{-1} \tag{10b}$$

in both cases taking $[\nu]$ = Mc/s and $[T]$ = °K.

A similar approximation exists for extremely high frequencies, i.e. for $h\nu/kT \gg 1$. Then exp $(h\nu/kT) - 1 \sim$ exp $(h\nu/kT)$, and Planck's Law becomes

$$B_\nu^{\mathrm{W}} = (2h\nu^3/c^2) \exp(-h\nu/kT), \tag{11}$$

which is called *Wien's approximation.*

The *brightness temperature* $T_b(\nu)$ of a source for a certain frequency is defined as the temperature a black body must have to emit an equal intensity at that frequency, so

$$I_\nu \equiv B(\nu, T_b). \tag{12}$$

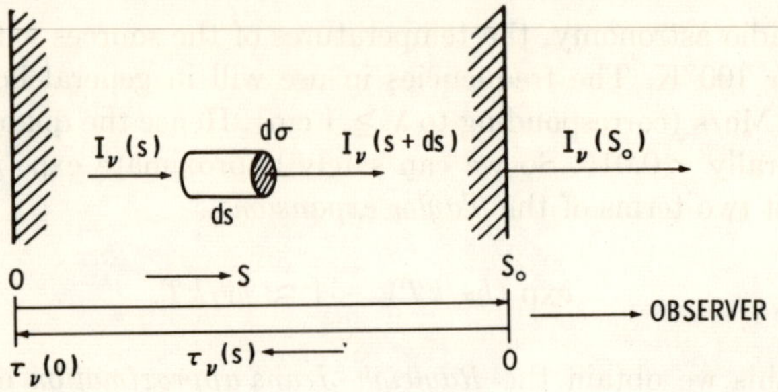

FIG. 2. Simplified radiative transfer problem. The observer is assumed in the direction perpendicular to the surface of the layer. s is the geometrical, τ_ν the optical thickness.

Here I_ν is the experimentally determined intensity. In radio astronomy, using the Rayleigh–Jeans approximation we find

$$T_b(\nu) = (c^2/2k)\ (I_\nu/\nu^2). \tag{13}$$

The brightness temperature corresponds only to the actual temperature if the source emits as a black body at that frequency; i.e., if the source is isothermal and if it absorbs all incident radiation. In all other cases the brightness temperature of a source will be a function of the frequency.

RADIATIVE TRANSFER

We will discuss here only a simplified case of radiative transfer applicable to most radio astronomical problems. A general treatment of the radiative transfer problem is given by Chandrasekhar.[4] We will calculate the intensity of the radiation emitted by a layer of matter of finite thickness. The surface of the layer is assumed to be perpendicular to the line of sight.

The *absorption coefficient* κ_ν is defined as the *fractional decrease* of the intensity, when the radiation passes through a layer of 1 cm

$$dI_\nu/I_\nu = -\kappa_\nu\, ds. \tag{14}$$

κ_ν is generally dependent upon the position of ds, and upon the frequency. It is $[\kappa_\nu] = \text{cm}^{-1}$. *The emission coefficient* ϵ_ν is defined as the amount of energy emitted by a unit volume per unit solid angle per second and per frequency band width of 1 c/s. Its dimensions are accordingly $[\epsilon_\nu] = \text{erg sec}^{-1}\ \text{cm}^{-3}\ (\text{c/s})^{-1}\ \text{ster}^{-1}$.

With the notation as explained in Fig. 2 we can write the equation of *radiative equilibrium*

$$I_\nu(s + ds)\, d\Omega\, dt\, d\sigma\, d\nu - I_\nu(s)\, d\Omega\, dt\, d\sigma\, d\nu$$
$$= -\kappa_\nu \cdot I_\nu(s)\, ds\, d\Omega\, dt\, d\sigma\, d\nu + \epsilon_\nu\, ds\, d\sigma\, dt\, d\Omega\, d\nu$$

or

$$dI_\nu = -\kappa_\nu I_\nu\, ds + \epsilon_\nu\, ds, \tag{15}$$

where I_ν, κ_ν, and ϵ_ν depend upon ν and in general on the position of ds.

For simplification, it is assumed that T is a function of s alone. Furthermore, it is presumed that all the radiation arriving at s comes from regions close enough to s to have their temperatures approximated by $T(s)$; that is, that changes of the temperature over the mean free path length of the radiation are small. This condition is called *local thermal equilibrium* (LTE). It is a very useful restriction, because then *Kirchhoff's Law* holds, saying that the ratio of the emission coefficient to the absorption coefficient is independent of the material and given by *Planck's function*:

$$\epsilon_\nu/\kappa_\nu = B_\nu[T(s)]. \tag{16}$$

This equation holds independent of the manner in which the LTE is established; whether it be by radiation, convection or conduction. Note that $B_\nu[T(s)]$ is independent of the direction of the radiation. The *optical thickness* τ_ν is defined by

$$d\tau_\nu \equiv -\kappa_\nu\, ds. \tag{17}$$

Clearly, we want τ_ν to be zero at $s = s_0$, assuming there is no absorption between the layer and the observer (see Fig. 2). And for $s < s_0$, we want $\tau_\nu(s) > 0$. Hence

$$\tau_\nu(s) = \int_s^{s_0} \kappa_\nu\, ds. \tag{18}$$

Note that τ_ν is a dimensionless quantity. Division of (15) by (17) and using (16) yields:

$$dI_\nu(s)/d\tau_\nu(s) = I_\nu(s) - [\epsilon_\nu(s)/\kappa_\nu(s)] = I_\nu(s) - B_\nu[T(s)]. \quad (19)$$

Multiplying (19) by $\exp[-\tau_\nu(s)]$

$$\exp[-\tau_\nu(s)][dI_\nu(s)/d\tau_\nu(s)] - \exp[-\tau_\nu(s)]I_\nu(s)$$
$$= -\exp[-\tau_\nu(s)]B_\nu[T(s)] \quad (19a)$$

and, integrating between the limits $\tau_\nu = 0$ and $\tau_\nu = \tau_\nu(0)$ (i.e. $s = s_0$ and $s = 0$) we find:

$$I_\nu(0)\exp[-\tau_\nu(0)] - I_\nu(s_0) = -\int_0^{\tau_\nu(0)} B_\nu(\tau_\nu)\exp(-\tau_\nu)\,d\tau_\nu. \quad (20)$$

Assuming an *isothermal* layer we have $B_\nu[T(s)] = B_\nu(T)$ and hence $B_\nu(T)$ independent of $\tau_\nu(s)$. Inserting this in (20) yields

$$I_\nu(0)\exp[-\tau_\nu(0)] - I_\nu(s_0) = B_\nu(T)\{\exp[-\tau_\nu(0)] - 1\}. \quad (20a)$$

Assuming that no radiation enters the layer from the rear side we then have the intensity of the radiation leaving the surface of an isothermal layer perpendicular to the surface:

$$I_\nu(s_0) = B_\nu(T)\{1 - \exp[-\tau_\nu(0)]\}. \quad (21)$$

If $\tau_\nu(0) \gg 1$ so that $\exp[-\tau_\nu(0)] \ll 1$, then the intensity for the emission from an optically thick layer ($\tau_\nu \gg 1$) is:

$$I_\nu = B_\nu = (2\nu^2/c^2)kT. \quad (21a)$$

Hence the brightness temperature of the outgoing radiation is equal to the temperature of the layer

$$T_b = T. \quad (21b)$$

If $\tau_\nu(0)$ is small ($\tau_\nu \ll 1$) we approximate $\exp[-\tau_\nu(0)]$ by $1 - \tau_\nu(0)$ and obtain the intensity of an emission from an optically thin layer

$$I_\nu = \tau_\nu(0) B_\nu = \tau_\nu(0)\ (2\nu^2/c^2) kT. \tag{21c}$$

Accordingly, the brightness temperature of the radiation is for $\tau_\nu(0) \ll 1$

$$T_b(\nu) = \tau_\nu(0)\, T. \tag{21d}$$

From (21) it also follows directly that in general

$$T_b(\nu) = T\{1 - \exp[-\tau_\nu(0)]\}. \tag{21e}$$

In an isothermal, homogeneous layer of ionized gas κ_ν is obviously independent of position and direction; thus we have

$$\tau_\nu(0) = \int_0^{s_0} \kappa_\nu\, ds = \kappa_\nu s_0 \tag{22}$$

and

$$I_\nu(s_0) = (2\nu^2 kT/c^2)[1 - \exp(-\kappa_\nu s_0)]. \tag{23}$$

From equation (23) it is clear that the spectrum of a radio source will be determined by the dependence of the absorption coefficient κ_ν on the frequency ν.

There are two known physical processes which are generally believed to cause the radio emission with continuous spectra from celestial objects. One is a *thermal process* (Coulomb-bremsstrahlung), the other a *non-thermal* one (synchrotron emission = magneto-bremsstrahlung of relativistic electrons).

1. The *Coulomb-bremsstrahlung*; i.e., the radiation emitted by electrons during Coulomb collisions, experienced when moving through the plasma with thermal (non-relativistic) velocities.

2. The *Magneto-bremsstrahlung*; i.e., the radiation emitted by electrons which are accelerated in a magnetic field, while moving with non-relativistic (cyclotron emission) or relativistic velocities (synchrotron emission).

We can compare observational data with theoretical curves and decide whether the source is *thermal* (Coulomb-bremsstrahlung) or *non-thermal* (Magneto-bremsstrahlung). All extragalactic radio sources (galaxies, radio galaxies and quasars) presently show spectra which can be interpreted as synchrotron emission spectra.

Magneto-Bremsstrahlung of Relativistic Electrons

Characteristic for magneto-bremsstrahlung of relativistic electrons spiraling in a magnetic field ("synchrotron emission") is a spectrum in the radio-frequency range which basically can be represented by

$$\mathrm{I} \sim \nu^{-\alpha} \tag{24}$$

where I is the intensity, ν is the frequency and α an exponent which often remains constant or nearly constant over a wide range of the radio-frequency range (10 Mc/s to 10^4 Mc/s).

The spectrum of a single relativistic electron circling in a magnetic field (H) is given by

$$P_\nu = \frac{\sqrt{3}\, e^3}{m_0\, c^2} H.\, F\left(\frac{\nu}{\nu_c}\right) \text{erg sec}^{-1}\ (\text{c/s})^{-1} \tag{25}$$

where m_0 is the rest mass of the electron. ν_c is the so called critical frequency of the emitted spectrum. It essentially determines the high frequency cut-off. For frequencies $\nu > \nu_c$ the emitted spectrum falls off exponentially. The critical frequency is given by

$$\nu_c = \frac{3}{4\pi} \frac{eH}{m_0\, c} \left(\frac{E}{m_0\, c^2}\right)^2 \tag{26}$$

where E is the energy of the electron. The function $F\left(\frac{\nu}{\nu_c}\right)$ can be approximated with sufficient accuracy (according to Wallis 1959) by:

$$F\left(\frac{\nu}{\nu_c}\right) \approx 1.78 \left(\frac{\nu}{\nu_c}\right)^{0.3} e^{-\frac{\nu}{\nu_c}} \tag{27}$$

The maximum of the emission occurs at

$$\nu_{\max} = 0.3\, \nu_c \tag{28}$$

If the energy distribution for the relativistic electrons and positrons follows a simple power law, then the contributions of the individual electrons to the emitted spectrum will superpose in such a way that an intensity distribution is obtained as given by formula (24) with a constant exponent α.

If the energy distribution of the electrons is represented by

$$N\,(E)\,dE = \text{const} \cdot E^{-g}\,dE \qquad (29)$$

then the spectral index α is given by

$$\alpha = \tfrac{1}{2}(g - 1). \qquad (30)$$

For a more detailed description of the synchrotron emission process, we refer the reader to our detailed report on "Extragalactic Radio Sources" NASA Technical Note D-2888.

RADIOGALAXIES AND QUASI-STELLAR RADIO SOURCES

General Description

Although radio waves from outer space were first detected by Jansky in 1932, it was not until 1946 that a *discrete* source was observed. The first tentative identifications with optically known objects were made by Bolton, Stanley, and Slee[5] who identified the radio sources Taurus A, Centaurus A, and Virgo A with the Crab nebula (NCC 1952), NGC 5128, and NGC 4486, respectively. But it was only in 1952, when Baade and Minkowski[2] identified the radio source Cyg A with a peculiar galaxy which seemed to consist of two galaxies in collision, and also studied several other objects extensively, that astronomers generally accepted the identifications. Presently there are 1159 sources listed by Mills, Slee and Hill[6], 328 in the revised 3rd Cambridge catalogue[7], 1219 in the 4th Cambridge catalogue[7a] and about 2000 in the Parkes catalogue[7b]. In this chapter only those extragalactic radio sources that have been identified with optically-observed objects will be discussed. Many radio sources will probably never be identified because their distances are too great for the associated object to be found on the photographic plates.

Normal galaxies [such as our own and the Andromeda Nebula (M31)] have a total power in the radio frequencies (between 10 and 10^4 Mc/s) between 10^{38} and 10^{40} erg sec^{-1}. Their radio luminosity is comparatively low and therefore difficult to observe at great distances. Of about 50 identified sources only 8 were found to be normal. The term *normal* is here defined to mean that the ratio of optical to radio emission is in the same range as for the closest neighbor galaxies. It is expected that the majority of all galaxies will be weak radio sources, i.e., normal galaxies, and that only a rather small group constitutes the radio galaxies, which are much easier to detect due to their extensive radio luminosity. The total power emitted from radio galaxies in the frequency range from 10 to 10^4 Mc/s extends up to 10^{45} erg sec^{-1}. Therefore they constitute the majority of the *observed* radio sources.

It is now generally accepted that the main process which generates radio emission with continuous spectra in the meter wavelength range in galaxies is *synchrotron emission.* Matthews, Morgan and Schmidt[8] investigated the manner in which identified radio sources are distributed among the different types of galaxies. They studied 52 objects. More recently, Bolton et al.[8a] investigated 211 sources. Their results on the percentage-distribution among the different types is summarized in Table I.

The *weak* sources are mostly spiral and irregular galaxies such as our galaxy, the Andromeda nebula, and the Magellanic Clouds. So far they are always single radio sources with a simple structure, centered in the galaxy, but often surrounded by an extensive *halo.*

The *strong* sources all seem to be related to certain types of galaxies; namely, to the ellipticals, the D galaxies, DE type galaxies, dumbbell galaxies, N galaxies and to the quasi-stellar objects which have much smaller dimensions than the bright galaxies.

D type galaxies are characterized by a bright, non-flattened nucleus, surrounded by an extensive envelope. The very large giant D galaxies are often the dominating objects of the clusters of galaxies. D galaxies frequently have diameters three to four times as large (40–80 kpc) as the other members. Their optical luminosities cover a range of a factor of 10, but their radio luminosities differ within a range of 10^4. Although all have the distinctive nucleus and the extended envelope, there are large differences in size, the nuclei sometimes consist of two or more condensations and frequently the cluster

TABLE I

Range of radio luminosity for different types of galaxies

Galaxy type	Percentage	Radio luminosity, erg sec^{-1}	Optical power
Spirals	4	10^{38}–10^{40}	Average galaxy 10^{42} erg sec^{-1}
Ellipticals (incl. DE type)	32	10^{40}–10^{43}	
D type	13	10^{42}–10^{45}	giant galaxy 10^{44} erg sec^{-1}
db type	7	$10^{41.7}$–10^{43}	
N type	5	10^{43}	
Quasars and possible Quasars	35	10^{44}–10^{45}	10^{46} erg sec^{-1}

of galaxies itself is elongated in the same manner as the envelope of the dominating D galaxy. Often absorbing dust bands are visible, revealing rotational symmetry. However, no optical feature indicates whether a giant D type galaxy should be a strong radio source or not.

The correlation between D galaxies which are radio sources and those which belong to clusters is interesting. Fifty percent of all D type radio sources belong to clusters, but only five percent of all D type galaxies in clusters are radio sources. Perhaps this might indicate that a D type galaxy becomes a strong radio emitter only for a rather short period; possibly because of some catastrophe occurring within its nucleus. Assuming the lifetime of a D type galaxy (or better, its nucleus as it characterizes the galaxy) to be 10^9 years as a reasonable estimate, we find that it would spend 5×10^7 years as a radio source.

The weakest of the D, DE type galaxies show a simple radio structure; the more luminous ones are identified with double, or more complex, sources.

The diameters of the D sources (or their components) are, as with the weak sources, comparable to the optical size of the parent galaxy. Only very large halo sources and very strong double or multiple sources extend much farther; the separations of the components of the double sources are approximately ten times larger than the size of the galaxy.

The DE type galaxies indicate a transition between the elliptical and the D galaxies; i.e., these include all variations of the nucleus slowly fading into the envelope. It is remarkable that galaxies of the DE type, closely resembling the D galaxies, have a higher absolute radio luminosity than the ones resembling the elliptical galaxies.

The db type galaxies (i.e., *dumbbell galaxies*) are very similar to the D type, the only difference being that there are *two* nearly equal, very distinct nuclei contained in the same envelope. Their radio luminosity range is about equal to that of *D* sources. Three of the radio sources identified with db galaxies are single and only two are double radio sources. This indicates a rather large proportion of single sources compared with the general case of radio galaxies where only five to ten percent are single radio sources. However, the statistical sample is still rather small.

The N type galaxies have very bright and small nuclei, the radiation whereof constitutes most of the optical luminosity. They have very faint nebulous envelopes, often just visible as a streak, and never far extended. Their radio luminosity is on the average higher than the luminosity of the D sources; however, this may be due to selection effects. They are closely related to the compact galaxies observed by Zwicky, characterized by a very small nucleus, only to be distinguished from stars by the lack of the diffraction pattern on the photographic plate. These compact galaxies seem to exist over a wide range of optical spectral types. Here, too, are no visible differences between the strong radio sources among them and the normal ones. In 1963 Zwicky suggested that the quasi-stellar radio sources are the most luminous, and are extreme cases of these compact galaxies.

The N sources probably have a complex radio structure, with several small components which seem to be rather long and narrow. Since only a few have been found, and their angular diameters are very small, more refined instruments are necessary to resolve their structure.

Quasi-stellar radio sources (“*quasars*”). The main optical features of the quasi-stellar radio sources are:

1) they can be distinguished from stars only by their unusual excessive brightness in the blue of the spectrum, extending into the ultraviolet.

2) their spectra show emission lines with similarities to the emis-

sion lines of planetary nebula and sometimes narrow absorption lines, which are slightly displaced towards the blue with respect to the corresponding emission lines. Both emission and absorption lines show an enormous redshift ranging up to 2.22.

3) occasionally very faint wisps or jets can be distinguished optically on the photographic plates.

4) their radio diameters are usually very small; about 1″. However, at least 13 quasars are known as double radio sources the separation of which extends up to 20″.

A more detailed description of the quasars shall be considered later in this chapter.

Description of Individual Objects

The following discussion applies to several of the more closely-studied radio galaxies:

Centaurus A, NGC 5128

The galaxy NGC 5128, identified with the radio source Centaurus A, is a DE type galaxy with a greatly extended envelope. The distance of Cen A is about 4.7 Mpc. The nucleus is elliptical with a broad absorbing band containing dust and gas in the equatorial region. This band is rotating about the major axis of the nucleus, like the nucleus itself, but with a greater velocity.

The weak extensions of the optical features along the axis of rotation are very peculiar; they reach out to more than 40 kpc and turn into the region of the radio emission. The only plausible explanation for the extension of matter along the rotational axis seems to be the ejection of material (ionized gas) along a magnetic field that might be aligned with the rotational axis.

The radio source shows a rather complex and very extended elongated structure. It extends over about 9 degrees on the sky. This corresponds to a diameter of 650 kpc. (The diameter of the Milky Way is about 30 kpc.) The detailed brightness distribution of Fig. 3 shows at least 3 double sources in the extended source and the so-called *inner source* close to the galaxy. This might indicate that these components were ejected by the galaxy during four successive events. This galaxy represents a special class of radio galaxies (another outstanding member of this class is Fornax A associated with the

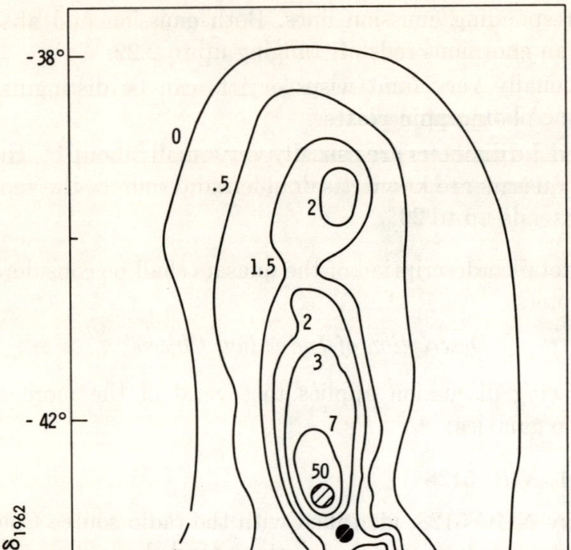

FIG. 3. Centaurus A (outer sources). The brightness distribution in Centaurus A at 21.3 cm. The small black dot in the center equals approximately the galaxy as shown in Fig. 4 (Cooper, 1962).

galaxy NGC 1316). These objects generally seem to have very large linear diameters in both their optical and their radio size. Their radio luminosities are in the range of 10^{42} erg sec^{-1}. Matthews, Morgan and Schmidt[8] give the following values

$$\text{Cen A}: L = 10^{41.87} \text{ erg sec}^{-1},$$

$$\text{For A}: L = 10^{41.27} \text{ erg sec}^{-1},$$

for the range 10 to 10^5 Mc/s.

Synchrotron radiation is plane polarized perpendicular to the magnetic field in the region of emission.

The *polarization* of the radio emission has been carefully studied, together with the *depolarization* effects occurring in the outer envelopes of Cen A or along the line of sight (Bracewell, Cooper, Cousins 1964; Gardner, Whiteoak 1964; Cooper, Price 1964). The inner source is highly polarized, in some wavelengths up to 13 percent, but the polarization is generally restricted to rather small areas. The two components are not polarized to the same degree, as can be seen from Fig. 4. Comparing the position angles of the axis of the nucleus, the

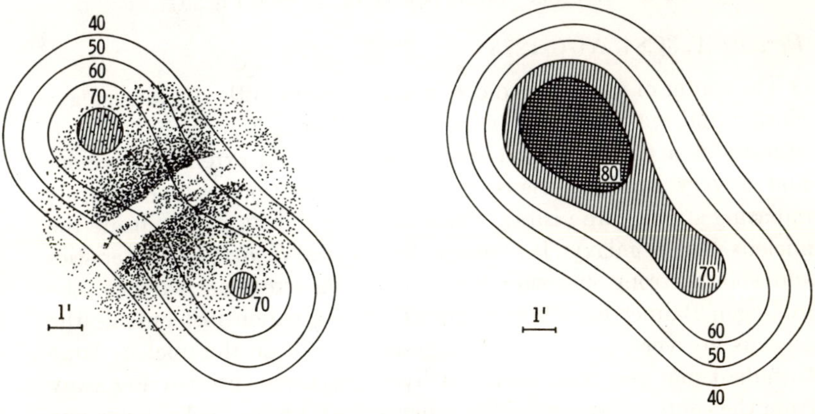

FIG. 4. Centaurus A (inner source). The central part of the radio source Centaurus A at a wavelength of 10 cm. The two drawings show the measured brightness distribution for two different polarization angles. In the left part the position angle of the electric vector is 25°, in the right part 115° (adapted from Bracewell, Cooper and Cousins, 1962).

double source closest to the nucleus and the magnetic field, we find differences which are confined within an angle of about 30°.

Fornax A, NGC 1316 (*D type galaxy*)

This source was identified by Mills. It is in many ways similar to Cen A. It also shows some absorption features across the nucleus that curve up on one side and down on the other. Arp was able to obtain photographical evidence of very faint extensions. At larger distances from the galaxy they curve towards the centers of the double radio source connected with it. The diameter of NGC 1316 as it is usually seen on photographic plates (see insert on Fig. 5) is of the order of 20 kpc while in the case of Cen A (NGC 5128) it is about 25 kpc.

Polarization experiments have shown that the magnetic field was not aligned with the line connecting the centers of the two radio components. It is now clear that the magnetic field follows closely the shape of the optical extensions. The linear size of the radio emission area of Fornax A can be expected to cover up to 200 × 300 kpc. Thus, it is comparable to Cen A. The main difference in the radio structure is that Cen A has several double sources and the inner source, which is very close to the nucleus, while For A consists apparently of one extended double source only. (see Fig. 5).

Perseus A, 3C84, *NGC* 1275

The center of a strong radio source was found in the Perseus galaxy cluster and was identified with the peculiar galaxy NGC 1275. It appears as a tightly-wound spiral, but it has an extended envelope and a very bright nucleus. Several condensations and outwardly radiating streaks are visible, which indicate violent events inside the nucleus of this galaxy. Its nucleus has typical *Seyfert* characteristics and shows strong emission lines exhibiting irregular features; the hydrogen Balmer lines being emitted from a region just outside the nucleus are very broad. This suggests that inside the nucleus large turbulent motions exist, and that hydrogen clouds are moving away from the center at speeds of the order of 3000 km sec^{-1}. As the galaxy is situated in a cluster, these high velocities might be due to a collision between two galaxies; however, no nucleus of a possible other galaxy colliding with NGC 1275 can be seen.

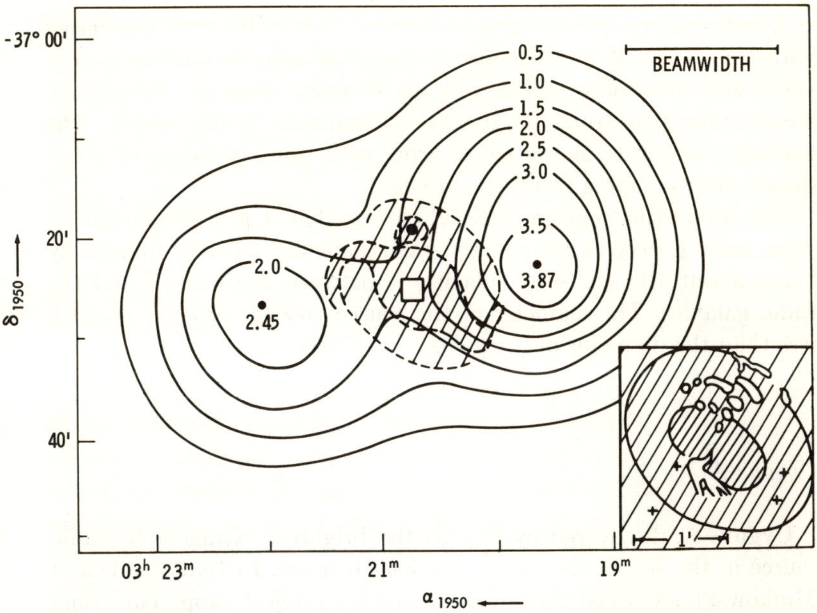

FIG. 5. Fornax A (NGC 1316). The radio contours of Fornax A are from Wade (1961). The large optical features are sketched from a photography published by Arp (1961). The square in the center contains the bright part of the galaxy NGC 1316. This part is given in the lower right corner ten times enlarged. There the brightness contours and the absorption patterns are sketched from the Cape Photographic Atlas of Southern Galaxies. The large optical features of NGC 1316 on the photography envelope the SO *p* galaxy NGC 1317, the position of which is indicated by a dot 6.'5 north of the center of NGC 1316.

The radio source Per A has a very small core being somewhat smaller than the galaxy (20 kpc), and a very extended halo of about 90 kpc, or perhaps much larger.[9] The central source accounts for 80–85 percent of the radio luminosities while the halo provides the remaining 15 percent.

The radio spectrum of Per A shows a very remarkable feature in the high frequency region (5000-10 000 Mc/s), where it curves upward. At 8000 Mc/s an increase in intensity of more than 10% per year has been observed by DENT (1966). This fluctuation is similar to the ones observed in the radio flux of several quasars, though less drastically.

A radio galaxy quite similar to NGC 1275 is NGC 1086, associated with the source 3C71. It is also a *Seyfert galaxy*; its nucleus is very small and extremely bright and the emission lines are broadened, which indicates strong, large-scale movements in the center. The extended structure shows spiral arms with some evidence of a hydrogen flare escaping from the nucleus.

The interesting feature of NGC 1086 is that it is a comparatively weak radio galaxy of exceedingly small diameter, its radio luminosity being about 10^{40} erg sec^{-1}, placing it between the normal and the radio galaxies. The diameter of the source (600 pc or less) is much less than the diameter of the galaxy ($\sim$15 kpc).

Cygnus A

Cygnus A, discovered by Hey, is the brightest extragalactic radio source in the sky in the meter wavelength range. In 1952 Baade and Minkowski[2] identified this source with a faint object (apparent visual magnitude 15^m). The redshift in the emission lines is 0.056, corresponding to a recession speed of 16,800 km/sec. Using a *Hubble constant* of 100 km sec^{-1} Mpc^{-1}, we obtain a distance of about 170 Mpc. The optical diameter is about 8 kpc. The optical spectrum of Cyg A is very similar to that of the Seyfert galaxies, although the broadening of the lines is not as large as in the case of NGC 1275 and the emission comes from a region larger than the emitting regions in other Seyfert galaxies. Its optical power in the emission lines is 2×10^{44} erg sec^{-1}; its total power is about 4×10^{44} erg sec^{-1}.[10]

The radio source is double with two nearly equal components approximately 34 kpc in diameter, and with a separation of 79 kpc. The components are extended along the major axis, on the center of which the galaxy is located. A polarization up to 8 percent has been measured. The direction of the magnetic field, however, is uncertain because of depolarization effects and *Faraday rotation* inside and outside the source. The radio luminosity is $10^{44.7}$ erg sec^{-1}. Cyg A is one of the brightest galaxies observed so far, in both the optical and the radio frequency range.

Virgo A, NGC 4486 (*M* 87)

Bolton, Stanley and Slee[5] identified Virgo A with the galaxy M87 (= NGC 4486), a member of the Virgo cluster. Baade and Minkowski[2] investigated M87 on the Mt. Palomar 200 inch plates and especially its optical jet which protrudes from the nucleus. The galaxy is surrounded by a very large number of globular clusters. The nucleus of the galaxy is bright and small ($\sim$100 pc). The total envelope also encloses the jet and has a diameter of 6 kpc. The distance between the outermost condensation in the jet and the center is 1100 pc. As Virgo A is a strong source, polarization measurements were carried out on the jet to find out if this peculiarity is associated with the source. The jet condensations were found to be optically highly polarized, but with slightly different directions. The optical spectrum of the jet shows a blue continuum, without emission lines, similar to the Crab Nebula. This gives strong evidence that the optical radiation from the jet is also due to synchrotron radiation. There is also ionized gas present in the nucleus since there appear strong emission lines of O II that show double features, indicating that large amounts of gas are moving away from the nucleus at speeds that are, however, still comparable to the random motion inside the nucleus.

The radio source has a rather complex structure which has been resolved into two components:

1) an unpolarized double core source with a separation of 2.5 kpc. The core source is elongated in the direction of the jet, another indication of the importance of the jet as part of the radio source.

2) a halo source emitting about 50 percent of the radio radiation. The halo is slightly polarized and has a diameter of about 20 kpc.

3C33

The peculiarity of this object is that it is a double radio source whose components have a very small size compared to their separation. It is a DE type galaxy with angular dimensions of $3''.3 \times 8''.4$. It also has an outer envelope of elliptical form ($22'' \times 9''$), but with its major axis aligned with the minor axis of the nucleus. The galaxy is probably rotating about the major axis of the envelope. The major axis of the double source forms an angle of 35° with the axis of the envelope. Polarization measurements of the radio source show that it is highly polarized, with the direction of the magnetic field in between the axes of the galaxy and the source. The separation of the components is 200 kpc (3′.8) and the size of the components, which are elongated along the axis, is 7 × 14 kpc (8″ × 16″). The strong polarization and the small size of the components show a highly-aligned magnetic field which is strong enough to confine the relativistic electrons to a rather small region.

Quasi-Stellar Radio Sources (Quasars)

It was stated in the introduction that for several years (from 1960 through 1962) the quasi-stellar radio sources were believed to be peculiar stars within our own galaxy. This was due to their perfect star-like appearance on photographic plates.

The main breakthrough in the identification came with Maarten Schmidt's paper[3] on the radio source 3C 273. The source coincides in position with a rather bright (13^m) star-like object. He was able to explain the optical spectrum when he accounted for a redshift of 0.158, corresponding to a nominal velocity of 47,400 km/sec. Following this discovery, 4 more sources belonging to the same class were soon reinvestigated; as they had already been identified with stellar-like objects. These are 3C 48, 147, 196, 286.

By December 1963 the number had increased to 9 objects. The additional 4 objects are 3C 47, 245, 9, and 216. 3C 47, however, has a larger radio diameter equal to about 1 minute of arc. At the third Texas Symposium on Relativistic Astrophysics in January 1967, Allan Sandage stated redshifts have been determined for 93 quasistellar sources out of a total of 150 objects known at that time.

The identification of the source 3C 273 was facilitated by the fact that the spectrum shows the Balmer series down to H epsilon. The appearance of 5 Balmer lines is quite unusual for objects of these types. A further peculiarity is that the forbidden O II line at 3727 A, which is usually very strong, here is extremely weak.

The star-like object is accompanied by a faint whisp, or jet. The jet has a width of 1″–2″ and extends away from the star in position angle 43°. It is not visible within a distance of 11″ from the *star* and ends abruptly at 20″ from the *star*.

Fortunately, this radio source was occulted by the moon three times in 1962; on April 15, August 5, and October 26. The occultations were observed with the 210 ft. radio telescope at Parkes, Australia, by Hazard et al.[11] at 136, 410 and 1420 Mc/s. This led not only to the most accurate determination of the position so far, but also to a determination of the inner structure of the source. It consists of two components, designated A and B. B is associated with the star-like object, A with the jet (Fig. 6). The difference in position be-

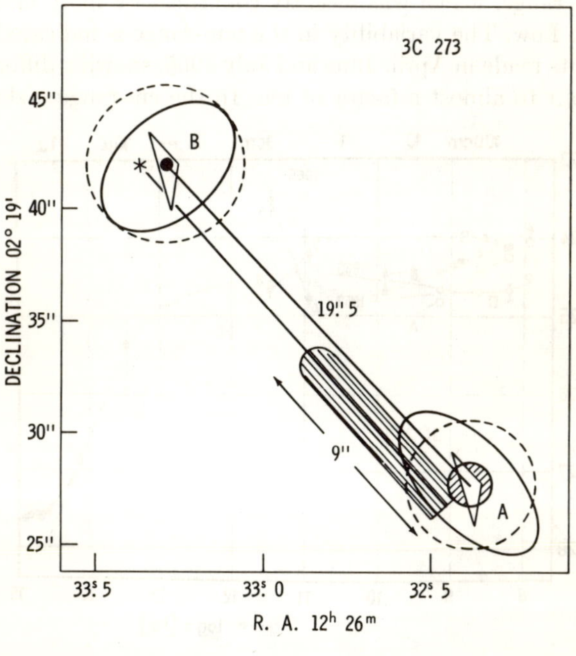

FIG. 6. Optical and radio features of 3C 273. The "star" designates the position of the optically bright object; the large hatched area, the faint jet. The circles (A, B) show the positions of the related radio sources each consisting of a luminous core (inner circle) and a halo of elliptical or spherical shape. The triangle gives the edges of the moon during the 3 occultations in 1962, determining the high accuracy of the radio positions (see Hazard et al.[11]).

tween the starlike object and the source was found to be less than 0.5″ and has been even more improved since. The deviation is so minute that one cannot doubt the identification, in particular since the source A coincides with the optical jet.

The source B consists of a very small core with a width of less than 0.04″. There are indications that the source B might in itself be double at frequencies below 1000 Mc/s, the separation being about 1″.

The source A is elongated along the optical jet with a maximum brightness close to the outer edge of the jet.

In Fig. 7 the entire spectrum is presented from the radio to the optical range. The most remarkable feature is the strong intensity in the infrared range, which was recently confirmed by more measurements by F. Low. The variability in the mm-range is indicated by 4 measurements made in April, June and July 1965, showing differences which amount to almost a factor of ten. In the cm-range a slow in-

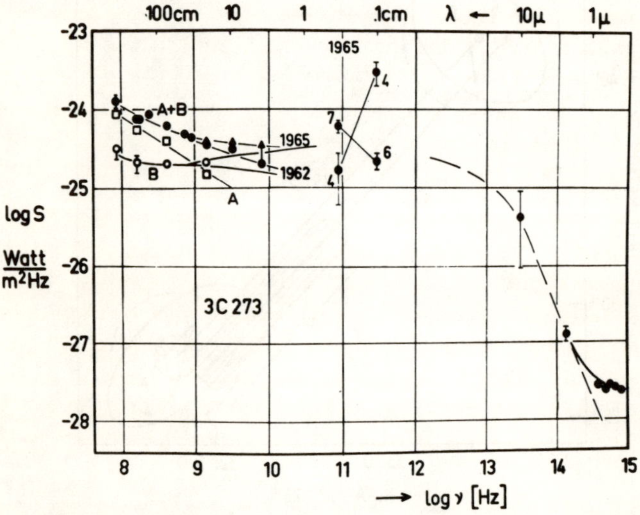

FIG. 7. The spectrum of 3C 273 for the frequency range from 10^8 to 10^{15} c/s. In the radio range (below 10^{10} c/s) the black dots represent the combined spectrum of the components A and B, showing the slow increase in intensity at the cm-wavelengths from 1962 to 1965. In the mm-range four measurements of 1965 are given, the numbers indicate the month of observation. All data for frequencies above 10^{10} Mc/s apply to the star-like object (source B).

crease of the intensity has been observed from 1962 to 1965. At 3 cm this amounts to a factor of two whereas at 30 cm variation remained undetectably small. The variations which have been found in the optical spectrum will be discussed below.

Immediately following Schmidt's finding of the large redshift in 3C 273, Greenstein[13] reexamined the spectrum of 3C 48. Although this spectrum does not show a Balmer series, by applying a redshift of 0.368, corresponding to an expansion velocity of 111,000 km/sec, Greenstein was able to identify the forbidden lines of O II, Ne III and Ne V. The strongest line in the spectrum then appeared at 3830 Å, which belongs to ionized Mg with an unshifted wavelength of 2800 Å.

By January 1967 redshifts for 93 quasars are known from emission line observations made at Mt. Palomar, Lick and Kitt Peak Observatories. They cover the range up to $z = 2.22$.

Figure 8 shows the schematic spectra of 20 star-like radio sources with identified redshifted spectra together with the spectra of the radio galaxies Cyg A and 3C 295. The peculiarity of 3C 273 is that the forbidden O II line (3727 A) is so weak that it can hardly be seen on the spectra and, therefore, was not listed in Schmidt's original paper. In 1964 Greenstein and Schmidt published an excellent detailed analysis of the radio sources 3C 48 and 3C 273.[15]

A few words should be said about the possibilities that the redshift is not purely cosmological, but rather gravitational. In that case, the quasars could be peculiar stars within the Milky Way, rather than distant extragalactic objects. A close examination of the implications of gravitational redshift, however, rules out this possibility. The gravitational redshift depends on the ratio of mass to radius. For 3C 48, with $z = 0.367$, Greenstein derived a radius R (in km) = $4\ M/M_\odot$ and a radius R (in km) = $9\ M/M_\odot$ for 3C 273, with $z =$ 0.158.

A star with about 1 solar mass and a reasonable surface brightness and a diameter of 10 km would have to be so close that its proper motion should have been observed. Jefferys, however, reported that there is no proper motion more than half the mean error of the determination. He used 14 plates from 1887 through 1963. The absolute proper motions which he obtained are about 0.001 ± 0.0025 sec of arc per year. Since this yields a lower limit for the distance of 2×10^4 pc, the object is outside the galaxy.

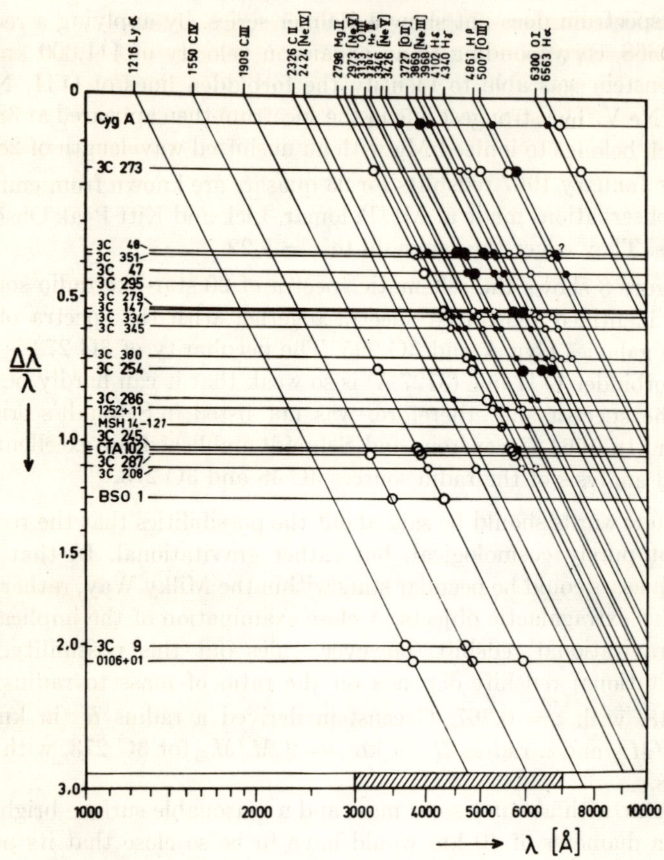

FIG. 8. Schematic spectra of 20 quasars with Cyg A and 3C 295. The ordinate is the redshift $\Delta\lambda/\lambda$. The scale is logarithmic in $1 + \Delta\lambda/\lambda$ (adapted from Greenstein, 1963). Forbidden lines are represented by black dots.

At the symposium on Gravitational Collapse[12] at Dallas in 1963, Williams reported on observations of the 21 cm line absorption of the galactic hydrogen situated between the source 3C 273 and the Earth. Unfortunately, the source is in the constellation Virgo at the galactic latitude of 64°. The line of sight, therefore, does not intersect dense hydrogen clouds. It appears, however, that all hydrogen, which can be observed in emission in the surrounding of the direction toward the source, appears in absorption when the telescope is pointed on the source. Nevertheless, the effect is rather marginal.

KOEHLER and ROBINSON discovered in 1966 an absorption profile in the radio continuum of 3C 273. This could be interpreted as the 21 cm line with a redshift equal to that of the Virgo cluster, which is close to 3C 273 in the sky. This would mean that 3C 273 is further away than the Virgo cluster, and therefore certainly at a cosmological distance. This observation, however, has not yet been confirmed.

A strong argument against gravitational redshift is the fact that there are strong forbidden lines which can originate only in highly-dilute gases, but barely in an atmosphere of a small object with extremely high density. A further argument is the fact that the spectral lines are comparatively sharp. The widths are in the order of 50A. In fact, if the lines originate in an object with a 10 km diameter the variation of the redshift within the layer, from which the lines are emitted, would create much broader lines.

Another possibility which recently has been discussed extensively is that quasars have been ejected during a gigantic explosion within our galaxy or another nearby galaxy, e.g. NGC 5128. In this case, the distances of the quasars would be of the order of 1 to 10 Mpc. It is, however, difficult to imagine that speeds of ejection up to more than 0.8 c (240 000 km/sec) can be obtained by an explosion. Many astronomers, therefore, favor the assumption of cosmological distances. Using these distances, it is found that these star-like objects are by far the brightest objects in the universe.

Table II gives the nominal speed, the observed visual magnitudes, m_v, and the absolute magnitudes, M_v, based on a Hubble constant of 100 km sec^{-1} Mpc^{-1} and an assumed linear relation between redshift and distance. Furthermore, two values are given for redshift-

TABLE II

Apparent and absolute magnitudes of 3 star-like radio sources and two radio galaxies. Two redshift corrections on the absolute magnitudes have been given, between which the true values can be supposed. The absolute magnitudes of the brightest elliptical galaxy and of our own galaxy are given for comparison. The second column gives the nominal recession velocity cz.

	10^3 km/s	m_v	M_v	$M_v-2(\Delta\lambda/\lambda)$	$M_v-5(\Delta\lambda/\lambda)$
3C 273	47.5	12.6	−25.6	−25.9	−26.4
3C 48	110	16.2	−24.5	−25.2	−26.3
3C 47	127	∼18	−23	−24	−25
Cyg A	16.8	14.1	−21.5	−21.6	−21.8
3C 295	138	20.9	−20.1	−21.0	−22.4
Brightest elliptical galaxy:				−22.7	
Our galaxy:				−21.0	

corrected luminosities. The first $(-2\Delta\lambda/\lambda)$ is the necessary minimum correction, the second $(-5\Delta\lambda/\lambda)$ applies for a spectral energy distribution in a normal galaxy. Since the star-like objects appear to have high intensities in the blue and UV range, the necessary correction would probably lie in between the two given values. For comparison, the corresponding data for the two giant radio galaxies Cyg A and 3C 295 are also given, as are the absolute magnitudes of the brightest observed elliptical galaxy and our own galaxy. It is evident that the star-like objects are about 100 times brighter than normal giant galaxies. Typical values for the total emission in the optical and the radio range were given in Table I.

In 1963 Matthews and Sandage[14] discovered light variations in 3C 48 and Smith and Hoffleit in 3C 273. In 3C 273 there appear to exist small cyclic variations with a period of about 13 years; also there are flashes lasting a week or a month during which the object is up to 1 mag brighter; and thirdly, in 1929 a sharp drop of $0^m.4$ occurred; following which the object restored to normal brightness in 1940. These variations impose a severe limit on the size in which they can occur. The 13-year cycle imposes no problem, if 3C 273 has only a diameter of about 7 light years according to the optical model by Greenstein and Schmidt (1964), but the shorter variations might occur only in local regions of the object, in which case the energy content reaches very high values. Although several processes have been suggested even a supernova event would not change the magnitude of the object appreciably.

The sudden drop in 1929 took place in 7 months and must have occurred through the whole object, thus limiting the size of the body, where the optical radiation originates, to 0.3 pc, and resulting in a very high electron density in that gas cloud.[15]

The so far most remarkable light variation has been observed on the object 3C 446 by Sandage and later on by Kinman et al.[15a] From October 1964 to July 1966 the luminosity increased by 3.2^m corresponding to a factor 20. The main portion of the raise occurred during the time from November 1965 to July 1966.

On the spectra obtained on July 13-15, 1966 the emission lines were hardly visible while they clearly stood out in September 1965. This fact can be explained if only the continuous part of the light emission increased while the line intensities remained so far unaffected. During the fall of 1966 there seem to be short-time variations superposed on a general decline of 0.2^m per day.

With respect to the lifetimes of these radio sources, only crude times (within several orders of magnitude) can be obtained. The outer edge of the jet in 3C 273 is 1.5×10^5 light years away from the star-like object. Since it is likely that the jet originated from the star-like object a lower limit is obtained for the lifetime of this object of 1.5×10^5 years.

If a lifetime of 10^7 years is assumed as a likely value, then the star-like objects have to provide a total energy in the form of relativistic electrons of 3×10^{58} erg. In the case of the largest radio galaxies, this value may even go up to 10^{60} erg. Since the efficiency of producing relativistic electrons cannot be very high, the actual requirements for the energy supply are still larger by at least two orders of magnitude. Since 2×10^{62} erg corresponds to the rest-mass energy of 10^8 solar masses, it is possible to appreciate the difficulty in explaining the energy supply for these strange objects.

Radio Sources and Clusters of Galaxies

The investigations by Pilkington[16] indicate that there exists a significant correlation in position between discrete radio sources and clusters. Using the new 178 Mc/s Cambridge survey and Abell's catalogue of rich clusters, taking an area containing 435 sources, he found that 22 sources had positions within 0.4 times the radius of the cluster. Only 3 out of the 22 can be expected to be due to chance. A similar investigation using the catalogue by Mills, Slee and Hill[6] yielded 21 more coincidences, within 0.6 times the cluster radius, of which 5 might be due to chance. The coincidences tend to be close to the center of the cluster, which is in agreement with the fact that many D galaxies, often being close to the centers of clusters, are strong radio sources. Another point of evidence that the coincidences are real is the fact that the apparent radio flux from sources coinciding with nearby galaxies is generally larger than from those coinciding with distant clusters. There was no evidence that the richness of the cluster affects the probability of strong radio emission. A quite different picture is presented when we use only the identified radio sources.[8] The results then obtained are: 50 percent of the parent galaxies occur in clusters with richness ≥ 0 and about 33 percent in clusters with richness ≥ 1, but only 20 percent do not appear to be associated with clusters of galaxies. At most, they might belong to very small groups. The quasars do not seem to belong to clusters, based on the poor statistical evidence. The stronger sources tend to

This shows very clearly that one has to be cautious in exercising statistics with these objects. On one hand many faint clusters are not included in Abell's list and, therefore, reduce Pilkington's coincidences. On the other hand, using only identified sources certainly does not present a homogeneous sample. It may be concluded, however, that beyond any doubt there is a correlation between radio sources and clusters, the degree of correlation has yet to be determined by avoiding selection effects as much as possible.

The Radio Spectra

The spectra of the radio galaxies and quasars will be discussed herein. The radio spectra of nearly all radio sources are very similar in shape and suggest strongly a great conformity regarding the processes governing the radiation of these sources in the radio frequencies. A source radiating by means of the synchrotron radiation process would have a spectrum represented by $F_\nu \sim \nu^{-\alpha_0} = \nu^{-(g_0-1)/2}$, provided that the energy spectrum of the relativistic electrons has the form $N_e(E)\,dE = A \exp(-g_0)\,dE$. Hence, plotting $\log F_\nu$ versus $\log \nu$ we expect to find a straight line. This is essentially confirmed by the majority of the spectra obtained presently. The average value of the spectral index is found to be 0.74. Departures from the straight line can be explained in several ways, e.g.: 1) the electron energy spectrum is not a power spectrum, 2) the radio spectrum may be the superposition of several components.

1) Because of energy losses due to synchrotron radiation, ionization, Coulomb-bremsstrahlung and inverse Compton-effect, the energy spectrum of the electrons may become quite different from the spectrum of the injected electrons (the source spectrum). Since these losses vary differently with energy for the different processes, the spectrum will be determined by the process responsible for the largest losses. If synchrotron radiation dominates, the spectral index will tend to $\alpha \rightarrow \alpha_0 + \frac{1}{2}$, if ionization is the main process, $\alpha \rightarrow \alpha_0 - \frac{1}{2}$, with all transitions possible. (Kellerman 1964). Hence, in sources with very large absolute power and high temperature brightness, where relatively strong magnetic fields might be expected, the spectra will be considerably steeper than for less strong sources, where other processes may take over, all of which tend to flatten the spectra.

Most spectral indices lie within the $\alpha_0 + 0.5$, $\alpha_0 - 0.5$ range (i.e., between 0.24 and 1.24). This might be interpreted as a confirmation of the dependence of the spectral index on the dominating process for energy losses of the electrons. This outline holds only for a process of continuous supply of electrons. If the supply of relativistic electrons ends the energy spectrum of the electrons will show a band which will move to lower energies with increasing time. This would result in a very steep part in the radio spectrum from a certain frequency onward (dependent upon the magnetic field and time). This frequency then, would also decrease with time.

2) One can easily see that for a complex source with unequal components with different spectral indices, the curvature of the integrated source will tend to be positive [compare Fig. 7 of the radio spectrum of 3C 273 (A and B)] even if the curvatures of the components are negative (for low frequencies the steep component will dominate, for high frequencies the flat component). This is in close agreement with observations by Howard, Dennis, Maran and Aller[17]; they found:

a) sources with positive curvature tend to be double, halo-core, or complex and of greater linear extent than sources with negative curvature. It is quite possible that all sources with positive curvature have a complex structure.

b) sources with negative curvatures tend to be smaller in linear size and in separation, and tend to have higher luminosities and surface brightnesses and seem to be favorable ones to look for intensity variations.

As only few sources with positive curvature were found, and the great majority of all sources is double or complex, the components are generally expected to have a similar physical structure, and to have the same origin.

Size and Separation of Radio Sources and Their Components

The radio features of about all identified strong sources show a double or more complex structure. Maltby and Moffet[18] studied 174 extragalactic sources. They were able to resolve 75, as follows:

13 did not have a complex structure,
15 had two equal components,
40 had two or more components of unequal intensity,
7 had a bright core source and a faint halo.

Correcting for the spatial distribution, which causes us to see several double-sources edge-on, we find that 80–85 percent of all resolved sources have double features; 10 percent are of the halo-core type, and only 5 to 10 percent are single. These percentages hold also for sources identified with radio galaxies.[8]

The double structure of nearly all strong sources is centered about the nucleus of the parent galaxy, but the components do have separations up to 360 kpc (Centaurus A). Maltby, Matthews and Moffet[9] suggested that due to some catastrophe occurring in the nucleus of the galaxy, a radio source of small diameter (1 kpc) had been ejected from the galaxy nucleus. While moving away from the galaxy the source might then expand to more than 100 kpc at a distance of 150 kpc from the nucleus.

Many of these galaxies have a rotational symmetry, recognizable by ring-like absorbing dust bands, and sometimes actual rotation has been observed. Where detailed optical observations were possible faint extensions in the direction of the axis of rotation have been found. These are very peculiar features in a rotating body, and can only be explained by the presence of a magnetic field aligned with the axis of rotation, and along which a plasma has been ejected. In many cases the main axis coincides approximately with the rotational axis of the galaxy and the components are extended in the direction of that axis.

Sometimes the components are connected by an emissive bridge and show rather sharp external edges. This might be associated with the shock front of a plasma moving through a magnetic field, away from the galactic plane.

Since it seems obvious that the magnetic field structure of the galaxies is decisive for the structure of the sources, many experiments have recently been carried out to detect any significant polarization in the radio region. As the main emissive process in these sources is believed to be synchrotron radiation, the polarization should give some information regarding the strength and the direction of magnetic fields. Though, some depolarisation occurs when the ray passes through our galaxy.

Optical and radio polarization experiments have been carried out on Messier 87. Linear polarization of both components was found. The difference between the angles of polarization of both components was generally smaller for sources with small separations, than for sources with larger separations. This is in agreement with the model of ejected matter from the galaxy and we expect the two components to behave rather independently from each other when they have moved a considerable distance from the galaxy. Close to the galaxy there will probably be a large interaction between the two components. There seems to be no simple relation between the angle of polarization and the position angle of the main axis of the double source, although there is a tendency toward a 90° difference. This would then give the result predicted by the model; an alignment of the magnetic field along the axis of the source, and approximately along the axis of rotation of the galaxy.

The double-source structure which is so abundant among the radio galaxies has also been found among the quasars. At present there are 13 quasars known which show double sources of their radio emission. If we accept the cosmological distances of the quasars as given by their redshifts, the separation of the sources is in the same order of magnitude as found for the radio galaxies. The largest separation observed so far is 188 kpc for the quasar 3C 47. From this point of view there is a striking similarity between quasars and radio galaxies.

Cosmological Implications of Radio Source Counts

A new discussion of objects of the 3C revised catalogue of radio sources has confirmed the earlier findings[19] that the relationship between the number of radio sources and their flux densities is steeper than one would expect from cosmological models, which are based on the assumption of homogenous distribution of radio galaxies in space without evolutionary effects.

Veron[19a] has shown that the number or sources plotted versus their observed flux in a double logarithmic scale can be represented by a straight line with a slope of $1.55 + 0.05$ for radio galaxies and $2.2 + 0.2$ for quasistellar sources, all these having a flux larger than $S = 9 \times 10^{-26}\ Wm^{-2}(c/s)^{-1}$ at 178 Mc/s.

The slope of 1.55 in the number-flux relation for radio galaxies is within its error limit in agreement with the maximum value derived from relativistic cosmology or steady-state cosmology, which all show a curved relation that starts at large flux densities with a slope of 1.5 and then bends over to smaller values.

The large value of 2.2 derived for quasi-stellar sources might still be influenced by various selection effects, but it is generally agreed upon that the slope will certainly be larger than 1.8, indicating that evolutionary effects have to be considered. In the following this will be done using a lower limit of 1.8.

The requirements for evolutionary effects in the radio sources have been investigated by Oort[23] and Davidson.[22] In order to represent the slope of 1.8 one can either assume that radio sources in the past were more numerous per unit volume of the universe or that their effective luminosity was considerably larger in the earlier evolution of the universe. Of course both possibilities can contribute in part.

From the formulas given by Priester[21] one can easily derive a lower limit for the requirements of the evolutionary effect if one requests obtaining a slope steeper than 1.5. Here we shall present the relation for the Einstein-de Sitter model with euclidean metric and a cosmological constant $\Lambda = 0$. In this model the expansion is proportional to $(t_0 - T)^{2/3}$, where t_0 is the present age of the universe as defined in this model and T the time counted from the present into the past. T is used to characterize the travel time for the radio waves from the source until they reach the observer.

The requirement for the time dependence of the number densities of radio galaxies per unit volume is

$$n = n_0[t_0/(t_0 - T)]^a, \tag{24}$$

where the exponent $a > 4$. For the luminosities the requirement is less severe

$$P = P_0[t_0/(t_0 - T)]^b, \tag{25}$$

where $b > \frac{8}{3}$, n_0 and P_0 are the present values for the number density and luminosity derived from the closest objects.

If we assume an age of the universe of 10^{10} years we find that for times of 3, 5, or 7×10^9 years ago, the required number density of radio galaxies was more than 4, 16, or 120 times larger than at present. The corresponding values for the required luminosities are 2.6, 6.3, or 30 times the present luminosity for 3, 5, or 7×10^9 years ago, respectively. Due to the singularity of the model for $T = t_0$ one is, of course, not permitted to extend this result up to that time. From the numbers given it is apparent that in particular for the luminosities, the required increase in the past appears to be reasonable, in particular since there is also the possibility that both the numbers and the luminosities were larger then. This, of course, would lessen the requirements on each of them correspondingly. In any case it must be kept in mind that the estimates given above can only provide some insight into the order of magnitude required for evolutionary effects.

If the interpretation of the large red-shifts as cosmological red-shifts is correct we can be hopeful that the statistics of quasars can contribute substantially to cosmology, in particular to the solution of the problem of the structure of the universe. For the near future, however, this will be limited to distinguishing between cosmological models which are based on the assumption that Einstein's cosmological constant Λ is zero or in other words that Newton's law of gravitation holds throughout the universe without modification.

Those models predict how the "observed" bolometric brightness of any object depends on the distance (or on the red-shift). If we assume that all quasars have the same intrinsic luminosity (or that in the statistics the average intrinsic luminosity of quasars remained the same during the evolution of the universe) we can calculate the brightness-redshift relation for different space-metrics. Those curves are presented in Fig. 9. The ordinate represents the bolometric magnitude in the usual astronomical scale, the abscissa is the red-shift $z = \Delta\lambda/\lambda$. The uppermost curve, labeled $q_0 = -1$, corresponds to the steady-state cosmology of Hoyle, Bondi and Gold. The other curves belong to models of the relativistic cosmology with hyperbolic metric ($0 < q_0 < \frac{1}{2}$), euclidean metric ($q_0 = \frac{1}{2}$) or spherical metric ($q_0 > \frac{1}{2}$).

The dots in Fig. 9 represent the quasars with observed redshifts. Since the bolometric brightness cannot be observed or calculated at present the observed visual magnitudes have been used. This implies a slight uncertainty in the ordinates of the dots. But at present, the statistics are, in any case, still so poor that only preliminary indications can be obtained and therefore the use of the visual magnitudes is still justified. From Fig. 9 one might conclude that the observed data from quasars seem to favor cosmological models with spherical or euclidean metric and to rule out the steady-state theory. One should, however, bear in mind that the selection effects in the observational evidence will also tend to favor the lower curves (i.e., the spherical metrics).

The models with spherical metric have the property that the expansion comes to a halt during the evolution of the universe in a finite

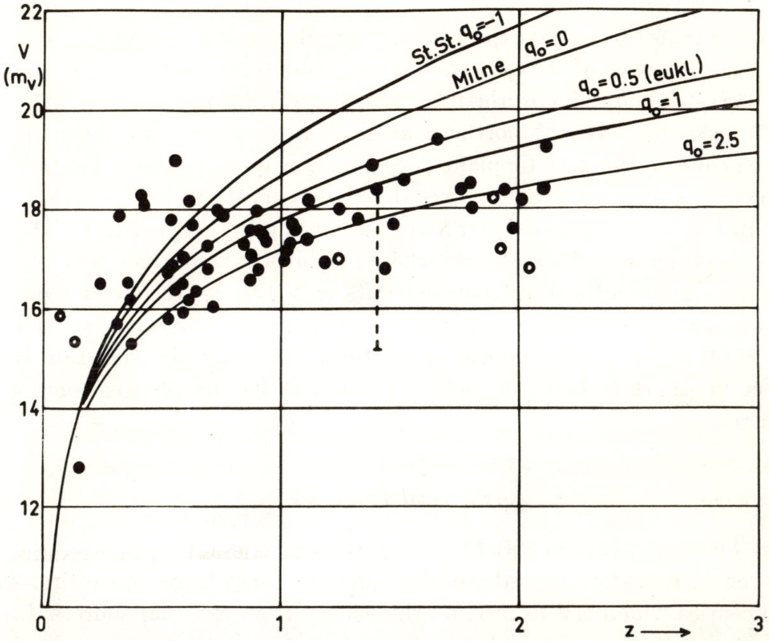

FIG. 9. *Relation between apparent brightness and redshift* $z = \Delta\lambda/\lambda$. The ordinate represents the bolometric magnitudes for the theoretical curves, belonging to different cosmological models as explained in the text. The dots represent the apparent visual magnitudes of those quasars where the redshift has been measured. The open circles represent radio-quiet quasars whose radio flux is below the present level of detectability. Since the dependence of the bolometric correction on the redshift is not yet known, some caution must be exercised in the comparison of the data with the theoretical curves.

The dotted line represents the range of variability of 3C 446 whose luminosity increased by a factor of 20 between November 1965 and July 1966.

time. Then a contraction follows into the same kind of singularity from which the expansion originally started. The time-lapse between the two singularities depends on the deceleration parameter q_0 and the present "radius" of the universe. In the euclidean universe the expansion would continue indefinitely, but with a decreasing speed which finally becomes zero.

The distribution of the identified quasars on the sky is sufficiently isotropic, as one would expect from extragalactic objects (see Fig. 10). On the sky map of the quasars the area around the southern celestial pole is still blank due to the fact, that indentifications are made from photographs taken at northern hemisphere observatories. Also the area close to the galactic plane does not show quasars due to the fact that the galactic absorption excludes the observations in this area.

In Fig. 10 the quasars are grouped according to their redshifts. The crosses represent those objects where redshifts have not yet been obtained. Apparently, the larger redshifts ($z > 1.0$) are relatively more abundant on southern galactic latitudes by 2:1 for the ratio between $z > 1.0$ and $z < 1.0$ whereas on northern latitudes the situation is vice versa. It is, however, believed that this has no physical significance.

Physical Processes

The main problem with the strong radio galaxies is that they require, even with modest estimates of their lifetimes, very large total energies. Theories which try to provide the energy from external sources, for instance from collisions between galaxies or even with large clouds of antimatter, are less favoured nowadays than those based on processes within the galaxy. All have to explain a total energy of the order of 10^{58} to 10^{60} ergs, supplied in the form of relativistic electrons. This value is obtained as follows: the lifetime of the relativistic electrons causing the radio frequency radiation is of the order of 10^7 years, if the magnetic field has a field strength of $H = 10^{-5}$ Gauss and if the initial energy of the relativistic electrons is of the order of 10^9 eV. As these sources emit between 10^{41} and 10^{45} erg sec^{-1} in the radio range and up to 10^{46} erg sec^{-1} in the optical range, the energy supplied in the form of relativistic electrons is of the order of 10^{59} erg, if the lifetime is correctly estimated. From this, one can conclude that the total energy for the largest sources is likely to exceed 10^{61} erg, since there will surely be only a small percentage of the total energy stored in relativistic electrons.

The energy requirements for quasars are of the same order of magnitude. While the lifetime of quasars may be smaller than those of

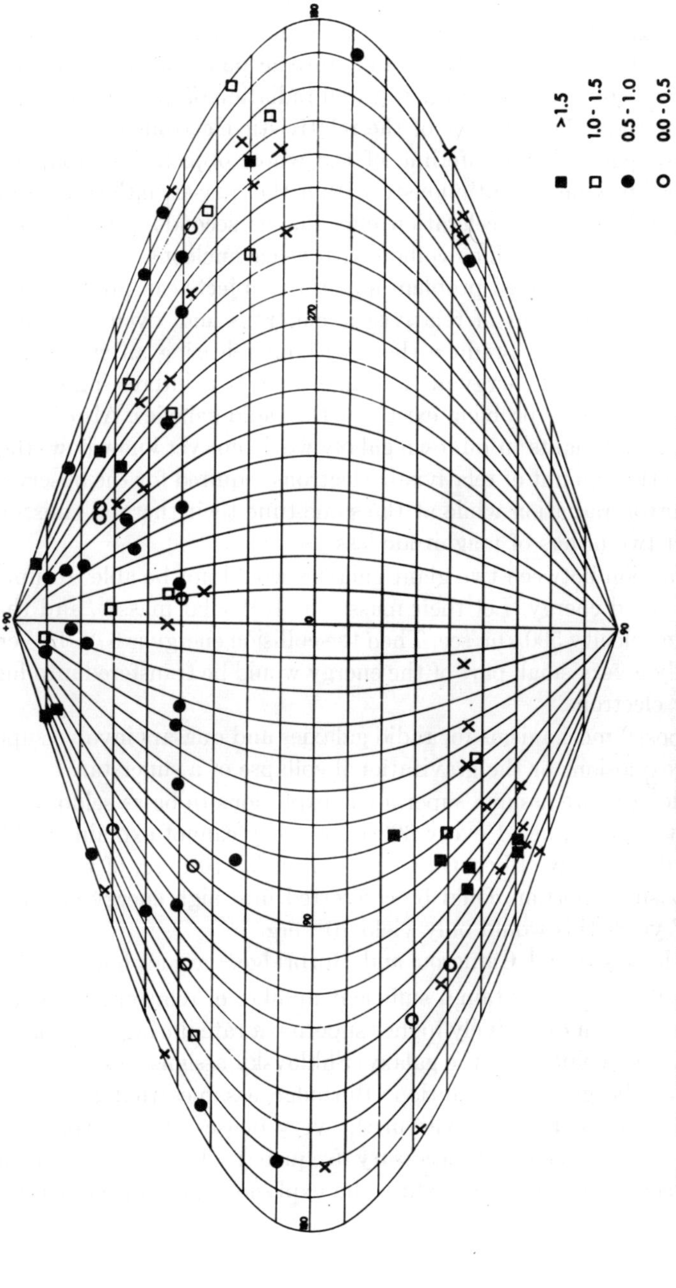

FIG. 10. *Distribution of quasars, classified by their redshift, on the sky in galactic coordinates*. The center of the figure is at $l^{II} = 0$, $b^{II} = 0$. Crosses: Quasars with as yet unknown redshifts.

radiogalaxies, the brightness is at least as high, and even greater by a factor of about 10 or more if we take into account that probably also the optical continuum and the infrared radiation of quasars is of non-thermal origin. The lifetime of the relativistic electrons in quasars is small compared to the lifetime of the whole object. Electrons are, therefore, produced continuously within the quasar, either by some not yet understood acceleration effect or as secondary products of high energy protons which collide with other nucleons.

It is still an unsolved problem how these objects can produce such a tremendous amount of relativisitic electrons and, even more puzzling, how they can produce them without releasing an even larger amount of thermal energy at the same time. It is interesting to note that a very similar problem exists for the radio-galaxies. Also for the supernova-remnants in our own galaxy we do not yet know how they produce the amount of relativistic electrons required for the observed synchrotron radiation while at the same time their thermal emission is about two orders of magnitude less.

A collision between two giant galaxies would not be able to supply the required energy. Let their masses be 10^{11} solar masses, and their relative velocity 500 km/sec. Then the collision energy is 6×10^{59} erg, but only a very small part of the energy would be transfered into high energy electrons.

Proposed mechanisms for radio galaxies and quasars involve supernovae explosions or the gravitational collapse of a superstar.

As long as we assume supernovae explosions to be as seldom as in our own galaxy, i.e., 1 every 300 years, they cannot explain the high amount of energy either. One explosion yields 10^{51} erg, of which only a very small portion could be converted into high energy electrons. In 10^7 years this would only yield 10^{54} erg.

Shklovsky[24] and Cameron and Burbridge[25] argue, however, that a plausible way to obtain a sufficient number of relativistic electrons is to assume an excessively higher supernova rate during some periods within the evolution of the galaxy. Shklovsky assigns this to an early stage of the galactic evolution. Burbidge assumes that a chain reaction of Type I supernovae in the very dense central regions of a galaxy can provide the necessary frequency of explosions. In this close-packed cluster of old stars the explosions are triggered by the shock waves from the exploding neighbor stars. It seems not implausible that the required very high star densities might be found in the nuclei of D galaxies, N galaxies, or perhaps in quasars.

Strong radio galaxies are usually of the D and N types, which can be considered as a special form of elliptical galaxies. This fact favors also the suggestion by Cameron[25] who assumes that intense supernova activity can arise in these galaxies during the star formation process in a large gas cloud. The more massive stars contract quickly and all spend about 3×10^6 years on the main sequence in the Hertzsprung-Russell diagram. Subsequent evolution will be rapid and will lead to Type II supernovae within a relatively short period (of the order of 10^7 years).

Hoyle and Fowler[26] try to derive the required energy from a gravitational collapse of an excessively massive superstar (10^6 to 10^8 solar masses). This suggestion is particularly important in connection with the quasars. The difficulties of this idea are that the contraction will lead to a rotational crisis; a faster and faster rotation which will break up the object into smaller objects. On the other hand, this provides a possibility for explaining the variations in the optical emissions, as Hoyle argued at the Symposium on Gravitational Collapse in Dallas 1963. If there are large numbers of subcondensations formed, which will reach their luminous stage statistically distributed, one might expect noise-like light variations.

A more detailed account of the ideas of explaining the physics of radio galaxies or quasars might be found in the proceedings of the "Symposia on Relativistic Astrophysics" which had been held in Dallas (1963), Austin (1964), Miami (1965) and New York (1967).

REFERENCES

1. H. Y. Chiu, Gravitational Collapse, Physics Today **17,** 21 (1964).
2. W. Baade and R. Minkowski, Identification of the Radio Sources in Cassiopeia, Cygnus A, and Puppis A, Astrophys. J. **119,** 206 (1954).
3. M. Schmidt, 3C 273: A Star-Like Object with Large Redshift, Nature **197,** 1040 (1963).
4. S. Chandrasekhar, Radiative Transfer, Clarendon Press, Oxford, 1950.
5. J. G. Bolton, G. J. Stanley, and O. B. Slee, Positions of Three Discrete Sources of Galactic Radio Frequency Radiation, Nature **164,** 101 (1949).

6. B. Y. Mills, O. B. Slee, and E. R. Hill, A Catalogue of Radio Sources Between Declination +10° and −20°, Austral. J. Phys. **11,** 360 (1958).
7. A. S. Bennet, The revised 3C Catalogue of Radio Sources, Mem. Roy. Astron. Soc. **686** (pt. 5), 163 (1962).
7a. J. D. H. Pilkington and P. F. Scott, A Survey of Radio Sources between Declinations 20° and 40°. Memoirs Roy. Astron. Soc. **69,** 183 (1965).
7b. G. A. Day, A. J. Shimmins, R. D. Ekers and D. J. Cole, The Parkes Catalogue of Radio Sources. Austral. J. Physics **19,** 35 (1966).
8. T. A. Matthews, W. W. Morgan, and M. Schmidt, A Discussion of Galaxies Identified with Radio Sources, Astrophys. J. **140,** 35 (1964).
8a. J. G. Bolton and J. Ekers, Identification of Strong Extragalactic Radio Sources. Austral. J. Physics **19,** 559 (1966).
9. P. Maltby, T. A. Mathews, and A. T. Moffet, Brightness Distribution in Discrete Radio Sources, Astrophysical J. **137,** 153 (1963).
10. G. R. Burbidge, E. M. Burbidge, and A. R. Sandage, Evidence for the Occurrence of Violent Events in the Nuclei of Galaxies, Rev. Mod. Phys. **35** 947 (1963).
11. C. Hazard, M. B. Mackey, and A. Shimmins, Investigation of the Radio Source 3C 273 by the Method of Lunar Occultations, Nature **197,** 1037 (1963).
12. I. Robinson, A. E. Schild, and E. L. Schücking (Eds.), Gravitational Collapse, Symposium Proceedings, Dallas, 1963, University of Chicago Press, in press, 1965.
12a. J. A. Koehler and B. J. Robinson, intergalactic atomic neutral hydrogen I and II, Astrophys. J. **146,** 488–516 (1966).
13. J. L. Greenstein, Quasi-stellar Radio Sources, Sci. Amer. **209,** 54 (1963).
14. T. A. Mathews and A. R. Sandage, Optical Identification of 3C 48, 3C 196, and 3C 286 with Stellar Objects, Astrophys. J. **138,** 30 (1963).
15. J. L. Greenstein and M. Schmidt, The Quasi-Stellar Radio Sources 3C 48 and 3C 273, Astrophys. J. **140,** 1 (1964).
15a. T. D. Kinman, E. Lamla and C. A. Wirtanen, The Optical Brightness Variations and Polarization of the Quasi-Stellar Radio Source 3C 446. Astrophys. J. **146,** 964 (1966).
16. J. D. H. Pilkington, Radio Sources and Rich Clusters of Galaxies, Monthly Notices **128,** 103 (1964).
17. W. E. Howard III, T. R. Dennis, S. P. Maran, and H. D. Aller, Curvature in the Spectra of Non-Thermal Radio Sources, Nature **202,** 862 (1964).
18. P. Maltby and A. T. Moffet, Brightness Distribution in Discrete Radio Sources, III, Astrophys. J. Suppl. **7,** 141 (1962).
19. P. F. Scott, M. Ryle, and A. Hewish, First Results of Radio Star Observations Using the Method of Aperture Synthesis, Monthly Notices **122,** 95 (1961).
19a. P. Véron, Comptages des Radiosources dans le catalogue 3C révisé. Ann Astrophysique **29,** 231 (1966).
20. G. C. McVittie, Counts of Extragalactic Radio Sources and Uniform Model Universes, Austral. J. Phys. **10,** 331 (1957).

21. W. Priester, Zur Statistik der Radioquellen in der Relativistischen Kosmologie, Z. Astrophys. **46,** 176 (1958).
22. W. Davidson, The Cosmological Implications of the Recent Counts of Radio Sources, II. An Evolutionary Model, Roy. Astron. Soc., Monthly Notices **124,** 79 (1962).
23. J. H. Oort, Some Considerations Concerning the Study of the Universe by Means of Large Radio Telescopes, OECD Symposium on Large Antennae for Radio Astronomy, Paris, 1961, p. 35.
24. I. S. Shklovsky, Radio Galaxies, Astron. Zh. **37,** 945 (1960).
25. A. G. W. Cameron and G. R. Burbidge, Star Formation in Elliptical Galaxies and Intense Radio Sources, Nature **194,** 963 (1962).
26. F. Hoyle and W. Fowler, On the Nature of Strong Radio Sources, Roy. Astron. Soc., Monthly Notices **125,** 169 (1963).

Chapter 23

NUCLEOSYNTHESIS

J. W. Truran and A. G. W. Cameron

It is only during the present century that it has been possible to argue that the elements must have had some point of origin in time. In order to conclude that such an origin is logically necessary, the naturally radioactive substances had to be discovered. It could then be argued that if the elements had had an infinite age, the daughter products of the naturally radioactive substances would have piled up enormous abundances, so that there should have become an infinite ratio between the abundances of the daughter products and the abundances of the naturally radioactive parent species. Since such infinite abundance ratios were not observed, it was immediately concluded that the naturally radioactive substances, and therefore presumably all of the elements, have a finite age.

It is interesting to dwell for a moment upon the evolution of ideas concerning the origin of the elements. During the last century, astrophysicists discovered that the solar spectrum contains characteristic absorption lines with the same wavelengths as those which are observed in laboratory spectra on the earth. To those scientists it was an exciting discovery to find that the sun contained the same elements as did the earth. Later it was discovered that the stars also contained the same elements as the sun and the earth, and eventually this discovery was extended to other galaxies. The natural qualitative assumption to make, based upon these discoveries, is the obvious generalization that the chemical composition of the universe is everywhere the same, apart from such easily understood variations in abundance as the lack of volatile substances on the earth.

This assumption of chemical homogeneity throughout the universe played an important role in defining the boundary conditions within which people tried to construct theories of the origin of the elements. It appeared that the universe also had a finite age, since the recession of the distant galaxies appeared to have started from some common point a few billion years ago. Hence it was natural to seek conditions during the early stages of the expansion of the universe in which the elements might have been formed.

To begin with, a variety of people put forward equilibrium theories of the formation of the elements. In these theories it was generally assumed that in the early stages of the history of the universe, when all the matter was squeezed tightly together, and was presumably at very high temperature, conditions might be ripe for producing the elements by a nuclear cooking process. Such theories suffered from the difficulty that no single set of conditions of temperature and pressure would give all of the elements with the observed relative abundances, so that it was necessary to make such theories extremely complicated, with varying sets of conditions assumed to be responsible for different regions of the periodic table of the elements.

As the complexity of the equilibrium theories increased and their plausibility diminished, other authors began to seek non-equilibrium mechanisms which might be responsible for the formation of the elements. In particular, Maria Mayer and Edward Teller[1] suggested that during the primordial stages of the universe there existed very large neutron masses, from which clumps of nuclei would split off, and subsequent fission and readjustment would take place. This theory was not developed very extensively. Also, Alpher, Bethe, and Gamow[2] proposed an elaborate theory in which the initial stages of the expansion of the universe contained matter which was largely a neutron gas in composition, so that as the expansion took place some of the neutrons would decay into protons, and these in turn were then expected to capture neutrons and in this way form all of the heavy elements. This theory also had weaknesses, among the greatest of which was the lack of stable nuclei at mass numbers 5 and 8 through which the neutron capture might progress.

However, a little more than a decade ago the astrophysicists started to measure the abundances of the elements in stars with good quantitative accuracy, and this changed the picture considerably. No longer was it a tenable assumption that the universe was chemically homogeneous. Two general types of abundance anomaly were found. On the one hand, it was found that the oldest stars in our galaxy, particularly those likely to be considerably older than the sun, had a deficiency of heavier elements relative to hydrogen as compared to the sun itself. This deficiency ranged from a factor of a hundred to a factor of a thousand in some extreme cases. The second type of anomaly concerned stars which in their overall composition might be quite similar to that of the sun, but in which specific elements could be seen to be overabundant. Thus certain stars were observed to be enriched in carbon and other stars to be enriched in certain heavy elements. Perhaps most significant, certain of the stars were observed to have a quite large abundance of the element *technetium*,[3] which contains no stable isotopes, and whose longest lived isotope has a half-life of just a little more than two million years.

These observations demonstrated that rather complex nuclear reactions must be going on in the interiors of stars in the galaxy today, and they also demonstrated that there was a tendency for the general abundance of the elements to increase in a star as the star under consideration was formed more and more recently. This naturally led to the general hypothesis that the galaxy may have been initially composed either entirely or nearly entirely of hydrogen gas, but that upon the initial star formation and stellar evolution, the first generations of stars to reach the end of their lifetimes would eject the products of nuclear reactions in their interiors back into the interstellar medium, and thus the interstellar medium would become continually enriched in such products.

It is to the latter ideas concerning nucleosynthesis in stars that we shall devote our attention in this chapter.[4–7] If one postulates that the products of nuclear reactions in the interior of a star are to become the elements that are formed in nature, then we wish to ascertain whether there is any resemblance between the abundances of these products and the observed relative abundances of the elements. Thus our first task must be to determine what nuclear reaction products are formed during the evolution of the star.

Chapter 21 has discussed the methods by which theoretical models of stars are constructed. Some of the basic nuclear reactions responsible for nuclear energy generation were discussed in that chapter. In the present chapter we shall be concerned very little with energy generation as such, and to a much greater extent with the changes in chemical composition that take place during stellar evolution. For this purpose we shall not discuss detailed models of stars at all, but we shall be content with considering matter that becomes more and more dense and hotter and hotter as its evolution progresses, so that after the exhaustion of one nuclear fuel the subsequent rise in temperature will be able to induce further nuclear reactions in the ashes.

For this purpose it is useful to consider a general result which bears on stellar evolution and is derived from the *Virial Theorem*. This result states that in the interior of a self-gravitating body, such as the sun or a star, and in which the ratio of the specific heats of the gas is 5/3, half of the released gravitational potential energy is stored as internal heat, and the other half has been lost into space by some means. The essential principles of stellar evolution are contained in this statement. When a star is first formed by contraction of gas out of the interstellar medium, some of the gravitational potential energy release must go into the ionization of the principal constituents, the hydrogen and helium, as has been discussed in the chapter on the origin of the solar system. The rate at which such a star can contract further is governed by the rate at which half the released gravitational potential energy can be transported to the surface and there radiated away into space. As this radiation occurs, a continual shrinkage of the star will take place, accompanied by a rise in the central temperature and density. When the central temperature becomes high enough to drive a thermonuclear reaction, the shrinkage of the star will cease as long as nuclear energy can supply the energy which the star radiates away into space. When the nuclear fuel is exhausted, shrinkage once again takes place with a continued rise of the central temperature and density. It should be emphasized that we are here considering what happens in the innermost part of the star, and the fact that the outer layers of the star may expand while the central shrinkage is taking place during certain stages of stellar evolution is an irrelevant detail which is not of interest in terms of the nuclear reactions that are taking place in the center. Thus the life story of a star consists of a number of periods of shrinkage, separated by plateaus during which nuclear reactions supply the energy which is emitted from the surface of the star. When all possibility of obtaining energy by nuclear transformations has been exhausted, then either the star will have found a way to settle down in the degenerate state as a white dwarf, or else it must face the prospect of still continued shrinkage and an eventual catastrophe. We shall return to this question later.

THERMONUCLEAR REACTION RATES

Before beginning a discussion of the various nuclear burning processes which contribute to the synthesis of the elements observed in nature, it is appropriate to outline briefly the methods employed in the determination of the relevant reaction rates. In general the number of reactions per unit volume per second, r, of two nuclear species with number densities n_1 and n_2 can be written

$$r = n_1 n_2 <\sigma v>$$

Here $<\sigma v>$ is an appropriate average of the product of the reaction cross section $\sigma(v)$ and the relative velocity of the nuclei, v

$$<\sigma v> = \frac{\int \sigma(v)\, v\, N(v) dv}{\int N(v) dv}$$

where $N(v)$ is the number density of nuclei having relative velocities v. In stellar interiors, the velocity distributions of the two species are well represented by the Maxwellian distribution, and the expression for $<\sigma v>$ takes the form

$$<\sigma v> = \left(\frac{8}{\pi\mu}\right)^{\frac{1}{2}} (kT)^{-3/2} \int_{E_t}^{\infty} E\, \sigma(E) \exp(-E/kT) dE$$

where k is the Boltzmann constant, T is the temperature, E is the kinetic energy of relative motion, E_t is the threshold energy and μ is the reduced mass

$$\mu = m_1 m_2/(m_1 + m_2)$$

Much of our discussion of nucleosynthesis during early stages of stellar evolution will be concerned with reactions proceeding on light nuclei at low energies ($kT = 861$ eV for $T = 10^7$ °K). At these energies, the nuclear excited states through which the reactions can proceed are few in number and widely spaced in energy. Under these conditions, the reaction rates may conveniently be expressed as sums over the contributions of the individual resonance states.

1.) Resonant reactions

In the vicinity of a resonance the reaction rate can be determined from a knowledge of the resonance parameters. The cross section at resonance is assumed to be well represented by the Breit-Wigner single level formula. Specifically we consider a resonance of energy E_r and spin J. Designating the widths for the entrance and exit channels by Γ_a and Γ_β respectively the cross section is given by:

$$\sigma_{a\beta}(E) = \pi\lambda_a^2 \, g \, \frac{\Gamma_a\Gamma_\beta}{(E\text{-}E_r)^2 + (\Gamma/2)^2}$$

In this expression λ_a is the reduced de Broglie wavelength for the incident channel, Γ is the total width for the decay of this compound nuclear state and g is the statistical factor

$$g = \frac{(2J + 1)}{(2s + 1)\,(2I + 1)}$$

where s is the spin of the bombarding particle and I is the spin of the target nucleus. In general Γ_a and Γ_β represent summations over all values of the orbital angular momentum l consistent with spin and parity restrictions.

Employing this form for the cross section, we obtain the following expression for $<\sigma v>$:

$$<\sigma v> = \left(\frac{8}{\pi\mu}\right)^{\frac{1}{2}} \frac{\pi g}{(kT)^{3/2}} \int_{E_t}^{\infty} \frac{\lambda_a^2\Gamma_a\Gamma_\beta}{(E - E_r)^2 + (\Gamma/2)^2} \, e^{-E/kT} \, E \, dE$$

We assume that the resonance is extremely sharp, and approximate those terms in the integrand which vary slowly with energy by their value at resonance, hence:

$$<\sigma v> = \left(\frac{8}{\pi\mu}\right)^{\frac{1}{2}} \frac{\pi g}{(kT)^{3/2}} \lambda_a^2(E_r)E_r e^{-E_r/kT}\, \Gamma_a\Gamma_\beta \int_{E_t}^{\infty} \frac{dE}{(E-E_r)^2+(\Gamma/2)^2}$$

As the contributions from the resonance tails are small we can extend the limits of integration from $-\infty$ to ∞ and evaluate the integral by the method of residues:

$$\int_{E_t}^{\infty} \frac{dE}{(E-E_r)^2+(\Gamma/2)^2} \sim \int_{-\infty}^{\infty} \frac{dE}{(E-E_r)^2+(\Gamma/2)^2} = \frac{2\pi}{\Gamma}$$

Thus the contribution to the rate at resonance is given by:

$$<\sigma v> = \left(\frac{32}{\mu}\right)^{\frac{1}{2}} \left(\frac{\pi}{kT}\right)^{3/2} \lambda_a^2(E_r)E_r e^{-E_r/kT} \left(\frac{g\Gamma_a\Gamma_\beta}{\Gamma}\right)_{E_r}$$

We see that resonance contributions to the rates can be determined from experimental values of $(g\Gamma_a\Gamma_\beta/\Gamma)$ at resonance. If there are many resonances occurring in the energy region of interest for which experimental values of the resonance parameters are available, the rates can be computed as a sum of these contributions.

2.) Non-resonant reactions

At low bombarding energies and far away from resonances the cross sections for nuclear reactions involving charged particles can be written in the form

$$\sigma(E) = \frac{S}{E}\exp(-2\pi\eta)$$

where S is effectively a constant and the Coulomb parameter, η can be expressed in terms of the reduced mass and the charges of the two species:

$$\eta = \frac{Z_1Z_2e^2}{\hbar}\left(\frac{\mu}{2E}\right)^{\frac{1}{2}}$$

This limiting expression for $\sigma(E)$ may be derived readily from a consideration of the energy dependence of the particle widths. In this approximation $<\sigma v>$ can be written as:

$$<\sigma v> = \left(\frac{8}{\pi\mu}\right)^{\frac{1}{2}} \frac{S}{(kT)^{3/2}} \int_{E_t}^{\infty} \exp\left(-\frac{E}{kT} - \frac{(2\mu)^{\frac{1}{2}}\pi Z_1 Z_2 e^2}{\hbar E^{\frac{1}{2}}}\right) dE$$

The factor, S, can be determined by a fit to the experimental reaction cross section at somewhat higher energies.

This expression is not integrable in its present form. We note that the integrand is a sharply peaked function: at low energies the Coulomb penetration probability increases rapidly while at high energies the Maxwell distribution falls off exponentially. The usual procedure[8] is to approximate the integrand by a Gaussian form, which is integrable. The maximum of the integrand defines the 'optimum bombarding energy'

$$E_m = 0.122\,[Z_1^2\, Z_2^2\, \mu\, T_9^2]^{1/3}\ \text{MeV}.$$

where μ is in amu and T_9 is the temperature in units of 10^9 °K. The full width at half maximum of the Gaussian peak is given by:

$$\Delta E_m = .198[Z_1^2\, Z_2^2\, \mu\, T_9^5]^{1/6}\ \text{MeV}.$$

We have thus defined the energy region for which we can expect the maximum contribution to the integrand.

These approximations are found to be extremely good for interactions involving light nuclei at low energies. They find wide application in astrophysics due to the fact that hydrogen and helium thermonuclear reactions predominate for most of the lifetime of a star. Evolutionary models have not yet been carried in any detail past the carbon burning phase.

Detailed studies of the modes of synthesis of the heavy elements demand accurate determinations of the appropriate reaction rates. For these nuclei, $A > 28$, there are many resonances in the energy range of interest for which experimental parameters are not available. From studies of the average properties of the relevant nuclear parameters, Truran *et al.*[9] have obtained values of $<\sigma v>$ in good agreement with experiment both for charged particle reactions on medium nuclei and for neutron-capture reactions proceeding on nuclei in the mass range $A > 60$.

HYDROGEN REACTIONS

The first nuclear reactions which will take place during this history of shrinkage consist of the reactions which destroy the light elements, deuterium, lithium, beryllium, and boron. Since, as we shall see, it is an interesting point of contention as to how the deuterium in the earth and in the meteorites was made, then at the present time we do not know how much deuterium to expect in the gas which contracts out of interstellar space. If the amount of deuterium in this gas bears the same ratio to hydrogen as does the deuterium on the earth, then the stage of deuterium burning can last long enough for the contraction of the star to be halted in the first plateau of the stellar evolution. However, the abundance of deuterium is in any case likely to be relatively small, so that the deuterium burning phase of stellar evolution cannot last too long. The abundances of lithium, beryllium, and boron are in any case so small that the nuclear reactions that ensue when these nuclei combine with hydrogen, in order to form various isotopes of helium, cannot supply very much energy, and no perceptible halt in the process of contraction will occur. The principal point to be noticed here is that the light elements that we have been discussing are completely destroyed by the thermonuclear reactions and are not manufactured. These elements have a relatively low abundance in the solar system, and this is consistent with the obvious requirement that somewhat unusual and evidently rather rare reactions must be invoked for their manufacture.

It will now be useful to discuss those nuclear reactions which can occur in a star which is initially composed of pure hydrogen. Such stars may have existed in the earliest history of the galaxy, and hence it will be useful to see just how elaborate a set of elements can be produced in such a first generation star. Later we shall come back and see what additional reactions will take place with the other minor constituents that may exist in the star when it is formed.

When the central temperature in a star like the sun rises to a little more than 10^7 °K it is possible for the hydrogen to start reacting with itself. The basic reaction is the so-called proton–proton reaction, as follows:

$$H^1(p, \beta^+\nu)D^2.$$

In this reaction two protons must come together and, while they are within the range of nuclear forces, a beta decay must take place in which one of the protons is converted into a neutron with the emission of a positron and a neutrino, and with the neutron captured to form part of the ground state of the deuterium atom. This is an exceedingly slow process. If hydrogen were able to react with itself in an ordinary nuclear reaction, that reaction would certainly take place at a temperature in the vicinity of 10^6 °K. Under such conditions it would only be the protons very far out in the tail of the Maxwell distribution of particle velocities which would be able to penetrate the mutual Coulomb potential barrier between the protons and thus to come within the range of the nuclear forces. Such collisions are extremely rare, because of the fact that there are so very few protons far out in the high energy tail of the Maxwell distribution. When the extreme slowness of the actual nuclear reaction involved, a beta decay rather than a particle or photon emission process, is also considered, then it turns out that the temperature has to be raised by an order of magnitude in order that the rate of collision between the protons, which brings the particles to within the range of the nuclear forces, is no longer a rare event, but may occur for particles which are much nearer the peak of the Maxwell energy distribution. Then it is only the slowness of the intrinsic beta decay process which governs the rate. The rate of an ordinary nuclear process under astrophysical conditions increases approximately as the 20th power of the temperature, but in the case of the proton–proton reaction such high temperatures, relative to those which would normally suffice, are required, that this is not true, and the rate of the proton–proton reaction varies more nearly as the fourth power of the temperature. The neutrino which is emitted in this reaction escapes easily from the interior of the star, and carries away with it part of the energy released by the reaction, but this reaction is still a source of energy for the interior of the star because of the energy carried out by the positron, followed by the energy release when the positron annihilates to form two photons, which do not go very far in the stellar interior before being absorbed.

The deuteron that has been formed in the hot environment of the stellar interior by the proton–proton reaction has, like the proton itself, only a single nuclear charge. Consequently, the combination of this deuteron with one of its neighboring protons is a process that can take place with very great rapidity, in about one second in the case of the solar interior, since the reaction is a straightforward capture with the emission of a gamma ray:

$$D^2(p, \gamma)He^3.$$

Since we are considering material that is initially all hydrogen nothing more than this will happen for some period of time. However, after the proton–proton reactions have been taking place for some time, enough He^3 will be built up in the interior of the star so that it can start reacting with itself. The reaction is:

$$He^3(He^3, 2p)He^4.$$

This completes the basic process of converting hydrogen into helium in the deep interior. These reactions by themselves would suffice.

However, things are actually a little more complicated than this. Again after some period of time, a significant amount of He^4 will accumulate, and then, at least in the more massive stars, a new reaction can take place which is somewhat faster than the one we have just written. This reaction is:

$$He^3(\alpha, \gamma)Be^7.$$

Ordinarily, in the laboratory, Be^7 is a naturally radioactive nucleus which decays by the capture of one of its electrons from the K-shell to produce Li^7. However, here we deal with the deep interior of a star, and such atoms as those of beryllium are stripped of their innermost electrons owing to the high temperatures present. Hence there are no K-shell electrons available to be captured. However, the nucleus is surrounded by a large number of free electrons, and these are just about as effective. The laboratory lifetime of Be^7 is 53 days, and under typical conditions in the interior of a star the Be^7 will capture a free electron with a mean time of the same order of magnitude. The reaction is:

$$Be^7(e^-, \nu)Li^7.$$

The Li^7 that is produced by this reaction is relatively rapidly destroyed in the interior by reactions with protons. The reactions consist of:

$$Li^7(p, \alpha)He^4.$$

We notice here that we get back the alpha particle that participated in the original reaction with the He^3, and now we have an additional alpha particle as well. Hence we can consider that the original alpha particle that reacted with the He^3 played the role of a catalyst.

At somewhat higher temperatures the proton capture lifetime for Be^7 will be comparable to the electron capture lifetime.

$$Be^7\ (p,\ \gamma)\ B^8$$

The B^8 nucleus formed in this manner will decay readily to an excited Be^8 nucleus via positron emission:

$$B^8\ (\beta^+\ \nu)\ Be^{8*}$$

This reaction sequence is terminated by the breakup of the unstable Be^{8*} to two alpha particles. Thus, as with the previous sequence, we can consider that the original alpha particle reacting with He^3 served as a catalyst, and the net result of this third proton-proton sequence is the fusion of four protons to one alpha-particle.

For stars more massive than $\sim$ 2 solar masses ($M_\odot$), the initial contraction will continue until the central temperature exceeds 1.7×10^7 °K. At these temperatures, hydrogen burning by the carbon-nitrogen-oxygen cycle may provide more energy generation than do the proton-proton chains. In these reactions, the carbon and nitrogen present in the medium act as catalysts for the conversion of hydrogen to helium, as illustrated in Table 1. Further, for stars of several solar masses, which burn their central hydrogen at temperatures of $\sim 3 \times 10^7$ °K, the isotopes of oxygen and F^{19} will be largely transformed into CNO-cycle nuclei. Calculations of the abundances of carbon, nitrogen and oxygen in equilibrium for the CNO cycle predict that N^{14} will be the most abundant nucleus.

This sequence of reactions is clearly not relevant to discussions of first generation pure hydrogen (or hydrogen and helium) stars. However, for stars possessing a solar abundance of carbon and nitrogen, energy generation by this process must be taken into account. Reeves[10] has given accurate expressions for the rates of energy generation for the proton-proton cycles and the CNO cycle. Employing these expressions we have plotted the energy generation rates as a function of temperature for a star containing solar concentrations of hydrogen, helium, carbon and nitrogen in Figure 1. The density was chosen to be 10 gm/cm^3.

TABLE I

Hydrogen burning reactions: the CNO cycle. (The arrows indicate a recycling of CNO nuclei.)

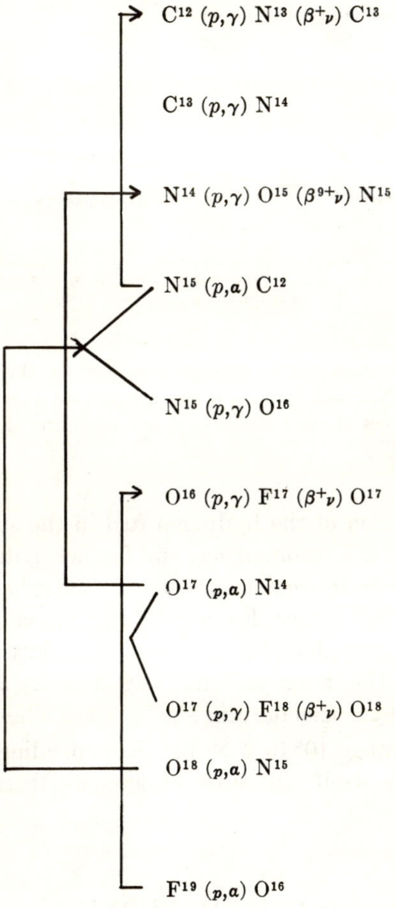

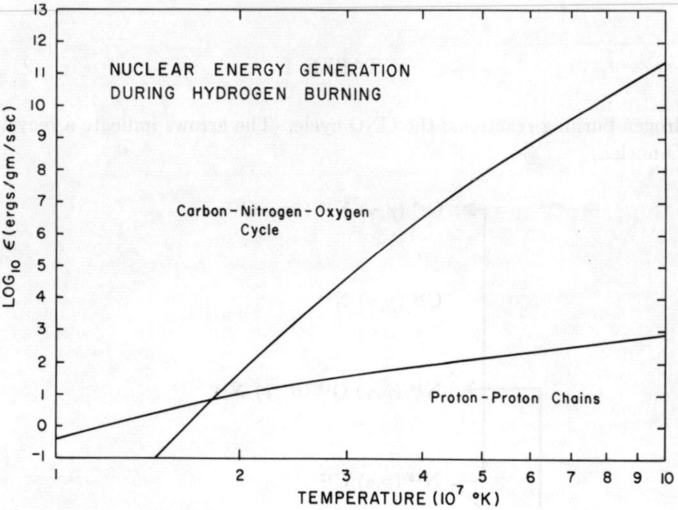

FIG. 1. The energy generation rates (ergs/gm/sec) for the Proton-Proton chains and for the Carbon-Nitrogen-Oxygen cycles are plotted as a function of temperature. Solar abundances of the elements are assumed, and the density is taken to be $\rho = 10$ gm/cm^3.

Following the exhaustion of the hydrogen fuel in the stellar core, the shrinkage of the star will recommence, and further hydrogen may be converted into helium in the region surrounding the exhausted core in a hydrogen burning shell source. Eventually the region of the central helium will be large enough and it can contract rapidly enough so that the temperature at the center will climb significantly above that which characterizes the hydrogen burning shell source. When this temperature climbs to the range 10^8 to 2×10^8 °K, depending upon the stellar mass, the helium itself will start undergoing thermonuclear reactions.

HELIUM AND CARBON BURNING

The basic helium reaction consists of the combination of three α-particles to form a C^{12} nucleus:

$$3He^4 \rightarrow C^{12}.$$

There is an interesting story behind this reaction. Nearly a decade ago Fred Hoyle[11] presented an argument in which he claimed that the above reaction should be greatly accelerated through the presence of a resonance in the C^{12} nucleus at an energy of 7.68 MeV. His argument was based upon the assumption that thermonuclear reactions in stellar interiors have indeed been responsible for the formation of the elements, and that the triple alpha reaction must have been the mechanism by which C^{12} was formed. However, once C^{12} is formed, it is vulnerable to destruction by the reaction:

$$C^{12}(\alpha, \gamma)O^{16}.$$

Hoyle noted that the formation of C^{12} from three α-particles was a two-step process in which the product of the intermediate step, Be^8, was an unstable nucleus. It decays back into two α-particles in an extremely small fraction of a second. Hence in effect very nearly a triple collision of the α-particles is required, and this must be an intrinsically slow process. On the other hand, the capture of an α-particle by C^{12} is a normal nuclear reaction which may be expected to go at a reasonably normal rate. Hence, if the triple alpha reaction were a normal non-resonant reaction, then the formation of C^{12} would, under ordinary conditions in stellar interiors, take place much less rapidly than its destruction by additional α-particles to form O^{16}. Since it is unreasonable to suppose that the rate of combination of α-particles with C^{12} is very much less than that of a normal reaction rate, it becomes logical to assume that the triple alpha reaction is greatly increased in rate by a resonant process. The energy that Hoyle suggested for the resonance is that which one gets by adjusting the resonance so that the rate of formation of C^{12} will be suitably rapid compared to the rate of its destruction.

This prediction of Hoyle's stimulated several people at the California Institute of Technology to determine whether there was in fact a resonance near 7.68 MeV. They determined without question that such a resonance exists, and that it can decay into three α-particles, and hence, from the principle of reciprocity, it can be formed from a combination of three α-particles. The energy measured for the resonance was 7.656 MeV, satisfactorily close to Hoyle's prediction.

In subsequent years a long series of ingenious experiments has been carried out to measure the nuclear properties of the resonance at 7.656 MeV in the C^{12} nucleus. Hence at the present time the rate of the triple alpha reaction is known to an accuracy of better than 50 percent. Unfortunately, however, the rate of the next reaction,

$$C^{12}(\alpha, \gamma)O^{16},$$

is not yet known. Various estimates have been made for this rate, but these estimates could still easily be incorrect by a factor of ten or more. Consequently, we still do not know with any degree of accuracy what the relative rates of formation and destruction of C^{12} will be in a stellar interior, other than that these rates are generally of the same order of magnitude. It seems likely that the helium burning process will leave some $C^{12}$and some O^{16} under typical stellar conditions. The greater sensitivity of the rate of the triple-alpha reaction to the abundance of helium, however, suggests that if helium burning goes to completion O^{16} should be the major product. The rate of capture of α-particles by O^{16} to form Ne^{20} has been shown to be negligible under ordinary stellar conditions.

Thus at the end of helium burning we may expect the center of the star to be composed predominantly of carbon and oxygen. We are now beyond the stage which has been explored by the construction of theoretical stellar models. We must proceed to discuss the subsequent evolution of the material from very general physical principles rather than from precise knowledge of models.

CARBON AND OXYGEN BURNING

After helium burning the central regions of a star will once again contract. In general the difficulty of penetrating nuclear Coulomb potential energy barriers results in the favoring of those reactions which can take place involving the smallest product of charges of the bombarding and target nucleus. This leads us to expect that the next reaction to take place will be that of C^{12} with itself. Indeed, this turns out to be the case. When C^{12} reacts with itself, two exothermic reactions are possible in which particles are emitted. These are:

$$C^{12} + C^{12} \rightarrow p + Na^{23}$$

$$\rightarrow \alpha + Ne^{20}.$$

These reactions take place at a temperature somewhat in excess of 6×10^8 °K, a point which will be discussed in more detail shortly.

There is also a weak endoergic branching to $Mg^{23} + n$, but the direct effects of this neutron emission are of minor importance at these temperatures.

At slightly higher temperatures ($T \sim 10^9$ °K) oxygen burning by $O^{16} + O^{16}$ is also possible. The dominant reactions in this case are

$$
\begin{aligned}
O^{16} + O^{16} &\rightarrow p + P^{31} \\
&\rightarrow n + S^{31} \\
&\rightarrow \alpha + Si^{28}
\end{aligned}
$$

where again the neutron branching is rather weak. Experimental data for these reactions is quite limited, but it is probable that the proton and alpha-particle yields are comparable.

The protons and α-particles that are released in these reactions find themselves in an environment which is very much hotter than that in which the hydrogen and helium was initially destroyed in the stellar interior. Consequently, the protons and α-particles will be very rapidly absorbed by some of the nuclei that are present. Two very obvious reactions would involve the formation of Mg^{24}:

$$Na^{23}(p, \gamma)Mg^{24}$$

and

$$Ne^{20}(\alpha, \gamma)Mg^{24}.$$

These reactions will be very important, but also the protons and α-particles may react with other nuclei present, and quite a wide variety of possible reaction products is actually possible. We shall not try to discuss the details of these here, other than to note that a number of nuclei in the mass range $24 \leq A \leq 32$ can be formed by the carbon and oxygen reactions.

Even neutrons can be formed in an indirect way by the carbon thermonuclear reactions. During the early stages of carbon burning, the protons that are produced find it easy to react with the unburned carbon:

$$C^{12}(p, \gamma)N^{13}.$$

The N^{13} nucleus is naturally radioactive, and, with a half-life of ten minutes, it emits a positron and a neutrino and becomes C^{13}:

$$N^{13}(\beta^{+}\nu)C^{13}.$$

The C^{13} can then react with some of the α particles that are produced by carbon burning to form a neutron:

$$C^{13}(\alpha, n)O^{16}.$$

Quite large numbers of neutrons are produced as long as the temperature at which carbon burning takes place is in the vicinity of 6 to 7 × 10^{8} °K.

However, if the temperature becomes 8 × 10^{8} °K or higher, a difficulty arises which interrupts the production of neutrons in this fashion. The N^{13} nucleus has only a very small binding energy for the proton which was captured by C^{12} to form it. At the same time the temperature is now so high that there are a significant number of photons at quite high energies in the tail of the Planck distribution. These photons can act to reverse the reaction which produced the N^{13}, namely to produce a photodisintegration process:

$$N^{13}(\gamma, p)C^{12}.$$

The higher the temperature, the more rapidly this reaction takes place. Already at 7 × 10^{8} °K. it is interfering significantly with the production of neutrons, and by 8 × 10^{8} °K it has virtually suppressed the whole process. This occurs because the capture of protons by C^{12} is not allowed to progress through to the formation of C^{13}, and the protons that are re-injected into the interior of the star by this *photodisintegration* process have a good chance of being captured by one of the other types of nuclei that are present. However, at the same time the direct production of neutrons by the endothermic process;

$$C^{12}(C^{12}, n)Mg^{23},$$

becomes possible. However, this can only take place providing the relative energy of the bombarding particles in the interior is greater than the threshold of 1.6 MeV, and even then only about one-tenth of the reactions between the C^{12} nuclei are expected to produce neutrons, so that this is not expected to be a very prolific source of such neutrons. We shall discuss what happens in general during the capture of these neutrons later. In the present discussion they would all be captured by the nuclei which are in the vicinity of Mg^{24}.

NEUTRINO PROCESSES

With the onset of carbon burning, the temperature in the interior of the star is sufficiently high that a probable new energy loss mechanism comes into play. This consists of the formation of pairs of neutrinos and antineutrinos deep in the interior of the star, which can escape into space without hindrance, and which carry with them large amounts of energy.

There are a variety of processes which are probably responsible for the creation of these *neutrino–antineutrino pairs*, but perhaps the simplest to explain is the process of *electron pair annihilation.* Just as the high energy tail of the Planck distribution of photons in the star under these conditions is intense enough to provide for rapid photodisintegration of the N^{13} nucleus, so also it is intense enough to create significant numbers of positron–electron pairs in the interior of the star. Ordinarily the positrons created as a part of these pairs re-annihilate with the creation of two photons:

$$e^+ + e^- \rightarrow 2\gamma.$$

However, the modern theory of beta decay and the associated theory of weak interactions now strongly suggest that an alternate mechanism of the annihilation of the positron with an electron to form a neutrino–antineutrino pair is possible:

$$e^+ + e^- \rightarrow \nu + \bar{\nu}.$$

The relative rates of these two types of annihilation process will be in approximately the ratio of the electromagnetic to the weak coupling constants. This means that the neutrino emission process will occur with the probability of roughly twenty orders of magnitude less than the probability for the annihilation of the electron–positron pair to form two photons. This very small probability for the formation of the neutrino–antineutrino pair does not, however, imply that such decay cannot have important astrophysical consequences. The photons which might be produced move only a centimeter or less in the deep stellar interior before being absorbed by the surroundings. On the other hand, the neutrinos escape immediately into outer space. An extremely large number of absorptions and re-emissions of photons is necessary to transport energy from the interior to the surface of the star. This very large number of absorption and re-emission processes is to be compared to the immediate escape of the neutrino, and when the temperature becomes high enough, the neutrino wins out. The rate of emission of energy in the form of neutrino–antineutrino pairs from the deep interior of the star will start to exceed the rate of radiation of electromagnetic quanta from the surface. This will have the effect in general of greatly increasing the rate at which the center of the star contracts, since the emission of the neutrino–antineutrino pairs from the deep interior does not actually cool the interior, but induces it to contract faster and faster, so that additional amounts of energy are stored as internal thermal energy and the interior becomes hotter and hotter. In general, any thermonuclear reaction taking place in the deep interior of the star must greatly speed up in order to supply, not only the energy that is being radiated from the surface in the form of protons, but also the energy from the deep interior which is being radiated as neutrino–antineutrino pairs. This will require relatively high temperatures for the thermonuclear reactions, and the time scale of such thermonuclear reactions will become quite short.

Actually there are a variety of other reactions which can also result in the emission of neutrino–antineutrino pairs. The rate of any quantum electrodynamic process which results in the emission of a photon can be calculated from the assumption that, during an extremely short interval of time, a virtual electron–positron pair is created which annihilates to form the photon emitted by the process. Hence any process which results in the emission of a photon has an analogous process in which a neutrino–antineutrino pair can be emitted. For our present considerations the most important of these processes is the photo-neutrino process:

$$\gamma + e^- \rightarrow e^- + \nu + \bar{\nu}.$$

During the stage of the carbon thermonuclear reactions, this is the reaction which gives the predominant energy loss by neutrino–

antineutrino emission. It will cause carbon burning to take place at a relatively rapid rate, and the temperature at which the carbon burning will take place may even exceed 10^9 °K.

Another process which can be important under these conditions is plasmon neutrino emission. While both classical radiation theory and quantum electrodynamics predict that a free photon cannot create electron-positron pairs, this is not the case for electromagnetic waves in a plasma. Thus, the reaction

$$\gamma \rightarrow \nu + \bar{\nu}$$

is theoretically possible.

In Figure 2, the regions of dominance of the three processes we have discussed are illustrated as a function of temperature and density.[12]

THE APPROACH TO THE IRON EQUILIBRIUM PEAK

It should be pointed out that by now we are dealing only with very massive stars, two or three or more times the mass of the sun. Stars of solar mass are unable to achieve central temperatures high enough

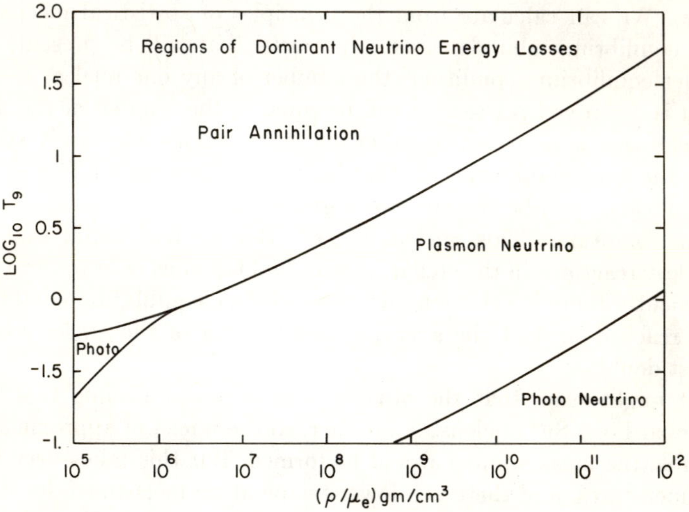

FIG. 2. The regions of dominance of the neutrino energy loss mechanisms we have considered are shown as a function of temperature and density.

to progress much beyond the carbon burning stage before settling down as white dwarf stars.

These more massive stars now evolve very quickly, owing to the enormous flood of radiation in the form of neutrino–antineutrino pairs which is streaming out of their deep interiors. There is some uncertainty, in lieu of this energy loss, as to whether the carbon and oxygen burning stages will go to completion. We will assume, for the moment, that carbon and oxygen burning continue uninterrupted to completion. We have found that, generally, the reaction products should be nuclei with masses in the range $20 \leq A \leq 32$, particularly the alpha-particle nuclei Mg^{24}, Si^{28} and S^{32}. Of the nuclei in this region, Si^{28} is the most stable against nuclear photodisintegrations. As the temperature is further increased, Si^{28} will be the last nucleus to photodisintegrate and will therefore accumulate. It can therefore be assumed that the material consists mainly of Si^{28} at temperatures $T \geq 2 \times 10^9$ °K.

In order to understand the next set of nuclear reactions that takes place, we must look forward a little bit. We are approaching a situation which is described as *nuclear statistical equilibrium*. Under equilibrium conditions, the rates at which nuclear reactions take place are extremely fast, but we are not directly concerned with these rates. We can calculate from the principles of statistical mechanics the equilibrium abundances of the nuclei that will be present, and under equilibrium conditions the number of any one nuclear species that is destroyed per second will be equal to the number of the same species that is reformed by other nuclear reactions. Generally speaking, the most stable nucleus, that having the greatest binding energy per nucleon, will be the nucleus of greatest abundance under equilibrium conditions. This nucleus is Fe^{56}. But we had found that the nuclear reactions in the stellar interior had taken us only to form Si^{28}. Evidently there is yet a long way to go before equilibrium conditions are reached; in fact the average mass number of the nuclei present must double.

It might seem that the obvious way of accomplishing this is to bombard one Si^{28} nucleus by another, so the nuclei of approximately double the mass number would be formed. But this takes very high temperatures, and there is a lower-temperature mechanism for doing it which is nevertheless quite a good deal more complicated. This mechanism involves the operation of a large number of photodisintegration reactions.

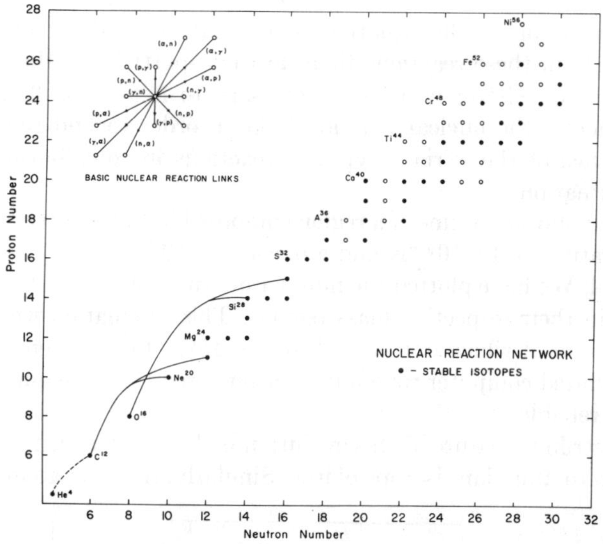

FIG. 3. The network of nuclear reaction links used in the calculation in which we followed the transformation of Si^{28} to Fe^{56}.

The situation is illustrated by Figure 3. This shows a section of a nuclide chart, with those nuclei which are stable in the laboratory indicated by closed circles and those which decay by beta emission to their neighbors indicated by open circles. A key to these nuclear reaction links is shown in the upper left portion of the figure. Each reaction link represents in fact two reactions: a forward reaction and its inverse. This reaction chart corresponds to that used in a calculation done by the writers[13] in which they followed the transformation of Si^{28} into iron peak nuclei.

The essentials of the transformation are as follows: Photodisintegration reactions must knock light nuclear particles out of about half of the Si^{28} nuclei that are present. Thus, for example, Si^{28} could lose an α-particle leading to Mg^{24}, this in turn could lose an α-particle leading to Ne^{20}, and a further loss of two α-particles would produce in turn O^{16} and then C^{12}. This in turn can be broken up to form three additional α-particles. These α-particles can be captured by the remaining Si^{28} nuclei to build them up toward the iron region.

This oversimplifies the situation, since neutrons and protons play a larger role in these reactions than do the α-particles in the build-up process from Si^{28} to iron. It is necessary to follow the flow of the nuclei across the nuclear reaction links in order to find the relative importance of the various types of reactions in contributing to the transformation.

The evolution in time of a region composed initially of pure Si^{28} at a temperature of 3×10^9 °K and a density of 10^6 gm cm^{-3} is shown in Figure 4. We have plotted the number densities of the most abundant nuclei in their respective mass regions. This calculation was carried to the point at which iron constituted $\sim 15\%$ of the system by mass; accumulated computer roundoff error rendered the subsequent evolution unreliable.

The striking feature of this evolution is that Fe^{54}, rather than Fe^{56}, is the most abundant isotope of iron. Similarly, the most abundant iso-

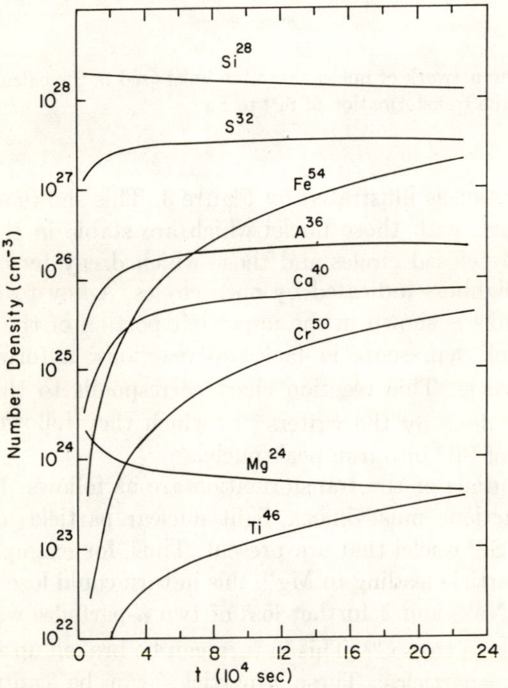

FIG. 4. The number densities of the more abundant isotopes are traced as a function of time during the Si^{28} to iron peak transformation.

tope of titanium and chromium, Ti^{46} and Cr^{50}, are not the most abundant isotopes observed in nature. This unusual abundance pattern can be understood in the following manner. Of the iron isotopes, Fe^{56} is favored by nuclear stability as it has the maximum binding energy per nucleon. However, we have started with a region composed completely of Si^{28}, for which the total numbers of protons and neutrons in the system are equal. In the absence of beta decays, this would favor the production of nuclei with $Z = N$ throughout the network: "for our conditions Ni^{56} should be the most abundant nucleus in equilibrium. However, in our calculations nuclear positron decays have resulted in an enhancement of neutrons relative to protons ($N/Z \approx 1.002$) sufficient to favor Fe^{54} in equilibrium."

There is a mechanism which might further enhance the number of neutrons relative to protons at somewhat higher temperatures—electron capture by free protons. Even for our present conditions ($T = 3 \times 10^9$ °K) the electron capture lifetime for free protons is of the order of the evolutionary time scale ($\sim 10^5$ seconds). However, the free proton number density was only $\sim 10^{-5}$ of the total particle number density and the total neutron to proton ratio was substantially unchanged. For a temperature of 5×10^9 °K, the free proton number density will be comparable to the total particle number density, but the relaxation time scale ($\sim 10^{-2}$ seconds) is several orders of magnitude less than the electron capture lifetime (>10 seconds). However, in the intermediate temperature region ($T \sim 3.5$–4×10^9 °K) the free proton number density will be substantially increased over that for $T = 3 \times 10^9$°K and the electron capture lifetime will be comparable to the silicon-to-iron conversion time scale—the enhancement of neutrons relative to protons under these conditions may be sufficient to result in an iron peak abundance pattern in better argeement with that observed in nature.

THE LATE STAGES OF EVOLUTION

After nuclear statistical equilibrium is reached in the deep interior of a star, no further nuclear energy release can take place, since the nuclei that are present already contain the greatest binding energy per nucleon of any that are available. Iron represents the bottom of the nuclear packing fraction curve; energy can be released either by building lighter nuclei up toward it or by fissioning heavier nuclei down toward it. But by this time the temperature in the deep interior has become quite high, 3 to 4 $\times 10^9$ °K, and the rate of energy loss by the emission of neutrino–antineutrino pairs is so great that the time scale associated with the shrinkage of the interior of the star has been reduced to about 1,000 seconds. And this rate of energy loss by neutrino emission will not be shut off just because the star has exhausted its nuclear energy sources. It will continue unabated and with an ever-increasing rate as the shrinkage of the stellar interior raises the central temperature. We are now fast approaching the catastrophe of dynamic instability at the center of the star.

Two types of instability are possible. It is apparent that one instability will trigger the other, and then both will contribute to the dynamical collapse of the star. The first of these types of instability is illustrated in Figure 5. This figure shows the nucleus which statistical mechanics predicts to have the largest abundance under equilibrium conditions. At lower temperatures and higher densities we see that the nucleus of greatest abundance is our old friend, Fe^{56}. However, at lower densities and higher temperatures statistical mechanics tells us that Fe^{56} is no longer the nucleus of greatest equilibrium abundance, even though it may have the greatest binding energy per nucleon. Higher temperatures favor nuclei of lower mass number, and eventually He^4 takes over as the nucleus of greatest equilibrium abundance. Actually, the region of transition from Fe^{56} to He^4 as the nucleus of maximum abundance is a rather sharp one. In Figure 5 the combinations of temperature and density at which the assembly of material contains the same masses of Fe^{56} and of He^4 is denoted by the line. The abundance ratio of these two substances changes rapidly as one crosses the line.

If the combinations of temperature and density which exist in the deep interior of the star should appear to the right of this line in the composition diagram of Figure 5, then statistical mechanics will require that the nucleus of greatest abundance in the interior change from Fe^{56} to He^4. However, in the preceding evolution of the star a great deal of energy has been released by the inverse transformations, in which He^4 has been evolved into Fe^{56}. We can understand, in terms of the nuclear reaction network diagram of Figure 3, how the breakdown of Fe^{56} into He^4 can occur. However, we must ask where the necessary energy would be obtained. The star has exhausted its nuclear energy resources, and hence the only source of energy which

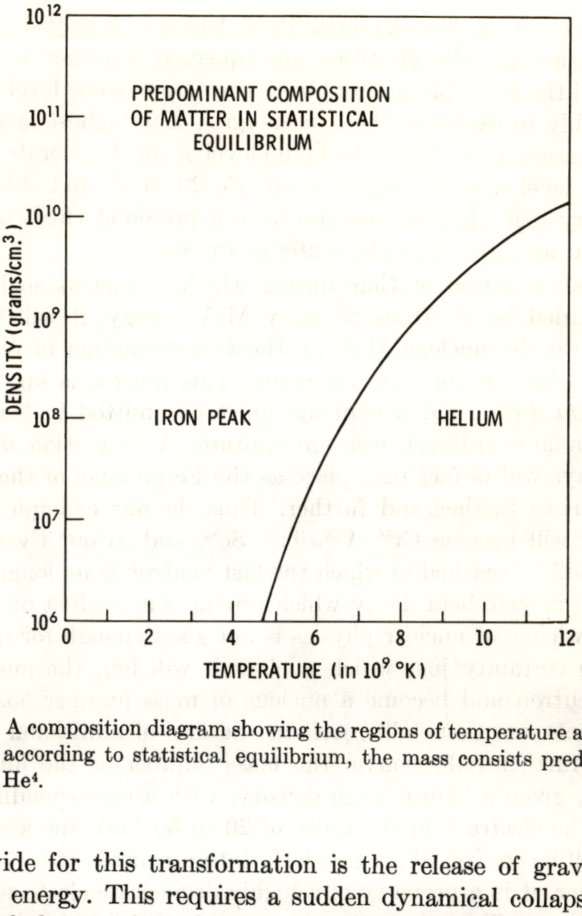

FIG. 5. A composition diagram showing the regions of temperature and density in which, according to statistical equilibrium, the mass consists predominantly of Fe^{56} or He^4.

can provide for this transformation is the release of gravitational potential energy. This requires a sudden dynamical collapse of the interior of the star in order to release the additional gravitational potential energy to effect the transformation.

The second type of instability is that which is associated with enormous density. As the center of the star shrinks and the central temperature rises, it turns out that the rate of energy loss by neutrino–antineutrino pair emission rises as a fairly large power of the temperature. Chiu[14] has recently shown that the central regions of the star will tend to become extremely dense under these conditions, and that the highest temperatures will be postponed as long as possible. In other words, a condensation of very high density forms at the center of the star. The material there becomes highly degenerate. This means that the electrons are squeezed together until they occupy all the available cells of phase space up to some level which is considerably in excess of thermal energies. The highest level of occupied phase space is called the Fermi level of the degenerate electron gas. This level mounts higher: 5, 10, 15, 20 MeV, and still higher. These very high electron energies have a profound effect upon the nuclei that are present at the center of the star.

After some period of time during which a nucleus such as Fe^{56} is bombarded by electrons of many MeV energy, it will be transformed into the nucleus Mn^{56} by the transformation of one of the protons in its interior into a neutron. This process is known as an *inverse beta decay*, and a neutrino must be emitted in the process just as in any ordinary electron capture. A succession of inverse beta decays will in fact take place as the Fermi level of the electron energies rises further and further. Thus, in our example, in turn the Mn^{56} will become Cr^{56}, V^{56}, Ti^{56}, Sc^{56}, and so on. Eventually a nucleus will be reached in which the last neutron is no longer bound. Upon the inverse beta decay which produces a product of this kind (our knowledge of nuclear physics is not good enough for us to predict with certainty just which nucleus it will be), the nucleus will lose a neutron and become a nucleus of mass number 55. Further inverse beta reactions will result in the loss of additional neutrons and the further reduction in the mass number of the nucleus. In this way, given a high enough density, with a corresponding Fermi level of the electrons in the range of 20 to 30 MeV the assembly of nuclei will be converted progressively into a neutron gas.

However, it is a property of a highly degenerate electron gas that the electrons will produce the greatest contribution to the internal pressure in the star. This is a consequence of the high velocities associated with their high energy, and the corresponding high transport of momentum. But we see that these electrons are being progressively removed by the inverse beta processes which transform the nuclei into neutrons. Consequently, the pressure at the center of the star will fail to remain high enough in order to hold up the overlying layers, and in this way, also, a dynamical collapse of the interior can begin. Such a collapse will indeed accelerate, since as the central region progresses to still higher densities, the rate of the inverse beta processes accelerates, and hence the electrons are removed at a faster and faster rate.

It is important to note at this point the implications of an equilibrium peak centered on Fe^{54}. This may be realized at somewhat higher temperatures ($T_9 \gtrsim 4$) in the absence of beta decay where it implies that the net silicon burning sequence is $2Si^{28} \rightarrow Fe^{54} + 2p$, which is endoergic by 1.3 MeV. In our calculation at $T_9 - 3$ the Fe^{54} resulted from a slight neutron enrichment due to positron decay in which a substantial fraction of the energy released may be carried by the neutrino. In either case, the silicon-to-iron reaction stage may not provide a sufficient energy source and the collapse of the core may again be triggered by the removal of electron degeneracy pressure.

The dynamical time scale associated with the collapse of the interior is exceedingly short, a matter of some milliseconds only. The collapse will result in the release of a great deal of gravitational potential energy. There is no possibility for central pressures high enough to withstand the weight of the overlying layers to once again be reached until the density has been raised beyond the density of matter in the ordinary interiors of nuclei. Normal nuclear density is about 3×10^{14} gm/cm^3, but sufficiently high pressures will not be attained again in the interior of the star to support the overlying layers until the density is raised to the region of 10^{15} gm/cm^3 or still higher. This is the region in which the neutron gas itself becomes degenerate. At this time the rise of the internal pressure will slow the rate of collapse of the star, thus releasing a great deal of thermal energy in the interior, and it is likely that this internal energy can go into the formation of an extremely intense shock wave. Indeed, recent investigations by Colgate and White[15] and by Arnett[16] suggest that the energy which goes into the shock wave is initially propagated out of the interior of the star by means of the emission and absorption of neutrinos and antineutrinos deep in the interior. We have now achieved matter densities so exceedingly high that the neutrinos can no longer freely escape from the center of the star, but will make several interactions on the way out.

We are not really concerned here with the ultimate fate of the material at the center of the star. One hypothetical possibility is that this material may form a new kind of degenerate configuration called a neutron star in which the internal pressures are provided by the degenerate gas of neutrons to which we have already alluded. However, there is an upper limit to the mass of a stable neutron star, which is uncertain but appears to lie in the range one to two times the mass of the sun. If the mass of the imploded remnant of the star should exceed this limit, then it appears that the central portion of the star must go into a configuration which will be indefinitely collapsing, without the possibility that a stable configuration will be reached. Such an object poses many fascinating problems to the theoretical physicist, but we shall not discuss it further here.

The outgoing shock wave in the envelope of the star has potentially an interesting role to play in nucleosynthesis. This shock wave will typically raise the temperature of the matter it traverses to a few times 10^9 °K, and in so doing it will accelerate the matter to velocities exceeding the velocity of escape from the star. Thus a large fraction of the mass of the star can be ejected in an event of this type, and it is evident that we are discussing a *supernova* explosion, in which such large mass ejections are observed to occur. At these high temperatures which take place in the gas behind the shock front which traverses the envelope material, proton and alpha particle thermonuclear reactions can take place quite quickly. Even carbon thermonuclear reactions take place rapidly. However, the basic proton–proton reaction and the triple alpha reaction will not be increased very much in rate during the passage of a shock wave, because of the fact that their rates are limited by weak interaction rates in the case of the proton–proton reaction, or by density effects in the case of the triple alpha reaction.

Studies of element synthesis in supernova shock waves[17] have revealed that regions composed initially of nuclei from C^{12} to Si^{28} can be processed readily to nuclei in the vicinity of the iron peak. The condition that the total numbers of protons and neutrons are equal still presents a difficulty, as both beta decay lifetimes and the electron capture lifetime for free protons are orders of magnitude longer than the nuclear reaction time scale at temperatures in the vicinity of 5×10^9 °K. This suggests that if an iron peak resembling that observed in nature is to be formed in supernova explosions, some neutron enrichment of the material must take place before the passage of the shock wave. This may take place in the regions immediately surrounding the core, if the material is not broken down to neutrons, protons and alpha-particles.

ABUNDANCE OF THE ELEMENTS

We have in the preceding discussion talked principally about the conditions at the center of the star, and the progressive increase in temperature and density, accompanied by more and more exotic nuclear reactions, which take place there. But it is evident that the surface of the star has remained relatively cool and has not been the site of thermonuclear reactions. Hence one can expect that there will be a gradation of temperature throughout the interior of the pre-supernova configuration along which occurs the complete range of compositions starting from hydrogen and going through Fe^{56}. The succession of compositions will correspond to those that we have discussed in connection with the succession of nuclear transformations. This set of compositions may receive certain modifications that result from the passage of the supernova shock wave, but then they will be ejected into space and subsequently will mix with the interstellar medium. Thus we are able to arrive at a general kind of prediction as to the type of nuclei which should be most abundant in nature. Hence at this point it is instructive to examine an abundance diagram of the nuclides as observed in nature.

Figure 6 shows a plot of the nuclide abundances as a function of their mass number. The reader should be warned that this diagram was produced by something of a circular argument, but nevertheless it is most useful for illustrative purposes. The source of the abundance data consisted of the abundances of elements as measured, mainly by radiochemical techniques, in stony meteorites, principally the chondritic meteorites, or when possible the carbonaceous chondritic meteorites. Some of the abundance data has come from measurements of abundances in the sun or other stars. Some of the points represent interpolations of the data (that for the volatile elements, notably Ar, Kr, and Xe) which have been made with specific mechanisms of nucleosynthesis in mind. This diagram shows a number of features related to processes of nucleosynthesis quite clearly, much more clearly than was present in the diagrams based on much smaller amounts of analytic and spectroscopic data available several years ago.

We notice first that the most abundant element or nucleus is hydrogen. About three quarters of the mass of the sun appears to be composed of hydrogen, and most of the remaining quarter of the mass of the sun is composed of He^4. As we discussed previously, the light elements deuterium (and also He^3), lithium, beryllium and boron are all extremely low in abundance; we have not found them to be formed as products in the stellar interior. Helium-burning

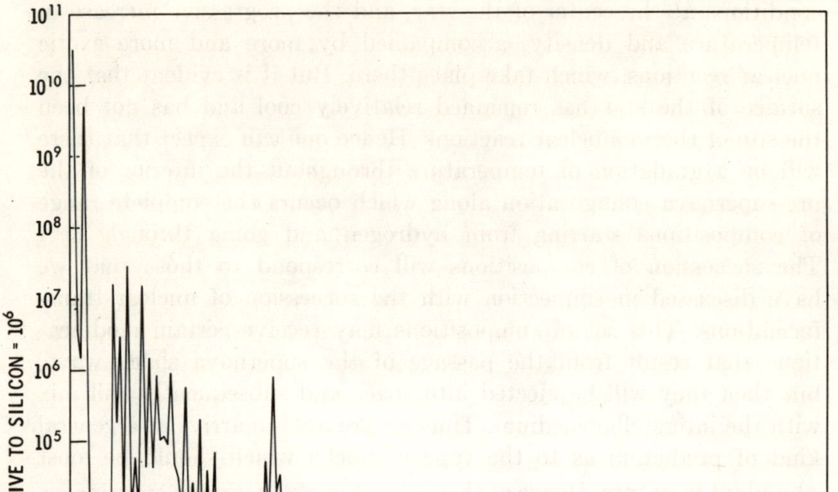

FIG. 6. The abundances of the nuclides plotted as a function of mass number.

makes C^{12} and O^{16}; we see that these are very abundant, and also the intermediate nuclide N^{14} has a rather high abundance; this, as we have seen, results from hydrogen burning by the CNO cycle. There then follows a gradual decline in the abundance as we go to higher and higher mass numbers and approach the condition of statistical equilibrium. This range includes Ne^{20}, Mg^{24}, S^{32}, and so on. Then there is a huge abundance peak centered about Fe^{56}. We associate this peak with the process of nuclear statistical equilibrium, which we shall discuss in more detail shortly. Essentially all of the features so far discussed were found to be suggested by the history of evolution of a star initially composed of pure hydrogen which we have gone through above.

But we see that the abundances of the elements do not drop to zero immediately beyond the iron peak. The general abundance level falls continually until about mass number 100 is reached, and above this point the level wanders up and down a bit but it is at roughly the same level until mass numbers greater than 200 are reached, and we reach a final peak at the positions of lead and bismuth. Then there is a gap at the positions of very short-lived radioactive species, and some naturally occurring longer-lived radioactive species in the form of isotopes of thorium and uranium. We have not found processes in the evolution of the first generation pure hydrogen star that will produce these heavy elements. Evidently we must look at the second and subsequent generations of stars to see if additional nuclear reactions can take place that will suffice.

Let us next look at the iron peak in more detail. Figure 7 shows the result of an important calculation originally carried out by Burbidge, Burbidge, Fowler, and Hoyle. The plot shows a calculation of the abundances of nuclides that would be produced under equilibrium conditions in the vicinity of Fe^{56}. This is compared to the abundances of the elements as measured in chondritic meteorites, with the exception of the abundance of iron which has been reduced from the meteoritic value to the approximate value that is found in the sun. The isotope ratios in each element are, of course, those which are measured for terrestrial elements. It may be seen that there is in general a striking agreement between the calculated and measured abundances. This agreement fails on the far shoulders of each peak, where additional processes of nucleosynthesis can produce more material than is predicted by the equilibrium process.

Now let us consider the heavier elements. We see in Figure 6 that the abundance pattern shows some striking regularities in the abundances of the heavier elements. Thus it may be noticed that there are sharp abundance peaks near mass numbers 90, 140, and 208. These are just the places where closed neutron shells occur along the

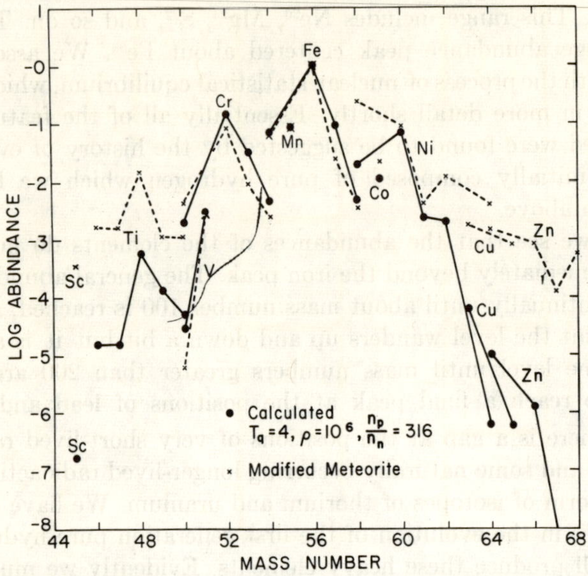

FIG. 7. The relative abundances of the nuclides in the equilibrium peak compared with those in nature, as deduced partly from meteoritic and partly from solar data.

valley of beta stability. These closed neutron shells consist of configurations of 50, 82, and 126 neutrons in the nucleus. Such configurations are exceedingly stable, and nuclei with nearly a closed shell of neutrons have a much reduced cross section or probability for capturing neutrons. This fact suggests that it must have been principally a neutron capture process which was responsible for producing these sharp abundance peaks. We notice also that preceding the two higher sharp peaks there are additional broader peaks. In these peaks the jagged odd–even character of the abundance variations is much reduced. We shall also attribute these features to a neutron capture process, but one which occurs much more rapidly than that which produces the sharp peaks, as will be shown below.

SYNTHESIS OF THE HEAVY ELEMENTS

A typical section of a nuclide chart is shown in Figure 8. Those nuclides which are stable against beta decay are outlined with

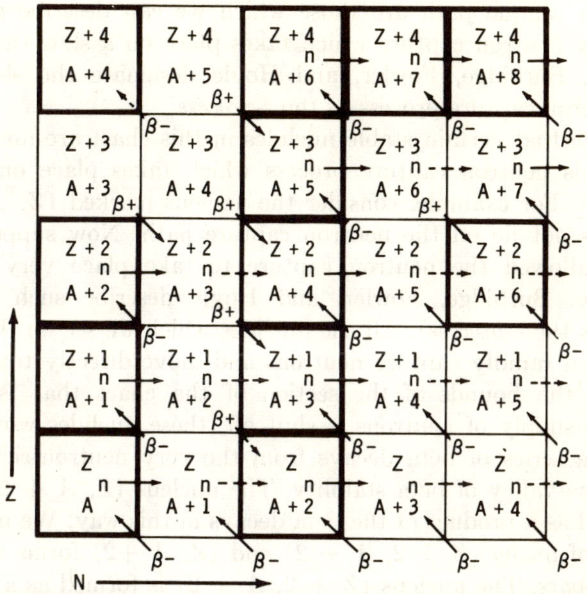

FIG. 8. A typical section of a nuclide chart. Nuclei which are stable against beta decay are indicated by heavy black borders.

borders of solid black. Let us suppose that we start in the lower left-hand corner of this diagram, and capture a neutron. The product following the neutron capture lies immediately to the right of the lower left hand square. Let us further suppose for the moment that this product nucleus, which we can see to be radioactive, has a half-life which is much shorter than the time required before an additional neutron is likely to be captured. Then the product nucleus will undergo a beta decay which will carry it diagonally toward the upper left and produce a product in the square immediately above the lower left hand square. We see that this nuclide is stable against beta decay. Consequently, after a while, an additional neutron will be captured in this nuclide, producing again a radioactive nucleus, which will once again decay toward the upper left. One could continue in this way and outline a unique capture path which threads its way through a specific set of the nuclides which are shown in this section of the chart. The stable nuclides which lie along this unique path are those which we can describe as being formed by neutron capture which takes place on a slow time scale. Burbidge, Burbidge, Fowler, and Hoyle designate the slow time scale neutron capture process as the *s-process*.

We note that certain stable nuclides in this chart are not formed during this neutron capture process which takes place on a slow time scale. For example, consider the nucleus marked $(Z, A + 2)$. This does not lie on the neutron capture path. Now suppose that we had allowed the neutron capture to take place very rapidly (Burbidge, Burbidge, Fowler, and Hoyle describe such a rapid process as the *r-process*). All the nuclides which we see on this chart would then rapidly capture neutrons and move directly toward the right off the bounds of the section of the chart that is shown. After the supply of neutrons is shut off, these nuclides would then undergo a series of beta decays from the very neutron-rich region toward the valley of beta stability. The nucleus $(Z, A + 2)$ would be formed as a product of the beta decays in this way. We note that the pair of nuclei $(Z + 2, A + 2)$ and $(Z, A + 2)$ forms a pair of stable isobars. The nucleus $(Z + 2, A + 2)$ is formed as a product only of neutron capture on the slow time scale, while the nucleus $(Z, A + 2)$ is formed only as a result of neutron capture which takes place on a fast time scale. The nuclei of the neighboring mass numbers, where there are not pairs of stable isobars, will be formed by a superposition of both processes.

We also notice in Figure 8 that the nucleus at the upper left-hand corner, $(Z + 4, A + 4)$, is not formed by a neutron process of any kind. We shall describe such a nucleus as a by-passed nucleus, and look for processes not involving neutron capture at all which may be responsible for its formation. We notice also that this nucleus and the nucleus $(Z + 2, A + 4)$ form another pair of stable isobars. The nucleus $(Z + 2, A + 4)$ is formed as a result of the superposition of neutron capture on both the slow and the fast time scales, and therefore, the nucleus $(Z + 4, A + 4)$ will give a measure of the abundance level of nuclei which are not formed in neutron capture processes. Thus we see that the existence of stable pairs of isobars allows us to separate the effects of various of the processes which have been responsible for building the heavy elements from one another in the heavy region.

Figure 9 shows how the abundance levels which have been produced by the various processes may be separated in this way. The various features shown by the different curves in Figure 9 have for the most part been separated out on the basis of the abundances of the isobars. However, in some places details are shown which have in addition been inferred from the abundances of nuclides which, while made by processes of neutron capture on both slow and fast time scales, are believed to be made predominantly by one or the other of these processes. This plot shows quite clearly that the products which are formed by neutron capture on a slow time scale are those which are associated with the sharp peaks near mass numbers 90, 140, and 208. The intervening abundance levels tend to be quite small. We have already indicated that the large abundances at these positions are strongly correlated with the smallness of the neutron capture cross sections at the same positions, and they are just what is to be expected to be produced by a slow process in which neutron capture builds up lighter nuclei into heavier nuclei, and where larger abundances must accumulate at regions of small cross section in order that a smooth overall rate of flow of the nuclei may be maintained.

The abundances of the nuclei that are formed by neutron capture on a fast time scale show now more clearly the rounded abundance peaks that lie just to the left of the two highest mass number peaks which are formed by neutron capture on a slow time scale. In order to see how these peaks may be formed, consider what will happen during a process in which some seed nuclei are exposed to an enormous flux of neutrons. These seed nuclei will very rapidly capture the neutrons, and as the product nuclei become more and more enriched in neutrons, the binding energies of the neutrons that are successively captured will progressively fall. Naturally, we expect some reasonably high temperature to be present in the region in which this occurs, in order to have created the neutrons in the first place; this temperature is probably slightly in excess of 10^9 °K. At these temperatures the Planck spectrum will contain fairly large numbers of high energy photons in its high energy tail, and these photons will be effective in inducing (γ, n) reactions in the products of the neutron capture when the neutron binding energy becomes small enough. Thus the seed nuclei will only be able to capture neutrons until they reach the region where the rates of photodisintegration become comparable to the rates of additional neutron cap-

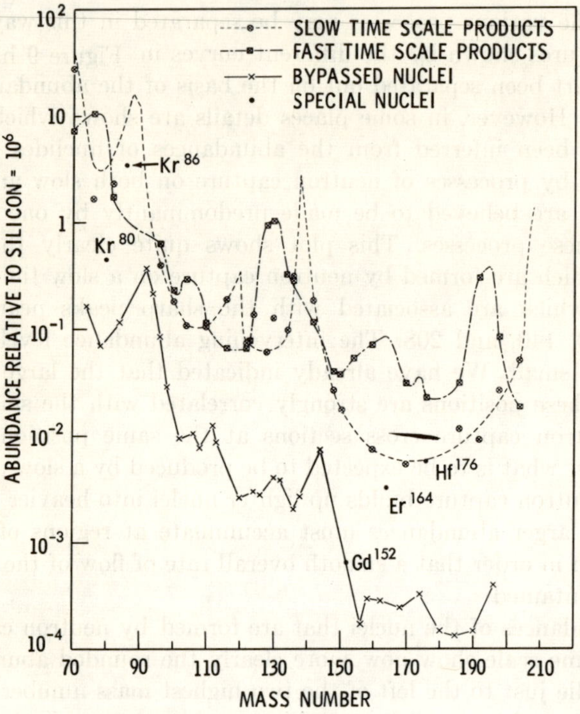

FIG. 9. Abundances of heavy nuclei separated according to the mode of synthesis (the "special nuclei" represent unusual conditions not discussed in this paper.) These curves are deduced mostly from isobaric abundance data.

ture. The nuclei will then wait in these regions until a beta decay can take place which increases the nuclear charge and allows further neutrons to be captured. This process is likely to continue until the region in which waiting occurs coincides with a closed-neutron shell. Beyond the closed neutron shell there will be an abrupt drop in neutron binding energy to a value much smaller than usual. A fairly long series of beta decays must then take place before neutron capture will be able to carry the seed nuclei beyond the closed neutron shell toward the much greater neutron rich region once again. During this series of beta decays, the seed nuclei have been brought closer to the valley of beta stability than is usual, and in so doing the beta decay energies have been reduced and the beta decay lifetimes of the nuclei have been correspondingly increased. Consequently, the nuclei must spend longer waiting times when they are at these closed shell positions than they otherwise do, and consequently their abundances in these regions will be necessarily increased in a constant flow-rate pattern. It is probable that the rounded peaks associated with the abundances of the products of neutron capture on a fast time scale are just the peaks that result from this enhanced waiting time at the closed neutron shells in the neutron rich region. We note that the rounded peaks correspond to the same closed shell numbers as the adjacent sharp peaks, but because the process takes place in the neutron rich region, then there are fewer protons associated with the rounded peaks, and this accounts for the lower mass numbers at which they are found. There are many other secondary features of the abundance curve associated with neutron capture on a fast time scale, but the interpretation of these is somewhat controversial, and hence it will not be pursued here.

Referring again to Figure 9, we see that the by-pased nuclei form a very smooth abundance curve, but one which lies at a much lower level then either of the curves associated with neutron capture. Thus the presence of minor contributions of the by-passed nuclei to all of the mass numbers does not interfere with the separation of the neutron capture effects between the two time scales.

Clayton and Fowler[17] have shown that if one chooses s-process nuclei and plots the product of the observed abundance times the cross section (σN) in the vicinity of 25 KeV. against the mass number one obtains a remarkably smooth curve. This behavior of σN can be understood in terms of the physical cnoditions under which slow neutron capture will take place. In the virtual absence of nuclear photodisintegrations, the rate of change of the abundance of a nucleus of mass A with time is given by

$$\frac{d\,N_a}{dt} = (\text{neutron flux})\,(\sigma_{A-1}\,N_{A-1} - \sigma_A\,N_A)$$

where σ_A is the neutron capture cross-section. If the flows are locally in equilibrium ($d\,N_A/dt = 0$) then we have

$$\sigma_{A-1}\,N_{A-1} = \sigma_A\,N_A$$

and analogous expressions relating neighboring nuclei. This predicts a constancy of the product σN with mass number for regions in which the flows have come into equilibrium.

The variation of σN with mass number A is shown in Figure 10. Only those nuclei for which experimental determinations of the cross sections are available have been included. A smooth variation of σN is evident in this figure. Superimposed on this trend are plateau structures, falling sharply at the positions of the closed neutron shells

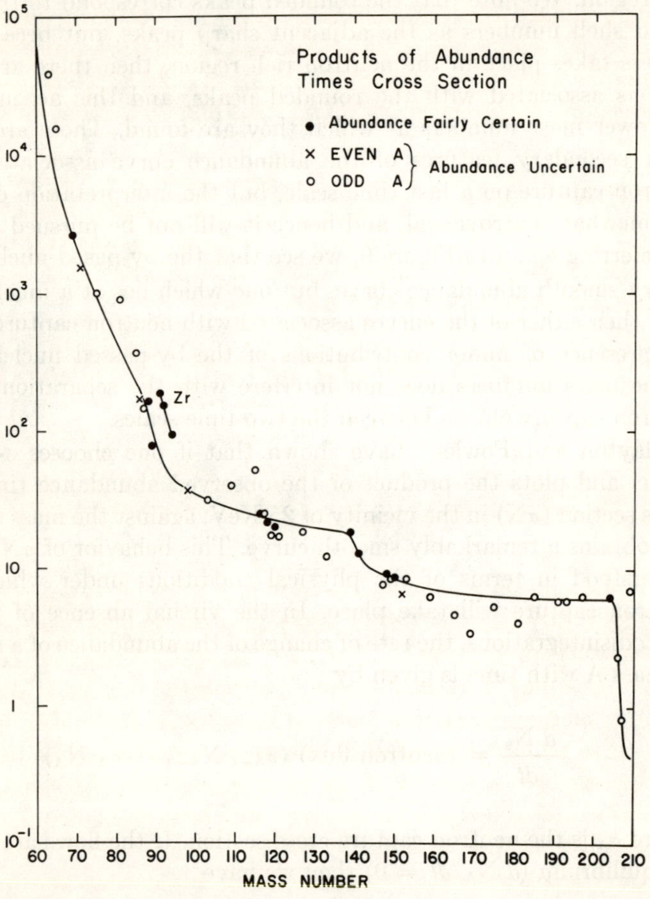

FIG. 10. The product of the cross section times the observed abundance for nuclei formed by slow neutron capture is plotted as a function of mass number.

($A \sim 138, 204$). At the closed shells, the neutron capture cross sections are substantially reduced. A larger neutron flux is demanded if equilibrium is to be realized. In so far as the detailed structure of this curve is a measure of the time integrated flux to which seed nuclei have been exposed, it contains a history of the s-processing of the solar system abundances. The smooth local variations of σN with mass number provides a means of estimating the relative contributions of r-process and s-process synthesis to the abundances of nuclei which can be formed by both processes.

Having defined the various modes of heavy element synthesis, it remains to determine the conditions under which these processes will proceed. We must first find suitable neutron sources which can produce neutron capture on each of the two time scales that have been discussed.

We have noted that stars which burn hydrogen via the CNO bi-cycle will convert the majority of the initial cargon, nitrogen and oxygen nuclei into N^{14}. When helium burning commences in these stars, the sequence of helium reactions which takes place starting with N^{14} can provide a neutron source as:

$$N^{14}(\alpha, \gamma)F^{18}(\beta^{+}\nu)O^{18}$$

$$O^{18}(\alpha, \gamma)Ne^{22}$$

$$Ne^{22}(\alpha, \gamma)Mg^{26}$$

$$Ne^{22}(\alpha, n)Mg^{25}.$$

Two alpha captures convert the N^{14} into Ne^{22}. A temperature of about 2×10^{8} degrees is required before the helium will combine with the Ne^{22} to destroy it. The $Ne^{22}(\alpha, n)Mg^{25}$ reaction is endothermic, and has a bombardment threshold of 0.5 MeV. But the optimum thermonuclear bombarding energies in the stellar interior under these temperature conditions are just such as to bombard the Ne^{22} nucleus in the vicinity of its neutron emission threshold. Thus it is quite probable that fairly large numbers of neutrons can be produced in this way, and at the present time it seems most likely that these neutrons are those which are responsible for building heavy elements by neutron capture on a slow time scale. The neutrons are initially captured by the nuclei of the iron peak and build these up along the unique path associated with neutron capture on a slow time scale toward the upper end at lead and bismuth. Neutron capture in bismuth leads to short-lived alpha particle radioactivity which merely forms lead and bismuth again, so that at the upper end the lead and bismuth simply accumulate the products of neutron capture on a slow time scale.

Next we need to find a source of neutrons which can produce neutron capture on a fast time scale.

Consider the behavior of the innermost ejected layers of the supernova. This material has been largely converted to neutrons during the implosion. Arnett[16] has found that many of the neutrons will be converted back to protons in the early stages of expansion by interaction with the positrons which are part of the electron pairs associated with the radiation field:

$$n + e^{+} \rightarrow p + \nu.$$

These protons will then be rapidly formed into alpha-particles when the temperature falls to $\sim 10^{10}$ °K.

There is much external evidence that the process of neutron capture on a fast time scale has frozen at a temperature of about $T \sim 10^{9}$ °K. During the expansion of the hot neutron gas the temperature will be several billion degrees when the density is $\sim 10^{7}$ gm/cm^{3}. This implies a neutron number density $n_n \sim 10^{31}$ cm^{-3} at an expansion time of $\sim 10^{-2}$ seconds.

We believe that this situation is very promising for the process of neutron capture on a fast time scale. Our estimates of beta decay half-lives for very neutron-rich nuclei are based on an explicit integration over realistic level densities in the daughter nucleus. Making allowance for the weakening of the beta decay matrix elements in the transitions to the more highly excited states, using a technique developed by Hansen,[12] we find half-lives $\sim 10^{-3}$ seconds. This is short enough to fit into the supernova expansion time scale.

One major question which we have not yet investigated quantitatively is the question of the seed nuclei for the process of neutron capture on a fast time scale. It is evident that nuclear processes building higher than He^{4} tend to be rather slow, so that the triple-alpha reaction, $3He^{4} \rightarrow C^{12}$, will convert only a small part of the He^{4} in the supernova expansion time scale. Because of the high-density neutron-rich medium, there may be other useful reaction chains, such as

$$He^4\ (2n,\ \gamma)\ He^6\ (\alpha,\ n)\ Be^9\ (n,\ \gamma)\ Be^{10}\ (\alpha,\ n)\ C^{13}$$

These have not yet been investigated. Nuclei more massive than carbon will be processed readily into the vicinity of the iron peak, providing the seed nuclei.

The bypassed nuclei are nuclei on the neutron deficient side of the valley of beta stability which cannot be produced by any neutron capture process. In principle these can be produced by photodisintegration processes or by proton capture processes, in each case operating on the products of previous neutron capture.

Consider first the abundance curve for the bypassed nuclei shown in Figure 9, There is remarkably little abundance variation between adjacent stable isobars of bypassed nuclei. This argues against production by photodisintegration, since photodisintegration rates will be proportional to exp—B/kT, where B is the appropriate neutron or proton binding energy. Since $kT \sim 0.5$ Mev, and the photodisintegration path should pass through nuclei having variations of B of ~ 1 Mev, we should expect abundance fluctuations of order e^2. Clearly these are not present.

However, it is very plausible that the bypassed nuclei should be made by proton capture. Note that there is a striking similarity between the abundance variations of the products of neutron capture on a slow time scale and those of the bypassed nuclei. The principal differences are that the ratio of bypassed to slow capture products decreases with increasing mass number, and that the abundance level of the bypassed nuclei stays high right up to the neutron closed shells, beyond which it drops abruptly. These are the effects that one would expect to have produced if the products of neutron capture on a slow time scale were raised to a temperature of a few billion degrees for a brief moment in a very proton rich medium.

We suggest that this happens when the supernova shock wave transverses the outermost unaltered layer of the pre-supernova star.

LIGHT ELEMENT SYNTHESIS

In this discussion we have found that the various features of the abundance distribution of the elements can be correlated with processes that we expect naturally to take place during the course of nuclear reactions in stellar interiors, with the exception of the light nuclei deuterium, lithium, beryllium, and boron. As has been mentioned in the chapter on the origin of the solar system, Fowler, Greenstein, and Hoyle have suggested that these lightest nuclei are produced by *spallation* reactions on boulders a few meters in dimension which latter collect to form the planets in the solar system. The writer believes that such light nuclei are more likely to be made by a variety of non-equilibrium reactions which can take place during the course of supernova explosions. The matter is very controversial and the appropriate reactions have not been worked out in detail as yet.

THE PROBLEM OF HELIUM

The primordial abundance of helium in the galaxy is currently a subject of considerable discussion. Its formation as the product of the first stage of nucleosynthesis in stellar interiors would seem to be consistent with its high abundance in nature. On this basis, it has been usual to assume that in its early stages the galaxy was composed of pure hydrogen. (It should be noted that the high helium abundance, 25–30 percent by mass, is difficult to obtain on the basis of the usual theory of stellar nucleosynthesis.[19,20])

Recent studies of stellar spectra[21–24] have yielded evidence for an abnormally low helium abundance in the atmospheres of old halo B stars. These observations would seem to support the view expressed above that the primordial helium abundance by mass in the galaxy, $Y = 0$.

There has recently been a revival of interest in the possibility of the cosmological synthesis of helium.[25] An isotopic background radiation in the region of centimeter microwaves has been discovered[26,27] which appears to have a blackbody spectrum corresponding to a temperature of about 3°K. This has been interpreted as the red-shifted radiation produced in a much hotter, more condensed phase of the universe, when the temperature was high enough to ionize hydrogen and hence to make space opaque to electromagnetic radiation. Such a condition has been described as the primordial fireball.

At a very early stage in the fireball, in the temperature range 10^9-10^{10} °K, a general equilibrium would be set up between neutrons, protons, neutrinos, antineutrinos, electrons, and positrons. This will establish a preponderance of protons over neutrons when the interactions break down as the temperature decreases. At temperatures near 10^9 °K the following reactions will occur which will convert most of the remaining neutrons into helium:

$$H^1\ (n,\ \gamma)\ D^2$$

$$D^2\ (n,\ \gamma)\ H^3$$

$$D^2\ (d,\ n)\ He^3$$

$$D^2\ (d,\ p)\ H^3$$

$$H^3\ (d,\ n)\ He^4$$

$$He^3\ (d,\ p)\ He^4$$

About 25 to 30 per cent of the mass of primordial matter is likely to be formed into helium in this way. Most celestial objects appear to have helium abundances of this order. However, if even one object could be found with essentially no helium in its composition, this would be a strong argument against the cosmological synthesis of helium.

Greenstein *et al.*[28] have offered an explanation of the low helium abundance in old halo B stars. They suggested that the surface helium has diffused below the photosphere, and that these stars may indeed possess a normal helium abundance in their interiors. Estimates of the diffusion velocities in these stars indicate that the helium diffuses downward one scale height in a time far less than the typical lifetime for the star on the horizontal branch. For main sequence B stars, the downward diffusion of helium will be opposed by meridional circulations induced by their rapid rotation—hence, studies of their spectra should reveal a normal helium abundance.

If Greenstein *et al.* are correct, then there is no evidence against the cosmological explanation of the helium abundance. Further research is clearly necessary on this subject.

It must be emphasized that this general discussion of the nuclear physics associated with nucleosynthesis is less controversial than the problem of finding the correct astrophysical location where the nuclear processes take place. That is to say, the stellar evolution calculations which result in constructing models of more and more advanced phases of stellar evolution have not yet reached the stage where it is possible to say precisely under what conditions the nuclear reactions can take place. It is evident that most future research in the area of nucleosynthesis must be devoted toward fitting the nuclear processes into the general context of stellar evolution.

BIBLIOGRAPHY

1. Mayer, M. G., and Teller, E., On the Origin of Elements, Phys. Rev. **76**, 1226 (1949).
2. Alpher, R. A., Bethe, H. A. and Gamow, G., The Origin of Chemical Elements, Phys. Rev. **73**, 803 (1948).
3. Merrill, P. W., Technetium in the Stars, Science **115**, 484 (1952).
4. Burbidge, E. M., Burbidge, G. R., Fowler, W. A. and Hoyle, F. Synthesis of the elements in Stars, Revs. Modern Phys. 29, 547 (1957).
5. Cameron, A. G. W., Stellar Evolution, Nuclear Astrophysics, and Nucleogenesis, Atomic Energy of Canada Report CRL-41.
6. Cameron, Nuclear Astrophysics, Annual Rev. Nucl. Sci. 8, 299 (1958).
7. Burbidge, G. R., Nuclear Astrophysics, Annual Rev. Nucl. Sci. 12, 507 (1962).
8. Salpeter, E. E., Nuclear Reactions in the Stars. I. Proton-Proton Chain, Phys. Rev. **88**, 547 (1952).
9. Truran, J. W., Hansen, C. J., Cameron, A. G. W. and Gilbert, A., Thermonuclear Reactions in Medium and Heavy Nuclei, Can. J. Phys. **44**, 151 (1966).
10. Reeves, H., Stellar Energy Sources, Chapter 2 in Stars and Stellar Systems, Vol. 8, Aller and McLaughlin (editors), Un. Chicago Press (1965).
11. Hoyle, F., On Nuclear Reactions in Very Hot Stars. I. The Synthesis of Elements from Carbon to Nickel, Astrophys. J. Suppl. **1**, 121 (1954).
12. Hansen, C. J., Neutrino Emission from Dense Stellar Interiors, Yale University thesis (1966).
13. Truran, J. W., Cameron, A. G. W. and Gilbert, A., The Approach to Nuclear Statistical Equilibrium, Can. J. Phys. **44**, 563 (1966).
14. Chiu, H. Y., Pre-Supernova Evolution, **In** Stellar Evolution, **edited** by R. F. Stein and A. G. W. Cameron (Plenum Press, N. Y., 1966).
15. Colgate, S. A. and White, R. H., The Hydrodynamic Behavior of Supernovae Explosions, Astrophys. J. **143**, 626 (1966).
16. Arnett, W. D., Mass Dependence in Gravitational Collapse of Stellar Cores, Preprint (1967).

17. Clayton, D. D. and Fowler, W. A., Abundances of Heavy Nuclides, Ann. Phys. **16,** 51 (1961).
18. Cameron, A. G. W., Truran, J. W. and Arnett, W. D., Formation of Heavy Nuclei in Nucleosynthesis, Arkiv för Fysik, (1967, in press).
19. Hoyle, F. and Taylor, R. J., The Mystery of the Cosmic Helium Abundance, Nature **203,** 1108 (1964).
20. Truran, J. W., Hansen, C. J. and Cameron, A. G. W., The Helium Content of the Galaxy, Can. J. Phys. **43,** 1616 (1965).
21. Searle, L. and Rodgers, A. W., The Horizontal-Branch Stars of the Globular Cluster NGC 6397, Astrophys. J. **143,** 809 (1966).
22. Greenstein, J. L., The Nature of the Faint Blue Stars, Astrophys. J. **144,** 496 (1966).
23. Sargent, W. L. W. and Searle, L., Spectroscopic Evidence on the Helium Abundance of Stars in the Galactic Halo, Astrophys. J. **145,** 652 (1966).
24. Greenstein, J. L. and Munch, G., The Weakness of Helium Lines in Globular Cluster and Halo B Stars, Astrophys. J. **146,** 618 (1966).
25. Peebles, P. J. E., Primordial Helium Abundance and the Primordial Fireball, Astrophys. J. **146,** 542 (1966).
26. Penzias, A. A. and Wilson, R. W., A Measurement of Excess Antenna Temperature at 4080 Mc/s, Astrophys. J. **142,** 420 (1965).
27. Dicke, R. H., Peebles, P. J. E., Roll, P. G. and Wilkinson, D. T., Cosmic Black-Body Radiation, Astrophys. J. **142,** 414 (1965).
28. Greenstein, G. S., Truran, J. W. and Cameron, A. G. W., The Helium Deficiency in Old Halo B Stars, Nature, in press (1967).

AUTHOR INDEX

Numbers in parentheses are reference numbers

SUBJECT INDEX